AF184139

Brazilian Steel Town

DISLOCATIONS

General Editors: August Carbonella, *Memorial University of Newfoundland,* Don Kalb, *University of Bergen & Utrecht University,* Linda Green, *University of Arizona*

The immense dislocations and suffering caused by neoliberal globalization, the retreat of the welfare state in the last decades of the twentieth century, and the heightened military imperialism at the turn of the twenty-first century have raised urgent questions about the temporal and spatial dimensions of power. Through stimulating critical perspectives and new and cross-disciplinary frameworks that reflect recent innovations in the social and human sciences, this series provides a forum for politically engaged, ethnographically informed and theoretically incisive responses.

Volume 27
Brazilian Steel Town: Machines, Land, Money and Commoning in the Making of the Working Class
Massimiliano Mollona

Volume 26
Claiming Homes: Confronting Domicide in Rural China
Charlotte Bruckermann

Volume 25
Democracy Struggles: NGOs and the Politics of Aid in Serbia
Theodora Vetta

Volume 24
Worldwide Mobilizations: Class Struggles and Urban Commoning
Edited by Don Kalb and Massimiliano Mollona

Volume 23
The Revolt of the Provinces: Anti-Gypsyism and Right-Wing Politics in Hungary
Kristóf Szombati

Volume 22
Frontiers of Civil Society: Government and Hegemony in Serbia
Marek Mikuš

Volume 21
The Partial Revolution: Labour, Social Movements and the Invisible Hand of Mao in Western Nepal
Michael Hoffmann

Volume 20
Indigenist Mobilization: Confronting Electoral Communism and Precarious Livelihoods in Post-Reform Kerala
Luisa Steur

Volume 19
Breaking Rocks: Music, Ideology and Economic Collapse, from Paris to Kinshasa
Joe Trapido

Volume 18
The Anthropology of Corporate Social Responsibility
Edited by Catherine Dolan and Dinah Rajak

For a full volume listing, please see the series page on our website: https://www.berghahnbooks.com/series/dislocations

Brazilian Steel Town

Machines, Land, Money and Commoning in the Making of the Working Class

Massimiliano Mollona

berghahn
NEW YORK • OXFORD
www.berghahnbooks.com

First published in 2020 by
Berghahn Books
www.berghahnbooks.com

© 2020 Massimiliano Mollona

All rights reserved. Except for the quotation of short passages for the purposes of criticism and review, no part of this book may be reproduced in any form or by any means, electronic or mechanical, including photocopying, recording, or any information storage and retrieval system now known or to be invented, without written permission of the publisher.

Library of Congress Cataloging-in-Publication Data

Names: Mollona, Massimiliano, 1969- author.
Title: Brazilian steel town : machines, land, money and commoning in the making of the working class / Massimiliano Mollona.
Description: New York : Berghahn, 2020. | Series: Dislocations ; volume 27 | Includes bibliographical references and index.
Identifiers: LCCN 2019037815 (print) | LCCN 2019037816 (ebook) | ISBN 9781789204339 (hardback) | ISBN 9781789204346 (ebook)
Subjects: LCSH: Companhia Siderúrgica Nacional--History. | Steel industry and trade--Brazil--Volta Redonda--History. | Volta Redonda (Brazil--Economic conditions. | Volta Redonda (Brazil)--Social conditions.
Classification: LCC HD9524.B84 C6255 2020 (print) | LCC HD9524.B84 (ebook) | DDC 330.981/53--dc23
LC record available at https://lccn.loc.gov/2019037815
LC ebook record available at https://lccn.loc.gov/2019037816

British Library Cataloguing in Publication Data

A catalogue record for this book is available from the British Library.

ISBN 978-1-78920-433-9 hardback
ISBN 978-1-78920-434-6 ebook

To my children Sofia, Rosa and Zeno with much love

Contents

Figures

Acknowledgements

This book benefitted a great deal from the conversations I had with colleagues and friends over time. I am especially grateful to Johnny Parry, Leo Panitch, Gavin Smith, Don Kalb, Sharryn Kasmir, Frances Pine, Victoria Goddard and to my Brazilian colleagues José Ricardo Ramalho, Marco Aurélio Santana, Raphael Lima, Sabrina Dias and Virginia Fontes. I am also deeply thankful to my friends in Volta Redonda, especially Stella, Vanderlei, Marco Aurélio, Renata and Gerardo and to Juliene de Paula for letting me use her wonderful photo for my book cover. This book would not exist without my students whose openness, intelligence and engagement always kept me inspired and hopeful for a better future.

Abbreviations

5S	Five Star (Sorting, Straightening, Systematic Cleaning, Standardising and Sustaining)
AAP-VR	Associação de Aposentados e Pensionistas de Volta Redonda (Association of Retirees and Pensioners of Volta Redonda)
ACIAP-VR	Associação Comercial, Industrial e Agropastoril de Volta Redonda (Commercial, Industrial and Agro-pastoral Association of Volta Redonda)
ACO	Ação Católica Operária (Catholic Workers Action)
BANERJ	Banco do Estado do Rio de Janeiro Rio de Janeiro (Bank of the State of Rio de Janeiro)
BNDES	Banco Nacional de Desenvolvimento Econômico e Social (National Bank for Economic and Social Development)
Bovespa	Bolsa de Valores de São Paulo (São Paulo Stock Exchange)
BVRJ	Bolsa de Valores do Rio de Janeiro (Rio de Janeiro Stock Exchange)
CBS-CSN	Caixa Beneficente dos Empregados da Companhia Siderúrgica Nacional (Employee Pension Fund of CSN)
CEB	Comunidade Eclesial de Base (Basic Ecclesiastic Communities)
CECISA	Imobiliária Santa Cecília (Real Estate Santa Cecília)
CGT	Comando Geral dos Trabalhadores (General Command of Workers)
CIA	Clube de Investimento Alternativo (Alternative Investment Club)
CIPA	Comissão Interna de Prevenção de Acidentes (Internal Commission of Accident Prevention)
CLT	Consolidação das Leis do Trabalho (Consolidation of Labour Laws)

CMDU	Conselho Municipal de Desenvolvimento Urbano (Municipal Council of Urban Development)
CODIN	Companhia de Desenvolvimento (Regional Development Agency)
Conlutas	Coordenação Nacional de Lutas (National Coordination of Struggles)
COHAB-VR	Companhia de Habitação de Volta Redonda (Housing Company of Volta Redonda)
COSEMA	Conselhos de Saúde Municipal dos Aposentados (Municipal Health Council for Retirees)
CP	certificate of privatisation
CSN	Companhia Siderúrgica Nacional (National Steel Company)
CTB	Central dos Trabalhadores e Trabalhadoras do Brasil (Central of Male and Female Workers of Brazil)
CUT	Central Única dos Trabalhadores (Unified Workers' Central)
CVRD	Companhia Vale do Rio Doce (Sweet River Valley Company) (now Vale)
DAS	Departamento de Assistência Social (Department of Social Assistance)
ETPC	Escola Técnica Pandiá Calógeras (Pandiá Calógeras Technical School)
FAT	Fundo de Amparo ao Trabalhador (Worker Support Fund)
FC	factory commission
FEEMA	Fundação Estadual de Engenhenaria do Meio Ambiente (State Foundation for Environmental Engineering)
FEM	Fábrica Estruturas Metalicas (Metal Structures Factory)
FGTS	Fundo de Garantia do Tempo de Serviço (Service Assurance Fund)
FS	Força Sindical (Union Force)
Funcef	Fundação dos Economiários Federais (Foundation of Federal Economists)
GGAF	Gerência Geral Alto Forno (smelting shop)
GGFM	Gerência Geral Folha Metálica (sheet metal)
GGLF	Gerência Geral de Laminados a Frio (cold strip mill)
GGLQ	Gerência Geral de Laminados a Quente (hot strip mill)

GGMA	Gerência Geral Metalurgia do Aço (steel metallurgy)
GSC	Gerência de Centro de Servico (Service Centre)
GT	Gerência Técnica (Technical Administration)
INSS	Instituto Nacional de Seguro Social (National Institute of Social Security)
IPPU	Instituto de Pesquisa e Planejamento Urbano (Institute for Urban Research and Planning)
IPTU	Imposto Predial e Territorial Urbano (Urban Land and Territorial Tax)
ISO	International Organization for Standardization
IURD	Igreja Universal do Reino de Deus (Universal Church of the Kingdom of God)
JIT	just-in-time
JOC	Juventude Operária Católica (Catholic Worker Youth)
MCMV	Minha Casa, Minha Vida (My House, My Life)
MDB	Movimento Democrático Brasileiro (Brazilian Democratic Movement)
MEP-VR	Movimento Ética na Política Volta Redonda (Movement for Ethics in Politics Volta Redonda)
MOP	mode of production
MST	Movimento dos Trabalhadores Rurais sem Terra (Landless Workers' Movement)
OAB	Ordem dos Advogados do Brasil (Order of Attorneys of Brazil)
PAC	Programa de Aceleração do Crescimento (Growth Acceleration Programme)
PCB	Partido Comunista Brasileiro (Brazilian Communist Party)
PDT	Partido Democrático Trabalhista (Democratic Labour Party)
PLR	Participação nos Lucros e Resultados (Participation in Profits and Results)
PND	Programa Nacional de Desestatização (National Privatization Program)
PO	Pastoral Operária (Pastoral Worker)
Previ	Caixa de Previdência dos Funcionários do Banco do Brasil (Bank of Brazil Employees' Pension Fund)
PSB	Partido Socialista Brasileiro (Brazilian Socialist Party)
PSDB	Partido da Social-Democracia Brasileira (Brazilian Social Democracy Party)

PSOL	Partido Socialismo e Liberdade (Socialism and Freedom Party)
PT	Partido dos Trabalhadores (Workers' Party)
PTB	Partido Trabalhista Brasileiro (Brazilian Labour Party)
SENGE-VR	Sindicato dos Engenheiros de Volta Redonda (Engineers' Union of Volta Redonda)
SIGMA	Sistema de Gestão de Manutencão (Maintenance Management System)
SINTRACO-NSMONPES	Sindicato dos Trabalhadores nas Indústrias da Construção Civil, Montagens Industriais e Construçao Pesada de Vota Redonda e Região (Union of the Civil Construction and Heavy Construction Workers of Volta Redonda and Region)
SMSF	Sindicato dos Metalúrgicos do Sul Fluminense (Metalworkers' Union of Sul Fluminense)
SMVR	Sindicato dos Metalúrgicos do Volta Redonda (Metalworkers' Union of Volta Redonda)
SOE	state-owned enterprise
TQM	total quality management
TRT	Tribunal Regional de Trabalho (Regional Labour Court)
TST	Tribunal Superior do Trabalho (Superior Labour Court)
UPV	Usina Presidente Vargas (President Vargas Power Plant)
Usiminas	Usinas Siderúrgicas de Minas Gerais (Steel Mills of Minas Gerais)
VW	Volkswagen

INTRODUCTION

Brazilian Steel Town and Companhia Siderúrgica Nacional

The first time I saw Companhia Siderúrgica Nacional (CSN) was at the end of a long bus journey from Rio de Janeiro. As the bus exited Rio de Janeiro's north station, it cut through interminable rows of favelas with the burnt-down National Ethnographic Museum in the background. Then, it wound through the beautiful Guanabara Bay – filled with plastic trash, industrial waste and broken fishermen's boats – and continued along Presidente Dutra Road through semi-rural favelas and mega shopping centres – some of which functioned, while others were empty. Leaving behind Duque de Caixas, the bus crossed yet a different landscape, mixing smaller urban sprawls, bits of tropical forest and semi-derelict motels with exotic names such as Dallas, Samurai and Middle Age Castle or just the name of their owners like Tony, Cassio and Carlos. Climbing the winding roads of the Serra da Mantiqueira range, small mud houses, water silos and parabolic TV antennas were barely visible through the tropical vegetation of the valley, and its residents occasionally appeared by the road selling bananas, papayas or coconuts. Finally, as the bus descended the mountain, the monstrous body of the Usina Presidente Vargas (President Vargas Plant – UPV) sat in the Paraíba Valley like a giant animal. With its smoky whirlpools, coloured fumes and coils of vapour, the plant seemed like an organic extension of the surrounding forest. Plants, fruit trees, tropical flowers, iron ore, pipes, rusting machinery and bright pools of chemicals formed a half-organic, half-industrial landscape. Wild capybaras and iguanas ran through the rubbish and chemical waste fermented on the banks of the Paraíba. Loud cricket chirps reverberated across

Figure 0.1. Companhia Siderúrgica Nacional (© Massimiliano Mollona).

electricity wires, stopping every time the siren of the furnace went off.

Such interpenetration between industry and nature reminded me of the modernist 'Manifesto da Poesia Pau-Brasil' (Manifesto of Brazilwood poetry) by Oswald de Andrade (1924). Influenced by the European avant-garde, Andrade pictured the early Brazilian working class through an essentialising mixture of indigenous identity and primitive communism. Brazilian modernism reflected the aristocratic fantasies of the cosmopolitan coffee elite (see Schwarz 1992), as it did Getúlio Vargas's dream to industrialise Brazil by building the huge CSN steel complex on the site of a half-dilapidated coffee fazenda (farm). Vargas's modernist fantasy projected onto Volta Redonda transpires from the speech he made when visiting the UPV in 1943:

> This industrial town will be the mark of our civilization, a monument to the potential of our people; the evidence that will disperse any doubt or anxiety about our future and will establish a new way of living and a new mentality . . . The country that was semi-colonial agrarian, importer of manufactured goods and exporter of raw materials will be able to assume the responsibility of an autonomous industrial life. (Morel 1989: 48)

Despite Vargas's hope that CSN would modernise the nation by making it industrially self-sufficient, the company remained under regionalists powers, was managed by an industrial bourgeoisie moulded in the authoritarian style of the rural elite and was considered as much a rural fazenda as an industrial business until it was privatised in 1992. Thus, the steel town's working class emerged from the entanglement of the forward-looking dream of industrial modernity and the backwards vision of rural elites, violent coronels and colonial and neocolonial powers. Indeed, to fully appreciate the lopsided history of Brazilian development, travellers ought to continue the bus ride behind Volta Redonda and stop at the majestic fazendas of the Fluminense region – in Vassouras, Barra de Piraí and Valença – which during the 'second slavery' were among the world's biggest coffee plantations, with hundreds of slaves and labourers, and today are run as tourist attractions by the same aristocratic families that once made up the elites of the Brazilian empire.

Brazilian Steel Town is solidly rooted in Marx's radical tradition of political economy, especially the critique of capitalist value, combined with an anthropological attunement to the cultural and phenomenological forms through which struggles between capital and labour take place in the everyday. I use steelmaking as a metaphor to explore the 'magmatic' relationships between capital and labour; the polymorph metamorphoses of value on the shop floor; the blurred boundaries between the factory, the township and the surrounding nature; and the ghostly existence of the subcontractors, cleaners and maintenance workers who move at the margins of the UPV. Although this book is situated in the disciplinary world of Western academia, its core concepts and issues originate from Brazilian sociology, anthropology and cultural theory. This book aims to offer a vision of twenty-first-century working-class life in Brazil, by threading together the micro- and the macro-political, the affective and the material and the ethnographic and the historical perspectives. Moreover, *Brazilian Steel Town* combines anthropology and political economy following a double movement – one of economicisation of politics and another of politicisation of the economy – in the tradition of Karl Marx, Peter Kropotkin and Rosa Luxemburg, who considered the economy as a 'total fact' (i.e. entangled in social relations); historically (paying attention to diverse histories and temporalities) and comparatively (i.e. in relation to the livelihood of often marginal 'others').

The political economy approach in anthropology (Roseberry 1988; Smith 1999; Wolf 1982) goes beyond 'mere' social analysis and sketches possible future scenarios of human-centred economies and even of utopian socialism (Clastres 1974; Graeber 2004). Inspired by Marx's critique of political economy, *Brazilian Steel Town* is conceived as an ethnographic critique of capitalist economic categories. Indeed, there is a mounting movement calling for the discipline of economics 'to get real', leave behind its abstract and disingenuous assumptions about human behaviour and engage in long-term, grounded, critical and human-centred political economy by looking at those value forms that transcend the commodity form – concretely negating the abstractions of capitalism – and contribute to the reproduction of life.

In other terms, there is increasing awareness that the 'realism of capitalism' is a political construct and that economic theory is itself a form of social reproduction. Inspired by Theodor Adorno's negative dialectics, the political theorist Frederick Harry Pitts (2018: 110) argues that a critical approach to capitalism rests 'in the refusal to accept at face value the economic objective forms taken by congealed social relations in capitalist society, whilst at once working within this face value as long as it is the value refused'. Taking a feminist and Althusserian perspective, J. K. Gibson-Graham (1996) argue orthodox economics is an ideological discourse revolving around 'capitalocentric' and normalising – patriarchal, xenophobic, homophobic and racist – assumptions. Using a queering, discoursive and ethnographic methodology, they map the 'diverse economies' – including co-operativism, informal labour and self-entrepreneurship – that coexist in the capitalist mode of production.

Similarly, the anthropologist Douglas Holmes (2013) describes economics as a top-down process of discoursive 'framing' of reality by powerful global financial institutions (central banks and the IMF), which become a normalising social force that shapes the expectation of the public-at-large and adjusts its behaviours. His *Economy of Words* describes ethnographically the 'perfomative apparatus' of technical reports, forecasting models, public statements and published policy deliberations, through which bankers and economists construct linguistically and communicatively 'economic truths'. Likewise, Mitchell Abolafia (2004) looks at the micro-politics of the US Federal Reserve by analysing its members' 'framing moves' – that is, the efforts to influence the interpretation of information – within the competing ideologies of Keynesianism and monetarism.

The fact that corporations increasingly rely on anthropologists for their latest market trend forecasts and technological innovation, or that global financial institutions manage expanded anthropological networks to capture cultural content for their economic models,

(Holmes 2016) shows anthropology's potential to create new vocabularies, practices and models of political economy grounded in people's everyday livelihood and accessed through the anthropological practice of fieldwork, itself as a potential form of revolutionary praxis of research intervention. Anthropology can both politicise the economy – by showing its situated, contested and rhetorical performative processes – and economicise politics, highlighting how ideologies and structures of imagination are grounded in the materiality of everyday life. Politics and economics cannot be separated, as they are two combined and uneven movements inherent in human development. Anthropology captures the human roots of capitalism intended as a process of cultural construction of alterity that sustains the imperialist project of occupation and destruction of life and annihilation of 'the other'.

CSN and Brazilian Extractivism

In this book, I discuss how Vargas's modernist dream – based on the entanglement between industrial modernity and agrarian backwardness and on the alliance between the *presidente* and the Brazilian 'people' – re-emerged under President Luiz Inácio Lula da Silva's neo-extractivist regime, and how this impacted a section of the Brazilian working class. I describe neo-extractivism as a particular form of *capitalism*, combining extensive commodity extraction, finance and large-scale state infrastructures, and associated with particular patterns of social relations of production, in the Marxist sense. I contend there has been a global shift in predominance from *exploitation* through wages of various forms to *extraction* through rents of various forms (Gago and Mezzadra 2017; Standing 2016). In this book, I look at how such a rentierist mode of production has materialised in Volta Redonda, one of Latin America's most important manufacturing and financial hubs.

Although my ethnography focuses on steelmaking, CSN's main source of income comes from iron ore extraction and logistics, so the firm is an engine of Brazil's neo-extractivist regime. Unlike other Latin American models of neo-extractivism, which are ultimately controlled by foreign capital (Veltmeyer and Petra 2014), the Brazilian model revolves around the alliance between the 'state investor', corporatist trade unions and the corporate sector. Using intensive commodity extraction to feed mass programmes of poverty reduction, the ruling Partido dos Trabalhadores (Workers' Party – PT) emancipated urban and rural masses while keeping alive the power of the landed and financial elites. I test this hypothesis by looking at the articulation of social relations of production on the ground – in the UPV and the city of Volta Redonda. I argue neo-extractivism in Volta Redonda takes a conglomerate form, encompassing global logistics, mining, steelmaking and urban land speculation. I also argue the 'internalisation of imperialism' (Oliveira 2005) that followed the transformation of the state from developer to financial investor (Musacchio and Lazzarini 2014) generated a new exploitative labour regime in CSN based on extreme deskilling, precarisation and the flexibilisation of the *re-internalised* workforce.

Funded in 1946 by Vargas, Companhia Siderúrgica Nacional (National Steel Company) was the first independent national industrial complex in Brazil and the first modern experiment of a mixed economic enterprise. In the first half of the twentieth century, the state radically expanded its role as 'entrepreneur' in the electricity, hydroelectric, logistics, banking, insurance and social welfare and pension sectors through mixed economic corporations called state-owned enterprises (SOEs) also known as *autarquias* (autarchies). These were under private law but relied on an administrative public infrastructure that allowed them to retain a public interest agenda. *Autarquias* were de facto decentralised public services providers – of credit, social welfare and logistical management – aimed at de-bureaucratising the state. They were autonomous juridical subjects with administrative and financial autonomy but under state control (Venancio Filho 1968: 36).

Akin to the modern-day public-private partnership, they constituted a radical departure from the existing Brazilian civic code, which had only three forms of public juridical person: the union, the state and the municipality. Like other SOEs that mixed public infrastructures and private management and finance, CSN embodied Vargas's 'organicist' view of the relationship between the state and the market, which defeated both classical liberal and Marxist conceptions. For Alfred Stepan (1988: 30), this mixture of planned economy, akin to communism, and capitalism is typical of Latin America:

> Policy makers in Latin American commonly commit themselves to an intermediate statist model that is neither communist nor capitalist by replacing private initiative with overall public regulation in economic life, and at the same time, retaining the marketplace as a basic mechanism for distributing goods and service. They retain a system that is heavily dependent on entrepreneurial initiative and market flows, whilst at the same time, undermining both.

Moreover, CSN embodied Vargas's regime of *trabalhismo* (workerism) based on a special relationship between the state and the working class achieved by linking citizenship rights to labour rights. In 1943, Vargas's Ministry of Labour and Employment created the Consolidação das Leis do Trabalho (Consolidation of Labour Laws – CLT), which continues to be one of the most comprehensive labour legislations in the world.[1] Among other things, the CLT establishes the workers' right to minimum wage, vacation, leave, professional training, housing, pensions and child benefits and regulates trade union affiliation, training and education. The sociologist Angela de Castro Gomes (1987) describes Vargas's workerist regime as 'occupational welfare', in which the working class achieved labour and civic rights within a framework of moral and legal dependency on the state. Moreover, she draws a stark contrast between the quiescence of the early Brazilian working class and how the early English working class emerged as a self-conscious political agent at the end of a long struggle led by different social strata in opposition to the state. The idea that the CLT produced a 'passive social revolution' is by shared by several Brazilian sociologists.[2]

Figure 0.2. Getúlio Vargas in CSN. Published with permission from the Instituto de Pesquisa e Planejamento Urbano (Institute for Urban Research and Planning).

Elsewhere, I argue this contrast between the early British and Brazilian working classes is based on an idealised version of working-class struggles in England[3] and a restricted reading of Brazilian working-class history (Mollona 2014). The quiescence of the working class had more to do with the fact that Brazil under Vargas was not an industrial democracy: it was still under the authoritarian rule of rural leaders, generals and the old coffee elite (French 2004). Moreover, the formal working class was, and continues to be, a small section of the working population mainly employed in the informal economy (Braga 2016; Singer 2018a). Unlike nineteenth-century capitalist England, the Brazilian labour regime mixed slavery and wage labour – a vast sub-proletariat and a small aristocracy of labour – so struggles for civic emancipation and struggles for economic redistribution went hand in hand.

In a country where 80 per cent of the population was below the poverty line, the 'regulated citizenship' promised by the CLT was 'utopian' (Cardoso 2010: 775). But such utopia could only be realised with the inclusion of rural citizens – 70 per cent of Brazil's population – two-thirds of whom worked informally or under a regime of semi-slavery. However, because of resistance from rural elites, the new labour legislation excluded foreign, agricultural, informal and unregistered workers. Thus, by making the urban industrial worker the exclusive recipient of citizens' rights based on the *carteira de trabalho* (work permit, on which occupational status was formally registered), the CLT de facto reproduced the previous regime of slavery for black, rural and informal workers. Indeed, without birth registration on their civic registers, rural workers were automatically excluded not only from formal employment but also from education, health benefits and the rights to vote and to property (Holston 2007).

For the sociologist Francisco de Oliveira (2005), Vargas's workerist legislation was meant to split the workforce between a small core of formal worker and a vast urban and rural lumpen, to disempower rather empower the working class and to decrease industrial wages. Indeed, since its establishment of the CLT, the minimum salary was constantly kept below subsistence levels. In other words, the pro-labour legislation of the *Estado Novo* (New State) forcedly implemented a capitalist labour market in Brazil not to develop a modern industry but rather to reproduce the previous system of slavery within the new industrial system (Singer 2018a: 21). Indeed, the immense scale of informal labour in contemporary Brazil continues to testify its peripheral condition vis-à-vis economies of the Global North and the marginality of its population, which, fluctuating in and out of employment, continues to lack of political, labour and welfare rights (Braga 2016: 60).

CSN was the crown jewel in Vargas's policy of state-led industrialisation in the steel, oil, mining and infrastructure industries together with other SOEs such as Companhia Vale do Rio Doce (now Vale) funded in 1942, and Petrobras funded in 1952. These SOEs were run by highly educated engineers and military elite[4] – trained in US-style military academies – and supported by the landowning class. This technocratic bureaucracy had penetrated deeply in the Brazilian economy and formed a separate, centrifugal force within the state (see Martins 1974), which Guillermo O'Donnell (1973) describes as 'bureaucratic authoritarianism'. One of these semi-independent 'empires' (Trebat 1983) was what is now the Banco Nacional de Desenvolvimento Econômico e Social (National Bank for Economic and Social Development – BNDES), set up in 1952 after the joint US-Brazil Commission called for the creation of a credit institution to finance the expansion of national infrastructure. In its early years, the BNDES focused on railroad modernisation and power generation and transmission. In the late 1950s, the bank began supporting the expansion of the steel and electricity sectors. From 1958 to 1969, half its funds went to the steel sector, so it became known as the National Steel Bank. In the 1960s and 1970s, the bank operated as a giant holding company for the steel industry.

Moreover, through its own investment bank, BNDESPAR, the BNDES had a double role of lender and controller of many national industries. Typically, the bank would start by financing minority portions of a company and later, through equity injections or convertible debt, become a majority shareholder. Through the BNDESPAR, the state become a minority – and, in several instances, majority – shareholder in several strategic industries, thus morphing from 'entrepreneur' to 'investor' (Musacchio and Lazzarini 2014). Since its reorganisation in the 1980s, the BNDES had moved from giving developmental support to public firms to implementing infrastructural projects, especially in the financial sector (Montero 1998: 34). Starting from this period, the bank functioned as the 'operational agent of the state' – but with autonomous budget and organisational structure – and as the main force behind the 'statitization of the Brazilian economy' (Musacchio and Lazzarini 2014: 89).

Through the BNDES, the state expanded control over powerful national conglomerates, such as the construction company Odebrecht,[5] and renationalised Petrobras and Oi (previously Brasil Telecom). During the PT administration, Petrobras became the fourth largest oil company in the world. Aldo Musacchio and Sergio Lazzarini (2014: 99) describe the Brazilian state as a 'leviathan'. In 2009, there were forty-seven public companies in which the federal government had majority shares, with assets worth $625 billion, and five of these were state-owned holding enterprises that controlled sixty-eight subsidiaries. As a result, 'the federal government in Brazil had almost three times the assets under management (AUM) of the Government Pension Fund of Norway, the world's largest sovereign wealth fund, with an AUM close to $500 million in 2009'. But, pace Musacchio and Lazzarini, such a model of state capitalism has nothing to do with socialism or with public control over private business. Instead, it reflects the interests of a narrow industrial and banking elite and of foreign capital that have occupied the state since the *República Velha* (Old Republic).

The role of industrial entrepreneur of the Brazilian state started with João VI (the Portuguese prince and future king of Brazil) who was keen to break Brazil's dependent development on coffee, sugar and rubber monocultures and commodity exports controlled by European capital. João supported the steel, iron and silk sectors through exemption duties on imports of raw material, high taxes on commodity exports (40 per cent) and direct subsidies. In the *Código Brasiliense* (Brazilian code) of 1811, João described these measures as 'free gifts to the capitalists from the state'. His vision of autarkic and independent national industry was vindicated by the crisis of the coffee economy in the 1870s, and the financial collapse of the landowning class sparked by the end of slavery.

By the mid-nineteenth century, there were more than sixty-four national industries in Brazil and a powerful industrial association founded as early as 1880. In the 1900s, the state withdrew support to coffee growers and increased support to the steel and mining industries – half of which were controlled by fazendeiros (Luz 1978: 48; Villela and Suzigan 1973: 124). It was under Getúlio Vargas's *Estado Novo* that the national industry boomed, associated with import substitution policies, so that Brazil could follow in the development path of 'the developed nations' and become economically self-sufficient. Vargas's industrialisation plan could not have succeeded without support from the United States and the landed elites, who made sure industrialisation would not turn into a fully-fledged bourgeois revolution (Fernandes 1975). Thus, the model of state-led industrialisation under the control of agrarian elites established in the early empire and reproduced under Vargas's *Estado Novo* turned out to be the blueprint of Brazilian development.

The turning point for the current model of state capitalism was the radical programme of privatisation of the early 1990s, including of CSN, which I describe extensively in chapter 5. In the privatisation process, the BNDES acted once again as the operational agent of the state and through the BNDESPAR purchased minority stakes in the privatised firms thus, putting the state in control, often through a system of expanded and invisible web of pyramidal and cross-participations, of several strategic steel firms (Musacchio and Lazzarini 2014). The privatisation of CSN in 1992 turned it from being a traditional steelmaker into a conglomerate operating in the finance, logistics and mining sectors. Paradoxically, it was with privatisation that the state gained control indirectly – that is, by controlling the Grupo Vicunha (Vicunha Group) of CSN – and turned into one of its extractivist arms.

The privatisation of CSN transformed it into a typical 'national champion' with a conglomerate and internationalised structure, vertically integrated and diversified into logistics, extraction, finance and steelmaking. Characteristic also of China, India and South Korea, conglomerates are considered the 'developmentalist' alternative to the flexible, flat, dispersed and highly financialised capitalism in the north (Goldstein and Ross Schneider 2004; Amman and Nixson 1999) Conglomerates are said to perform particularly well during economic crises because they provide internal credit and mechanisms of internal transfer of the workforce instead of lay-offs. On the other hand, Latin American conglomerates are said to be hierarchic and 'despotic' (Schneider and Soskice 2009). Indeed, the hyper-centralised and verticalised post-privatisation structure of CSN reflected the despotic regime of the new owner, a magnate whose main business is the textile industry.

Moreover, privatisation led to what Oliveira (2005) calls 'the internalisation of imperialism' (via the BNDES) leading to the metamorphosis of national debt dependency on foreign companies into working-class debt dependency on the state. As for other contexts (from post-communist Poland to Chile in the 1980s), privatisation in Brazil was branded as a form of 'people's capitalism'. The minister of finance of the time, the famous world dependency theorist Fernando Henrique Cardoso, considered privatisation a tool of 'emancipation of the periphery'. I show in chapter 5 that privatisation in fact used mostly working-class savings – in the form of pension funds and the Fundo de Amparo ao Trabalhador (Worker Support Fund – FAT) – to reproduce existing monopolistic powers while fending off foreign investors. Thus, the privatisation of CSN not only broke the workers' movement in Volta Redonda and nationally (Graciolli 2007). It also used working-class money to subsidise the banking and industrial bourgeoisie.

The FAT was created with the 1988 Constitution of the Federative Republic of Brazil, which earmarked 60 per cent of it for unemployment insurance and the remaining 40 per cent for the BNDES to finance development programmes. By 2006, the FAT was responsible for 67 per cent of the disbursements of the BNDES, amounting to more than $100 billion in 2010 (Zibechi 2014: 47). Pension funds were created in 1977 by the military regime. In 2001, President F. H. Cardoso passed Law 108, allowing union members to manage the funds. Cardoso believed workers' participation in the management of pension funds would boost the economy and open it up to popular forces. But it was under the Lula administration that the privatised system of pension funds and the FAT came to dominate the Brazilian economy.

The popularisation of financial capitalism via pension funds was already part of Lula's first electoral manifesto. From 2003 onwards, unionists, especially those associated with the confederation Central Única dos Trabalhadores (Unified Workers' Central – CUT), created their own pension funds, began to frequent business milieus and even joined the board of the Bolsa de Valores de São Paulo (São Paulo Stock Exchange). Raúl Zibechi (2014: 41) estimates the 2010 pension fund assets in Brazil reached $300 billion, 16 per cent of Brazil's GDP, becoming the largest institutional investor of the country. In 2010, the PT controlled fifteen pension funds, including Previ, Petros and Funcef, which are among the world's biggest. For instance, Previ is ranked twenty-fifth in the world, controls seventy Brazilian companies and owns shares, some of which are majority, in Vale, Embraer, Petrobras, Usiminas, and Gerdau.

In Lula's second administration (2006–2010), the link between the government, the banking elite and the industrial elite and the BNDES strengthened further so that 'financial accumulation now occurred mainly in the state realm'. The state-led financialisation of the economy turned ex-workers – 'the hard core of the PT' (Oliveira 2006: 23, 25) – into pension fund operators and powerful public managers in control of private conglomerates such as Vale. These ex-labour-activists-turned-public-fund-managers constituted a new financial class (Zibechi 2014: 147; Boito 2007: 117). The ruling PT also strengthened further the influence of the BNDES, which today is the biggest development bank in the world with investments representing 7 per cent of Brazil's GDP. The BNDES and pension funds will account for 40 per cent of Brazil's GDP by 2024.

Moreover, Lula consolidated his model of 'developmentalist extractivism' based on, according to Maristella Svampa, 'large-scale enterprises, a focus on exportation and a tendency for mono-production or monoculture. Its emblematic figures are strip mining, the expansion of the petroleum and energy frontiers and the generalization of the agribusiness model (soy and biofuel)'. According to Svampa, neo-extractivism is in need of continuous expansion and hence is tied to citizen consumption, overdevelopment of state infrastructures and 'an imperialist mode of living' (2017: 66, 67). In 2008, Lula launched the Programa de Aceleração do Crescimento (Growth Acceleration Programme – PAC) a four-year investment programme with a budget of R$291 billion (23 per cent of the GDP) concentrated in the infrastructure and energy sectors. The second phase (PAC 2) included the construction of fifty-four hydroelectric plants, with dams to be built mostly in the Amazon basin.

CSN has a central role in the neo-extractivist developmentalism of the Brazilian state. The firm controls important mines, forests and logistical nodes, such as the Port of Sepetiba, which provides access to the global maritime trade. Besides, CSN owns the operator of the Transnordestina railway – a 1,700-kilometer line connecting the state of Rio de Janeiro with the northern states of Ceará, Maranhão and Piauí. Once completed, the line will connect all the major ports and oil refineries in the states of Rio de Janeiro and Espírito Santo and carry commodities such as soy, corn, iron ore and gypsum from the Piauí north to the Port of Pecém and east to the Port of Suape, and from there be shipped to China, Brazil's main export market.

The Transnordestina railway displaced millions of small farmers in Brazil's notoriously poor Região Nordeste (Northeast Region). But at its launch in 2006, Lula described the R$11.2 billion ($3.53 billion) Transnordestina project as a miraculous tool of development for the rural masses. Although CSN agreed to inject R$3.6 billion in the project in 2016, the railway is largely funded with BNDES and federal money. In turn, CSN greatly benefits from the infrastructural expansion of the state. For instance, PACs boost the internal market for steel, which in 2010 constituted 80 per cent of the steel revenues of the company. Moreover, the government's funding of the mining industry through the BNDES and the softening of environmental restrictions[6] boosted CSN mining revenue, which in 2010 was R$1.2 billion, the highest income generated in the group.

Linked to this extractivist model is the internationalisation of Brazilian companies, which Virginia Fontes and Ana Garcia (2014) describe as 'imperial capitalism'. As for other emerging economies, the internationalisation of Brazilian capital and the dominance of Brazilian firms were boosted by the economic crisis of 2007–2008. According to the Boston Consulting Group (Bhattacharya et al. 2013), Brazil has thirteen companies in the 'global challengers' list, behind only China and India, in addition to its already established global conglomerates such as Vale and Petrobras (Fontes and Garcia 2014: 209). The main regional focus of the 'sub-imperialist expansion' (Zibechi 2014) of Brazilian conglomerates in the oil, gas, mining, hydroelectric, ethanol, cellulose and construction sectors are the northern and Amazonian regions and Latin America, where corporate expansion has taken the form of brutal 'expulsion', land grabbing and paramilitary violence (Sassen 2014: 99). Indeed, Lula's renationalisation of Petrobras was largely driven by his intention to turn the energy complex into a global ethanol producer and protect this strategic sector from competition from the United States.

In this book, I discuss how the workers of Volta Redonda negotiate the social and economic regimes associated with the extractivist state. I argue there is a structural coupling between Brazil's dependency in the global economy, in the car, steel and IT sectors, and the strategies of accumulation by dispossession by CSN based on rent seeking, commodity export (mainly iron ore) and a precarised working regime. In 2008, the metalworkers of Volta Redonda thought of themselves as capitalists. But during my fieldwork, they struggled with unemployment, debts and declining purchasing power.[7] With a powerless Sindicato dos Metalúrgicos do Sul Fluminense (Metalworkers' Union of Sul Fluminense – SMSF), conservative evangelical organisations and loose business-citizens' platforms increasingly took up the struggles of the *metalúrgicos* (metalworkers). Seeing their bourgeois lifestyle waning, the metalworkers were resentful towards the PT and the upwardly mobile poor with precarious jobs in retail, IT and service sectors or in the tertiarised manufacturing industry. They would often blame these 'new rich' for 'milking the system' either by having jobs on the side while also receiving benefits from Bolsa Família (Family Allowance) or by applying to the Minha Casa, Minha Vida (My House, My Life – MCMV) subsidised housing programme without qualifying for it. Given the racially segmented occupational structure of Volta Redonda – the black community is employed in precarious sectors or as subcontractors in the manufacturing industry – these off-the-cuff remarks about the new rich were intrinsically racist.

Paradoxically, right-wing President Jair Bolsonaro had a big following among the evangelical community in Volta Redonda, rooted predominantly in poor and black bairros (neighbourhoods). But his racists and homophobic remarks during the 2018 electoral campaign appealed mainly to the lower middle class of Volta Redonda – entrepreneurs, commercial families and the working class disillusioned with the PT – which voted for him (Mollona 2018b). Another section of the working class – informal and tertiarised workers in the service and building industries – faced a similarly harsh employer: the municipality of Volta Redonda and illegal contractors, who exploited their labour in the desperate attempt to survive in the shadow of the UPV. A third section of the working class – tertiarised industrial workers, IT workers and autoworkers –experienced first-hand the harsh tertiarisation regime by big national and foreign corporations. Their struggles against labour deregulation took the form of regional alliances (characterised by a mixture of business pragmatism) and social justice activism, strengthened by the legal principle of subsidiary responsibility.

Moreover, I show the trade union's factory-based struggles, the land activism of civic coalitions and the legal and the business activism of the new working class were different strategies of labour struggle happening at different state levels and reflecting historically and geographically diverse trajectories of capitalist development. These different working-class constituencies – the *metalúrgicos*, autoworkers, pensioners, cleaners, builders and other disenfranchised workers – were part of the same undifferentiated and magmatic centre ('Class C') that brought together conservative and progressive forces thus, complicating the traditional opposition between labour and capital, and creating a fragile political equilibrium that collapsed with the economic crisis.

Lulismo

At the beginning of my fieldwork in October 2008, I was reading *The Economist* at Heathrow Airport waiting to embark on the flight to Rio de Janeiro. The magazine's front cover showed the *Cristo Redentor* statue shooting into the sky like a rocket and the title 'Brazil is taking off'. A few months later, the news of the momentous financial crisis in the United States reached Brazil as some exotic epidemic the country seemed immunised from. Amidst global recession, the country was growing at a rate of 8 per cent, and the real currency was getting stronger, thus allowing millions of Brazilians to buy foreign white goods, travel worldwide and embrace the middle-class way of life.

In the G20 meeting in Washington, DC, that year, President Lula da Silva (2008) addressed the finance ministers and central governors from the Global North with a damning assessment:

> The crisis started in advanced economies. It is a result of the blind belief in the market's self-regulation capacity and, by and large, of the lack of control of the activities of financial agents. For many years, speculators have made excessive profits, investing money they did not have in miraculous business. We are all paying for this adventure. This system, just like a card castle, collapsed and killed the dogmatic faith in the principle of no government intervention in economies.

Indeed, because of state regulation, Brazilian pension funds had been relatively untouched by the financial crisis. In addition, the Lula government had instituted a 2 per cent tax on capital inflows and passed a law that earmarked 20 per cent of pension funds towards productive investments, especially infrastructure. Brazil had just won the bid to host the World Cup in 2014, and photos of Lula holding the trophy went viral on social media and in national newspapers. No doubt about it – I had arrived in Brazil at the apex of *lulismo* (Lulism).

Indeed, the working-class experience in Volta Redonda cannot be reduced to the political economy of neo-extractivism. The charismatic personality of ex-president Lula had a huge impact on working-class politics, and not a single day passed during my fieldwork without Lula being mentioned by friends and acquaintances. André Singer (2012: 14) describes *lulismo* as the encounter between a style of government (i.e. redistributive populism) and a class formation, the sub-proletariat. Lula's redistributive populism empowered the sub-proletariat and the rural masses – the historical social formation that emerged from slavery – while maintaining the power of the industrial and financial elites. Economically, it combined high interest rates, overvalued currency, programmes of poverty reduction and wage increases for the lowest section of the working class.

The PT's ascent to power and its consolidation up until 2013 was largely the result of two political shifts: (1) the overwhelming support by the poor communities in the rural Northeast and (2) the transformation of the top echelons of the trade union into a new financial class. The lumpen's overwhelming support reversed Brazil's historical trend. In the 1990s, the poorest section of society supported the neo-liberal policies of Cardoso and voted against Lula – being suspicious of the Left, particularly its disruptive strikes and liberal values, for instance, on abortion or women's and blacks' rights. Then, it was students, public functionaries and the educated middle class in the South that supported Lula. In the 2006 elections, the situation had reversed – 55 per cent of the super-poor and those up to two minimum wages had voted for Lula, whereas 65 per cent of the richest section of the population voted for the centre-right candidate of the Partido da Social-Democracia Brasileira (Brazilian Social Democracy Party – PSDB). Relying on the 'extremes' – the extremely poor and the financial elite – *lulismo* was intrinsically precarious (Braga and Purdy 2018; Singer 2012) and even 'schizophrenic' (Braga 2016).

Oliveira (2005) describes the incorporation of the financial elite within the state controlled by a working-class party 'hegemony in reverse'. He argues such a contradictory power bloc reflects Brazil's peripheral capitalism – a mixture of technologically advanced and precarised IT sector, outdated manufacturing industry and vast informal economy and precarious labour – which he describes as a 'duckbilled platypus' (half-mammal, half-bird). During his two administrations, Lula massively increased the minimum wage;[8] set up Bolsa Família, a radical programme of poverty reduction; democratised credit and popularised and 'moralised' stock market capitalism (Jardim 2009) developing financial and housing markets for the poor (Ricci 2010: 35–42).

Because of these anti-poverty and pro-labour economic policies, the population below the poverty line decreased from 28.5 per cent in 2003 to 16 per cent in 2008 (Fishlow 2011: 105), and the Brazilian middle class, or Class C, boomed.[9] In a 2012 interview with the *Financial Times*, Dilma Rousseff claimed: 'This, I think, is a very important gain for Brazil – that is, to transform itself into a middle-class population. We want this; we want a middle-class Brazil' (Leahy 2012). Indeed, since Brazil's PT took power in 2003, more than forty million Brazilians joined the middle class, including, in income terms, the working class.[10] As of 2015, Class C Brazilians accounted for 53 per cent of the country's population (Pearson 2015).

Yet, for Marcio Pochmann, the notion of 'Class C is a myth' (2012: 25). His statistical analysis shows the Brazilian 'middle class' barely makes it above the poverty threshold. Its occupations are as precarious as those of informal workers and the poor newly entered in the job market. For Pochmann, the Brazilian Class C is the increasingly marginalised and precarised working class – employed in call centres, fast food restaurants, banks or supermarkets with flexible and precarious contracts – which expanded the most during the second Lula administration. The precarisation of the middle-class is a global phenomenon (Banerjee and Duflo 2008), but according to Marcello Neri (2011), there is a special porosity between the lower middle class and the working class in Brazil. For instance, a 10 per cent increase in retail prices in 2013 hit the working class hard. Most working-class families struggled to pay for mortgages and for food and basic services. But the middle class was hit even more violently by the rise in the costs of services. The series of demonstrations that shook the Rousseff government in 2013 had a strong anti-inflation component shared by both the middle and working classes.

During the devastating economic crisis of 2015, 3.7 million people were displaced from Class C and entered the ranks of the new poor (Singer 2018a: 30). Today, the wealthier classes A and B – 33 per cent of Brazil's population – account for 60 per cent of spending growth in Brazil, while Class C accounts for only 33 per cent (Pearson 2015). In his first term, while pushing pro-labour policies, Lula continued to appease the financial community by keeping interest rates among the highest in the world (Braga and Purdy 2018)[11] and massively appreciating[12] the real, which increased industrial unemployment and boosted mass consumption among the poorer classes also facilitated by new forms of cheap credit (*crédito consignado*). Because of Lula's early pro-labour policies, the social state spending rose to 23 per cent of the GDP, which led him to cut welfare expenses and deregulate the labour market in his second term. From 2003 to 2005, the minimum salary grew by 6.8 per cent but by only 1.5 per cent from 2015 to 2016 (Carvalho 2018: 20). Moreover, about 94 per cent of the jobs created during the Lula and Rousseff administrations are low income (below 1.5 minimum wages) and concentrated in nonindustrial sectors such as retail, construction work, transport and general service (Pochmann 2012); 60 per cent of these involve young people (18–24 years old), mainly women and ethnic minorities. In 2012, low-income and casualised labour and informal and unremunerated work made up 55 per cent of the labour market.

The sociologist Ruy Braga (2016) describes *lulismo* in terms of a 'schizophrenic' dualism between policies that formalised employment and decreased occupational precariousness for informal workers and the lumpen on the one hand, and policies that increased the flexibilisation and precarisation of formal labour on the other hand. The precarisation of wage labour had a deleterious impact on the traditional working-class occupations, as demonstrated by the exponential rise of turnover rates and industrial accidents in traditional industries.[13] In terms of party structure, Lula put an end to the PT's consensus-based and horizontal structure, centralising it, bureaucratising it and excluding the grassroots movements that propelled it to power, especially the Movimento dos Trabalhadores Rurais sem Terra

(Landless Workers' Movement – MST) (Wainwright and Branford 2006) and transformed the trade union movement into a 'bourgeois state formation' (Braga 2016: 70) involved in private-public partnerships, pension fund management and corporate banking. Moreover, *lulismo* involved a specific discoursive shift. In his first term, Lula addressed his people (*povo*) as 'working class'. In his second term, he took the distance from his image of working-class leader and presented himself as a champion of the poor. Thus, discoursively, he turned the class conflict between left and right (the PT's traditional focus) into a conflict between poor and rich mediated through his personal charisma. Lula's transformation into a populist leader alienated the base of the PT, which traditionally opposed populism (Singer 2012).

Yet, in a country where the formal working class constitutes a small minority within a much broader dispossessed population, populism does not imply the veering away from class issues. Instead, it may mark the emergence of a new form of class articulation that extends to the more marginal constituencies (Singer 2018a: 20). In this sense, the historical legacy of Brazilian populism reflects the entanglement between industrial development and backwardness. Indeed, against the common claim that the Brazilian political system is extremely fragmented (Mainwaring 2018), Singer (2018a: 23) argues that at least since Vargas's *Estado Novo*, Brazilian politics has been characterised by a clear-cut bipartisanism between a conservative party representing a small middle-class elite and a liberal party representing Brazil's vast rural and urban underclass with a clientelist party of the rural elites acting as mediator. In the past forty years, such polarisation was between the PSDB and the socialist PT mediated by the clientelist centre-right party Movimento Democratico Brasileiro (Brazilian Democratic Movement – MDB). By incorporating the MDB inside his governmental coalition and luring the financial class, Lula depolarised his administration and stepped into the electoral ground of the 'the party of the rich'.

Starting from his personal trajectory – a poor *nordestino* (northeasterner) and ex-metalworker who became Brazil's president – Lula perfectly incarnated such polarisation between wealth and poverty, as well as the contradictions and limitations of any attempt to go beyond it. At the time of my fieldwork, inequality was decreasing nationally, and labour incomes were converging towards a median just above the poverty line. But such convergence hid a deep political instability. The upwardly mobile poor overwhelmingly supported the government, whereas the downwardly mobile casualised working class I describe in this book, and the impoverished precarious public workers (lecturers, bankers) and service workers (teleworkers) – the lower Class C – strongly opposed it.

In chapter 2, I show the wageworkers of CSN were largely disaffected with *lulismo* because they were badly hit by the labour deregulation and wage decline during the PT's second administration. A big portion of line leaders, operative workers and the affluent *mineiros*[14] of the operative departments, considered themselves Class C, had an individualistic and middle-class mentality that prioritised lifestyle over occupational identity and were strongly critical of the PT's turnaround from being a working-class party to being aligned with the financial elites. Among them, second-generation workers – now close to retirement – had supported Cardoso over Lula in the 1990s and were now affiliated with the conservative MDB and the Força Sindical (Union Force – FS) trade union confederation. These disaffected *metalúrgicos* had developed an anti-*lulista* stance and voted Bolsonaro against Fernando Haddad in the 2018 elections. Surprisingly, the leadership of the Metalworkers' Union supported Bolsonaro, too.

A second group of employees, consisting mostly of unskilled workers, identified with the populist tradition of the Partido Democrático Trabalhista (Democratic Labour Party – PDT), the working-class party funded by Leonel Brizola, to revive the legacy of Vargas's Partido Trabalhista Brasileiro (Brazilian Labour Party) particularly, its workerist, developmental and corporatist vision. Some skilled engineers and operatives identified with this vision, too, which they felt constituted the real core of the PT and which was endorsed more forcibly by Rousseff and Minister of Finance Guido Mantega after 2010. Third, younger workers who entered the company under the current 'authoritarian' regime were more politically militant against the company and the government and affiliated with the radical Partido Socialismo e Liberdade (Socialism and Liberty Party – PSOL), which was gaining support among precarised wageworkers, including university and schoolteachers, IT, service and retail workers. In 2005, disaffected with the PT's first term in power, the PSOL and the Partido Comunista Brasileiro (Brazilian Communist Party) broke away from the PT (Wainwright and Branford 2006).

The only working-class constituencies that supported the PT were the cleaners, informal workers and builders I describe in chapter 6 who, coming from extremely poor backgrounds, benefited from Lula's Bolsa Família and the regulation of outsourced labour pushed by the PT in the 2000s. It helped that the PT in Volta Redonda was represented by Cida Diogo, a charismatic general practitioner who led important environmental struggles against CSN in poor neighbourhoods for more than thirty years. Many top CSN managers supported Lula, too, for his imperialist expansion in Latin America, the militarisation of the national soil against foreign investors and the financialisation of the steel industry through the BNDES.

These diverse attitudes towards Lula – the *metalúrgicos* and middle managers criticising him, and precarious workers and top managers supporting him – reflected the broader economic scenario. CSN was losing money as a steelmaker – its losses were offset by profits in the mining division in Congonhas (Minas Gerais) and by financial returns– and cutting down investments and personnel in Volta Redonda. In parallel to the trajectory of deindustrialisation, new jobs were being created every day in the shopping centres, retail shops, private medical clinics, IT centres and call centres spreading all over steel city. These jobs were extremely precarious and short term, especially after President Michel Temer introduced the new 'interim' contract of employment in 2017. Paradoxically, the booming service and leisure industries were being fed with the household debts of the steelworkers. For sometimes it seemed precarious service workers had become affluent Class C, too. Endorsing the middle-class lifestyle of the metalworkers, they bought heavily subsidised cars, televisions and even homes.

Towards the end of my fieldwork, when the first signs of the collapse of the commodity boom emerged and CSN dismissed workers en masse, cleaners, tertiarised workers and residents of poor bairros were hit so hard by spiralling household debts, house repossessions and unemployment that they finally turned against the PT and the Metalworkers' Union, too. This is when the hegemony of the PT finally collapsed. At the national level, such re-articulation of impoverished lumpen and casualised and precarised wageworkers led to the extraordinary resurgence of working-class militancy in 2013, which combined 'the *political struggles* of the urban precariat in defence of their social right, and the *economic struggles* of the formal working class for better working conditions and salaries' (Braga 2016: 63; Mollona 2018a). In Volta Redonda, it was the most impoverished section of the working class – women cleaners, garbage collectors and tertiarised service workers – that led this struggle.

In chapter 3, I show the spatial dimension of class struggles in the context of the broader dynamics of financialisation and decentralisation of the economy started with democratic constitution in 1988, when municipalities became both indebted economic subjects and autonomous political subjects. The class struggles of the 1980s combined factory and urban activism around the 'land issue'. In the 1990s, these struggles polarised in the particularistic factory-based actions of the metalworkers on the one hand, and the Right to the City movement of the Volta Redonda municipality and marginal constituencies on the other. When the financialisation of the Brazilian economy hit the ground in Volta Redonda (from the 2000s onwards), the focus of class struggle shifted from the factory and the city into the regionalist economic and environmental platforms involving municipalities, trade unions, regional businesses, wealthy landowners, global carmakers and the *metalúrgicos*. In this context, the Sul Fluminense regional alliance acted both as a powerful anti-systemic force against the São Paulo region controlled by the ruling PT and as a new global capitalist network.

In chapter 4, I show how the Metalworkers' Union embodied the multifaceted dimension and deep contradictions of *lulismo*. Politically, most CSN workers and Metalworkers' Union leaders considered themselves 'more left-wing' than the ruling PT and openly criticised Lula for endorsing neo-liberalism. Yet, the union endorsed financial capitalism, too, operating like a financial and legal service provider while also being involved in social unionism and local activism. Subscribing to Lula's ideology of people's capitalism, the central branch of the Metalworkers' Union functioned as a transmission belt of national policies of democratisation of credit and financialisation of working-class households typically, through MCMV. Moreover, reflecting the depolarisation of class struggle taking place within the governmental coalition, the SMSF incorporated the centrist and populist PDT, MDB and FS inside its leadership and cut links with grassroots urban movements (neighbourhood associations, women's organisations, the MST) and with the precarised automotive, civic construction, service and informal sectors workers. In discoursive terms, trade union leaders followed *lulismo* in 'moralising capitalism' using the same workerist Christian language Lula inherited from his early Catholic militancy.

In chapter 5, I show how a large portion of CSN workers, especially those who did not belong to the aristocracy of labour, endorsed Cardoso's and Lula's vision of 'stock market capitalism' as an instrument of democratisation of the economy and bought stocks of the company, private pensions and bonds, thinking of themselves as capitalists rather than workers. On the other hand, those CSN employees who opposed privatisation and worker-led financial capitalism retreated into an increasingly productivist, sectionalist and factory-based stance. Throughout this book, I will explore these multiple dimensions of the working-class experience and fragmentation – between *lulismo* and anti-*lulismo*; anti-capitalism and imperialism; neighbourhood-based and factory activism; productive and speculative class identities and between nationalism and regionalism.

Fieldwork

The idea of doing fieldwork in Volta Redonda came while discussing my Sheffield research with a Brazilian colleague. We were both intrigued by the comparison between the two steel towns. Deindustrialisation in Sheffield had been so radical because Prime Minister Margaret Thatcher's programme of privatisation and industrial closures was meant to destroy the British working class, which the English bourgeoisie always held in great contempt, also because of its Darwinian view of the market economy as a healthy process of destruction of weaker forms (i.e. in its view, labour) and survival of stronger ones (i.e. capital). During my fieldwork in Sheffield, mainstream media and MPs described the ongoing closures of the steel plants with a fitting botanical metaphor – as a process of pruning of dead brunches and of mutilation for the purpose of social and economic regeneration aimed at developing a healthier post-industrial landscape.

My hypothesis was that Brazil lacked such extreme market and anti-labour ideologies because of the country's strong tradition of developmentalism and *trabalhismo*, as well as its already overextended labour reserve army, which could not be stretched further without breaking the system. Besides, because of the structural entanglement between finance, industry and landownership, Brazil's transition into post-Fordism was highly unlikely. But my comparison was not intended to highlight the differences between post-Fordism in the core and developmentalism in the periphery. Instead, I wanted to understand their articulations in space and time. As the world's capital of finance, the United Kingdom could afford to close down its less profitable and more polluting industries and buying commodities and semi-manufactured goods on the world market, whereas Brazil is stuck in its historical role of exporter of raw material and primary and semi-finished goods and as provider of cheap labour and land to foreign multinationals localised in the country.

Indeed, Brazil's relationship with dependent development vis-à-vis Britain started in 1808, when the British Navy escorted Prince Regent

João from Lisbon to Bahia, helping the Portuguese royal family to establish its court in the tropics, and continued through British control of the Brazilian cotton trade in early nineteenth century, the coffee and iron industries in the twentieth century, the national electricity infrastructure in the 1940s and through various forms of direct and indirect investments today. Besides, British Protestantism and the Victorian bourgeois values of self-interest and individualism were central in the 're-Europeanisation' of Brazil during the Old Republic (J. Souza 2017: 62). On the other hand, Sheffield and Volta Redonda can also be considered two similar 'world peripheries' where industry always remained subordinated to the interests of landed and financial elites. Indeed, after the global economic crisis of 2008, Sheffield was more of a world periphery than Volta Redonda – a forgotten land, abandoned by both finance and industry.

In 2008, when I arrived in Brazil, it seemed Lula's 'social-developmental' formula (Bresser Pereira 2017; Singer 2015, 2018a) – re-nationalisation of industries, controlled financialisation and social redistribution – constituted a tangible alternative to the deregulated capitalism in the Global North, which appeared to be on the verge of collapsing. Barack Obama's election to the US presidency, which I witnessed together with jubilant Brazilian colleagues at the Universidade Federal do Rio de Janeiro (Federal University of Rio de Janeiro), reinforced the impression that a new, post-neo-liberal order was emerging. The consolidation of the so-called pink tide in Latin America seemed to suggest this new global order had its epicentre in the Global South.

In Volta Redonda, I lived in bairro 60, a historical working-class bairro funded by CSN in 1946. My house was in road 60/2, bordering the vast Cicuta Forest and surrounded by lush hills where local residents hiding from CSN private police cultivated subsistence food (tomato, maize, courgettes, farofa) that made up a substantial part of their daily diet. This particular stretch of the road was poorer than the rest of the bairro and was inhabited for most of the day by children and women, while men nomadically travelled to and from their jobs. I hoped I could do in-depth ethnographic research inside the plant and even work there as I did in Sheffield. I hoped that in the tradition of the Italian operaist working-class enquiries, I could look at the relations between capital and labour at the point of production (Burawoy 1985). But, upon insistence of the management, my research inside the plant was initially limited to qualitative interviews with employees and managers twice a week, eight hours per day, mainly in the firm's productive departments. Towards the end of

Figure 0.3. The City Centre (© Massimiliano Mollona).

the fieldwork, I spent more time in informal conversations with the workers in the canteens, offices and break rooms.

During the rest of the week, I interviewed social activists, representatives from metalworkers' unions, cleaners and civic construction workers, community leaders, priests activists, labour lawyers, labour judges, council workers, teachers and leaders of bairros and residents associations. Subsequently, as my informants became friends, I was actively involved in legal campaigns, pamphleting and direct activism. Besides, I occasionally lectured in primary and secondary schools in poor bairros and favelas and organised weekly meetings with workers during which we discussed books, poetry and newspapers articles. I also met workers at the swimming pool of the workers' club, at Samba nights and at feijoadas (a traditional stew of beans and pork) at my friends' homes on Sundays.

Ten months into my fieldwork, I inexplicably started receiving death threats. In the following months, my mobility was severely restricted. In December 2009, a year into my fieldwork, following the advice of Brazilian friends and colleagues, I left Volta Redonda. When I left Brazil, the economy was booming. President Lula had just finished his mandate with an 83 per cent approval rating – the highest in Brazilian political history (Marinho 2010). Yet, the political crisis of *lulismo* had been tangible throughout my fieldwork. During my subsequent visits to Volta Redonda in 2011 and 2012, I experienced the slow onset of the economic crisis, as CSN started to restructure and lay-off people and unemployment grew across all sectors. But I could not imagine the scale of the crash in 2015 when the economy shrunk by 3.8 per cent; unemployment rose to 9.5 per cent (38 per cent up from the previous year), wiping out three million jobs from the labour market; and informal labour peaked from 40 per cent to 51 per cent. Neither could I fathom the coup against Rousseff in 2016, the jailing of Lula just before the start of the 2018 presidential elections and the election of Bolsonaro to the presidency.

My last visit to Volta Redonda took place in November 2018 on the day Bolsonaro won. During the visit, I was deeply troubled by the extent of the political and economic crisis in Volta Redonda. For instance, back in 2008, my fifty-five-year-old friend Gerardo; his wife, Lara; and their son, Cassio, led a comfortable lifestyle – which included two cars, Cassio's expensive motocross hobby, Gerardo's regular fishing trips to Mato Grosso and Lara's weekly physiotherapies – out of the pension he received from CSN, while Cassio apprenticed for a subcontractor of CSN and his fiancée, Jane, worked at a homeopathy clinic. In 2018, they had moved out of Volta Redonda (which had become too expensive) to the 'dormitory town'[15] of Pinheiral. Cassio had been unemployed for three years and had married Jane, who quit her job because of the high costs of commuting. The family now lived off the extra income generated by renting out the flat in Volta Redonda and off Gerardo's pension. Unsurprisingly, the whole family had shifted its political support from the PT to Bolsonaro in the 2018 presidential election.

Now under a conservative leadership, the Metalworkers' Union had supported Bolsonaro, too. Some of the historical leadership of social movements kept on struggling. For instance, Zezinho, the leader of the Movimento Ética na Política – Volta Redonda (Movement for Ethics in Politics – Volta Redonda) and of the Pastoral Operária (Workers' Pastoral) continued to organise weekly meetings in poor communities and to campaign against CSN on issues of employment, working conditions, industrial accidents and environmental pollution both in his parish and through his community radio. Beth, the leader of the women's movement, continued to raise consciousness among young women of poor bairros and to campaign against domestic violence; Renata, a nurse in the city council, kept on working on empowering young black women; Lincoln, the militant president of the planning department and a PT councillor in the 1990s, continued to be actively involved in legal proceedings against CSN's ownership of 25 per cent of urban land.

University lecturers continued to lead militant research on human rights, working conditions and labour activism against CSN and schoolteachers in poor bairros continued to use Paulo Freire's pedagogy despite its criminalisation under the new proposed Escola sem Partido (Nonpartisan School) legal framework.[16] A week after Bolsonaro became president, I took part in the event 'Thirty Years after the 1988 Strike' organised by the Department of Anthropology and Sociology of the Universidade Federal Fluminense (Federal Fluminense University). In the introductory speech in the packed lecture theatre, one of the lecturers said: 'We must remember the strike of 1988 not as we remember objects that end up in museums. We must take the memory of the strike with us every day and use it to understand what is happening in the present. In the light of the recent political election, this active sense of history is all the more important.' Given the heavy censorship that was falling upon Brazilian universities around the time of the elections, the lecturer's combative speech and lucid sense of history touched me.

Ten years earlier, in 2008, I had taken part in the twentieth commemoration of the strike, in a theatre in the Volta Redonda city centre. Back then, Brazil's economy and *lulismo* were riding high. But the mood of the commemoration was subdued and melancholic, as if the object of commemoration indeed pertained to a forever lost past. Why was the mood so combative and positive today? It has been argued that the military coup of 1964 was a reaction of the Brazilian bourgeoisie to the rise of labour strikes and activism that followed the formation of the *ligas camponesas* (peasant leagues) in 1961 and the Comando Geral dos Trabalhadores (General Command of Workers) in 1962, and that the 'new trade unionism' of the 1980s was a form of labour re-articulation against the military-industrial complex solidified during dictatorship. Likewise, *bolsonarismo* arguably marked the resurgence of such military-industrial complex in response to the labour conquests achieved (for better or for worse) under the PT administration.

Can it be that the new combative spirit I witnessed in Volta Redonda reflected the beginning of yet another working-class articulation? Can it be that labour is already organising itself, already coalescing around a new form after its violent crushing by capital? These broader questions opened up at the end of my fieldwork. Indeed, how does one relate the experience of fieldwork to the structural articulation of capitalism in space and time? With what temporal delay do global and national political and economic events (crises, coups, inflation hikes, electoral successes and radical privatisations) impact everyday life, and in turn, what is the cumulative effect of local actions on the broader context? What can one's fieldwork say or do about capitalism, the relationships between core and peripheries and the future of labour struggles?

Was the political and economic crisis I observed in Volta Redonda in 2009 the consequence of the global financial crisis of 2008? Or was it a local process of deindustrialisation triggered by regional extractivist forces? Or was I observing yet a different process – namely, the political crisis of *lulismo*? And was this a national trend? Or was it part of the regional crisis of the pink tide? Or of the global collapse of 'progressive neoliberalism'? (Brenner and Fraser 2017). Doing fieldwork involves juggling several different dimensions – seeing behind the visual evidence, thinking across different temporal and geographical scales and continuously testing and reifying one's theory, in my case Marxism, vis-à-vis the complex social and political entanglements of everyday life. For instance, it was easy to dismiss the anti-Lula talks of the Metalworkers' Union directors as mere rhetorical exercises of a bourgeoisified section of the working class that was itself a state agent. But missing from that judgement were the invisible tensions between national and local union branches, between the Metalworkers' Union and the CUT confederation, between social movements and the PT and between Metalworkers' Union's active role in factory negotiations and its corporatist dependence upon the state.

Likewise, from a cursory glance at CSN and considering its dramatic visual impact on the surrounding city, it was easy enough to describe the firm as an obsolete and polluting steelmaker and Volta Redonda as a typical 'rust-belt city'. But such observation missed the vast and invisible economic ramifications of CSN in the city – its army of subcontractors, tertiarised workers and suppliers – and its national and international linkages, interdependencies and cross-participations across mining, logistics, finance, infrastructure and construction that make Volta Redonda both a traditional industrial steel town and a global financial and logistical hub. Moreover, beyond the 'evident' environmental and cultural differences – the lush nature surrounding the plant, the mellow vibe of political demonstrations or the dilapidated architecture of poor bairros – is the working class of Volta Redonda not the same as the working class in Sheffield, with its everyday struggles, precarious existence and enduring solidarities?

Doing fieldwork through a political economy framework entails a constant suspension of belief and of sensory experience, in the light of what happened 'before' and 'elsewhere' and for reorienting our actions and social relations in the present. In such combination of physical and analytical displacement, relational openness and self-discovery fieldwork, more than a methodology, is a political praxis. Unlike in Sheffield, where my fieldwork was immersive and I managed to 'blend in', I was always perceived, for better or worse, as a foreign researcher in Volta Redonda, perhaps also because of the defensive posture of the Brazilian working class at the time. I also struggled to get used to the claustrophobic atmosphere of the city and to the tentacular extensions of the plant – the fumes, the noises, the CCTV cameras and the private police – which seemed to follow me everywhere.

Yet, I was slowly captured by and immersed in the magical and surreal experience of living in the Brazilian steel town, walking and cycling in the shadow of the gigantic silhouette of the UPV, playing football by the carcinogenic mountains of coal dust, breathing benzene while jogging across misty yellow smokes, fishing diseased fish, shopping for groceries amidst loud furnace explosions and watchful CCTV cameras and drinking coffee in homes with no roofs, electricity or sanitation. It is this resilience of the families of Volta Redonda that my ethnography seeks to interrogate and to celebrate as a way to explore the condition of the working class in the twenty-first century. In Volta Redonda, I became absorbed in the reading of *Brazil: Land of the Future,* written by the Jewish Viennese poet and writer Stefan Zweig (1941) and published in Brazil the year of CSN's founding.

In the book, Zweig describes Brazil and Europe as being at the opposite ends of time and history. With its concern with history, tradition and authenticity, Europe is stuck in the past. In Brazil there is no history, origin or tradition – only an 'ongoing transformation of everything'. Transformation is so ubiquitous that there is no standard time, and 'hours have more minutes than in Europe' (1941: 57). The future constantly leaks into the present. Moreover, unlike European and American capitalism, which create 'dark, homogenous and unnumbered masses' and where the cultural and political homogenisation of the working class generates class conflict, the Brazilian working class is fluid, colourful and heterogeneous – 'it expands outside the factories to include the *caboclos* of the Amazon, the *seringueiros* in the forest, the *raqueiros* in the plains and the *indios* in the impenetrable jungle' (133).

Zweig and Vargas met several times during Zweig's stay in Brazil. From their diaries, we know they liked each other, for they both considered the *Estado Novo* the future of decaying Europe. Vargas was particularly obsessed by the spectre of European communism and Zweig by that of Nazism. Their shared vision reveals the entanglement between industrial modernity in the Global South and the crisis and revival of the European project, played out in the phantasmatic space where exilic intellectuals, authoritarian dictators and the spectres/fetishes of the communist proletariat meet. With the growth of National Socialism, Zweig left Austria in the 1930s and, after a long stay in England in 1940, settled down in Petrópolis (not far from Volta Redonda), where Pedro II had established the royal court in the nineteenth century. In Petrópolis, Zweig (1943) wrote *The World of Yesterday*, an emotional memoir of the collapse of the Hapsburg Empire and of *Mitteleuropa*, with its cosmopolitan intellectuals, open cafe culture and international solidarities, because of the ascent of Nazism and the prospect of the Second World War. In Petrópolis, haunted by depression, Zweig committed suicide with his second wife, Lotte Altmann, in 1942.

As my fieldwork proceeded, I was struck by how Zweig 's exilic life resonated with my own experience. I too had abandoned old Europe at a moment of deep economic and political crisis, which in Italy had triggered violent right-wing movements, the ascent to power of Berlusconi and the multiplication of Mafia terrorist attacks on civilians. I too had moved first to the United Kingdom, lured by the hopes generated by the New Labour party of Tony Blair, and then to Brazil, following the trail of the 'new' Latin American socialism, embodied in the charismatic figure of Lula da Silva – whom several of my interlocutors called 'the new Vargas'. Like Zweig, I too was hunted down and exhausted from chasing a future that never materialised and from the disappointment and depression generated by these crushed hopes. But as my fieldwork unfolded, I appreciated that the melancholic exilic space I was experiencing was not neutral. Instead, in it, my feelings and expectations for Brazil were partially implicated with a neocolonial gaze projecting Western ways of thinking and feeling onto the South. Aware of the affective drives and contradictory positionality in which it was produced, *Brazilian Steel Town* is also a silent reflection on my own past and on my hopes for the future.

Notes

1. Despite the changes introduced under the Temer administration in 2017.
2. Viana (1955: 58) argues Vargas's workerist legislation led to a peaceful social revolution by offering three types of inclusions: within the firm, through the promise of stable employment; within the state, through compulsory union participation and corporatist decision-making structures; and within the social welfare.
3. The sociologist Ching Kwan Lee (2007: 12) makes a similar case in her analysis of labour protest in contemporary China.
4. For a discussion of the performance of military-trained managers in Brazil, see Musacchio and Lazzarini (2014: 133–140).
5. In 2017, the construction firm was at the centre of a major corruption scandal which led to the impeachment of both Brazilian President Dilma Rousseff and of Peruvian President Pedro Pablo Kuczynski and reverberated throughout Latin America
6. 'A committee of the federal senate in Brazil approved a Constitutional Amendment Proposal (PEC 65/2012) stating that no project should be suspected or cancelled once an [Environmental Impact Assessment] has been submitted. The proposal has been returned to the committee, but it could be released for a vote of the full senate at any time' (Ritter et al. 2017: 165). If approved, major infrastructure projects will be able to proceed regardless of their impacts on biodiversity, indigenous areas, traditional communities and conservation areas (Rocha 2016).
7. While the minimum wages at the low-end of the occupational scale (up to five minimum wages) had doubled during the Lula administration, theirs barely kept above inflation levels.
8. The minimum wage rose by 23 per cent during the first Lula administration. Rising wages accounted for 60 per cent of the fall in the Gini coefficient from 2001 to 2011. According to the Institute of Applied Economic Research (IPEA 2012), the Gini coefficient fell to 0.527 in 2011 from 0.595 in 2001. From 2001 to 2011, the income of the poorest 10 per cent of the population grew by 91 per cent, while that of the richest 10 per cent by just 16.6 per cent.
9. In Brazil, social classes are marked by letters and based on the household's gross monthly income. For instance, the Instituto Brasileiro de Geografica e Estatística (Brazilian Institute of Geography and Statistics) defines Class A as above R$15,760;

Class B, R$7,880; Class C, R$3,152; Class D, R$1,576; and Class E as below R$1,576 (Fujikawa Nes 2016). The relationship between income and classes is also represented in terms of the national minimum salary, which is currently R$998 per month (as of April 2019).

10. Occupationally, Class A is composed of bankers, investors, business owners, major landowners, etc.; Class C, people who provide services directly to the wealthier groups (teachers, managers, mechanics, electricians, nurses); Class D, people who provide services to Class C (housemaids, bartenders, bricklayers, civil construction workers, small store owners, low-paid drivers); Class E, people who earn minimum salaries (cleaners, street sweepers, unemployed people).
11. Almost half the revenue was used to pay off national debt, and Brazil's four-largest banks earned profits in 2013 that surpassed the GDP of eighty-three countries.
12. R$3.95 to $1 in December 2005; R$1.65 in December 2010 (Bresser-Pereira 2017: 312).
13. Braga (2016: 75) estimates the rate of tertiarisation of labour increased by 13 per cent per year from 1996 to 2010. In 2013, 12.7 million workers were tertiarised.
14. This appellate specifically refers to workers with provenience from the state of Minas Gerais.
15. The residents consist mainly of retired CSN employees who were laid off in the 1990s.
16. If passed, the law will illegalise discussing 'political subjects' in primary schools and universities.

– Chapter 1 –

Capital Enclosures, Labour Abstraction and the Struggle over Value Forms

During my first days of fieldwork in Volta Redonda, I joined a guided tour of the Usina Presidente Vargas. While our group of visitors walked through the smelting shop, a smelter opened the furnace door. We all stopped and stared at the red magma boiling inside it – thick, formless, multicoloured and sinuously moving. 'It's magic, isn't it?' the tour leader said. Mesmerised, we all agreed in silence. That sight reminded me of how, for the philosopher Castoriadis, the magma is the metaphor for human collectives, which are interconnected and, at the same time, autonomously self-organised. Unlike the autonomy of non-human species, which is based on closed self-organisations, human autonomy involves an ontological opening and the potential for radical ruptures. In fact, in their quest for freedom and autonomy, humans 'continuously question their own institutions and significations' (Castoriadis 1994: 23) and can destroy anytime the principles governing their lives. Castoriadis argues: 'Democracy is the only tragic political regime – it is the only regime that *takes risks*, that faces openly the possibility of self-destruction. But human autonomy has 'always been occulted and covered by the representation, itself highly instituted, of an extra-social force (the gods, the ancestors, or "reason", "Nature")' (27). These institutionalised ideologies lock humans in a state of permanent closure.

For Castoriadis, the metaphor of the magma prefigures revolutionary societies – based on forms that are also formless, closures that remain open, collectives sharing autonomy and a movement that never ceases or disperses. The magma is the threshold between movement and stasis – between the flow that erases differences and liberates people, and the institutions that protect and occupy their spaces and freeze their development. The philosopher George Simmel (1918) describes the human condition as a process of 'life transcendence' involving a similar magmatic process. Non-human mediums – written texts, images or money – allow us to step outside the physical and mental boundaries of our bodies, to see ourselves from the outside and hence to experience finitude. But with time, these non-human mediums become ossified substitutes for social relations, circulating independently from them and blocking the flow of life. Thus, for Simmel, the process of self-transcendence also implies the construction of new boundaries and the abstraction, objectification and externalisation of life into autonomous and rigid forms. From the human point of view, such a movement of abstraction is experienced as progressive dematerialisation, stasis and impotence.

Under normal circumstances, life breaks all human-made moulds and remerges as flow. But under capitalism, according to Simmel, the dematerialisation and ossification of life reaches its apex because of the social technology of money. Instead of being a simple mediator of value, capitalist money is valuable in itself and circulates independently from social relations taking up a life of its own. While money is subjectified and made alive, humans are in turn objectified, abstracted, frozen and turned into formal equivalences of wealth and hollow vessels of labour. Consequently, for Simmel, modern capitalist metropoles have an eerie quality, as they are inhabited by ghost people who, following the flow of money, circulate in the form of flat cinematic images, flickering surfaces and fragmented, isolated and competitive individuals. Thus, as I was staring at the furnace during the guided tour, I asked myself: 'Is steelmaking so mesmeric because the animacy of steel and its internal tensions between flow and form prefigure, like Castoriadis's magma, human autonomy and freedom? Is not the labour of the smelters so magical because of how it manipulates liquid and formless matter to produce the solid objects and infrastructures that make up our everyday environment?' Indeed, the mesmeric, hypnotic, spirited and enchanting movements of steel in the furnace seemed to me to embody the movement of labour and prefigure its potential for emancipation.

In a polemical exchange with Toni Negri, the philosopher Giorgio Agamben (2005) argues that left-wing scholars' (ab)use of the term 'movement' betrays their lack of trust in democratic institutions including the role of 'the people'. Likewise, this book looks at the tensions and contradictions between capitalist institutions – including states, trade unions, factories, money and finance – and the movement of labour. On one hand, capital aims at freezing labour in the fetish forms of land, machines and money. Confronting it, labour resists reification, occupation and enclosures and strives to overflow them and reproduce life. I am interested in introducing the notion of movement into political economy. What can the animacy of steel tell us about work under capitalism and life beyond it? How can it describe the unevenness of labour (Kasmir and Gill 2016) and its motion – around, across and beyond the spatial, temporal and institutional fixes and enclosures of capital?

Capital as Enclosure

The edifice of capitalism rests on the institution of private property – a state-sanctioned enclosure of common goods (Polanyi [1944] 2001) that produces 'scarcity', and the market system of valorisation and distribution. For Rosa Luxemburg, capitalist enclosures are always associated with imperialist processes of dispossession and annihilation of 'the other'. Capitalists tend to justify their appropriation of common goods and of labour through the ideologies of equality, self-interest, freedom and social progress. Yet, central to capitalism is the bourgeoisie's sense of entitlement to lead a parasitical life at the expenses of the marginal classes. Indeed, for the American sociologist Immanuel Wallerstein (1983), the historical function of the bourgeoisie is not to increase profits but rather to live off rents and enclosures, like the aristocrats did before it. It is the process of working-class proletarianisation – achieved despite and not because of the bourgeoisie – that forced the parasitical classes to act like business people. Here, I will develop an anthropological framework for the study of contemporary capitalism intended as a form of enclosure – of land, knowledge and labour. I use the Marxian mode of production (MOP) framework adopted by the anthropologist William Roseberry to focus on capitalism's historical and contemporary entanglement with non-capitalist social relations such as kinship, slavery and state despotism.

Modes of production have two components: factors of production (land, labour and tools) are the material support structure for humans' actions, and relations of production are the specific social forms (slavery, serfdom, wagework) taken by these actions. MOPs are historically specific forms of labour organisation and corresponding forms of consciousness – they show the historical interplay between material existence (infrastructure) and forms of imagination (superstructure). Marx's MOP framework has been criticised for being Eurocentric and dismissing the cultural specificity of subaltern struggles (Chakrabarty 2000; Chalcraft 2005: 15).[1] At the other extreme, idealist anthropologists and anarchist scholars have used MOP analysis to describe the 'original' communism and affluence of indigenous and non-Western communities (Clastres 1974; Wengrow and Graeber 2015; Overing Kaplan 1975; Sahlins 1972). Unlike these scholars, I used, in line with the anthropological structural-Marxism of the 1970s, the MOP framework comparatively to look at the articulations between capitalism and non-capitalism. For instance, Emanuel Terray (1972: 22) shows that kinship societies are stratified along lines of class, as elder kinsmen control the reproductive labour of women, and junior kinsmen through various kinship obligations (virilocal residence, patrilineality, exogamy) rather than in monetary terms.[2]

Claude Meillassoux goes even further by arguing that capitalist and domestic MOPs – productive and reproductive labour – are structurally entangled both among traditional pastoral communities and in contemporary Western societies. In both contexts, wage contracts deprive workers of their means of subsistence and make them dependent both on domestic labour and on commodity markets. After the French colonisation of Guroland (Ivory Coast), foreign food and agricultural companies broke the local self-subsistence agriculture, employing local peasants as wageworkers and underpaying them, thus forcing them to rely on the domestic labour of their kin for survival. As well as incorporating the family system into the capitalist system, colonial industrialisation created masses of dispossessed peasant proletarians who migrated circularly between factories and villages to survive. For Meillassoux (1964: 58) the forced circular migration of the Guro people is akin to that of the impoverished rural workers in early capitalist Europe. For the French anthropologist, labour mobility is a central prerequisite for capitalist accumulation in the centre, too. For instance, he presents ethnographic evidence of how a big share of the workforce of a French multinational corporation consisted of underpaid African workers on short-term and precarious contracts and surviving on domestic and informal labour.

Moreover, anticipating the reality of contemporary flexible capitalism, Meillassoux (1981: 92) highlights how South-to-North circular migration of wageworkers is central to the functioning of capitalism. North African migrant workers employed in the French company were paid three times less in indirect wages than French workers, and this wage differential was justified through racialised categories of skill and seniority. Because of their marginal position in the labour market, impoverished ethnic workers returned to their villages periodically for survival. Indeed, the management of transnational circular migration, through regularised temporary programmes that feed the demand for flexible labour in developed countries, is a central mechanism of accumulation in late capitalism (Kalm 2010: 37). In other words, Meillassoux reframes Marx's argument of the fragmentation between wageworkers and lumpen, as a tension between productive and reproductive labour within the broader context of dependent development. Combining anthropology and political economy, Meillassoux shows capitalism cannot be separated form imperialism, intended as brutal movement of colonisation and occupation of the sphere of reproduction, and in fact is structurally but unevenly related to underdevelopment in those parts of the world where extraction of reproductive labour is possible.

Thus, Meillassoux reworks anthropologically Trotsky's notion of 'combined and uneven development', unveiling ethnographically the historical, spatial and ideological oscillatory entanglements between production and reproduction. Moreover, in his book on African slavery, Meillassoux (1986) shows how slavery, capitalism and war form a unique and interconnected economic system. Thus, the French articulationist school shows that, in addition to class relations being reproduced in non-capitalist societies, capitalism is *internally* constructed through the non-capitalist relations of kinship, war and slavery. Adopting the structuralist perspective of Claude Lévi Strauss, this Marxist anthropological school considers the tension between infrastructure and superstructure – production and social reproduction – both historically and diachronically, that is, through cosmologies, myths and ideologies associated with specific forms of livelihood and recurring across cultures (see Lévi-Strauss et al. 1976). Indeed, I will argue the fetish forms of machines, land and money cut across different geographical and historical contexts.

Lastly, French structural Marxism put an end to the long dispute between formalist and substantivist anthropology around the issue of 'economic rationality' – the former arguing that rationality and self-interest are universal human features, and the latter that non-capitalist economies are socially embedded and geared towards the basic reproduction of life rather than towards growth and profit. Reworking the notion of 'bounded rationality' formulated by the Nobel Prize–winning economist Herbert Simon, the anthropologist Maurice Godelier (1966) argues all societies – whether ruled by kinship, state violence or economics – mix conscious and unconscious, selfish and social, and rational and irrational drives, evaluations and motivations. Such a focus on the entanglement between social reproduction and capitalism is central to feminist theories of labour value, too. Silvia Federici (2012) in particular argues the reproductive and affective labour – vastly performed by women – is systematically devalued and exploited under capitalism and that the sphere of social reproduction is one of the many human commons that is constantly under threat of occupation by capital (De Angelis 2018).

The economic historian Thorstein Veblen shows that industrial capitalism marked a shift in the form of capital's enclosures – from land to labour. Veblen ([1904] 2005: 56) describes modern business enterprises as legalised forms of predation, theft, sabotage and waste that combine the speculative logic of capitalism and the predatory logic of feudalism. In economic terms, business enterprises do not produce wealth by maximising outputs, as per traditional economic theory. Instead, they are moneymaking endeavours that profit from the enclosure and waste of human labour. In this light, the wage contract is a form of rent extraction achieved by enclosing and commodifying labour. In particular, modern factories (1) deskill workers and externalise their knowledge into machines, thus making them dependent on the market for their reproduction, and (2) embed the logic of speculation and finance within the production system. In other words, the abstraction of labour under capitalism happens first by externalising human knowledge and skills into the technological apparatus of the factory and then by creating a market for it, independent from human value. In experiential terms, business capitalism is predatory because it dispossesses humans of their 'instincts of workmanship' ('to turn things into human use') and 'care for the other', and forces them into permanent competition and emulation of the wealthy 'leisure classes' (Veblen [1914] 2006a: 23).

Moreover, Veblen anticipates current debates on how finance is parasitical on industry (Marazzi 2009; Vercellone 2013). The bigger and more complex industrial production, the more human knowledge is externalised into the firms' technological system controlled by a new class of 'technical experts'. As technical systems become more complex and valuable, technicians morph into 'consulting engineers' who audit the value of machines and advise the bankers involved in the underwriting of the company (Veblen [1904] 2005: 68). The main objective of the new financial class – consulting engineers, business managers and corporate bankers – is to inflate the monetary value of firms rather than make them productive. Under such a regime, corporate profits come not from production but from financial speculation and leverage on the assets of firms mainly machines. In fact, the financial classes increase their profits by systematically obstructing production and deskilling workers (Veblen [1899] 1994: 10). But with the deskilling of workers, the knowledge of how to run the machines is lost, so business capitalists are 'out of touch', 'incompetent' and 'absentees' (15). As the money value of the firm is invested in the financial market and returns on investments overshadow industrial profits, the capitalisation of the firm becomes independent from the productive cycle ([1904] 2005: 27). The more the industrial system expands, the more financial capital spreads and multiply itself, in the form of loans, securities and collateral debts, capturing the main state infrastructures – railways, iron mills and main roads (10).

As a result, the financial class controls industrial monopolies and state infrastructures.

In other words, finance accelerates the repression, occupation and waste associated with capitalist markets[3] and entangles it within the state – making it work on a logistical and infrastructural level. Besides, the economy of war of modern nation-states creates new forms of finance and speculative bubbles that redistribute wealth from the industrial to the financial class accentuating existing inequalities. Exponents of new Marxian labour theory argue abstract labour is a transmutation of the money form, which hence both predates and follows production (Bellofiore 2009; Pitts 2018). With Veblen, we can appreciate how such a transmutation of labour into the money form takes place through the technological apparatus of production. Lenin ([1910] 2010) and Hobson ([1904] 2011) – contemporaries of Veblen – also famously discussed the relationship between finance, industry and imperialism. But unlike them, Veblen emphasises the internal and national dimension of imperialism. For instance, the early German state systematically opposed the emergence of capitalist markets and private property, as these undermined its monopoly over the economy. Following the 'warlike logic' of the Prussian empire, the German state systematically sabotaged and inhibited national economic development to make citizens dependent on it (Veblen [1915] 2006b: 262). The German nation-state controlled its 'community' not only through corporate plunder but also through forms of taxation and redistribution that required heavy and efficient bureaucracies. This predatory form of state imperialism is central to my analysis of the Brazilian state.

Veblen's argument that states and firms are part of the same infrastructure of capitalism runs counter to the bourgeois ideology that the economy – both as field of knowledge and as form of human behaviour – exists independently from social relations. In fact, the fictitious separation between the economical and the political functions of the state started with the legal invention of the 'corporate person' in Europe in the twelfth and thirteenth centuries. Corporate persons were organisations such as charities, religious communities and universities, to which the Crown granted by royal charter semi-independent status, privileges and immunities (Hodgson 2015: 225). Initially used by the Crown to raise revenues by selling privileges to specific groups, the chartered form eventually became a way to regulate the economic behaviour of state subjects through highly moralised and codified agreements. In the sixteenth and seventeenth centuries, European monarchs used chartered companies such as the East India Company (1600), the Royal African Company (1660) and the Hudson's Bay Company (1670) to plunder and police the territories of the colonies and freely operate in the Atlantic slave trade. In these territories, corporations acquired extra-legal status being legally allowed to use force to maintain economic supremacy.

As the exploitation of the colonies intensified, the need for further capital concentration and separation between economies and states was achieved with the new legal form of the joint-stock company. The Limited Liability Act of 1855 granted full juridical personality to corporations, de facto granting them political immunity. Thus, the modern corporation emerged as extra-legal extension of state power, which allowed modern sovereigns to discipline their economic subjects through a mixture of violence, loyalty and the impersonal law of the market. For the legal historian Joshua Barkan (2013), this progressive personification of the corporation runs in parallel with the depersonalisation of the Crown and the birth of the Hobbesian state as 'social contract'. Thus, modern corporations emerged from the historical transformation of colonial mercantilism into industrial capitalism fuelled by the imperialist movements of nation states and from the entanglement of slavery, state violence and market economy.[4] Indeed, state bureaucracies and corporations are 'two sides' of the same national infrastructure of dispossession, occupation and destruction of physical and human resources underpinning the condition of 'generalised competition' of capitalist markets.

Indeed, for institutionalist economics, *bureaucracies* and *corporations* are similar forms of organisation of human labour, which involve different 'transactions costs' from market-based forms of human organisations (Coase 1937; Williamson 1981). *Markets* are the most efficient forms to organise standardised labour whose value is easy to establish – for instance, end-of-the-line packing. But those activities that are difficult to assess – for instance, an engineer's report –are best kept inside bureaucracies, where efficient mechanisms of monitoring, assessment and sanctioning exist (Williamson 1981: 567). In other words, corporations emerge in response to opaque environments, actions that are difficult to evaluate and work that cannot be easily coordinated. In the market, human actions have a clear economic value attached to them, and there is no need for co-operation. Market relations are impersonal, free and often immediate or 'spot-like', such as when we buy produce in grocery stores. They involve competitive relationships between employers (the agent) and workers (the contractors) that can be entirely mediated by the price mechanism. Unlike competitive market relationships, bureaucracies (and firms) are based on workers' co-operation. Co-operation can be achieved either through coercion and direct control or by generating a common corporate culture and loyalty towards the firm. In both cases, co-operation is costly to set up and maintain.

According to the institutionalist economist Albert Hirschman (1972), bureaucracies and markets are mutually entangled. As firms grow, they develop monopolistic powers and internal labour markets, which allows them to cut labour costs, fire and replace workers. But the more firms grow, the greater their employees' bargaining power – and hence their ability to *voice* their discontent towards the company and eventually *exit* resorting to external labour markets. Thus, monopolies have high political costs, and efficient labour markets eventually force firms to downsize (Coase 1937). Indeed, in early corporate debates, markets were considered democratic because they provided an *exit* option against the vicious circle of labour subordination, deskilling and despotism of big corporations and states (see also Anderson 2016; Hayek 1944). The economist William Ouchi (1980: 130) adds a third institutional mechanism of labour coordination: the *clan*. When labour is extremely complex, skilled and difficult to assess and the objectives of the employers and those of the employees are congruent, both market and hierarchies are inadequate. The best way to coordinate skilled workers is to allow them to self-organise according to their own 'culture' and through non-hierarchical and informal mechanisms. High-tech workers and highly paid professionals especially tend to develop fluid and egalitarian organisational structures that run along the principles of trust and solidarity and where 'the individual' and 'the collective' are blurred, as in 'primitive societies' or among 'utopian communities' (140).

Ouchi's clan culture is the same 'informal shop floor culture' Michael Burawoy (1985) identified in the Allied engineering shop floor in the 1980s, consisting of informal knowledge and 'games of production', which are not part of the wage contract but through which the workers internalise the wage relation. The sociologists John Meyer and Brian Rowan (1977: 351) go even further. They describe firms as creators of corporate myths, ceremonial rules and production rituals and argue that, especially in highly skilled and competitive sectors, firms survive through such mythopoeic activities. In fact, management scholars in the 1980s veered away from the rational economy paradigm and described firms as 'anarchies' (Weick 1976), 'rationally bounded' (Simon 1991), self-referential (Tversky and Kahneman 1986) and driven by a statistical wisdom that 'has the same function that magic and divination play in primitive society' (Devons 1961: 135).

Such recognition within the management science of the 1980s – that firms are driven by irrational, informal, ritualistic, symbolical and even magical world views and operate in a largely unknown environment – reflected the global reorganisation of production, also known as Toyotism, that took place in the 1980s when the classical Fordist production line was replaced by the model of flexible production based on extreme subcontracting. Subcontracting complicated the traditional economic wisdom of the time. It is a space located in between the market and the firm, so it transcends the logics of both. The economist Robert Eccles (1981: 10) describes organisations that rely on extreme subcontracting – for instance, in the construction and high-tech industries – as 'quasi-firms'. The services exchanged by these organisations are too complex and interdependent to be entirely left to the market yet too spread out and diversified to be dealt with internally.

Quasi-firms are inherently unstable. Main contractors struggle to find the right outsourcing partners and to remain independent from them by not losing expertise. Moreover, embedded in complex webs of interdependent activities, they find it difficult to delimit their core business. On the one hand, subcontracting commodifies and makes labour abstract, impersonal and precarious. But, on the other hand, it shifts the valorisation process away from the direct control of the capitalists. Thus, although subcontracting is normally considered a form of precarious labour, it can also be seen as a form of precarious capital. In chapter 6, I discuss an instance in which subcontracting generates precarious capital and enhances the political struggles of a marginal category of workers located 'at the periphery' of the wage system. In other words, the discovery of the creative, collective, unconscious and irrational dimension of the economy reflected capital's temporary loss of control over labour after Fordism's crisis and the multiplication of labour generated by flexible production and extreme subcontracting. Instead of working through its mix of repressive bureaucracies and free market (like the Veblen's business enterprise), post-Fordism co-opts the ritualistic, horizontal, affective, informal and collective relationships into its new factory regime – which Burawoy (1985) calls 'hegemonic' and Luc Boltanski and Eve Chiapello (2007) describe as the 'new spirit of capitalism' – mimicking the logic of the 'non-economic' institutions of families, communities and clans. But this latest form of labour co-optation has thrown capital into a precarious condition, too.

In the same way early capitalism enclosed land and commodities and turned them into abstract value, the wage system of industrial capitalism enclosed, abstracted and speculated on human co-operation, collective knowledge and labour commons. The collapse of Fordism and the colonisation of life by capital – the emergence of the social factory (Negri 1991) – increased the permeability and co-dependency between capital and human co-operation or commons – both the processes of capital's capture of commons, and the insurgence of commons against it. With the collapse of the coercive production apparatus of the factory and the ideology of free market, it has become evident that capitalism cannot function without human co-operation. Indeed, the current trend of labour re-internalisation, economic protectionism and state developmentalism shows the terminal crisis of the Polanyian compass between profit and social redistribution (Streeck 2016) and the return of state violence and coercion as the main mechanisms of economic coordination. Economists tend to study corporations in isolation from the state or, at best, in terms of their 'isomorphism' to it, but anthropologists rarely consider the intersection of states and corporations. In this book, I look at the capitalist mode of production as a 'total social fact' – that is, in the uneven and heterogeneous entanglements between state bureaucracies, corporate structures, kinship, market ideologies and unfree labour.

In relation to the Marxist framework I have laid out, if capital/labour relations overflow the boundaries of factories and cut across families, communities, states and markets – taking up productive, distributive and redistributive functions – then what counts as labour? Where does its value originate? Where do we draw the line, for instance, between productive, unproductive and reproductive labour or between wagework, bonded and co-operative labour? Reframing the above institutionalist literature through a Marxist framework,

I suggest thinking about modes of production as composite forms of livelihood combining despotic, capitalist and communist modes, each associated with specific practical, ideological and affective constructs but open to change and contestation. Within such an entanglement of capitalism and non-capitalism, processes of value creation span across different domains, kinds of actions and institutions. It is precisely in this contestation, negotiation and 'social validation' (Heinrich 2012) of the meanings and forms of human actions and of the boundaries of institutions – through force, economics or the law – that the terrain of struggle between capital and labour takes place.

Abstract Labour

Adopting a Gramscian framework, the historian Dipesh Chakrabarty (2000) argues the Marxian notion of 'abstract labour' is grounded in Western teleological history and in the Eurocentric notion of personhood – as homogenous, self-contained and individualistic – and misses the diverse, heterogeneous and living essence of labour that always exists in between 'bare life' and abstraction, and hence resists commodification, homogenisation and subsumption. Likewise, in this work, I emphasise the 'liveness' of labour, which more often than not exists outside the realm of wage labour (Van der Linden and Lucassen 2012: 6) and is engaged in 'minor' forms of resistance and struggles against abstraction. These struggles are as much cultural as they are material. To emphasise the liveness and agency of labour means to acknowledge that capital does not always make people. People make capital, too. Capital emerges from the stories of 'big men' (Godelier 1986), the rituals of enrichment of the peasants of the Cauca Valley (Taussig 1980), the anti-establishment attitude of working-class kids (Willis 1978) or the 'games of production' of wage-workers (Burawoy 1985). People produce capital every time they turn women into wives, oceans into farms and machines into dangerous monsters that take away their labour. Besides, because of people, some things or persons cannot be commodified and alienated – wives (Dalton 1965), ritual work (Parry 1986), gifts (Mauss [1950] 2002), heirlooms (Weiner 1992), land (Smith 1991) or the labour of children and enslaved people (Miers and Kopytoff 1977).

In this book, I show 'capital' in Volta Redonda is made by the nationalism and patriarchy of metalworkers, the sanitised ecological activism of middle-class professionals, the business corporatism of 'the new working class', the legalism and managerialism of trade union cadres and the narratives of spiritual salvation of evangelical leaders. At the same time, capital is 'unmade' by the practices of solidarity, commoning and sharing of women, cleaners and subcontracted workers. This unmaking of capital by labour happens not through acts of cultural refusal but through theories of action emerging from material struggles and historically contingent relations of production. The struggle between capital and labour is first a struggle around the definition of value starting from that of labour. The notion of labour applies to virtually all human activities – material and immaterial – through which societies reproduce themselves: the physical labour of steelworkers or of parents who raise their children; the immaterial labour of employees in call centres and IT firms; the production of symbols, stories and ideologies by priests, shamans and politicians; and the 'affective' labour of nannies, nurses and maids.

Human actions can take two distinctive value forms: one productive and one reproductive. Reproductive labour includes those kinds of actions that go into the reproduction of people in the present and in the afterlife (food, childcare, clothing, education, mourning). Productive labour includes those actions that produce objects and wealth – which includes war, oratory or lush rituals of gift exchange.[5] The different orders of morality associated with these spheres of labour create inequalities between those who perform them. The handling of money, wages or gifts – or the involvement in war, government and diplomacy – is called 'work' and performed, often ostentatiously, by people (usually men) with high status and rank in public. The activities that go into the reproduction of people are normally performed by women and children in the domestic or informal realms, are considered less valuable and defined 'labour'. Under capitalism, *work*, on the one hand, acquires nearly a supernatural value and is considered a magic and transformative force possessed by talented individuals or an inalienable property that gives to their 'owners' the right to citizenship or the dignity of personhood. On the other hand, *labour* is considered a source of alienation, material dependence and social exclusion. This is evident in the way we often make connections between black, unregulated and illegal labour and foreigners or marginal people.

This tension between gestures of permanence and of impermanence – between reproductive and productive labour – is at the very core of 'the human condition' (Arendt 1958). Nonetheless, non-capitalist societies cope with this tension by keeping these two spheres of labour separated into different moral orders and with rituals that reconcile them (Parry and Bloch 1989). In Malaysia, the cash produced through the labour of fishermen can be used for the household only after women purify it by cooking it on the household hearth (Carsten 1989); in Benares, India, the Brahmin priests performing death rituals accept money only after an elaborate ritual of refusal performed to avoid spiritual pollution (Parry 1989). Under capitalism, the sphere of reproduction is subsumed and incorporated into the sphere of production, leading to the progressive abstraction and commodification of life. Throughout history, reproductive labour has been systematically devalued, and women – the holders of the ultimate power to give life – have been constantly marginalised and persecuted. But under capitalism, the exploitation of reproductive labour is the very engine of economic profit.

Looking at labour relationally means to appreciate that productive and reproductive labour are two faces of the same capitalist coin – that the wage economy always entails incomplete labour commodification and semi-proletarianisation (Denning 2010) and the geographical and temporal 'unevenness' (Kasmir and Gill 2016) of capital, as the surplus value generated in capitalism's core institutions (the market or the factory) relies on unpaid labour in the periphery (the household or the informal economy) enforced through imperialist relationships between Global North and Global South and systematic processes of social exclusion such as sexism, homophobia and racism (Balibar and Wallerstein 1991; Federici 2004; Meillassoux 1981; Wallerstein 1983). In the struggle over value, in one camp is the alienated and divided world of capitalism, moulded on the commodity form and striving towards the *concrete* abstraction and homogenisation of labour through price equivalences, market standards and 'synthetical time'.[6] On the other hand, labour opposes capitalism's abstraction through everyday acts of reconnection, recomposition and occupation of capital's enclosures.

This struggle emerges, first, at the point of production and, subsequently, at critical junctures along processes of exchange, consumption and circulation. It is on the capitalist shop floor that labour leaves behind its spectral quality – its existence as pure and abstract productive potential – and becomes 'real' and productive, adopting the material and illusory appearance of money and technology, a state Fredrick Harry Pitts (2018: 118) describes as 'abstract concretedness'. Within the wage relationship, the living agency of labour – its power to reproduce life – is systematically obscured, neutralised and annihilated by the forces of capital. Indeed, commodity fetishism inverts the relationship between humans and things so that unanimated objects become living beings, endorsed with the power of changing human destinies. Capital – in the fetish forms of money, land and machines – is the invisible force that puts things in motion. Human labour, on the other hand, is emptied out of life, cut out from social relations, and only exists in the abstract and lifeless commodity form. The living machines, commodities and money of the capitalist cosmos push people and nature violently into the background as they forcedly move to the fore. Alienation is the state of consciousness where objects, machines and money come at humans with overbearing force – making them motionless, impotent and abstract.

This splitting between a frozen humanity and the moving capitalist cosmos is never resolved. Existing in between hollowed-out existences and enchanting virtual realities, the capitalist 'individual' oscillates between depression and aggression; hard work and overconsumption; acceleration, impotence and stasis. The more capital embeds itself in the material texture of life, the more extreme is its movement of objectification, dehumanisation, expulsion and abstraction, in the forms of xenophobia, sexism and homophobia. In my ethnography of Volta Redonda, I discuss capital's three fetishes – machines, money and land – and the processes of company reorganisation, expulsion and privatisation conjured up by these fetishes. First, on the CSN shop floor, the systematic dehumanisation of workers and misrecognition of their labour takes place through the technological system. Technological fetishism, the belief that technology is a godly force and the 'materialized body of capital' (Marx [1858] 1973) overpowering humans, is a central mechanism of class reproduction.[7] In the Usina Presidente Vargas of CSN, this fetishism emerges in the form of lopsided relationships between labour and other agencies that embody the 'animacy'[8] of steel. The capital-intensive operations of making and forging steel are considered highly skilled because they seem to embody, as in sympathetic magic, the animacy of steel. Unlike these, the labour intensive tasks at the end of the line are considered repetitive, boring and, ultimately, unskilled.

Moreover, skilled workers value labour as a form of embodied knowledge – unalienable and non-commodifiable – whereas line operators and unskilled workers, whose knowledge is externalised into machines, value it following standard measures of 'sales', 'quality' and 'productivity'. In chapter 4, I show how these different notions of labour value also underpinned different attitudes towards the privatisation of CSN. Skilled workers saw it as an attack to their skills, and as a form of 'theft' and externalisation of their knowledge. Unskilled workers saw it as a way of 'making money'. Technological fetishism also underscores the 'dependent' industrialisation of Volta

Redonda. The UPV was built with US money and with the assistance of the engineering consulting company Arthur G. McKee & Co., which also built MagnitogorskI Iron and Steel Works in the middle of the Russian steppe in 1928, the Reichswerke Hermanm Göring industrial complex in Germany in 1937 and United States Steel Corporation in Gary, Indiana, in 1908 – three plants and regimes (Stalinism, Nazism and capitalism) that embodied the spirit of industrial modernity of the time. The rolling mills and furnaces imported from the United States moulded the gestures of generations of Brazilian workers on the principles of scientific organisation, accounting and human relations that gringo managers and engineers taught to the local workforce. Technological dependency on American and European technology, managerial knowledge and operating systems continues to shape the firm's life.

Indeed, most of the machines in the UPV come from the United States, Japan and Germany and were purchased during different reorganisations and expansions of the plant. The first expansion (in 1974) was made predominantly with US technology, the second (in the 1980s) predominantly with Japanese technology and the most recent in the 2000s predominantly with German one. After each expansion, Brazilian engineers and skilled workers had to undo machines, assemble pieces from different machines and make different national standards compatible. The operations of Brazilian capital were largely constrained and limited by this foreign technology. The purchase of the long billet mill from China with the turnkey contract in 2008 limited the operation of Brazilian capital even more, as the contract puts suppliers in charge of the instalment and servicing of the technology throughout the mill's life.

With machines, money is the second capitalist fetish that obscures the value of labour on the CSN shop floor. The younger and unskilled workers affiliated with Union Force – a collaborationist and pro-management union historically from outside the traditional labour movement – value labour through quantifiable and monetary measures, especially after the introduction of the Sistema de Gestão de Manutencão (Maintenance Management System – SIGMA), which automated, financialised and outsourced labour and introduced 'the market onto the line'. Unlike these, the older electricians, engineers and mechanics mostly affiliated with the traditional labour movement consider labour unalienable and resist its commodification, externalisation, disembodiment and quantification by the SIGMA. These traditional workers, all too often dismissed as 'productivist', are in fact 'reproductivist' because they consider labour an action contributing to the long-term reproduction of the plant and oppose its intensification associated with short-term profit. The implementation of the SIGMA aimed at commodifying and monetising labour, way beyond the wage contract, pitted these two radically different views of labour – one speculative and short term, the other reproductive and long term – against each other.

The third and final capitalist proxy for human productivity is the productivity of the land. A core ideology of Western capitalism is the separation between land and industry, and the progressive devaluation of the former vis-à-vis the latter. The anthropologist Fernando Coronil (1997) argues this European model of capitalist valorisation does not apply to world peripheries like Venezuela whose extractivist state fetishises the magical fertility of the land while systematically devaluing human labour. Likewise, in Brazil, because of the historical entanglement of extractivism and industrialisation and the alliance between the landowning class and the industrialist elite, the productivity of labour and the productivity of land are considered mutually dependent rather than opposing forces (Coronil 1997; Villela and Suzigan 1973). During the Old Republic,[9] the ultranationalist rural elite considered state-led industrialisation an 'artificial' deviation from Brazil's natural economies of coffee and mineral extraction. This extreme ruralist nationalist elite (*sertanista*) opposed all non-agricultural developments – including in the foreign-controlled banking, commerce and insurance sectors – through an ecological discourse revolving around the fertility of the land. The words of Alberto Torres (1938: 93), the leader of the rural movement, have a strong environmental undertone:

> The savages tricked by the intelligence of the sailors exchanged precious stones and gold for mirrors, shiny objects and ostentatious objects. Similarly, we are blinded to the fact that our wealth is exploited by strangers by cars, palaces, fashion, jewellery and those appearances that we perceive as progress and which bewitches us with their smell and theatrical luminosity and which hide the fact that our real Brazil, the Brazil of the forest virgin and of the mines has been spoiled, stolen, pulverised.

This ecological and anti-productivist discourse was shared by the slave owners, coffee planters, financial brokers and merchants of semi-manufactured good, mostly the members of liberal classes, who advocated supporting the 'natural economy' of coffee extraction. Indeed, according to Jacob Gorender (1978), the extractivist economy of the planters in Brazil's 'colonial mode production' formed an organic whole with the subsistence economy led by formerly enslaved people and dispossessed rural workers, which prevented industrialisation and proletarianisation in Brazil for long time.

On the other hand, many planters-turned-industrialists after the abolition of slavery (especially in São Paulo) advocated the development of the mining and steel industries because these relied on 'autochthonous' raw material – iron, coal and labour extracted from the national soil. Thanks to the state-sponsored programme of mass immigration of European labourers, this class of fazendeiros-industrialists was able to reproduce the previous slavery system within the new workerist regime of the *Estado Novo*. In the UPV, too, nationalistic narratives revolving around the productivity and fertility of the Brazilian soil or the inequality between the high-energy-consuming centres and the highly polluted and low-energy-consuming peripheries[10] are shared by Trotskyist activists, trade union representatives, engineers and top managers alike. The fetishisation of the value of the national land cuts across classes and makes opaque the state's intensive extraction of sugar, coffee, oil, ethanol (in connivance with ruthless landowners, ranchers, loggers and global companies); the mass construction of dams, roads and other state infrastructures; and Brazil's sub-imperialist expansion in the Americas (Marini 1977). In fact, many consider the commodification of the national soil 'a necessary evil' for the consolidation of Brazil's global political hegemony.

Class and Culture

In this book, I want to rethink Marxian class analysis from an anthropological and humanist philosophical perspective – as opposed to Louis Althusser's anti-humanism or recent post-humanist philosophy. In the past thirty years, failing to read the new class articulation – uneven, heterogeneous, affective and in flux – associated with contemporary capitalism, sociologists and political scientists have argued the working-class has either disappeared, under the so-called embourgeoisement theory, or become so fragmented along lines of gender, race and cultural values that it is sociologically irrelevant. Because of this assumption, scholars veered away from class analysis and retreated into a fetishised identity politics. This failure to understand new forms of class composition has led to the collapse of the socialist project and the bizarre resurrection of the working-class as political subject if only discoursively, by the hands of right-wing populist parties. The fragmentation, dispersion and multiplication of labour that followed the collapse of the Fordist factory, the deregulation of finance and the systematic attack against the working class in the past thirty years have shifted the struggle between labour and capital outside the traditional factory context and embedded it within the micro-texture of human life. Hence, a humanist – but not anthropocentric – anthropology is fundamental to contemporary class analysis.

The work of anthropologists is often assumed to show the role culture plays in complicating, contextualising and relativising universal economic and social categories and trends. In this sense, Marxism, with its universalistic and structuralist notions of 'class', 'capitalism', 'inequality' and 'exploitation', is often assumed incompatible with 'proper' anthropology. In fact, in the past forty years, anthropologists either have been uninterested in Marxism or have subscribed its postcolonial critique as Eurocentric, materialistic and in denial of the local histories, discourses and cultures that underpin subaltern forms of class struggles at the margins. Against this trend, there has been a recent resurgence of anthropological studies of class reconnected to the Marxist tradition (Carrier and Kalb 2014; Kalb 1998; Kalb and Mollona 2018; Kasmir and Carbonella 2008). In *The Communist Manifesto*, Marx and Engels ([1848] 2012) describe human history as a history of class struggle. In capitalist societies, because of the continuous drive to increase profits and commodify life, this struggle reaches the apex. Masses of impoverished and disenfranchised workers are 'crowded into factories, organised like soldiers . . . and capitalist society as a whole is more and more split up into two great hostile camps directly facing each other: the Bourgeois and Proletariat'. Moreover, capitalism, unlike pre-capitalist or non-capitalist societies, where masters and kin have the duty to keep enslaved people alive, leaves people to starve. In a world dominated by pure economic interest and productive actions, the activities of care and support for the other are valueless. By forcing workers into misery and starvation, capitalism is ultimately breeding a revolutionary class and 'producing its own grave diggers' (Marx and Engels [1835–1895] 1987: 14.)

Marx's prediction – that the sharp polarisation, under capitalism, between bourgeoisie and proletarians would lead to the revolution of the proletariat – failed to materialise. In fact, the mass proletariat described by Marx did not constitute a unitary front but was internally divided between the aristocracy of labour and precarious workers,[11] between men and women (the former in formal work, the latter in unpaid domestic labour)[12] and between the workers of the empire and those in the colonies. This internal fragmentation of the working- class between wagework and precarious labour – and between productive and reproductive labour — continues to prevent the formation of a unitary front of working-class struggles.[13] To a certain extent, class fragmentation reflects the fact that labour is experienced differently by different subjects, according to their gender, sexuality, age and race – these intended not as abstract sociological categories but as uneven and embodied experiential fields. That is to say, using a Gramscian framework, class is experienced through the lenses of culture. In fact, these diverse experiential dimensions of labour are subsumed under the point of view of the dominant class – as abstract labour – which therefore becomes hegemonic (Gramsci 1971: 181). The hegemonic culture of capitalism abstracts labour through identitarian discourses that naturalise and objectify external bodily appearances – skin colour, age or sexuality – as both 'authentic' and standardised markers of personhood, morally and politically loaded, which legitimise labour exploitation. Because the systematic exploitation and devaluation of reproductive labour are central to capitalism, the naturalisation, objectification and moral stigmatisation of women – of their gender identity and sexuality – is central to it.

Thus, my anthropology is 'against culture' intended as a bourgeois hegemonic social construct that fetishises identities and follows instead Marx's same drive to understand the universal human impulses, needs and relationships – cross-culturally, historically and relationally– that contribute to the well-being of societies (Chibber 2013) and how such generalised well-being can be achieved first by the hands of those who mostly provide for it – that is, the working class. At the same time, with Gramsci (1971: 10–13), I am interested in how culture can transcend specific economic and material conditions and empathically connect with the point of view of 'the other'. That is to say, I am interested in culture as a *political* movement away from *economically* bound locations – a shift from base to superstructure (181–182). Seeking connections, mending ruptures and separations and experiencing the point of view of 'the other' – bridging economic divides and cutting across classes – culture can be a revolutionary weapon of the working class. In fact, class and culture are two central and antagonistic dimensions of the experience of capitalism. Historically, industrialisation, nationalism, mass production and state centralisation contributed to the emergence of a homogeneous working-class culture cutting across the dimensions of gender, race and age. The financialisation of the economy, the flexibilisation of production, the deregulation of labour and the dissolution of the wage contract within the global context of uneven development contributed to the decline of class politics and the proliferation of identitarian, nativist and cultural forms of political identification.

Intended to valorise those human relations that are not contained within the rigid realm of wagework, identity politics naturalised personhood within the rigid and timeless boundaries of 'cultures' while at the same time reproducing the logic of capital – through medicalising discourses, biotechnology or sheer racism – within the very micro-texture of the human body. Despite its apparent democracy, identity politics reduces the uniqueness, mutability and relationality of individuals to flat, anonymous and standardised stereotypes. 'Culture' and 'identity' constituted new terrains of class struggle in Europe and the United States in the 1980s. The emergence of culture as an alternative political field to that of class in the United Kingdom coincided with the crisis of the welfare state, the collapse of the nationalised industrial economy, the crisis of the trade union movement and the onset of the capital deregulation and the financial revolution that swept the city of London in the 1980s. Indeed, Boltanski and Chiapello (2007) claim the cultural critique of capitalism of the 1960s – its celebration of cultural diversity, identity politics, horizontalism, imagination and creativity against the materialistic, homogenising and one-dimensional class politics of Western social democracies – was incorporated into the post-Fordist capitalist regime of the 1980s, which they call the 'new spirit of capitalism'.

It was under Margaret Thatcher that the most powerful cultural critique of Marxism and of class politics emerged. In the 1970s, the sociologists John Goldthorpe and colleagues (1968) argued the British working class had high salaries, secure pensions, comfortable homes and incomes and therefore had internalised the values of the petty bourgeoisies. Particularly, the affluent autoworkers in Southern England had an instrumental and individualistic attitude that differed from the culture of solidarity of traditional working-class communities in the North, such as those of miners. Goldthorpe argued the affluent workers were 'bourgeoisified', that is, they experienced their class position through 'a system of practices, meanings, values and structures of feeling' (Williams [1961] 1984)[14] akin to those of the middle class. Sociologists from the Marxist tradition attacked the study as 'culturalist' and argued perceptions of work cannot be assumed to be homogenous within the same industry and that they vary even within the same factory (Beynon and Blackburn 1972). Marxist sociologists working in the anthropological tradition of the Manchester School led exceptional industrial ethnographies showing how workers' motivations and consciousness are largely the effect of managerial ideologies (Lupton 1963), relations *in* production (Burawoy 1985) and shop-floor organisation (Beynon 1973). But even Marxist sociologists eventually conceded the industrial working-class in the North was sectionalist (Beynon 1973), 'particularist' (Hayter and Harvey 1993), patriarchal (Rose 1992; Valenze 1995) and racist (Hall 1970, 1980b).

It was becoming clear that culture played a central role in both reproducing and transcending class relations. In 1975, the Jamaican scholar Stuart Hall in *Resistance through Rituals* (Hall and Jefferson 1975) showed there was a clash within the British working class between youth subcultures and the petty bourgeois racism and patriarchy of the white working class. In a controversial article written in *Marxism Today* just a few months before Thatcher became the leader of the Tory Party, Hall (1980b) denounced the 'moral panic' of the white working class about youth revolt, union activism and black immigration. But racism was not the only value the working class shared with the petty bourgeoisie. Both were deeply disaffected with the interventionist and centralised state under the ruling Labour Party. Thatcher would soon tap into this generalised culture of fear and discontent and *discoursively* frame her opposition to the Labour Party as a struggle of 'the people' against a technocratic, materialist and centralist state.

Referring to Gramsci's notion of 'passive revolution', Hall describes capitalism as a project of 'regressive modernization' that combines both 'economical' and 'cultural' aspects. For instance, Thatcher understood that by constructing a new economic narrative, she could trigger a radical political change and that monetarism, the extreme version of free market economy proposed by Milton Friedman, could be used as an ideology for radical change. Already at the beginning of the 1980s, when few people in the United Kingdom could predict the seismic shifts caused by the 'Big Bang'[15] in the City of London, Thatcher understood the ideology of monetarism would play a central political role – in breaking the trade unions, legitimising privatisation, attracting global finance – and built a new popular imaginary that wove together the contradictory positions and components of Toryism (family, authority and patriarchalism) with those of neo-liberalism (individualism, mobility, success, anti-statism, economic deregulation and global finance). Using the example of Thatcher's ideological appropriation of monetarism, Hall takes the Gramscian view that political economy is both a political and cultural construct – which caused him to clash with some of his comrades within the New Left Review group (Jessop 1985). After all, was not the idea that economics is just a technical matter at the heart of the right-wing ideology of Thatcherism?

For Perry Anderson, 'Hall's fundamental contribution to the political debate of the time, was to argue that the Left had to embrace a broad socialist project which both valorised and encompassed cultural difference, including differences of gender, race, kinship, sexuality, education, consumption and leisure' (2016: 76). In other words, culture was a new terrain of class struggle. Hall's cultural approach to class analysis was influenced by his close conversations with the Argentinian philosopher Ernesto Laclau, who, like Hall, was using Gramsci to build a postcolonial critique of orthodox Marxism. In *Hegemony and Socialist Strategy*, written with the philosopher Chantal Mouffe, Laclau criticises Marxian class analysis for its reductive economicism and positivism and calls for an anti-materialist, psychoanalytic and discoursive approach to political economy (Laclau and Mouffe 1985). Laclau's (2005) *On Populist Reason* veers further away from Marxism and brings together discourse analysis, psychoanalysis and political theory. To the notion of class, Laclau opposes the notion of 'the people', which he claims has a long history – from Rousseau to Le Bon, Lenin and Peron –that transcends and encompasses the formation of the industrial working class and the Eurocentric logic of industrial democracies.

Analysing the phenomenon of populism, Laclau describes politics in terms of open-ended *form* rather than as material demands arising from specific class positions. In fact, populism brings together diverse and contradictory ideological positions and political demands. Working through chains of equivalences, it reconciles conflicting frameworks – for instance, xenophobia and working-class solidarity and left-wing and right-wing demands – within a unified social subjectivity antagonistic to the status quo and coalescing around empty signifiers such as 'the people', 'the elites' or 'capitalism'. Laclau also distinguishes between democratic demands (e.g. for increasing social welfare or against inequality) and popular demands (e.g. the undifferentiated rage against the government) and argues that globalised capitalism generates incremental leakages between the two. Moreover, Laclau's Freudian and Lacanian description of the leader, whom people identify with emotionally and instinctively – sometimes as a father and sometimes as a peer or friend – constituted a challenge to the Marxian notion of class interest and its theorisation of relationship between the party and the masses.

The most scathing criticism of the Marxian class framework came from the Subaltern Study Group gathered around the journal

Peasant Studies, which brought together the Gramscian and the postcolonial frameworks. According to such a perspective, Marx's class framework is universalising and ethnocentric because it discounts the local discourses, cultures and histories through which anti-capitalism articulates itself at the margins. To argue the working class is moved by material interests reflects a 'bourgeois rationality' (Chakrabarty 1989: 70) and the Eurocentric notion of 'autonomous, self-bounded, rational individual' (Escobar 1999). Among them, the Indian historian Ranajit Guha (1999) looked at how popular culture and intellectual leadership in India shaped the everyday struggles of subaltern classes – workers, peasants, non-industrial urban poor and the impoverished bourgeoisie. Guha's description of subaltern forms of peasant insurgency constituted a powerful counter-narrative to the 'grand narrative' of class struggle developed by European Marxism, one in which culture was responsible for power structures that went beyond class exploitation or direct coercion. For instance, Guha showed the cultural hegemony of the postcolonial elites can take the form of Western-style education, modern infrastructures, Orientalist heritage projects or the Hindu doctrine of Dharma.

Like Guha, the anthropologist Rajnarayan Chandavarkar's (1994) fascinating historical analysis of the labour movement in Bombay shows how problematic it is to apply Western notions of class to the Indian context. Despite their political activism, the Bombay mill-workers were internally differentiated along lines of caste, ethnicity and kinship and did not conform to Marx's idea of homogenous proletariat. But for Chandavarkar, unlike Guha, the entanglement of labour, caste and kinship and the fragmented class consciousness of the Indian workers was the consequence of capitalist development rather than of their 'primordial loyalties', 'personalism' or 'propensity to violence' as in Chakrabarty's culturalist view.[16] Indeed, in specific postcolonial contexts, class and culture are part of the same anti-capitalist struggle and reinforce each other, as in the case of the Bolivian tin miners' struggles against the privatisation of the mines described by the anthropologist June Nash (1979). The core of the miners' identity and political activism was in their strong cosmological belief in Pachamama and Supay, mythical gods that existed long before the Spanish conquest. Reacting to the privatisation of the mines, which fragmented and mechanised work, the miners made offers (*challas*) of cigarettes, liquor and coca to the devil Tio (the Spanish incarnation of Supay) and received back from Tio the power to challenge labour reorganisation by the management.

At the same time, under the protection from the goddess Pachamama, women intensified their subsistence activities and organised political rallies by the Church of the Virgin Mary (the incarnation of Pachamama). Through their workplace and community activism, and the joint efforts of men and women, the Oruro miners stopped the reorganisation of the company. Nash argues the Oruro miners are politically conscious and militant, but they do not think about themselves solely in terms of class. Rather, their consciousness is a mixture of socialism, pre-conquest beliefs, Catholic theology, and narratives of ancestors kept alive in oral traditions. Similarly, Michael Kearney (1996) discusses the articulation of class and identity in the mobilisation of the people of San Jerónimo in Oaxaca, Mexico. The migrant workers of San Jerónimo had fluid identities, being at the same time peasants in their town (where they practiced forms of farming based on ancient environmental and ethnobotanical knowledge), informal labourers in Californian agro-businesses, urban migrants in the Mexican shantytown and proletarians in the cities along the US/Mexico border. Kearney argues the Oaxaca peasants had multiple political subjectivities and forms of consciousness – indigenous, class-based, informed by human rights or ecological awareness – and that this multidimensionality challenged the one-dimensional logic of capitalism and hence halted the development of agro-businesses in the region.

Nash and Kearney ask, 'How do industrial workers develop a common political consciousness, given their social fragmentation in the capitalist labour process?' Their answer is that 'class for itself' can emerge only within a wider cultural consciousness – religious, ethnic or racial – that both encompasses and expands class identity. Their argument is that political movements purely based on economistic notions of class are doomed to fail and that successful workers' mobilisations must bring together class and identity politics. Indeed, the main theme of much recent international labour studies literature, particularly on 'new trade unionism',[17] focuses on the intersectionality of labour movements. It tends to stress the divide between the working-class in the South – cross-sectional, socially diversified and combining class and identity struggles – and the working class in the North – sectionalist, class-oriented, factory-based, white and male. For instance, the Congress of South African Trade Unions and the Brazilian CUT confederation of the 1990s were instances of 'new trade unionism' in the way they brought together social forces outside the traditional labour movement including church organisations, women, squatters, intellectuals and human rights activists protesting the military rule.[18]

Likewise, for Huw Beynon and José Ramalho (2000: 230), the Brazilian Workers' Party (PT) first operated as 'a novel political conduit for social movements associated with the rights for women, cultural and ethnic minorities, and the environment – all without losing its identity as a party of the working-class'. The sociologist Ronaldo Munck (2011) goes as far as to consider trade union and labour activism in the South as a new global form of class struggle. But the mutual entanglement of class activism and social movement activism in the South is as historically contingent as the sectionalism and materialism of the working class in the North. Indeed, I discuss in this book the consequences of the PT severing its links with grassroots organisations and turning into a centralised, vertical and business-oriented governmental party. In the North, since the 'post-industrial' economic turn, it is assumed the working class has disappeared – incorporated into the middle class, gentrified, turned into white-collar workers or precarised and politically defeated by neo-liberalism. Guy Standing (2011) compellingly makes this argument in *The Precariat*, a bold analysis of the experience of precarisation of the Anglo-American working class linked to processes of financialisation, privatisation and 'workfarisation' of labour. Standing identifies a five-tiered class structure, with a plutocratic class of ultra-rich at the top and an expanded precariat class at the bottom, including the ex-working class, the young and educated unemployed and the 'denizen' (migrants, long-term unemployed, 'illegal' workers).

Unlike the old proletariat, the precariat is employed in insecure intellectual jobs – on unpaid labour, zero-hour contracts, crowd labour or freelancing – and is entirely wage dependent (i.e. it has no welfare or social rights). The mentality of the precariat is very different from that of the old proletariat. The precariat is usually young, has family background in insecure employment and does not identify itself with the prospect of having permanent jobs or actively seek work. On the contrary, it strives to be free from work or to work autonomously. Moreover, disaffected with politics and oppressed by workfare, the precariat does not seek protection from the state and wants less direct intervention and more redistribution from it. Standing argues the proletariat, in its violent disaffection with mainstream politics and in its broad demands, often overlaps with ultra-right movements. Standing's description of the 'precariat' captures well the global army of impoverished and disenfranchised wageworkers, in the United Kingdom, the United States or Brazil, who are turning to populist and right-wing leaders to voice their discontent with the establishment. For instance, the notion of the precariat applies well to the condition of service and retail workers and of the precarised working class in Volta Redonda. Yet, as noted by the anthropologist Jan Breman (2013), Standing's notion of the precariat is Eurocentric and discounts the general condition of accumulation by dispossession in the South, where most people rarely make it into the privileged rank of wage labourers. But my Brazilian case shows how the precarised wageworker and the flexible service workers are two sides of the same global 'precariat'.

Against the argument that the working class is disappeared in the post-industrial North, Marxist sociologists have argued sometimes that white-collar workers are subject to the same process of labour standardisation, deskilling and alienation as the Fordist manual workers (Braverman 1974) and that workers in the creative industries (Lazzarato 1996), service and affective economies (Narotsky and Besnier 2014), IT industry (Braga 2016; Huws 2003, 2014); art sector (Sholette 2011) and even Wall Street (Ross 2004; Martin 2015) are the new lumpenproletariat. The collapse of Fordism and the financialisation of the economy dislocated the male, white and affluent manual worker and further 'multiplied labour' (Mezzadra and Neilson 2013) and generated workforces composed predominantly of young women from minority ethnic backgrounds and with little previous experience of work or trade union activism. This new capital configuration brought to the fore issues of ethnicity, gender and kinship, accentuating even more the tensions between class and identity in the contemporary experience of labour. A big part of the value generated in the post-industrial factories in the North consists in the affective and reproductive labour of caring, cleaning and looking after people – mostly in the service and leisure industries – that Fordism consistently devalued. Yet, such a commodification of reproductive labour is associated with greater job insecurity and with increased misogyny, racism and homophobia so that the struggles of women and of racial, ethnic and sexual minorities are even more central than before to class struggle more generally.

The tension between class and identity – between 'redistribution' and 'recognition' (Fraser 1995) – is a central concern of feminist politics. According to Marxist feminism, capitalism is intrinsically patriarchal in that it naturalises gender and sexual identities into markers of alterity that legitimise the exploitation of women. For instance, Sallie Westwood's (1987) ethnography of an English Midlands hosiery factory looks at how relations of production are shaped by gender roles among a largely female workforce. The capitalist labour process is based on a repressive gender ideology enforcing the rigid divide between 'home' and 'workplace'. The women of the hosiery factory challenged this ideology by 'domesticating' the shop floor – filling it with personal objects such as mugs, personalised aprons, family photos and house slippers – and organising informal exchanges and 'selling parties' that disrupted the Taylorist regime and the formal economy of the factory. But this 'culture of femininity' – often expressed through romantic idioms of love and betrayal – ended up reinforcing the dominant patriarchal ideology of the shop floor, which considers women 'domestic' and 'unproductive' subjects.[19] In turn, the work regime is reproduced at home. Exhausted after a long working day, women must perform household chores because of the supposed 'femininity' associated with the domestic realm.

Westwood sees capitalism and patriarchy as neither autonomous nor reducible one to the other but rather complementary, with race adding another dimension, as Asian women are the most exposed to the violence of managers and husbands. Similarly, Ching Kwan Lee's (1998) ethnography of labour in China shows how women's exploitation in the shop floor is enforced through the idioms of kinship and ethnicity. Because capitalism systematically devalues and exploits reproductive labour, the condition of women and their struggles are central for class struggle in general. Kathi Weeks (2007) argues that instead of pushing for the monetisation of reproductive labour or for more and better-paid wagework, women should campaign for their liberation (and that of men) from work. Indeed, the rise of feminism as identity politics coincided with the post-Fordist transformation of capitalism, deindustrialisation, the collapse of the socialist project and the extreme individualisation and atomisation of social relations under late capitalism (see also Fraser 2013; Watkins 2018). To be sure, women are often as class militant as men (if not more) in socialist contexts (Rofel 1999) or in the context of socialist struggles against capitalist expansion such as in the struggles of miners in post-Perestroika Russia[20] or in the UK miners' strike in 1984.[21] Likewise, Silvia Federici (2012) shows how central to the anti-capitalist struggle are the valorisation of reproductive labour and the acts of commoning mostly by the hands of women workers.

Finally, anthropologists have stressed the entanglement between capitalism and religion, showing how narratives of spiritual enlightenment (Rudnyckyj 2011), personal salvation (Comaroff and Comaroff 2000) and moral pollution (Pinney 2000) reproduce existing class relations or, conversely, how religion and magic can open up new forms of contestation against capital, through acts of mimicry (Taussig 1980), possession (Ong 1988) or shared performative ethical judgements (Lambek 2013). The anthropologist Aiwa Ong (1988) focuses on three Japanese semiconductor factories located in the free trade zone of Kuala Langat in southern Malaysia. There, women's employment in industry conflicts with the values of their traditional Muslim communities (kampung), where men embody reason and self-knowledge and women are said to be influenced by carnal desires and spiritually polluted. When young women find employment in the new multinational corporations, these traditional values are undermined. Becoming economically independent, they leave their family homes, refuse arranged marriages and embrace the new urban lifestyle associated with foreign capitalism. But this freedom has a price. Emancipated from their fathers, they become the 'factory daughters' of male managers who use the same male kampung discourse to exploit them on the shop floor, forcing them to perform wearing and under-remunerated tasks.

In turn, women challenge such capitalist discipline though performative acts of spirit possession. Chased by ghosts, they walk into forbidden spaces – toilets, prayer rooms, supervisors' break rooms – speak loudly and in incomprehensible language and witness the Kampung cosmology coming alive in their support: tall spirits lick sanitary towels in the toilet, black tigers run free in the prayer room and old men appear on the lenses of the microscopes. At the time of Ong's fieldwork, militant Islamic organisations were mounting a moral campaign against foreign capitalism, claiming it was spreading immoral values in rural communities. In this context, Ong shows spirit possession ultimately reinforces the patriarchal morality of both community leaders and factory managers. Rudnyckyj also reflects on the intersection between religion and capitalism, by looking at how the management of Krakatau Steel in Banten (Western Java) implements spiritual training sessions for the employees – combining group prayers and multimedia presentations – to forge 'a more disciplined, less corrupt, corporate employee' (2011: 64). Here, it is the mixture of the Islamic notion of piety and Euro-American management jargon – embodied in the empty signifiers of 'proactive subjects', 'total action' and 'self-control' – that construes the 'embodied dispositions' and 'affects' of the workforce. The emotional impact of these training sessions is so strong that employees literally weep on the shop floor. Lurking in the background of such blending of 'corporate competitiveness and other-worldly salvation' is the reorganisation of the company, which followed a long wave of national privatisations starting in the 1990s and coinciding with the national Islamic spiritual reform (*reformasi*).

Lastly, Comaroff and Comaroff (2000: 316) describe the neo-Pentecostalist evangelical Universal Church of the Kingdom of God as an epiphenomenon of 'millennial capitalism' whereby the magical and invisible forces of global finance are locally appropriated through cargo cults and rituals that, mimicking the logic of capitalism, seek immediate and short-term returns. Indeed, religion has a great impact on class relations in Volta Redonda, where about half the population is evangelical, most of whom are affiliated with the Igreja Universal do Reino de Déus (Universal Church of the Kingdom of God – IURD). The IURD is a powerful autonomous political force in Brazil with shifting political allegiances. For instance, having supported Fernando Henrique Cardoso over Lula da Silva in the 1998 elections, the leader of the Brazilian IURD, Edir Macedo, supported Lula in 2002, allowing him to woo the poor evangelical communities who were suspicious of PT's left-wing liberalism, especially its take on abortion. In 2018, Macedo's support of the right-wing candidate Jair Bolsonaro of the Partido Social Liberal (Social Liberal Party) was central to his electoral victory (see Mollona 2018b). At the national level, corrupted politicians use the IURD to buy the vote of black, women and young constituencies and make electoral inroads into favelas and shantytowns.

In Volta Redonda, a big section of the workforce is neo-Pentecostal or Kardecist.[22] These beliefs do not provide a short-term, spiritual or moral form of resistance vis-à-vis the rational and materialist logic of capitalism as for Ong. Neither do they reproduce spiritual and mystical forms of neo-liberal corporate governance, as for the Comaroffs and Rudnyckyj. Rather, neo-Pentecostalism constitutes a bottom-up force that embodies the *lulista* ideology of social improvement and entrepreneurialism in poor neighbourhoods and favelas and propels the political career of local corrupted politicians and leaders. The enforcers of the capitalist discipline on the CSN shop floor – the team leaders and supervisors – are often evangelical pastors affiliated with the IURD and from extremely poor bairros. Their Pentecostalism lacks metaphysical flare and is directly invested in intensifying labour, fostering individualist practices of stock market investment, self-entrepreneurship and cost-oriented behaviour and subservience to the corporate hierarchy. But *Brazilian Steel Town* veers away from those culturalist approaches (see Buarque de Holanda 1936; DaMatta 1983; Freyre 1987) that emphasise Brazil's primordial and pre-capitalist mentality – a mixture of gentleness, trickery, patriarchy, personalism and deference towards power – which is supposedly responsible for the corruption of the patrimonialist state, the lack of public consciousness, the populist politics, the familism and the economic backwardness. According to Jessé Souza (2017), these culturalist accounts of Brazilian society are both racist – they forget the role of the white Global North in the reproduction of slavery and underdevelopment in the Global South – and elitist because they reinforce the hatred for the state, the public and the poor held by the Brazilian bourgeoisie.

Instead, I look at culture as a central component of working-class struggle. On the one hand, culture is the local form through which the abstraction of capital emerges and is constituted on the shop floor and outside it – the bourgeois construct that fragments and objectifies the working class. On the other hand, culture is the common sense that opposes the money form, speaking through everyday gestures and raising above divisions, barriers and enclosures – temporarily surveying the unknown and painstakingly weaving together different perspectives, beliefs and self-perceptions into a collective that is moved by its seams, differences and imperfections. Working-class culture is the long-term process through which labour articulates its needs, desires, and hopes in the light of contingent resources, the history and memories of past struggles and the momentous structural forces of capital steering in directions that are difficult to predict or fathom from the workers' singular and particular locations. Lastly, working-class culture is the cumulative effect of local actions, how they fall back on enduring power relations – between slaves and masters, capitalists and workers, states and subjects – but also undermine and change them, generating counter-feedback loops with indeterminate and uneven temporal and geographical effects and which hence haphazardly affect the wider political and economic spectrum.

This book aims at capturing the working-class culture of Volta Redonda in the structural context characterised by a state that both follows and betrays the socialist ideal; a national economy that is heavily industrial, progressively immaterial and under the grip of neocolonial powers; a company that is at the same time an old-fashioned steelmaker, a landowner and a global financial conglomerate; and a labour movement that is both co-opted and repressed by the state. I am interested in capturing the historically and geographically contingent cultural forms through which the struggles between labour and capital and negotiations around value emerge in the shop floor and in the city. Because culture is both a capitalist enclosure and an escape route from it, this struggle is always contradictory and unpredictable. As well as being a piece of materialist history and ethnography, *Brazilian Steel Town* is a journey through the capitalist imagination whereby human actions, knowledge and nature are enclosed, compartmentalised, institutionalised and commodified inside factories, machines, markets and precarious housing developments through processes of state violence and cultural and economic imperialism. Confronting the material and phantasmatic enclosures of capital, the forces of labour strive towards change, life transcendence and self-transformation, moved by solidarities and fragmentations and compelled to build the world anew all day, every day.

Notes

1. For a response to such critique, see Chibber (2013).
2. Meillassoux challenges Terray's argument of class exploitation on the grounds of the continuity between production and reproduction within the domestic unit.
3. "Sabotage" comes from the French *sabot*, which means clogs, dragging the feet, slowing down, withdrawing outputs, misdirecting or under-employing.
4. For a nuanced historical materialist historiography, see Tomba (2013: 409).
5. Lambek (2013) also distinguishes between two orders of actions. But he distinguishes between actions (with ethical value) and labour, whereas I consider all actions as imbued with both ethical and material values.
6. According to Sohn-Rethel (1978: 22), this is the concrete experience of socialised time under capitalism.
7. See also in Marx and Engels ([1835–1895] 1987); Marx ([1867] 1990, [1867] 1992).
8. The anthropologist Mel Chen (2012) uses the notion of animacy – and the related ideas of liveness, freedom and agency – to explore contemporary race, sexual and class politics.
9. From 1889 to 1930, when Vargas instituted the *Estado Novo*.
10. Mitchell (2011) also argues that the fossil-fuel-driven economies in the north continue to extract agricultural labour and commodities from the solar-based economies in the south.
11. For an account of the internal stratifications within the English working class at the beginning of twentieth century, see Roberts (1971). Marx revises this simplified class structure in his discussion of the industrial reserve army in *The Capital, Vol. 1* ([1867] 1990) and the 'The Eighteenth Brumaire of Louis Bonaparte' ([1852] 1994: 259).
12. The exclusion of women from wagework happened through progressive factory reforms supported by liberal legislators and the trade unions (Fraser 2016).
13. According to Parry (2018), the fragmentation between core workers and casual and temporary labourers continues to constitute the main dimension of the experience of labour under neo-liberalism.
14. Following Gramsci, Williams ([1961] 1984) calls these 'common sense'.
15. The deregulation of the credit market.
16. For Chakrabarty (1989: 218), primordial caste loyalties prevented the formation of the working class.
17. For the debate on 'new trade unionism', see Kilusang Mayo Uno trade union in the Philippines highlighted by Scopes (1992), or Seidman's (1994) work on new unionism in Brazil and South Africa.

18. For CUT, see Ramalho and Santana (2001b).
19. Likewise, Ong (1988) shows how spirit possession among Malay women workers reinforces the patriarchal images of them as 'factory daughters'.
20. For instance, Ashwin (1999) argues women were more militant than men in opposing the privatisation of the mines in post-Perestroika Russia based on their greater sense of belonging to the sphere of work, and Dunn (2004) shows how women workers in a Polish baby food factory resisted privatisation because they saw their working identity as continuous with that of carers and mothers.
21. Massey and Wainwright (1985) show how the miners' strike in 1984 – which mainstream media represented as 'typically' male, white and productivist – combined traditional forms of labour activism (strikes, pickets and lockouts) led by the mining community with the open campaigning style and grassroots networks of women, the CND and ethnic urban movements.
22. A form of spiritism introduced into Brazil by the French priest and educator Allan Kardec in the nineteenth century. Unlike neo-Pentecostalism, Kardecism is often contaminated by Afro-religious strands such as Umbanda.

– Chapter 2 –

Cyclopes at Work

Capital as Technology

Ode to the Cyclopes
(To the Workers of Volta Redonda)

Chant II 'On the Promised Land'
In the old station with rotten wood falling into pieces, the cyclopes arrived
They dismounted. Suspicious of the armed soldiers giving orders
Left and right in the grey morning. An unusually grey morning for the Eldorado
Giant tractors cut the land that discharges dump
People here. People there
Blocks of concrete getting shaped
Sledge hammer feathering the edges of irons until mermaids touch the end
Until the end of the day and the beginning of the next and many other days
Until the end of the agony, an unusual agony for those who came in the hope of meeting the Eldorado.

In the planned chaos came the machines. So big to be scary.
In the meantime, in the old Europe, in Asia and Africa, the world was dying.
But these indigenous cyclopes, gathered from all corners
From Mina Gerais, São Paulo. With all their saints leading the way
In the Vulcan, the workers forged the earth
In the breast of the earth, the workers forged the new land of Eldorado!
In the meantime, the world was an open wound
An unusual insanity and madness for those who built the Eldorado.

—Waldyr Bedê (2004), local historian

Cyclops
CHORUS LEADER
What's the matter, Cyclops?
CYCLOPS
I'm done for.
CHORUS LEADER
You look terrible.
CYCLOPS
I feel terrible.
CHORUS LEADER
Did you get so drunk you fell in the fire?
CYCLOPS
Nobody wounded me.
CHORUS LEADER
Then you are not hurt.
CYCLOPS
Nobody blinded me.
CHORUS LEADER
Then you are not blind.
CYCLOPS
Blind as you.
CHORUS LEADER
How could nobody make you blind?
CYCLOPS
You mock me. Where is Nobody?
CHORUS LEADER
Nowhere.

—Euripides (408 BCE)

During the guided tour of the Usina Presidente Vargas, I was surprised most of my fellow 'tourists' were current employees with their families. I recall asking, 'Why did you come here with your families? Is it not enough to be in this harsh environment for five days a week?' Looking amused, a man answered, 'Because this is our family.' He then told me a local saying, dating back to the Vargas era: 'Volta Redonda is a steel family. Vargas is the father, CSN the mother and the steelworkers the children'. The tour of the plant was exhausting. In its massive scale, the shop floor felt like a stage for the mythological struggle between humans and Gods. As we walked through a 2.5-kilometre-long balcony suspended above the hot rolling mill, the tour guide went through production processes and statistics. From the 85,000 tonnes of crude steel produced in 1946, the plant currently produces 5.6 million tonnes of crude steel and 5.1 tonnes of laminated steel. It employs twelve thousand direct and five thousand tertiarised workers. Overall, CSN employs twenty-two thousand direct workers and seventeen thousand subcontractors and is the largest fully integrated steel producer in Brazil and Latin America.

During the nearly hour-long walk across that shop floor, the scale of those numbers suddenly hit me. How do these workers put up with a factory that produces the equivalent of eight hundred Eiffel Towers per year – nearly three every day? How do they manage to exist, act and be heard inside such a gigantic mechanical system where everything is connected, synchronised, in movement and planned from above? And are the workers moving the machines, or the machines the workers? 'The scale of this plant is fit for the Gods', the tour guide exclaimed as we walked. From where I stood, this corporate myth seemed true. Machines looked like giant tools moved by the hands of Gods, in turn putting workers in motion. But as my fieldwork unfolded and I got closer to the labour process, it became more evident that, indeed, the workers' actions put the technical system in motion and that the technological system is fragile and unreliable, whereas workers are resilient and reliable.

On our way back, we walked through the smelting shop. As we watched the magma boil, our tour guide said: 'Steel is alive. Its crystalline structure is porous and filled with interfaces and holes that make it quivering, unsettled and vibrant. It has an internal vitality. It is contagious. It can change and morph from solid to liquid and from hard to soft . . . one moment it is living fire . . . and the next it is dead like stone!' He explained the magic of steelmaking consists in its circular transformations – organic matter (ores and carbon) becomes living magma and then a dead, cold and solid commodity circulating in faraway markets. But at the end of their life cycles, steel-made commodities are reused and melted in electric arc furnaces[1] – old construction frames, beer cans and gardening furniture find a new life in the form of railway sleepers, kitchen sinks and cars.

The circularity of steel is reflected in the working patterns. First, the skilled and slow performance of the smelters and the long wait for the steel to 'cook' inside the furnace, punctuated by the frantic activities of helpers and cleaners around them. Then comes the climax of the tapping of steel. For a few seconds, everybody stands still. Towards the cold end of the production process, human activities become quicker, syncopated and precise – billets and rods are reshaped, reformed and cut down into increasingly neater and smaller pieces, following market specifications. They are measured, tested and accounted for as they slowly move to the end of the line and towards the market. In each downstream stage of the production process steel is reheated and made formless again so that shorter production cycles start anew.

Close to the fire, human movements are gracious, self-conscious and circular, taking the form of a reciprocal exchange between humans and matter. As the production process moves away from it, they become increasingly fast, short and disembodied until the point of near stillness at the end of the packing line – when labour is fastest. As the production process unfolds, the transformation of steel from hot to cold is mirrored in a perceived coldness, detachment and immobility of people. As for sympathetic magic, the quality of steel seems to be transferred to humans. In the hot part of the shop floor, work is embodied in the gracious movements of the smelters; in the cold part of the shop floor, it is machines and objects that move while workers appear to stand still. Compared with the artisanship of smelting, the end-of-the-line operations look like tedious office work.

In Western and Eastern cosmologies, as well as being procreators and warriors, smiths and smelters embody the demiurgic skills of the worker – of one who makes objects. Like the medieval alchemists, they are the emblems of artisanship. They are mysterious and dangerous creatures whose mastery of the fire brings them close to the work of the devil. Because of their sympathy with the fire, they are often filled with furore, anger and extreme heat emanating from their bodies. They are deformed and one-eyed like Cyclops, or the Japanese *ame no ma-hitotsou no kami* ('one-eyed god of the sky') or the one-eyed Nordic god Odin, whose phantom horse also is one-eyed. Smiths live underground with dangerous mobs of metalworkers. Among some Australian and Siberian communities, smelters are shamans who feed themselves with iron crystals fallen from the heavenly vaults, can see spirits and souls and can fly.

It is the celestial origin of iron that gives smelters and smiths their mysterious power. For a long time, humanity used tools and artefacts made of meteoritic iron fallen from the sky. When Cortez asked the Aztec chiefs where they obtained their knives, they simply pointed to the sky (Eliade 1978: 21). In many ancient religions and alchemic philosophies, metallurgy is a form of manipulation of the sexuality and fertility of Mother Earth. Iron ores are embryos growing at a different pace from those of humans, animals and vegetal organisms. By extracting metal from the bowels of the earth, smiths and smelters supersede the work of nature and precipitate the geological tempo into human time. Their furnaces are new matrixes – artificial uteruses where the ore completes its gestation. For the Arabian scholar and mystic Ibn Sina (980–1037), iron is animated by romantic love, and caves and mines are the vagina of Mother Earth.

In parts of West Africa, smiths are high-caste members of sacred societies, civilising heroes and procreators of primordial foetuses from embryonic, formless and chaotic primary matter. It is by stealing the fire from the women's vaginas that they acquired such procreative power. Smelters are warriors associated with the Iron Age, when agriculture and new solar gods – the male fertilisers and spouses of the great terrestrial mothers – displaced the minor gods and the economy of hunter-gatherers. Yet, the beginning of the Iron Age was a time of intense philosophical and religious ferment,[2] which saw the birth of the major world religions – Zoroastrianism, Judaism, Buddhism, Jainism, Hinduism, Confucianism, Taoism, Christianity and Islam. It was during the early Iron Age that the philosophies of idealism and materialism were brought together through the invention of modern money (Graeber 2013: 200). Smelters incarnated this contradictory world. They were irascible, authoritarian and patriarchal, but also nomads and wise guardians of ancient metallurgical mysteries and rites.

The smelting shop was the attraction of our tour. Like the other visitors, I was taken by its magical aura. The immense shop floor was covered in thick, white smoke punctuated by sharp cones of light coming from the punctured aluminium roof. In the semi-darkness, the 'cyclopes' (as smelters were often called) looked like grey, tall ghosts cautiously moving around these light-cones and occasionally hitting them. In the light, they looked like old-fashion sci-fi characters – staring behind old goggles, holding long tools with oversized aluminium gloves, wearing heavy boots and overalls patched with layers of self-made clothing protections. After the tapping, we got mesmerised again. The cyclopes stopped for a moment, lifted their goggles and helmets, wiped the sweat from their faces and cleaned the dust from their eyes. Only then could their black skin be seen. Since the first run in furnace blaster 1, in 1946, attended by Vargas himself, CSN went through several stages of diversification and expansion. Much of this expansion took place in the 1970s and 1980s, when the UPV underwent three major developments. The first, completed in 1974, increased annual output capacity to 1.6 million tonnes of crude steel and diversified its production. The second, completed in 1977, expanded output capacity to 2.4 million tonnes, and the third, finished in 1989, added new coke and sinter plants, a basic oxygen shop, a third blast furnace, new hot strip and cold strip mills, two continuous annealing lines and two electrolytic tinning lines.

The latter two expansions took place while CSN was part of Siderbrás, the holding company that controlled most of the federally owned steel companies. The idea of reorganising the governance of the SOE steel sector around a holding company started in the second half of the 1960s, with the first Plano Siderúrgico Nacional (National Steel Plan). The holding company centralised the ownership of SOEs, standardised their practices and created economies of scale among them and with the transport and energy sectors (Schneider 1991: 102). In 1990, the Fernando Collor de Mello government put Siderbrás in liquidation and listed its subsidiaries, including CSN, to be privatised.[3] Even though the firm ceased to be an SOE in 1992, CSN still performs important state functions such as education, through the local Escola Técnica Pandiá Calógeras (Pandiá Calógeras Technical School – ETPC); the policing of the local working class; and the consolidation of state infrastructures. The post-privatisation system of corporate governance is more transparent but less democratic than the previous public regime. Before privatisation, CSN was heavily controlled by Brazil's president and the Ministry of Planning, Budget and Management, which appointed the five executives who run the company, while the board of directors had only nominal power.[4]

Under the new private regime, the directors are autonomous from the executive and under the sole supervision of the CEO, who is directly under the firm's owner-president. This vertical structure was reinforced by the operative system's total quality management (TQM) and the Maintenance Management System (SIGMA), which disempowered if not disposed of the middle management and decentralised control into the line. The quality system was introduced in the 1980s as a way to fragment the workforce between a skilled core – whose wages and working conditions were above the national average – and a periphery – precarised and outsourced. The auditors of the Superintendência Geral de Controle de Qualidade (General Superintendence of Quality Control) were put in charge of implementing the Toyotist principles within the Fordist structure of CSN. On the line, they observed the workflow and conversed with shift leaders and supervisors to establish productive standards on the base of which the production process could be rationalised. Militant workers resisted the auditors' attempts to acquire their knowledge. To break the workers' resistance and acquire their knowledge, the CSN directory set up quality control circles, a form of shop-floor brainstorming based on the format of the militant shop-floor assemblies led by trade union activists at the time. The International Organization for Standardization (ISO) certification requirements on the best quality management system practices fragmented and standardised the production process further (Caruso 2009: 125).

Thus, the post-privatisation structure of the company is transparent, vertical and split between top management and line workers, all remunerated through performance-related systems, unlike the previous system, which was based on fixed wages.[5] This new hierarchical corporate structure reflects the financialisation of the firm by the current owner – Benjamin Steinbruch, whose main business is in textiles[6] – whereby steelmaking is only a small portion of his conglomerate business, which also includes logistics, telecommunication, energy, transport and commodities. It was through the privatisation of CSN that Steinbruch took control of its immense public infrastructure complex including of ports, railways and telecommunications – as well as of 70 per cent of the land in Volta Redonda. Moreover, privatisation moved CSN further under the influence of the BNDES. Today, CSN operates like a 'typical' Brazilian conglomerate (Goldstein and Schneider 2004), which have highly vertically integrated operations, producing large shares of its primary and intermediate inputs, with each input functioning as a separate profit centre. After steelmaking, mining is CSN's most profitable business. All iron ore consumed by CSN is extracted and processed in its Casa de Pedra mine, located in Congonhas (Minas Gerais). In addition to extracting fifty million tonnes of iron ore per year, CSN also extracts limestone, dolomite and manganese in Arcos and Conselheiro Lafaiete. In 2014, CSN exported thirty-two million tonnes of iron ore[7] worth R$4 billion – nearly half the total corporate income (R$11.5 billion) in the same year (CSN 2015).

After privatisation, CSN also internalised much of its energy and logistics production. Currently, the company owns the old federal line connecting Mina Gerais, Rio de Janeiro and São Paulo; the 4,500-kilometre-long Notheastern line connecting seven Notheastern states;[8] and a container terminal in the Port of Sepetiba, Brazil's second largest, from where it exports forty-five million tonnes of steel per year, which generates an annual income of R$2 billion. The UPV is energy self-sufficient, and CSN has majority shares in two major hydroelectric facilities in Ita and Igarapava (São Paulo). To balance its currency deficit, the company expanded internationally, purchasing the galvanising plant in Portugal and a billet mill in Unterwellenborn, Germany. CSN also has the laminating plants Galvasud in Porto Real and CSN Paraná. The strong over-evaluation of the real decreased the competitiveness of Brazilian exports, so 65 per cent of the conglomerate's income comes from the domestic sale of steel and from mining and logistics, representing 25 per cent of the company's income, equally distributed between internal and external markets. This internal market expansion is also driven by the expansion of the construction and housing sectors linked to the government's flagship infrastructural development Growth Acceleration Programme (PAC) and the housing programme Minha Casa, Minha Vida. In fact, the billet mill, which was opened in 2008, is totally dedicated to the construction sector.

Thanks to its conglomerate structure, CSN went through the 2008 global recession relatively unscathed by offsetting financial losses and decline in global steel demands with sales of cement and iron ore. But, by 2012, the strong currency had hit the firm's finances hard; the iron ore market sharply contracted after China's slowdown, and the political turmoil around President Rousseff led to the suspension of PACs, the flight of foreign carmakers from the region and hikes in labour costs, which had already risen exponentially in the last five years as an effect of Lula pro-labour policies. This was the beginning of the crisis that would terminate with a full-blown recession in 2015. The smelting shop (Gerência Geral Alto Forno – GGAF) was hit the hardest in all the company's reorganisations. The privatisation of CSN in 1993 cut its workforce by a third and eliminated the departmental director. But this new flat structure generated conflicts among the middle management so that it was reorganised again. Supervisors were flown to Japan, and Japanese consultants implemented the new TQM and Five Star (5S) systems. The Japanisation of the furnace cut its workforce by another half. In 2004, the two blast furnaces were put under different directors. Older and more politicised workers were replaced with younger ones. During the 2008 crisis, the second shop (AF2) closed down for three months, and all furnace workers were made redundant. Most of them were re-employed when the crisis was over. Today, the GGAF has three furnaces and seven hundred employees, half of whom are subcontractors. The number of subcontractors doubles during the seasonal shutdown, when armies of workers go underground to empty the furnace tanks of toxic grease, waste and chemicals.

In December 2015, 1,500 workers were laid off. In January 2016, CSN announced the indefinite closure of AF2 to adjust production to weak demand. The furnace, CSN's oldest, produced about five million tonnes of steel per year, about 30 per cent of the firm's total output, and employed three hundred workers, who were indefinitely laid off. However, a few months after the announcement of the closure of AF2, the company celebrated AF3's production record of one hundred thousand tonnes of steel, making it the most productive furnace in Brazil.[9] AF3 employs only 190 direct people and 100 subcontractors – technicians, engineers, operative, mechanics and electricians. AF3's production records followed a series of improvements led by maintenance firms, whose relevance I discuss in chapter 6. AF3 opened in 1976 and was refurnished by a Japanese firm in 1984 when its daily production capacity was expanded from 6,000 to 7,200 tonnes. In 2001, a UK maintenance firm expanded its daily production capacity to 9,400 tonnes. At the time of my fieldwork, the company was cutting down on subcontracting and consolidating an internal labour market of low-level operatives whose working conditions for instance in terms of hazardousness, and wages were often lower than those of the subcontractors.

This regime of 'internalisation of precarious labour' was achieved in several ways. First, through the SIGMA, which deskilled line workers and implanted a regime of continuous cost control. Second, with the 'up or out' career system, whereby recruits who failed to progress to level 3 after two years, were laid off. Because progress depended on the employees gaining some additional self-funded degrees in higher education, the system de facto linked professional advancement to personal wealth and reinforced existing inequalities within the workforce. Third, CSN wages were extremely low compared to the national average. Skilled engineers earned R$2,100 per month (50 per cent less than the engineers in other local steelmaking firms), and non-qualified operatives earned R$700 (less than two minimum wages at the time, just above the poverty threshold) against a national average of R$1,200. For instance, the employees in the invoice department – where twenty-three thousand letters are signed every day (forty-seven every minute) and seventeen thousand tonnes of paperwork are processed every year – would normally progress from a starting salary of $R700 to a maximum of R$1,100. Basic jobs such as security guards, train drivers, postal workers and cleaners were paid just the minimum wage.

Fourth, employees were obliged to accept being permanently relocated to different company branches, which included the faraway Northeast, and to join the company's transfers, for instance, in China, Canada or India, which often lasted several months. These transfers and relocations were used to threaten and discipline the workforce. Fifth, because of under-investment and lack of skilled maintenance workers (who were made redundant with the SIGMA), there was a very high incidence[10] of, often deadly, work accidents. These were often blamed on the employees, as breaches of the company's disciplinary policy, and punished with dismissal. It was not uncommon for workers to hide work-related injuries for fear of being dismissed. Lastly, there was a strict moral code in place. Employees who 'spoke in a derogatory way of the company in public', 'made fun of or offend a superior', 'discussed confidential corporate issues in public' or 'failed to destroy confidential documents' were punishable with immediate dismissal. Because of this harsh labour system, the CSN workforce was both extremely young – roughly half the employees were eighteen to twenty-four years old – and very short term.[11]

This system of internalised precarious labour focused on the furnace and the hot strip mill (Gerência Geral de Laminados a Quente – GGLQ) departments, the most capital-intensive in the firm. In these departments, CSN was not concerned with recruiting well-qualified line workers. Its main concern was to improve labour flexibility. Ronaldo, a production engineer in charge of recruitment, claims:

> We are not too bothered to recruit skilled line operatives. The competitive advantage of our Brazilian workforce is in not in their technical skills, but in their *malandragem* [trickery] and *jeitinho* [improvisation]. Western companies are stuck in a rigid methodological approach. They try to improve labour productivity through suggestion boxes, formal meetings or corporate reports. Our line workers are generous and creative. They like to improvise. You can see it everywhere on the shop floor: pipes patched up with cellotape, makeshift electric wire extensions and handmade tools. These are the small tricks that make our labour force productive!

Top managers like Ronaldo were not bothered by the operatives' exceptionally high turnover. In fact, his dream was to have a flexible workforce. But at the same time, the SIGMA drained the company of precious technical knowledge, so CSN's main problem was to recruit badly needed skilled engineers. Ronaldo is one of the many engineers who teach engineering at the ETPC, which after privatisation went under the control of CSN and is now private. Through the ETPC, CSN recruits a vast number of seventeen-year-old trainees on short-term contracts, which fits with the firm's high turnover policy. Ronaldo argues the opposite: high labour turnover is the consequence of recruiting subcontracted workers, and in fact, since the firm followed the new policy of recruiting children of employees, labour turnover in his department from fell from 5 per cent to 2 per cent.

Raissa is one of these bright, young ETPC graduates who has worked at the furnace for one year. She loves the hard-wearing work on the line but fears that, being a woman, she will end up in the control room at the end of her training. Her father is an unskilled mechanic for a subcontractor of Volkswagen and Peugeot, and her two brothers work for rival steelmakers Votorantim and Gerdau. At every meal, the siblings argue about who is the best employer. Raissa is positive about the future:

> Working on the line is not at all hard except for the inspections, when I have to climb eight hundred steep steps up to a bridge fifty metres above the ground. I am confident that here I can progress to a managerial position. Overall, employers prefer not to have women in operative jobs. But for me, being the only woman in 'a man's job' is an advantage. Last year I was the top performer at the technical exam because of the great support and training I got from my *padrinho* [supporter]. I work on night shifts, so during day I can attend the classes for my BA in engineering. My only concern is job security. Last year we had a terrible explosion at the furnace. It was a miracle nobody was killed.

CSN's strategy of internalising *flexible* labour and cutting down on subcontracted labour created a stratification within the workforce between simple line operators with declining wages and working conditions, and well-paid engineers like Raissa. These young engineers are less politicised than the older engineers who were dismissed because of the implementation of the SIGMA. In between them are the leaders who have slight better wages than the operatives but more precarious roles and positions than engineers. The working conditions and wages of the operatives depend on the 'relations in production' (Burawoy 1985) in each department (Figure 2.1).

The Steelmaking Production Process

Smelting (GGAF)

After going through the sintering department, the mixture of coke, pellets and lump ore is transported to the smelting shop. Here, blast furnaces 2 and 3 run continuously with ten or eleven heats per day, each heat lasting two to three hours, and produce 10,000 and 4,100 tonnes of pig iron per day, respectively. Cyclopes call furnaces 'old ladies' because of their anthropomorphic shape – slim at the top (the 'throat') and wide at the bottom (the 'belly'). Ladies are said to be strong and temperamental. Their thirty-metre-high steel armoured bodies continuously grumble, scream and burst loudly on the shop floor. Besides, 'ladies' are said to be cold-blooded and greedy because their 'heart' (where ignition valves are located) is below their bellies. Routine furnace operations are performed by eight contractors and

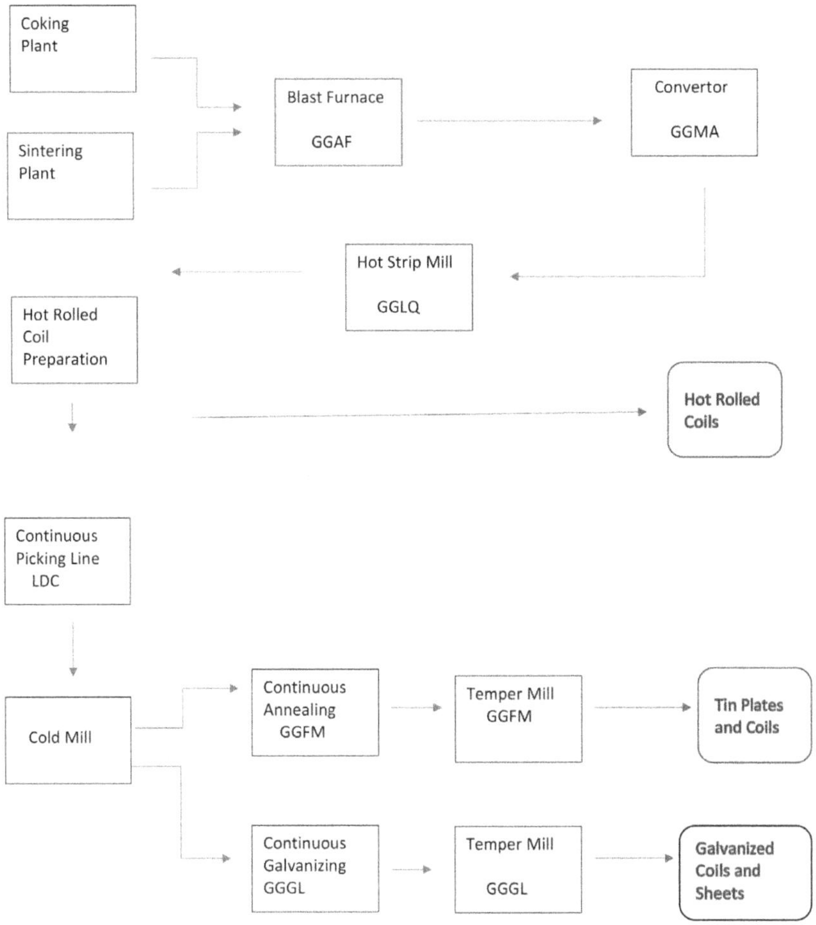

Figure 2.1. Usina Presidente Vargas production flow. Figure created by Massimiliano Mollona.

eight operatives. Contractors have the harshest tasks: they put refractory moulds on the furnace's door,[12] break moulds of slag and clean gas, minerals and ores after every heat. The operatives' tasks are similarly wearing: they measure the heat of the furnace, manage the gas outflow from the pipes, prepare the raw material and direct the flow of pig iron into the torpedo. Every time they open the furnace, a 2,000°C blow of air hits them. By the furnace, the temperature reaches 60°C. Thick yellow smoke burns eyes and throats. Barely visible in the haze and the heat, giant red quartz letters address the workers with football-like cheers: 'We support you, you are our champion!' or 'Well done, you have scored another production goal!'

The operations at the furnace are as follows. First, the subcontractors feed the 'burden' (a mixture of ores, pellets, coke and fluxes) inside the furnace's throat. As the burden sinks into deeper and hotter zones, individual layers of ores, pellets, coke and fluxes go through different chemical reactions. At 500°C, the burden loses water and reduces in mass. Meanwhile, residual gas is rerouted outside the throat into the power-generating system of the plant. In the belly, the burden reaches 1,000°C and expands again, becoming spongy and semisolid. Then, as the temperature reaches 1,500°C, the burden loses carbon and lands at the bottom of the heart in a liquid state. Oxygen is continuously blasted at immense speed though small copper valves inside the heart. Matter is propelled up by these blasts and falls to the bottom of the belly in circular loops. Although these phases are automated, smelters check them continuously by opening the door of the belly. Smelters can determine the temperature and thickness of the steel by its colour and smells and make small adjustments based on these sensory observations. When the hot metal reaches 1,600°C, the smelter taps the molten metal. Running through channels across the shop floor, the metal falls into the torpedo car below the furnace. In this part of the shop floor, the room temperature is 40°C. Operators move across thick columns of vapour released from the hot metal running across the shop floor. With long spears, they control the flow of metal by breaking the moulds of slag to ease it. The hot slag runs through the shop floor inside a refractory channel and then free-falls into an outdoor pool, while it is sprayed with blue coolant. The falling red slag and the sizzling blue coolant make a permanent rainbow outside the windows of the department.

Refining (GGMA)

After the steel is smelted, torpedo cars transport the hot metal to the steel metallurgy department (Gerência Geral Metalúrgia do Aço – GGMA), following the ancient railway line that crosses the plant's main courtyard and connects the plant to the city. A loud siren and a strong heat wave precede the arrival the torpedo car. Pedestrians stop and stare at its slow movements. The GGMA is sandwiched between the furnaces and the mills and connected to these departments by a maze of half-hidden stairs, bridges and flyovers floating mid-air. In the refining department, hot metal is transformed into liquid steel by reducing its carbon and slag content and achieving the specified alloy composition. The process of reduction has two phases. In the primary process, pig iron is mixed with basic materials (lime, magnesium, carbon and silicium) in the LD-KGC and blown with oxygen at 1,660°C. This process lasts only eighteen minutes. It starts with a deafening and iridescent explosion of oxygen and terminates with a violent release of 230 tonnes of steel and 20 tonnes of waste in the ladles below the converter.

During the secondary process, called *metalurgia de panela* (pot metallurgy), the quality of the steel is fine-tuned mainly through micro-increments in aluminium content to increase the harshness of the steel, achieved by detracting oxygen and sulphur from the melt or adding affine elements such as soda, magnesium or lime. There is also a station that produces ultra-low carbon content steel for the automotive industry through a process called degassing. At the end of the refining process, continuous casting begins. Liquid steel is released from the ladle into copper moulds. The mould passes through circular arcs and is transformed into semisolid strands that run along the caster and solidify along the journey. Water is continuously sprayed on them to prevent them from solidifying too rapidly, which would cause them to explode. The slabs now roughly twenty-five millimetres wide go into the hot strip mill, where the forming process begins.

Forming

To obtain the shape, size and characteristic required by consumers, the steel must be reheated, rolled and cut to size. Steel is made of crystals with specific space lattices.[13] The rolling process breaks the crystals and shifts the atoms along the glides of the lattices. When the destruction of crystals has reached the limit, atoms stop gliding and reach a steady state. The metal is now permanently deformed. But even in a steady state, the metal is under stress – that is, it strives to return to its previous crystalline structure. This tension is responsible for the hardness of the metal. Crystals form again when steel is reheated between 800°C and 1,150°C and reaches yet another permanent configuration. Thus, steel is reshaped either by regenerating crystals through reheating or by crushing them at cold temperatures. These two different processes are performed respectively in the hot and cold strip mills.

Hot Strip Mill (GGLQ)

In the hot strip mill, slabs are preheated in the oven (*forno*); in the rolling mill, they are 'thinned' (*desbaste*) into sheets between 38 and 250 millimetres thick; in the coiling mill, they are turned into coils (*bobinas*). Thicker ordinary coils are used for car chassis, tyres and industrial pipes; thinner ones are used for metallic structures, gas pipes and packaging. These coils and sheets are sent to the Port of Sepetiba and from there are shipped to national and international markets. Other sheets are sent to the cold mill, where they are transformed into tinplate. Roughly half[14] the coils go to continuous pickling line (*linha de decapagem continua*), where iron oxide is removed from their surface with nitric acid and they gain the degree of flexibility and high quality required by the automotive and white goods industries.

Sheet Metal (GGFM)

The Gerência Geral Folha Metálica (GGFM) produces metallic sheets and coils for the food, automotive and durable industries. The department's most successful product is tinplate for the milk, soft drink and meat industries, which generates 20 per cent of the company's revenue. The GGFM has 880 employees, divided in four departments, each responsible for a phase of the production process: the cold mill, the temper mill, the cold strip mill and the service centre. The production process is as follows: (a) sheets are rolled into a thickness of 0.14 to 2.25 millimetres (the process hardens and smooths the sheets surface but alters their original crystalline structure); (b) in the annealing process, the sheets are reheated and cleaned, and their original crystalline structure restored (this process can be continuous or, for achieving greater effect, performed in small batches); (c) in the temper mill, the sheets that go into cars, refrigerators and gas cookers are further hardened and thickened; (d) rolls are cut to costumers' specs and checked for quality (the material is then inspected, generally by women workers, and classified in terms of quality); (e) they are packaged as either sheets or coils.

Cold Strip Mill (GGLF)

The cold strip mill (Gerência Geral Laminados a Frio – GGLF) produces galvanised coils, sheets and rolls for the car, construction and white goods[15] industries. The cold sheets produced by the GGLF constitute 38 per cent of the firm's total output. From walking around the department, it is clear the company invested heavily in it. Walls are freshly painted, most machines are in good condition and its main two ovens, which cost R$80 million, are entirely computerised. Among workers, there is a perception that the GGLF technical process is more skilled than in other departments, mainly because its processes of galvanisation and coating sheets and coils are made bespoken. Coating makes tins resistant to erosion by external agents and hence suitable for food preservation. In addition, it makes customised labelling and branding possible. Galvanisation hardens the steel by immersing it in electrolytic baths of molten coating metals – mainly zinc and chrome – and a mixture of caustic sodium and sulphuric acid at 450°C.

The Workforce

The Smelting Shop (GGAF)

In the job description, smelters are considered the least skilled in the company.[16] They are on average between forty-two and forty-five years old and have high seniority in the company. Leaders (as supervisors are now called) and indirect workers earn higher salaries than smelters. This labour stratification is reflected in the workers' residential patterns.[17] In AF2, unlike in the main furnace, 50 per cent of the workforce has been in CSN for less than four years. The furnace department is also known as 'the heart' of the plant. Furnaces must work continuously. If they stop, they get damaged and the production process can be seriously disrupted. In the past, the firm's 'heart' was said to be Trotskyite and Leninist because it was traditionally controlled by the most militant wing of the Metalworkers' Union (SMSF). In fact, the smelting shop was the first department occupied by the workers in the strike of 1988 and became the stage of the violent confrontations between the SMSF and the army that led to the killing of three workers. During privatisation, the older and militant smelters were replaced with younger (second-generation) workers recruited mainly from outside the city, especially from former coffee towns with a long tradition of slavery and indentured labour.

Osvaldo, the leader of AF2, is one of these second-generation workers. He tells me of his life before coming to Volta Redonda:

> I worked in a fazenda in Goiás for twenty years before I came to CSN. The fazendeiro was a German guy. He was a nice bloke. My first job was to cut through five thousand acres of wild forest and fence up the whole property. I had sixty-seven 'hands'[18] under me. I turned them into proper workers. The fazenda is now a thriving cattle ranching business. I become the owner's right-hand. I had a big income – double my current salary at CSN. But I felt too isolated there. There was nothing I could spend my wages on. Here in Volta Redonda is the opposite. There is no money, but plenty of commodities

> and goods of consumption. When I arrived here I was already forty years old; I am fifty now.

Osvaldo is sceptical of both the management and the Metalworkers' Union:

> I preferred the hierarchical regime of the fazenda. Here, politics is a game between the trade union and the management, especially around the time of collective agreements when they play the game of presenting totally opposite arguments. They play out a fictional class struggle in front of the workforce. In reality, there is no struggle because the company overpowers the union. The trade union pretends to be assertive just to increase its membership; some trade union representatives are said to have even accepted bribes from the management.

José, another furnace leader, is also from rural background. He comes from the neighbouring town of Vassouras, a leading centre of the region's coffee economy. He is the first wageworker in a family with a long history of slave and casual labour. José started working at the furnace when he was sixteen – just a few days after his father had left the family in search of a job and never returned. A few months into José's job, CSN was privatised:

> Privatisation was a 'political school' for me. It led me to think about myself as an entrepreneur or investor, rather than a worker. I bought company shares and private health care, and I set up a family business [which was recently liquidated]. Today, my wage and my wife's wage as a primary schoolteacher combined together are barely enough to pay for the mortgage. By I am still proud of myself. My ninety-four-year-old grandmother was a plantation worker. She went through unbearable suffering. I do not want to depend on anybody – not even on my company.

José's view of himself as an 'investor' and 'entrepreneur' was not uncommon in CSN, especially among second-generation workers.

Line workers are more politicised than leaders are. Anderson, a second-generation line worker, lost both parents when he was sixteen and worked as a street vendor until he was offered a job at CSN. He worked for thirteen years as a furnace operator before he was promoted to his current position in the control room (also called 'the brain'). His job is to stabilise the smelting operations by intervening on three stock variables – temperature, time and chemical composition of the burden – using complex computer software. Even though his control operations deal with advanced chemical and engineering processes, they are classified as low skilled. Anderson considers himself lucky to have been promoted without having a technical degree, which is a strict prerequisite for trainees' career advancement. CSN's few bursaries cover just 30 per cent of the cost of the degree (50 per cent if related to metallurgy), so only the more affluent workers have a chance to progress professionally. In addition, young trainees from relatively wealthy backgrounds tend to live with their parents until they get married, whereas poorer employees become economically autonomous very early on. According to Anderson, this penalises the poorer workers:

> It is not so much the inequality between managers and operatives that worries me but that between operatives. For instance, level 3 workers earn nearly twice as much as the operatives on level 2. These, in turn, have wages that are double the ones of the subcontractors. Have you ever done some underground gas checks or repaired high-voltage circuit repairs or removed chemical waste under the furnace? This is slavery, man; it is not wage work. Even within the same wage category, some workers are illiterate or poor or orphans. They do not have family money to pay for a higher degree and do not stand a chance to progress to level 3. They get fired after a few months.

Anderson reflects on how the CSN workforce is deeply polarised in terms of family background and education, and how these social inequalities are reproduced on the shop floor. Those colleagues of his who are better educated and come from wealthier families end up in managerial or supervisory positions, leaving their working-class lifestyle and political militancy behind and 'becoming middle class'. Sadly, the Metalworkers' Union does not tackle the issue of wealth inequality within the working class.[19]

Antonio, a black smelter in his fifties, is much worse off than Anderson. Despite having worked at the furnace for thirty years, he is still level 2. His salary is at the bottom of the operative scale. Normally, workers in hazardous areas like him have a retirement age of fifty-five. But his job was recently reclassified as non-hazardous, so his retirement age was moved to sixty. When I asked why he was being discriminated against, he pointed to his skin and said, 'It is not a mystery, is it?' The Paraíba Valley was the epicentre of the world's coffee economy until the late nineteenth century, and according to many of my friends, the legacy of slavery was still strong in Volta Redonda, in terms of racially segmented labour markets and working conditions. Laureano, a retired black worker and president of the Palmares Club[20] (a recreational group for black CSN employees), comes from a Minas Gerais village and started working as a shoemaker with his mother when he was twelve. One day, while he was at shoe factory finishing the last pair of shoes that had to be delivered by the next day, his brother showed up and told him there was a letter for him at home. The letter invited him to travel to Rio de Janeiro and play for the Madureira football team. He was stunned by the offer:

> I looked to all the pairs of shoes I had made on that day and I thought to myself, 'I am going to leave all these shoes behind!' When I went home, everybody was crying because I am the fifth son in a family of nine children. My parents did not want to let me go, whereas my sisters encouraged me to go.

Laureano ended up playing for Madureira until he was eighteen. In 1954, he moved to Volta Redonda and continued to play for Cruzeiro (the local team). He became very popular. After an injury at twenty-three, he started working at the UPV as attendant (*servente*). Most black workers were *serventes* like him. They performed the worst tasks in the company: unloading stone, cement and sand. Moreover, to progress professionally, employees needed at least two years of secondary education, which was unusual for a black person. But Laureano went back to study, and after ten years of technical school, he became a supervisor. It was then that his problem started:

> It was very difficult to perform the function of supervisor [*chefe*]. Even if my wage improved, I still faced discrimination. Brazil is a racist country. As long as I was *servente*, my white bosses treated me well. The problems started as soon as I became a *chefe*. I understood very quickly that a black person has no authority over white workers.

Discrimination continued outside the shop floor in terms of residential segregation. White managers lived in the rich bairros of Santa Cecilia or Laranjal; whereas Laureano lived for nine years in a bairro made up only of wooden shacks. As for other black residents, he was forbidden access to the employees' club, the aeronautic club or the sailing club. Laureano continues:

> The white working class was not officially banned from these clubs, but it was definitively not welcomed either. These were for the elites of white workers. We, black, still carry the image of slavery. We have conquered some spaces that were not allowed to black people before. But we remain at the bottom of the social pyramid.

Yet, shifting from his personal history to the broader national history, Laureano reproduced the modernist and *workerist* narrative that casts industrialisation as a process of social emancipation, including of black workers:

> Before Vargas, Brazil was a colony. The CLT was an immense gift to the workers – in particular, black workers, who for the first time in history started to have rights, including the right to retirement! It was through Vargas's

> industrialisation that we, black, started to rise socially. CSN was a foundational company in the movement for industrialisation of the country. The company was important, especially for us black people, because we were those who performed manual labour in the country.

Refining (GGMA)

The GGMA has a small workforce of mainly low-ranking managers and supervisors living in high-income working-class neighbourhoods (such as Conforto or Sessenta) and describing themselves as middle class, or Class C. Originally, most GGMA workers come from Minas Gerais. In the early days, the CSN workforce consisted mainly of rural migrants from Minas Gerais. Unlike migrant workers from São Paulo and Rio de Janeiro, European migrants or descendants of enslaved or formerly enslaved people, who had experience in political activism and were affiliated with the Brazilian Communist Party (PCB), *mineiros* were mostly illiterate peasants with little experience in political militancy. Moreover, several top CSN managers were from Minas Gerais and maintained strong ties with the *coronéis* (generals) of the region. Indeed, the *mineiro* ethnicity defines the 'authentic' working class of Volta Redonda. *Mineiros* are highly valued by the company, given the best jobs and considered 'the true sons of CSN'. Refining is nearly entirely mechanised and, compared with smelting, requires less manual skill; in the words of some smelters, 'It is like an office job.' But refining workers describe it as an extremely skilled job that requires sophisticated alchemical knowledge. José, a line operator, explains:

> The 'refining station' [*estação de aborbulhamento*] is not as dramatic as work at the convertors, where workers deal with explosions and oxygen blown up at very high temperature. We work on low-temperature, micro-variation and micro-alchemical reactions based on affinities, for instance, between sulphur and lime, or repulsion, for instance, between aluminium and oxygen. Our work is old fashioned, a bit like cooking. We know when steel is ready to tap from the way it boils, the purplish colour of the slag floating on the surface or how the pungent smell of sulphur gives way to a pleasant aroma of lime.

Thus, in the steelmaking department, there is a polarisation between the smelting shop and the refining department. Since privatisation, the smelting shop has undergone several reorganisations, which disposed of its militant workforce and made work more intensive and hazardous. In terms of working conditions, career prospects and salary, the older and predominantly black workforce of the smelting shop is much worse off than the younger and better-paid Class C in the refining department, even though smelting jobs are more skilled and hazardous than those in the refining department.

Hot Strip Mill (GGLQ)

Entering the GGLQ, visitors are overwhelmed by the vast scale of the technology and by the harsh environment. The nearly two-kilometre-long mill is under a thick veil of steam coming from the cooling system underneath it. Because of the mill's immense scale and the thick chemical fog hovering over it, the department at first looks populated just by machines. Occasionally, some human silhouettes appear through the fog, illuminated by the fleeting orange lights of the running coils. They then quickly recede in the background and disappear. In fact, the department has the greatest concentration of employees in the plant – 150 direct workers in the strip mill and 150 workers in the pickling section. The room temperature is 30°C on the floor and 40°C on the passageway above the mill, where the operators stand and walk most of the time. As for the other productive departments, the GGLQ's general director is directly under the CEO and the president. Below the general director are four managers for the general process, maintenance, the acid tank and the cylinder section. The average GGLQ employee is forty years old, married and with young children, with low formal education (secondary school or technical degree) and twenty years of seniority at CSN. Many GGLQ employees are second-generation workers hired during the company's privatisation.

Displayed in the cavernous entrance of the GGLQ are the department's official mission, values and philosophy. The mission is 'to provide the internal and external clients with the amount, quality and kind of coil agreed with the lowest possible cost and the highest degree of physical security for the workers and the machines'. The main values are humility, honesty, loyalty, empathy, teamwork, enthusiasm, discipline and knowledge. The philosophy revolves around the three principles: 'being focused on the orders coming from the top management', 'caring for security, productivity and quality' and 'having a preventive mind'. Despite these self-professed values and philosophies, the department's main concern – from the helpers to the general manager – is productivity. According to Octavio, the general manager, the mill produces a profit of R$200 per second.

Octavio has an MA and a PhD in metallurgy from the prestigious Federal Fluminense University. He had a rapid career at CSN. He started in 1982 as a process engineer in the steelmaking department, and in 2007, he was already a general manager at the GGLQ. Octavio says he does not 'think like an engineer'. Engineers focus on the long term and on quality improvements, whereas he focuses on short-term productivity and cost efficiency. As a general manager, he introduced two major organisational changes to boost productivity. First, he implemented the just-in-time (JIT) system, which introduces the market principle into the line (see also Mollona 2009) so that coils are produced only when a client – internal or external – orders them. Roughly two-thirds of these clients are internal – the pickling line and the cold mill – and one-third external. Second, Octavio introduced the cell systems, transforming the different stations of the hot strip mill into separate 'productive cells' with dedicated supervisors. Cellular manufacturing is a workplace design developed in Japan in the 1970s, and with the JIT system, it continues to shape the organisational structures of corporations across the globe. Unlike traditional functional organisations, where similar machines produce functionally separated outputs, different machines produce similar outputs in the cell system. The main aim of cellular management is to 'multi-skill' operators by having them work on different machines and tasks.

This managerial system dates back to the 'job enrichment' agenda of US management science in the 1970s. Indeed, the first cell system was introduced in CSN in 1974, under the military regime, by the US consulting firm Arthur D. Little, which led the factory expansion known as plan D. Based on the analysis of more than 848 tasks and 3,370 occupation workflows, the US consultant radically changed the job descriptions and salary structures. Peripheral workers (cleaners, builders and other general service workers) were dismissed or encouraged to take early retirement, and 'strategic workers' were 'multi-skilled'. The new job description and salary structure was kept secret, spreading uncertainty and competition on the shop floor (Caruso 2009: 130). After the latest round of cellular reorganisation led by Octavio, the turnover of the GGLQ reached a stunning 60 per cent. The reorganisation was badly timed, as it happened just when foreign steel and car corporations moved to the Fluminense region, offering wages well above those of CSN and hence generating a mass exodus of employees. Of GGLQ employees, 80 per cent are under thirty years old. According to Octavio, these young employees are 'wage hunters' with no corporate loyalty or career vision. Nonetheless, he is not worried about high turnover and in fact considers it a blessing in disguise:

> I know that management theory considers turnover as a cost in terms of training and human capital. But in fact, in my opinion, it is more expensive for us to have long-term wage relations and fixed contracts. High labour turnover can be a blessing as long as the system is entirely standardised. My ideal case scenario would be to have an entirely mechanised workplace with entirely replaceable labour.

Octavio's ideal of JIT labour, where workers are totally substitutable and employed only with short-term contracts, reflects CSN's strategy of the time of building an unskilled, short-term and flexible internal labour market. Such flexibilisation and casualisation of the permanent workforce creates a monstrous a hybrid between Taylorism and Toyotism.

Octavio is not worried about the human consequences of work intensification either. He sees labour as infinitely elastic and flexible. But he is very concerned about the fragility of the machines, especially the strip mill; 80 per cent of the GGLQ machines were installed in 1982 and were programmed for a maximum yearly level of production of 3 million tonnes. But they currently produce 5.1 million tonnes, and the top management is planning to increase production further and process all 5.4 million tonnes of slabs produced in CSN. Octavio comments:

> I am worried the 'old guy' [strip mill] will not be able to keep up with such levels of production. It is already thirty years old. You can see. Its body is rusted and bent. Look at the rolling stands. They are like crooked bones. If we push it too far, it will snap and break. There are several furnaces and continuous casting machines but only one hot strip mill. This department is like a funnel: everything passes through it. This mill is the nodal point for the whole production process and where greatest value is produced. If the old guy dies, this whole firm will shut down.

Concerns for the hot strip mill also dominate the workers' narratives. For instance, according to José: 'The old guy generates a value of R$200 per second. People from other departments take for granted that it will last forever. But when it will stop for good, they will understand how harshly we all have treated it.' Together with productivity, the other central narrative of the department is that of 'success'. Indeed, most GGLQ employees, especially the unskilled ones, have experienced upward mobility. Raised in poor bairros in the city's periphery, they now live in affluent working-class neighbourhoods with good sanitation, public services, including schools and medical centres.

There is a strong regional dimension to this trajectory of upward mobility. The better-off *mineiros* workers of the GGLQ continue to refer to the company as their mother and link their professional success to 'her' generosity and the nurturing and caring attitude of the managers, rather than in terms of meritocracy. Marcos was orphan since the age of five. He worked as a street vendor until he was seventeen, when he became a subcontractor in the UPV smelting shop. After five years as a subcontractor, he applied for a job at CSN. A few months after applying and not hearing back from HR, he showed up at the office of Juvenal Osorio, CSN's president at the time. He describes the event as follows:

> I reached the top floor of the company's headquarters. A security guard stopped me. Right then, Osorio was passing by and invited me into his office. I told him I wanted to be given a chance in life. I wanted to be part of the CSN family. I wanted to work hard and to earn enough money to buy a house and make my own family. Two days later, on 1 December 1987, I got a job at the GGLQ, where I made it 'all the way up' to level 3.

Narratives of success are plentiful. CSN adopted Dudú when he lost his parents at fifteen. He started working as a message boy and 'made it to the top' to his current position of shift supervisor (level 3). José, his workmate, describes himself as a 'high achiever'. He started as a 'servant' twenty-five years ago when only eighteen. In quick succession, he was promoted to the role of helper, operator and supervisor – a role he still covers. He says his career, thanks to the generous support by his *padrinho* (godfather), was 'stellar'.

Experienced workers become sponsors (*padrinhos*) of young recruits, training them in their first year of employment and supporting them throughout their careers. Candidates who apply for promotions are assessed by a technical commission (*banca técnica*) composed of area managers, supervisors and *padrinhos*. During the job interview, *padrinhos* stress the candidates' strengths and fitness for the new position. Indeed, job positions such as mill operator, oven operator, panel operator or floor assistant do not have formal descriptions except for the generic progressive classifications of levels 1–3. With these generic job profiles, the oratory skills of one's *padrinhos* are fundamental to progress. Employees with strong *padrinhos* are likely to make it 'all the way to level 3. Failure to progress to level 3 after the first three years of training normally leads to dismissal. *Padrinhos* must be treated 'like family'. They must be showered with gifts, favours and some free labour – on the shop floor or in their household – and visited during festivities and special occasions. The institution of the *padrinho* dates back to the Catholic institution of *compadrio*, which was re-appropriated at the time of slavery in the Paraíba Valley. Fugitive or manumitted slaves needed powerful *compadres* – mostly planters but sometimes older freed slaves – to win back the trust of their masters or intercede for them in legal manumission proceedings (Chalhoub 2006; Mattoso 1986; Reis 2008).

Indeed, 'kinship', both real and fictional, is a central organising principle in the GGLQ, where nearly every employee has siblings or children working in the department or in departments nearby. Even though all perspective employees must go through a formal job interview, kinship remains the central recruitment mechanism. In fact, against the general manager's dream of JIT labour, HR prioritises the recruitment of current employees' children to keep turnover low and to increase loyalty towards the company. Kinship is also a route for progression. Workers with parents or relatives already in the company have access to influential *padrinhos* and well-predisposed technical commission. Moreover, kinship functions as an informal supervision mechanism, which sometimes creates intergenerational conflicts. For instance, Jorge keeps a constant eye on his son Danilo: 'If the quality of the slabs that Danilo processes on the hot mill is low, I call him on the walkie-talkie and tell him, "Hey son, what are you doing? This stuff is rubbish!"' Older production workers discourage their children from working at CSN and push them to go into the media or IT industries now booming in Volta Redonda. Top managers and high-level engineers push their children to stay 'in the steel family' and put them on training schemes inside the company from early on. But the children of unskilled employees more often than not end up becoming steelworkers, whereas the children of managers go into the media or service sectors. More than father-children relationships, these career paths reflect structural factors – managerial jobs in the steel sector are low-income and precarious, whereas managerial jobs in the media sectors are better paid and have higher status.

In addition to the support of *padrinhos*, workers need the backing of the departmental leader to progress professionally. The position of leader, introduced during the company reorganisation of 2004, replaced the former position of line supervisors. Leaders do not have the same status of *padrinhos*, who de facto control the internal labour market, or the formal power of shift supervisors, who are in charge of production targets. Their main role is pastoral care, exercised especially during the 'flash meetings' (*reunião relâmpago*) held fifteen minutes before the start of each shift. These meetings focus on the employees' family financial and health circumstances rather than on production. Leaders chair these meetings informally by reading excerpts from the Bible or from the local newspaper or telling stories, anecdotes and gossips. Flash meetings work as a subliminal form of corporate communication, especially on issues of remuneration, complaints and internal security. With its dense networks of kin, leaders and *padrinhos*, the GGLQ perfectly incarnates Vargas's old corporate ideology of the *família siderúrgica*. But family ties and regional solidarities of the department are overridden by an invisible corporate hierarchy running along lines of income and race. For black and poorer workers, success is a chimera. They are predestined to stay at the bottom of the career ladder.

Poor workers who cannot afford to pay for university degrees[21] do not progress to level 2 and, because of the company's 'up or out' policy, are laid off within three years of employment. In addition, the careers of leaders and workers end at level 3. They cannot progress to managerial positions. Nevertheless, the truly stellar ascension to power of Enéas – who became the company's CEO from being a simple production engineer – keeps the corporate dream of mobility alive. Parallel to these dreams of professional success especially among the more affluent *mineiro* Class C are narratives of rural nostalgia and the striving to return home. Aldo, leader at the rolling mill furnace, tells me:

> Both my parents are *mineiros* and come from the countryside. They grew up in farms. When I was young, they would tell me wonderful stories about places and people of that rural world. I could nearly smell the grass and animals in their stories. I have a plot of land here. When I work on it, I feel sad and melancholic. I feel I want to live there forever. I feel sick and tired of industry.

Despite all the talks of 'success', 'stellar carriers' and upward mobility, the GGLQ offers little prospect for social advances. Together with family men, the department is full of younger 'wage hunters' with a contractual and detached attitude towards the company and who consider employment in the UPV in strict monetary terms and as a first step towards finding better employment. This working-class perception of employment as 'money' fit well with the JIT labour vision of the top management. Such instrumental and materialistic world views by younger and more precarious workers and the availability of easy credit (*crédito consignado*) by local lenders led them to spend well above their available income. For some time, this precarious working class had a consumption style comparable to that of the affluent Class C. But the economic crisis hit their households hard, generating bankruptcies and mortgage defaults and bringing to the surface pre-existing inequalities within the workforce.

Cold Strip Mill (GGFM)

Because the GGFM holds the national monopoly of tinplate and produces tins and cans for prestigious customers such as Coca-Cola and Nestlé, the top management of the department has great power in the company. For instance, unlike other departments, it has equal say on specifications, branding and packaging with the purchasing department in São Paulo. Unlike the GGLQ, where the organising principle is 'productivity', the organising principle of the GGFM is 'sale maximisation'. Most line operations are mechanised, and pressures to increase productivity and sales fall mainly on the mill operators and packers located at the end of the line. When production is intensified, the twenty tonnes of steel sheets that run every day along the 350-metre-long rolling mill regularly build up bottlenecks, leading to small explosions that stop the line for some time. Under pressure to minimise production slowdowns, line operators quickly unwind, cut and reset the sheets into the rolling trains by hand. To an external observer, these packing operations look ordinary, but sheets are as sharp as razors, and even when handled with gloves, they can cut, bruise, and mutilate – indeed, the GGFM has the highest rate of work accidents in the plant.

Moreover, consisting mostly of packing and handling operations, the GGFM is considered an unskilled department. Historically, the GGFM was the only productive department to employ women, pejoratively called *vira latas* (can turners). Today the department continues to employ more women, unskilled and black workers than any other department in the firm. Overall, the employees identify with the prestigious global brands they serve and proudly consider themselves the drivers of the firm's sales and profits. But such market-based ethics is unevenly distributed among the workforce. Workers at the top end of the line and on automated tasks go along with labour intensification. Those on alienating and hazardous packing tasks at the bottom end – mainly women, black and young – resist it. Take Douglas for instance:

> My father often tells me to pay attention to the 5S. He is quite emphatic about it. But the system is too rigid. In the past, I had just to put the piece of plastic around the coil, cover the coil and tape it. Now we have to load it on the forklift, lift the coil . . . This 5S is madness. The management innovate continuously.

Unlike the workers on automated tasks who live in affluent areas, end-of-the-line workers live in extremely poor neighbourhoods,

Figure 2.2. Women in the cold strip mill. Published with permission from the Instituto de Pesquisa e Planejamento Urbano (Institute for Urban Research and Planning).

where 20 per cent of residents earn less than one minimum salary and 15 per cent are without income. Overall, the age and socio-economic profile of the department is well below the corporate average. Thus, it is not surprising that, in the GGFM, employees, especially young ones like Douglas, are sympathisers with the Socialism and Freedom Party (PSOL) and the PCB, the two radical working-class parties – left of the PT – with a great following also among precarised school and university teachers, IT workers and retail workers. In Volta Redonda, the PSOL leader is the federal MP Marias das Dores Mota (Dodora), the charismatic woman leader of the Sindicato dos Educadores e Profissionais da Educação (Union of Educators and Education Professionals) who sparked the university movement in the 1980s and ran for mayor in 2008.

The recently introduced cell system tried to break the resistance of end-of-the-line workers. By putting line workers in charge of other line workers, the cell system shifted the traditional power struggle between workers and management onto the very line. Arnaldo, the GGFM general manager, had to convince 320 employees of his department to endorse the cell system:

> In the cell, there is no hierarchy. Instead of the old shift supervisors, we now have cell leaders. This system simply formalises the previous pattern in which the most experienced workers informally trained and assessed their workmates. The aim of the cell system is to give workers ownership over the production process – make the production line self-sufficient and self-directed. Every day I remind employees they are the ones who own the equipment – not me or the company. At the core of the notion of the cell is the very principle of self-management. The cell is a vision of men and machines coming together – of humans becoming machines and workers becoming owners. Yet, several workers opposed the introduction of the cell system. I had to lay them off.

José Geraldo, a top manager in the GGFM with a background in engineering and business administration, was in charge of implementing the cell system from 2001 to 2003. He encountered a lot of resistance, especially from the skilled workers who controlled the labour process. He argues:

> The cell system created a new job profile that curbed the informal power of older and more skilled workers. Now 80 per cent of the jobs are 'routine'. Our new operators are less qualified, have low career expectations and do not mind moving across functions, departments and even firms. They like routine jobs, yet they are also flexible and independent. They see the wage contract as a form of exchange: 'I give to the company good productivity and CSN gives me back its expertise.'

The creation of this flexible internal labour market allows the company to downsize rapidly during economic crises and rehire when the firm's finances improve.[22] Leaders generally share the managers' enthusiasm for the cell system. José, a thirty-six-year-old leader, is totally behind the philosophy of the cell. He entered CSN in 1990 as operator 1 and was made supervisor in 1997, a position he still covers but under the new title of leader. He comes from a family of unskilled workers from Brazil's poor Northeast Region and lives with his parents in a small house in the bairro 9 April. Like Arnaldo, he considers the cell system one of self-management:

> The cell is a system in which everybody can embody the figure of the leader, where each employee thinks like the superior: the operator thinks like a leader, the leader like a supervisor and the supervisor like a manager. Self-management would work perfectly in a non-human, mechanical world. But the human systems are not like that. They are imperfect because every human being is unique. I talk and transmit a message, but you understand it in your own way. I have to put myself in your chair, where you are sitting now, and imagine what is going on in your mind. The role of the cell leader is to use empathy to build a group of people who think alike and act as one man – one machine.

Leaders are great believers in the cell system, which they argue fosters an egalitarian, empathic and co-operative culture when, ironically, the cell system destroys these very human attitudes. Yet, these notions have a strong appeal among the many spiritualists, evangelicals and Kardecists who make up a big portion of the unskilled CSN workforce.

For instance, Getúlio Jr., a twenty-eight-year-old unskilled line operator is fervently evangelical and pro-management. He was named after Getúlio Vargas. His father, Getúlio Sr., was born in Valença, a centre of the coffee economy, and his mother died giving birth to him. Getúlio's uncles raised him and carried him with them when they moved to Volta Redonda in search for a job at CSN in the 1940s. He was born with a twisted spine and had three major surgical operations in the last ten years. Petrobras turned down his job application because of his invalidity. But faith in God made him persevere, and CSN offered him a job in 1993. He spent seven years on the line and then was promoted leader. He says:

> My co-workers criticise me for being too pro-management. But for me the world of production is like the world of God, for every action follows a pattern. No human action should be wasted. For instance, look at *rendimento metálico* [metallic productivity – RM, an operational target based on waste and residual material]. The average RM target is 98.8 per cent, which means 1.2 per cent of the material is lost in waste. I ask my workers for an RM of 100 per cent. If they truly have faith in God, their actions will be wasteless.

The forty-two-year-old line leader Claudio is an evangelical minister based in the bairro of Santa Cruz. When he entered CSN in 1989 as subcontracted furnace cleaner (the most unskilled job in plant), he was a fervent Communist. But in the 1988 strike, the army killed a close friend of his, and he was left traumatised and deeply changed by the event. He quit the Metalworkers' Union and enrolled in a BA in theology. Then, he had a religious epiphany and decided to give up politics. Every evening, Claudio preaches at his local evangelical church with his wife, also an evangelical minister. He has many followers inside and outside the company – among managers, workers and neighbours. He speaks with an idiosyncratic mix of managerial and Christian jargon:

> I see myself more as a spiritual leader than a technical expert. I teach to those who are physically or intellectually disadvantaged to dominate their machines – to become aware of their role in the grater production chain. Workforces should be rewarded professionally and not just in monetary terms. If you have high wages but you are not happy in your job, this is not

> good money. In the Bible is written that you must do everything as if you were doing it for God. When someone starts complaining about the boss, I do not defend the company: I talk about biblical principles and attitudes. I tell them they are blessed to be here. Many men do not have a job to support their families – no matter if the wage is good or bad. They have no money to support their house.

Marcelo, a twenty-seven-year-old fresh recruit, has a much more pragmatic view of religion. Marcelo works in the toxic and hazardous continuous annealing process. He travelled all over Brazil, including to the extreme regions of Pará and the Amazon, following his father, an unskilled painter now working as a subcontractor for CSN. He became religious after he witnessed the extreme poverty of the favela Paraiso, located between bairro California and Barra Mansa. The favela has 380 families living in about 160 barracks with two to three families per barrack. Each barrack is fifteen square metres. Most children in the favela are starving and seriously ill. With the community of the Santa Maria Aparecida Church in the bairro Conforto, Marcelo puts together monthly provisions of flour, rice and sugar and organises Christmas lunches. He is now trying to bring a team of GPs in the favela to vaccinate the children. Most residents have never seen a doctor in their entire life. He says:

> I do not like Volta Redonda. It is a small town but with the problems of a big town, including violence, pollution and extreme poverty. When I mentioned the favela Paraiso to the departmental manager, he was shocked. 'I did not know there were favelas in Volta Redonda', he told me. How can they be so blind? The whole city is surrounded by favelas. Why in such a rich city and with such a profitable company are so many people starving? It was the same in Macaé, the oil city where my father was an oil rig worker. Poverty and industry seem to go hand in hand.

Arnaldo, the general manager, describes the GGFM employees as 'young and highly qualified'. In fact, they are much less qualified than most employees in the firm[23] coming from the army, security firms or farming rather than from the usual recruitment channels such as the ETPC, the Serviço Nacional de Aprendizagem Comercial (National Apprenticeship System) or internal subcontractors. As well as women, the department has several black and mestizo workers from rural background. On the shop floor, there are many stories of past slavery and of present-day forms of labour exploitation akin to that of the Paraíba Valley fazendas. But the black GGFM employees from rural backgrounds employed are less antagonistic vis-à-vis the management than the black smelting shop workers, who have a long history of industrial labour and political activism. Being the first wageworkers in a long family tradition of indentured and slave labour, they strive to integrate in the company and avoid antagonistic postures.

Likewise, the younger and better-qualified female employees of the GGFM do not articulate their narratives around class or gender. Discussions about gender and class discrimination are more frequent among older women, perhaps because most of them had one or more *vira latas* in the family. For instance, Maria, a black quality inspector, lives in Siderlândia, one of the poorest bairros in Volta Redonda. She has worked fifteen years in the GGFM and is still on level 3:

> I enjoy being a process inspector, but relations with men are very difficult. Male supervisors do not like women to have the last word. Women constantly have to prove themselves. They are always cast as physically weak . . . They constantly have to show they are strong. In fact, the production process is not wearing at all. Even a child could do it. Women end up producing much more than men just to prove themselves. You need to work more, a lot more, to be respected on a professional level. If you are a man and you make a mistake, that's fine. But if you are a woman and you make a mistake, that's because you are a woman. For a very long time, women were only allowed to work in administration, or as secretaries or in the sorting room. They were treated inhumanly, like slaves or machines. Under the old public regime, this shop floor was full of women, but only in administration or as *vira latas*.

It was the interim president, Maria Silvia Bastos, who, during the privatisation of the company, opened the doors of the shop floor to women. Then, Maria moved from the sorting room to the production line:

> Being discriminated on both counts of being black and a woman made me a more compassionate and responsible leader. If you ask me about class, I do not know what to answer. I think it is actually more important to have women in managerial positions than struggle against the management. I encouraged my two daughters to study engineering at the Federal Fluminense University. They are about to graduate. I hope they will become top managers in CSN and improve working conditions for all workers. Look at Lula. He was just a poor black worker from the Northeast and went on to become Brazil's president!

The fact that the department with the highest number of black, female and poor workers is also the most productive and profitable in the company should not come as a surprise. Younger workers tend to resist the continuous labour reorganisations and the intensification and precarisation of their work through slowdowns, health and safety campaigns or absenteeism. Indeed, the Volkswagen, Peugeot and Thyssenkrupp CSA factories in the region provide alternative sources of employment and hence increase their chances of exiting the wage contract with CSN.

But most employees in the department seem to subscribe to their leaders' ideologies of self-empowerment, self-management and moral or religious enlightenment such as Arnaldo's idea of the cell as a form of self-management, Maria's career-based vision of female empowerment or Claudio's evangelical perception of work as spiritual self-empowerment. Moreover, several GGFM workers are involved in informal economy, especially subsistence agriculture, commerce and service, for instance, as clerks in shopping malls and retails shops, car salespeople, street vendors, lottery operators and illegal parking attendants. Such self-entrepreneurship makes employees less dependent on the exploitative work in the GGFM.

Cold Strip Mill (GGLF)

When asked about their labour, GGLF workers often describe it as a 'more noble' version of the GGFM, where sheets and coils are not coated or galvanised. Unlike the cheap, light, disposable cans and tins produced in the GGFM, the hard sheets produced in the GGLF go into expensive durables and prestigious cars such as Fiat, Ford, General Motors and Mercedes. José explains to me that the expensiveness of the final product and the nobleness of the material go hand in hand. The more noble materials go into the motor industry, and the less noble ones into cheaper industries. Line 3, the least noble line in the department, produces sheets for the white goods industry up to a thickness of 2.1 millimetres. The process is poor in alchemical terms because it involves no hardening of crystals. Line 1 produces sheets for tubes, drums and silos in the civic construction industry with a thickness of 2.84 to 3 millimetres. These sheets contain lead crystals and hence are slightly 'nobler' and harder than the one produced in line three. Line 2, the 'most noble' line, serves the automobile industry. It uses antimony crystals, which are purer and more resistant to oxidation than lead. Antimony is a grey, silvery crystal. José says Egyptian princesses used it as eye cosmetics in 2000 BCE and that metal sheets containing antimony are smooth, translucent and beautiful to look at.

The revenues of the GGLF are higher than in all other departments, and the outputs are twice as much as those of the GGFM. Yet, the GGLF is much less profitable than the GGFM and struggles to break even because of fluctuations in the global steel market. In 2010, the combined effects of the re-evaluation of the real and the economic slowdown nearly halved the sales of galvanised sheets in the international market, and 23 per cent of the GGLF workforce was laid off. The international branches Lusosider (Portugal) and Heartland Steel (US) partly improved the firm's currency deficit. But the position of CSN in the galvanised and hard sheets markets improved radically only after the firm focused on the internal markets at a time when these were being heavily subsidised by the government. The 'less noble' products going into the civic construction industry saved the department from bankruptcy. Production is now back to pre-crisis levels, and the department started to re-internalise labour. The nobility of the steel processed in the GGLF does not square with the profile of its workforce, which, like that of the GGFM, comes from families of poor and unskilled labourers (painters, carpenters, cleaners) who travelled from former coffee towns to Volta Redonda searching for temporary work at CSN. But in residential terms, the GGLF workers are considerably better off than those of the GGFM. Yet, memories of harsh labour in coffee plantations, cattle ranches and rural fazendas are still vivid on this shop floor.

Wagner, a twenty-eight-year-old crane driver, is one of them. He comes from a family of formerly enslaved people from Espírito Santo. He has two brothers – one is a stonemason, and the other is unemployed. His wife works as a part-time cashier at the local shopping centre. Wagner started working at CSN in 1998 as a subcontracted furnace cleaner, one of the most dangerous and unskilled jobs in the plant. He was permanently hired in 2006, dismissed during the economic crisis in 2008 and rehired eight months later. Because he was fired twice at CSN, he is still officially classified as a trainee. Since he restarted at the GGLF, he has been working on continuous twelve-hour shifts in recent months to substitute a more senior driver who has been absent for several weeks:

> I never got used to spending twelve hours all alone in this crane suspended fifty metres above the ground. Can you see? The people down there are small and unreal. They move slowly and are folded in smoke and vapour. When I am here, I keep on thinking of my son. I miss him very much. I also often think about the stories my grandmother told me about working in a coffee plantation. She saw one of her sons killed by the fazendeiros. Our family is still scarred by that memory.

The recent re-internalisation of labour led to more precarious working conditions in the GGLF. Workers on level 2 are either dismissed or 'flexibilised' – made to rotate on different jobs or put on short-term contracts. These flexible and short-term internal contracts allow the firm to adjust its manning levels to market demand without the costs of hiring and firing associated with the traditional wage contract. Moreover, overtime by unskilled workers or young trainees is regularly used to cover for those senior workers who are absentees or have second jobs.

Rolando, the general manager, makes no secret that the firm's training system allows CSN to adjust manning levels easily and without big political costs. By law, apprentices have a forty-day period of training under the supervision of a godfather. At the end of the period, a committee assesses them. If they pass, they are hired permanently. Apprentices are under the 'up or out' rule. They are expected to proceed from level 1 to level 3 in two years. If they do not, they are out. Rolando acknowledges the system is harsh on the less wealthy trainees:

> The 'up or out' rule is important. They cannot grow old and start families on trainees' wages. But those without technical qualification need to study part-time and get a technical diploma. Otherwise, they cannot pass. I am worried for these new lads who come without education. To get a technical diploma while working is extremely difficult. Some of them even failed primary school!

Through lay-offs, internal and external transfers, and other forms of flexibility, the firm has developed an internal 'quasi-market' based on contracts that are in between wagework and part-time contracts. 'Quasi-markets' allow those departments of CSN that are more exposed to the global economy to change manning levels without resorting to firing or subcontracting – both coming with high political and economic costs. At the firm level, such combination of precarisation and internalisation allows the company to operate in the international steel market – highly profitable but volatile and hence with unpredictable staffing requirements – and in the domestic market – less profitable but more predictable and with fixed manning requirements.

Managers, White Collars and Leaders

The offices of the middle management are in modular units in the courtyard to the back of the furnace. At regular intervals, their walls and windows are shaken by the electric discharges of the furnace. The rooms are overheated; inside them, desks, computers and windows are coated with grey dust. Managers shout at each other at a short distance. The courtyard that connects the offices of the middle management, the furnace and the top managers' quarters is an unusual ecosystem in which nature and industry are oddly entangled. Chemical vapours flow upwards from the cracks of broken pipes leaving trails of green grass and wild flowers along the exhaust system. Cooling water falling from the coal tower gathers in bright blue circular pools surrounded by mango trees and tropical flowers. Hummingbirds from the nearby forest fly over silvery coal dust carpets. Rusty water drops from discarded train wagons and metal scrap assembles in orange pools half-solidified in dark crystal formations. In the courtyard, the primal elements of steelmaking – iron ore, water and coal dust – reappear in the form of industrial waste.

The flat and hierarchical post-privatisation structure of CSN disempowered the middle managers and precarised their jobs. In 2004, the reorganisation of AF2 and AF3 into independent production centres led their directors to clash. The rivalry between the furnace directors ricocheted on the middle management. The situation deteriorated so much that, after refusing the company's proposal of a 30 per cent salary increase,[24] the AF2 managers and engineers left to join the rival firm Thyssenkrupp CSA, bringing with them the entire furnace's production team. Steel companies that have recently moved into the region offer higher salaries than CSN – in some instances, 50 per cent higher. In 2015, the turnover rate of the middle management was 30 per cent. Fighting lay-offs on a daily basis, middle managers have become 'wage hunters', too, like the younger operative workers, with whom they often ally against the top management. For instance, Jonas says:

> After privatisation, the middle management was totally balkanised and squeezed between the top management and the line. With my degree and an MSc in metallurgy, I am more of a burden than a resource for the company. [The top managers] need yes-men like the leaders who are illiterate, unskilled and docile.

Inside the offices of the top management are small landscaped gardens with *bromélias* and cacti, birds in small cages hanging from the ceilings and broad skylights pouring natural light on desks filled with paper and dust. In the firm's vertical organisational structure, top managers have full control over their productive departments and unlimited power – second only to that of the CEO. Unlike the middle managerial rank, top managers identify totally with the company. They tend to have a middle-class family backgrounds – they are children of lawyers, architects or teachers – and are technically competent and well qualified; most have executive MBAs from prestigious Brazilian universities. They all share an American-inflected business jargon and a cosmopolitan, developmental, globalist and imperialist outlook.

Fabian, the forty-year-old black director of the GGAF wearing a Rolex, Ray-Bans and a polo T-shirt and speaking in a thick American accent, is exemplary of such cosmopolitan and well-qualified top management. He started at the ETPC when he was seventeen and in 1988, at the age of twenty, began working in the coke department. Studying at night, he graduated in engineering at the Federal Fluminense University in 1992 and, after an MBA in management, moved to the GGAF, where he progressed from team leader to general manager. Fabian supported the privatisation of CSN, which he experienced first-hand as a young recruit, because in his opinion it improved the efficiency of the UPV and made it independent from foreign capital – unlike privatisations in other Latin American countries, which imposed upon these countries a 'Western regime of austerity and monetarism'. Fabian's views are shaped more by international relations than by the politics of the shop floor. The international commodity market and Brazil's trade deals in the Mercosul heavily influence Fabian's daily decisions. Raw material (coal, coke and iron ore) constitutes 60 per cent of the GGAF's production costs and 50 per cent of CSN's overall production costs. Production-wise, not much can be done to cut these costs. But according to Fabian, 'it is important to have a sound geopolitical strategy'.

Up until 2007 CSN imported 100 per cent of its coal (450,000 tonnes per year) from China. After the economic crisis, the Chinese government set up export tariffs and the price of coal rose by 500 per cent. So, CSN switched to Colombian suppliers and such switch improved radically CSN's finances. Fabian says: 'Since Rousseff went into power, Brazil has a dominant trade position in South America and this allows us to access raw material cheaply from our Latin American friends. If we want to continue to be competitive as a company, Brazil must continue to improve its geopolitical power.' More generally, the top managers' sense of loyalty, technocratic excellence, discipline and self-perception as 'civil servants' is in line with the tradition of authoritarian technocracy of the Vargas era, which persists in big public conglomerates such as Petrobras, Vale and the BNDES.[25] Indeed, top CSN managers tend to have a statist mentality and admire Lula's strong presidentialism and interventionist stance in international relations. For instance, Jaime, another top manager, says:

> You know, I will surprise you, but I really respect Lula. He has brought order and discipline in Brazil. He is taking back *our* Latin America and militarising the country against foreign capital. He does not let the North bully him. Have you seen how he chastised the developed nations at the G20 in Davos recently, blaming their economic crisis on their irresponsible capitalism? He is not one of these 'soft lefties'.

The white collars of the Gerência Técnica (Technical Administration – GT) service the general direction and the main productive departments. During my fieldwork, the GT had just been created as part of the restructuring connected to the implementation of the SIGMA. On my arrival, the department was still under refurbishment. About thirty employees were cramped inside a big room, sitting in small desks in parallel rows and wearing earplugs to protect them from the loud construction noises from the adjacent building.

All GT employees are engineers or highly qualified technicians with a strong educational background[26] who live in affluent areas of the city. Most of them are extremely knowledgeable about the production process. They know the exact quantity of pig iron and special steel that goes into each furnace, the technical specifications of the steel of each final product – cars, metallic structures, laminates and billets – and the organisational structure of the company – from the top management to the subcontractors. They tend to be very critical of the management, mainly because of the SIGMA, which deskilled them and drained technical knowledge away from the firm. As with the GGAF engineers, the turnover rate of the GT engineers is very high (30 per cent), and their migrations to rival steel companies are frequent. Vanderlei, the head of the GT, considers the SIGMA part of the broader process of industrial downsizing of CSN and of its transformation into a financial conglomerate. Vanderlei is a talented engineer. He invented a coating material for furnaces that saved the company millions of reals. He is also an extremely competent accountant with an MBA from a prestigious international university. But he is not committed to the company, and instead of seeking internal promotion, he pursued a parallel career as a technical consultant. Now forty-eight and close to retirement, he intends to fully pursue this consulting career.

The CSN engineers have a long history of independent mindedness. In the 1980s, the Sindicato dos Engenheiros de Volta Redonda (Engineers' Union of Volta Redonda), affiliated with the PCB, played a central role in the anti-dictatorship movement and supported the anti-capitalist struggles of the radical *metalúrgicos*, especially through legal struggles focused on workers' right. In recent years, the union has veered away from labour activism and focused instead on the environmental activism. Privatisation radically diminished the engineers' status and power by replacing them with subcontractors or with the automated SIGMA. Vanderlei argues:

> Paradoxically, it was the currency crisis of the 1990s that started the process of financialisation of CSN and diverted it from its core business of steelmaking. At that time, during the Cardoso administration, the ideology that supported financialisation was that of free market. Now, under Lula, financialisation is presented as a form of social inclusion.

After the discovery of the Tupi oil field (later renamed the Lula oil field), President Lula had announced the creation of an oil sovereign fund (a state-owned capital reserve) to be used on anti-poverty and educational programs and on the PAC, a series of ambitious infrastructural projects.[27] According to Vanderlei, mayors and governors promised grand infrastructural developments to boost their electoral campaigns. But after elections, these projects failed to materialise. In fact, the PAC's completion rate is very low. Moreover, according to Vanderlei, PACs in Volta Redonda are monopolised by ruthless entrepreneurs who put subcontracted workers under extremely hazardous working conditions. Indeed, a week before our conversation, ten subcontractors had died while working on a PAC in Minas Gerais.

Moreover, even when these projects are realised, they are too short-term and low-tech to have any economic impact. Vanderlei points out how the Chinese government uses sovereign funds to boost exports and R&D and to create trade barriers to foreign capital. As for most countries in the Global South, the Brazilian steel industry is technologically obsolete and low-value: 44 per cent of the costs of steel are concentrated in the steelmaking process, which produces low-quality and low-value steel slabs and is heavily pollutive. However, steelmaking in the Global North is not pollutive because it focuses on end processes. Instead of using the oil fund for social redistribution, according to Vanderlei, the state should boost industrial trade and invest in new technologies, rebalancing the technological gap between North and South. Vanderlei considers the owner of the firm a 'speculator' and a member of the global financial elite that is destroying industry. His view is shared by the younger engineers of the GT, for whom the relationship between precarious working conditions and global economic and environmental inequality is central.

For instance, Beto, a metallurgical engineer in the coke department, is developing an applied project for his environmental management

MA that aims at reducing the incidence of volatile material in the coke reduction process by improving the seals of the blast furnace's doors. He also set up the CSN 'green helpline' for residents to denounce instances of CSN-related pollution. Beto opens the window of his office and tells me:

> Can you see that white, thick cloud of smoke? That's how we turn coal into coke. It is a very backwards process – nearly primitive. The residents of Volta Redonda never got used to the amount of carbon released into the city from the coke cavern. Pollution is increasing. The top management of the company is careless about the environment.

Beto relates such a careless attitude towards the environment to the firm's cost-accounting system, which considers the productive department an autonomous unit of profit and put it under heavy pressure to cut costs. For instance, under the current system, iron ore is acquired from the mining department at a market price rather than the internal transfer prices, and this is financially straining for the coke department, which in turn squeezes salaries and cuts investments first on pollution-reducing technologies and processes. Tadeu, a metallurgic engineer who worked at the quality lab for twenty-one years and supervises 180 employees, agrees with Beto. He is in charge of checking the quality of the steel in the primary refinement process – when it changes from pig iron to finished steel. Tadeu is very critical of the recent company reorganisation:

> For all its talks on TQM and cell management, the top management is just focused on cost cutting. Each year, I attend process-improvement training at the Nippon Steel Corporation in Japan and America. They truly have a holistic view of the production process whereby the cost dimension is entangled with the human, environmental and quality dimensions. This holistic view also existed when CSN was under state ownership. Privatisation destroyed it. It created conflict, competition and insecurity.

Samuel's work experience is different. Unlike most of the highly educated employees in the GT, his only qualification is a degree in mechanical turnery. Forty years old, married and with three children, Samuel has been a CSN lab analyst for fifteen years. For him, employment at CSN carries with it little status, especially compared to, for instance, civil service work, where the starting monthly salary is R$4,000 instead of the R$2,500 in CSN. On Samuel's desk is the paperwork related to the legal case he will discuss in labour court later today. Two years ago, Samuel was about to quit his job and train as a full-time labour lawyer. But, not wanting to lose his pension fund, he agreed with the management to work flexibly. He gave up his position at the radiology lab and moved to the photo lab, where he earns 30 per cent less. In exchange, the company allowed him to work half-time for two days a week and to be on leave after a set quota of overtime. After the apprenticeship, Samuel intends to be a labour lawyer specialising on subcontracting.

The white-collar workers of the GT had great occupational fluidity. Several employees had parallel careers or are retraining for different jobs. At the top end, Vanderlei juggled the jobs of top manager, external researcher and corporate consultant. At the bottom, Samuel, without postgraduate qualification, worked part-time while training as a labour lawyer. This occupational fluidity reflected the precarious conditions of the engineers and technical workers whose knowledge was externalised and disposed of after privatisation. Unlike the 'particularist' opposition to the firm by the middle management, or the cosmopolitan identification with it by the top management, the white-collar workers and the engineers of the GT opposed the company from a postcolonial and developmentalist point of view. The engineers' militancy reflects their historical alliance with skilled production workers. But militant production workers read their precarious working conditions through Marxist lenses – because of the deskilling and precarisation of their labour by the capitalist owner. Unlike them, engineers framed it in terms of 'uneven development', financialisation of the economy and anti-industrialism by parasitical and unproductive national and international financial classes.

They considered Benjamin Steinbruch part of such a financial elite, which was parasitically profiting from the decline of industry. As an alternative to the current system, they endorsed a kind of social neo-developmental model based on active state intervention including currency re-evaluation, the slashing of interest rates, industrial subsidies and economic renationalisation. This neo-developmentalist view partly betrayed a nostalgic longing for the policies of national autarky implemented by Vargas in the 1950s. Such industrial nostalgia was widespread among technical workers, engineers, some skilled workers and trade union representatives – which explains the popularity of the *varguista* Brazilian Labour Party in Volta Redonda. At the bottom of the corporate hierarchy were the leaders, who acted as transmission belts between the management and the workforce. Leaders were 'second-generation workers', circa fifty years old, who entered the company at the time of the 'big strike' and the privatisation of CSN, thus experiencing early state violence, reorganisation and mass redundancies. Most of them described their early employment in CSN as traumatic and frightening. Some claimed to have spent their first years in the company in survival mode. Leaders were more senior than the average employee and yet more vulnerable.

Being of poor family backgrounds, they had no savings to fall back on in case of redundancy. In terms of personal assets, second-generation workers were worse off than the previous generation of employees who purchased homes in the central bairros of Sessenta or Conforto when the housing stock of CSN was privatised. Just after privatisation, second-generation workers purchased homes in the bairro Vila Rica, which the company had just built, by using their newly acquired company stocks. At the time advertised as suburban and affluent working-class neighbourhood, Vila Rica had become a very poor bairro by the time of my fieldwork. Often of neo-Pentecostal background, leaders saw their role as one of spiritual mentoring or support in the workers' difficult circumstances. Empathy was their strength.

Unlike the more formal and authoritarian relations between employees and *padrinhos*, leaders had a quasi-peer relationship with their workers. Unlike *padrinhos*, who were the official representatives of the corporate family, leaders were not formally associated with the company hierarchy, yet their role was to enforce production standards, and they were co-responsible with their workers on productive targets, absenteeism and those technical breakdowns due to employees' negligence. Leaders incarnated the most the divergence between CSN's ideology of success and the reality of its hierarchic and punitive structure. They described their careers as 'phenomenal' or 'stellar'. But their salaries and job security were the same (if not worse) as those of most line workers. Career-wise, they had no chance of becoming managers or finding alternative employment.

The Enchantments of Technology and of the State

Marx describes commodity fetishism as a lopsided relationship between subjects and objects, whereby commodities take the appearance of powerful living entities and humans are progressively turned into things. On the UPV shop floor, technological fetishism involves a similar lopsided relationship between people and objects, whereby machines are seen as living gods animating and empowering humans, and humans as inert by-products of broader technological and economic forces. Indeed, technological fetishism is a central mechanism of class reproduction on the CSN shop floor, where the magical vitality of steel – which seems to transcend human agency – is responsible for the lopsided relationship between labour and the super-human forces of capital embodied in the empty signifiers of 'depreciation', 'productivity' and 'profit'. These social technologies of capitalism are enchanting, mesmerising and deceptive: their power is aesthetic as much as it is material. Particularly, the metallurgical and alchemical framing of labour, which relates the workers' skills and abilities to detect the microscopic and crystalline structures and the invisible properties of steel, depoliticises and naturalises labour hierarchies.[28] Workers in the capital-intensive operations of making and forging steel are considered highly skilled, whereas those performing the labour intensive tasks in the 'cold' end of the line are considered unskilled.

Smelter-alchemists are said to put the steel in motion through their secret, performative and embodied knowledge, and control the unpredictable 'mood swings' and bursts of the capricious furnaces. But the workers of the refining department claim to be the real metallurgists. They can see the inner movements of steel – the infinitesimal and slow unfolding of cracks, fractures, tensions, expansions and realignments in the crystalline structure down to the molecular level. These ancient metallurgists make up the affluent Class C of Volta Redonda.

Unlike them, line operators in labour-intensive departments such as the cold strip mills (GGLF and GGFM) are at the mercy of the economic logic of costs, sales and profits. In these departments, the movements of the machines are fast, rhythmic and continuous and systematically outcompete the movements of handlers and packers, whose actions appear clumsy and who are constantly falling behind or catching up. Moreover, the technological and political fields are mutually entangled. Skilled workers value labour as a form of embodied knowledge, unalienable, un-transferable, non-commodifiable and external to the technological system. Unlike them, line operators and unskilled workers, experience labour through measures of sales and productivity embedded in the technological system. The former experience work through their bodies, and the latter through money and the market. The former resist the money form, the latter embody it (Pitts 2018: 93). Indeed for line worker José Maria da Silva:

> Some workers due to their class consciousness resist the externalisation of their knowledge. They do not tell the management all they know. They are aware that by passing information about their work to the capitalists they sowed the seeds for their own dismissal. The say, 'I own this knowledge, I am not going to pass it to those who will replace me.' (Caruso 2009: 138)

These different perceptions of labour-value underpin different attitudes towards the privatisation of the company. Skilled workers saw it as a direct attack to their work. Unskilled workers saw it as an opportunity to become 'investors' and to improve their finances.

Indeed, the implementation of the automated SIGMA deskilled workers and put them under a regime of permanent attention and vigilance of technological breakdowns – which both immobilised them and entangled their lives with the broader movements of finance. Such logic of permanent maintenance turned the labour of line operators into a commodity – a monetary measure of the value of machines – whose value originated from 'elsewhere' and 'outside' and fluctuated according to future projections. The workers' deep anxiety for the lives of machines that I encountered in most departments of the UPV reminded me of Heidegger's argument that technology, especially artisanal tools, has a poetic function as it provides a sensorial, existential and empathic entry point into how it feels to be used, consumed and discarded by 'the other'. In Graham Harman's (2002) more extreme version, humans are 'tool-beings' kept alive by an invisible technological support structure – both human and non-human – which becomes evident only in moments of breakdown.

Heidegger also argues that modern industrial machines are rational and 'anti-poetic', in that they destroy human empathy and desacralise and de-spiritualise work. Indeed, for Mircea Eliade (1978: 65–66), the invention of steel obtained by mixing iron and oxygen in the blast furnace during the industrial revolution annihilated the *poetic* skills of the medieval alchemists. Alchemists mastered time and operated within a cosmological frame, whereas industrial workers and modern steelmakers are prisoners of time and, consequently, have lost their sacred aura. Heidegger's romantic critique of industrial work was taken up by Marcuse and by the anti-capitalist cultural movements in the 1960s and was later incorporated into 'the new spirit of capitalism' in the 1980s (Boltanski and Chiapello 2007).

But the secularisation of work of industrial capitalism never happened. Capitalist shop floors continue to be magic and animistic. Every day, line operators witness obsolete, broken and exhausted machines being magically put in motion by the invisible and mighty forces of finance. This cognitive dissonance between the limitations and vulnerability of technology and the limitless possibilities of finance transports the workers into a mental zone of hope, anxiety and empathy towards machines whose movements and breakdowns dominate life on the shop floor. Trapped in algorithmic webs of finance, human relation, production and markets and lost in the real-time integration and capillary circulation of these money forms, workers become abstract 'machine fragments' and ghostly materialisation of the commodity form. The workers' obsessive empathy towards technology, which Marx calls 'fetishism', turns machines into powerful, domineering and capricious agencies. In fact, the situations described by Harman (2002), whereby the value of tools becomes evident only when they break down, in the CSN shop floor is inverted. Here, tools are continuously broken, and humans are those who are taken for granted. The faulty nature of machines constantly obscures the workers' reproductive acts of mending, patching up and reconnecting. It is only when humans break down and life on the shop floor is disrupted by redundancies, deadly accidents or retirements that the value of humans, if only shortly, is publicly acknowledged.

This regime of production based on continuous imaginative labour defeats standard descriptions of alienated manual labour. David Graeber, for instance, describes it as follows: 'Laborers are relegated to mind-numbing, boring mechanical jobs, and only a small elite is allowed to indulge in imaginative labor, leaving the feeling, on the part of the workers that they are alienated from their own labor' (2015: 95).[29] Instead, in my shop floor ethnography, workers are forced to perform continuous exhausting interpretative labour, a big part of which has to do with the maintenance of machines. Maintenance operations are simultaneously material, affective and reproductive, whereby the care of 'the other' is continuously at stake. Against the stereotypical image of the traditional working class as productivist, skilled production workers are in fact reproductivist, because they value labour as a kind of action that contributes to the long-term reproduction of the plant and the survival of vulnerable machines whose lives they are deeply entangled with.

For skilled workers, maintenance cannot be disentangled from productive labour, and this cannot be detached from their sense of self-worth. Their control over technology, which encompasses material, affective and cognitive dimensions, prevents the abstraction and alienation of labour normally associated with the wage contract. Unlike them, the workforce of labour-intensive departments experiences work in terms of money, income, credit or market depreciation. Their sense of self-worth is based on their self-identification with the money form – in terms of monetary value of their factory work inside the plant and of their consumption and investments outside it. Thus, the technological system is central in the social construction of labour value and in the relations *in* production in the shop floor.

The second form of enchantment on the shop floor is linked to the state and its 'effects'. Politically, most CSN workers consider themselves 'more left-wing' than the ruling PT and openly criticise Lula for his endorsement of neo-liberalism. Yet, they strongly identify with the nationalist and imperialist culture of *lulismo* fully embodied by CSN, which in this sense functions as a transmission belt of state capitalism. Black and women employees of CSN are deeply aware of the racism and patriarchy that cut across the workforce. At times, women workers and subcontractors join forces against the management. But even marginalised women and black workers tend to identify with the imperialist and nationalist ideology of *lulismo*. National pride first emerges in relation to Brazil's extractivist model of development. Workers and managers alike talk proudly about Brazil's immense rural wealth and its world-leading role in the agricultural and energy sectors, and express support for Lula's policy of militarisation of the Amazon, the levying of agricultural tariffs and barriers against foreign agribusinesses and his sub-imperialist role in the Mercosul or in the União de Nações Sul-Americanas (Union of South American Nations), which they call 'our land'. Moreover, instead of linking environmental pollution to CSN's exploitative and cost-cutting operations, engineers and young employees blame the degradation of the national soil on the energy and technological inequality between Brazil and the Global North. They emphasise how CSN, because of its technological obsolescence, is forced to focus on the highly polluting iron-reduction process.

The second national narrative consists in the claim that Brazil is a racial democracy, a myth that dates back to the debates on race that emerged after the abolition of slavery in 1888. Under the Old Republic intellectuals, politicians and doctors described racial miscegenation as a dangerous social problem leading to criminality, deviance and even madness (Schwarz 1992). In the early twentieth century, intellectuals, liberal politicians and the modernist artistic avant-garde radically challenged such racist medical ideology, arguing instead that miscegenation between African and Brazilians was a factor of national strength (Schwarcz and Starling 2018: 381). Strongly rooted in Western Marxism, as well as in French surrealism and Freudian psychoanalysis, Brazilian modernism cast the working class as the embodiment of racial miscegenation. The myth of the industrial proletariat as the embodiment of racial miscegenation was also central to Vargas's modernising project.

But Vargas's *Estado Novo* provided an ethnicised version of it, which both reproduced and disguised the previous slavery system. The *mineiro* peasant was the new ideal proletarian subject – docile and reliable because they were neither entirely white like the fractious European migrants employed by industrialists in São Paulo, nor entirely black like formerly enslaved people. This ideological ethnicisation – or 'whitening' of race – runs against the fact that Minas Gerais is the state with the largest black population in Brazil.[30] The affluent *mineiro* employees – the descendants from impoverished peasants-turned-modern-industrial-proletarians – were central to the corporate narrative that cast CSN as a 'crucible of ethnicities'. This narrative of social mobility – from fields to factories – was mirrored in the personal trajectory of President Lula, an ex-metalworker from the impoverished rural Northeast Region. In fact, the ethnic crucible of the firm is strongly racially stratified with black workers and workers from poor family backgrounds (often descendants from enslaved people) in labour-intensive and precarious jobs and the 'peasants' from Minas Gerais in skilled production jobs.

Third is the historical narrative that cast Brazil as a young nation with a bright future. During my fieldwork, this sense of a present suspended in an immanent and ever-shifting future and overshadowed by an overbearing sense of historical backwardness, haunted the plant. Employees keep repeating: 'Brazil has everything it needs to grow', 'is following the path to prosperity', 'will soon become a First World nation', 'is the land of the future', 'but is not quite there yet'. Some employees frame this sense of incomplete development through a postcolonial reading: 'We are still a rural and colonial nation'; 'We had an incomplete industrial revolution'; 'In Brazil, industry has created more slums than progress.' Others frame it through the middle-class narrative revolving around the 'corrupted', 'ignorant' and 'thief' state. Indeed, the workforce often depicts Brazilian society as being atavistically patrimonial and personalist and lacking of a sense of the public and with an endemically corrupted state. This culturalist narrative, which violently remerged with the *Operação Lava Jato* corruption scandal, ultimately reinforced the point of view the middle-class elites and their pro-privatisation and anti-state stance.

Moreover, the workforce's general hostility towards the Global North does not translate into South-to-South solidarity. On the contrary, workers and managers often cast the 'Brazilian people' against the 'people' of other BRIC nations and showed resentment especially towards China. Take, for instance, the remarks of a top manager:

> China is the big world power but has no democracy. If you do an investment there now, you do not know what is going happen to it tomorrow. They are pretty ruthless in business. They kept on overpricing the coal they sold us. Besides, the Chinese state is more imperialist than the United States. They have bought half of Brazil, especially our natural resources and more profitable industries.

This nationalist-populist world view – based on shared narratives of racial democracy, rural identity, fertility of the land, future prosperity and military expansion – forms the backbone of the corporate ideology of CSN shared by the working and the middle classes. But when the economic crisis hit Brazil, this national populist consensus turned against the PT. Nationalism was reinforced by the old patriarchal and religious culture of the company. Even though formal education remains central to the firm's career and recruitment policies, kinship – in the form of blood relations, patronage or patriarchal authority of managers and leaders – continued to dominate recruitment and promotion patterns that relegated women, black and young workers to unskilled and underpaid jobs and put older white workers on skilled production tasks.

Moreover, in line with the organicist and spiritual workerist tradition of the *Estado Novo*, the wage relation continued to be framed in terms of gift, reciprocity, empathy and co-operation within the *família siderúrgica* rather than in economic terms. In Volta Redonda in the 1980s, the top echelons of the Catholic Church supported the military regime, but the grassroots Catholic movement 'liberation theology' joined the anti-dictatorship struggle. Under dictatorship, the PCB had been disbanded, the plant was occupied by the military and the trade union had been repressed and reduced to a corporatist extension of the state. Political militancy happened through grassroots youth Catholic organisations operating in favelas and working-class neighbourhoods and led by the charismatic Bishop Waldyr Calheiros Novaes. These religious organisations brought together diverse movements such as women's and black organisations, the movement for the occupation of universities and the Landless Workers' Movement. This same Catholic coalition bringing together labour and social movements led the anti-privatisation movement in the 1990s.

But these historical militant Catholic movements are radically different from the Kardecism or the neo-Pentecostalism of the Evangelical Universal Church of the Kingdom of God that were spreading in Volta Redonda at the time of my fieldwork and whose entrepreneurial, profit-oriented and materialistic vision translated well into the corporate 'ethics' of CSN. In the words of Coco, a Kardecist line leader who holds an MA in theology:

> Capitalism is not about money and selfishness when it is made in the name of God. God is a force that permeates the plant and move workers to act well. To act well means to be like tools or implements of profit. Good leaders are like experienced healers who harness spiritual forces for the benefit of the company.

Charismatic leaders like Coco preach the corporate gospel through the many informal religious activities that take place in the UPV during the workday. This entanglement between sacred gospel and production standards finds echoes in the way managers describe labour reorganisation system such as the cell system and the SIGMA by using mystical and spiritualist terms such as empathy, holism, self-ownership and self-realisation.[31]

At the end of my visit to the UPV, the tour guide escorted us to the exit. It was sunset. We followed an outdoor path with colourful and geometric signals for pedestrians whose meanings were mysterious to me. I felt as if we were walking in an alien city. As we walked, a loud siren went off. A giant torpedo train carrying molten steel to the mill passed by. We stopped. The torpedo moved very slowly. Hundreds of workers all around waited motionless. Enfolded in orange steam, they seemed alien and abstract. This atmosphere of emotional suspension and mental displacement made me think of a conversation I had with Samuel, the quality control worker. Every half hour in the quality room, Samuel cuts a section of laminated steel, photographs it and analyses it on contact sheets. The minuscule brown dots on the grey contact sheets are blown-up versions of steel impurities only a few microns big. From these dotted patterns, he judges the quality of the samples. He does not know what specific sample he is looking at. They are classified in four-digit numbers. Describing me his job, he told me: 'Outside the plant, we are in the world of kilograms. Inside the plant, we are in the world of tonnes.'

This expression is normally used by smelters and mill workers who deal with large-scale machines and processes, and it sounded odd when told by Samuel, who works on patterns of dots only a few microns big. In fact, the infinitely small scale of Samuel's work is difficult to grasp. To a lay observer, the dots on the contact sheets look like vast and distant constellations. On the shop floor throughout my fieldwork, I too was haunted by Samuel's cognitive confusion between the infinitely big, abstract and distant movements of capital and the infinitively small actions of the workers. In such a reified atmosphere, it was easy to forget that labour was real and, indeed, carried with it real risks and dangers. One evening while I was walking home, I witnessed a terrifying explosion that plunged first the plant and then the whole city into total darkness. Residents gathered in the main square, standing still and watching the vortex of red flames enveloping the plant, with red glows on their faces.

First, they expressed concern for the broken machines and described the explosion of the oxygen plant as a 'lament of a wounded animal' and 'a cry for help'. Then, they emphasised the superhuman and Cyclops-like resilience of the workers and their heroism. This narrative of steelmaking as a heroic male productive activity in fact obscured the workers' more valuable reproductive work of fitting pipes, fixing valves, screwing bolts and patching up, remaking and recycling mechanical parts to keep the plant going. Indeed, as skilled fitters, engineers and maintenance workers were progressively outsourced and substituted by the SIGMA, bottlenecks, explosions and deadly accidents intensified.

In the poem at the beginning of this chapter, the local historian Waldyr Bedê describes the migrant labourers who arrived in Volta Redonda to work in CSN in the 1940s as cyclopes lost in an inhospitable land while seeking the Eldorado. Old Europe was dying, and a new Brazil was being cast in the forge of Vulcan through the sacrifice of these indigenous cyclopes. Unlike them, Euripides's Cyclops is not a force of progress but one of backwardness. Tragically, he stumbled upon a new force, embodied in the modern hero Odysseus, that he could not understand, name or even see. These two forces and embodiments of human industry – one material and productive, the other invisible and speculative – are also two versions of the same capitalist tale that continues to enfold Volta Redonda.

With their overwhelming sensory assault to the body, oversized and super-human scale of operations, systematic separation, cutting and fragmentation of social relations and violent forms of labour extraction, capitalist factories destroy human bodies, negate human creativity and create a permanent state of cognitive confusion and magical enchantment where the abstract gods of the economy and of the state come alive and are felt on the body. In this magical zone, worker-citizens come to realise that economic progress cannot exist without strong states, and strong states cannot survive without the forces of capital. In such material and sensuous state of abstraction and false consciousness, labour forgets its own creativity. Indeed, mirroring the reproductive labour of mending, fixing and repairing machines is the workers' imaginative labour of weaving, mending and reconnecting their fragmented, isolated, out-of-scale and overloaded selves into the greater social fabric and name it class, workforce or union. This poetic function of labour, pace Heidegger, makes capitalism both possible and vulnerable.

Notes

1. Small electric furnaces that use steel scrap as raw material.
2. On the spirituality of the axial age, see Jaspers (1953: 1–25).
3. CSN was privatised through a series of auctions held in 1993 and 1994 (see chapter 4).
4. Article 14 of CSN's bylaws (before privatisation) said, 'The boards of directors and of executive officers will manage the company, respecting the guidelines and norms defined by the Supervising Ministry, as the representative of the Federal Government's interests in the steel enterprises and related activities' (quoted in Pinheiro 2002: 10).
5. The managers' wages were capped to four times the average employee's remuneration.
6. The textile group Vicunha acquired majority control, and Steinbruch became the president of CSN in 1997.
7. CSN is currently second biggest exporter of iron ore in Brazil.
8. Maranhão, Piauí, Ceará, Rio Grande do Norte, Paraíba and Pernambuco e Alagoas. CSN is strongly investing in expanding this line, aiming to cover traffic for the entire Northeast Region.
9. The furnace is the second biggest in Brazil terms of production capacity.
10. During my fieldwork, there were at least three serious accidents per month and five deaths, which is well above the fatal injury rate in the United Kingdom, according to the statistics of the Health and Safety Executive. From 2011 to 2016, there was an average of one death per year.
11. The average length of employment was five years, and in departments such as the GGAF, 80 per cent of the workers had less than one year of employment.
12. In informal conversations, the supervisors described to me the task of 'moulding' as a 'slave job' and argued it was eliminated when the furnace was partly mechanised during privatisation.
13. In crystallography, a lattice is an array of points repeating themselves in three dimensions.
14. In 2011, 2.5 million tonnes of pickled coils were produced out of 5 million tonnes of coils.
15. Refrigerators, gas cookers and other domestic appliances.
16. For instance, unlike most other jobs, they require only 'second degree' and 'confident reading' – 'no other additional skill'.
17. About 40 per cent of the workforce comes from the more affluent bairro 60 and Barra Mansa, and the rest come from the poor bairros Santa Rita, Santa Lúcia, Vila Rica and Açude. In the former bairros, people earning less than one minimum salary constitute only 5 per cent of the local population; only 3 per cent are out of employment, and most residents earn between three to ten minimum salaries. In the less affluent bairros people earning less than one minimum salary on average constitute 20 to 25 per cent of the local population; 10 per cent are out of employment, and only 25 per cent of residents earn three to ten minimum salaries (IPPU/VR 2013).
18. A derogatory term, which in the past was used for enslaved people.

19. The sociologist Armando Boito Jr. (1999: 203) calls this fragmentation within the working class 'regressive hegemony'.
20. Palmares, or Quilombos dos Palmares, was a fugitive community of people who escaped enslavement in colonial Brazil that developed from 1605 until its suppression in 1694.
21. Fees are a minimum of two minimum wages per month, which is two-thirds of the average pay at CSN.
22. After the partial closure of two furnaces and heavy downsizing in 2008, the company rehired when the economy picked up.
23. Having completed the secondary school or with a technical course but without a bachelor's degree.
24. These rival firms offer, on average, 50 per cent higher salaries.
25. Schneider (1991) discusses in detail the profile of Brazilian state managers.
26. Most of them have postgraduate degrees in engineering.
27. Launched in 2007 with an initial investment of $4.2 billion, PAC is used as an umbrella term for infrastructural projects aimed at fighting poverty in the country, such as building new homes and roads and developing water, electricity and sanitation in areas of urban decay.
28. Chen (2012: 10–12) shows how claims of animacy reproduce class, race and gender relations in the United States.
29. Braverman (1974) famously criticised the opposition between manual and intellectual labour reproduced by Graeber.
30. I thank Danilo Caruso for making this point to me.
31. Rudnyckyj (2011) describes a similar dynamics in an Indonesian steel plant.

– Chapter 3 –

Old and New Land Questions

Capital as Land

> Losing its way, a spider monkey strays from the Cicuta Forest and ends up in the municipal zoo by the monkey cage. The wild monkey says to the captive monkey, 'Don't worry, I will steal the keys from the porter and set you free.' The captive monkey replies: 'I don't want to be free. I like it here. There are baboons from Africa, tigers from India and unspoiled nature. Outside the trees are sick, the air is dirty and the big tractors of the company are destroying the forest. Soon you will have no home.' Having heard this, the wild monkey steals the keys from the porter, locks itself in the cage and throws the key into the forest. 'I like it here, too', it says.

An old man told me this story at the municipal zoo at the beginning of my fieldwork. As the fieldwork progressed, the story of the spider monkey became for me an allegory of the deadlock in which the working class of Volta Redonda finds itself. The global economic slowdown has undermined the city's manufacturing base and working-class identity and brought to the fore the fragility of the post-industrial economy on which it now relies. The privatisation of CSN cut its workforce by 40 per cent and relocated its core businesses away from the city. During my fieldwork, industry employed only seven thousand direct workers, less than 30 per cent of the population, and jobs were concentrated in the service, commerce and public sectors. Yet, the company owned 25 per cent of the municipal land and continued to control the local economy, especially through the extensive use of outsourced workers, which amounted to eleven thousand in Volta Redonda and the Fluminense region, where unemployment was at 19 per cent – well above the national average of 15 per cent (CSN 2010: 257).

In this chapter, I show how the municipality, local entrepreneurs and civic organisations try to halt the expansion of CSN in the municipal land through a mixture of civic activism and economic reconversion, in the tourist and leisure sectors. But the local working class is too impoverished to spend on culture or leisure, with the exception of going to the cinema and the shopping mall. Scenes of working-class families living in dark homes to save on electricity bills reminded me of similar scenes among working-class families in Sheffield. Expenditures on service and leisure come mainly from pensioners – the wealthiest and most influential social group in Volta Redonda – and the wealthier working-class families employed by the global carmakers and steelmakers that recently settled in the region. These autoworkers and steelworkers are younger and less politicised than the *metalúrgicos* and tend to challenge the role of the Metalworkers' Union of Volta Redonda and their narrow focus on factory-based activism by being actively engaged in regional cross-sectional alliances with progressive municipalities, civic movements and businesses aimed at countering the power of the multinationals in the region. In this chapter, I will assess these new forms of labour and civic activism within the broader historical framework of land and municipal struggles against CSN in Volta Redonda.

The story of the monkey was told to me in 2009, a few days after CSN announced that, because of the economic downturn, it would lay off 1,500 employees for one month, return to eight-hour shifts and reduce wages across the board to avoid compulsory redundancies. After the Metalworkers' Union of Sul Fluminense (SMSF) rejected these conditions, CSN made 1,200 redundancies. The public outcry that followed led to the creation of the Fórum Demissão Zero (Forum for Zero Redundancies) – a coalition of civic movements led by the fourteen mayors of the region and consisting of the Movimento Ética na Política de Volta Redonda (Movement for Ethics in Politics – Volta Redonda – MEP-VR), the Movimento Resgate da Paz (Movement for Peace), industrial and civil servants' unions and the powerful Ordem dos Advogados do Brasil (Order of Attorneys of Brazil – OAB). The forum paralysed the city for days and reached out to the governor of Rio de Janeiro; the Ministry of Industry, Foreign Trade and Services; and even President Rousseff.

But with the slowdown of the national economy and deindustrialisation in the region, this regional alliance lost momentum. Today, 'Steel Town' (*Cidade do Aço*) is simultaneously industrial, post-industrial and rentierist; made up of proletarians, pensioners and IT workers; and divided between environmentalists who would like to see the plant close, those who would like it to increase production and those, especially pensioners, who heavily rely on its service and retail industries. There is a new sprawling favelisation of the town; CSN violently evicts residents from its land; and levels of industrial pollution, and related benzene poisoning, and blood and skin cancer are rising. The thick, yellow clouds of carbon monoxide generated in the smelting process hover over the city like ghosts that people have learned to ignore.

Even though the industry employs only 30 per cent residents, the city could not exist without the plant, as it accounts for 60 per cent of the municipality's revenue in the form of urban rents and taxes. As my fieldwork unfolded, the story of the spider monkey came to symbolise this dramatic situation. The monkeys are fleeing deforestation and favelisation to find refuge in an artificial and manufactured forest. The cage represents the post-industrial economy, a new capitalist enclosure emerging from the ashes of deindustrialisation. Located between the city centre and the Cicuta Forest – one of Brazil's biggest urban tropical forests – the civic zoo of Volta Redonda embodies the spatial paradoxes of late globalisation. Under dictatorship, when the zoo was built, steelworkers entering it must have felt as if they were stepping into a space of both urban leisure and wilderness. But today, in the context of tropical deforestation, sprawling favelisation and demographic and economic decline, the fault lines between the forest, the city and the plant are vanishing. The zoo is a liminal zone where leisure and industry, urbanisation and deforestation, and work and unemployment meet.

Recent studies of deindustrialised cities and rust-belt regions in the United States and China have highlighted how political decentralisation, both regionalism and municipalism, may foster new forms of participatory democracy and cross-sectional alliances between the traditional working-class and civic movements, including middle-class and employers' organisations. Indeed, in Brazil the anti-dictatorship movement developed cross-sectional alliances between municipalities, employers and trade unions, especially in the industrial region of São Paulo, where the Workers' Party (PT) had its stronghold. But the economic geography of Volta Redonda does not match the classical profile of industrial rust belts. CSN is simultaneously a global financial operator with a globally dispersed production process, a declining Fordist factory, a powerful landowner and a logistical state infrastructure. Even in such a strongly mono-industrial company town, deindustrialisation can only be partial – based on a mixture of underemployment, low wages and favelisation – and industry continues to dominate the local economy. In this chapter, I show how such coexistence of industrial capitalism, global finance and land rentierism produces an idiosyncratic urban landscape, in which the factory is both invisible and omnipresent.

Seeing through the Eyes of a Dictator: The Steel Town in the Fazenda

In 1941, building an industrial city from scratch must have been a daunting prospect for Vargas. Having to choose where to draw a line on the vast stretches of his empty land, he must have felt like a painter facing an empty canvas. With respite, he must have noticed on the map of the newly founded Cartography Institute[1] that, in the middle of the Paraíba Valley, at the intersection of the states of Rio de Janeiro, Minas Gerais and São Paulo, a small rural community was surrounded by an extended railway network (Estrada de Ferro Central do Brasil), a hydroelectric plant (LIGHT Rio de Janeiro) and abundant coal and iron reserves. Like all dictators, Vargas employed the most famous urban architects and planners for personal propaganda. In Rio de Janeiro, to break with the old, colonial, urban landscape, he had his architects destroy the imperial skyline and build grid-like patterns and minimalist buildings. Like all dictators, Vargas liked monuments, perhaps because they froze human movement into the solid and unchangeable shape of history. But this time was different. Vargas wanted Volta Redonda to be 'a living monument' (F. Lopes 1986) – made of real people and animated by the modern force of industry.

Taylorism fascinated Vargas, perhaps because it offered a rationale and a methodology to build such living monuments. So, for his steel complex, Vargas hired the best industrial consultancy in the world: the US engineering firm Arthur McKee, which had already built the Reichswerke Hermann Göring steel complex in Germany, the United States Steel Corporation in Pittsburgh and the Magnitogorsk steel and iron complex in Russia. Yet, being a man of rural and military origin, Vargas did not share with Hitler, Stalin and Roosevelt the view of industry as a force of 'creative destruction'.[2] He worried industrialisation would attract communists from all over Europe and spark new civic and labour movements, as it happened in São Paulo and Rio de Janeiro. Moreover, as a member of Brazil's rural elite, Vargas was anxious about the prospect of Brazil breaking with its colonial past and becoming industrially dependent on the United

Figure 3.1. Vargas's statue in Brazil Square (© Massimiliano Mollona).

States. In fact, in choosing Attilio Corrêa Lima as the chief planner of Volta Redonda, Vargas distanced himself from the mainstream, US company-town approach to industrialisation.

Corrêa Lima had studied at the Sorbonne under the direction of the Saint-Simonian architect Toni Garnier. Unlike what Taylorism and Stalinism would later effect by way the urbanisation of the countryside and the mechanisation of farming, the utopian approach of Saint-Simon was based on small-scale and self-sufficient communities relying on a combination of industry, farming and energy production.[3] Garnier's (1917) *Industrial City* (*Une Cité Industrielle*) was the first modern book on urban planning based on Saint-Simon's ideas. For Garnier, industrial cities had to be located in the proximity of coalmines, water and grazing land for food and energy self-sufficiency. Moreover, they had to be functionally divided into residential, commercial, leisure and industrial zones. Machinery and land would be collectively owned, and committees of workers and capitalists would collectively run the city. This democratic setup required no police or central administration.

Vargas was no socialist, but this small-scale, self-sufficient and semi-rural model of industrialisation was closer to his colonial worldview than the revolutionary industrialism of the European and

US company towns. Vargas's chosen location for the steel complex was the Santa Cecília farm in the community of Santo Antônio, a 2,300-hectare colonial house owned by an impoverished coffee planter and located in the Paraíba Valley, at the intersection of the states of Minas Gerais, Rio de Janeiro and São Paulo. The Paraíba Valley had witnessed several colonial economies. It was occupied by slave hunters, bandits and Jesuits in the in the seventeenth century; depopulated by the gold and diamond rush in the eighteenth century; and the epicentre of the world's coffee economy in the nineteenth century. The Portuguese aristocracy established itself in the area of Volta Redonda in 1744 through *sesmarias* granted personally by José I of Portugal. In return, the Portuguese court demanded from it to convert to Christianity and, if necessary, exterminate the local *índios* – Purí and Coroado – whose belligerent behaviour and nomadic lifestyle was documented by Jesuit travellers and who proved to be an unproductive workforce to the local fazendeiros. By the beginning of the nineteenth century, most of the local Indian population had been exterminated.

Because of the phenomenal expansion of the coffee economy, a second slave trade boomed in Brazil after the abolition of the Atlantic slave trade in 1838. From 1840 to 1851, at least 371,615 were smuggled into Brazil, peaking in 1848 when 60,000 new African slaves arrived in Rio de Janeiro (Stein 1957: 65), most of whom would be purchased by the fazendeiros in the Paraíba Valley (see also Marquese and Tomich 2015; Tomich 2011). Wealthy coffee barons ruled these vast fazendas with slave quarters and satellite worker villages for resident sharecroppers, putting self-sufficient farmers and smaller planters out of business and turning the landscape into an intensive coffee monoculture. The legalisation of squatting in 1850 led to the marginalisation of both legal (*sesmarias*)[4] and illegal squatters (*posseiros*), a sharp decline of smallholdings and increased land concentration. But the abolition of slavery and land exhaustion put an end to the valley's golden era.

As the slave trade ended in the 1850s, eighty thousand immigrants moved to the Paraíba Valley to work under the highly over-extractive *parceria* system of indentured labour, and by the end of the century, the valley was depleted. Those landowners prescient enough to see the end of the coffee economy made immense fortunes by selling their remaining indentured labourers, whose value had tripled since the abolition of slavery. The coffee barons who stuck to their old ways were accused of 'routinism' and soon went bankrupt because of the diminishing returns on the labour of their aging formerly enslaved people and the lack of cash and credit.[5] From the ranks of planters, overseers and slave merchants, a new commercial elite emerged integrated, via the recent national railway and road network, with the national industrial economy. The shrinking coffee economy was increasingly controlled by foreign capital and hence extremely volatile. In 1920, the value of Brazilian coffee collapsed on the New York Stock Exchange, wiping out the local economy. The value of the land dropped from R$700,000 to R$40,000, settlers left en masse and fazendeiros merged and formed extensive dairy farming and beef cattle businesses. Overgrazing and slash-and-burn agriculture made the land even more barren.

The formal abolition of slavery and the decline of the coffee economy did not automatically lead to a regime of free labour. Many slaves continued to live in their formers quarters and work for their masters but now as daily 'hands'. Cash-stripped landowners resorted to various forms of free labour. They hired gang labourers (formerly enslaved people or *posseiros*) by the day, paying them in kind with food, tools and accommodation; employed subcontractors (*empreiteiros*), who hired their own gangs and paid flat rates on specific jobs; practiced sharecropping; rented land to *parceiros* and *colonos* (who avoided dispersion of their large holdings) and encouraged their *colonos* to help each other for free (*mutirão*) and to co-work. In the mega coffee plantations of the Paraíba Valley, *colonos* made 50 per cent of the workforce (Stolcke 1988: 29). Because they were allowed to cultivate and keep agricultural produces, planters systematically kept wages below subsistence levels.

Bringing together the basic institutions of colonialism (indentured labour, large land ownership and a subsistence agriculture that kept industrial wages low) and a modern urban infrastructure in the nearby town of Barra Mansa,[6] Volta Redonda was the perfect location for Vargas's industrial utopia. Besides the Santa Cecília fazenda, Vargas also confiscated, as specified in Law 237 of March 1941:

> 39 brick houses for settlers; 2 guesthouses; 1 house for the administrator; 9 storage houses; shelters for chicks and pigs; barns, sheds and stables; 35,000 metres of reinforced barbed wires; a 700-metre orangery; electricity, water and telephone connections; 200 hectares of forest; and approximately 542 cows, 145 calves, 25 bulls, 51 oxen for carts, 417 oxen for reproduction, 27 donkeys, 4 mares, 3 wild horses and 2 riding horses. (Costa 1978: 40)

After a long negotiation with Hitler and Roosevelt, Vargas agreed with US Steel a loan of $20 million from the Export-Import Bank of the United States in 1941 to purchase US equipment, machinery and services for the construction of CSN. Roosevelt was strongly invested in this collaboration, considering Brazil an important ally in the looming war.

The team of Corrêa Lima in Rio de Janeiro, Arthur McKee in Cleveland and US Steel in Pittsburgh worked together on the architectural model they presented to the valley's two thousand residents – impoverished landowners, shopkeepers, and settlers – who were anxiously waiting for Brazil's first industrial revolution to unfold on their barren land. Completed, the model was so big that it filled the entire municipal hall. Unlike the architectural models of the time, which aimed for a naturalistic effect and included human figurines, vehicles and trees, this one was modern and abstract. Residential zones and industrial and public infrastructures were arranged in parallel grids and painted in uniform grey. The model contained no people. In fact, how could planners, architects and engineers have possibly rendered those modern men – half proletarians, half peasants – who so far existed only in the imagination of a dictator?

Two diagonal halves cut across the balsawood landscape, with the plant on top and the town underneath. Inside the perimeter of the

Figure 3.2. The team of Attilio Corrêa Lima stands by the architectural model of Volta Redonda. Published with permission from the Instituto de Pesquisa e Planejamento Urbano (Institute for Urban Research and Planning).

plant, long stretches of the Paraíba Valley and Paraíba River were enclosed. Surveying the model, the residents must have learned how neatly the company's hierarchy was to be inscribed in the social texture of the city. The Laranjal and Bela Vista neighbourhoods for the directors stood north on top of a hill. To the south, the Santa Cecília neighbourhood for the middle management sat on the edge of the valley. West of the company and inside the valley were the working-class neighbourhoods: those of the maintenance and skilled workers on the valley's outer edge, and those of the unskilled labourers deeply buried inside it. Moving from east to west, the dimension of houses decreased and their density increased – the mansions and villas of middle-class neighbourhoods sparsely scattered on the edge of the tropical forest, the working-class districts densely patterned with little houses. That architectural model must have been experienced as the embodiment of modernity.

But in the model, Garnier's utopian principles had already disappeared, overtaken by the productivist logic of US managers and engineers (A. Lopes 1992: 73). Instead of collective workers' residences and socially mixed neighbourhoods, there were American-style single-family units and segregated garden city-style suburbs

Figure 3.3. Detail of the architectural model. Published with permission from the Instituto de Pesquisa e Planejamento Urbano (Institute for Urban Research and Planning).

(*jardim*) for the middle classes. The planners' original uniform, bare-concrete, grid-like structures were intended to prevent the formation of social hierarchies. But the architecture ended up reproducing them by having the houses of wealthy residents done with expensive autochthonous materials and personalised architecture in the 'folk' style of the time. Moreover, instead of keeping residential, industrial, and commercial zones separate, several key places were located in close proximity to the plant: CSN's administrative offices, the police station, the new commercial centre, the hospital, the nursery, the playground and a school for one thousand pupils. The architectural model conveyed a fundamentally confusing message. The inside area of the plant, aside from its basic components (the oxygen plant, coke oven, furnace and mills) consisted mainly of empty land. Empty land also extended to the north of the plant – the 'old' Volta Redonda controlled by landowners and speculators. This corporate enclosure of empty land contrasted with the urban density of the European and US steel cities. Was the company a real industrial employer, or was it just another colonial rentier?

Moreover, US engineers and consultants did not endorse Vargas's dream of industrial self-sufficiency. Although self-sufficient in terms of electricity, the plant was not productively integrated either upstream with agriculture or downstream with the car sector. The real Volta Redonda turned out to be very different from its architectural model mainly because the executive commission spent the entirety of the first US loan on the plant. No money was spent on the township. Only in 1943, after a second loan of $25 million from the United States, did the commission built accommodations, access roads and basic welfare infrastructure for some residents. In these two transitional years, the township was made nearly entirely of wooden shacks. The company invested in technology rather than housing because it had no plans to permanently house the builders, constructions workers and stonemasons whom it would make redundant after the construction phase. Already in 1941, CSN's housing stock was privatised, and up until 1967, the company maintained a residential deficit of 30 to 40 per cent of its workforce.

Aside from housing permanent workers, CSN used its land mostly for real estate speculation. The extreme housing deficit led to the rapid favelisation of the periphery. In 1942, there were only 662 houses and 3 hotels, mostly for foreigner managers, for a total of 6,160 residents. In 1946, at the end of its first expansion (known as Plan A),[7] the company reduced its workforce from 11,719 to 8,177. Many of these 3,542 unemployed and homeless workers occupied

Figure 3.4. Vila dos Índio favela. Published with permission from the Instituto de Pesquisa e Planejamento Urbano (Institute for Urban Research and Planning).

land in the northern part of the city, founding the favelas do Morro dos Atrevidos and Vila dos Índio, now Bairro Sessenta (A. Lopes 1992: 111). Other unemployed workers became *posseiros* (owners of private, often illegal, developments) or moved to rented dwellings, mostly irregular, owned by private developers. In 1948, only 34 per cent of workers lived in company housing (Piquet 1985: 97).

During the city's early developments, Corrêa Lima had already warned against the creation of 'two separate Volta Redondas: the old, unplanned, dangerous and backward periphery and the new, clean, planned and forward-looking township' (Costa 1978: 40). Since CSN's early days, these areas were under different economic regimes. Private developments, legalised land (*áreas de posse*) and leased land were controlled by private speculators and big fazendeiros and traded at market values; company homes were subsidised by CSN[8] and under the company's strict surveillance.[9] Even though Volta Redonda was planned by Brazil's best architects, the UPV was designed by the world's most renowned engineering firm and the steel town was Vargas's crown jewel, the city underwent the same process of slummification described by Mike Davis (2006) in relation to dependent industrialisation and predatory development of megalopolises in the South. The first land occupations by unemployed workers date back to the 1940s, even before the plant became operative. In fact, favelisation was also the unintended consequence of the logistical nightmare of building a steel complex in the middle of a dilapidated coffee valley.

For a start, the valley was uneven, flooded and corroded by ants,[10] and US engineers had to heavily level and tamper it to lay the plant's foundations. Moreover, because of the Second World War, the delivery of machinery, tools and services from the United States was halted several times, leading to long periods of unemployment and inactivity. In addition, by 1945, Vargas had lost support in the National Congress of Brazil, which led to cuts in public subsidies to Volta Redonda. But the gap between urbanisation and industrialisation in Volta Redonda was caused not only by the housing policy of CSN, bad urban planning or international politics. Workers themselves fled the city en masse to escape the plant's harsh labour regime. Construction workers spent whole days underground laying the plant's foundation or suspended fifty metres aboveground on makeshift wooden cranes and ropes assembling vast blast furnaces, coke ovens and thermoelectric plants. During the construction years, a huge number of them were crushed and died from falling beams, rolls and bars; poisoned by benzene, lead and tetanus; or killed by typhoid or malaria. Moreover, slavery in the Paraíba Valley was just one generation away, and its legacy was still strong. Apart from a small core of about three hundred office staff, drafters and engineers, the vast majority of employees were paid by the hour (Dinius 2011: 49) and had no labour rights. Under Vargas's workerist regime, only formally registered and literate industrial workers were granted political and welfare rights. Unskilled, precarious and rural workers had no rights – not even the right to vote.

But for the 'migratory birds' (*arigós*),[11] as these workers were called, running away was not easy. Under the presidency of General Edmundo de Macedo Soares da Silva (Vargas's Minister of War), the plant was under a regime of martial law and run as a military installation.[12] Draftees and 'free labourers' worked side by side on ten-hour shifts under the supervision of the industrial police; the criminal police unit (made up of military personnel and ex-convicts) and the US intelligence service. Ordinary workers had restricted freedom of movement and were not allowed voluntary dismissals. Those who left the company were persecuted as deserters and punishable with four years in prison. Labour turnover was extremely high: from 1941 to 1946,

46,838 migrant workers came to Volta Redonda, and 35,140 had left (Dinius 2011: 51). Macedo Soares had developed strong links with the top management of US Steel through the National Steel Commission, which he chaired. Fascinated by the American way of life, he famously compared the grid-like structure of Volta Redonda with that of New York City. But as much as he celebrated American modernity, he acted as a colonial landlord. Workers remembered him on horseback, always ready to strike with his whip. Like Vargas, he treated the working class with a mixture of repression and paternalism, describing Volta Redonda as a steel family (*família siderúrgica*) with CSN as the mother, Vargas the father and the steelworkers the sons.

Between Old Land Struggles and New Working-Class Utopias

For the families of permanent workers, Volta Redonda was a dream city. It had spacious, clean family homes, state-of-the-art infrastructures and modern urban spaces, which included public parks, tropical forests and private gardens. The following extract from CSN's journal *O Lingote* captures this image of urban working-class paradise:

> After the hustle and bustle of the plant, getting away from the heat of its furnaces and ovens and from the noise of its mills and powerful machines, the visitors meet on the other side of the train track a well-designed city, with a smell of new things, full of gardens and trees, flowery, and appropriate. It is the Steel Town, which emerged from refined planning and was made up of human communities artificially created. (Dinius 2011: 83)

This feat of social engineering was achieved through the company's various welfare departments. The Departamento de Assistência Social (Department of Social Assistance – DAS) organised all the city's leisure activities. It sponsored and sat on boards of social clubs, educational and cultural associations, grassroots Catholic associations – some of which would later become powerful anti-dictatorship subjects, such as the Juventude Operária Católica (Catholic Worker Youth – JOC) – carnival balls, folkloric festivals and the town's soccer league. DAS's crown jewel was the Workers' Club, a luxurious sports complex with an Olympic-size swimming pool, three tennis courts, an Olympic gymnasium and a three-thousand-seat auditorium.

The DAS had a monopoly on the city's leisure time. New civic associations had to register with it to legally operate. But the Departamento de Saúde (Department of Health) was the most important tool of

Figure 3.5. Bairro Laranjal. Published with permission from the Instituto de Pesquisa e Planejamento Urbano (Institute for Urban Research and Planning).

corporate propaganda. It ran the town's hospital with state-of-the-art machines, spacious and sanitised consulting rooms and an ultra-modern child welfare centre, which included a training centre for nurses. Medical services were charged to the employees and their families at differential rates depending on their salaries (with subsidised rates still unaffordable for unskilled workers). The Departamento de Administração do Serviço Público (Urban Department) was in charge of urban infrastructure,[13] firefighters, the police force and the town's musical band. CSN's welfare provisions even exceeded the generous allowances established by the Consolidation of Labour Laws (CLT), on which industrial unions based their claims and which included a funerary fund, a five-day break after minor accidents, sickness insurance and coffee and lunch breaks for night-shift workers. Moreover, CSN established the first scheme in the world of workers' participation in corporate profits, long before any European company.

CSN invested in welfare and housing only in the 'new city' (*cidade nova*) located to the right of the river and including the bairros Laranjal, Vila Santa Cecília, Conforto, Jardim Paraíba and Sessenta. The 'old city' (*cidade velha*) located to the left of the river, which included the Niteroi community in Belmonte e Retiro, was left in the hands of speculator and landowners. Indeed, the gap between 'the two Volta Redondas' was widening. The township became a gated community for the aristocracy of labour, which excluded unemployed, unskilled workers and non-employees. The old city was a sprawling rural camp made up of precarious and unsafe dwellings in continuous transformation due to self-construction, forced relocations and speculation. The old city was an 'in-between city' (Davis 2006) suspended between the urban and the rural or, to use the term coined by the radical Brazilian geographer Milton Santos (2000: 82), a space of 'tropical flexibility' – a blend of deprivation, underdevelopment and progressive politics. In fact, it was in the old city that an anti-corporate coalition of professionals, fazendeiros and merchants under the umbrella of the Masonic organisation Loja Maçônica Independência e Luz (Masonic Organisation for Independence and Light) emerged for the town's political and administrative emancipation.

In 1954, the coalition gained control of the newly formed municipality and initiated a series of improvements in the old city and of legal actions against the expansion of CSN. In 1956, the new mayor, a wealthy landowner, regularised the illegal private developments in the old city and upgraded them with public money (Cabral de

Figure 3.6. Proletarian houses. Published with permission from the Instituto de Pesquisa e Planejamento Urbano (Institute for Urban Research and Planning).

Souza 1992: 60). This flow of municipal subsidies towards private landowners mirrored the way CSN's was being developed with state finances. The old aristocracy and commercial classes that controlled the municipality created a power block against the state-controlled CSN. Thus, the old colonialist block and the new capitalist state were struggling over the monopoly of urban rents and land. CSN's policy was to keep its residential stocks low and maximise the profit over the rentable land. At the end of its second stage of expansion in the 1960s, the residential stock of CSN covered only 40 per cent of the workforce, and the firm was making huge profits from the sale of residential land (111). The formal working class was spared from the market. Their houses either continued to be managed by CSN or were sold to existing tenants at heavily discounted rates. The informal working class and the unemployed were driven into poverty by unaffordable housing prices. When the municipality pressured CSN to increase its housing stock for the working class, CSN demolished individual family homes and replaced them with high-rise buildings to maximise returns on rents.

The municipality started to gain control of the urban land by resorting to civic law. For instance, it passed a law – the Lei Orgânica dos Municípios (Organic Law of Municipalities) – in 1948 that required new private development to be approved by the municipality and revoked CSN's status of fiscal exemption. But the company continued to dodge taxes through legal loopholes. The federal government and the state governor backed CSN, arguing that as a strategic supplier to global manufacturers and the national industry, especially the car industry, the company should be tax-free. Yet, in 1963, because of pressures from local MPs, the Metalworkers' Union and neighbourhood associations, CSN was forced to retroactively pay seventy million cruzeiros in taxes to the municipality. To be sure, the coalition controlling the municipality was hardly revolutionary. Like all 'ghost mayors'[14] of Volta Redonda, the new mayor was a figurehead of the federal government affiliated with the Partido Social Democrata (Social Democratic Party) – a conservative alliance of the landed classes opposed to Vargas's pro-labour stance.

The Brazilian Labour Party – Vargas' populist party – in Volta Redonda was pressing for reforms on urban land use as part of its broader agenda of national democratisation and agrarian reforms. But defying the clear-cut spatial and political divides between the old and new Volta Redondas, the residents of favelas and those of the township joined forces against both CSN and the landowners. This grassroots and cross-sectional Right to the City alliance brought together the unionised working class; neighbourhood associations; farmers; women's associations; co-operatives of cleaners, builders and service workers; grassroots Catholic organisations; and lawyers and GPs – two professional classes active in poor bairros. In 1959, the residents of the favela Morro dos Atrevidos demanded water and electricity infrastructures (CMVR 2015: 63) and at the same time petitioned with the working class of Conforto against the pollution created by the concrete factory owned by CSN (80).

After the 1964 military coup, CSN management put an end to the paternalistic relationship between CSN and the town and pushed towards greater civic autonomy. The military regime inaugurated a new era of predatory expansion, speculation and intensive favelisation. It crushed all civic movements and put CSN under direct control of the federal government. Under the direction of the new president general, Oswaldo Pinto da Veiga (Vargas's Minister of Raw Material), CSN was vertically integrated with the mining sector upstream and the car industry downstream. Marking the end of CSN's paternalistic regime, the company sold its urban infrastructure to the municipality of Volta Redonda, which purchased it by subscribing CSN stocks. Welfare and public services previously covered by the company were externalised to the municipality.[15]

Moreover, following the new administrative reform under the Humberto Castelo Branco government, CSN set up the real estate holding Imobiliária Santa Cecília (CECISA) – funded jointly by the workers' pension funds and CSN – to build and commercialise new residential developments and privatise the existing housing stock. The land confiscated by CSN in 1941 pertained to the municipality of Barra Mansa (registered as lot 3045) and not to Volta Redonda, which would become a municipality only in 1954. The housing stock privatised in 1967 was transferred to the land register of Volta Redonda through ad hoc agreements between the municipality and CSN, but the rest of lot 3045 remains in the land registry of Barra Mansa. (I will discuss why this has great political implications for the current land debate.) The municipality consolidated its autonomy, too. Mayor Sávio Cotta de Almeida Gama, a wealthy landowner, created two municipal autarchies[16]: the residential Companhia de Habitação de Volta Redonda (Housing Company of Volta Redonda – COHAB-VR) and Serviço de Águas e Esgotos (Water and Sewage Service), servicing 98 per cent of households. With CECISA speculating in the township and the COHAB-VR developing popular housing in the periphery, the political economies of the new and old Volta Redondas were now inverted. The township was now being wildly privatised, and the old city was put under a regime of redistributive welfare.

Indeed, the link between industrialisation, privatisation and favelisation became even more evident with Plan D, which led to fifty thousand redundancies[17] of unskilled workers and simultaneously increased the number of engineers and middle-income workers. These wealthier new residents bought CECISA's newly privatised housing stock, pushing low-income, working-class families – especially former construction workers – and the unemployed to the margin of the city, in favelas and *áreas de posse*. By the end of the 1980s, the township was nearly entirely privatised; there were eighty favelas in Volta Redonda (where 12 per cent of the urban population lived),[18] and 69 per cent of the dwellings were made of wood and had no running water. Even after the COHAB-VR built new houses for low-income workers and regularised urban favelas, the housing situation was dramatic. The municipality responded by urbanising some favelas but without the involvement of the local communities and without any social intervention. With the labour movement silenced by the military, a new coalition of urban squatters[19] and grassroots Catholic organisations[20] led a second wave of anti-corporate campaigns focused on 'rights to the city' and spatial justice, including regularisation of favelas and poor working-class neighbourhoods and the right to land ownership. The movement opposed commercialisation of cities, real estate speculation, the privatisation of public services and urban segregation,[21] adopting a new political language in which access to the city constituted the new ground of popular democracy.

The army violently repressed this Right to the City coalition, but the newly established Instituto de Pesquisa e Planejamento Urbano (Urban Research and Planning – IPPU) in the municipality supported it and set up a department of participatory planning that included squatters and local residents. The squatters' movement was also supported by the Metalworkers' Union, Catholic organisations, women's movements and eighty-five neighbourhood associations. In 1983, the eighty families living in the favela Três Poços asked the IPPU for the right to purchase the unused private land they occupied for a symbolical price. The right was refused, setting the ground for an escalating confrontation. In 1985, the central government passed the Lei dos Posseiros Urbanos Municipal Law of Urban Squatters – a landmark victory for the squatters' movement. The law legalised squats, created a municipal fund for public land management and guaranteed the squatters' right to take part in discussions, elaboration and implementation of urban plan and infrastructure.[22] The Right to the City movement continued to grow, incorporating NGOs, social movements, trade unions, architects, engineers and favelas and neighbourhood associations. The most influential activists of this urban movement took the leadership of the Metalworkers' Union, turning it into a 'new union' (see chapter 4).

In 1984, the Metalworkers' Union organised the first general strike in the town's history. The workers occupied the plant and set up a camp by its main entrance, where they organised assemblies, discussions and public events with residents and civic organisations. It was in this new camp, symbolically connecting the inside and outside of the plant – labour and urban movements – that the 'new trade unionism' started to emerge. The triumphal mayoral election of the GP Marino Clinger Toledo Netto (PDT), the preferred candidate of Governor Leonel Brizola, sealed this cross-sectional alliance between the municipality, civil society and the labour movement. But, paradoxically, the advent of democracy and the privatisation of CSN that followed shortly after coincided with the end to this extraordinary period of political mobilisation. In 1988, with the principles of decentralisation and federalism enshrined in the new democratic constitution, Volta Redonda gained municipal autonomy. Administrative and financial autonomy meant municipalities now controlled their urban plans, policies and budgets and were able to make strategic alliances with the private sector and civic coalitions against the federal government. On the other hand, autonomy reduced state and federal funding and expanded payroll expenditures and inherited debts, creating immense municipal deficits (C. Souza 1997). To leverage these debts, municipalities issued municipal bonds and turned to the financial market and to real estate. Like other municipalities under the democratic regime, Volta Redonda found itself constantly over-indebted and on the brink of bankruptcy.

For Franco Abrucio and Márcia Soares (2001), federalism's other effect was to make municipalities politically autonomous vis-à-vis regional and national politics and develop strong ties between these and urban social movement, a process the authors describe as *novo municipalismo* (new municipalism). This was the case in Volta Redonda, which nonetheless remained under the strong influence of national politics. Ever since the establishment of free elections in 1989, reflecting national politics, the Partido Socialista Brasileiro (Brazilian Socialist Party – PSB) and the PT greatly influenced mayoral elections in Volta Redonda. The presence of the PT in Volta Redonda consolidated further after the formation of the Congresso

Nacional da Classe Trabalhadora (National Congress of the Working Class) in 1981 and of the Unified Workers' Central (CUT) in 1983. The Democratic Labour Party (PDT), funded by the charismatic Brizola, the other working-class presidential candidate, also gained strength both nationally and in Volta Redonda.

Throughout the Lula administration – based on the alliance between the PT and the Brazilian Democratic Movement (MDB) – the mayors of Volta Redonda came mainly from the ranks of the MDB with support of the PT. Since 2000, the power of the MDB in Volta Redonda had increased because of the immense power of Former Rio de Janeiro Governor Sérgio Cabral Filho (MDB). Cabral Filho used Volta Redonda as electoral springboard for his candidacy to the presidency and showered the *Cidade do Aço* with PACs, structural funds and state funding for special projects that never materialised, such as the new bicycle lane; the high-speed train between Volta Redonda, São Paulo and Rio de Janeiro; or the Steel City airport, which was initiated in 2003 with a series of land confiscation in peripheral bairros. Moreover, the federal state continues to have strong political influence over Volta Redonda, especially through its structural investments (PACs), founded jointly by the Central Bank of Brazil and the BNDES. For instance, in 2008, Volta Redonda was given R$124 million – R$52 from the BNDES – for a PAC aimed at improving water and housing infrastructures, upgrading the water treatment system and reducing the pollution of the Paraíba River.

After the incarceration of Cabral Filho and the impeachment of Rousseff, Volta Redonda ceased to have privileged regional status, and President Michel Temer scrapped the regional mega-projects started in Volta Redonda by Cabral Filho. The new democratic constitution of 1988 reflected the central demands made by the urban movements of the 1980s, such as urban justice, direct democracy and property as a fundamentally public and collective right. But in Volta Redonda, the process of democratic opening coincided with a dramatic increase in wealth inequality, and the urban dispossession that followed the privatisation of CSN (see chapter 5) amounted to a full-blown privatisation of the city. The privatisation of 1992 violently split the Metalworkers' Union (Graciolli 2007) but led to a strong civic articulation between Catholic organisations, the CUT and the PT, leading to the creation of the Comitê Popular Contra a Privatização Popular Committee against Privatisation in the 1990s. Bishop Waldyr Calheiros Novaes was central to this anti-privatisation movement. In an interview given to the *Folha de São Paulo* shortly after the privatisation of CSN, he says:

> Private capital is going to turn the city into a factory, with an entrance in Presidente Dutra Road and the other in Lúcio Meira and the master's house in the central offices of the company. I bet the owner of CSN is going to be the local mayor soon. Citizenship is going to disappear. (Lima 2018: 40)

In 1992, the left-wing coalition in power, the Popular Front (the PSB, PT, PV and PCB) led by Paulo Baltazar, a charismatic GP with a background in grassroots activism in favelas, directly confronted CSN on the land question. First, Baltazar ordered the expropriation of the unproductive land owned by CSN, including stretches of the Cicuta Forest, and forced CSN to pay environmental taxes in line with the new Programa Ambiental Compensatório (Compensatory Environmental Programme). Second, the mayor established the municipal body for participatory planning Conselho Municipal de Desenvolvimento Urbano (Municipal Council of Urban Development – CMDU) involving squatters, militant Catholic organisations and women's and civic movements to tackle the issue of spatial justice in the city. Baltazar also created the Fundo Urbano Comunitário (Community Urban Fund), a municipal body in charge of regularising favelas and occupations.

For the first time in the city's history, the municipality was in charge of drawing its own urban plan independently from CSN's expansion plans. Moreover, the mayor sought alliances with municipalities, social movements, trade unions and businesses in the Fluminense region against CSN and the global carmakers that had just moved into the region, which had an equally harsh regime of labour. These grassroots initiatives did not come unilaterally from the major but were the result of a strong pressure made on the major by grassroots movements in the city. In the 1980s, when the ABC Region of São Paulo was being heavily industrialised, the PT and the CUT developed the famous 'participatory budgets' – collective and neighbourhood-based processes of economic decision-making involving municipalities, the civil society, industrial unions and the private sector. Participatory budgets were the embodiments of the intersectional, horizontal and participatory democracy that the PT advocated for Brazil.

Baltazar tried to replicate this form of participatory planning in the Fluminense region with the support of CUT experts, economists and urbanists who had masterminded the experience in São Paulo. Yet, for labour lawyer Vanderlei Barcelos, these experiments marked the beginning of the institutionalisation of urban struggles in Volta Redonda and of the co-optation of its social movements (Lima 2016: 104). CSN's privatisation in 1992 (see chapter 5) divided the local labour movement and silenced the most militant neighbourhood associations.[23] The company cut 30 per cent of its workforce and started a process of delocalisation and diversification from its core business of steelmaking. The municipality had acquired from CSN the main hospital and urban infrastructure. But CSN still owned two-thirds of urban land[24] and controlled the most important leisure, cultural and educational organisations of the city.[25] In 1996, CSN employed only a few hundred more workers than the public sector and a third less than the public and commercial sectors combined. In 1996, Mayor Antônio Francisco Neto from the MDB incorporated the participatory budget and its dialogue with the social movements inside the CMDU, thus institutionalising them. The Neto administration marks a second post-privatisation strategy by the municipality informed by economic pragmatism rather than social activism. Neto's ambition was to diversify the economy of Volta Redonda by developing the service, leisure and tourist industries. Rebranding Steel City as 'the city where the river bends', Neto built a water park, sports centres, cultural landmarks celebrating the city's multicultural heritage[26] and re-landscaped the Paraíba River, making it accessible only to cyclers and joggers.

By 1996, Neto had implemented 1,300 infrastructural projects and generated a regional and federal income stream of R$11,400 per month. He described his entrepreneurialism as a way of boosting the self-confidence of Volta Redonda that was destroyed by CSN – a goal embodied in his electoral slogan: 'I believe in Volta Redonda, you must believe in it, too'. In fact, much of the funding came from corrupted Governor Cabral Filho, who was very close to Neto, and reflected the broader trend of financialisation of the Brazilian state and its entanglement with the private sector especially major construction companies, whose illegal practices of bribery and illegal tendering were unveiled with the *Operação Lava Jato* investigation. Moreover, Neto's top-down urban regeneration – from industry to the leisure and tourism industries – followed the well-rehearsed path of forced economic reconversion implemented in post-industrial European cities like Sheffield, where the steel industry was replaced by the cultural industry. But unlike Sheffield and other typical rust-belt cities, Volta Redonda continues to depend on a steel complex whose production continues at full blast.[27]

In 2002, when CSN partially closed the Civic Construction Department (see chapter 5) and downsized the research department, Neto formed the Companhia de Desenvolvimento (Development Company – CODIN) aimed at bringing together industrial trade unions and the private sector to jointly develop a regional economic plan. Neto's ambition was to create a regional industrial district, based on the European model, using the existing network of dispersed, diversified and interdependent small subcontractors and suppliers of CSN. During privatisation, back in 1992, Governor Brizola had attempted to form a regional consortium of banks, trade unions, small businesses and municipalities with majority control of CSN. After acquiring CSN, Benjamin Steinbruch also embraced regionalisation. Following the advice of the US consulting firm McKinsey (see MGI 1993), in 1997 he split the company into regional centres of profit – logistics, transport, steel, power, mining and finance – under the direction of the corporate centre based in São Paulo. For some scholars (Arbix 1996; Ramalho 2005, 2012) the Fluminense regionalism that emerged in the 1990s created a strong cross-sectional counter-hegemonic block – composed of businesses, municipalities and the labour movement – against CSN and the global carmakers that had moved into the region.

I argue the opposite. Regionalism was the form taken by the new global capitalist reorganisation, based on the territorialisation and decentralisation of finance, which turned municipalities into economic actors burdened by debt and unemployment and having to balance the interests of capital and labour. This new form of capitalist reorganisation coincided with the democratic opening, the privatisation of national companies such as CSN and the second wave of industrialisation that brought global carmakers into the Fluminense region.[28] Displacing São Paulo as the most sought-after Brazilian regional district, the Sul Fluminense region developed an 'automotive labour regime', to lure in global carmakers and steelmakers. The synchronisation between the financialisation and downsizing of CSN and the rapid industrialisation of the region had three major consequences. First, the privatisation of CSN dramatically reduced the number of its direct and indirect workers, subcontractors and suppliers[29] in the region. The outsourcing of production to companies not based in the region broke the Sul Fluminense rust belt but also paved the way for the service industry – thus, changing the region's profile, from mono-industrial to a dispersed and diversified industrial district – and released cheap labour for the automotive sector.

Second, regionalism unleashed a 'fiscal war' (Arbix 2000) between states and municipalities to attract foreign investors, by offering cheap labour, fiscal and environmental incentives and free land and public infrastructure. For instance, Volkswagen opened its first global 'modular factory' in Resende in 1996 thanks to the generous concessions and contributions by the state of Rio de Janeiro,[30] and PSA Peugeot Citröen opened its first Brazilian factory in Porto Real in 2001, also with the support of the municipality[31] and the state of Rio de Janeiro, which owns 30 per cent of the factory's stocks. Moreover, this business-oriented regionalism reinforced the power of local elites. For instance, the sugar baron Renato Monteiro da Costa was behind the recent political activism of the municipalities of Resende and Porto Real. Since the 1960s, the family patriarch Renato had supported the development of the automotive sector to counter the region's excessive dependency on CSN. As the biggest landowner in the region, the Monteiro da Costa family was personally involved in, and benefited immensely from, the development of the industrial pole and became an influential political figure in Resende, Porto Real and the CODIN. Indeed, these tripartite regional consultations, even when involved broad constituencies,[32] ended up subsidising the automotive, tourism and leisure sectors[33] or reinforcing the ties between Neto and his corrupted ally Governor Cabral Filho[34] with no substantial gains for labour.

On the other hand, new working-class struggles were taking place in these newly industrialised towns. In 2002, the municipalities of Resende and Porto Real set up 'industrial chambers' and 'municipal employment commissions', with trade unions, employers' federations and neighbourhood and other civic associations, whose role was to monitor the decisions taken by VW and Peugeot that affected the local community.[35] These municipal bodies became new institutional frameworks that limited the power of the global corporations in the region. Moreover, Resende set up the Departmanento de Movimentos Sociais (Department of Social Movement) in charge of participatory planning and land redistribution, which added to the traditional corporatist alliance new civic subjects including the Catholic Church, the Landless Workers' Movement (MST), neighbourhood associations, the black community, and the MEP-VR. This new institutional opening at the regional level was made possible by Lula's establishment of the Ministry of Cities in 2000, which allowed politically active municipalities such as Resende and Porto Real to pursue their Right to the City agenda against private developers and companies.

But unlike Resende and Porto Real, which tried to contain the power of global carmakers by engaging with diverse civic constituencies, Neto was mainly interested in attracting foreign investors,[36] offering credit, fiscal incentives, free land and even advanced purchase of merchandise. But industries have stayed clear of Steel Town because its local economy was still haunted by the overpowering presence of CSN – in terms of monopoly both over the local labour market and over the land – and by its heavily unionised workforce. Volta Redonda's attempts to lead regional coalitions failed because of various factors: the SMSF's loss of power and its failure to develop links with the autoworkers, other municipalities gaining power vis-à-vis Volta Redonda and lack of interest in the city by global investors.

Between Expulsion and Gentrification

In 2008, when I arrived in Volta Redonda, a fight broke in my bairro between the private police of CSN and local residents. The police were fencing a community allotment that local people used to grow vegetables and graze animals. It was late at night, and I was struck by the fact that mainly women and children were confronting the police. Fights and confrontations between local residents and the company were intensifying because, according to local residents, the municipality had adopted a reconciliatory stance towards CSN. In fact, even if Neto had retreated into an unfruitful business pragmatism, political constituencies within the municipality associated with the PT continued to fight CSN on the land issue. For instance, the CMDU had just passed the city's first independent urban plan, which was drafted in 1992 but had been blocked by CSN ever since. Forbidding any future residential and infrastructural expansions by CSN, the plan was a huge blow to the company.

The head of the IPPU, Lincoln da Cunha Pereira Filho, had campaigned for the city plan since the 1970s. His first job at the municipality was to campaign against CSN's plan to relocate all the residential and commercial units in the old city outside the urban zone and expand the plant on the vacated urban land. In the 1980s, Cunha supported the movement for land occupation, which he claimed was destroyed by privatisation. In a personal conversation, he told me:

> The Left missed the fact that the real asset that was being privatised was not the plant but the land that went under the ownership of CSN. You see? All the land south of the river is now owned by CSN and controlled by the federal police. You see? [He points to the map.] It is empty. Occupations are concentrated in the northern part of the city. Article 783 of the democratic constitution states that areas that are unused, under-exploited or without construction have to be put to use or be liable to progressive taxation for five years – and at the end of the five years, it can be expropriated by the state. We are challenging CSN on this ground. But after privatisation, Volta Redonda

> has become an empty city: 12 per cent of its homes are empty, against a national average of 6 per cent. The poor people are leaving the town. Only the rich and the favelados are staying.

At the time of my fieldwork, Cunha was helping a civic collective for land restitution to claim back the land CSN had acquired during the privatisation process. The legal argument Cunha makes is that the land CSN confiscated in 1954 was never regularised and transferred to the register of Volta Redonda after privatisation and hence was still under the public jurisdiction of the Barra Mansa municipality. Moreover, according to Brazilian law, land can only be expropriated to perform some social function, which was the case when the state originally expropriated the land to create CSN. But after privatisation, the social function of the expropriated land ceased, so it should be returned to the municipality. In response to the municipality's lawsuit, CSN intensified its occupation of the city. In bairro 60, where I was living, the company, from one day to the next, fenced off soccer fields, playgrounds and vegetable allotments. Irregular homes and building extensions were knocked down (see Benson 2005).

Farming and animal grazing were important parts of the bairro's informal economy. Farming on urban land between April and November, most families were self-sufficient on vegetables and legumes throughout the year, with small urban farms producing up to seventy tonnes of vegetables per day in the peak season. So, land expropriation by CSN pushed most families into starvation. Women had little connection with the trade unions, neighbourhood associations or local politicians. When the private police of the company or the municipal police stormed the neighbourhood to evict local residents, it was women and children who confronted them. These police raids affected the local residents unevenly depending on their ownership status. The 'affluent' residents in the upper side of the bairro had regularised their houses and confronted the police with property deeds and proofs of regularisations of planning extensions, claiming their rights to property as taxpayers and homeowners. The renters, poor landlords and squatters in the lower side of the bairro could only claim their constitutional right to live in the land left unproductive by the company. But the police consistently dismissed their claims and forcefully relocated them to the next favela.

The anthropologist James Holston (2007: 180) describes the acts of self-construction of homes, regularisation of ownership and claims to 'contributors' rights' by the residents of the bairros Jardim de Camelias and Lar Nacional in São Paulo in the 1990s as acts of 'insurgent citizenship'. Today, these acts reflect the logic of popular ownership and the financialisation of poverty that split bairro 60 between a class of landlords and a class of homeless.[37] This split was visible in the very architecture of the bairro. The original 1940s one-story, white-stucco family homes in the affluent north side were personalised with colourful and baroque extensions or, mirroring the managers' bairros, were designed in 'tropical Palm Springs' style' (Ruggeri 2007).[38] Some even mimicked the luxurious style of the villas in the managerial bairro Laranjal, with minimalist concrete volumes, organic landscapes à la Roberto Burle Marx[39] and American-style swimming pools and open patios. In my bairro, a wealthy friend had his house built as a 'new age' Egyptian temple with an underground grotto for rituals and prayers and a waterfall in the living room. Unlike these, the houses in the south expanded horizontally in makeshift huts, roofless structures, courtyards and semi-derelict brick houses. Lastly, the rural favelas at the edge of the city consisted mainly of mud huts dispersed and semi-hidden in the forest.

In the past ten years, microfinance and affordable housing have spread in the more impoverished bairros and in surrounding favelas. The bottom-up processes of home construction and self-directed property claims described by Holston were replaced by microloans and federal programmes of social housing. For instance, My House, My Life (MCMV) offers state subsidies to low-income families through lease-to-own agreements in which the poorest families contributed as little as 5 per cent of their monthly income. Under the MCMV scheme, municipalities relocated low-income families from bairros with no running water, electricity or sanitation – but often located in proximity to the city centre – to 'modern' homes in more peripheral bairros. In this way, they boosted the value of municipal land and pushed poor families further into the periphery. The scheme also involves great profits for constructors, who are entirely subsidised by the government, but cut on production costs, especially labour wages and safety.

Thus, according to Ruy Braga (2016: 26), MCMV generates social apartheid benefitting only middle-income families (Camboriú 2013) and corrupt intermediaries within municipalities while spatially segregating poor families and pushing them into indebtedness and (often indentured) building work. During my fieldwork, the municipality legalised several squats and informal settlements, with pompous ceremonies in which property deeds were officially given to favelados. These regularisations of favelas were the first steps towards their occupation by the housing market and their co-optation into popular credit and compulsory municipal taxes. Important grassroots struggles for a decent livelihood continue to be fought in poor neighbourhoods. But the residents' insurgent acts of reclamation of the city – squatting, auto-construction and legalisation – are constantly captured by an expanding real estate market and state programmes leading to the financialisation of poverty. As my fieldwork unfolded, it became clear that both regionalism and trade union activism were failing the unemployed, informal workers and squatters of my bairro. The ethical turn I describe next reframed the land question through the lens of environmental and fiscal activism.

The Ethical Turn

In Volta Redonda, the ethical critique of capitalism has been a central trope of the anti-corporate struggle since the 1980s, when militant Catholic grassroots organisations, inspired by liberation theology, catalysed the anti-dictatorship movement. But the ethical discourse I witnessed during my fieldwork came from a more mainstream Catholic tradition. When, after the global economic crisis, CSN, Peugeot and VW announced a wave of mass redundancies, Dom João Maria Messi, the new Catholic Bishop of Volta Redonda, organised the Forum for Zero Redundancies. With employers, trade unions, civic and commercial associations and grassroots Catholic organisations, the forum also involved neighbourhood associations, the MST and black and women's organisations. In 2008, the forum organised public talks, events and demonstrations that reflected on the social and economic consequences of the lay-offs on communities, families and small businesses. Partly, the forum replicated the horizontal, cross-sectional and business-oriented regionalist alliances promoted by the municipality of Volta Redonda that emerged after the privatisation of CSN, bringing together social movements, municipalities and regional and municipal businesses and entrepreneurial associations.

But the forum, unlike these earlier entrepreneurial platforms, had a stronger anti-corporate orientation, replicated the modus operandi of the Catholic anti-dictatorship movement and were led by two influential grassroots Catholic organisations linked to these earlier struggles: the MEP-VR and the Movement for Peace. The MEP-VR was funded by José Maria da Silva (Zezinho) a CSN metalworker and *cepista* (health and safety representative) who led several political campaigns with the PT in the 1980s and was dismissed by the firm in 1996. The MEP-VR brought together the historical network of militant

Catholic organisations such as the JOC, Comunidade Eclesial de Base (Basic Ecclesial Communities – CEB) and Ação Católica Operária (Catholic Worker Action), social movements and professionals, especially lawyers and judges,[40] around an explicit 'post-materialist' agenda revolving mainly on environmental, pedagogical, housing and anti-corruption issues. For instance, the MEP-VR monitored corruption practices in the municipal chamber, sat in economic and environmental commissions and led the Projeto Político Pedagógico (Political Pedagogical Project), a self-organised network of primary and secondary schoolteachers providing basic education in poor bairros.

The Movement for Peace was more 'hands on' and operated in low-income neighbourhoods affected by redundancies and unemployment, domestic violence and drug-related violent crime. Father Juarez Sampaio, the CSN-worker-turned-priest who founded the Movement for Peace, acknowledged the legacies of early trade unionism on the organisation:

> I worked in CSN until 1992. That was the golden era of trade unionism. The trade union was very combative. But with privatisation, my department was closed down and I had to leave. I started my priesthood because I wanted to be part of a church engaged with the working class and with the struggles of poor people. (Lima 2010: 87)

Perhaps because of Sampaio's own experience of redundancy in 1992, the Movement for Peace operated mainly in neighbourhoods heavily hit by the privatisation of CSN such as Santa Cruz, where six thousand residents were made redundant as a result of privatisation, or Parque Maíra, a semi-rural favela at the periphery of the city. Sampaio is a charismatic activist. It was mainly thanks to him that the first lawsuit against CSN for environmental damages was successful. Because of the strong influence of Bishop Calheiros Novaes, who was in close terms with the Metalworkers' Union, the Movement for Peace was able to partly connect to the working-class base, especially with its less affluent strata. The mixture of socialism and Catholicism of the Movement for Peace resonated with the political language of trade union leaders like Renato Soares – who were politically formed among grassroots socialist catholic organisations – and captured the atmosphere of creeping anxiety about the region's progressive deindustrialisation and the dark prospects of a future without steel or car industries.

Together, the Pentecostal and the Catholic congregations represent more than half the Volta Redonda electorate. The former are associated with the business community and the MDB and operates mainly on anti-poverty projects in poor neighbourhoods, whereas the Catholic Church has a clear anti-corporate agenda and works in liaison with the PT and the SMSF. On the Feast of Corpus Christi, the political relevance of these two religious communities becomes evident.[41] For the Pentecostal community, the Feast of Corpus Christi is a magical day when the congregation literally meets the Holy Spirit. Composed of various churches, but under the lead of the powerful Universal Church of the Kingdom of God, the evangelical community led a 'march for Jesus' through the city with coloured banners, T-shirts and balloons, accompanied by collective gospels. Consisting of roughly five thousand demonstrators, mainly black, young[42] and from poor neighbourhoods, the demonstration was joyous and festive. It paused by CSN, then by the Zumbi memorial and made a final stop in the bairro Retiro, where hip-hop concerts and rap battles continued through the night. In their speeches, religious and neighbourhood leaders mentioned unemployment and the economic crisis only in passing, focusing heavily on racism and urban violence instead.

On the same day, Bishop Messi – the successor of Calheiros Novaes – organised a mass at a sports centre on the island of São João,

Figure 3.7. Evangelical parade (© Massimiliano Mollona).

a small lush islet in the middle of the Paraíba River. The scenery was evocative. An estimated seven thousand people stood under the altar, surrounded by the tropical forest. From behind the altar, the sun projected the Cross's long shadow onto the believers. Most of the congregation were well dressed and middle aged. There were well-known representatives from the traditional Left, and civic organisations such as the MEP-VR and the Movement for Peace. Messi's talk for the Holy Communion took place at sunset, under a sky on fire, and with the audience listening in total silence:

> Through this Holy Communion, I ask you to appreciate the value of communion not only with God but also between yourselves and against the values of competition, individualism and self-enrichment. Let this Holy Communion remind us of the importance to struggle for justice. The salaries of the politicians, the mayors, the lawyers and the managers always go up. But the salaries of the workers and of 'the people', those who sweat for our community and who should earn more, get smaller and smaller. Through this Holy Communion, we must learn to fight for justice and social equality.

The silence after the sermon was broken by a rapturous samba version of *Gloria*. An endless line of believers slowly crawled towards the altar, dancing on the spot and collectively singing a samba-style gospel: 'For the working class and the labourers, for the fighting people, we offer our struggle to you God . . . glory to God'. Then, there was a collection for the Landless Workers' Movement in a nearby favela, followed by another talk that emphatically denounced capitalism, its unjust profits and the poverty it generated. At dusk, Messi left, walking through the congregation on a carpet of sand and flowers, followed by a long procession of priests and altar boys and accompanied by a celestial hymn – like a vision.

The 'ethical turn' arguably represented a subdued, depoliticised and 'post-materialist' form of working-class activism that reflected the weakness of the Metalworkers' Union vis-à-vis the global deregulation of capital and the financialisation and statalisation of the economy led by the leadership of the PT and the trade union movement. In 2010, as the economic crisis slowly crept in, the Forum for Zero Redundancies changed direction, too, becoming a business think tank devoted to improving the local economy and attracting new companies through various form of land and service tax reductions, credit and purchasing consortia. Even Renato Soares committed himself to buying new VW cars for the Metalworkers' Union. The change in the forum's direction reflected the fact that the city, like

Figure 3.8. Waiting for mass on the island of São João (© Massimiliano Mollona).

other municipalities, now acted as a global financial subject seeking to attract foreign capital. The slow financialisation of the city slowly aligned the municipality, CSN and the Metalworkers' Union.

The ethical turn also comprised the new wave of environmental activism, by lawyers, GPs and architects, that started during the privatisation of CSN, when environmental organisations campaigned for and obtained the establishment of the measure of 'environmental liability', which opened the possibility of CSN being charged for environmental damages. In 1993, for the first time in history, the company was charged R$60 million in environmental fines – a sum well above the municipality's yearly budget of R$50 million (Lima 2010) – and was forced to pay the Imposto Predial e Territorial Urbano (Urban Land and Territorial Tax – IPTU) and Imposto Sobre Serviços (Services Tax). This convergence between labour, legal activism and civil rights movements seemed to replicate the successful formula of the new unionism of the 1980s. But the emphasis on corporate issues such as 'corporate social responsibility', 'environmental sustainability' and 'green balance sheets', endorsed by the Metalworkers' Union in the context of the forum, often overshadowed the framework of class struggle and the issues of declining wages, unemployment and precarious working conditions.

As well as reigniting urban activism, environmentalism empowered federal and regional agencies vis-à-vis the company. For instance, in 1992, the Regional Secretary and the Ministry of the Environment jointly increased CSN's estimated environmental liability and forced the company to invest R$108 million in pollution-preventive measures. Moreover, since then, the company's transfers to the municipality are strictly monitored by BNDES, which is also the company's main credit supplier and now acts as enforcer of environmental regulations in Volta Redonda. But, as in other working-class towns,[43] environmentalism and health activism were slow to take off in Volta Redonda. The first legal environmental framework was set up by the military in 1981 with the creation of the federal Secretaria de Meio Ambiente (Secretary of the Environment), the national environmental administrative body Instituto Brasileiro do Meio Ambiente e dos Recursos Naturais Renováveis (Brazilian Institute of the Environment and Renewable Natural Resources) and the new form of civic responsibility for damages to the environment. But for a long time, CSN censored open debates on the conditions of leukopenia[44] and leukaemia – both related to benzene poisoning. Fearing being made redundant, affected workers did not speak out. The Metalworkers' Union remained silent, too. Only in 1985, when the charismatic GP Cida Diogo became the health director of the SMSF, was the issue of leukopenia openly addressed and became a central issue of contention between the Metalworkers' Union and CSN.

The debate had a huge effect on the labour movement and the city more broadly. Residents were told what they must have feared all along: for a long time, they had been poisoned by benzene fumes released from the UPV. According to Diogo:

> Only then did people start to worry that the condition might not be limited to the factory and that pollution might expand outside the factory – into the city. Of course, the gas did not stop at the factory's gates. Volta Redonda grew around CSN. The factory's walls are the homes of its residents. It was the first time the trade union raised this issue in public. The community got involved. The workers organised massive strikes, occupying the central office of CSN and squatting by the factory gates. At that time, the company was still denying the condition. (J. Lopes 2004: 106)

Only then did the residents realise their health and the health of CSN workers was deeply entangled and that CSN was harming not only the working class but the whole city. The Metalworkers' Union developed an extensive Programa de Saúde do Trabalhador (Workers' Health Programme) in coordination with GPs, the regional Secretary of Health and the Ministry of Labour and Employment, thus forcing CSN to put in place an extensive programme of monitoring and prevention of environmental pollution and heavily investing in new research on leukopenia. In 2000, in coordination with federal and regional environmental agencies, the BNDES halted a R$10 million loan to CSN, resuming it only after the company emitted R$80 million worth of bonds guaranteeing its plan to invest in pollution-reduction measures. Yet, during my fieldwork, more disturbing revelations emerged. An inspection by the regional environmental commission of the Assembléia Legislativa do Estado do Rio de Janeiro (Legislative Assembly of the State of Rio de Janeiro) showed levels of benzene in the Paraíba River were 400 per cent above normal levels. Fishes had horrendous deformities and were dying of cancer.

Soon after the publication of a damning report by the World Bank highlighting the exorbitant levels of pollution in Volta Redonda, CSN was heavily fined for environmental damage and river pollution. The BNDES suspended all loans to CSN for breach of anti-pollution regulations. Unlike the grassroots and labour-based cross-sectional alliances gathered around the 'land issue' in the 1980s, the environmental coalition was based on a broader institutional and class spectrum, which together with the SMSF included the Fundação Estadual de Engenhenaria do Meio Ambiente (State Foundation for Environmental Engineering – FEEMA), the Conselho Municipal de Defesa do Meio Ambiente (Municipal Council of the Defence of the Environment), neighbourhood associations,[45] the SENGE-VR, the Partido Verde (Green Party) and different ecological organisations.[46] These organisations had very different environmental agendas. Neighbourhood associations based in favelas and poor working-class neighbourhoods campaigned against industrial dumping and pollution and articulated their campaigns in terms of rights to the city and urban democracy. Environmental associations, with mainly a middle-class base, had a conservationist agenda and focused especially on preserving the Cicuta Forest. Together with the municipality, this environmental coalition set up Agenda 21, an ambitious programme against pollution, dumping, land degradation and cruelty against animals, which included a 'green hotline' to call to denounce environmental breaches by CSN. In Agenda 21, the fight against waste and pollution was framed as part of a broader process of improvement of working-class life, which included also better working conditions, education and culture.

Some argue the 'post-materialist' approach of such civic movements as the Forum for Zero Redundancies or Agenda 21 lacked of a broader political vision (Vanderlei Barcelos in Lima 2010: 78). But, these struggles for health, employment and the environment are as much struggles for a decent working-class livelihood as factory-based struggles. Under the strong leadership of Bishop Messi and its grassroots base, the forum led important environmental struggles in poor neighbourhoods, bringing together the labour movement, civic organisations and the middle class. For instance, in 2003, it forced CSN to compensate more than one hundred resident families in Volta Grande IV, a neighbourhood built by the SMSF at the beginning of 1990s on land where the company systematically dumped chemical waste. Analyses by the chemical engineers attached to the union showed the company's chemical waste had corroded the houses of local residents and was responsible for the leukaemia and cancer spreading in the community (Ramalho et al. 2013: 184). Living in a lethal environment, the residents of Volta Grande IV are still awaiting for relocation.

For José Sergio Leite Lopes, the workers' health mutated from being a labour issue to be discussed at the factory level to becoming a 'public' issue to be discussed at the level of the municipality and the state. Lopes describes this change as the 'environmentalisation of social conflicts' in Brazil. Yet, when Lopes's study was made, the state struggled against powerful corporations such as CSN. In the word of a functionary of FEEMA: 'FEEMA cannot stop CSN because the state nowadays is weak. It needs CSN. Volta Redonda needs CSN. There if this fundamental commitment' (J. Lopes 2004: 106). But under the current neo-extractivist regime, CSN and the state share the same understanding of nature as a commodity to be exploited, extracted and distributed en masse for the benefit of the Brazilian nation. Thus, it is not surprising the environmental activism of the state against CSN dwindled under the PT administration.[47]

Indeed, environmentalism in Volta Redonda seems to be an oxymoron. The city and the plant are so entangled into each other that even under strict environmental control, everyday life is intrinsically unhealthy. People continue to put up with unbearable levels of sound and air pollution[48] as if CSN does not exist. Even though 64.5 per cent of the residents of Volta Redonda declared in 2004 to have pollution-related respiratory problems, only 1.4 per cent of them partook in environmental activism, while 25 per cent were involved in sport associations (Lopes 2004: 141). On the other hand, the regional dispersion and geographical diversification of CSN created new opportunities for labour reorganisation. I have mentioned the global commodity boom turned the iron mine division of CSN located in Casa de Pedra (Minas Gerais) into the main profit generator of the company. In 2008, CSN's mining subsidiary, Nacional Minérios, employed 64 per cent of the residents of Congonhas (Minas Gerais), where the Casa de Pedra mine is located, and transferred to the municipality R$24 million in annual taxes. Work at the Casa de Pedra mine is notoriously harsh. The Metabase Inconfidentes (Miners' Trade Union) has no representation in the factory council or health and safety representatives (*cipistas*),[49] and wage levels and working conditions are the worst in the group.

But the profits generated by Nacional Minérios strengthened the bargaining power of the Miners' Trade Union. In 2009, the union opposed the R$12 million reorganisation plan for the plant, which included mass redundancies. The opposition was strengthened by a report produced by the Grupo Rede de Congonhas (Congonhas Network Group) – a pro-labour coalition of engineers, lawyers, judges, GPs, civic organisations and the municipality – showing that the proposed expansion of the company would damage the river system and the UNESCO-protected Santuário de Bom Jesus do Matosinhos (Ramalho et al. 2013). This broad coalition continued to oppose the development. In 2011, mineworkers led three days of strikes, which paralysed the company's entire operations. In the following year, after a three-year litigation between the municipality, the union, the Grupo Rede de Congonhas and CSN, the development was finally scrapped. In this instance, the workers' opposition to CSN succeeded because it was supported by an environmental coalition of lawyers, environmental engineers and local MPs who framed the mobilisation in environmental and ethical terms. Besides, the struggles of the Minas Gerais mineworkers also empowered the subcontractors in Volta Redonda, who in that year won important struggles (see chapter 5).

In Brazil, regionalism – and the fiscal and political autonomy of municipalities – was enshrined in the 1988 constitution as a fundamental tool of decentralisation of the state and a step towards participatory democracy. Indeed, the regional coalitions between municipalities, grassroots movements, trade unions and local industrialists formed in the early 1990s opposed the neoliberal stance of President Fernando Henrique Cardoso. In the industrial region of São Paulo, these cross-sectional alliances between municipalities, employers, civic movements and trade unions against global carmakers were led by the militant autoworkers (Humphrey 1982; Ramalho 2005; Silver 2003). According to José Ricardo Ramalho and colleagues (2009), the cross-sectionalist alliances of the Sul Fluminense region are similar anti-systemic forces against a progressively spatially fragmented capital (see also R. Santos 2006). For instance, by diversifying regionally, CSN became vulnerable to the reorganisation of labour along the nodes of its value chain, most recently in the Casa de Pedra mines.

In reality, these cross-sectional alliances can do very little against the violent movement of capital accumulation in the Fluminense region. In the late 1990s, Volta Redonda Mayor Neto embraced a business-oriented posture in the hope of securing regional funds to reconvert the city into a leisure and tourist destination and, in so doing, attract global investors. At the same time, the more militant regionalism of municipalities and autoworkers of Resende and Porto Real was systematically undermined by the constant threat of relocation, extreme subcontracting, systematic breaches of labour regulation and sheer financial power of global carmakers. In this context, the seemingly short-term militant particularism of the Volta Redonda steelworkers was vindicated. Indeed, the only real 'spatial fix' of capital in the region remains the UPV.

In this chapter, I have shown how the internalisation of the national debt and the spatial reorganisation of national and foreign capital impacted grassroots politics. The cross-sectional alliances that have taken place in the Sul Fluminense region in the past ten years are radically different from the grassroots favela activism of the 1970s, the movement of land occupation of the 1980s and the militant regionalism of the PT in São Paulo in the early 1990s. They are hopeless attempts by cash-stripped municipalities to resist the regime of internal debt imposed upon them by a state-turned-financial capitalist and to counter the ruthless regime of accumulation implemented by global carmakers in their 'ghost factories'.[50] Thus, the 'land issue' marks different historical articulations of the conflicts between labour and capital – and between CSN and the city – involving different social constituencies (the working class, trade unions, municipalities and regional agencies, the middle class, the church and the poor), narratives (ecological, ethical, of class or social marginality) and scales of political action – spanning from the global, the federal, the regional and the neighbourhood. Initially, the struggle between CSN and the city reflected the struggle between the new state capitalist and the old feudal landlords, and between the public economic regime of the former and the private rentierism of the latter. Thus, the early municipalism against CSN was an attempt to restore the old economic order.

Dictatorship inverted these positions. CSN was reorganised along criteria of economic efficiency, and its welfare and social facilities were sold to the municipality or privatised. Now it was the municipality's duty to perform welfare and redistribution functions. In the 1980s, the Volta Redonda municipality aligned itself with the labour movement and the powerful anti-dictatorship coalition against CSN. It was through the active involvement of its IPPU that the struggles of the SMSF, squatters and civic organisations gained strength and the notions of rights to the city and urban justice gained visibility. In 1985 the Municipal Law of Urban Squatters gave the Volta Redonda municipality further ammunition against CSN. This was now a struggle between a repressive privatising regime pursued by a national company, and a new grassroots movement of urban commons supported by a radical municipality. Throughout the 1990s, the political actions of the municipality against CSN and in support of the commons with grassroots civic coalitions and the labour movement was further enhanced by the principle of municipal autonomy enshrined in the new democratic constitution.

But the process of democratic decentralisation coincided with Brazil's opening to the world economy and with the first wave of national privatisations, starting with that of CSN. Privatisations led to the internalisation of the national debt – described by Francisco de Oliveira (2005), as 'internalisation of imperialism' – and transformed the Brazilian state from 'economic regulator' to 'economic investor' operating through its capillary network of regions and municipalities. In chapter 5, I discuss in detail the nexus between the privatisation of CSN, the privatisation of the city, the financialisation of working-class households and the transformation of CSN employees, especially unskilled workers, into 'modern investors'. Thus, as well as a process of democratisation, regionalism can be understood as one of spatial reorganisation of capital, which spread relationships of debt dependency from the state downwards towards cities and neighbourhoods and captured the impoverished working class and favela residents into the financial and real estate markets. Moreover, with its principle of municipal accountability, regionalisation turned municipalities into nodes of a decentralised and dispersed national financial-logistical infrastructure open but also vulnerable to the speculations of global capital.

In this context, Volta Redonda, burdened by debts and rising unemployment, gave up on its struggle against CSN and co-opted the Metalworkers' Union and civic associations in the hopeless attempt to lure local businesses and international capital against an increasingly deterritorialised and financially dispersed CSN. Unlike the discourse of Baltazar in the 1980s, which mapped the struggle between CSN and the municipality as a one between state capitalism and urban commons, Neto framed the struggle between CSN and Volta Redonda as a struggle between the state and the market. Neto's business-oriented stance mimicked that of Governor Cabral Filho, who was behind him and whose charisma and dealings with the business community had made him the most likely successor to Lula to the presidency. Indeed, as a conglomerate combining land and iron ore extraction, steelmaking, finance and logistics, CSN embodies and reproduces the neo-extractivist model of the state. The functioning of CSN as an agent of state capitalism is particularly striking. Volta Redonda can be considered the historical laboratory for the modernisation of Brazil along the visions of Cardoso and Lula da Silva, starting from the privatisation of CSN and ending with the financialisation of poverty and the 'proletarianisation of capitalism'.

Indeed, on the one hand, CSN is dis-investing from industry and investing in the service sector – deindustrialising Volta Redonda and financialising it – reflecting the 'schizophrenic posture' of the PT government. In 2015, it scrapped plans to develop a new furnace (AF4) and proposed instead the development of an immense commercial area on its land – including a shopping centre, a supermarket, a university, a hotel and a conference centre – partly funded with the company's pensions funds. But as a powerful landowner and industrialist operating through the old logic of accumulation by dispossession, CSN is also a force 'against progress', especially in terms of its extensive use of outsourced and precarised labour, which prevents the full transformation of Volta Redonda into a frictionless and post-industrial hub – a logistical extension of the state or a 'truly' national champion. In fact, the BNDES and other federal agencies continuously monitor and discipline CSN through punitive interventions – taxes, environmental fees or conditional funding such as PACs and other federal programmes.

CSN sits uncomfortably in this spatial and historical divide between the 'progressive' forces of state-backed financial capitalism and the ancient rentierist block including colonial elites, progressive municipalities and the new working class. The company is simultaneously a global financial operator, with a production process globally dispersed, a Fordist employer and a powerful landowner. It currently employs only 30.6 per cent of the local population[51] but owns a third of the municipal land and controls the local economy. Moreover, most of the municipality's revenue come from CSN in the form of urban rents, taxes and environmental fines. Unlike most rentierist post-industrial cities (Harvey 2012), Volta Redonda is both deindustrialised and mono-industrial, and its urban land is at the same time overcrowded with slums and empty, including the Cicuta Forest.

In the *Cidade do Aço*, the rentierist logic of late capitalism described by David Harvey (2012) takes a disorganised form, encompassing global logistics, mining, informal and indentured labour, steelmaking and land and housing speculation. In this 'postcolonial space of capital' (Mezzadra 2011) – where heavy industry, cattle grazing, informal and tertiarised labour, and the media and tourist sectors blend into one other – local residents continue to jog, walk their dogs, play futsal and bike along the Paraíba River, seemingly ignoring the yellow smoke, violent explosions, dark clouds and acidic smell hovering over the city. Vargas wanted Volta Redonda to be a model of industrial democracy for the whole nation. In fact, Volta Redonda replicates Brazil's pattern of extreme inequality.[52] The history of land struggle in Volta Redonda is at the crossroad of three structural forces: state-backed dependent industrialisation, associated with favelisation and accumulation by dispossession; market speculation and financialisation both at the state and the municipal levels; and the commoning and urban struggles by social movements and marginal urban constituencies. While systematically opposing CSN, the municipality aligned itself with the cross-sectional and marginal constituencies that reclaimed the land as part of a redefinition of the urban commons in the 1980s, whereas both in the early times and recently, as result of democratic decentralisation, it aligned itself with market forces.

The struggle over the urban land I have described in this chapter is ultimately a struggle over the meaning and forms of livelihood. CSN and the local elite consistently see the land of Volta Redonda as capital to be maximised *despite* the need of the people – of decent housing and education, basic food and sociability – either through the state or the market, whereas in the 1980s, the cross-sectional civic coalition revolving around the SMSF saw the land as a *commons* to be managed collectively, horizontally and democratically, through housing co-operatives, squatting and collaborations with the municipality. More recently, with the economic crisis hitting Brazil, CSN, the Metalworkers' Union and the municipality are joining forces against the state. In 2018, Benjamin Steinbruch actively supported Ciro Gomes (PDT) based on an explicit 'anti-Lulista' manifesto.[53]

In this chapter, I have also considered the argument that municipalities and regions are new sites of labour reorganisation and activism. Against this argument, I have claimed the regionalisation and democratisation of the Brazilian state and the administrative and financial autonomy of municipalities was largely a form of reorganisation of state capitalism that turned municipalities into financial units, democratised credit (pushing working-class families into debt and stratifying them along housing property lines) and forced the Metalworkers' Union to oscillate between a hopeless business pragmatism and militant particularism. This story is not new: Brazil is not the only developing nation where the advent of democracy coincided with economic deregulation and the triumph of neoliberalism.

The first question emerging from this chapter is whether the legacy of the Right to the City, squatters and civic movements developed in Brazil against the authoritarian state in the 1980s can be revived by the Brazilian working class in the twenty-first century. Indeed, Right to the City movements such as the Movimento Passe Livre (Free Fare Movement) led the exceptional surge of political activism in 2013 at the national level (Braga 2016; Zibechi 2014) and in Volta Redonda.[54] The Coletivo Terras de Volta (Collective for Land Restitution) fighting for the restitution of the 147 areas of public interest appropriated by CSN with the privatisation is another hopeful recent development in Volta Redonda. Formed in 2015, the collective brings together various social movements, the Party for Socialism and Freedom and the powerful juridical body Ministério Público Federal (Federal Public Prosecutor's Office) but has failed to capture Volta Redonda's working-class base (Lima 2018).

The second question is how to reconcile struggles for reproduction and for a decent livelihood including health, housing and the environment with factory struggles – instead of considering these two forms of struggle as mutually exclusive. Lastly, this chapter asks whether the global workforce of neo-extractivist conglomerates can form an anti-systemic force and hit capital in its core logistical nodes, as happened in the case of the Casa de Pedra mine. The extreme power of global carmakers and the militant particularism of the Volta Redonda *metalúrgicos* point to the opposite direction. Moreover, in the current context of economic protectionism, nationalisation and monopoly capitalism in the Global North, cross-sectional struggles and solidarities are even more unlikely, although a regional labour block against the neo-extractivist consensus seems to be emerging in Latin America.

Notes

1. In the first years of his dictatorship, he passed laws for forests, water, land and mines concession and funded national geography, demography and statistics institutes.
2. The economist Joseph Schumpeter (1942) describes economic development as a process of creative destruction – that is, of replacement of old factors and forms of production with new ones.
3. The anarchist political thinker Peter Kropotkin (1919) also supported this small-scale and localised view of economic development in opposition to Marx's endorsement of mass industrialisation and urbanisation as necessary historical forces.
4. Settlers who received grants of land from the Portuguese Crown.
5. Banks were reluctant to make loans without the collateral guarantee of slave property, and new national banking elites were emerging (Stein 1957: 113).
6. Including a post office, police station, cemetery, slaughterhouse, bakery, chemist, school, railway station and water and electricity (Costa 1978: 16). The municipality of Volta Redonda was founded only in 1954. Until then, CSN was part of the Barra Mansa municipality.
7. The company went through four stages of expansion: 1941–1948 (Plan A), 1950–1954 (Plan B), 1956–1960 (Plan C) and 1960–1986 (Plan D).
8. Their rent was fixed at 20 per cent of workers' salaries.
9. CSN could expropriate them at only thirty days' notice.
10. Hence, its characteristic name 'anthill'.
11. This was the name given to early migrants who moved across workplaces without settling permanently.
12. Decree Law 11.087 (10 December 1942) was made just after Brazil entered the Second World War.
13. Including roads, streets, parks and commons; the electricity, water, sewage and waste systems; public transport; and reforestation. *Boletim de Serviço de Volta Redonda* 2 (7 January 1948): 1280.
14. Because of its strategic importance, Volta Redonda was considered an 'area of national security' and was tightly controlled by the federal government. Its mayors were appointed directly by Brazil's president and hence were called 'ghost mayors'.
15. *O Lingote* 222, no. 18 (1971).
16. As I discuss in the introduction, public autarchies were financially and administratively autonomous municipal bodies that Vargas established to improve public services.
17. In 1975, 4,200 workers were made redundant, 12,400 in 1978 and 30,000 in 1984, on completion of the works on AF3 (Lima 2016: 64).
18. Compared to 5 per cent in 1966 in (Cabral de Souza 1992: 128).
19. Movimento dos Posseiros Urbanos (Movement of Urban Squatters). See Lima (2010: 56).
20. Worker-priests, CEB and the JOC.
21. In the 1980s, the Right to the City movement was spreading all over Brazil, laying the conceptual foundations for the urban reform of the constituent assembly of 1988.
22. Prefeitura Municipal de Volta Redonda (Documento do FURBAN).
23. See the interview with Isaque Fonseca in chapter 5.
24. Including residential developments, gardens, grazing land, courtyards, fields, unregistered streets, the Santa Cecília farm and the vast Cicuta Forest, which consists of 131 hectares of UNESCO-protected urban tropical forest.
25. Namely, the ETPC, the workers' and employers' clubs and the Hotel Bela Vista.
26. The most important of which is the memorial of Zumbi, a symbol of black community. Zumbi do Palmares was the last of the leaders of the Quilombo dos Palmares, a fugitive settlement in Alagoas.

27. Between 50 per cent and 60 per cent (R$10 million) of its yearly income consisted of IPTU and environmental taxes from CSN.
28. Aimed at attracting foreign carmakers, restructuring the existing national car sector and consolidating Brazil's strategic industries in the Mercosul.
29. According to the Associação Comercial, Industrial e Agropastoril de Volta Redonda (Commercial, Industrial and Agro-pastoral Association of Volta Redonda – ACIAP-VR), reducing direct employment in the rust belt from 3,000 to 1,300.
30. The state gave to VW an electric station for the sole use of the plant worth $4 million, a gas network worth $7 million and a $2 million grant to the Resende municipality for the construction of electricity, water and transport facilities to the plant, inaugurating a new institutional arrangement.
31. The municipality granted the company exemption from corporate taxes and gave three million square kilometres of industrial land.
32. For instance, in 2002 the Federação das Indústrias do Estado do Rio de Janeiro (Industry Federation of the State of Rio de Janeiro State) led an economic consultation with twelve municipalities; the ACIAP-VR; the Sindicato das Indústrias Metalúrgicas, Mecânicas, Automotivas, de Informática e de Material Eletro-Eletrônico (Union of Metallurgical, Mechanical, Automotive, Computer and Electro-Electronic Industries); the SENGE-VR; the SMSF; the local Fluminense Federal University; and the Confederação Nacional da Industria (National Confederation of Industry – CNI).
33. E.g., housing, hotels, restaurants, cleaning, security or transport.
34. Cabral Filho was especially influential in the 2008 creation of the Agência de Desenvolvimento Regional do Médio Paraíba (Regional Development Agency of Middle Paraíba), a consortium fourteen municipalities, banks and federal universities in charge of developing regional economic district.
35. For a detailed analysis of these factories' organisation of labour, see chapter 5.
36. With initiatives such as *Investing in Volta Redonda* and *Rethinking Volta Redonda*, organised by the newly created Pólo Industrial de Volta Redonda (Industrial Hub of Volta Redonda).
37. For a similar conclusion on the effect of regularisation of housing in poor neighbourhoods, see Taschner (1995).
38. Ruggeri (2007) describes the small replicas of South California in the gated communities of Hong Kong.
39. Brazil's most celebrated architect and landscape artist.
40. Namely, the OAB, ministers for justice, public attorneys and members of the Comissão Brasileira de Justiça e Paz (Brazilian Commission for Justice and Peace) and the Conferência Nacional dos Bispos do Brasil (National Conference of Bishops of Brazil).
41. Celebrated in May, the Feast of Corpus Christi celebrates the suffering death and resurrection of Christ.
42. According to an estimate by an organiser of the church, 70 per cent of the demonstrators were below twenty-five years of age.
43. I describe workers' attitude to health and safety in the United Kingdom in Mollona (2009).
44. Involving dizziness, headache, nausea and fainting.
45. The Federação das Associações de Moradores (Federation of Resident Associations) and the Conselho das Associações de Moradores (Council of Resident Associations).
46. The Associação Ecológica de Volta Redonda (Ecological Association of Volta Redonda) and the Associação dos Defensores da Mata da Cicuta (Association for the Protection of the Cicuta Forest).
47. However, this is not the case for regional environmental agencies. For instance, in December 2017, following pressures from regional environmental agencies, Governor

Luiz Fernando Pezão nearly closed CSN because it had not complied with environmental regulations.

48. Located in proximity of employees' dwellings, CSN's White Martins works on eighty-six decibels against a permitted threshold of seventy.
49. Comissão Interna de Prevenção de Acidentes (Internal Commission of Accident Prevention).
50. These factories are described as 'ghost' because their workforces are nearly entirely tertiarised (see chapter 5).
51. Fifty per cent of the residents are employed in commerce and service, and 10 per cent in public administration.
52. *O Globo*, 4 April 2005, based on the statistics of the Brazilian Institute of Geography and Statistics. About 25 per cent richer residents of Volta Redonda earn twenty times more than the 40 per cent poorer ones.
53. Gomes was the CSN director until the 2016 and served as president of Transnordestina Logística. Steinbruch put himself forwards as Gomes's vice presidential running mate.
54. On 20 June 2013, thirty-nine thousand people marched in the Volta Redonda city centre in protest of the Rousseff administration.

– Chapter 4 –

Of Ants and Steelworkers

Capital as Labour

You can kill an ant but you cannot get rid of the whole ant nest.
—SMSF President José Juarez Antunes

The central branch of the Metalworkers' Union of Sul Fluminense (SMSF) is in the 'old city', the town's commercial centre, filled with 1950s-era beauty salons, fashion retailers, bicycles shops, Internet points, spiritualist stores, legal firms, post offices, stationeries and building supply shops. Despite the high unionisation rates in CSN, workers have mixed feelings about the Metalworkers' Union. Several times, I heard from workers on the shop floor that the SMSF has just a 'performative' or 'ritualistic' role. One of these performances I witnessed was the ballot for the proposal to change the current three eight-hour shift regime to a four six-hour shift regime. Union ballots are not legally binding. They are used to generate public awareness on specific labour campaigns and influence the management on contested labour issues.

In preparation for the ballot, the Metalworkers' Union's bulletin *9 de Novembro* discussed the benefits of the change: a national increase of 2.8 million jobs 'for your sons and grandchildren', an increase in hourly and overtime labour rates and more time for study and leisure. Direct salaries would not be reduced. There would be only a slight reduction in the Fundo de Garantia do Tempo de Serviço (Service Assurance Fund)[1] and the thirteenth salary, which the Metalworkers' Union estimated around R$100 per year. Moreover, the bulletin spelled out the historical legacy of the struggle over the reduction of the workday – from the first struggles at the beginning of the century to reduce the workday from fifteen hours to ten, to the constitutional law of 1988 that established the legal work week as forty-four hours. The article concluded: 'The reduction of the workday is not a struggle against unemployment but a struggle for the workers' right. This was the central demand of the strike of 1988, which deeply signed the history of the city and culminated with the death of three workers of CSN' (SMVR 2008). But there was a widespread suspicion among the workforce that the Metalworkers' Union was pushing the change in order to minimise the lay-offs CSN would implement in case the much-feared economic recession hit Brazil and Volta Redonda.

During the ballot, the workers reclaimed, once again, 'their space', cordoning it off from traffic. From 6 a.m. to 6 p.m., CSN employees queued patiently to register at the union stall and then vote in the wooden cabins located by the workers' memorial. Meanwhile, the eight SMSF directors gave speeches in turn, standing on top of the union van. They spoke energetically, emphatically and at length even when the square was semi-empty. The crowd came in and out, unevenly, depending on working shifts and lunch breaks. The health director took the centre stage, emphasising the physical and mental benefits of a reduced workday and of a healthy work-life balance. The ballot closed at 6 p.m., coinciding with the end of the afternoon shift. Four SMSF representatives and four CSN representatives stepped inside the fenced space where the boxes containing the electoral forms had been placed and started the count. A crowd gathered around that space to witness the counting. The union organisers and CSN representatives who were counting communicated through hand gestures with colleagues outside the fence. Forms were counted once, divided in bundles of one hundred and secured by rubber bands; then, all the representatives counted them again. The CSN witnesses gathered with anxious and threatening expressions. Tension and excitement mounted as the results informally circulated. At sunset, it was declared that nearly two-thirds of the workers voted in support of the six-hour shift. The legal coordinator exulted, 'This is direct democracy in action . . . a triumph!' A crowd of workers remained in the square cheering, joking and drinking as darkness fell. CSN agreed to start consultations with the SMSF on the reduced shift regime.

A few days later, it emerged that the workers supported the six-hour shift but opposed the proportional reduction in their benefits. The negotiations that followed were a fiasco. After a short pilot scheme, the six-hour regime was scrapped and taken up again by the company a year later in the midst of the economic crisis and lay-offs, but this time with a radical salary reduction. Some workers asked: 'Was the Metalworkers' Union supporting the reduced shift one year earlier because it knew the company was about to downsize?' 'If so, why were the workers not made aware of this?' Indeed, during my fieldwork in Sheffield, I witnessed several instances of management and unions pushing for supposedly pro-labour, less hazardous or more environmentally friendly changes, framed through narratives of 'work-life balance' and mental well-being, which ended up cutting labour or reorganising shop floors. Hence, the scepticism that manual workers tend to have towards environmentalism and health and safety activism comes from their experience of how these post-materialist issues have legitimated labour reorganisations in the past.

For Stella the 'fighter' (*lutadora*), the charismatic lawyer of the SMSF, the ballot fiasco and the steelworkers' rejection of the proposal to reduce working hours in unhealthy areas and their endorsement instead of hazard-related bonuses showed their short-term and materialistic mentality and lack of class consciousness:

> The Metalworkers' Union has become a corporation, like CSN. And the steelworkers are like bankers. When they come to the *sede* [headquarters], they are only concerned about how to make more wages and about their *cesta básica* [basic basket]. They can only think about 'the now' . . . about their own production quotas and wages . . . always in the present moment! They do not worry about the other workers or even about themselves in ten years' time . . . They are really narrow-minded . . . They lack class consciousness![2]

In 2008, at the beginning of my fieldwork, Beth, the union's communication officer, claimed it was a good moment for labour. Brazil and CSN were growing economically, and the Metalworkers' Union thus was increasing its bargaining power vis-à-vis CSN. Renato Soares, the new Communist leader of the Metalworkers' Union, was re-energising the labour movement. With a background in journalism, Beth rehearsed a well-known narrative, connecting the SMSF to the history of the Brazilian labour movement, especially the 'new unionism' of the 1980s (Santana 2006; Seidman 1994). According to this narrative, Soares was getting rid of the hierarchical and bureaucratic structures of the previous Union Force (FS) leadership and liaising with grassroots organisations (including women's and black organisations, pensioners, and informal workers), thus bridging factory-based and community-based activism.

For instance, the Metalworkers' Union regularly organised the 'Zumbi day'[3] in collaboration with the black community on the

20th of November, supported struggling primary schools and had established the Secretariado de Mulheres (Women's Secretariat) in association with the women's association led by Beth Lobo, an organisation with an impressive record in women's struggle. The new SMSF secretariat had already commissioned a questionnaire on the condition of women metalworkers in Volta Redonda. Run by Beth and César, two young journalists with degrees in media and communication, the communication department of the Metalworkers' Union actively promoted its brand of 'new unionism' among local media, universities and its members through press releases, conferences and community workshops, as well as through its weekly bulletin *9 de Novembro*. Given the defensive posture of the Metalworkers' Union in the last few years, most members brushed off such a discoursive identification with the new unionism of the 1980s as the latest exercise in corporate branding.

In fact, outside the opening hours of the legal office, not much was going on in the Metalworkers' Union House. People would walk into the courtyard to seek refuge from the heat of the day. They would pick up the bulletin, sit and discuss the latest collective agreement or contentious labour issue. Is the rising cost of oil and beans in the *cesta básica* in line with current levels of inflation? Are dental costs included in the free medical package? Some queued by the three computers located in the courtyard to check their inboxes or browse the Web. In the courtyard of that 1950s black-marbled modernist building, a question came up in my mind: 'Is this all there is to Brazil's new unionism?' The controversy surrounding the ballot on the six-hour shift revealed the materialistic outlook of the metalworkers, which was reflected in the SMSF leadership. The union was mainly active in factory-based negotiations around economic and work-related issues such as basic food allowance (*cesta básica*), health and safety, Participação nos Lucros e Resultados (Participation in Profits and Results – PLR) and health and educational provisions and did not seek cross-sectional alliances or campaigned on 'post-materialist' issues like social or community unions.

In discussing the sources of workers' power, Eric Olin Wright (2000: 962) distinguishes between *associational* and *structural* power. The workers' associational power is linked to their collective organisation such as trade union and political parties. Structural power is based on the workers' location in the economic system and can be further divided into *market bargaining* power, associated with the conditions of the labour market, and *workplace bargaining* power, which results from the workers' strategic location within key industries. For

Beverley Silver (2003: 13), marketplace power refers to specific skills needed by the employers, or the workers having an extra source of income; workplace bargaining power is linked to the workers' position in the production process, including inter-firms' relationships and broader value chains. In chapter 2, I discussed some shop-floor determinants of the power of the CSN workers. I have described how the heavy reorganisation that followed the privatisation of CSN deskilled the strategic and militant workforce – smelters and maintenance workers – and created a 'mean and lean' system based on the division between core workers and subcontracted workers.

But in CSN, the classical 'mean and lean' Japanese model was nearly inverted. The wages and working conditions of core workers were not better than those of subcontracted workers – in fact, they were worse at times, also because of the subcontractors' strong bargaining power vis-à-vis the company. The situation was also affected by the influx of new steelmakers in the region, offering higher wages and better working conditions than CSN and draining skilled workers and engineers away from the company. These circumstances explain the firm's back sourcing of labour from 2009 onwards. Thus, buoyant economic conditions, the haemorrhage of skilled labour and increased inter-firm competition strengthened the bargaining power of the Metalworkers' Union and shifted the epicentre of the workers' struggles back to the factory level. In this chapter, I assess the regional and national determinants of labour struggles, including those of the workers of the multinational carmakers that settled in the region in the 1990s, which are also represented by the Metalworkers' Union and partake in CSN's same global value chain. In particular, I assess the Metalworkers' Union's role of regional broker among different working-class constituencies in the light of its structural dependency on the state and on national trade unions.

I argue the Metalworkers' Union only 'ritualistically' supported civic forces outside the traditional labour movement, such as black and women's organisations, schoolteachers and neighbourhood associations, as a form of media propaganda and recruitment strategy. In reality, the union and its base were firmly focused on traditional wage- and factory-based struggles and dissociated from the struggles of the new working class, subcontractors and precarious workers such as civic and construction workers, cleaners, maids and garbage pickers. In their commodified vision of *labour* as *wage* and as *capital*, and in their belief in the *workerist* social contract between the working class and the state, the metalworkers end up reproducing at the local level the broader neo-extractivist consensus. Unlike them, builders and marginal workers consider labour as commons (see chapter 6). In the builders' union, workers often talk about themselves as 'ant nests' (*formigueiros*) to describe their acts of gathering, coming together and acting collectively and in unison and oppose their communal ethics to the rigid and centralised structures and behaviour of the *metalúrgicos*.

The Metalworkers' Union of the Sul Fluminense

The Brazilian state has a major influence on the labour movement through the labour legislation. The Consolidation of Labour Laws (CLT) was created by Vargas's Ministry of Labour and Employment in 1943 and continues to be one of the most comprehensive labour legislations in the world. Among other things, the CLT establishes the right to minimum wage, vacation, leave, professional training, housing, pensions and child benefits and regulates trade union affiliation, training and education. The labour code made unionisation de facto compulsory by establishing that only unionised workers were eligible for social benefits and nominated workers' representatives in charge of collecting trade union dues on behalf of the state. In the same year, Vargas's constitutional assembly created the Brazilian Labour Party (PTB), the biggest workers' party in Latin America. The CLT's generous social benefits were given based on the workers' affiliation to state-controlled trade unions and the PTB rather than based on their demands. As I argued in the introduction, only the formal working class was entitled to the generous civic rights of the *Estado Novo*, whereas foreign, agricultural and informal and precarious workers were excluded from them.

Hence, the sociologist Angela de Castro Gomes (1987: 73) describes the *Estado Novo* as an 'occupational welfare state' where the working class gained political emancipation through concessions from above rather than grassroots activism (see also Vianna 1978; Weffort 1978). To be sure, non-governmental parties such as the Brazilian Communist Party (PCB) – the biggest and strongest in Latin America at the time– had a radical anti-state and anti-capitalist agenda and was violently targeted and outlawed by the workerist state. But overall, the formal working class had high occupational status and little class consciousness and was socially emancipated without being politically so. Castro Gomes draws a stark contrast between the early Brazilian working class and the 'making' of the working class in England. The English working class emerged as a self-conscious political agent at the end of a long struggle led by different social strata in opposition to the state and the capitalists, whereas the Brazilian working class achieved labour rights within a framework of moral and legal dependency on the state. In Britain, the workers were conscious of their class status, whereas in Brazil they were not. Britain was a capitalist society driven by class interests and conflicts. Brazil was a workerist society driven by asymmetrical reciprocity.

Elsewhere (Mollona 2014), I argue this contrast between the early British and Brazilian working class is too stark and based on an idealised version of working-class trajectory in England. Besides, the quiescence of the Brazilian working class had more to do with the fact that, under Vargas, the country was under the authoritarian rule of rural leaders, generals and the old coffee elite rather than being an industrial democracy on the European model (French 2004: 142). The formal working class was historically a very small section of the working population, which remained mainly employed in the informal and subsistence economy. Unlike nineteenth-century capitalist England, the Brazilian labour regime mixed slavery and wage labour – a vast sub-proletariat and a small aristocracy of labour – so struggles for civic emancipation and struggles for economic redistribution went hand in hand. Indeed, according to Sidney Chalhoub (2011, 2012) and Marcelo Badaró Mattos (2008), the seeds of working-class formation in Brazil are to be found in the struggles for civic recognition of manumitted slaves and informal workers in urban centres in the nineteenth century.

For the Labour historian John French (2004), although the CLT was designed to produce quiescence, in practice it politicised the labour movement by providing a political language through which workers' rights were acknowledged. In a country still dominated by indentured labour, legal activism allowed for the articulation of a new class consciousness (see also J. Lopes 1988; Sigaud 1980). Similarly, the labour scholar Regina Morel (1989) shows how, in the 1950s, the Metalworkers' Union was under a strong Communist leadership and won the rights to paid overtime, night shifts and recreation and health and safety provisions by using the CLT to challenge the private legal framework of the employees' regulation, rather than through top-down concessions. The list of successful legal struggles by the SMSF is long. In 1952, under the charismatic leadership of the Communist Allan Cruz, the union famously won a lawsuit against CSN for breach of CLT Article 71 on mandatory break times and rest. In 1955, Director Nestor Lima obtained a 50 per cent increase on workers' participation on profits based on CLT Article 30. In 1957, under the leadership of Othon Reis, a former personnel director of

CSN, the SMSF had individual labour agreements between the HR of CSN and employees repealed.

Reis was a close friend of Luís Carlos Prestes,[4] the famous leader of the PCB and member of the Communist International in Moscow. This deeply scared the management of CSN and the national political establishment. Indeed, the combativeness of the Metalworkers' Union of Volta Redonda (later the SMSF) reflected the broader political effervescence that followed the creation of the peasant leagues in 1961 and of the General Command of Workers (CGT) in 1962 and which would soon be repressed by the military coup. Moreover, the workerist ethic of the *Estado Novo* was informed by the European tradition of Christian paternalism, especially by Pope Leo XIII's famous encyclical *Rerum Novarum*, which addressed the social question of working-class poverty at a time of great political turmoil in Europe. In his emphatic public speeches, Vargas called Brazil his 'family', the workers his 'children' and the CLT a generous 'gift' (*dádiva*) to the 'people' (*povo*). Paternalism and nepotism spread among state-affiliated trade unions, too, which during dictatorship became bureaucratic, technocratic and corporatist bodies and more preoccupied with social redistribution than with class politics (Weffort 1978).

The success of these early legal struggles was also because of the independence of labour courts from the executive. Heavily repressed during dictatorship, the labour justice system was reinforced by the 1988 constitution and heavily cut back again by President Fernando Henrique Cardoso, who was determined to break the power of trade unions and the labour courts and 'put an end to the Vargas era' (see Sorj et al. 2011). When Lula was elected president, he passed Amendment 45 to extend labour law to informal labour, which had peaked during Cardoso's presidency. On the other hand, the heavy regulatory function of the state under the Workers' Party (PT) weakened the power of regional labour courts and trade unions, especially in relation to collective negotiation. I discuss the issue of legal activism in chapter 5. Here it is important to stress that while compulsory union dues were abolished under Michel Temer, compulsory affiliation and state's approval of new unions are still in place (although currently under revision) thus making trade unions heavily dependent on the state.

The sociologist André Singer (2012) argues the CLT continues to shape the PT paternalist approach to industrial relations and its 'privileged' relationship with the industrial working class (see also Vianna 2004, 2007). In Volta Redonda, the pro-labour nature of the CLT does increase the bargaining power of the metalworkers, prevents their precarisation against the national trend of labour deregulation and supports pro-labour legal activism. The fact that the first important constitutional amendment proposed by Temer was the dismantling of the CLT and the passing of Law 4331 shows how such pro-labour legislation is against the interests of the conservative classes. Founded in 1943 in Barra Mansa, the Metalworkers' Union of Volta Redonda (SMVR) was jointly run by CSN employees and company representatives, thus becoming the symbol of the workerist social contract between the working class, the state and CSN. Even under the early Communist leadership of Cruz, the SMVR had a strong *getulistas* (pro-Vargas), pro-governmental and nationalist stance it shared with the management.

The state elected the union's board members and enforced union dues, and the union functioned as provider of legal, financial, educational, housing and medical services. The SMVR delivered these services in the schools, hospitals, workers' associations and banks owned by CSN (but controlled by the state) functioning de facto as a redistributor of public welfare within a private/public corporate structure. The developmentalist and autarchic economic model promoted by Vargas seemed to benefit both the local management and the local working class, whose wages, career prospects, pensions and welfare entitlements were among the best in the country. Industrial production also benefited the local impoverished landowners and former coffee planters now working as steel retailers, iron ore merchants, developers or financial brokers. President Juscelino Kubitschek's goal of 'fifty years of development in five' implied a veering away from Vargas's workerism. To please those foreign carmakers who depended on Brazilian steel, Kubitschek put pressure on the CSN management to increase production, lower wages and cut steel prices. In 1956, after Kubitschek visited the plant with US Vice President Richard Nixon and a delegation of American managers, CSN underwent a radical reorganisation. The factory regime changed, from paternalist to Taylorist. In the 1950s, as a consequence of the 'new US style of management' enthusiastically promoted in the company's journal *O Lingote*, real wages had declined by a third.

But the Metalworkers' Union of Volta Redonda was under a strong Communist leadership and strongly opposed the regime of scientific management and the new career structure established by US consultants. The strength of SMVR was reflected in the fact that, in 1956, the first National Conference of the Metalworkers of Brazil was organised in Volta Redonda. The military coup in 1964 and the economic recession ended this corporatist model and started an era of violent labour repression. Volta Redonda was brutally targeted by the military. A coalition formed by the CNS's top management, the army and the right-wing União Democrática Nacional (National Democratic Union) controlled the municipality. Rights to vote and be voted in trade union elections, to hold political demonstrations and to meet in public were suppressed. All labour rights and civil liberties attached to them were abolished. In the first stages of dictatorship, corrupt union leaders colluded with the municipality, and CSN controlled by the army.

In the 1980s, a new generation of labour activists replaced the existing corrupted leadership. The old union leaders were from Communist and rural backgrounds, trained in the local, prestigious technical school[5] and employed for life in public companies (SOEs) run like semi-military organisations. Their identity was mainly occupational (rooted in the notions of 'skill', 'category' and 'wage band'), their activism was sectional (revolving mainly around legal negotiations within the existing CLT framework) and their logic was bureaucratic, productivist and corporate. Unlike them, the new leaders were well educated, from the urban, lower middle class and involved in grassroots Catholic militancy. For instance, the union leaders of the ABC Region of São Paulo – the historical core of the PT – were mainly precarious and unskilled workers of multinational carmakers with ruthless labour regimes. Based on their precarious labour, they described themselves as 'unsophisticated' (*peões*), subaltern and poor (*pobre*). Unlike the old sectional and nationalistic guard, they were internationalist, identitarian and open to horizontal solidarities with Catholic, black and women's groups; teachers; lawyers; GPs; and employers' associations. Their politics was a peculiar blend of Communism, liberation theology, plant activism and urban guerrilla.

The new Brazilian unionism is traditionally associated with the struggles of the autoworkers of the ABC Region led by the young and charismatic Lula. International scholars, with a few notable exceptions (Beynon and Ramalho 2000), have overlooked the new unionism of Rio de Janeiro.[6] In Volta Redonda, unlike in the ABC Region, the city and the factory were historically entangled in a typically 'hegemonic' regime of production (Burawoy 1985). CSN controlled workers through family-based forms of recruitment, training and supervision, paternalistic management and corporate rituals that fostered a sense of 'steel family' among the workers, some of which continue today.[7] Outside the plant, the company controlled the workers through the municipality's basic welfare infrastructures, as well as its neighbours' associations, housing co-operatives, parishes,

Figure 4.1. The army in the Usina Presidente Vargas. Published with permission from the Sindicato dos Metalúrgicos do Sul Fluminense.

workers' clubs, cinemas, schools, samba clubs and leisure organisations. Up until dictatorship, the Metalworkers' Union was in charge of these leisure and welfare corporate infrastructures and hence functioned as a hegemonic extension of CSN into the city. But when the dictatorship violently severed ties between 'mother' and 'children', this urban corporate infrastructure suddenly turned into an invisible labour network (Santana and Mollona 2013).

Indeed, for Morel (1989), the absence of a clear spatial divide between the company and the city was a strong political asset for the labour movement during dictatorship. With the army stationed in the plant and the company's internal police targeting, torturing and jailing worker-activists, neighbourhoods were the epicentres of workers' resistance. Women organised anti-police urban guerrilla actions, led grassroots movements in favelas and paralysed nurseries, hospitals and schools. With the youth, they occupied the local university and schools, and rallied squatters and slum dwellers against the municipality. In November 1988, a few months before the democratic presidential elections, the Metalworkers' Union mobilised more than eight thousand workers in a strike against CSN. The federal army and the company police violently repressed the strike, leading to the murder of three workers.

Figure 4.2. Funeral of Valmir Freitas Monteiro. Published with permission from the Sindicato dos Metalúrgicos do Sul Fluminense.

In response to the repression, the newly constituted Popular Front – an alliance between the metalworkers' union, the civic construction union and religious, women's, black and student movements – organised the 'hug' (*abraço*), a cordon of thirteen thousand people holding hands around perimeter of the twenty-five square kilometre plant. Moreover, during the *abraço*, the working class of Volta Redonda shouted 'we own the plant' (*a usina é nossa*) to let President José Sarney know it opposed the privatisation of CSN, which was being discussed at the national level. The *abraço* shook the nation, becoming a powerful symbol of Brazil's democracy and of the new working-class power. Popular unrest followed the strike. José Juarez Antunes, the president of the Metalworkers' Union, became the mayor of Volta Redonda. For the first time in Brazil's history, a trade union leader was in charge of a public office. Only a month after becoming mayor, Antunes was killed in a road accident orchestrated by the military.

The privatisation of CSN split the Metalworkers' Union into two factions – one associated with the conservative FS, which supported it, and the other with the Unified Workers' Central (CUT), which opposed it. In chapter 5, I describe how the leader of the Metalworkers' Union Luiz Antônio de Oliveira ('Luizinho') lobbied in favour of the privatisation of CSN among the base and kept in

Figure 4.3. The *abraço*. Published with permission from the Sindicato dos Metalúrgicos do Sul Fluminense.

daily contact with the privatisation team of the BNDES and with President Itamar Franco to facilitate the privatisation of CSN. Expelled from the SMVR, at the time under CUT leadership, the group around Luizinho named itself *formigueiro* and formed the separate FS. The dispute between the CUT and the FS escalated in 1998, when the proposal to fund the new Sindicato dos Siderúrgicos de Volta Redonda (Steelmakers' Union of Volta Redonda) formed by Carlos Henrique Perrut and other FS members, challenged the hegemony of the Metalworkers' Union. Initially legalised by the Ministry of Labour and Employment, the Steelmakers' Union was later declared unconstitutional. Privatisation also divided the CUT between the mainstream Articulacão Sindical and the Communist Corrente Classista, leading to the creation of the spin-off radical confederation Coordenação Nacional de Lutas (National Coordination of Struggles – Conlutas). Reflecting the new capitalist regionalist trend I described in the chapter 4, the Metalworkers' Union allied itself again with the reformist FS in 2011.

Union Confederations

During the anti-dictatorship movement, union confederations developed as coordinating platforms for the struggles taking place in factories and communities across the country. At the time, most confederations wanted to break away from state control over unions and from Vargas's paternalistic system of labour relation based on mandatory union taxes, state-controlled labour courts and legalisation of unions. The proposal for a national confederation of all Brazilian workers emerged in the first Conferência Nacional da Classe Trabalhadora (National Conference of the Working Class) in 1981, when Brazil was still under dictatorship. Alongside demands for wage increases across the board, the conference set some ambitious goals such as trade union autonomy from the state, land reform and political democratisation of the country. The conference was organised by the new union leaders affiliated with the PT who called themselves the 'independent' or 'authentic militants' and by delegates from Unidade Sindical (Union Unity) affiliated with the existing unions including those close to the PCB (Barros 1999).

The group of 'authentic militants' consisted of Trotskyites, Marxist-Leninists and left-wing Church activists with an anti-state stance along with a philosophy of rank-and-file activism and direct bargaining. Against these, Union Unity defended the existing corporatist framework. In 1983, the 'authentic militants' founded the CUT,[8] and in 1986 some Union Unity delegates founded the CGT, which continued to favour union taxes and normative labour courts. In 1991, Luiz Antônio de Medeiros, president of the biggest workers' union in Latin America, the Metalworkers' Union of São Paulo, founded the FS. In opposition to the CUT's dogmatic stance, the FS cast itself as a pragmatic confederation. Medeiros had a populist, charismatic and managerial style of presidency. His business unionism appealed to many steelworkers in Volta Redonda – those who had bought into the post-privatisation narrative of workers' ownership but also those left-wing workers who advocated for factory-based rather than state-controlled representation. In this respect, the FS had a democratising spirit. It was the first union confederation to establish shop steward elections in factories.

Throughout the 1980s, the CUT led the union movement through a mixture of revolutionary agenda and grassroots activism and was deeply connected to the PT. In the 1990s, privatisations, labour deregulation and radical organisational restructuring led by foreign companies under the Collor and Cardoso governments broke the labour movement and led to a reconfiguration of power within the existing union confederations (Santana 2000b). The CUT assumed a strong anti-state stance. At the macro level, the CUT opposed privatisation and labour deregulation. At the factory level, the CUT questioned the ideology of Toyotism and the related practices of multi-skilling, just-in-time, tertiarisation and mechanisation introduced by foreign multinational companies. But, defeated by the powerful lobby of carmakers and industrialists of São Paulo in the 1990s, the CUT had to turn conciliatory and endorse the capitalist logic. It supported the official Programa Nacional de Qualificação Profissional (National Program for Professional Qualification), sponsored by employers and implemented by the Worker Support Fund (FAT), took the distance from the struggles of precarious, outsourced and informal workers, and participated in the Sectoral Chambers of the Automotive Industry, getting closer to the powerful business leaders' union the Federação das Indústrias do Estado de São Paulo (Federation of Industries of the State of São Paulo). By the end of the 1990s, the CUT had come to terms with capitalism, seeking to change it 'from the inside' and make it more democratic.

The charismatic leader of the FS, Medeiros, went even further, fully endorsing the new flexible labour regime. He believed workers had to proactively support labour reorganisations rather than oppose them like the CUT and the CGT, working with the management to improve the profitability of industry. Medeiros believed opposition to the new regime of flexible capitalism would in the end lead to predatory capitalism, widespread labour precarisation and unemployment. In the 1990s, some global carmakers relocated from São Paulo to the Sul Fluminense region and developed a new model of extreme lean production and Toyotism that radically changed the existing system of labour relations. The SMSF fiercely opposed such a transplant of the Japanese model in the Fluminense region, whereas the FS supported it. In a proposal for trade union action, the FS (1993) argued the new flexible production model in the long term would benefit the workers as well as the capitalists through their PLR. Besides, according to the pamphlet, the revitalisation of the trade union movement depended on its financial and political autonomy from the state and greater involvement in regional consultation processes with managements and local municipalities combined with a rolling back of state interventionism and trade protectionism.

The FS stance towards lean production mirrored that towards privatisation. Unskilled workers affiliated with the FS supported the vision of 'workers' capitalism' through which the BNDES and

Cardoso were selling the privatisation process to the working class. In 1998, autoworkers and unskilled steelworkers were part of that precarious working class that, according to Singer (2012), voted Cardoso over Lula. These workers were not historically affiliated with the PT or the CUT and distrusted Lula both for his connections with the social movements and for how it rubbed shoulders with rich bankers and pension fund managers. In CSN, most 'second-generation workers' who supported privatisation saw themselves as 'investors' but ended up in precarious job positions. When Lula was elected in 2002, the CUT underwent a deep reorganisation, becoming hierarchical, centralised and professionalised (Ricci 2010; Wainwright and Branford 2006). Currently, CUT leaders are trained in independent trade union schools and taught by well-established union professionals. The CUT's consensus-based decisional system was substituted with a centralised system of strategic planning, originally developed by the bankers' union. The CUT's bureaucratisation mirrored the one taking place within the PT, where decision-making is now centralised in the hierarchical national executive committee and detached from the grassroots party constituencies. Contradicting the principle of autonomy enshrined in its founding charter, the CUT never abolished the union tax or state monopoly of representation. Instead, it forged closer alliances with the executive to the extent of being incorporated into the state (Zibechi 2014).

According to the sociologist Rudá Ricci (2010: 35), trade unionists in 2008 covered approximately thirty thousand state managerial positions in health, education and social assistance. In the first Lula administration, 26 per cent of the ministers came from the union, and during the second, 16 per cent.[9] The consolidation of the union oligarchy within the executive paralleled the emergence of a new governmental elite predominantly male, white and well educated.[10] The union's incorporation into the state happened through its involvement in the management of the FAT and pension funds, which play a central role in Brazil's economy. The FAT was created with the 1988 constitution, which earmarked 60 per cent of it for unemployment insurance and the remaining 40 per cent for the BNDES in order to finance development programs. By 2006, the FAT was responsible for 67 per cent of the disbursements of the BNDES, amounting to more than $100 billion in 2010 (Zibechi 2014: 47). Pension funds were created in 1977 by the military regime. In 2001, Cardoso passed Law 108, allowing union members to manage the funds. Cardoso believed workers' participation in pension funds management would boost and open up the economy to popular forces. Like him, Lula used pension funds to accelerate economic growth and promote 'popular capitalism'. Indeed, the popularisation of financial capitalism, via pension funds, is arguably a central component of *lulismo*.

From 2003 on, unionists created their own pension funds, began frequenting business milieus, went into pension fund management courses organised by the CUT and even joined the board of the São Paulo Stock Exchange. Raúl Zibechi (2014: 41) estimates the pension funds assets in Brazil reached $300 billion in 2010, 16 per cent of Brazil's GDP, becoming the largest institutional investor of the country. In 2010, the PT controlled fifteen pensions funds, including Previ, Petros and Funcef, which are among the world's biggest. For instance, Previ is ranked twenty-fifth in the world, controls seventy Brazilian companies and owns shares, some of which are majority, in Vale, Embraer, Petrobras, Usiminas and Gerdau. In Lula's second administration, the links between the government, the banking elite, the BNDES and the corporate world strengthened further so that, according to Francisco de Oliveira (2006: 23), 'financial accumulation now occurs mainly in the state realm'. According to Oliveira, state control over the financial economy happens nearly entirely through ex-workers-turned-pension-fund-managers and technocrats within the state. The PT also has a strong hold on the BNDES, the biggest development bank in the world with investments representing 7 per cent of Brazil's GDP. Despite the latest radical political shift within the state, the current union/banking/financial nexus likely will grow on its own, independently from the political affiliation of the ruling class. The BNDES and pension funds are estimated to account for 40 per cent of Brazil's GDP by 2024.

But not all the labour movements subscribed to the CUT's endorsement of pension fund capitalism. In 2004, two new confederations, Conlutas and Intersindical, split from the mainstream Articulação Sindical of the CUT, and in 2007, affiliates of the PCB founded the Central dos Trabalhadores e Trabalhadoras do Brasil (Central of Male and Female Workers of Brazil – CTB) with close links to those movements that the PT and Articulação Sindical had marginalised, including Comissão Pastoral da Terra (Pastoral Land Commission), the Landless Workers' Movement (MST), the Movimento dos Trabalhadores sem Teto (Homeless Workers' Movement), and women's and black organisations. Renato Soares is a fervent critic of *lulismo* and Articulação Sindical, and a founder of the CTB. In 2011, with CSN making mass redundancies, the SMSF broke further away from the CUT and merged two factions – one associated with the militant splinter groups Conlutas and CTB, and the other with the conservative FS. This alliance with the FS was surprising considering Soares's affiliation with the radical PCB and CTB. It showed that, as the economic crisis advanced, the SMSF was rapidly losing bargaining power and was forced to ally with conservative forces, including with the Brazilian Democratic Movement (MDB) and Democratic Labour Party (PDT) governmental parties.

The PDT was founded in 1982 by the charismatic Rio de Janeiro Governor Leonel Brizola as an alternative to Vargas's old PTB, which continues to exist. Brizola cast himself as the alternative working-class candidate to the presidency and the representative of the working class of Rio de Janeiro, against Lula's PT, which was rooted in the ABC Region. In fact, Brizola, like Vargas, was a populist politician with great appeal among the working-class. During the tumultuous 1960s, Brizola, as governor of Rio Grande do Sul, catalysed popular support around the proposal to expropriate land from rich landowners and distribute it to landless farmers (*agricultores sem terras*), led the takeover of a US electricity company and rallied one hundred thousand civilians and soldiers against the advancing militia of the interim president preventing the military coup in 1961 (Schwarcz and Starling 2018: 495). In the 1990s, Brizola first energetically campaigned against privatisation, famously describing it as a neo-liberal stunt imposed upon Brazil by the IMF, but he later supported the regional consortium set up by the SMSF and including business representatives, banks and municipalities that bid for the control package of CSN.

As the economic crisis deepened, the Metalworkers' Union lost power. In 2015, after CSN announced four hundred redundancies, the new president of the union, Silvio Campos, called a general strike. In 2016, CSN confirmed the lay-off of three thousand workers announced in November of the previous year and tabled a proposal for the collective agreement that included halving the remuneration for night work and overtime, no adjustment to inflation, and the cessation of the *cesta basica* and company scholarships.[11] Rejecting the proposal as 'offensive' Campos travelled to Rio de Janeiro to pressure the governor and the labour minister to intervene in the negotiation in support of the Metalworkers' Union. The delegation was unsuccessful. In January 2016, CSN confirmed the lay-off of three thousand workers. The news came after heavy redundancies by Volkswagen had violent knock-on effects on the region's economy. By March of the same year, four workers had died in separate accidents. The mayors of the region pleaded with the SMSF to liaise with the autoworkers and form a united block against deindustrialisation in the region.

Campos and Soares changed strategy. Instead of directly confronting CSN, they put together a regional delegation of federal and state MPs and the mayors of Volta Redonda, Resende and Porto Real petitioning the Ministry of Employment and Labour to pressurise the BNDES to release credit to the struggling CSN. This was a radical departure from Soares's strategy of direct confrontation or from the regional alliances of the 1990s and early 2000s against CSN. This time, the delegation of Soares and Campos was led by a CSN director, Ciro Gomes – the PDT candidate in the 2018 presidential elections. Thus, on the one hand, the Metalworkers' Union partakes in the regional cross-sectional block, which acts as a counter-hegemonic force vis-à-vis the financial-industrial complex of São Paulo, controlled by the governmental PT and the CUT. But at the local level, the union is tied in with the governmental party coalition (consisting of the PT, MDB, and PDT) and the conservative confederations CUT and FS against radical left-wing forces such as the PSOL and PCB.

The Trade Union Family

Labour activism entails skills passed down from one generation to the next and produced through specific career paths that are as much integral to working-class culture as to the productive apparatus of CSN. The first trait of a good union leader is technical expertise. Leaders are normally skilled electrical or mechanical maintenance workers, trained at the prestigious Pandiá Calógeras Technical School (ETPC) or the Escola de Engenharia Industrial Metalúrgica de Volta Redonda (School of Industrial Metallurgical Engineering of Volta Redonda). But, unlike the old productivist union leaders, the maintenance skills of the 'new' leaders involve human empathy – a special gift for caring and looking after machines. Empathy is also important to become a *cipista* (safety representative), the second fundamental profile of the good union leader, for whom the life and well-being of fellow workers are central. This was the case of 'Renatinho' Soares whose parents, Seu José and Dona Marta, were respected Catholic leaders in the community of Conforto. Like his father, Soares attended the ETPC before entering CSN and was a *cipista* in CSN before becoming union leader. Soares continuously stresses his identity of unskilled workers 'born and bred' (*peão nato*) or his 'simple' job position as maintenance electrician, echoing the way the leaders of the 1980s, unlike the old *varguista* leaders, identified with the subalterns and the poor and downplayed their professional status.

Rhetoric is another fundamental skill of labour leadership. In Volta Redonda, this refers back to the specific form of speech of the worker-priests and militants in the Catholic Workers Action (ACO) or the Catholic Worker Youth (JOC) of the 1980s. Bringing together liberation theology and Marxism, these speeches are a blend of mysticism, revolutionary socialism and Christian ethics. In the 1980s, such language catalysed a wide political spectrum, including Trotskyites, Marxist-Leninist unionists, PT activists and favelados within the same anti-dictatorship network.[12] Soares has inherited the rhetorical gift of the anti-dictatorship labour leaders. Take, for instance, this personal conversation:

> Capitalism is an immoral and brutal force that oppresses people in a very subtle way. Young people are lured by easy money and luxurious lifestyles. They lack an ethical compass. They are alienated, disengaged and fearful of their bosses. Capital is the essence of all human beings. Like all human beings, it continuously transforms itself . . . Now we are witnessing another period of transformation: the end of the neo-liberal consensus and the beginning of violence and mass killing. The anti-democratic regime initiated by the United States in Iraq will continue in Latin America. Trade union representatives, local leaders and workers, and all the people against capital will be killed. Brazil may become like a new Ireland. There may be some terrible religious and civic conflicts ahead.

Soares was worried about the climate of fear hovering over CSN shop floor and lamented that CSN forbade the creation of factory councils and had recently incorporated the *cipa* – the only form of workers' representation allowed in the factory – into the managerial board. As a consequence of the *cipa* losing autonomy, work accidents were growing steeply, and Soares felt 'a truly horrible accident' was going to happen in the near future.

In 2009, after a long period of collaboration with the management, the Metalworkers' Union organised the first strike against CSN in seventeen years after the firing of six hundred workers. Soares was angry for the way the ruling PT was colluding with multinational corporations, closing an eye on their mass redundancies and authoritarianism while simultaneously supporting its own 'national champions'. He claimed the base was disillusioned with the PT and Lula, whom he considered cynical:

> Lula took a cynical turn towards neo-liberalism. My father used to say that the more educated, the more stupid people are. This is true when we compare Fernando Henrique Cardoso and Lula. Lula did not even finish primary school; Cardoso is a sociology professor. But Lula is more intelligent. More

> than Fernando, he has embraced financial capitalism and betrayed the working class. But he has great popular support.

The two famous ex-leaders of the SMSF, Vagner and Vanderlei Barcelos, are also children of community activists in the bairro Retiro. Their father, Emidio, also a CSN employee, was a pastor and a labour leader. Their mother, Dona Jandira was a charismatic community organiser. The Barcelos brothers studied at the ETPC and became skilled maintenance workers. In 1983, Vanderlei and Emidio put together an independent union list that challenged the reformist FS of José Juarez Antunes. Two years later, Vanderlei and Vagner became directors of the Metalworkers' Union and fiercely fought President Antunes for his support of the privatisation of CSN. Antunes was a friend of Emidio and condescendingly called Vanderlei and Vagner 'his little children'.

Marginalised by the FS leadership after privatisation, the Barcelos brothers quit the union, following radically different career paths. Vagner became a local councillor and Vanderlei a legal activist, married to Stella 'the fighter'. Thus, when trade union affiliates say 'the SMSF is like a family', they refer to the specific career paths, family backgrounds and rhetorical skills of leaders like Soares and the Barcelos brothers, rooted in the new unionism of the 1980s. Unlike these, Luizinho, the charismatic leader of the SMSF who led the pro-privatisation consortium in 1992 and ran for the SMSF presidency against Soares in 2006, did not belong to 'the trade union family'. He has a much poorer family background than the average trade union leader. His father, Sebastião, worked in a rural fazenda in Barra Mansa before moving to Volta Redonda to take up a job at CSN in the 1940s. Like other ex-peasants, Sebastião supported Vargas's *Estado Novo* and considered industrial work a means of social mobility. But when Luizinho was only six years old, CSN dismissed Sebastião, and the family had to relocate from the working-class bairro Conforto to one of the poorest favelas in the city. His dad's humiliating fall, from the prestigious rank of steelworker to that of poor slum dweller, changed Luizinho forever.

After his father's death, the family went through extreme poverty. Nicknamed 'already dead' (*já morreu*) for his fragile physical completion, Luizinho spent most of his childhood as street vendor. He was the family's only breadwinner, supporting his mother and six siblings. Illiterate, he began his career as a building subcontractor. Early on, he got involved in neighbourhood activism with the Catholic organisation Pastoral Operária (Pastoral Worker – PO). Unlike the

ACO, based in Retiro, which was affiliated with the PT, the PO was not affiliated with any political party. Luizinho led important political campaigns, including for free vaccinations of children and the sanitation of poor bairros. Moreover, in 1984, he was an organiser of the civil construction workers' strike, the first strike in the company's history, called 'the strike of the marginals', which stopped CSN's second big planned expansion. In the 1990s, Luizinho entered CSN and became an SMSF director under Antunes's leadership. People from his bairro considered his shift from grassroots activism to trade union directorship a betrayal of his 'true' working-class origins.

Luizinho's second betrayal came in 1992 when he supported the privatisation CSN. As I discuss in the chapter 5, CUT-affiliated workers opposed privatisation and called those who supported it *pelegos* (traitors). But Luizinho did not belong to the labour aristocracy. He believed in Cardoso's 'people's capitalism'. Like other unskilled workers, he saw the privatisation of CSN as a means to escape the labour hierarchy and seniority system reproduced both by CSN and by most steelworkers. As I argued in chapter 2, unskilled workers like Luizinho had a short-term, speculative and capital-oriented view of their labour and saw privatisation as a way to increase its value by investing in pension funds and company stocks. Moreover, like other FS directors, Luizinho believed that by becoming stockholders, workers would get access to the board of directors and, in so doing, increase their control over the labour process and their participation to the company's profits. The idea that trade unions should be involved in finance was controversial among CUT and PT affiliates at the time. Lula himself opposed it when he was the leader of the metalworkers but endorsed it as president.

Luizinho understood the privatisation of CSN marked the beginning of a new populist alliance between the state, the financial and industrial elites and the working class, which would last well beyond the Cardoso administration and, indeed, would come to fruition only under Lula. In fact, several union activists of the time went on to become pension fund managers or businessmen, like Luiz Albano, another charismatic ex-leader of the SMSF. Unlike Soares, who framed his politics in anti-capitalist and moral terms, Luizinho endorsed capitalism from a 'realist' point of view replicating, Lula's business pragmatism (Ricci 2010). Moreover, like Lula, Luizinho had a miraculous 'rags to riches' story to tell. Like Lula, he escaped poverty and starvation thanks to self-determination and ambition, and climbed up the social ladder, from being a poor favelado to becoming the leader of the Metalworkers' Union.

As well as indicating a specific trajectory of leadership, the saying 'the SMSF is like a family' emerges when people recall their traumatic experiences of militancy under dictatorship. After the *golpe* of 1964, the entire leadership of the Metalworkers' Union and the militant section of the working class were laid off, imprisoned and tortured.[13] As Bishop Waldyr Calheiros Novaes's involvement in the anti-dictatorship movement grew, activists from the JOC were also jailed and persecuted. For instance, Edir, a JOC militant, recalls:

> We paid the price twice. We were persecuted twice. They called us the Communist Worker Youth instead of Catholic Worker Youth. I was captured only three months after I got married. In the room where I was being held, the army soldier came to me and told me: 'I am an expert. I learned torture techniques in the United States.' When the torture finished, I felt that the flesh of my body had only bones. Everything was broken; everything ached after these electric shocks. (Felício 2008: 62)

CSN, the army and the secret Inquérito Policial Militar (Military Police Inquiry), under the directorate of CSN, formed a ruthless and mighty 'military-entrepreneurial apparatus' responsible for human rights violations in the shop floor and in the city (Estevez and Lima 2015). Informants from the secret police infiltrated lunch breaks and the more militant departments – the coke oven, smelting shop and Fábrica de Estruturas Metalicas (Metal Structures Factory) to gather incriminating evidence against activists, spreading fear and suspicion in the plant. Raids continuously went on in the working-class neighbourhoods of Retiro and Niteroi.

Workers went underground, spending their days in buses or at the cinema and sleeping rough. In this context of terror and underground existence, the union provided support that biological families could not offer. Marcelo Felício, the president of the Metalworkers' Union during the famous 1988 strike, tells me:

> We would go into hiding all together and live together in the forest surrounding the city or in São Paulo. We would disappear all of a sudden. We did not even have the time to tell our families. They were just told we had gone into hiding. Then, all contacts were cut. When captured, we were kept in small cells with no bathroom together. We were forced to defecate and urinate right there, in front of each other. They tortured and humiliated us in groups. They burned our hands, put candles in our anuses and made us sing, 'We are all going to die together.' We lived through these traumatic experiences together. We never told of these experiences to anybody, not even to our families. We did not even share them among ourselves. But we knew what we went through together, and we stuck together. We were like family. Like a body of people. A body they managed to crush. But our women were left alone, to feed large families.

A whole generation of children of trade union militants who were tortured or imprisoned experienced first-hand their parents' traumas. Most 'grandchildren of CSN' (young CSN workers) struggled with copying with the absence, pain and stigmatisation of their parents by the local community. Some drifted into drugs, alcoholism and unemployment.[14] When men went underground, women were left struggling against the military police on their own while also having to look after their children, often without support from the union. Male trade union representatives tend to downplay women's role in the anti-dictatorship struggle or in the 1988 strike and mainly focus on their household struggles.

For instance, there is little mention that in May 1980, four hundred women from the public teachers' union (SEPE), led by the charismatic Dodora, organised a series of strikes and walkouts against the municipality and kick-started the university lecturers and students uprising, which was a central component of the anti-dictatorship movement. Moreover, there is little mention of the women who led the Ação Libertadora Nacional (National Liberation Action)[15] or the armed guerrilla in Volta Redonda, such as Jessie Jane Vieira de Sousa, who in 1970 kidnapped an airplane with thirty-five passengers at the Galeão International Airport, asking the release of thirty-five political prisoners, an action that cost her eight years in prison. And few now remember Rosalice Fernandes, who was only fifteen years old when she was arrested with her father, Othon Reis, for occupying the Radio Siderúrgica Nacional on the day of the military coup, 1 April 1964. After the arrest, Fernandes became a prominent labour activist and campaigner against the privatisation of CSN homes and for the rights of women political prisoners and founded the Communist pamphlet *DT INFORMA*, documenting the struggles of factory councils.

In May 1976, during dictator President Ernesto Geisel's visit to Volta Redonda, Fernandes distributed an inflammatory issue of *DT INFORMA* entitled 'Two Words Regarding Your Struggle' in which she called for the coming together of all the factory committees of Volta Redonda against the dictatorship. The pamphlet caused panic among the armed forces in Volta Redonda about an imminent workers' uprising. Fernandes was captured, taken to the Departamento de Ordem Política e Social (Department of Political and Social Order) in Rio and tortured for ten days. The army was particularly ruthless with militant women, who, according to Fernandes 'transgressed twice: first, as women and then against the dictatorship'. The testimony of a general of the junta confirms her claim: 'Women are much more ferocious than men. This is my experience. They are much more cruel and ruthless than men. Many of them were terrorists and had leading roles in the movement. They had to be punished harshly' (D'Araujo and Castro, in Lelis 2008: 28). Fernandes testified to the Comissão Nacional da Verdade (National Truth Commission)[16] of how she avoided being raped by Captain of the Federal Police Francisco Borges:

> I am looking at the son of bitch who is in charge of the barrack. He comes close and lifts my skirt and says he is going to rape me . . . and much more. I went crazy. He charges violently at me and I charge violently at him. There was blood everywhere. It was a grotesque scene. I talked with other women comrades who were sexually abused. The issue was always the same. Either they did what I did or they got frozen and paralysed. (CMV 2015: 417)

Indeed, despite being a central force in the labour movement, women never made it into the directorship or the presidency of the Metalworkers' Union. Historically, women's militancy consisted in community activism, on anti-poverty, health and abortion issues often in liaison with Catholic organisations. Male labour activists were typically factory-based and tended to look down on community activism (Perrot 1988), when, in fact, the two forms of activisms were linked. Thus, when people talk about the 'trade union family', they also refer to a specifically male and skilled job profile – fitter, maintenance worker or skilled operator – with high salaries and purchasing power and with senior positions within the SMSF.

In fact, it seems the only movement within the anti-dictatorship coalition – including squatters' co-operatives, neighbourhood associations and black organisations – that resisted the neo-liberal turn of the 1990s is the women's movement. For instance, the newly instituted Women's Secretariat, led by the charismatic Beth Lobo, is an important link between the Metalworkers' Union and the civic movements of Volta Redonda. Beth has a strong background in neighbourhood activism, especially in the Movimento Contra a Carestia (Movement against Hunger)[17] led by the grassroots Basic Ecclesiastic Communities (CEB) in the 1980s, which combined labour and gender activism. In the 1980s, she funded the Organização Popular das Mulheres (Popular Organisation of Women) and was a founder of the PT and the CUT. In the Metalworkers' Union, she campaigns for women's rights together with the union's more radical section Articulação Sindical, and in the CUT, she leads the Comissão Especial da Mulher Trabalhadora (Special Committee on Working Women), a permanent trade union body for (and run by) women. According to Beth:

> Here in Brazil, we do have the tradition of trade unions empowering women and seeking to integrate them within the movement. In Volta Redonda, trade unions are composed mainly by men. But the Metalworkers' Union was always exceptional in the way it built bridges with social movements and with the worker's wives. In Volta Redonda, there are very few working women. In CSN, only one productive department employs women, the GGFM, where they are still derogatorily called *vira latas*. Women work mostly in the central offices and in administrative jobs. In the technical departments, that is, the most qualified sectors, women are absent. Until ten years ago, the ETPC did not accept women. When young girls were first allowed in the ETPC, they didn't even have women's toilets!

Beth led a study commissioned by the Metalworkers' Union on the conditions of women workers in Volta Redonda, showing that women continue to be excluded from the most qualified jobs and have mostly secretarial and administrative jobs. Moreover, they are burdened by an excessive workload, having to combine domestic work and underpaid wagework. More worryingly, one in three women in Volta Redonda has been a victim of sexual violence. Beth says:

> The 'steel culture' of the city and of CSN is very *machista*. We work a lot with abused women, to strengthen their self-esteem and help them recognise themselves as citizens. Of course it is also a class issue. Thirty per cent of the women who look for counselling are from a middle-class and upper-class origin. The other 70 per cent are from poor neighbourhoods. But that 30 per cent of women have academic qualifications and degrees – psychologists or lawyers – and are on paid employment outside the household. Still, they are victims of domestic violence.

Being a skilled organiser, Beth has connected the Women's Secretariat to a wider network of institutions and organisations, including universities, NGOs specialising in women's labour rights, neighbourhood associations in poor bairros (the most active on women's rights), the local police, the Secretaria Nacional de Políticas para as Mulheres (National Secretariat for Women's Policies) and the regional Fórum de Mulheres (Forum for Women)in the municipality, where she also works. In addition, Beth set up a special police force within the municipality dealing with women's issues, a shelter for at-risk women and an agreement with the local law university for a Labour Law for Women course led by Linda Beto, a well-known labour judge. Moreover, in 1993, Beth set up the Casa da Mulher Bertha Lutz, a women's association that is now in charge of providing services for women within the municipality. With Beth, other charismatic women activists make systematic connections between the labour movement and civic organisations – for instance, the regional MP Dodora (PSOL), who led the struggles of precarious public university lecturers and schoolteachers in the 1980s, and Cida Diogo (PT), a charismatic GP, ecological activist and former health director of the Metalworkers' Union who set up the famous ecological platform Agenda 21 in the 1990s. Dodora and Diogo ran in the mayoral elections in 2008 but were defeated by the corrupted Antônio Francisco Neto.

'New unionism' arguably broke with the previous professionalised and corporate *varguista* system but in turn reproduced the patriarchal and hierarchical system of CSN, whereby *padrinhos*, *cipistas* and skilled maintenance workers take up the leadership of the SMSF. Younger steelworkers and autoworkers do not buy into the SMSF's narrative of the union family (hence the older steelworkers' claim that they are apolitical) in the same way they do not feel they belong to CSN's 'steel family'. The 'grandchildren of CSN' see themselves as precarious employees, low-income earners, trade union clients or single-issue militants – in health, human rights or ecological issues – rather than loyal and permanent members of CSN or the Metalworkers' Union. Such distrust of traditional labour institutions is not necessarily a sign of political apathy: quite the contrary, it may reveal a more radical posture. But the spirit of the 'new unionism' of the 1980s is not entirely disappeared or 'brandified' for propaganda's sake. Militant women such as Beth and Cida continue to connect civic and factory unionism – struggles for recognition and struggles for retribution – going against the particularistic, factory-based and male-dominated logic of CSN and the SMSF.

Central Branch: Learning to Be Investors

At the time of my fieldwork, the Metalworkers' Union had a base of sixty thousand members and forty executive members, including President Campos, Vice President Soares, eight executive directors[18] and the directors of the sub-branches of Resende and Retiro. The president supervises two decisional bodies: the Fiscal Council and the Council of Delegates, both composed of three active members and three substitutes. The FS controlled the Metalworkers' Union from the year of the privatisation, 1993, until 2006 and from 2012 onwards. Thus, with the exception of the short communist leadership of Soares, the Metalworkers' Union has been under FS leadership since the confederation was founded.

The core functions of the main branch of the Metalworkers' Union are financial services and legal representation. Financial services are

Figure 4.4. The central building of the Sindicato dos Metalúrgicos do Sul Fluminense (© Massimiliano Mollona).

delivered by trade union office workers in impersonal and bureaucratic interactions framed in terms of a client-provider relationship. Indeed, most metalworkers – especially second-generation workers – go to the *sede* seeking financial advice – enquiring on the returns of their CSN stocks, pensions funds, municipal bonds, foreign currencies and the CSN private insurance scheme – Caixa Beneficente dos Empregados da Companhia Siderúrgica Nacional (Employee Pension Fund of CSN). The Metalworkers' Union also runs very popular workshops on financial literacy and household management that deal with mortgage repayments and interest rates on credit cards, cars and white goods, as well as how to make balance sheets and plan important family investments. In Brazil, interest rates on credit cards, mortgages and car loans are extremely variable, depending on the bank or retailer issuing the loan. For instance, credit card interest rates vary from 70 to 700 per cent – compared to the international benchmark interest rate of 14.25 per cent – which makes them even by the economists' standards a form of usury (see Rapoza 2015).

In February 2018, the percentage of Brazilians officially registered as 'in default' for having overdue bills totalled 40.5 per cent of the population aged eighteen to ninety, an estimated 61.7 million people. The Metalworkers' Union is not the only institution to offer financial training to the masses in Volta Redonda. Casas Bahia – Brazil's main furniture and white goods retailer – has its own in-house financing operations and offers its customers one-to-one financial training sessions on how to repay interests and avoid defaulting. Seventy per cent of Casa Bahia's customers are either in informal employment or earn below the minimum income (Prahalad 2004: 209). Nonetheless, the retailer offers loans on purchases without credit checks with interests in the region of 40 per cent. Moreover, like other local grocery stores and supermarkets, Casa Bahia uses the 'passbook' system that allows customers to make small instalment payments, with increasing interests, ranging from one to fifteen months. By promoting financial literacy and facilitating workers' access to credit and financial investments, the *sede* is one of the nodes of working-class financialisation I discussed in chapter 3. Indeed, at times, top trade union representatives from São Paulo visited the central branch to deliver training in participatory budgeting, pension fund management or advanced financial and statistical analysis.

The second important function of the *sede* is legal consultation. The legal drop-in centre run by full-time officers and lawyers opens daily from 9 a.m. to noon. The rows of plastic chairs facing the officers' desks fill up immediately as the office opens. People, often looking tense and frightened, inquire about unfair dismissals, harassment and underpayment. Some are illiterate and come to have letters read out to them by the trade union officers. Even workers from other categories – cleaners, maids or cooks – queue up for legal advice. But the drop-in centre can only advise on economic issues, such as PLR, payments, returns on pensions, and investment and overtimes rates. Legal issues that are more contentious, such as work hazards, unfair dismissal or unpaid salaries, are discussed in one-to-one appointments with the union's lawyers. Moreover, the SMSF legal director, João Nery Campanário, was in charge of the collective negotiation with CSN. People joked that Campanário was the real *presidente*. Born in a powerful family of Niteroi, he was a militant Communist during the dictatorship and the legal officer of several rural trade unions, including the Confederação Nacional dos Trabalhadores na Agricultura (National Confederation of Agricultural Workers). In 1983, Campanário was the legal councillor in Juarez's municipality and in 1994 ran for Congress for the Brazilian Social Democracy Party. At the time of my fieldwork, he was in charge of all the collective negotiations in the Sul Fluminense region – including with PSA Peugeot Citröen, Volkswagen and Usiminas. As he was often seen in public with powerful politicians and entrepreneurs, it was rumoured he would run for Congress.

Soares and Campanário spent a lot of time together and made an unlikely couple. Soares belonged to the PCB, while Campanário was affiliated with the populist and *varguista* PDT. Soares was critical of how the state controlled the unions, particularly of the system of compulsive union dues and the labour court system, while Campanário supported them. Soares criticised Lula's neo-liberalism; Campanário supported his mix of capital and labour deregulation and social redistribution, particularly through the newly established sovereign fund. Soares spoke the language of new unionism and civic participation; Campanário, the corporatist language of wage bargains, collective negotiations, public-private partnership and people's capitalism. Moreover, the impersonal and corporate logic of the financial department and the workerist, corporatist and personalist logic of the legal department make a strange mix. The Metalworkers' Union was suspended in a time warp – between the SMSF of the Vargas era, and the CUT of the Lula era.

When the legal office closed at 2 p.m., the directors left for their private practices, and the building suddenly became silent and deserted, except for the president's office. Stepping inside it from the black-marbled, neon-lit, empty corridors was like entering a different time zone. The Communist 'old guard' close to Soares was always there, wearing baseball caps, T-shirts and backpacks, surrounded by photos of Fidel Castro and Che Guevara, chatting animatedly with their feet on the long corporate table and pulling drinks from the 'presidential fridge'. They looked like squatters ready to leave. These secret gatherings at the presidency lasted several hours, and Soares often left halfway through a secret backdoor. Despite these informal meetings, the central branch of the Metalworkers' Union functioned

Figure 4.5. Renato Soares (© Massimiliano Mollona).

mainly as a bureaucratically run legal service provider and financial adviser on wages, basic welfare provisions, PLR or returns on investments. Even inside the president's room, issues not strictly related to the steelworkers' wages or the UPV – for instance, the precarious and hazardous labour of subcontractors or the condition of the autoworkers – were seldom discussed.

Moreover, despite the rhetoric of new unionism, most discussions in the central branch focused on economic issues, as traditional business unions do. For instance, it addressed how, in the most recent collective negotiation, the union had obtained a twofold increase in food checks, the extension of free medical care to children of CSN workers from eighteen and twenty-one and nursery support of R$300 per month – or how it successfully campaigned outside the offices of the Instituto Nacional de Seguro Social (National Institute of Social Security – INSS),[19] because of its delay in paying unemployment and disability benefits, forced some families to the edge of starvation. According to Soares, this long delay was not because of the institute's insolvency but was a retaliation against the Metalworkers' Union, which in the latest collective negotiation with CSN obtained a real adjustment on employers' INSS contributions from 1.5 to 2 per cent.

Symbolically, the space of the central branch extended until Antunes Square, located opposite the entrance to the UPV. Named after José Juarez Antunes, the square was said to belong to the steelworkers. In the centre of the square is the *9 November* memorial designed by the famous architect Oscar Niemeyer in 1989 in remembrance of the three workers killed by the army in the 1988 strike. Only three days after its inauguration, the military blew it up. The memorial is still half-destroyed. Instead of repairing it, the steelworkers' families decided to leave it as it was after the explosion – with deep cracks cutting through the central image of the three murdered workers – as a reminder of state violence. The square's layout has been an ongoing struggle between the workers and the city council. Until privatisation, the steelworkers treated the square as an extension of the plant and gathered there for leisurely conversations, concerts, political discussions, demonstrations and ballots. After privatisation, the city council redirected the city main's central avenue through the square. Now the workers enter the factory through a long metallic overhead tunnel above the square. The workers' previous meeting point has been replaced by a series of bus stops.

The Branch of Retiro: Empowering the Real Base

Retiro is a working-class suburb with a predominantly black, unemployed, elderly population and with powerful commercial and landowning families connected to the early emancipatory movement. In the 1940s, the area of Retiro was occupied by squatters and ex-workers who were laid off after the initial phase of construction of the company. In the 1960s, some employees purchased CSN homes or regularised their properties, but Retiro continued to grow unevenly mainly through irregular squats and occupations. In the 1970s, with the support of militant catholic organisations, Retiro formed a neighbourhood association, one of the first in Volta Redonda, campaigning around issues of spatial justice – including the residents' right to food, health and land ownership – wages, labour rights and human rights abuse by the military. As a result, the neighbourhood was regularised in 1979. Under the influence of Bishop Calheiros Novaes, the local church of São Sebastião became an epicentre of the anti-dictatorship movement and the main meeting point for Catholic grassroots organisations (e.g. CEB, the ACO and the PO), women's groups and workers.

It was in the São Sebastião church of Retiro that the famous leaders of the 1980s emerged: the worker-priest Father Jacques and the community activist Jandira Barcelos, along with her sons, Vagner and Vanderlei (SMSF), and Diogo (PT). Today, the church is mainly visited by pensioners, the unemployed and elderly representatives of the ancient elite. The bairro has lost its political edge. It is extremely poor (80 per cent of its resident earn less than two minimum wages) (IPPU-VR 2013: 121) and very densely populated by a socially heterogeneous mix of people – the ageing and impoverished Class C; the socially mobile street vendors, retailers and petty urban farmers; and builders, lorry drivers and informal workers. Locally produced cheese, meat, poultry, vegetables, sweaters, leather shoes, bags and synthetic blankets, plastic toys and acrylic leggings 'made in China' are sold in street stalls, car booths and retail shops dotted along the main road.

People say the meat of Retiro has a peculiarly bitter taste because the local animals breathe the benzene coming from CSN's oxygen tank, blown by the nasty easterly wind. Residents often describe Retiro as 'a toxic beauty'. Indeed, the hilly configuration of the bairro creates a sensory confusion between the faraway plant and the local urban environment.

For instance, from the top of the bairro, the bright yellow flames coming from the UPV appear to spring up from the pavement, and the bright red coming from the furnaces' fires glows on people's faces like warm sunlight. With the company on one side of the bairro and the favelas on the hills on the opposite side, Retiro is a place of extreme visual contrasts. The main road connecting the plant with the regional highway cuts through the neighbourhood. The car density in the bairro even exceeds the exorbitant urban average.[20] Trucks, lorries, four-by-fours and 1950s Chevrolets speed towards the highway, dodging the bicycles, donkey carts and motorbikes slowly going by the local streets and, at times, violently colliding with them.

The union branch (*subsede*) of Retiro occupies a 1940s low-rise building located in the middle of this urban chaos. The SMSF vice president is in charge of the *subsede*, together with a full-time union organiser and two part-timers. The earthy colour of the small central building and the hand-painted blazon of the union – a blue crucible crossed by a yellow thunder – on the front gate make the building look like a colonial fazenda. Inside, the building is vast. It has three offices, a two hundred-seat cinema and an open-air theatre used for parties, barbecues, meetings and rallies. In this branch in the 1980s, labour leaders and workers' priests discussed Jesus' gospel and Marx's *Capital*, organised grassroots actions to better the sanitation of the surrounding favelas and trained workers and trade union activists in urban guerrilla. Today, the Retiro branch functions mainly as a medical centre and meeting point for the elderly and the pensioners, who represent one-fifth of the Volta Redonda electorate and are politically very influential.

The Associação de Aposentados e Pensionistas de Volta Redonda (Association of Retirees and Pensioners of Volta Redonda – AAP-VR) has forty thousand members. Because workers in unhealthy areas are pensionable after twenty-five years, most of the current pensioners in Volta Redonda are young – below fifty years of age. Founded in 1973, the AAP-VR grew exponentially after the waves of dismissals and early retirements that followed the reorganisation of CSN in the 1980s and has kept its anti-corporate stance ever since. In Brazil, women can retire on full benefits after thirty years of work, and men after thirty-five. The average retirement age is fifty-four, and the average pension is 70 per cent of the final salary, although several CSN ex-employees are on full salary pensions. In 2010, only 11 per cent of Brazilians were over sixty-five years old, yet pensions constituted 12 per cent of the GDP. Moreover, bereaved spouses are allowed to collect almost the full pension of their deceased partners, and pensions are automatically raised in line with the minimum wage. According to the World Bank, Brazil's demographic bonus[21] is running out, and the working-age population is quickly shrinking. According to some economists, the current pension system benefits the rich at the expenses of the worse off and children (Leahy 2016; Reid 2014: 272)[22] and heavily weighs on public deficit. They consider the pension issue in Brazil a 'time bomb'. Unsurprisingly, these economists support an alternative, partly privatised pension system that would increase further social inequalities.[23]

At the time of my fieldwork, the PT government was already cutting down on pension allowances, which outraged the AAP-VR. First, pensioners claimed their minimum wage was fixed to the 1988 index rather than the current one. For the workers who retired in the 1980s, pensions could be as low as R$226 (i.e. below the poverty line). Second, the pensions' adjustment to inflation was only 5 per cent compared to the 10 per cent used for salaries. Moreover, pensioners opposed the government's proposal to increase the pensionable age to fifty-five for women and sixty for men, arguing that early retirement would penalise the younger generations. But in Volta Redonda, in the context of raising unemployment, pensions are fundamental for the survival of entire households. One in three unemployed depends on their parents' pensions to survive, and 50 per cent of pensioners are the main breadwinners of large working-class families. The

AAP-VR has strong connections with regional and national politicians in Brasilia, where its delegates often travel for lobbying. The Metalworkers' Union and the city council constantly try to please the AAP-VR. The last two city majors built new facilities for pensioners, including leisure parks, swimming pools, gyms and walking tracks, as well as free Internet, dance and PE classes. Early in the morning, pensioners power-walk along the Paraíba Valley following their personal trainers in T-shirts with the municipal logo.

Jorge, the main part-time union officer of the *subsede*, showed me his business card of 'professional clinic psychologist', a profile he carved out from his degrees in biology, psychology and geography. Before going into hypnosis, Jorge worked in the city council's Department of Social Services for ten years. He offered three services: sermons, hypnosis and legal advice. In Retiro, he did not practice hypnosis but offered mainly legal and health advice. He was said to be one of Volta Redonda's top hypnotists. His private practice downtown was very popular. Jorge was also a talented speaker. His sermons were a mix of mysticism, radical anti-capitalism and environmentalism. He described his activity of pastor in the poor rural bairro where he lived:

> Many pensioners and young people are depressed. But for me, mental health is not a 'hard science.' Depression is not genetic. It is an unbalance of forces, a loss of soul, which in my bairro are mostly related to long working hours and unhealthy working conditions. In my career as a psychologist, I have realised how depression is often related to unbalances between people's souls and the earth's soul. In fact, the earth has a soul, too. The recent earthquake in China was the earth's soul rebelling against exploitation and ruthless extraction of mineral resources. These violent seismic shifts are the voice of the earth rebelling against us humans. In my hypnoses, I tune together the soul of people and the soul of the environment.

Although he worked closely with the São Sebastião parish, Jorge saw himself as a spiritualist rather than Catholic. Indeed, Jorge was very important for the Metalworkers' Union, because his mixture of evangelism and spiritualism appealed to the youth:

> Spiritualism does not position itself outside our contemporary time and politics. It works within the time and logic of our modernity. It is pragmatic rather than dogmatic – closer to ethics than to religion. For instance, my spiritualism is holistic – a mixture of psychology, geography and sociological class theory. The current depression of the working class reflects an historical moment. It is becoming clear that money is polluting the soul of the working class rather than opening the doors to happiness. There are many depressed young people here in Volta Redonda. I make them feel better.

Elisa, the other part-timer, helped the branch director, Artur Gomes de Lima, organise the pensioners' parties and workshops. Recently, with the communication department, she organised the workshop 'Memories of the 1988 Strike', which invited elderly people to give oral testimonies on the strike based on the original black-and-white photos taken by workers and activists at the time, which were displayed in blown-up sizes in the cinema. Among these were the iconic photos of the bodies of the murdered workers lying in open coffins with crying relatives all around, army tanks and colonels standing by the smelting shop ready to strike and the aerial photo of the thirteen thousand 'ants'[24] holding hands around the twenty-five square kilometres of the plant. The workshop participants were of different ages and backgrounds, but despite twenty years having passed since that traumatic event, they were all deeply moved by the photos. Touchingly, the elderly shared their traumatic memories – their anger and sorrow – with the younger participants. Elisa studied dentistry part-time at a local college, where she was learning to make dental prostheses and implants. Dental services are subsidised for pensioners, but she was campaigning for free dental care.

The director of Retiro's union branch was Artur Gomes de Lima – fifty-five years old, married and with five children. Artur was a self-made man. At thirteen, he travelled from Minas Gerais to Volta Redonda on the back of a lorry transporting chickens. He started as an unskilled builder and got a job at CSN when he was eighteen. In 2007 , after thirty-three years of employment, the company forced him to take early retirement. According to Artur, he was pushed out because of his union activism:

> The HR manager told me I should take early retirement to help the younger generations to find employment. It is true, older unionised workers like me are a nuisance to the company. With my salary of R$5,000, it can hire eight twenty-year-old workers full of dreams and hopes. These youngsters are afraid and inexperienced. During their first years of employment, they never get involved in trade union politics, and many of them are sacked before they become aware of their rights to a decent wage and to secure employment.

For twelve years, Artur chaired the Comissão de Trabalho (Working Committee), a public reconciliation body for labour disputes made up of trade union and employers' representatives to settle complaints before they go to court. The process is straightforward: workers fill in complaint forms and send them to the committee, which makes a decision within a week. If the committee unions and employers disagree on the president's decision, then the case goes to the Tribunal

Regional de Trabalho (Regional Labour Court). Most complaints are related to the workers' PLR, which are difficult to solve consensually. For Artur, the PLR is a tool of managerial co-optation:

> You will earn much more fraternising with the guys than bashing them on their heads. By giving up 10 per cent of their profits, employers have much greater returns on labour. The PLR make employees docile. They take much greater care of the company's machines, save on material and energy and abide to health and safety regulations, which normally are disguised forms of labour intensification. Because the minimum wage just covers for basic living expenditures [water, electricity and transport]and food consumption, nearly all employees subscribe to the CSN health plan. The PLR is the component of the wage that raises the workers' lives just a little above the subsistence level. There was always a lot of stress in the negotiations of the commission I chaired . . . I cried a lot, especially when my decisions were overturned.

Retirement hit Artur's finances hard. His income shrank from R$5,000 to R$2,000. Luckily, his five children were employed – the boys in CSN and the army, and the girls in retail. But he still paid towards their home mortgages and rents. So, he struggled to keep his head above the water. To make ends meet, he set up a small diary business in Jardim Belmonte, the bairro where he lived, employing his wife, his daughter and two temporary labourers. The business generated

Figure 4.6. Artur Gomes de Lima (© Massimiliano Mollona).

an extra income of R$2,000, which nearly took him back to his income from the previous year. But the business was harsh and wearing, and created much tension in the family.

Artur also supervised a few contracted farmers working the small plot of land at the back of the *subsede*, which provided the SMSF with some extra food for meals and parties. He belongs to the *brizolista* PDT and the reformist FS, and he is critical of the PT, especially under the Lula leadership:

> Lula's support to the poor and his redistributive program Bolsa Família is just a demagogical gesture. In fact, he always worked in the interests of the corporate sector and of 'first-class citizens' at the expenses of the working class and the 'middle section'. Besides, he says he is socialist, but he operates like an old fashion general. He is a man of power.

Artur ran with the MDB in 2004 and with the FS against Soares in the 2006 union elections. He lost, but Soares confirmed him as the union's delegate for the elderly, a post he still holds. Artur considers the two mainstream left-wing parties within the Metalworkers' Union – the PT and the PCB – too dogmatic and centralised. His main task in the *subsede* is to create links with local pensioners and powerful commercial families by offering medical services and organising leisure activities including games, weddings, balls, parties, discussions, conferences, workshops and gym, yoga and Internet classes.[25] As president of the Clube Náutico (Nautical Club) council, director of entertainment at the Aero Club and a social columnist for the pensioners' journal, he is a strong voice in the entertainment business. His parties are always a great success.

The Last Day of the Month Party

The last Friday of the month pensioners' party I attended in Retiro had been sold out for weeks. About two hundred people sat in the main cinema, which was decorated with balloons, workers' photos and Brazilian flags. Artur started the event by welcoming the audience and introducing the guests: Soares, the vice president and four directors of the Metalworkers' Union; the president of the regional AAP-VR; and an independent health consultant. Artur was used to large crowds. He talked at the microphone confidently, ending sentences with long and vibrating utterances, like a sportscaster. The party took off with Jorge's short sermon, which started discussing the economic crisis and went on comparing the passions of Saint

Matthew with those of President Soares. Jorge told how he was a marginal figure in the trade union movement, a simple security guard in CSN, until he ran into João Soares, Renato's father, giving a political speech-sermon to a group of workers. That speech changed his life. Thanks to their beloved João (everyone present, including Renato, crossed themselves), he became a labour activist. Jorge concluded emphatically: 'João was an exemplary instance of how to combine the world of the brain with the world of God. I tell you. This is the essence of human psychology: faith and the lucidity that comes from following ones' ideals.'

Renato Soares himself appeared at the party, and before leaving for an urgent meeting, he briefly enumerated the SMSF's recent successes, including a 3 per cent increase in pensions, and argued pensioners were at the core of the 'union family'. Next, Sandra, the health consultant, sketched the history of pensioners' activism. She described how the pensioners' movement came from the new trade unionism of the 1980s and consolidated in the 1990s with the creation of the Conselho Nacional de Previdência Complementar (National Complementary Pension Council) and the AAP-VR. In 2004, Volta Redonda created the Conselhos de Saúde Municipal dos Aposentados (Municipal Health Council for Retired Workers), a body with legal, executive and fiscal power composed of members from the city council and from civil society. Regrettably, the Metalworkers' Union had no representatives in the retirees' council, which, according to Sandra, showed the lack of collaboration between unions and pensioners. Then, she quoted a worrying statistic about abuses against the elderly. In 2008, there had been 170 cases of violence against pensioners by younger relatives, most of them in affluent neighbourhoods. Sandra argued that with the increase in family breakups and domestic violence, pensioners suffered increasing isolation, fragmentation and alienation. The SMSF, Sandra argued, ought to become a second family for them. Next, the charismatic president of the regional AAP-VR spoke:

> Retired comrades and pensioners, I urge you to rebel against the current corrupt political system! Let me give you some data and anecdotes. Last year, the families of MPs got R$3,000 in funerary allowances – their coffins are wrapped in gold! – and pensioners had a monthly income of R$250 compared to the R$415 of workers! We are a growing political force, with a strong presence in the parliament. The Metalworkers' Union must reckon this reality. Workers are losing ground. Pensioners are becoming stronger. You know why? Because wagework comes and goes, but sooner or later everybody becomes a pensioner.

Then it was Stella the *lutadora*'s turn 'I get very nervous when vulnerable people are abused. I get very nervous when workers, pensioners, women and black people are exploited. I get very nervous indeed!' After a long applause, Stella offered practical and legal advice to the audience, patiently and exhaustively, for nearly an hour. Then, she concluded with a poem she had written for the occasion:

> Today, I can mourn for the rain or for being without money
> or because I am bored of my housework or tired in the factory.
> But is up to me to decide
> what kind of day is going to be tomorrow.

Artur announced a raffle would begin. Directors extracted numbers in turn. People from the audience shouted, 'It's mine!' and rushed to the front table to fetch their prizes. These were expensive durables – coffee machines, duvets, pressure cookers and even fridges – that few in the audience, with the exception of wealthier affiliates, could afford buying. Most people attending the party were clearly from poor backgrounds, with the exception of a woman sitting in the first row, whose hands were covered in gold rings. She was rumoured to be the offspring of a big pharmaceutical family – a generous union donor. The woman had bought several raffle tickets and was on a lucky streak. For every prize she won, the audience booed angrily. To resolve the impasse, one of the directors took the microphone and said: 'I do not think this is fair. Everybody should have the same chance to win.' The audience cheered. The woman begrudgingly returned all her presents. The audience cheered again and clapped. More than twenty presents were distributed – nearly one for every four people. At the end of the raffle, a man in the crowd took the microphone and said the rich woman deserved her gifts after all. The audience cheered again. The secretary returned her prizes.

Artur then introduced some eminent figures in the audience – presidents of neighbourhood associations, former activists and community leaders. Salerno, the leader of Jardim Belmonte, spoke of his close connection with the sports secretary of Volta Redonda and promised free access to all sports facilities for all pensioners. Another wealthy-looking resident praised the important work the SMSF was doing in the local community. The health director spelled out the union's provisions to pensioners, which were much more comprehensive than CSN's health plan.[26] He said how the union was pressing CSN to lower managerial bonuses, which depressed employers' contribution and pensions and confirmed the rumour that the local hospital was about to close down, addressing the community's concerns about it. The SMSF vice president concluded the day saying pensioners made up the core of Volta Redonda's working class. Artur announced that the garçons would serve dinner, and the crowd migrated to the vast ballroom adjacent to the cinema. A live orchestra started to play *forró*, and the waiters served beer and churrasco, elegantly walking around the decorated tables and following Elisa's strict instructions. Huge S-M-S-F letters made of coloured balloons floated above the stage. People got drunk, danced and sung until late at night.

The success of the pensioners' party was the result of a coordinated effort of several full-time trade union officers. If the legalistic, impersonal and service-oriented interactions in the *sede* reflects the atomised corporate worldview of CSN, then Retiro's interactions with the local community show a different sets of power relations. Here, the union increases its associational power by liaising with powerful commercial families, the pensioners' community and poorer sections of the working class, following the logic of patronage, generosity and gift giving. The political support and cash donations from local landowners, commercial elites and pensioners are fundamental for the survival of the local branch. Moreover, the Metalworkers' Union in Retiro brings together the traditional Christian community of São Sebastião and younger evangelists through the charismatic presence of former worker-priests, like Jorge, whose language spun across politics, religion, ethics, psychology and environmentalism. Thus, the function of the Retiro *subsede* is to strengthen the workers' associational power by revitalising the memories of struggle, leisure patterns and forms of reciprocity (including free services, gifts and cash transfers) within the traditional working-class community, as well as consolidating the ancient rentierist block – constituted by local commercial families, landowners and subaltern workers – against CSN.

The Branch of Resende and Porto Real: The Rebirth of 'New' Unionism?

For Bartolomeu Citelli, the director of the Resende and Porto Real branches (*subsedes*), 'life blessed him'. He was born in the year of the Cuban Revolution. His father, a metalworker at CSN, died at thirty-three, when he was only nine. Citelli was the oldest of eight siblings. His struggle for life started very early on. When he was thirteen, he became militant in the grassroots ACO and then joined the Metalworkers' Union, becoming a leader of the anti-dictatorship movement. As I discussed earlier, the privatisation of CSN broke the radical leadership of the SMSF and put the union under the conservative FS. The privatisation of CSN destroyed Citelli's career. He was expelled from the SMSF, sacked by CSN, blacklisted and persecuted. For many years, nobody would employ him. He went into hiding and even begged in the streets. He was still alive thanks to the generosity of his close friends. Citelli said, 'Union people help each other; they are like family'. He was not married. His family was the union. When the new automotive regime moved to the Sul Fluminense region, the landscape of labour activism changed radically. The 'new working class' (Ramalho and Santana 2006; Santana 2006) of autoworkers is young, well educated, relatively well off[27] and very different from the Volta Redonda *metalúrgicos*. These younger autoworkers tend to identify more with their employer than with the SMSF, which they see as a mediator rather than a political organisation.

The global carmakers leading the 'third wave' of Brazilian industrialisation came up with a model of extreme outsourcing that completely overhauled previous models of labour relations. The most extreme version is the Volkswagen 'dream factory' in Resende, which is totally outsourced (see chapter 5). Unlike the traditional Japanese model, where core activities remain internalised, all operations in the dream factory – both direct (e.g. production and assembly) and indirect (e.g. cleaning, transport, food, health service, data processing and logistics) – are externalised to independent subcontractors operating in situ. The only core functions retained by Volkswagen are brand development and quality control. In 1996, when the factory opened, 1,300 out of 1,500 workers were subcontracted. This form of subcontracting of core productive activities was illegal in Brazil at the time of my fieldwork.[28] By establishing an exclusive legal relationship between the main contractor and the subcontractor, this legal form negates juridical personality and hence the labour rights of the employees of the subcontractors. I discuss in detail this form of subcontracting in chapter 5.

The Metalworkers' Union initially struggled with the modular factory, having to negotiate separately with each subcontracted firm in the factory. But in 1999, the charismatic director of the Resende *subsede*, Isaac Moares, organised a strike that paralysed the firm, obtaining the establishment of a factory commission (FC) inside the plant. FCs are composed of workers' representatives from the main company and from the seven companies of the consortium, dealing with workers' representation in a unified framework, unlike in the previous system. But FCs have no legal status and hence cannot represent workers in collective negotiations. It is the SMSF that discusses wages, PLR and working conditions with the management of the seven firms of the consortium. Nonetheless, FCs are extremely important because they coordinate workers' actions across modules, unofficially organise the tertiarised workers and feed information from the shop floors to the central direction of the SMSF. For instance, FCs organised a series of high-profile campaigns and legal actions against tertiarisation in Volkswagen, such as the national 'strike festivals' (*festivals da greve*) in 2005 and 2007 (Francisco 2015), participated in cross-sectional negotiations in the Cámaras Sectorial (Sectoral Chamber) and the Commissão Municipal de Emprego (Municipal Employment Commission) and created the national and world committees of VW workers.

Moreover, FCs organised anti-corporate campaigns, forums and public debates in collaboration with the Departmanento de Movimentos Sociais (Department of Social Movements of Resende)– a body made up of various organisations including the Comunidade Negra (Black Community), the MST, the Federation of Resident Associations, the Catholic Church and the Movement for Ethics in Politics. This kind of regional, cross-sectional and business-oriented activism of the autoworkers replicated the militant regionalism that grassroots movements and municipalities waged against CSN and foreign corporations in the 1990s. Then, horizontal relations were forged between the labour movement and civic organisations within a unitary anti-capitalist framework. Unlike it, the horizontalism of the autoworkers was premised on the collaboration between workers and the capitalist class. Indeed, for Zibechi (2014: 41), the combination of service-oriented and leisure-based mobilisation (e.g. fiestas, parties, carnival) of FCs and the CUT were forms of 'citizens' unionism' akin to the social unionism of the 1990s, which weakened class-based forms of struggle. Unlike Zibechi, Elaine Marlova Francisco (2015) argues the grassroots approach of FCs bridged factory and community activism, challenged the verticalism of the Metalworkers' Union and hence constituted an anti-systemic force within the traditional labour movement.

Citelli argued he learned a lot from dealing with Volkswagen and PSA Peugeot Citröen, the other global carmaker in the region. He negotiated with powerful international corporate lawyers and simultaneously with eight different companies, took part in cross-sectional forums with businesses and municipalities and co-organised strike festivals and other forms of community activism. Indeed, the Metalworkers' Union benefited from the autoworkers' struggles, if anything because it increased its bargaining power vis-à-vis the management of CSN. Yet, Citelli considered young autoworkers apolitical and sectionalist. In fact, the modular system fragmented the workers along different brand lines, each one represented in separate collective negotiations, and made it difficult for them to be aware and supportive of the struggles of other categories of workers. For instance, autoworkers showed little sense of solidarity with ancillary workers, who were experiencing lay-offs and reorganisation at the time of my fieldwork. Indeed, the rate of sindicalisation in Volkswagen declined from 62 per cent in 2001 to 27 per cent in 2009. But the decline seemed to be related more to the lack of integration between the factory councils and the Metalworkers' Union, which weakened the mechanism of workers' representation in collective negotiation, than to the workers' political inclinations (Ramalho et al. 2009).

Despite being part of the same global commodity chain and the same regional economic space, there was a disconnect between the autoworkers and the steelworkers of Sul Fluminense. Since 2006, representatives from Resende FCs were allowed to run for the SMSF presidency. After losing the election in 2014, some FC leaders, with the backing of the CUT, attempted to found an independent autoworkers' union, which was opposed by the SMSF and blocked by the electoral tribunal. Citelli was sceptical about the business-oriented and cross-sectional regional alliances of the 'new working class' and the CUT. The SMSF's focus continued to be 'particularist', that is, focused on wages and working conditions in the UPV (Rodriguez 2010).

For Bazerra (2017: 371) the SMSF's disconnect from the autoworkers and its *varguista* corporatism is the consequence of how neoliberal capital puts different nodes of the same global commodity chain – in this case the steel and car industries – in competition with each other. Yet, the *subsede* of Porto Real and Resende functioned as a coordinating node between autoworkers, global corporations, business-oriented municipalities, the Volta Redonda *metalúrgicos*, the Casa de Pedra miners and the autoworkers of Porto Real and Resende. This role of regional broker across the mining, steelmaking and car industries considerably increased the SMSF's bargaining power.

Moreover, the extreme Toyotism of Volkswagen and PSA Peugeot Citröen, the lack of legal rights of their employees, the maturity of the national car market and the threat of relocation to new free enterprise zones, combined with the absence of a tradition of militancy in the region, heavily limited the 'forces of labour' of the autoworkers – whereas the *metalúrgicos* benefited from the legal protection of the CLT, the strong tradition of labour militancy in the city and the legal activism of local judges and lawyers. Moreover, unlike the modular and flexible car factories, which could disappear and relocate overnight, the giant body of CSN was a permanent spatial fix in the Paraíba Valley, which prevented any capital mobility. As for other aspects of Brazilian politics, the past and the present – new and old – blur when trade union activism is considered. The Metalworkers' Union combined the 'new unionism' of the 1980s – horizontal, urban and grassroots – in Retiro; the 'new' unionism of the autoworkers – cross-sectional, regional and entrepreneurial – in Resende and Porto Real; and traditional *varguismo* – factory-based and corporatist – in the Metalworkers' Union House. These different forms of unionism can be conceptualised as different 'state effects' materialising in different spaces of labour.

In this chapter, I have argued CSN – with its operations spanning across finance, logistics, steelmaking and land speculation – is an extension of the neo-extractivist state. I have also argued the state controls CSN indirectly, mainly through the BNDES and the public bank system and through national infrastructural and welfare programmes. I have shown the Metalworkers' Union operates within the same extractivist logic, but from the antagonistic point of view of labour. Overseeing the regional alliances of autoworkers, the daily struggles of pensioners and precarious workers in Retiro and the financial and legal interest of the 'affluent' metalworkers, the SMSF brings together local, regional and global interests – subjects and powers spanning across finance, industry, households, logistics and urban planning.

Moreover, despite Soares's militancy in the PCB and affiliation with the CTB, the Metalworkers' Union is internally controlled by governmental parties (the PT, MDB and PDT) and the mainstream confederations FS and CUT. Benefitting from the legal protection of the CLT, the city's strong working-class tradition and its network of labour lawyers, judges and legal activists, metalworkers have greater bargaining and associational power than autoworkers or precarious and informal workers. But the union is losing support, especially from the youth and women, who do not feel represented by the traditional *família siderúrgica* and are now militant in more radical left-wing parties. Moreover, populism is a central feature of Brazilian labour politics, especially for those leaders like Vargas and Lula who claimed to have a special relationship with the working class, which they called the 'people' or the 'poor'. In the same tradition, Luizinho and Soares cast themselves as workingmen from humble origins, conflate 'working class' and 'poor' and propose cross-sectional alliances of the 'poor' against the 'rich'. In fact, they address different class constituencies. Soares represents those skilled production workers (Class C) who feel betrayed by Lula's shift towards financial capitalism and advocate for the return to developmentalism and reindustrialisation and for higher labour income. Unlike Soares, Luizinho represented the unskilled and informal workers who have been historically excluded from the labour aristocracy and endorsed financial capitalism in order to escape traditional labour hierarchies.

These workers are different articulations of the 'people'. The workers invoked by Soares are more like the industrial workers of the Vargas era than the rural lumpen empowered by Lula and Rousseff. The workers invoked by Luizinho are like the stockholder capitalists imagined by Cardoso in the 1990s, except they never quite became pension fund managers like some PT affiliates. The actually existing working class of Volta Redonda seems to be located somewhere in between these two abstract constituencies. The three pillars of the populism of the Metalworkers' Union are the moral critique of capitalism, the notion that the union is like a family and the view of labour as a gift. Drawing on their Catholic background, the leaders of the Metalworkers' Union, first of all Soares, frame capitalism as a dystopic, apocalyptic and violent force – responsible for wars, mass displacements, depression and drug use – tapping into the anxiety and fear of local residents about urban violence, especially after the recent migration of drug gangs from Rio de Janeiro.

This left-wing, apocalyptic, anti-capitalist discourse coincided with the *ethical* activist turn in Volta Redonda. This, I argue, provides an alternative framework to that of evangelical groups and CSN, which capitalise on widespread fear and paranoia to break the labour movement (Birman 1996; see also Ramalho 2012). More importantly, the moral critique of capitalism is part of a broader anti-state discourse that brings together the Left and the Right in Brazil in a magmatic and unstable centre. In the 2018 electoral campaign, President Jair Bolsonaro managed to turn such a moral critique of capitalism into a highly effective anti-PT electoral strategy. The idiom of the family reflects the old social pact between CSN and its 'children' rooted in the patriarchal and personalistic relations in production I described in chapter 2. As for the manual skills necessary to be a 'proper' steelworker, being a labour activist requires skills – from community organising to public speaking – passed down from one generation to the next and that follow a long genealogy of founders of the movement. Vargas's workerism and *lulismo* both cast the working-class as the privileged interlocutor of the state. Indeed, minimum wage increases and the social ascent of the traditional working class were the hallmarks of both *lulismo* and *varguismo*.

Despite the social and economic improvement for wage labourers in the past fifteen years, Brazil's economic growth still relies on the direct exploitation of its vast mass of unemployed and precarious workers – or, to put it like Singer, 'labour's backwardness is productive for capital' (2018a: 22) – and work still feels like a true *gift*. This is especially the case for second-generation CSN workers, with their traumatic employment histories, and for some black and women workers. In the current context of crisis and mass lay-offs, the trade union endorsed the idea that work is also a gift. Take the remarks of a union rep:

> There are so many people who are unemployed and poor in Volta Redonda. I am lucky to have a job. The company was always generous with me. To be able to work for twenty years in the company has been a gift. Thanks to my labour and to the salary of the company, I was able to build a family and buy a house and be who I am. I have to thank the company for all that I have and all that I am. I owe my life to the company.

We have seen how the state-driven financialisation of the economy impacts on the shop floor in terms of labour deskilling and intensification and internal class fragmentation. In 2008, wageworkers like Bobo thought of themselves as capitalists. Today, they struggle with unemployment, debts and declining purchasing power. Once the economic crisis hit Volta Redonda, the SMSF's support to the financial economy and to the dream of people's capitalism – the idea that 'labour is capital' – lost legitimacy, leaving way to an old-fashion and particularist encroachment on the wages and working condition of productive and direct workers

Today, it is conservative evangelical organisations and loose business citizens' platforms that take up the struggles of the *metalúrgicos*. Another section of the working class – informal and tertiarised workers in the service and building industries – faces a similarly harsh employer: the municipality of Volta Redonda. But through forms of labour commoning and co-operativism, they manage to stay afloat in an increasingly hostile social and economic environment. A third section of the working class, made up of subcontractors and autoworkers, is faring better than the metalworkers. Less formally politicised than the former, autoworkers, subcontractors and a growing number of IT workers have experienced first-hand the extreme tertiarisation practiced by foreign corporations. Their militant struggles against labour deregulation and regional alliances oscillates between business pragmatism and legal and civic activism (see chapter 6).

Moreover, the trade union's factory-based struggles, the neighbourhood activism of civic coalitions and the legal and business activism of the new working class are different strategies of labour struggle happening at different state levels and reflecting historically and geographically diverse trajectories of capitalist development loosely clustered around the Metalworkers' Union. Up until the recent crisis, these three sections of the working class – impoverished wageworkers, upwardly mobile urban poor and tertiarised workers – formed a magmatic and internally divided 'centre' oscillating left and right and kept together by various programmes of urban development, wage indexation, poverty reduction, home ownership and household financialisation. The economic crisis broke this magmatic consensus. The SMSF changed strategy and aligned itself with the industrialists of the Fluminense region, including CSN, against the financial-union-governmental complex of São Paolo and the mainstream confederation CUT. This new convergence between industrial capitalists and the working class against the state in Volta Redonda reflects the broader shift in Brazilian politics that ultimately led to the dramatic fall of the Workers' Party.

Notes

1. The fund supports employees who have been unfairly dismissed.
2. In chapter 5, I describe in detail the conditions of informal and precarious workers such as cleaners and builders and the more militant leadership of the builders' union.
3. Also called Black Awareness Day or Black Consciousness Day. Zumbi was the leader of the Quilombo dos Palmares, a runaway slave community emerged in the seventeenth century.
4. Prestes led the Aliança Nacional Libertadora (National Liberation Alliance) against Vargas in 1935 and was imprisoned and exiled during the dictatorship.
5. Most CSN workers underwent training at the company's technical school, ETPC.
6. Brazilian scholarly contributions to this issue include Antunes and Santana (2014); Mangabeira (1993); Morel (1989); Ramalho and Santana (2001c); Santana (2000a, 2010, 2013). For the autoworkers, see Ramalho (1989); for the ship workers of Rio de Janeiro, see Pessanha (2001); Pessanha and Morel (1991); for CSN, see Morel (1989).
7. Sport competitions, concerts and other leisure events in its sport and social clubs; a social column in the company's magazine; a Christmas party and other festive celebrations.
8. For the history of CUT, see Santana (2000b).
9. The average percentage in all post-dictatorship government is 11 per cent (Zibechi 2014: 42).

10. According to Maria D'Araujo Soares (2009), there are eighty thousand politically appointed positions in Brazil, forty-seven thousands of which are appointed directly by the executive. During the two Lula administrations, only 20 per cent of these posts went to women, 87 per cent to white people and 95 per cent to people with college educations or postgraduates mainly in economics, engineering and law.
11. In chapter 1, I show how these are fundamental for career progression.
12. For an analysis of the language of the Landless Workers' Movement, see Issa (2007).
13. In CSN, ninety-seven workers were laid off. Personal communication, Marcelo Felício (former SMSF president), 20 May 2008.
14. Personal conversation with Marcelo Felício, 20 May 2008.
15. A Brazilian Communist guerrilla movement founded in 1968 to fight the military regime.
16. Launched in late 2011 under the orders of President Rousseff to investigate the murders and disappearances of civilians under Brazil's military regime.
17. This movement has existed in Brazil since at least the beginning of the twentieth century. There was a Demonstration against Hunger in Rio de Janeiro in 1913, which then spread to other cities.
18. In the departments of health, pensioners, legal, finance, communication, organisation, treasury and small enterprises.
19. The INSS gives benefits for disability and unemployment, accidents, *salário família*, maternity, pension for death and dismissal without just cause.
20. Volta Redonda has the second-biggest concentration of cars in Brazil after Brasília. Brazil's total car park consists of 382 vehicles per 1,000 inhabitants, compared to the global average of 143 vehicles.
21. In which the working-age population is at its largest in relation to dependents.
22. At the time of writing, the Temer administration was discussing a radical plan of pension reform as a keystone of the president's cuts to public spending.
23. The reform proposed by Paulo Guedes, the current Minister of the Economy in the Bolsonaro administration, goes into this direction.
24. This is what labour activists called themselves after José Juarez Antunes famously claimed, 'YOU can kill an ant but you cannot get rid of the whole ant nest'. Antunes himself was called the 'ant leader' (*formigueiro*).
25. For a R$28 monthly subscription, SMVR members have unlimited Internet access, against the R$48 monthly rate of private providers.
26. The union plan cost R$200, compared to CSN's R$400. It offers free psychology and blood pressure consultations; for an extra R$8, it offers consultations by qualified GPs, prostate cancer tests and ECGs; for an extra R$15, it offers comprehensive dental care.
27. About 70 per cent of them are at the most thirty-four years old, 60 per cent are married with children, 95 per cent have a permanent job contract, 65 per cent gain three to five *salários mínimos* (the highest wage level in the region) and 69 per cent own a home, 59 per cent of whom have already extinguished their mortgages (Ramalho and Santana 2006).
28. The Temer administration passed the co-called Outsourcing Law (13,429/17), which legalised the externalisation of core activities.

– Chapter 5 –

Capital as Money and the Invention of People's Capitalism

On 28 February 1986, a few days after the fall of the military regime, President José Sarney made a dramatic appearance on national television, announcing he would implement a monetary shock at midnight. All contracts and wages would be frozen, and the new currency, cruzado, introduced. Sarney spoke of the ghost of hyperinflation hovering over Brazil's economy:

> Hyperinflation has been poisoning the economic system for a long time, invisibly and systematically. It penetrated the mind-set of the Brazilian people [*povo*]. Money is losing value as soon as it is issued . . . We want the Brazilian people to be agents of privatisation. We want the Brazilian people to be agents of the state. No hustling, bargaining or brokering will be tolerated anymore . . . The monetary shock will erase people's inflationary memory[1] – like an instant photo.

Inflation and debt dependency are as old as the story of the Brazilian Empire. In 1808, after establishing his court in Rio de Janeiro, Prince Regent João VI funded the Banco do Brasil, the first modern bank in Latin America (and still in existence), to finance the government's deficit with Portuguese money. But in the 1980s, the spectre of inflation haunted Brazil in the crucial moment of its transition into democracy. In 1979, the international crisis had devalued the national currency, the Cruzeiro, by 30 per cent. Inflation had peaked. Despite pressures from the IMF, João Figueiredo's administration struggled to rein it in. On mainstream media, money was described as 'alive' (*dinheiro vivo*), and with the nearly magical power of self-reproduction and inflation was visualised as a dragon mounting a deadly attack against the state (Sansi-Roca 2007). Sarney's monetary shock aimed at drastically reducing the liquidity in the system by materially reducing the volume of money in circulation. In fact, the animacy of money and its power for growth, self-accumulation and circulation proved to be untameable by the state. For three consecutive administrations, inflation was out of control.

Hyperinflation massively increased the already existing social inequalities. People with savings accounts were rewarded with high interest rates and had monthly corrections indexed with inflation. They used checks instead of cash, anticipating that prices would rise before their checks were cashed. To avoid handling cash, shops and businesses paid into their accounts once or twice daily. But poor people had no bank accounts; they were stuck with cash money (*dinheiro vivo*). Within the working class, hyperinflation hit especially hard informal and precarious workers – cleaners, builders or maids – who had no *carteira* and were paid in cash. The standard narrative describes Sarney's radical monetary measures taken after the fall of dictatorship as the usual story of a peripheral country pushed into structural adjustment by powerful global financial institutions led by the United States and at a time of global capital deregulation (the Baker Plan of 1985). While this story can be hardly denied, in this chapter, following the economist Albert Hirschman,[2] I will discuss the process of monetary stabilisation and the privatisation of CSN that took place from 1986 to 1993 as forms of statecraft through which the populist alliance between the state and the working class – the central component of workerism –was resuscitated in the new democratic era. In fact, money does not just have an economic value: it is also a symbol of the state (Hart 1986) and the social base of citizenship under capitalism. Thus, Sarney understood that the monetary consolidation of the Brazilian state and its victory over 'inflationary agents' were fundamental to build a new political consensus.

In this chapter, I describe first the series of monetary measures of stabilisation aimed at turning the working class into the 'proper citizen' of the new democratic state. Then, I describe the privatisation of CSN as a process of incorporation of the working class into the new regime of 'people-led' financial capitalism associated with democracy. I conclude by arguing such a monetary incorporation of the working class was both economically and politically disruptive for the Volta Redonda steelworkers. The young generation of Brazilian economists who supervised Sarney's and Cardoso's monetary policies in the 1990s had graduated from prestigious US universities – Princeton, Harvard and MIT – but shared Brazil's strong developmentalist 'counterculture', rooted in the tradition of dependency theory (Evans 1979) and economic neo-institutionalism (Hirschman 1977: 178), and were sceptical of the Western economic models being transplanted to other Latin American countries.[3] Monetarist and Keynesian economists treated inflation as a purely economic phenomenon: the former argued inflation was caused by excess of money; the latter, by production bottlenecks. Moreover, monetarist economists blamed inflation on the 'monetary illusion' of irrational economic agents who undervalued inflation in their everyday economic decisions. Unlike them, the heterodox economists working for Sarney considered inflation a political phenomenon – that is, the consequence of power struggles between different economic agents trying to maintain their real incomes.

The Brazilian state systematically used inflation and indexing to regulate power relations between different social constituencies and avoid social conflicts. For instance, in the 1950s, inflation and overvaluation of the currency diverted income from primary exporters to import-substituting industrialists without generating hostility from the coffee elite (Hirschman 1981: 181). Moreover, up until the 1980s, generalised inflation allowed the Brazilian state to redistribute income between the traditional working class and the bourgeoisie through oscillatory movements, conceptualised as a rotating credit – although of a competitive form rather than as generalised reciprocity (Bresser-Pereira 1986: 10; Hirschman 1981: 189). Moreover, the price indexation of contracts, wages, rents and commodities sheltered the private sector (especially finance) from the negative effects of national debt at the expenses of the rural and urban underclasses. In other words, inflation functioned as a mechanism of harmonisation of the tensions between labour and capital. But the phenomenon of hyperinflation showed that such an equilibrium between labour and capital was in crisis and people had lost trust in the monetary authority of the state – governors and mayors freely issued quasi-money to further their local interests,[4] rural oligopolies set prices independently form the state, financial assets were converted daily and wages every six months.

Inflation peaked especially during wage adjustments, because of workers wanting to defend their purchasing power. Reflecting the power of left-wing political forces within the government – the Brazilian Democratic Movement Party (MDB) – and outside – the Workers' Party (PT) and the Unified Workers' Central (CUT) – Sarney's stabilisation plan aimed at redistributing income towards the industrial sector, especially the working class, at the expense of banks, middle-class investors and rural producers, which had so far benefited from the economy of inflation. But the labour movement did not trust the MDB – a populist party emerged from dictatorship and which at the time was seeking re-election. Sarney implemented the monetary shock in a top-down manner, in the Brazilian tradition of authoritarian technocracy (O'Donnell 1973). The team of young economists seized the Banco Central do Brasil (Central Bank of Brazil), the treasury, the ministries for industry and planning and worked closely with the attorney general excluding the Congress and the relevant ministries (including the Ministry of Labour) (Sola 1993: 163). The shock aimed to restore the monetary authority of the state by using an automatic price and wage adjustment systems, which bypassed political negotiations with core social constituencies. Inspired by the principle of 'distributive neutrality', the heterodox shock eliminated three forms of state control over the economy: (1) intermediate prices (while it maintained control over the prices of final goods), (2) wage negotiations and (3) financial assets.

In fact, it contained many redistributive measures, including an across-the-board 8 per cent wage increase,[5] the introduction of unemployment benefits, the devaluation of the prices of staple foods and the drastic reduction of interest rates. To protect working class purchasing power, the new wage-escalation system guaranteed an automatic wage increase on every 20 per cent increase of the consumer price index. The intervention was so biased towards labour that subsequent wage adjustments – also called 'wage triggers' – provoked cyclical waves of hyperinflation. In the end, the shock combined monetarism and Keynesianism, trying to limit the liquidity in the system through forced price administration by the state. Sarney's monetary measures encountered widespread opposition. De-indexation was experienced as an expansion of the market economy at the expenses of the negotiated sphere of the economy generating losses in real value (Sola 1991: 189). Middle-class investors withdrew en masse their savings accounts and purchased stocks, houses and durables. Firms dodged price controls by withholding supplies, cheating official price tables, 'dressing up' their products or refusing to sell unless buyers paid premiums. Car and meat producers showed their strength, increasing their *ágios* (mark-ups) and prices, which put them on a collision course with the state. In addition to these price adjustments, money printing, speculations and a sudden expansion of the informal economy and the black currency market showed an array of non-institutional economic subjects resisting incorporation into the monetary apparatus of the state.

Even the trade unions, the main recipients of the monetary measures, feared state-guaranteed, upwards wage adjustments would take away their bargaining power. Thus, they bargained on the top of the official wage rates, causing a demand shock that, in addition to the existing production and consumption boom, overheated the economy and derailed the plan.[6] After a sharp drop from 235 to 60 per cent, inflation hiked to 416 per cent in a few weeks and steadily grew, reaching 1,783 per cent in 1994. In 1990s, after the failure of the Cruzado Plan II, the new President Fernando Collor implemented an even more radical anti-inflationary plan, freezing wages, prices and 80 per cent of the nation's saving; introducing a new currency; substantially increasing taxes; and putting a levy on all financial transactions. Collor's plan also failed, following his impeachment for corruption charges. Only in 1993 did Brazil achieve monetary stability as a consequence of the implementation of the Real Plan by Minister of Finance Henrique Fernando Cardoso. Cardoso's successful stabilisation of the real was linked to his radical deregulation and privatisation of the Brazilian economy, which put it in synch with the global economy. I will describe further the link between money and statecraft in the context of the privatisation of CSN which entangled working-class lives – and the *dinheiro vivo* of marginal constituencies – within the circuit of state money and turned Brazilian workers, as per Sarney's wish, into 'economic agents of the state'.

The fundamental mechanism of privatisation was the securitisation of public debt – that is, the conversion of credits against the state into 'privatisation money' – which led to a perverse entanglement between the state and the private sector. Because of the immense value of the credits that the BNDES held against the public companies, the state – via the bank – ended up with majority and minority control over the privatised companies. Moreover, such control was obtained by converting debts of the BNDES into 'privatisation money', which were considered substandard and hence came to be known as 'rotten money' (*moeda podre*). Moreover, because of the lack of national capital and of protectionism against foreign capital, the only form of liquid money (*moeda viva*) for privatisations came from the workers' pension funds. Thus, the involvement of the working class into the regime of 'popular capitalism' sketched by Cardoso and later implemented by Lula begun with the privatisations of the 1990s, starting from that of CSN. Even though the workers' pension fund constituted the bulk of the cash flow generated by the privatisation, the workers' consortium ended up with only 20 per cent of shares of CSN. This was not quite the form of workers' capitalism that was initially fantasised.

CSN: A Political Laboratory for Rebuilding the Nation

CSN was privatised in 1992. But the restructuring in preparation of the privatisation started in 1989 with the last stage of the so-called Plan D, consisting of a series of investments, expansions and reorganisations that radically improved the company's profile and laid off more than six thousand workers out of twenty-three thousand.[7] During the implementation of Plan D, Japanese management consultants were invited to Volta Redonda to implement the TQM and 5S systems in UPV. Walter, an engineer in CSN, describes the changes brought in by the 5S on the shop floor:

> The 5S was like this. You are on the shop floor and select what you need and what you don't need. Then you order what you do not find on the shop floor. Then you preventively clean the shop floor so that the new order does not get dirty. In this way, you educate the worker. When he understands the work environment needs to be clean, he starts cleaning himself. He starts tucking his shirt inside the trousers and polishing his shoes. This process of hygiene becomes a process of self-discipline, which continues at home.

Plan D turned CSN into one of the most advanced steelmakers in the world. At the same time, it increased the firm's financial exposure, which became the strongest argument for privatising it. Against these reorganisation measures in November 1988, more than eight thousand workers took part in a strike against CSN – the first strike in the company's history, which led SMSF President José Juarez Antunes to become mayor of Volta Redonda.

If the circumstances of Antunes's death are mysterious, its legacy is clear. It marked the birth of the reformist Union Force (FS) and of the 'workers' capitalism' that was being supported by the progressive forces within the new democratic government. The standard economic narrative describes the process of privatisation in 1990s in Brazil as a historical moment of democratisation of the economy that followed the country's transition into democracy. In fact, privatisation broke the labour movement, reinforced existing industrial and bank monopolies, financialised the economy and put it under state control thus, turning it from 'regulator' to 'investor' (Musacchio and Lazzarini 2014: 150). Among the one hundred thousand people attending Antunes's funeral were Leonel Brizola and Lula da Silva, the two main working-class presidential candidates. Few people in the crowd could have imagined that, many years later, Lula would perfect the project of financialisation of industry started with Plan D. The death of Antunes immediately sparked a leadership context within the Metalworkers' Union won by the radical left-wing group CUT Pela Base led by the Barcelos brothers.

New CSN President Lima Neto and the reformist wing of the SMSF supported privatisation, whereas the SMSF leadership, militant Catholic labour organisations, social movements and the Engineers' Union of Volta Redonda (SENGE-VR) opposed it. After a series of strikes, demonstrations and a public referendum showing that 80 per cent Volta Redonda residents were against the privatisation of CSN (Lima 2016: 100), the anti-privatisation coalition put together a proposal for restructuring the firm. The proposal reflected the point of view of the engineers and of skilled workers who led the coalition. It argued CSN suffered from structural dependency on companies in the Global North rather than being inefficient or its labour too expensive.

Reflecting the productivist view of the leadership of the Metalworkers' Union, the proposal suggested some measures to fix the company: (1) investing in new technology to upgrade the low-value-added and polluting production process and implement a continuous production system; (2) cutting bureaucracy and debt – which had on average doubled labour costs since the early 1980s and represented 98 per cent of production costs in 1989; (3) implementing a new market regime for the steel going to the automotive sectors, which was undervalued by 40 per cent; and (4) making sales more transparent. In fact, steel distribution was controlled by corrupted former army officers, local gentries and politicians who traded steel and its by-products (nickel, cement and benzene) and even new machines in second-hand markets.[8] Neto, an engineer close to the SENGE-VR and the Metalworkers' Union, dismissed the diagnostic report and reiterated his support to the privatisation of the company. The *formigueiros* who supported privatisation were expelled from the Metalworkers' Union and went on to fund the FS.

Unlike the engineers and skilled workers who led the Metalworkers' Union, the FS saw privatisation as a chance to implement the 'people's capitalism' that was being championed by the Minister of Finance Cardoso. This, they believed, would improve the economic and social position of marginal workers vis-à-vis the dominant aristocracy of labour. The *formigueiros* put together a shareholders' consortium based on the workers' pension funds of CSN (CBS-CSN). The consortium gained support from Neto, the BNDES, several regional business associations, the Banco do Estado do Rio de Janeiro Rio de Janeiro (Bank of the State of Rio de Janeiro – BANERJ) and the charismatic Rio de Janeiro Governor Leonel Brizola. Initially,

Brizola opposed privatisations. In a famous TV appearance, he denounced Collor's privatisation plan as 'dictatorial' and paying lip service to global financial interests and the IMF. By supporting the workers' consortium under the Democratic Labour Party (PDT) and FS umbrella, Brizola wanted to create a centrist working-class coalition based in the state of Rio de Janeiro that challenged the radical CUT-PT working-class coalition based in the industrial region of São Paulo.

A few months after passing the Programa Nacional de Desestatização (National Privatisation Program – PND), Collor was embroiled in a corruption scandal, put under impeachment and forced out of the presidency. The new President Itamar Franco froze the PND and dissolved the privatisation commission. The commission was a supposedly independent consultative body composed by public figures close to the president. But it de facto ran the privatisation programme. The new presidential circle was strongly anti-privatisation, fearing it would put the country under foreign control. Moreover, a strong middle-class constituency feared it would drain the public administration and force it to disinvest, especially in social welfare. Indeed, this was an atypical case. Most of the financial exposure of CSN was towards not the private sector rather the BNDES and state banks, and the strongest impulse to privatise came from within the top management of CSN. The president of the bank, the economist Antônio Barros de Castro, dissolved the privatisation group but did not fire Antonio Perrone, the economist of the bank in charge of privatisations. Perrone describes that moment as follows:

> This was a typically Brazilian moment. I was in charge of privatisation in the BNDES. They dissolved the team, but nobody told me I should stop. So, I went on. I looked for all possible alliances. I lobbied with the vice president of the bank. He fixed me up with José de Castro (Brazil's attorney general)[9] and Franco's personal lawyer. De Castro agreed to set up a meeting with President Franco. At the meeting, President Franco told me: 'The privatisation of CSN is ready to go. It's all in place. We cannot stop the process. There is no point in restructuring CSN and make it national now. I am not going to stop you now.' Franco did not agree with me. But he did not stop me. So, I went on. If you ask me how privatisation happened, I can't say. But Franco and I agreed on two things. First, we both disliked Brizola. We both thought he was a dangerous populist and a ruthless *caudillo* [strongman], an Argentinian who despised the Brazilian people and a populist, constantly addressing the 'working class' but without being a genuine Marxist. Second, we both understood that to solve the fiscal crisis of the state, we needed the support of the trade union and that CSN had such support.[10]

Moreover, both Franco and Perrone knew the Metalworkers' Union was deeply split between FS and the CUT and that the privatisation process would divide further the steelworkers of Volta Redonda, a militant section of the working class that Franco deeply feared. After the meeting with the head of privatisation, Franco passed Law 8,301, single-handedly resuming the privatisation plan.

The new board fixed the minimum prices of the firms to be privatised and introduced a new regime in which any prospective buyer had to pay upfront an amount of 'living money' (*moeda viva*) equal to the value of its assets and profits. With a clear break with the past, this emphasis on living money was meant to attract national and foreign groups[11] and exclude minority shareholders, especially the workers. In fact, Franco was determined to hijack the workers' consortium proposed by the FS (Graciolli 2007: 266). This sudden change at the top reverberated downwards. With no secure prospect of control over the company, soon-to-be-privatised Neto and the top management of CSN now opposed privatisation. Neto was dismissed, and a new pro-privatisation director was put in place. Albano and Luizinho, the leaders of the FS, challenged the exclusion of minority shareholders and put together a workers' investment consortium called the Investment Club. Several regional employees, mayors, business associations (including the powerful Commercial, Industrial and Agro-pastoral Association of Volta Redonda) and the BANERJ joined the consortium,[12] which ended up representing 52 per cent of CSN shares.

Thanks to the determination of Albano and Luizinho, the process of privatisation was put in motion again. The two trade union leaders and the government were in continuous conversation. Perrone and his team travelled several times to Volta Redonda to explain the details of the deal to the workers. These had been put off by the obscure economic language of the privatisation dossier the BNDES had circulated, which glossed over the most important issues such as the conversion rates of pension entitlements to shares, rates of interest and discount and modes of actualisations. Employees were more interested in these economic issues than in the issue of co-management, which for the FS leadership was the essence of the deal. To translate the technical details of the BNDES report to illiterate workers, Albano and Luizinho even put together an illustrated booklet explaining the impact of the privatisation of CSN on its employees. In public meetings, Albano and Luizinho attacked the privatisation team of the bank and its president. But in private, they collaborated tightly with them to seal the deal. Perrone updated President Franco on every move. Despite their dealings with the state, Albano and Luizinho's heavily anti-state narrative, especially against corrupted state managers, resonated with part of the working-class base.

Albano and Luizinho have personal histories of extreme poverty and precarious jobs. They do not belong to the 'new' trade union movement of the 1980s made up of educated, Catholic and skilled workers and engineers but instead come from grassroots activism in favelas and poor neighbourhoods. Being outsiders within the labour movement, they believed privatisation and 'people's capitalism' would close the social and economic gap between *peões* like themselves and the aristocracy of labour. Luizinho framed his pro-privatisation stance as a form of opposition to the left-wing aristocracy of labour of Volta Redonda – an anti-establishment posture:

> Volta Redonda was a left-wing city. You did not have space for any other political position. It was an ideologised city, and a small one, which was an additional problem. For the workers, it was easier to say they were against the privatisation rather than admit they were not left-wing. (Lima 2010: 116)

William Reddy (1987) describes the idea that money is a social equaliser as a Western liberal illusion. Unskilled workers like Luizinho believed in such liberal illusions that by delinking incomes from wages and attaching them to money and capital investments, they would improve their status and wealth. Moreover, they hoped the extra money coming from their stocks on the top of their wages would allow them to access the exclusive goods – homes, cars and domestic appliances – that so far were accessible only to the affluent working class. This consumption gap between affluent and peripheral working class was addressed later, under the Lula and the Rousseff administrations, which boosted consumption for the marginal working class through policies of popularisation of credit and financialisation of poor households I described earlier. But the proposal of Brizola, Albano and Luizinho came with a price of one thousand new redundancies,[13] longer shifts and labour intensification through the TQM system.

In fact, the FS consortium was much more than an economic project. It was a move towards the creation of a new regional coalition, at a time when the new democratic state was both decentralising and opening up to global capitalism. Global car manufacturers were relocating from São Paulo to Rio de Janeiro, looking for cheaper and non-unionised labour and lured in by municipalities engaged in 'fiscal wars' against each other. The cross-sectional alliance between unions, municipalities and private businesses set up by Brizola and the FS was meant to turn the state of Rio de Janeiro into the owner of the most profitable steel complex in Brazil, with the FS as the unlikely co-manager. Brizola described the consortium as a form of co-management between the state and the working class, but its ultimate logic was to turn the Sul Fluminense region into a global capital hub controlled by the state of Rio de Janeiro. Unlike it, the cross-sectional regionalism led by the PT in São Paulo reflected the principle of democratic decentralisation through forms of direct democracy and collective participation by employees, municipalities, civic associations and local entrepreneurs.

In 1992, 98 per cent of CSN employees supported the FS consortium. But, in the build-up to the auction, scheduled on 1 April 1993, the regional coalition started to disintegrate. The president of the Brazilian Social Democracy Party, possibly following Franco's directives, filed a lawsuit against the Investment Club. When Franco confirmed the restrictions on privatisation money, the BANERJ pulled out. It turned out the bank had joined the consortium in the desperate attempt to save itself from bankruptcy. The SENGE-VR put together a Clube de Investimento Alternativo (Alternative Investment Club – CIA), breaking the mass support to the Investment Club. In November 1993, accused of fraud, the CSN president resigned. Brizola asked for the impeachment of President Franco for his dealing with the privatisation of CSN.

The two consortiums of consultants who set the minimum price went under public investigation. The Conselho Regional de Economia (Regional Economic Council) of Rio de Janeiro estimated a 25 per cent discrepancy between their evaluations.[14] Moreover, it transpired that the minimum price of CSN was set lower than the lowest evaluation by the consultants (Graciolli 2007: 130). In fact, the minimum price was even well below the value of two rolling mills that had been purchased just two years earlier.[15] According to a reliable estimate,[16] CSN was sold for 14 per cent of its real value. Moreover, 95 per cent of the value was paid in poor money (*moeda podre*). The main buyers of CSN were the mining company Vale (CVRD), the textile group Vicunha (17 per cent) and the domestic banks Bradesco and Itaú (6 per cent) and Bamerindus (2.81 per cent). After a long legal contentious and thanks to a generous buyout from the BNDES, Vale withdrew from the control group. Vicunha, owned by Benjamin Steinbruch, acquired control of CSN. With CSN, Vicunha also acquired majority stakes in lucrative iron ore mines, national logistical infrastructures (including energy, ports and railways) and a vast portion of land and leisure and welfare facilities in Volta Redonda.

The post-privatisation structure of CSN is more vertical and centralised than the previous public one. In the new post-privatisation financial regime, the company is fragmented in independent business units with income from the mining, logistical and financial sectors, systematically offsetting losses in the main steelmaking business. Going against the principle of 'market transparency' that supposedly inspired the privatisation of national companies in Brazil, the privatisation of CSN turned Vicunha into an oligopolistic power,[17] directly and indirectly controlling companies from across the private and public sectors and entangled within the national logistical infrastructure of roads, ports, electricity and railways. Moreover, the privatisation of CSN put the national steel industry under the financial control of the BNDES, public pension funds and banks, laying the ground for the new logic of financial 'governmentality'.

Privatisation: Finance as Working-Class Money

In the introduction, I mentioned the early empire and the Brazilian state relied on state-owned enterprises – semi-private bodies with autonomous budgets – to deliver important public services from utility to education. Gaining immense financial power in the Vargas era, SOEs were downsized and privatised under the military regime in 1979. The 1988 constitution inverted the trend, establishing public monopolies in telecommunications, oil and gas distribution and setting up barriers to foreign ownership in mining and electricity. The return to state-led industrialisation and economic protectionism after the democratic opening was a reaction against the monetarist economic dogma that was forcing radical processes of privatisation in those Latin American and Eastern European countries that were transitioning into democracy. Despite an existing law that restricted foreign capital to 40 per cent of shares in national companies, the Brazilian elite was worried privatisation would put foreign capital in control over the Brazilian economy. Moreover, with the trade unions and the business sector, the wider public opposed the privatisation of basic industries, fearing the public services would be next in line. Yet, the worsening of state finances and the growing hyperinflation forced the president to restart the PND, which reached its apex with the Collor presidency in 1990s.

Sarney was personally opposed to privatisation, but as hyperinflation reached record levels, he saw it as the last resort against the collapse of the state. He delegated the process to a group of engineers and economists of the BNDES. Unlike the heterodox economists who devised the monetary plan, the BNDES privatisation team had a private-sector background and came from the banking sector. Indeed, the major stimulus to privatise the steel sector came from the fact that the bank was heavily exposed towards insolvent national steel companies. Under the new private regime, the bank could cash its debts[18] and restore profitability. The BNDES convinced the treasury to convert its certificates of public external debts into titles that could be used to buy shares in the soon-to-be-privatised companies. Because public debt is so central to the Brazilian economy, the conversion of state debts into 'privatisation money' took the form of mass securitisation. With a single gesture of creative accounting, the entire failing public industry would be recapitalised. But because there was no liquidity in the Brazilian economy, and no national or international private groups willing to pay in cash, the recapitalisation process had to be set in motion with the *moeda viva* of the working class. But how could the state convince workers to invest their savings and pensions in shares of yet-to-exist companies?

Criticism of the Western models of privatisation transplanted to Latin America and Eastern Europe dominated the early discussions on privatisation in Brazil. Licínio Velasco, one of the architects of the early privatisations,[19] believed the British market model, which Chile had just followed, would not work in Brazil. In the United Kingdom and Chile, the state and the managerial class believed in the economic benefits of privatisation, such as the devolution of the managerial functions to more efficient private entrepreneurs or the reduction of public deficit. Moreover, in the United Kingdom and Chile, there was an explicit neo-liberal consensus behind privatisation which was considered a tool to weaken the trade unions and downsize and deregulate industry. Although there was a mounting fear for the ascent of the militant CUT in Brazil, trade unions were still under state control and there was no clear support for state downsizing or free market. Moreover, Brazilian industrialists, especially steelmakers, considered neo-liberalism a self-defeating Western dogma. Consequently, instead of seeing privatisation as a tool of labour and capital deregulation and downsizing of the national industry, the Brazilian steel industrialists and the BNDES saw it as an opportunity to regulate, upgrade and consolidate the national industry under increased state control (Amann et al. 2004; Montero 1998). After all, because of the country's tradition of mixed economy, public and private economy clearly would remain deeply entangled.

Moreover, according to Velasco, the Brazilian economy lacked of liquidity, and the UK market-based model of privatisation would have not worked. Velasco and his team ruled out the French-state-led and consensus-based model of privatisation, too, because both Sarney and Collor lacked of broad political support. Their right-wing, anti-statist constituencies had no root in civil society or parliamentary politics so that they lacked of a neo-liberal mandate on the ground of which to sell privatisations as a political project. Thus, Brazil had nowhere to turn for inspiration. The architects of the early privatisations wanted to make the economy 'more democratic'. But there are as many ideas of democracy as models of privatisation. For the privatisation of British Telecom, British Gas and British Steel in the 1980s, the Thatcher government adopted a model of large floatation on the stock market based on competitive bids of ascending prices. Such an economic model reflected Thatcher's idea of democracy consisting of utility-maximising and competitive 'possessive individualists'. For Velasco, Brazil 'was not ready' for such a competitive model of market democracy. Instead, he and the other experts put together a model of privatisation called Usiminas,[20] based on consensus and openness to non-institutional investors such as public and private banks, NGOs, foundations and workers' pensions funds – none of which in a dominant position.

The auction was divided into three blocks: an auction of common shares (representing the majority control), an auction of non-voting shares and a public offering of heavily discounted shares exclusively for employees. The model was inspired by two main principles: openness and just price. Bids were open. Anyone could bid. There were no technical prequalification or minimum thresholds for bidders, provided they were registered with the stock exchange and had a National Insurance number. 'Just price' meant the price of shares had to meet the requirements of different constituencies, especially marginal ones. Technically, the just price was achieved though the mechanisms of the minimum price and debt securitisation. Unlike the English auction system based on open ascending price, the minimum price matches the purchase power of different investors who bid for different price bands. Besides, when prices are difficult to set because companies are not floated on the stock exchange, the minimum price allows substantial incremental market adjustments and discourages undervaluation. Second, the mechanism of securitisation converted debts papers – usually held against insolvent state firms – into privatisation money to be used to purchase stocks in the soon-to-be-privatised firms. The conversion followed non-market criteria, with the treasury filling the gap between the nominal and the market value of these debts.

The overall impression was that the state was injecting new money into the system. But because a market for privatisation money did not exist yet, the treasury renegotiated the initial debts on its own terms. More importantly, securitisation generated a cash flow for the state before its companies were sold. But apart from the virtual money coming from securitised debts, the only cash in the system consisted of working-class savings. To lure the working class, the BNDES president of the time, Eduardo Modiano offered to convert the government's debt to the Worker Support Fund (FAT) – the workers' biggest forced saving scheme (the FGTS) – at the discretion of workers, to certificates of privatisation (CPs) subsidised by the bank. Modiano presented his proposal as a historical opening towards the working class, the beginning of a new kind of workers' capitalism that would 'moralise' the Brazilian economy. Modiano's description of stock market capitalism as a moralising force was repeated nearly verbatim in a speech given by President Lula da Silva at the São Paulo Stock Exchange ten years later. The moral component of Modiano's offer was that by having their savings converted into privatisation currency ('social money' – *moeda social*) worker-investors would be allowed to sit on the boards of private companies and take strategic decisions with the management.

But at the time, the two main parties with substantial working-class constituencies – the PT of Lula and the PDT of Brizola – opposed privatisation and rejected the proposal. Modiano raised the stakes. He would now convert the workers' saving into CPs at a 70 per cent discount with ten-year loans and subsidised interest on additional purchases. This was the biggest discount rate offered to employees in the history of the privatisation worldwide. This second proposal convinced many workers and trade unions, and split the labour movement (see chapter 4). Among economists and in governmental circles, privatisation money was considered *moeda pobre* because it originated from outside the market, from insolvent state debts and marginal social constituencies, including the working class – the equivalent of contemporary subprime mortgages. This 'rotten money' had different origins and degrees of risk – personal deposits, debts with international financial institutions (such as the IMF and World Bank), people's mortgages, frozen currencies, currency swaps between multinationals and workers' pensions funds.[21]

The pooling into the same national currency of debts originating from totally different realms – spanning from poor households to the World Bank – was a massively ambitious state project. The BNDES team intuited such an ambitious operation would fail if pushed unilaterally by the executive. As Sarney hinted, it had to be endorsed by 'the Brazilian people'.[22] First, the bank promoted grassroots pro-privatisation campaigns among the working class with the help of union leaders, such as the one I have just described. More importantly, it structured the auction openly. According to Velasco, such openness was achieved thanks to a system of computerised transactions set up in the Bolsa de Valores do Rio de Janeiro (Rio de Janeiro Stock Exchange – BVRJ):

> I went to the president of the BVRJ and said: 'We need your help . . . we do not want a model that concentrates ownership, because we will not have political support for that. We can't sell the company to one buyer.' A group of state engineers came up with the idea of creating a computer that would coordinate bids from all the main stock exchanges across the country with the only exception of São Paulo, which opposed the project.[23] This is how the computerised Sistema Eletrônico de Negociação Nacional (Electronic System of National Negotiation, SENN) was born! SENN was small. The size of a big tape recorder . . . it did not look particularly high-tech. But it had a huge memory.

During the auctions, SENN stored and processed in real time bids from all sorts of investors – corporations, workers, trade union representatives and even housewives. First, control room operators checked the validity of each bid. Then, the SENN cleared out the bids for each price. Finally, the BVRJ director moved onto the next price, cutting out the outstanding bidders. There was still scope for discretional processes, especially during the residual bids that followed the initial sell-off of 65 per cent of the capital. These started on the highest price and quickly moved down becoming increasingly more chaotic.

On the day the bid was opened, a cordon of three hundred soldiers in riot gear surrounded the stock exchange. Moreover, the federal judge Aldo Rebelo from the Communist section of the Order of Attorneys of Brazil and the municipality of Volta Redonda tried to stop the bid heavily delaying its start. Started, the bid was stopped again after just fifty-seven minutes because of lack of buyers. When the bid resumed, the procedural chaos continued. For instance, quotas on foreign, national and public capital exceeding the legal limit would be cleared by the computer but stopped by the director much later. Moreover, as well as through the SENN, bids were sent in by telephone and even in writing. The SENN computer become the symbol of the transparency of the privatisation process and of the high-tech financial infrastructure of the new democratic state. In fact, some people argue the SENN rendered invisible the manipulations of the BVRJ director, who was said to follow the instructions of the privatisation team (Montero 1998: 47). But, following the principle of 'procedural transparency', the bank externalised the most controversial decisions – such as the setting of the minimum price and the patrimonial, technological and juridical audits – to external foreign auditors and law firms, which protected the BNDES from criticism on the process.

In a widely circulated report, Modiano made it clear the BNDES was leading the privatisation process because it was struggling economically and was heavily exposed to failing national companies. He described privatisation as a form of economic reorganisation of the bank and such reorganisation as a first step towards the financial restructuring of the Brazilian state. Thus, the bank presented itself not as a state agent but as a competent and profit-oriented economic agent determinate to restore the state finances. Even the managers of the public steel companies to be privatised saw the BNDES as a competent state manager and supported its role in the privatisation process (Schneider 1992: 25). But the BNDES's greatest achievement was to win the workers' trust, which guaranteed not only the success of the privatisation process but also the survival of the bank. In addition to the 40 per cent of the FAT automatically transferred to the BNDES by constitutional decree,[24] the securitisation of pensions funds recapitalised the bank and allowed it to invest in more productive sectors (Pinheiro and Giambiagi 1997; Schneider 1992).

It turned out that securitised debts (*moeda podre*) could be used only to purchase the control package and hence by richer investors. Only 19 per cent of the purchases were made in cash, that is, with the *moeda viva* of mostly marginal investors such as cash stripped investors or workers' funds, who had no debts to cash or enough capital to go for the main auction. The living money generated by these residual bids, mainly from the workers' pockets, generated twice as much financial liquidity for the state than the control package. Because of the extended use of 'rotten money', the PND generated with very little liquidity from 1990 to 1994 (Resende 1985: 128). Yet, the harsh reality of workers' capitalism became evident only in the moment of actualisation. CPs nearly halved their value only a few weeks after the auction, and most debts proved difficult to actualise.[25] Thus, privatisation money, like the 'primitive money' of traditional societies (Dalton 1965) circulated in separate spheres of exchanges and according to different social status of the economic agents. 'Rotten money' circulated within the rich industrial bourgeoisie and banking elites according to the logic of the gift and of reciprocal value conversions.

'Living money' circulated within marginal constituencies – the informal and precarious working class – according to the logic of the market. As a consequence of the magical monetary power of the state, the bourgeoisie saw their debts turned into living money, and the working class had its living money turned into tokens of credit in the failing national economy.

It is widely acknowledged that Brazilian privatisations restructured and recapitalised the industrial sector (Pinheiro and Giambiagi 1997). Yet, the privatisation of the steel industry created four major drawbacks for labour. First, it increased social inequality. From 1991 to 1997, privatisations led to a 30 per cent reduction of waged employment and a 10 per cent increase of informal labour mainly among women and black workers (DIEESE 2000). Second, instead of bringing capitalism to the people, it increased concentration[26] not by the existing steel groups but by private conglomerates (e.g. CVRD), public banks (typically, Banco Bozano Simonsen and Bradesco), pension funds (especially Previ) and steel distributors,[27] none of which wished to be directly involved in steel production. In the 1990s, Brazil's largest banks (Bamerindus, Itaú and Unibanco) had purchased CPs and were keen to unload them so that privatisations increased their control over the steel national industry, laying the ground for the emergence of industrial-financial complexes that now dominate Brazilian's economy (e.g. Bozano Simonsen and CVRD). As a result, the steel industry remained under-capitalised, oligopolistic and under a new financial elite consisting of public companies and pension funds.[28] These oligopolistic powers came to control the immense privatised pension fund, the CBS-CSN, which is also one of CSN's main stockholders.

Third, as it happened in Eastern Europe, privatisation increased rather than reduced the state's control over the industry, both as provider of equity capital during privatisations and as long-term investor after it. But unlike in Eastern Europe, where privatisations bred hypertrophic technical, juridical and administrative state apparatuses (Haggard and Kaufman 1995) or in Chile and Britain, where they reinforced the authoritarian power of executives (Goldstein 1999: 679), privatisation in Brazil empowered the state *economically* – through cross-participations and minority control packages managed by the BNDES – but not *politically*, triggering instead a process of internal fragmentation between ministries, autarchies and the bureaucratic technocracies run by economists and bankers. Fourth, privatisations created unstable and intricate patterns of pyramidal corporate governance, with small numbers of investors from different sectors, clustered together either directly or through holding companies. A typically complex case was the cross-ownership between CSN and Vale (CVRD),[29] which was solved with BNDES subsiding the Vicunha group. Today, these pyramidal forms of corporate governance allow dominant shareholders to control firms even with a very small share of their total capital and to forge inter-firm relationships in different and even competing companies.

Like the transition 'from socialism to clan' sparked by the privatisations in Eastern Europe (Verdery 1996), Brazilian privatisations spun a complex web of cross-participations, inter-firm relationship and 'nested hierarchies' controlled by big conglomerates, banks and pensions funds, and cutting across the private and the public sectors.[30] Moreover, privatisation turned the Brazilian state into a 'leviathan' that uses pension funds (the workers' 'living money') and public banks such as the BNDES[31] to control immense industrial and financial complexes and acting simultaneously as economic investor and regulator. Through a deeper penetration into the economy via the equity market, the Brazilian state transformed itself into 'pension fund manager', pushing relations of debt and dependency deeper into Brazilian society. Described as a new revolutionary form of 'people's capitalism' – bringing together the state, the capitalists and the workers – privatisation in fact captured the liquid money of the working class to the advantage of those bureaucrats, industrialists and bankers who had recycled themselves into the new apparatus of the democratic state.

Excluded from the control group of the CBS-CSN, most employees agreed to use their stocks to purchase houses that CECISA – CSN's construction company – was developing in the bairro Vila Rica and which the company advertised as the new luxury residencies for Class C. Twenty years later, Vila Rica is one of the poorest bairros of Volta Redonda,[32] inhabited by unemployed and precarious/informal workers.[33] At the macro level, such monetary magic – the transformation of 'poor money' into state money – was made possible by the hybrid privatisation model implemented by the BNDES, which combined top-down state control and market deregulation and entangled together finance and industry – the public and the private. On the shop floor, the monetary magic was achieved through managerial systems such as the SIGMA, which deskilled line workers, put the shop floor under a market regime and empowered those younger and unskilled workers who were keen to buy into the dream of working-class capitalism. The political process set in motion with the privatisation of CSN – that is, the transformation of the working class into middle class and of the trade union into financial class – would come to full fruition only later with Lula.

The idea that 'privatisation money' – the new currency created with the privatisation process – would reduce inequality and benefit low-income workers reflects the 'western liberal illusion' (Reddy 1987) that money under capitalism functions independently form social relations. In fact, in the privatisation process I describe, class relations determined how money functioned. Low-income workers were cash providers *to* the state, whereas oligopolistic capitalists such as Vicunha were credit recipients *from* the state. Like 'primitive money', 'privatisation money' circulated within separate spheres of exchange according to the status of the people involved. Against the prevailing anthropological trend of studying money and finance as separated from production (LiPuma and Lee 2004; Maurer 2005),[34] this chapter emphasises how relations *in* production impact on the circulation of state money and vice versa. The support to the privatisation of CSN coming from the marginal working class can be understood only by looking at the power relations between such a marginal working class and the aristocracy of labour on the shop floor, and how the managerial reorganisation that accompanied privatisation changed such power relations, which I explain in chapter 6. The SIGMA radically restructured relations *in* production, subcontracting maintenance jobs, deskilling the aristocracy of labour (skilled operatives and engineers) and empowering those marginal workers whose political support and 'social money' were vital for privatisation to succeed.

Moreover, against the well-rehearsed argument that finance and circulation deterritorialise and fragment the forces of industry and the state (Lee and LiPuma 2002), my ethnography shows how finance did in fact restructure, coalesce and reterritorialise class relations around the new democratic state by weaving together different forms of debt and regimes of values (Ortiz 2013). In other words, monetary regimes are not ideologies existing in a cultural vacuum but are enforced and negotiated through the material organisation of work and life and in their entanglement with the forces of production. The more unskilled and precarious the workers, the more they supported privatisation and identified with the model of 'worker-investor' presented to them by the BNDES. For these marginal constituencies, the stock market offered the hope of social mobility and of breaking away from existing occupational hierarchies. Overall, labour purchased $290 million in shares that by 1995 were worth more than $1 billion.[35] But 'worker-investors' were allowed to buy only up to $4,000 worth of CSN stocks so that they did not make it into the ranks of the real capitalists. It was mainly top union leaders or PT executives who ended up becoming executives or pension funds managers in the top echelons of the state.

The privatisation process also both catalysed and pitted against each other different state agencies and reconstructed a fragmented, delegitimised and indebted state around a new national-popular economic project. On Modiano's insistence, the Directorate for the Privatisation, the body in charge of the privatisation plan, stayed independent from the executive and hence was able to build cross-sectional alliances with trade unions, workers, bankers and industrial oligarchs. Because of the 'embedded autonomy' of the BNDES (Evans 1979), the privatisation plan continued and was successful despite the executive being internally divided about it. In fact, privatisation was a process of statecraft orchestrated by banking elites and the BNDES, which is both a public development bank and one of the world's most powerful state agencies and independent financial capitalist organisation. Moreover, the post-privatisation entanglement of governance, finance and industry at the state level is reflected in the post-privatisation structure of CSN and in the way such a private conglomerate complex controls strategic public resources and infrastructures. The success of the privatisation process can also be ascribed to the 'imaginative framing' (Abolafia 2004)[36] of the experts of the BNDES who set up the process as open, grassroots and democratic and brought together different regimes of value from across the country – working-class households, banks, corporations, municipalities, trade unions, hospitals and pension funds – through the high-tech SENN.

In the end, the privatisations of the 1990s deeply divided the working class between the traditional labour movement and the more marginal workers who identified with the 'people's capitalism' proposed by Cardoso.[37] The traditional working class associated with the PT and the CUT, which opposed privatisation, learned from the defeat. From 2003 onwards, the top echelon of this traditional working class opened up to the world of finance, becoming pension fund managers or technical experts for the ministries in the PT government, while the rest of the working class went through progressive precarisation. Since his first mandate, Lula accelerated the 'monetisation' of the popular classes through programmes of public-private partnership, national microcredit, the Bank of the People and the incorporation of the workers' pensions funds into the stock market in the form of the Bolsa de Valores Socioambientais (Socio-environmental Stock

Exchange). In a speech given in 2003 during his visit to the São Paulo Stock Exchange (Bovespa), Lula claimed:

> Never in the history of this country the issue of the pension funds will be discussed as much as we are discussing it now. The people must understand how these funds can help to solve many unsolved problems in our country . . . from now on we must think the function of the pension fund will be much more noble, from the social point of view, than it has ever been. (In Jardim: 2009: 130)

A few months after the talk, Selim, the president of the General Command of Workers – the second largest labour confederation in Brazil – was elected president of Bovespa's board of administration. Raymundo Magliano, the president of Bovespa at the time, described this development as follows:

> The worker was not born just to work. He also was born to earn money. What we want to do with the stock exchange is social inclusion. It is important that the worker acquires an asset; controls some firms and has his own rent. We needed a trade union leader in the stock exchange in order to consolidate the alliance between capital and labour. It is important that Selim is with us in order to consolidate our strategy. (*Exame*, 14 January 2005, in Zibechi 2014: 130)

The state vision of ordinary citizens acting as state agents sketched by Sarney and perfected by Cardoso started to materialise only on 1 January 2003, when the Brazilian nation woke up under the presidency of an ex-metalworker from Brazil's poor Northeast. Starting his presidency like all great populist leaders, in the 'Open Letter to the Brazilian People' ('Carta ao povo Brasileiro') of June 2002, Lula addressed 'the people' as a new economic subject, able to embrace the forward-looking and moralising forces of stock market capitalism and leave behind the backwards logic of industrial democracy. This convergence between 'left' and 'right' political agendas around the financialisation of the working class followed in the footprint opened by the 'New Lefts' in the United States and Europe. In this sense, the main story of this chapter is not so much how Brazil was forced to endorse the financial economy by conservative international financial institutions such as the IMF, the World Bank or neocolonial powers. Instead, the chapter tells the story of a long-term process of institutionalisation of inequality achieved through monetary inclusion and perfected by a working-class party of a global periphery, which with the advent of democracy found itself engulfed in the global forces of finance.

Notes

1. The economist Francisco Lopes (1986) coined this term in his well-known book *O choque heterodoxo: Combate á inflação e reforma monetária.*
2. In *The Passions and the Interests*, Hirschman (1977) argues the rise of Western economics in the seventeenth and eighteenth centuries reflected a shift in governmentality in which state violence was exerted through the technical apparatus of economics rather than the army. In a later work on inflation, Hirschman (1981) argues monetary policies in post-authoritarian Latin America were forms of state redistribution that functioned as a mechanism for co-opting radical political movements.
3. For instance, they advised the government not to follow the 'shock therapy' adopted by Argentina to stabilise inflation the year before and which was influenced by monetarism (Sola 1993: 169). Neiburg (2010) discusses the political and cultural construction of the notion of 'monetary instability' in relation to the monetary policies of Brazil and Argentina.
4. For instance, the Banco do Brasil was allowed to create base money through an open central bank discount facility (see Baer and Beckerman 1987: 41)
5. Fifteen per cent for those on minimum wage (42 per cent of the population) (Sola 1991: 164).
6. In February, labour had a 4.4 per cent income increase 'from one day to the next' (Baer and Beckerman 1987: 46).
7. By the time BNDES officials placed CSN on the auction block in 1992, its total debt had been reduced by 73 per cent ($1.9 billions of $2.6 billion) (Montero 1998 : 48).
8. Machines were left to blacken and depreciate outside the furnace department to be later sold at their real market value.
9. The authority in charge of the legal advising of the executive branch.
10. Fernando Perrone, personal communication, 10 June 2008.
11. Franco expanded from 20 to 100 per cent the control share allowed to foreign groups.
12. The proposed capital composition was BANERJ: 10 per cent; FGTS: 30 per cent; and CBS-CSN: 9 per cent.
13. In 1991 alone, one thousand employers were laid off – 5 per cent of the workforce (Graciolli 2007: 166).
14. Including underestimated production values and overestimated environmental liabilities.
15. Respectively, $1.462 billion and $2.3 billion.
16. According to the consultant Jaako Poyry, the establishment of a steel company equivalent to CSN would cost $11.2 billion.
17. For Goldstein and Schneider (2004), this structure is typically Latin American.
18. Debts against public companies are not executable, because they are underwritten by the treasury, whereas debts against private firms are executable.
19. This section is based on personal conversations with Licínio Velasco, Estela Almeida and Fernando Perrone, who were in charge of the privatisation team of the BNDES during the period I discuss. See also Velasco (1997, 1999 and 2010)
20. From the name of the first company to be privatised.
21. More specifically, 'rotten money' included deposit facility agreements , CPs, thirty-six billion cruzados novos that the Collor Plan had frozen in the Central Bank as part of its macroeconomic stabilisation, overdue credits against public companies (e.g. Siderbras debentures), mortgage notes, deposits of foreign currencies and various external debt credits. Collateral Debt Obligations (CDO) included multi-year deposit facility agreements, parallel financing agreements and co-financing agreements.

22. Under Collor, many decisions were taken without even the approval of the legislature. For instance, in one day, the Collor administration liberated more than $50 billion in Central Bank accounts (Montero 1998: 42).
23. Unlike the powerful São Paulo Stock Exchange (Bovespa) – which monopolised 85 per cent of total capital transactions and specialised in private stocks – the SENN integrated seven regional stock exchanges and specialised in state bonds and smaller institutional investors. The BVRJ was abolished in 2002 and subsumed to the Bovespa.
24. According to Giambiagi (2005: 215), 40 per cent of the current cash flow of the bank comes from the FAT, 70 per cent at the time of Sarney. By 2006, the FAT was responsible for 67 per cent of the disbursements of BNDES, amounting to more than $100 billion in 2010 (Zibechi 2014).
25. Foreign debts were discounted by 25 per cent.
26. 'By September 1994, six industrial and financial groups dominated the Brazilian steel sector: Usiminas-Cosipa, CSN, CST, Gerdau (Piratini), Mendes Jr.-Agominas, and Acesita. Usiminas and Cosipa alone controlled 61 percent of the market for flat, unalloyed steel plate used by the major producers in Brazil's extensive automobile manufacturing complex. CSN commanded the other 39 percent' (Montero 1998: 21).
27. In Volta Redonda, steel distributors were old merchant and industrial families with great local power.
28. At the time, SOEs accounted for 12.2 per cent and public pension funds for another 15 per cent of privatisation capital (Amann et al. 2004: 15).
29. CVRD held 9.9 per cent of CSN. The latter controlled 25.2 per cent of the holding company Valepar, which had a 52.2 per cent stake in CSN itself.
30. Schneider and Soskice (2009) describes such Brazilian model 'hierarchical capitalism'.
31. Since the 1960s, the BNDES is known as the 'steel bank'. Its finances to the steel industry increased from 9.7 per cent in 1989, to 23.2 per cent in 1991 and to more than 45 per cent in 2000–2001.
32. Fifty per cent of the residents earn less than two minimum wages and 10 per cent have no formal income.
33. For instance, the unskilled line operators and smelters of the GGAF live in the bairro.
34. Unlike them, Meillassoux (1972) and Dupré and Rey (1973) contextualise money within relations of production.
35. 'By June 1995, more than 105,000 workers in the public sector had purchased shares with *moedas socials* averaging 10 percent of total shares for each public firm set for auction' (Montero 1998: 48).
36. Abolafia (2004) describes economic policies as processes of cognitive and linguistic framing.
37. This is consistent with Singer's (2012: 80) argument that the marginal working class preferred Cardoso over Lula in the 1998 election.

– Chapter 6 –

Labour as Commons

I remember my grandparents used to tell me, 'A bird outside the group is an easy prey.' If you act as a group, you will win and not lose. With animals, we see this happening. I don't know if you ever noticed this? Do you know what is a *biquinho de laque*? It is a very small bird, like a sparrow but even smaller. It has a little yellow beak. One day we were having a meeting to put the electoral list together. We were in a room in the building of the Engineers' Union. We were waiting for people to arrive and I was looking at the back of the house. Someone was throwing food. The birds assembled around it. The *biquinho de laque* were coming in pairs to eat, but bigger birds would scare them away. The little birds did not get a chance to eat. They formed a big group and they returned as a group and they scared the big birds away. So, they were able to eat. In that moment, I called the others to show them what was happening. It was a lesson. If among animals there is such an instinct for co-operation, why shouldn't it exist among humans? I always give this advice: 'You need to unite comrades because this is the only way you can conquer rights and respect.'

—Builder, Sindicato dos Trabalhadores nas Indústrias da Construção Civil, Montagens Industriais e Construçao Pesada de Vota Redonda e Região (Union of the Civic Construction and Heavy Construction Workers of Volta Redonda and Region[1]) SINTRACONSMONPES

In 2005, outsourced maintenance workers led a strike in Volta Redonda that passed relatively unnoticed but whose implications turned out to be great for the direct workers and the management of CSN. They were striking against their employers and CSN, which they held jointly responsible for their decreasing wages[2] and benefits,[3] inhumane working conditions (such as lack of air conditioning and fresh water) and stigmatisation by direct workers. In that year, outsourced workers continued to undermine CSN through slowdowns, absenteeism and boycotts. Moreover, they led strikes in 2006 and 2007. In response, CSN back sourced several maintenance jobs, starting a trend of back sourcing that continues today. The trajectory of outsourced workers I describe in this chapter goes against standard narratives of class struggle and flexible capitalism. In fact, the resurgence of working-class activism in Volta Redonda came by the hands of outsourced workers, notoriously the weakest link of the labour movement. How did such a traditionally unskilled and politically fragmented section of the working class become the political vanguard of the labour movement, overshadowing, both at the local and national level, the leadership of traditional trade unions?[4] But also, how do we explain the recent trend of back sourcing not only in the South but also in Europe and the United States? How can we explain the recomposition of the Fordist factory in our era of lean production?[5]

To be sure, outsourced workers always came across as autonomous and independent during my fieldwork. On my first day in the UPV, I walked into the maintenance office of the CSN steel plant, interrupting the lunch break of a group of outsourced maintenance workers. It was at the height of the financial crisis. The company had just laid off the whole maintenance department. When I mentioned this to the subcontractors, one of them said:

> This comes as no surprise to us. They were just supervisors. They knew nothing of the company. The technical knowledge of the firm was lost when they privatised the company and introduced the SIGMA [an automated control system] and the engineers and technicians were fired. We are the only ones left who know how these machines work and how to fix them. I fit the refractory lining of the furnaces. This is more an art form than an industrial skill. Refractory material is very temperamental. It tends to break easily. There is no easy way of laying it inside the furnace. It is boiling down there. My skills? It all comes down to intuition and expert touch.

Subcontracting and outsourcing are spaces located in between the *market* and the *firm*. On the one hand, these forms further commodify and make labour abstract, impersonal and precarious. But they also shift the valorisation process away from the direct control of the capitalists. Here, I will describe how in this grey area located between firms and markets, and where labour is only partially commodified, workers have some agency against capital. Moreover, I discuss the role of the law, a force emanating from the state, in blurring the boundaries between wage labour, outsourcing and informal and co-operative labour, as well as the consequence this has for labour.

The privatisation process I have described divided the labour movement but did not fully deregulate labour because of the strong pro-labour nature of the Consolidation of Labour Laws (CLT). President Fernando Henrique Cardoso came into office famously determined to break the power of trade unions, deregulate the labour market and 'put an end to the Vargas era'. The legal professions and the trade union movement opposed Cardoso's proposed reforms of the legal labour system. When Lula da Silva was elected president, he passed Amendment 45, which expanded the scope of labour law from 'employment' to 'work' in order to cover informal labour, which had peaked during the Cardoso presidency. On the other hand, the PT's heavy regulatory labour framework restricts the scope of regional labour courts and trade union action, especially in relation to collective negotiation. I have already stressed that 94 per cent of the jobs created during the Lula and Rousseff administrations are low income (below 1.5 minimum wages)[6] and in non-industrial sectors, such as clerks, construction work, transport and general service (Pochmann 2012), and 60 per cent of these involve young people (between eighteen and twenty-four), mainly women and ethnic minorities. Moreover, low-income and casualised labour and informal and unremunerated work currently make up 38 per cent of the labour market and are exempt from the labour rights contained in the CLT. In this chapter, I discuss how subcontracting blurs the labour forms of wagework, informal work and entrepreneurship and in so doing, as well as precarising labour, introduce zones of uncertainty for capital.

Outsourcing in Brazil

First, it is important to distinguish between informalisation, subcontracting and outsourcing. Informalisation is the condition of incomplete proletarianisation associated with dependent industrialisation in the colonies, such as that described by Keith Hart (1973) in Ghana. I discussed the wave of informalisation and favelisation of Volta Redonda associated with the early development of CSN in chapter 3. Subcontracting refers to the practice of hiring an outside company or provider to perform specific parts of a business contract or project and is normally temporary. Subcontracting is a process of externalisation of labour that started at the end of the 1970s in Brazil and increased exponentially with the transition to democracy and the deregulation of labour and capital initiated by Cardoso. Typically, subcontracting concentrated in the building and cleaning industries. In that context, subcontracting reflected the subsumption of labour to foreign capitalists (under neocolonial regimes) rather than to foreign states as for informalisation. Outsourcing generally refers to processes that could be performed by a company's internal staff but are contracted to outside providers working independently. A typical example would be to contract an outside provider to manage internal technology. This model of externalisation of noncore economic activities is normally associated with short-term, cost-cutting, radical reorganisations – from vertical to horizontal structures – and labour deregulation. Outsourcing is a form of long-term precarisation of the formal workforce and is a more recent phenomenon that started with the expansion of the service and IT sectors in the late 1990s, the new frontiers of expansion of foreign capital.

The economists Michael Piore and Charles Sabel (1984) give a positive reading of outsourcing as the motor of the 'flexible specialisation' model, based on co-operative relationships, shared knowledge and functional interdependence among small firms. In practice, subcontracting and outsourcing often blur into each other, and the situation in Brazil is complicated even more by the huge scale of the informal economy, which overlaps with both although they are different economic processes. Overall, processes of labour outsourcing in Brazil led to the reduction of salaries and direct benefits, the worsening of labour conditions in terms of health and safety and labour deskilling. According to a study recently published by the CUT, 26.8 per cent of the Brazilian formal labour market is outsourced. Most outsourced jobs are concentrated in the lowest income band (up to two minimum wages), have wages 23 per cent lower than those of direct workers and involve longer working hours, on average three hours more per week. And in 2013, 90 per cent of slave labour in Brazil was outsourced, mainly in the building sector (CUT 2014).

There are different kinds of outsourcing entailing different degrees of vulnerability and precarity. Contracts between workers and subcontractors can be permanent or temporary. There are three kinds of temporary contract recognised in the CLT: the training (*experiência*) lasts ninety days and is not renewable, (2) the temporary contract (*determinado*) lasts two years and (3) the project (*obra certa*) is permanent. Outsourced workers in permanent employment relations experience a contradiction: while their contract with outsourced firm is permanent and stable, the contract between their employers and the main firm is temporary, and typically lasts one to four years. Moreover, the relationship between the main contractor and outsourced workers can be formal or informal. In Brazil, there are various forms of outsourcing: hiring autonomous workers for domestic work; hiring companies for supplying products, pieces or machinery; subcontracting auxiliary and support services to specialised companies; subcontracting services within the central production area to autonomous professionals; subcontracting to subcontractors (quarterisation – *quarterizacão*) or co-operatives. Moreover, Brazil has no legal or administrative distinction between subcontracting and outsourcing, so they blur into each other.[7]

Until 2017, there was no law on subcontracting in Brazil. The main regulation on subcontracted work was Article 331 of the Tribunal Superior do Trabalho (Superior Labour Court – TST) now incorporated into the CLT. Article 331 is a *sumula,* which means it is not legally binding but represents the best practices supported by labour judges. Article 331 establishes two principles. First, the contract of outsourcing takes place between two employers (the contractor and the subcontractor) and not between employer and employee (the contractor and the subcontracted workers). Unlike direct workers, outsourced workers are not legally subordinated. They have no physical personality and hence no labour rights (Dias 2010: 13). They are 'juridical persons' with only civic rights. For instance, in case of their employers' bankruptcy, they do not have the right to their outstanding wages and pensions. Thus, subcontracting establishes a state of exception within the labour law, which is active only within hierarchical relations.

Second, the article establishes that the externalisation of core (*end*) activities is illegal whereas it legalises the externalisation of marginal (*means*) activities. This distinction between core and marginal activities, inherited from the corporate world, is easy to circumvent and is contested even among labour judges. According to Sabrina Dias (2015: 7), it is precisely because of its fuzzy nature that Article 331 was used both to increase and to curb outsourcing. For instance, CSN uses contract workers for the core maintenance operations of waste extraction, cleaning and sealing the pig iron channels and the furnace, which were performed by internal workers before their reclassification as 'not core'. Article 331 had the merit of reducing the illegal and informal subcontracting of cleaners, maintenance workers and builders that spread in Brazil in the late 1980s. On the other hand, the legalisation of outsourcing led to an intensification of its use from the late 1990s onwards (Druck 1999).

The issue of subcontracted work was one of the triggers of the public moral outrage against Rousseff. The object of contention was the proposed Law 4330, which legalised the outsourcing of core activities. On the other hand, the proposed law would have given legal rights (including overtime pay) to marginal workers such as domestic workers and hence radically overhauled Brazil's established labour hierarchy. President Temer passed the 'outsourcing law' in 2017 (Law 13,429). Some argue it was the result of pressures made on the government by global agency firms and call centres to lower labour costs and expand outsourcing opportunities. Now the IT, health and government sectors can be radically reorganised and run entirely on a reserve pool of outsourced workers hired by global hiring agencies. The model of the 'ghost factory' adopted by some global carmakers in the region, where all core operations are outsourced, may become the norm. Outsourcing is a global phenomenon. It follows the flows of global finance and commodity chains and overflows national boundaries. Yet, the legal system in Brazil has a central role in shaping the institutional boundaries of firms and relationships between labour and capital. Indeed, Dias (2015: 23) suggests considering 'outsourcing' in Brazil a fluid and mutating reality, involving different degrees of precariousness and multiple histories of struggles.

Outsourcing in CSN

This chapter focuses on outsourced maintenance workers with permanent contracts.[8] These workers are based at CSN, formally registered (so, unlike informal workers, they are covered by the CLT) and perform skilled tasks such as mechanical and electric repairs, welding and refractory work, often working in parallel to direct workers. In this chapter, I show the company resorted to labour outsourcing during privatisation to purge the militant and skilled workforce but, because of this policy, became heavily dependent on external labour thus, losing bargaining power vis-à-vis subcontracted workers. Since the beginning, CSN resorted to various forms of legal and illegal outsourcing, especially of cleaners and builders. In the late 1980s, the company started extensively outsourcing maintenance workers into a separate Metal Structures Factory (FEM)[9] a subsidiary of the main company. The FEM consisted of the same workers previously employed by CSN who hence described the reorganisation as a 'simple change of shirts'. Being skilled mechanics and electricians, they continued to control the production line and in 1989 led the famous occupation of the factory (chapter 4).

The privatisation of CSN and the reorganisation of the company in 1992 cut the workforce by two-thirds. First went the many unskilled builders, carpenters and bricklayers who had been recently employed for Plan D. Then, more than half the maintenance workforce and engineers, the firm's most militant section, known as the 'strike force' (Dias 2015: 150) was outsourced and dispersed into several external firms. This mass outsourcing was made possible by Article 311, which was approved only a few months after CSN was privatised. Subcontracting of maintenance jobs took a pyramidal form. CSN subcontracted via FEM to a 'smaller FEM' – *feinha*.[10] The *feinha*, in turn, outsourced to smaller subcontractors – a structure called *quinteirizacão* ('quinterisation'). The FEM maintained only the functions of supervision and cost accounting. The few maintenance workers who stayed in the FEM were deskilled and turned into line managers in charge of supervising a vast and often underqualified external workforce. For instance, in the coke oven, there were more than two hundred outsourced maintenance workers. Initially, outsourced workers lost their labour rights and rights to pensions, and suffered a radical deterioration of their working conditions and status. In 2002, the FEM was closed, and all maintenance operations were outsourced to two multinational firms – the Italian Comau and the Japanese Sankyu. CSN kept only the functions of supervision, marketing, distribution and extraordinary maintenance led by a small team of maintenance workers. Outsourced workers gained permanent positions, but their wages and working conditions worsened further. In addition, CSN regularly employed short-term subcontractors for specific maintenance jobs.

Permanent workers would describe outsourced maintenance workers in derogatory and racist terms. But the jobs of these very permanent workers were precarious and deskilled because of the Maintenance Management System (SIGMA) that standardised maintenance and embedded supervisory control into the line. The SIGMA is a complex, layered automated maintenance system. 'Corrective' maintenance happens after faults occur and is relatively easy and costless to implement. But most faults tend to occur in opaque circumstances and are difficult to detect. Moreover, corrections can be 'curative' – when they restore the correct functioning of machines – or 'palliative' – when they delay the final breakdowns. Palliative maintenance is the most profitable for the company because it adds value to economic assets that are depreciated. For instance, the prolongation of the life of AF2 improved the financial position of the firm by several million reals. Preventive maintenance takes place before faults occur. Its value is much harder to quantify, especially when it refers to hidden faults. Unlike 'time-based' and 'condition-based' faults, which reflect, respectively, the behaviour and the physical properties of the machines, hidden faults are those that may already exist but cannot be detected through observation or measurement. Hidden faults can only be detected speculatively, through expert guessing or retroactively – that is, through evidence captured in the electronic logs of the workers' interventions and observations. Thus, there is a potentially endless amount of information on which retroactive interventions can be made.

This maintenance framework pushes the workers into a regime of continuous attention. Because hidden faults exist only in potentia – as imaginary states – workers must detect and correct them before they happen, and these speculative interventions are themselves retroactive and thus unsettle the existing field of human-machines interactions. This regime of continuous attention generated by the SIGMA embeds and interiorises control within the cognitive and affective apparatus of line workers. It does so by setting up speculative correlations between work operations (a productive target) and the depreciation of machines (a financial target) generated against the backdrop of real and imagined maintenance tasks. This regime of permanent attention embedded in the standardised maintenance system is extremely distressing for line operators. For instance, Nenê, a mill operator, claims:

> Unlike, say, the packers down the line, machine operators like me do not control their labour. Neither do we understand what goes on inside our machines. We are not engineers who can look inside them, like shrinks who can look inside people's minds! We can only speculate, guess and hope. This can be emotionally draining. Two years ago, 'the old guy' [the billet mill] broke down while I was in the control cabin. It was a traumatic experience. I felt guilty and anxious about the financial implications of my actions, about my own life and about the future of the company. For long time, I felt my whole body was paralysed, too. I lived in a state of suspension.

The sensuous connections that workers, especially in capital-intensive departments, draw between themselves and the mechanical system are an important dimension of their work identity. In capital-intensive departments such as the furnace, working knowledge is embodied, implicit and enacted as a reciprocal exchange with the technical system (Carrier 1992; Ingold 2000; Mollona 2005). In labour-intensive departments, knowledge is disembodied, abstract and externalised into the technical apparatus.

Workers experience the externalisation of knowledge into the technical system in terms of loss of control and impotence, especially when knowledge is externalised into obsolete machines as the weakness of machines is perceived as a symptom of their own physical weakness. For instance, the semi-automated machines of the tempering and cold mills are more than sixty years old and have long passed their life expectancy. They look as if they are about to collapse. Such vulnerability of machines generates a constant regime of attention, care and anxiety for Nenê:

> My job is to care for the old guy [he points to the mill], especially his crooked bones [he points to the transmission bars]. Caring for machines has a strong emotional impact on me . . . also because it has great financial implications for the company. For each of my maintenance operations increases the economic value of the company, each fault that I let slide, destroys it.

The lives of the workers and those of machines are experienced as monetised assets – as abstract and volatile as finance. Mechanical breakdowns, financial meltdowns and personal crises are constantly around the corner. Trying to make sense of the value of their work, employees are eaten by doubt and self-monitoring. Marx's notion of fetishism, the process whereby money and machines are humanised and overvalued and workers are devalued and objectivised, is stretched even further by the SIGMA, which values human labour only as long as it prolongs the lives of machines. Moreover, amidst the broader financialisation of the economy, where short-term financial returns are more important than long-term industrial profits, the SIGMA actualises the financial logic onto the shop floor in a Veblenian fashion – that is, by establishing a regime where preventing depreciation (increasing the value of assets) is more important than boosting productivity.

On top of the unpredictability of finance (and of the machines) comes the unpredictability of the state. In a bizarre move, the state privatised the company right after it completed the mega expansion Plan D, which brought a new furnace and oxygen shop and two laminating mills onto the shop floor, boosting the production capacity of the plant from five million tonnes to nearly five million tonnes. Workers remember Plan D as the company's 'golden era', which attracted skilled workers from all over Brazil and turned Volta Redonda into a world-leading steelmaking centre. But they also remember the same period as the dark era of privatisation. Many 'second-generation' workers[11] hired during the firm's expansion were traumatised in witnessing mass redundancies of colleagues, friends and relatives. A big portion of the labour force made redundant – especially black and women workers – ended up in informal and domestic work. Indeed, for many workers, it was difficult to understand what was really going on during privatisation. The general feeling was that the firm was making new investments and expanding production capacity rather than cutting costs. Workers greeted the arrival of Japanese and German machines on their shop floor with pride and excitement. José, a retired engineer, describes to me that moment:

> Entire rolling mills and ovens travelled in pieces from Japan, Germany, China or the United States to Volta Redonda. Some of them were already old, rusted and with missing parts. Our typically Brazilian 'inventiveness' [*jetinho*] made them work: we handmade missing parts; bought them in the black market and even smuggled them from Europe! We were proud of these foreign machines; we felt that, despite all their faults, they were taking us into a new stage of industrial development. None of us [maintenance workers] saw it [the privatisation] coming. Nobody guessed those machines would make us redundant.

The SIGMA had a catastrophic impact, especially on the Smelting Shop (GGAF), where a third of the workforce was cut, the position of the general director eliminated and the three smelting shops made into independent centres or profits, generating conflicts between their directors who acted – in the words of one smelter – as 'authoritarian rulers of independent kingdoms feuding against each other'. During my fieldwork, half of the GGAF's seven hundred employees were subcontractors. In the GGAF, subcontractors had the harshest jobs, such as putting the moulds on the furnace door or breaking the slabs.

But the operations of low-level permanent workers were also extremely wearying: checking the heat inside the furnace, managing gas outflows, cleaning the halogen tanks and directing the flow of pig iron into the torpedo car. Indeed, the working conditions of permanent workers were often worse than those of the subcontractors. For instance, during the economic crisis of 2009, all the workers of AF2 were immediately laid off for three months, and only some were reemployed when the crisis was over. For a long time, confusion reigned. Many permanent employees end up as contractors, and contractors were hired permanently. But at the time of the privatisation, it must have been impossible to foresee that the modernising forces of foreign technology and Japanese management – total quality management, quality control circles and the 5S model – would bring disruption and loss. In fact, workers witnessed a dramatic turnaround of the company that, from a net loss of $749 million in 1990, went to a net profit of $110 million in 1995. Some workers bought into the management's narrative that these improvements were the result of the 'productive investments' in foreign technology, rather than of the labour cuts that followed them.

Indeed, workers' recollections of this period are fuzzy. For instance, Bobo, a leader, got his job at CSN three days before the general strike of 1988. When the strike paralysed the factory, his contract was frozen. He was officially registered on the *carteira de trabalho* (working card) only one year into his job. A few years later, during privatisation, he saw the workforce being cut from twenty-two thousand to fourteen thousand. It was another life-changing event:

> Under public management, CSN was running out of money. It kept delaying payments to suppliers and employees . . . It was even delaying purchases of raw material. It was like a zombie, running out of blood and about to collapse . . . Privatisation came as a shock. Volta Redonda entered in a state of collective hysteria. Initially, I was worried, too, but then the SIGMA totally changed me. First, it made me appreciate the importance of money in the shop floor – monetary costs of mechanical breakdowns and monetary gains of preventive operations and correct set ups. Then, I had an epiphany: privatisation had made employees more vulnerable as workers but more powerful as investors. I bought company shares and invested in the pension scheme. Now I think about myself in terms of ownership – ownership over my life – rather than in terms of my wage. The state and the company do not own me anymore, I am the sole responsible for my pensions, housing and education. If this is capitalism, then I am a capitalist.

Second-generation permanent workers like Bobo accepted privatisation from a position of weakness and uncertainty. They went through an historical transformation, comparable to the transitions from socialism to capitalism, which they did not fully understand. Today, they are 'stuck' in between classes and generations. They are not as skilled or politically connected as the older generation, nor they are as formally educated and contractual as the younger one. They neither oppose capitalism, like the former, nor endorse it like the younger 'wage hunters'.[12] During my fieldwork, I was always struck by their docility vis-à-vis the management and their lack of identification with their job, reflected in how they talked about themselves as 'passers-by', a 'fleeting workforce' or 'migratory birds' (*arigós*), even if they had permanent contracts.

Unlike them, those engineers who had been displaced under the SIGMA system strongly opposed the company. For instance, Vanderlei, an engineer, was strongly critical of the management. The son of a lorry driver, he enrolled at the Industrial Gymnasium at the age of ten and, with a CSN scholarship, completed an MA and a PhD in engineering. Even if he supported the privatisation of CSN, Vanderley is very critical of the current financial regime:

> Under public regime, the interests of the company were subordinated to those of the nation. CSN was forced to undersell for inflation-control reasons and to subsidise secondary industries, mainly the automotive sector. Privatisation was good because it introduced a real market for steel. But the current owner subordinates the interest of the company to his private interests. He does not invest or try to maximise profits. He mainly focuses on increasing the financial and commercial returns of his group. The current PT government, despite being left-wing, plays this speculative game, too. With one hand, it keeps interests rate high to please the financial capitalists. With the other hand, it pays dear for redistributive programmes such as the Bolsa Família. I support the old developmental vision that the state should foster industrial development especially in R&D technology. What Brazil needs is to close its technological deficit[13] vis-à-vis Western steelmakers.

In contrast to permanent second generation workers like Bobo, the tertiarised maintenance workers of CSN have a strong occupational identity and are politically militant. From 2005, for three consecutive years, they confronted CSN with strikes, slowdowns, sabotages and absenteeism. These struggles by tertiarised workers were not easy. Many of them were dismissed and blacklisted. José, one of them, describes how these struggles evolved:

> Back then, we were considered the untouchables in the plant. By law, we had to be confined in 'subcontractor only' spaces, and these were cordoned off as in a crime scene. We had no canteen or locker room. We ate and got undressed by the machines. CSN employees would pass by, stare at us and call us 'faggots' or 'pussies'. One of us was beaten so severely that he spent one month in hospital recovering. They said we were stealing their jobs. They were scared. After the first strike, in 2005, me and several of my co-workers were dismissed and blacklisted. For a whole year, we remained unemployed. Once you have CSN against you, you are done for! Then, we started to get together and share our experiences. Initially, these were very informal meetings, at the local *churrascarias* [steakhouses]. We realised that while most of us were struggling to be on the employment ladder, others were making real fortunes as independent contractors. Some of these were our own employers! They were basically acting as 'pimps' between us and CSN. We said: 'As long as we stick together, we don't need them. We can negotiate directly with CSN.' In fact, we are the ones who know how to care for and look after the machines . . . If we all withdraw our labour, then the firm goes bust. We started a process of collective protest. As the blacklisting grew, we noticed a change in the attitudes of the employers. They softened up. They had realised they needed us . . . they needed to keep us in the firm.

Some direct workers realised the struggles of tertiarised workers were important for their job security, too. In 2007, direct workers and tertiarised workers set up the Campanha Salarial Unificada (Unified Salary Campaign). The former agreed to support the outsourced workers' demands for higher wages and better working conditions and to join them in a series of slowdowns and production line halts. As a result of their militancy, the contract workers' wages grew more rapidly than those of direct workers. In fact, the salary level established in collective agreements for outsourced workers became the benchmark for qualified direct workers, too, which meant they were on the same wage level of skilled engineers and technicians.[14]

Having won the right to have their collective negotiations jointly with those of direct workers, their company-related health[15] and pensions plans, participation to profits, monthly food provisions and salary levels are in line with those of direct workers. Their health and safety standards and working environment improved radically, too. Maintenance workers are also in high demand by the global carmakers and steel companies who have just moved to the region. Former Metalworkers' Union President Renato Soares said as the working conditions and wages of direct workers worsened and those of outsourced workers improved, the two categories ended up roughly in the same position (Dias 2015: 39). Such coexistence of precarised wageworkers and socially mobile outsourced workers within CSN shop floor, mirrors the 'schizophrenic' combination of wage precarisation and formalisation of precarious labour led under the Lula administration (Braga 2016).

As for the Chinese 'sunset workers' described by Ching Kwan Lee (2007), the 'insurgent identity' of maintenance workers consists in their citizens' right to legal justice based on a recent change in labour law, which was achieved through the activism of left-wing judges and lawyers. Going against the logic of Article 331, the new law establishes the principle of 'subsidiary liability' (*responsabilidade subsidiária*), which makes the main contractors jointly responsible with their subcontractors for accidents and underpayment of wages and benefits to their employees. But political and legal activism aside, outsourced maintenance workers won these important concessions because of their shop-floor-related bargaining power, as they are the only ones left in CSN who can run the company.

When the SIGMA was introduced, engineers and skilled maintenance workers were laid off, so the internal memory and knowledge of the technical system was entirely wiped out. The few internal engineers and maintenance workers left were turned into cost accountants and supervisors. But in order for the plant to run smoothly, the management needs workers who know about refractory material, the chemistry and alchemy of steel and how to set up inductions ovens, cooling and heating machines and mills. Moreover, dismissed maintenance workers know how to navigate through different international technological standards and to operate the foreign technology CSN continues to depend on. In the past, the centralised engineering maintenance department was in charge of mending, repairing and adapting the Japanese, US or German machinery to the existing machines, often copying them and remaking them anew.

At the time of my fieldwork, the company depended on those outsourced maintenance workers, often former CSN workers who, having been involved in Plan D, had such knowledge. Hence, it continued to renew contracts with the same maintenance firms in the hope of winning the trust of its skilled ex-employees. Everyday groups of subcontracted builders, mechanics and electricians burst onto the shop floor, commanding both respect and resentment from the direct workforce. Some small subcontractors had worked in CSN for decades and were considered quasi-employees. Yet, they showed no attachment to it. In fact, they considered themselves 'outsiders', and because they had skills highly needed by CSN, they indulged in absenteeism, absences and even sabotage against the company. Some skilled former CSN maintenance workers even set up their own businesses with higher remunerations and better working conditions that the internal workers.

Claudio, the director of a small maintenance firm, was one of them. Claudio even acquired a mythical status. He was said to be able to guess the heating speed above which the water system clogs, generating a thermal shock that can crack the refractory bricks of the annealing ovens, or tell the critical temperature of the clay pots in which zinc and aluminium fuse and beyond which molten zinc leaks into the rolls of the galvanising oven and clogs them. Not even the advanced thermo-graphic texts of the laboratory were able to detect these micro-events before it was too late. Whenever Claudio put his hands on some machine, the productivity of the company peaked. Claudio was proud of his economic independence:

> My salary is three times higher than the wages I got when I was a worker here – I am the boss now, you see. I get salaries, not wages. But I do exactly the same operations. And on the top of it, I am treated with much more respect. I have no attachment to the company. Since they fired me my relationship with it is merely contractual. It is they who need me now.

To retain subcontractors like Claudio, CSN had to concede greater profit margins or roll back on its operation targets. Moreover, both internal subcontractors and independent firms were expensive to monitor. Direct workers cost about 150 to 200 per cent more than outsourced workers but, as I have shown, come with fewer political risks. Because of their political activism, slowdowns, stoppages and absenteeism – and their shop-floor bargaining power vis-à-vis the company – outsourced workers end up being more expensive than direct workers. Outsourced workers use their bargaining power to increase their salaries and freeriding against CSN. So, it is not surprising that in 2007 CSN started a process of back sourcing of maintenance workers, as well of security and transportation workers, in the newly created general maintenance unit.[16] In 2010, CSN back sourced two-thirds of the maintenance workers previously employed by its main subcontractor, Comau. Maintenance workers 'changed their shirts back again'.

The maintenance workers' actions of caring for and looking after the machines are reproductive and long-term, rather than productive or cost-saving in the short term. They cannot be performed by standardised managerial systems. In fact, the expulsion of maintenance workers and skilled engineers due to the SIGMA shows the narrow-mindedness of capitalism, which systematically devalues and undermines the most valuable labour – the reproductive one – and over-valorises instead short-term productive actions. But the drain of skilled workers and the drop in profits associated with the SIGMA show that such narrow-mindedness does backfire. Outsourced maintenance workers were successful in their labour activism because they partially exited the firm and negotiated with it as external and collective subjects, having gone through a process of commoning of personal experiences and knowledge and of mutual support in times of adversity.

Other Forms of Tertiarisation

In the 1990s, the global carmakers PSA Peugeot Citröen and Volkswagen moved to the Sul Fluminense region, lured by the tax exemptions, free land and cheap labour offered by cash-strapped municipalities and supported by regional alliances (chapter 3). Initially, these regional alliances emerged from grassroots movements[17] and were modelled on the cross-sectional coalitions with employers and civic movements led by the Metalworkers' Union[18] in the industrial ABC Region of São Paulo in the 1980s. But in the Fluminense region, which displaced São Paulo as a new regional industrialisation pole, these regional blocks slowly became mere business alliances between municipalities, new labour unions and local entrepreneurs aimed at attracting foreign investors with cheap labour, fiscal and environmental incentives and infrastructures. Cash-strapped municipalities waged ruthless 'fiscal wars' (Arbix 2000) against each other. For instance, Volkswagen opened its first global modular factory in Resende in 1996 with generous concessions and contributions by the state of Rio de Janeiro,[19] and PSA Peugeot Citröen opened its first Brazilian factory in Porto Real in 2001 also with the support of the municipality,[20] as well as the state of Rio de Janeiro ,which owns 32 per cent of the factory's stocks.

These global carmakers leading Brazil's 'third wave' of industrialisation operate with a model of extreme outsourcing. The model replicates the form of 'industrial condominiums' – in which suppliers are clustered around the main assembly plant – typical of the reorganisation of global carmakers in the Global South (Abreu et al. 2000: 266; Salerno 1997). But the Volkswagen 'dream factory' in Resende is an extreme version of this model, as it is based on totally outsourced work (Ramalho and Santana 2001b, 2004). This 'factory of the future' pushes the Japanese lean model to the extreme. In the dream factory, unlike the traditional Japanese model, where core activities are internalised, all operations – both direct, such as production and assembly, and indirect, such as cleaning, transport, food, health service, data processing and logistics – are externalised to seven independent companies operating in situ. Unlike previous post-Fordist models, where component suppliers sent their components to the main factory, here subcontractors take part in the main production process and Volkswagen – the main corporation assembler – remains outside it. The only core functions retained by Volkswagen are brand development and quality control. Essentially, the firm is like a merchant capitalist operating through a spatially concentrated putting-out network. The dream factory is both a market and a firm. Its extreme marketisation is disguised by the fact that subcontractors are under the same roof and share the same HR management and Volkswagen uniform – although with a fine-printed logo of their company on the front. This total subcontracting was against Article 331.

In 1996, when the factory opened, 1,300 out 1,500 workers were subcontracted. Until 2004, the modular factory had two thousand employees, half of whom were tertiarised – three hundred from Volkswagen and seven hundred from the subcontractors. The situation complicated further in 2008, when the German MAN AG bought the Volkswagen factory and logistical operations were externalised to a separate firm located to the rear of the firm, adding a third level of subcontracting. In 2011, the plant had six thousand workers, three thousand from the modular factory – only nine hundred of whom belonged to MAN AG – and three thousand from outsourced suppliers. Thus, 80 per cent of labour was subcontracted. Most of these suppliers are located in the region of São Paulo, which means the 'dream factory' is both spatially concentrated and deterritorialised. Following a wave of mass redundancies in 2014, the plant now operates with 4,500 workers on just two shifts – from the previous four-shift regime.

Under the new MAN AG regime, the company transformed itself from a production centre to a logistical centre coordinating just-in-time hundreds of suppliers scattered across different regions. After in initial period in which each subcontractor held separated collective negotiations with the Metalworkers' Union, the employer agreed in 2009 all subcontractors should be in the same collective negotiations between the MAN AG and the Metalworkers' Union. Yet, tertiarised ancillary workers (e.g. cleaners and transport workers) who represent the majority of the workforce and externalised logistical workers continue not to be represented in Factory Councils or in collective negotiations. The management argues that such a model of nearly total outsourcing is legal because the company's sole business is in brand management and quality control, the only two internalised departments, and all the other activities, both direct and indirect, can be legally outsourced. The outsourced workers of the modular factory are different from the outsourced CSN workers.

First, they are internally fragmented between ancillary workers and line workers. Ancillary workers have low educational backgrounds, little experience of formal employment and of political activism and an instrumental attitude to work (Pereira 2007). Despite representing more than half the factory's total workforce, they are not represented in Factory Councils or in collective negotiations between the Volkswagen and the Metalworkers' Union and have marginal status vis-à-vis the more affluent[21] car workers. Unlike them, the direct workers of MAN AG[22] constitute a 'new working class' radically different from the *metalúrgicos* (Ramalho and Santana 2006; Santana 2006). They are young, well educated, relatively well off, non-unionised and loyal to their employers and 'brands' – with whom they develop trusting and long-term relationships. Their rate of unionisation is low.[23] But, this could be the consequence of the SMSF's a failure to represent them rather than of their political apathy, as for Bartolomeu Citelli, the SMSF director in charge of negotiations in Resende and Porto Real.

Despite being less explicitly political than the outsourced workers employed by CSN, the autoworkers of Resende and Porto Real kept their real wages intact and avoided lay-offs in a context of declining prices and profits in the industry (Ramalho et al. 2009) by entering tripartite negotiations with municipalities and management[24] and reviving the early experiments of the Workers' Party (PT) in São Paulo. Moreover, according to José Ramalho (2005), autoworkers rescaled their action from the factory to the region and created cross-sectional solidarities with municipalities, other social constituencies and workers in other industries, setting in motion a new process of contestation. Unlike them, steelworkers are focused on factory-based struggles concerning wages, participation to the company's profits and working conditions and are sceptical about the cross-sectional regionalism of the car workers. This conflict between the factory-based activism of steelworkers and the regional activism of car workers must be considered in relation to the ongoing tensions between the municipality and CSN.

Between Outsourcing and Informalisation

The process of privatisation of national companies initiated in the 1990s led to a peak in informal employment at both national and local levels.[25] Under the PT, formal employment rose exponentially, mainly in low-income jobs and for women, black and rural workers[26] traditionally outside the formal working class. These newly created low-income jobs bordered with the informal economy, which still made up 40 per cent of the GDP in 2011. In fact, most jobs created by the PT administration in the past ten years have pertained to this grey area between tertiarisation and informalisation. Brazil's formal employment system is based on the *carteira de trabalho*. Workers with a *carteira* are entitled to pensions, social contributions and labour rights, while those without have no welfare or labour entitlements. For employers, one obvious way to informalise wage labour is to not register the workers on the *carteira* or to register them after they have started working for the company. It is not unusual for companies to register their employees many years after their start date or just before they go bust.

Workers lose the rights to social contributions and additional salaries for the period in which they are unregistered. They can have their salaries and contribution back only if they can demonstrate in the labour court that they have been working for the company while unregistered, which is both difficult and dangerous, as workers in contentious with the company are by default fired and blacklisted. Moreover, to dodge taxes, companies keep nominal wages lower than real ones, pressurising workers to declare minimum wage in exchange for top-ups in cash. Employees do not know that by doing so, they lose social contributions. It is not unusual for them to realise it only at the time of retirement. A third way to informalise the wage relation is to underpay overtime and hazardous work. The law establishes three levels of hazardousness with related pay increases: minimum (10 per cent), medium (20 per cent) and maximum (40 per cent). Companies get around that. For instance, an electrician working in a high-voltage area for two hours per day is paid the extra 30 per cent only in proportion to those hours, instead of the legally due 30 per cent daily salary increase. Companies also frequently outsource work to ghost companies or fake co-operatives, which they then close down without notice, leaving the workers without wages and social contributions.

At the lower end of the labour market cleaners, domestic workers, builders and garbage pickers work without a *carteira,* sixteen hours per day and on a daily salary of R$5. In 2011, domestic workers made up an astonishing 27 per cent of the national population – nearly seven million people, or a fifth of the active female population (Pochmann 2012). Eighty per cent of them were undocumented and hence had no rights to social contributions. A big portion of them work in a regime of indentured labour also because of the strong racial component of domestic labour.[27] The presence of maids, cleaners, nannies, chauffeurs, personal trainers and security guards in the hyper-affluent families of Rio de Janeiro or São Paulo or in the large landowning families of the rural North is not as surprising as the sight of domestic servants in the households of the steelworkers of Volta Redonda.

Most domestic workers are employed as day labourers (*diaristas*) paid in cash at the end of the day. Two years ago, the Sindicato das Empregadas Domésticas do Sul Fluminense (Union of Domestic Workers of Sul Fluminense) won an unprecedented victory. As part of the broader debate over Law 4331, the TST passed a constitutional amendment extending workers' rights to domestic workers on daily contracts. But very few domestic workers applied to legalise their jobs, fearing they would be laid off by their employers. The informal jobs of builders and garbage pickers are even more precarious. Some garbage pickers set up co-operatives or small businesses on the base of formal contracts – and hence registered in the *carteira* – with the municipality, housing associations or condominiums. But most of them barely survive. Ronaldo, a garbage picker, describes to me his normal day:

> I leave the bairro with my cart at four in the morning. In the morning, [the cart] weighs only twenty kilos. The walk to town takes me two hours. On the way, I pick up things: paper, cans, plastic, clothes, food – whatever is thrown in the street. I put plastic, food and cans in separate baskets hanging underneath the cart, and paper and clothes on the top of the cart. My cart is like in an office. Everything has its own place, and I need to keep it tidy. When I hit the town at 6 a.m., the dustbins are overflowing with garbage. That's when I have my first meal of the day. There is always some discarded food in the rubbish. We have divided up the city. I cover three bairros. But even so, it is often the case that some lunatics come to me with a knife or gun and tells me that that's their dustbin. Halfway through my day, the cart is almost entirely filled up. I climb on the piles of paper and cloth and tie them with ropes. Now the cart weights fifty to sixty kilos, and I walk very slowly. Along the way, I sell clothes and paper to merchants. At the end of the day, I go to the co-operatives and sell all that is left. I earn twenty-three cents per kilo of paper. On an average day, I make R$5. On a really good day, I earn R$10 for fifty kilos of paper. Returning home takes me a little longer because it is pitch black. The road is dangerous. Many friends of mine were killed on their way home.

Figure 6.1. Garbage picker (© Massimiliano Mollona).

Several garbage co-operatives – in fact small firms dressed up like co-operatives – sell to the municipality and recycling companies the garbage they buy at much lower prices from independent garbage pickers.

The recent construction industry boom, sparked by federal housing and infrastructural programmes (e.g. the PAC and MCMV) had boosted even more the informal work in the building sector, where 40 per cent of the workforce is informal. The biggest employer of informal workers is the municipality, which regularly tenders to illegal subcontractors working with unregulated and indentured labour. By municipal building sites at night, it was not uncommon to stumble into groups of informal workers sleeping rough surrounded by baskets half-filled with rubble and by rows of donkeys still steaming and panting. Indeed, after CSN, the municipality is the biggest labour offender in the municipal labour tribunal. Because of the legal principle of 'subsidiary responsibility', the municipality is often summoned in labour courts on the count of exploiting cleaners, builders and garbage workers. Most of these trials lead to compensation, but rarely do they entail legal charges against the municipality.

In the 1980s and 1990s, the section of the working class made up of unemployed, informal workers, rural squatters and community leaders led the Right to the City struggle against CSN. During my fieldwork, these marginal constituencies were instead captured by evangelical organisations or neighbourhood associations through programmes of grassroots evangelisation, poverty reduction (Bolsa Família), popular homeownership (MCMV) or cultural entrepreneurship (Pontos de Cultura). But builders and civic construction workers continue to be a very militant section of the Volta Redonda working class. They were the central force behind the demonstration of 20 June 2013, when, following the national wave of demonstrations, about thirty-nine thousand people marched through Volta Redonda to protest the Rousseff government.[28]

The main branch of the Builders' Union used to be in a rural villa in the bairro Retiro, surrounded by lush trees and street markets.[29] The offices open onto a central courtyard where barbecues and folk music constantly go on. Despite a permanent long queue that starts from the street, coils around the central courtyard and climbs up the stairs leading to the legal and health offices, the atmosphere is always friendly and relaxed. Union clerks are busy running adult classes, training shop stewards, printing political pamphlets and cooking – activities that bigger unions like the SMSF externalise to private service providers. According to Dejair Martins de Oliveira, the president of the Builders' Union at the time:

> The Metalworkers' Union is very different from us. It is very bureaucratic and economically minded. First, it protects the interests of the firm, and then the interests of the workers. We are different. I first take the side of the workers. Besides, the Metalworkers' Union considers itself as an extension of the Brazilian Communist Party (PCB). For me, the union must make use of the party in defence of the workers' rights but remain autonomous from it. The PCB is very hierarchical and centralising. It has great influence over the management of the union.

The civil construction union is an important historical force of labour. In 1979, it led the first strike in the history of Volta Redonda called 'the strike of the unskilled'. In 1988, Dejair became the president of the CUT in the Fluminense region. The Builders' Union is affiliated with the left-wing current of the PT and the CUT but does not support the PT's national alliance with the conservative Brazilian Democratic Movement. The union has a small board of directors and a very large base, representing twenty-three categories[30] in seven cities of the region[31] with approximately eight thousand affiliated. As discussed earlier, public PAC programmes and infrastructural development funds benefitted mainly construction companies close to the PT, often global conglomerates such as Odebrecht, operating through a dispersed subcontracting network of illegal and highly exploitative firms. The main political focus of the Builders' Union is to look after the conditions of the precarious builders employed by the illegal subcontractors working for the municipality by making frequent inspections in factories and building sites and working closely with the Ministry of Work. In the past ten years, as federal PACs have financed new municipal buildings, tourist facilities and council flats, the labour conditions of the builders have dramatically worsened.

Builders and steelworkers used to support each other. In 1985, the Builders' Union and the Metalworkers' Union led a historical joint strike against CSN, which paralysed the city for days. CSN never allowed the Builders' Union inside the plant, so the Metalworkers' Union led informal negotiations with the management on behalf of the builders and gathered information for them. But current trade union regulations forbid steelworkers from supporting actions taken by other categories, and the Builders' Union has shifted the focus of its activities away from CSN. According to Dejair, builders and steelworkers now have very different work identities:

Figure 6.2. Dejair (© Massimiliano Mollona).

> Steelworkers consider themselves the elite of the industrial workforce. Unlike them, builders identify themselves with workers at the bottom of the social spectrum – cleaners, nannies and other domestic workers. In Brazil, we have a culture of slavery. Here, everybody has nannies or domestic workers who do not receive wages. They are unpaid or paid with food or soap. Builders are treated the same. We have more solidarity with these categories of exploited workers than with the steelworkers.

Co-operativism

Because even the militant unions are often too bureaucratic or institutionalised to look after precarious and informal labourers, these have organised themselves in the form of co-operativism. Brazil has a long history of co-operativism. Introduced by Jesuits fathers at the beginning of the twentieth century, the Brazilian co-operative movement emerged as a predominantly rural phenomenon. Urban co-operatives, mainly housing ones, developed after the economic crisis of 1980, when President Sarney radically cut public housing provisions and implemented the 'waterfall income policy',[32] which led to mass mortgage defaults and repossessions. Unlike associations, which are non-profit, co-operatives generate profits to be shared among their members,[33] which makes them akin to microenterprises. The 1988 constitution blurred the lines between co-operatives, microenterprises and NGOs further, by allowing co-operatives to make up their own statute as long as they met basic requisites.[34] Besides, co-operatives are exempt from taxes and the compulsory labour contributions established by the CLT, which makes them appealing to exploitative petty capitalists and tax dodgers. Nonetheless, with the decline of formal employment that followed the privatisations in the 1990s, the co-operative movement turned away from entrepreneurship and became a powerful force of civic activism.

'Popular co-operativism' was especially focused on actions of poverty and violence reduction in poor neighbourhoods. Lula made popular co-operativism a centrepiece of his first administration, founding the Secretaria Nacional de Economia Solidária (National Secretariat for Solidarity Economy), a national body consisting of civic, entrepreneurial and public organisations, in charge of developing solidarity economy and co-operativism. In chapter 3, we saw how housing co-operatives were pivotal in the anti-dictatorship movement and facilitated the alliance between the labour movement and civic organisations against CSN. In Volta Redonda, the most powerful co-operative was the housing co-operative of Volta Grande, which in the late 1990s had eleven thousand members and famously forced CSN to give a monthly housing subsidy of R$55 to each of its residents.

But during the second PT administration, housing co-operatives were incorporated inside neighbourhood associations and subject to bureaucratic requirements – for instance, the obligation to produce balance sheets and formalise their decision-making process. In Volta Redonda, the incorporation of housing co-operatives into the municipal system of neighbourhood associations alienated grassroots constituencies and put them under the control of politically ambitious leaders and the conservative Organisation of Brazilian Co-operatives. During my fieldwork, there was a consensus among my friends that housing co-operatives had lost their anti-establishment role and had become tools for furthering the political careers of ambitious neighbourhood leaders and local politicians.

But for Isaque Fonseca, a charismatic labour leader in the 1980s and currently an active player in the solidarity movement, co-operatives continue to be important means of self-organisation of those marginal workers – cleaners, recycling and domestic workers – who have no formal representation in the CLT and no support from the traditional labour movement. Isaque was born in Brazil's interior, one of twelve children of poor and illiterate *índios*. When he was thirteen, he was recruited by the special repression units of the military police. He went through two years of extremely violent training in urban guerrilla, at the end of which he quit the police force and went to Volta Redonda. While working at the GGMA in CSN, he read Marx and became politicised, joining a clandestine radical communist group. With Vanderlei Barcelos and other leaders of the anti-dictatorship movement, he founded the Associação Nacional dos Anistiados Políticos (National Association of Political Amnesties) and campaigned to free jailed comrades. He was also a founder of the national CUT and the PT in Volta Redonda and a leader in the 1988 strike. Isaque recalls the strike with an anecdote:

> I became a leader of the movement because I knew a lot about urban guerrilla. The secret police had trained me to think like a state. During the strike, we looked at the soldiers straight in the eyes, and we saw they were afraid of us. They were like Tarzan losing the liana! When an army tank invaded the furnace department, someone quickly climbed onto the overhead crane and lifted the tank with the big magnet normally used to transport scrap metal. Everybody laughed. The soldiers looked terrified and silly. We all screamed: 'Who is going to save the army? The workers!'

From 1988 to 1990, Isaque was the director of the SMSF under the presidency of Barcelos, a close friend. But he grew disillusioned with the trade union movement:

> In CSN, workers talked about the revolution of the proletariat, but they did not lift a finger. They had a physiological understanding of class but no class consciousness. The majority kept their head low. Moreover, steelworkers are very corporative, they think: 'A steelworker is a steelworker, a miner is a miner and a public worker is a public worker.' Not even a bus driver is a manual worker for them. Together with oil workers, they are the aristocracy of labour here in Brazil. Trade unionism here in Brazil is corporatist. It was not made by the workers but given to the workers by grace of the state. The base of the CLT is Mussolini's *Carta del Lavoro*: it has a fascist origin. Trade union dues are made compulsory by the state. But co-operatives do not have compulsory dues because they are constituted by people, not by the state. A worker in a steel factory is in a capitalist system. A worker in a co-operative is not.

Isaque quit the Metalworkers' Union and travelled to rural Cuba, meeting various co-operative leaders. The Cuban experience, where the state is strongly embedded in the co-operative movement, led him to found together with Sandra Vega, his lifelong partner, the first Fórum Brasileiro de Economia Solidária (Brazilian Forum for Solidarity Economy) in Rio de Janeiro, a regional umbrella of co-operatives, public-private partnerships, research hubs and Right to the City organisations. But in the end, the forum also alienated Isaque for whom these regional and federal platforms and public-private partnerships were forms of state control over the co-operative movement.

Evidence of how the state incorporated co-operativism in its neoliberal logic came to me during a visit to the Department of Solidarity Economy of the National Bank for Economic and Social Development (BNDES). Run by Carlos, a young and talented economist with a PhD in economic sociology and a background in the aerospace industry, the department is an unlikely blend of state developmentalism, micro-entrepreneurship and corporate management. The bank runs three lines of co-operative credit: non-repayable credits to set up worker-run companies aimed at low-income communities; credits to individual entrepreneurs normally in the informal sector; and micro-credit. Carlos makes it clear the programme's goal is to develop entrepreneurship and has nothing to do with 'old-fashioned co-operative principles'. Moreover, according to Carlos, the role of the BNDES is that of co-manager and not merely of redistributor:

> We only help those co-operatives that meet some basic rules: that the persons applying own the assets they are applying for, that they have environmental licenses and that they meet the juridical minimum for small enterprises [i.e. have a director and at least four dependents]. The BNDES looks at the different steps in the value chain rather than at individual co-operatives. The bank does not just give money. We finance them not directly but indirectly. We deal with local financial agents who give credit to entrepreneurs. In general, this is the way the BNDES tends to operate. Fifty per cent of our financial profits are indirect, and only 1 per cent comes from direct finance. We have moved from an approach based on income to an approach based on collaboration. Our idea of poverty is based not on income but on the notion of 'entitlement' coined by the Indian economist Amartya Sen. It is not about money but about the ability to pay. Technical help is central. Our experts are sent over to help. The BNDES co-manages production, logistics and commercialisation. The phase of commercialisation is the most difficult because of social pressures. Sometimes the market is not there. Sometimes there are dominant actors who try to block the development of the market.

The BNDES follows a territorial approach, developing partnerships with local states, municipalities and companies. For instance, in Porto Real, the bank is in partnership with the steel company Votorantim, the second biggest steel company in the region, on sustaining and developing a big co-operative network. In addition to finance, the

BNDES and Votorantim offer to the co-operatives of the network various business services including managerial training. According to Carlos, these co-operative programmes have been very successful in creating a transparent market system and eradicating the informal economy.

According to Isaque, on the other hand, by making mandatory the principle of private property, generating a divide between technical experts and non-skilled co-operative members and formalising decision-making against long-established consensus-based practices, these programmes run by the BNDES transformed co-operatives into small and medium-sized enterprises. Having helped coal burners, cleaners, recyclers and rural artisans set up co-operatives for nearly twenty years, Isaque is critical of entrepreneurial approaches.

> Look at the language they use: 'profitability', 'competition', 'excellence', 'sustainability' and 'growth'. What on earth has competition to do with co-operatives? There is no quick fix to learn how to co-operate. My seminars run for several months. We play, discuss politics and find new ways of working together. I think about my job as one of raising consciousness rather than teaching how to run a business. One of my biggest successes was to set up a co-operative among índio coal burners in the state of Espírito Santo. When I arrived in the village, a young boy welcomed me holding a severed human arm. It turned out it was the arm of a gringo that an old woman had cut off with a machete in response to his sexual advances. The gringos treated these *índios* like slaves. The woman was the only person who reacted. I came in and helped them organise. The gringos left, and the leader of the co-operative is now a municipal councillor. The co-operative has more than fifty affiliates.

The co-operative of women cleaners, the Abraço (Hug), is another example of how co-operativism provides tools for self-organisation and mutual support to marginal categories of workers. Maria founded the Abraço nearly twenty years ago in response to a traumatic life experience. Only fifteen years old, while working as cleaner for an affluent Volta Redonda family, Maria was repeatedly sexually abused by her employer. At the time, there was no cleaners' union, so she went to a local women's association. The lawyer of the association told her it would be impossible to bring her employer to court because he was a prominent member of the community and 'it was her word against his'.

Unsatisfied with the lawyer's answer, Maria used all her savings to pay for a private solicitor. The solicitor confirmed the opinion of the first lawyer. Speaking to her friends, she realised hers was not a separate instance. Several women in the group had also been repeatedly physically and morally abused by their employers. Moreover, during these informal meetings, Maria and her friends realised their employers breached all sorts of employment regulations. Most of them worked a minimum of twelve hours per day, including on weekends; were unregistered and hence did not qualify for any social benefits or paid leave; and did hazardous and unsafe manual tasks such as clearing gardens during the rainy season (thus often being bitten by the deadly mosquito dengue or poisonous snakes or spiders), working on high ladders or unblocking pipes or drainages with bare hands. Those who lodged in-house lived in shacks or tiny rooms in extremely unsanitary and hazardous conditions. One of them was electrocuted by an open wire and lost the use of an arm. The non-lodgers had exhausting and dangerous commutes, walking home after midnight or to work at 4 a.m. During these commutes, they were often assaulted by tramps or stray dogs. Some were raped.

So, the cleaners' co-operative initially functioned as a space for sharing work experiences and finding common strategies to fight back against their employers. After the co-operative was legally set up, the initial subscription money was used to hire a labour lawyer permanently, which gave sudden visibility to the Abraço. Maria described to me that moment:

> Initially, we were just a support group borne out of desperation. We would meet in the small room in the public leisure centre and cry from beginning to end. We were really scared someone would find us and report us to our employers. We were all young and emotionally vulnerable. We were all attached to our *dons*. We considered them our families – especially the lodgers. We were treated like members of the family – the ugly or poor members. You know what I mean? Initially, we felt guilty to meet in such secrecy and talk behind their back. But by sharing our experiences, we came to realise they were all abusing and exploiting us the same way. It was mostly the men. But the wives went along with them. We all had gone through similar experiences. We were all underpaid, overworked and with very low self-esteem.

When Maria hired a labour lawyer, long queues of women seeking legal advice started to form outside the door to the centre. Today, the Abraço has nearly one hundred members – mostly, but not exclusively, women. The offices of the co-operative are in a side street just opposite the UPV. Leading me out, Maria points towards CSN and says:

> There is where patriarchy begins, in the top offices of the firm. There, white men order people about and think of themselves as superior beings. A long time ago, on the very site of this plant, there was a coffee plantation where male planters acted exactly the same. History repeats itself. White men are in charge of the lives of black people. Humanity seems to be unable to learn from its past mistakes.

Legal Activism

From my field notes:

> The building of the Regional Labour Court [Tribunal Regional de Trabalho – TRT] is in the heart of the 'old city'. Its postmodern architecture – a dark-marbled Greek temple with electric blue windows and a pink entrance door – sits awkwardly within the austere modernism of the city. Outside the building, people celebrate, nervously wait, pompously arrive or walk hurriedly carrying heavy stacks of documents. Expensive cars and bicycles are parked. Rows of taxis wait. Popcorn vendors, shoe shiners and lottery sellers advertise their services. Workers meet their lawyers for the very first time. Inside the building is packed. Lawyers, workers and their families rehearse their testimonies in preparation for the hearings. People start to fill the courtroom at 9 a.m.
>
> The hearing starts at 10 a.m. when Linda Brandão, the president of the labour court of Volta Redonda, solemnly climbs the stairs to a high platform and sits at her desk. Below her, workers and relatives sit in religious silence. Above her is the Brazilian flag. Linda reads out a personal reflection: 'Simplicity is the companion of truth as much as humbleness is the companion of knowledge. Justice is not complicated. It is simple. It's people who complicate it. Simplicity creates its own truth.' Most cases relate to infringement of the labour rights of subcontractors. People stare and nod in silence. The first audience is a case of unfair dismissal against a local small firm. The defence lawyer, the public persecutor and the worker sit at the desk below Linda. The entrepreneur does not show up. Circumstances are discussed; supporting evidence presented. Linda says the worker must be re-employed and his lost salary paid. A dactylographer types. The whole conversation appears on a big plasma screen behind Linda.
>
> In the second case João Batista, a black subcontractor is suing a small engineering firm for loss of wages. He was paid only for half of the days he worked. For the other half he was paid in canteen vouchers. The woman owner of the firms says João Battista worked only intermittently. Stella asks if the company made any contract with him. The woman answers, 'Yes, but it was an improvised contract. Only for small and intermittent jobs.' People giggle. Stella asks that the entrepreneur formalises the contract and pays the arrears. Linda goes along with her request.
>
> The third case is against CSN. The company formally registered a worker eight months after the start of his employment. Stella, the lawyer of the Metalworkers' Union, walks onto the stage. She talks with confidence and charisma. She is very popular with the audience. A witness confirms the workers' testimony. Stella and Linda agree on evidences and consequences. Linda recommends the company retroactively register the worker. Stella solemnly states, 'Justice has been done.' The crowd cheers. The CSN lawyer, a tall, muscular man in his fifties, smiles scornfully.
>
> Linda concludes the morning session with a Zen sentence from Siddhartha concerning the futility of human sufferance. The crowd dissipates and the

building fills up with a new crowd of people queuing for documents at the legal office that just opened. I end up with Linda, Stella and a small group of friends at a local *churrascaria*. After a heavy meal of roasted pork and a few shots of cachaça [a liquor made from sugar cane], we return to the court. This time, Linda invites me to sit by her. Feeling tipsy, I climb up the stairs and sit at the high desk with her. Linda introduces me to the public as 'the Italian scholar.' People in the audience stare at me respectfully. A little boy sticks his tongue out at me.

In Brazil, there are three levels of labour courts: the local court (*vara*), the TRT and the TST. Brazil's president appoints directly the presidents of these courts according to the criteria of meritocracy and seniority. Before they are formally in charge of the courts, judges must go through a two-year apprenticeship as 'substitute judges'. Labour judges have an unusual profile within the legal profession. Half of them are under forty years old, and 43 per cent of *varistas* and 37 per cent of TRT and TST judges are women, compared with the national average of twenty 5 per cent (Morel and Pessanha 2006, 2007).

Moreover, unlike the traditional middle-class backgrounds of most legal professions, labour judges tend to be from working-class backgrounds.[35] This socially mixed composition of labour courts reflects the social inclusiveness of the 1988 democratic constitution, whose Amendment 45 increased the quota of women judges from 5 to 20 per cent and opened law faculties to women and lower social constituencies. Moreover, since the 1990s, the increased opposition to labour precarisation by the trade union movement led to the proliferation of labour courts. Despite such democratisation of labour courts, only 1.2 per cent of labour judges are black. Labour judges recently split from the all-powerful Associação dos Magistrados Brasileiros (Association of Brazilian Magistrates) and formed the independent Associação Nacional dos Magistrados da Justiça do Trabalho (Association of Labour Justice Magistrates), which represents them in the TST and the Justiça do Trabhalho do Brasil (Labour Justice of Brazil). But legal activism and labour politics do not easily map onto each other. In fact, there is a political divide between, on the one hand, civic and penal lawyers and magistrates who tend to be middle class and aligned with the traditional labour movement and, on the other hand, labour judges who tend to align themselves with marginal labour constituencies.

Partly, this conflict between the traditional leftists and labour judges arises from the different rationales attached to collective negotiations (led by unions) and the labour right system (led by judges). The judges' priority is to fight labour precarisation, whereas trade unions and their lawyers focus on protecting the interests of wageworkers ('the base') in collective negotiations. For the former, the compelling element of the law comes from the public nature of court hearing; for the latter, it comes from the compulsive nature of the private agreements between workers and employers. Because of their alignment with precarious workers, labour judges and lawyers were often on a collision course with both company managers and trade union representatives. Moreover, because of the pro-casualisation stance of some parts of the trade union within the PT government, labour judges fighting against precarisation are penalised career-wise and rarely progress to regional or federal positions. Indeed, one of the explicit aims of the CLT reform passed by Temer in 2017 was to disempower labour judges and cut down on labour lawsuits[36] de facto leading to the privatisation of capital/labour relations.

Judges aside, labour law is very different from penal and civil law. Brazilian labour law is customary, unlike the rest of the national legal system. When a judgment is used repetitively in courts, the supreme court declares it a jurisprudential orientation (JO), which means all cases that fall within the JO's remit are judged according to it. With time, a JO becomes normative orientation and then formal law (*sumula*). Dozens of *sumulas* are made every year, so the labour legal system is continuously transforming. Moreover, the CLT is extremely flexible. For instance, Article 8 allows judges, in absence of appropriate existing legislation, to use any legislation that is not included in the CLT, in any legal code – in any country in the world. Moreover, in line with the principle of analogy, judgements can be extended to circumstances that have no legal precedents. On this basis, for instance, the rule of ten-minute breaks every two hours was extended from typewriters to IT workers.

Finally, labour law is personalistic. In Brazil, there is no corporate mechanism for labour grievances resolution. The process starts when 'individual grievances' (*dissídios individual*) are made to the labour commission – a reconciliation body made up of employers' and employees' representatives. If the commission fails to reconcile the parties, then the petition enters the court system. For John French (2004: 100), this emphasis on individual grievances stems from the unequal and personalistic nature of the wage contract in Brazil and the protectionist stance of the CLT, which discourages the discussion of grievances in collective negotiations. The open, flexible and individualistic practices of labour judges and lawyers radically differ from the technocratic knowledge and distant impartiality of the traditional legal profession. During my fieldwork in Volta Redonda, a consensus was already mounting that collective negotiations were better than lawsuits in labour courts. According to Linda Brandão, the president of the Volta Redonda *vara* at the time, this push towards private collective negotiations was against the principle of democratisation of labour rights implicit in the democratic constitution and reflected the increased professionalisation of the Metalworkers' Union I discussed earlier.

Linda strongly opposed such professionalisation and privatisation of labour activism. Her *vara* processed thirty lawsuits per day, approximately five thousand per year,[37] mainly on unfair dismissal, unpaid social contributions and thirteenth salaries[38] against firms and co-operatives. Linda was part of an influential group of women judges who recently rose to power in the Fluminense region, including Maria das Graças Cabral Viegas Paranhos, the president of the Regional Labour Court of Rio de Janeiro. Speculations about why women make such good labour judges tend to rely on gender stereotypes. Labour judges and trade union representatives tended to assume the technocratic and rational logic of civic and commercial law was more suitable to the 'rigid' male mentality and that, unlike these, labour law required the 'traditionally female' quality of empathy. According to Linda, these stereotypes had a grain of truth.

> Labour law is based on informality, reconciliation and flexibility, and men are not good in any of these things. When a lawsuit is filed, judges must first establish the formal validity of the petition. Many petitions are formally invalid because workers are illiterate and their lawyers underqualified, and as a result, the information filed is patently false. For instance, in some petitions, you have people working twenty-six hours per day, or working on three machines at the same time. But, instead of rejecting these petitions, I try to regularise them before the first hearing [*audiencia inaugural*] so that the petitions are not declared invalid and the process is not stalled. Companies do not mind delays. But these hit workers very hard!

To keep the justice process going, Brandão adopted informal approaches, showing up on shop floors with labour engineers to check machines, ventilation systems and sound levels or gathering testimonies in situ – tasks that should be done by trade union lawyers or the Internal Commission of Accident Prevention (CIPA). Moreover, workers often do not show up at hearings, because they do not receive notifications, or cannot read them or afford to miss a day of work. Instead of following the protocol and 'condemning them by default', Linda would ring them up and tell them to get to the court and even send her assistant to fetch them at home or in their workplaces.

For Linda, the second 'feminine' quality of labour judges was that of being peacemakers. She tried to solve most cases through pacification between petitioners and defendants – often, respectively, the employee and the employer – in the first hearing:

> It is difficult to remain impartial when employers are consistently abusive towards the workers – especially CSN, which is a 'factory of accidents'. But because most labour cases are straightforward abuses against the workers, reconciliation always favours the workers. Unions are often too ideological and confrontational. This goes against the interests of the workers. The less conflict the better. The only good justice for the workers is a fast one. I took advantage of Article 8 in all possible ways. I even used French labour law in a case of unfair dismissal. But I never abused my power. There are many corrupted labour judges colluding with politicians in Volta Redonda.

Recently, at conference on labour law, an influential male MP said labour judges abuse their power and act like 'Batmans'. Confronting him, Linda responded that MPs and civic and penal judges are often part of a corrupted system. She mentioned the case of a famous MP who had been just suspended and trialled for bribing a penal judge.

When she was twelve, Linda started working in a shoe factory with her mother but received the *carteira* only at fourteen. She changed jobs many times. She worked as a *peão* (unskilled worker) in factories, retail companies, service companies and accounting firms. She became the first woman president of the CIPA in the region. While working, she graduated in law and specialised in labour law. Before her position of president of the Volta Redonda *vara*, she was a federal judge and law professor. She has been a working mother. She married and divorced twice. She claims: 'Despite being a judge, I have a *peão* background. I know very well the workers' conditions and how to communicate with them. This can make lawyers and politicians uncomfortable.' More than a gender issue, the professionalism and strong empathy of women labour judges like Linda are linked to their working-class background. As a working-class woman who rose to power in a traditionally male profession, a labour activist critical of the legalism of traditional trade unions and the entrepreneurialism of some co-operatives and a person from humble origin who uses the law as a tool for workers' self-help rather than a privileged field of knowledge, Linda was a typical anti-establishment figure. When I returned to Volta Redonda in 2012, I was not too surprised she had been suspended from the presidency and banned from the legal profession precisely for that altercation with the MP.

The two most active and respected labour lawyers in Volta Redonda are Stella and Vanderlei Barcelos – two prominent labour leaders in the 1980s. I have already discussed that Vanderlei led the coalition between the SMSF and the student, women's and black movements in the 1980s. From 1988 to 1990, he was the general secretary of the Metalworkers Union. But privatisation left him disillusioned. He stepped down from the leadership of the union when he realised that with the ruling PT, the historical opposition had been institutionalised:

> Why should I do trade union activism? If I am not allowed to think and behave differently? If I can only operate within the boundaries of what is allowed? Allowed by whom? By the father who prevents the son to wander in dangerous places? The trade union today only does what it is allowed to do. Before, the opposition operated outside the state. Now it operates inside the state. I am active as lawyer. I do not do penal, civic or marriage law. I only do labour law for the workers.

Vanderlei's legal activism operated at the margins of powerful institutions – the state and the trade union – and was mainly in support of marginal workers, especially builders, cleaners and domestic workers. In his view, metalworkers had become part of the state apparatus. They were uninterested, if not hostile, to the struggles of marginal workers.

Vanderlei's first action as legal activist was to challenge the blacklisting system of CSN. He convinced all the outsourced workers of CSN who had received offences by their employers to sue both their employers and CSN. CSN had to pay millions of reals in compensations. Subsequently, CSN dismissed and blacklisted every worker involved in the legal dispute. Consequently, the firm ran out of employees and was forced to issue a general amnesty to all the blacklisted workers. Vanderlei is a great supporter of the CLT and is deeply worried about the current attempts to flexibilise and restructure it:

> CLT is a strong asset for the working class because it has a strong pro-workers stance. The CLT's core principle is *dubio pro operario* [in Latin, 'in doubt for the worker'], meaning that when a contention between worker and employer arises, the worker is assumed to be right. Rather than saying everybody has the same right, as for civil and penal laws, the CLT makes it clear that in the wage relation, workers and employers are not equal. Workers are more vulnerable than the capitalists.

Stella, Vanderlei's lifelong partner, is the most renowned labour lawyer in town. She works for the SMVR, the Builders' Union, the

Sindicato dos Empregados em Empresas de Asseio, Conservação e Limpeza Urbana (Union of Employees in Cleaning, Urban Cleaning and Sanitation Companies) and the Sindicato dos Trabalhadores em Postos de Combustível (Union of Gas Station Workers). Despite these unions having radically different bases and agendas, Stella is able to represent them all. Her main goal is to fight labour precarisation, especially of builders and cleaners, and reduce working hours across all categories, an agenda that clashes with that of the Metalworkers' Union. When she advised the union to replace the salary supplement for hazardousness with a proportional reduction of working time, most workers refused, 'preferring money over good health'. For Stella, steelworkers think of themselves as bankers and have little political consciousness and sense of solidarity towards other categories of workers.

With a gift for self-transformation, Stella fluidly moves across the corporate offices of the Metalworkers' Union and the informal offices of the sub-branch of Retiro, where she is nicknamed 'the revolutionary'. But it is clear she feels most at home in the Builders' Union, where she shows great empathy and care towards her mostly vulnerable clients. Stella grew up in Buenos Aires, the daughter of a 'very charismatic and handsome' director of the General Command of Workers (CGT)[39] of Southern Italian origin. She would follow him to trade union meetings, negotiations and demonstrations. The CGT became her family. At the union, she learned to talk, swim and ski. She liked to go to the cinema with her dad. They shared a passion for Italian neorealist cinema, which kept them connected to their roots. When Juan Perón was overthrown and forced into exile in 1955, the CGT leadership was purged, including Stella's father, who was jailed and tortured.

The government tried to 'buy him', giving him the option to go into exile to the United States. But he refused. After studying chemistry for one year, Stella switched to law at the University of Córdoba, one of the most prestigious law faculties in Latin America. Becoming a labour lawyer, she fulfilled her father's dream. She got involved in the radical Peronist Youth Movement,[40] which was violently persecuted under the dictatorship. In the late 1970s, Stella left Argentina for Volta Redonda where she got involved in the clandestine labour movement and met Vanderlei, her lifelong partner and *companheiro* (comrade). Stella and Vanderlei have a law firm in partnership with Andreia, a young lawyer who specialises in workers' pensions and environmental rights. Andreia is leading a legal campaign with the Workers' Association for Leucopoenia to promote awareness of the illness and reduce levels of benzene in the city. The workers' demand to have leukopenia officially recognised for legal compensation was turned down by the regional justice system in the 1990s (chapter 3). Until Cida Diogo became the health officer, the SMSF joined CSN in denying the condition, withdrawing support to the leukopenia association. Andreia is fighting to bring the issue of leukopenia back into the political agenda.

As an environmental lawyer, Andreia also campaigns against CSN's planned development of a new billet mill in the Cicuta Forest, which was not opposed by the civic and environmental associations affiliated with Agenda 21. According to Andreia, when Gotardo – a GP representing the Green Party and close to CSN – became mayor in 2005, environmental organisations suddenly became silent about the ongoing environmental degradation and health hazards caused by CSN. Andreia is also active in animals' rights, campaigning against the abuse of horses and donkeys on construction sites where they used to transport transport heavy-duty building material. Andreia is not the standard lawyer type:

> I do not have a great relationship with my family. I left it ten years ago, when my mum died. I loved her very much. She was a powerful woman and a great witch doctor. She cured people in the bairro, and the bairro respected her. After she died, she reincarnated three times – once in my cousin and twice in me. In fact, she is still with me now. Two years ago, I discovered I had brain cancer. I refused to go through the torture of Western medicine, and I cured myself with medical herbs and the help of a candomblé priestess. Now I practice candomblé, too. Now I am a healer and white magic practitioner when I don't practice as lawyer. I do not miss my biological family. My real family are Vanderlei and Stella now. But I continue to be the only breadwinner for my father, brother and boyfriend – who are all unemployed. I also have seven dogs and fifteen cats.

Vanderlei, Stella and Andreia often met on weekends and were often joined by Isaque and Sandra. They would stay in a small wooden house Isaque and Vanderlei built with salvaged wood in the thick forest just outside Volta Redonda. Their political views – for instance, on the CLT or on co-operativism – were different. But the exilic character of their meetings brought them together. Fear of being followed or bugged by the security service of CSN or the federal police was always present in these meetings. But it seemed that being together made them feel safe. In different forms, they were all fighting for the rights of precarious workers against CSN and against the Metalworkers' Union – two institutions that belonged to the state in contradictory ways and at the same time were entangled with local and global trajectories of capitalism. Moreover, they were also all committed to the socialist project that Lula and the ruling PT both betrayed and continued to represent.

It is in these autonomous, exilic and self-enclosed gatherings of subcontractors, cleaners and labour lawyers I have described in this chapter that the practices of solidarity and *commoning* – intended at the same time as individual actions, collective experiential spaces and long-term embedded processes – emerge and take an explicit anti-capitalist character. From the perspective of labour, back sourcing – the return of the firm – marks a new repressive turn. The political voice of direct workers is being muted. Their economic conditions are worsening. Unlike outsourced workers, they can be dismissed at any time, heavily fined, punished or transferred to faraway subsidiaries – as far as China. Such re-internalisation of labour is the outcome of the increase in market bargaining power outsourced workers gained after they exited the firm, disentangling themselves from the asymmetrical wage relation, and acquiring a new collective voice, based on practices of mutual support and commoning.

Thus, labour subcontracting, internalisation and commoning – or, using the terms of neo-institutional economics, the market, the firm and the clan – are not discrete and competing institutional mechanisms of capital organisation. Rather, they are hybrid and overlapping political and economic spaces where the shifting boundaries and tensions between capital and labour are both concealed and revealed. The privatisation of CSN led to the externalisation of the technical knowledge of the firm into the SIGMA with all the effects we have seen in terms of overpowering labour. As the short-term cost-benefits of externalisation waned, the management realised nobody knew how to run the plant and that the company totally depended on subcontractors. These used their increased market bargaining power to improve their wages and working conditions and came together as a workers' collective. The company reacted by re-internalising work and making wage relations more authoritarian.

The national legal framework is fundamental in shaping the strategies of capital and labour and shifting the institutional frameworks of firms, market and commons, and we can say, without exaggeration, Law 331 was passed to allow firms like CSN to start massive programmes of outsourcing after privatisation. Likewise, the political victories of the subcontracted workers were made possible by the opposition to Law 331 led by left-wing lawyers and judges. Against this consolidation of labour rights, the law passed by Temer in 2017 erodes the rights and wages of the formal working class, especially in the IT and automotive sectors mostly controlled by multinational corporations, and makes the model of extreme outsourcing I have described a permanent feature. Will such legalisation of outsourcing put an end to the old flexible capitalism, deregulating and deterritorialising labour even further, as in the cases of the ghost factories? Or will it create a permanent state of precarious and exploited internalised labour as in the case of CSN?

Within this ever-shifting legal context, labour forms become as important as their content. The content of maintenance jobs – repairing refractory bricks, changing an electric valve, fixing a machine – is the same if performed in streets, workshops or factories. But its value changes according to the legal framework through which maintenance tasks are valorised by capital; in other words, they are institutionalised. The CLT continues to be one of the most advanced labour legislations in the world, but it addresses only a small elite of industrial workers and leaves most working people in a condition of 'structural precariousness'. In the realm of informal labour, the flexible and open-ended nature of the CLT paved the way for constant redefinitions and downwards shifts in labour/capital relations: wage labour becomes subcontracting; co-operatives turn into small businesses, small businesses into informal sweatshops and informal labour into slavery.

Wage Work/Slavery

In fact, the CLT must be read against Brazil's historical legacy of slavery. Up until the end of the nineteenth century, the Paraíba Valley was one of the world centres for the slave trade and coffee economy. From 1840 to 1851, at least 371,615 slaves were smuggled into Brazil (Stein 1957: 26) peaking in 1848 when 60,000 new enslaved African people arrived in Rio de Janeiro (65) most of whom would be purchased by the *fazendeiros* of the Paraíba Valley. The formal abolition of slavery in 1888 did not automatically institute a local regime of wage labour. The forms of indentured and precarious labour used in fazendas – unpaid work, work paid by the day (in cash or kind) and subcontracted work (*empreiteiros*) – were replicated in the industries of the Paraíba Valley, including in CSN whose workforce was partly recruited from local former coffee fazendas.[41]

The historian Sidney Chalhoub argues that because of the legacy of slavery, freedom in Brazil is a condition of structural precariousness. Brazil had unusually high rates of enslaved people's manumissions and freedom, but liberties were restricted by law and systematically revoked based on the shaky moral ground established by the master. Moreover, free and freed black people were systematically profiled and imprisoned by the police under suspicion of being slaves. As a result, 'in Brazil more than in any other slave society in the Americas slave emancipation occurred concomitantly with the continuation of slavery' (Chalhoub 2011: 409). Moreover, Brazilian labour scholars have emphasised the continuity between slavery and informal and precarious work in Brazil and between slavery and outsourcing.[42] The growing precarisation and informalisation of labour in contemporary Brazil have their historical roots in the structural precariousness described by Chalhoub and in Brazil's historical dependency on the Global North (Braga 2016).

I argue the back sourcing of labour internalises labour's structural precariousness within the corporate realm. For Ben Schneider and David Soskice (2009), the authoritarian, family-based and hierarchic structure of Brazilian corporations is 'isomorph' to the authoritarianism of the Brazilian state (see also Pargendler 2013: 813). Instead, I argue the state and the corporate sector, rather than being 'isomorphs', are mutually entangled in the same system of 'corporate sovereignty' (Barkan 2013) and that paradoxically, under the PT, such entanglement between corporations and the state is tighter because of the incorporation of the trade union within the state. Particularly striking is how Brazilian commercial law allows SOEs to operate like private firms and at the same time be under state control. For instance, the BNDESPAR – the wholly owned subsidiary of the BNDES operating as its equity arm – has equity stakes in seventy-four closely owned companies and in seventy-one publicly traded companies, which accounts for 60 per cent of Brazil's total market capitalisation. Consequently, according to Mariana Pargendler (2013: 830), the state controls a vast section of the private corporate sector.

Moreover, the Brazilian corporate legal regime reflects the 'symbiotic relations' between the state and corporate oligarchies (Pargendler 2012: 2923). Such symbiosis was reinforced in 1997 when Cardoso reformed the corporate law eliminating protections towards minority shareholders and further enhancing the power of business dynasties, banks and pension funds. At the same time, the corporate legal regime of privatisation, especially Law 331, deregulated labour and legalised outsourcing, leading to the totally externalised 'ghost factory' regime I describe earlier that denies legal personality to outsourced employees. Consequently, Brazil's commercial legal framework simultaneously allows extreme labour deregulation, extreme private capital concentration and extreme state control. This model is not the outcome of colonial transplant (Pargendler 2013) but reflects Brazil's trajectory of dependent industrialisation led by the landed elite, the military classes and the technocratic bureaucracy and based on the deep social division between state elites and private oligarchies on the one hand, and masses of precarious and informal workers on the other hand, with a weak industrial bourgeoisie floating between these two extremes.

Indeed, the Brazilian industrial bourgeoisie irrationally aligned itself with the landed and financial classes and rejected Rousseff's developmentalist model aimed at supporting and improving precisely the industrial sector. In the end, because of fear of being associated with working-class interests, because of its weakness vis-à-vis the financial class or because of its historical entanglement with it, the Brazilian industrial bourgeoisie decided to actively contribute to the fall of the developmental state (Singer 2018a). Contributing to this fall was the mounting anti-statism shared across the political spectrum. The scandal of Petrobras was a case in point. After being elected president, Lula reversed three decades of tertiarisation and partial privatisation of Petrobras, dismissing foreign subcontractors and re-internalising labour. From 2005 to 2008, Petrobras made thirteen thousand new assumptions, becoming one of the most powerful public oil employers in the world. Thus, while encouraging foreign companies to outsource labour – for instance, through the proposed Law 4330 – the state re-internalised it and put it under a stricter control regime. In fact, the popular rage against Rousseff that emerged after the Petrobras scandal can be read as a popular protest, this time coming *also* from the working class, against this form of state-led capitalism.

Francisco Oliveira (2005) describes the financialisation of the Brazilian state in the late 1990s as 'internalisation of imperialism'. The phenomenon of extreme precarisation of wage labour I describe in this book is the outcome of such internalised imperialism and of how Brazil's extractivist capitalism rests upon a strong power block consisting of state managers, trade unions representative and private corporations. The haunting structural precariousness of Brazilian labour relations in this chapter offers a reflexive tool for 'provincialising' not only wage labour but also, and more poignantly, capital.[43] The structural precariousness of labour is is the consequence of both capital accumulation and of the limits to its growth – with slavery constituting the extreme limit of capitalism and of the otherwise open-ended wage contract. The acts of solidarity, commoning and reciprocal support and care by subcontracted workers, cleaners and legal activists I have described are glimpses into a non-capitalist and human-centred economy produced in existential conditions of life precariousness, social exclusion and economic dispossession. Moreover, these struggles show how capital is deeply undermined by the threat of collective withdrawal of labour from the wage relation, and that this is more effective when workers are not entirely bound to the wage contract as for outsourced workers.

The experiences of marginalisation and exclusion of outsourced workers and their exit from the wage relation produced new, shared experiences and meanings of labour and opened up new legal and political spaces of struggle. Besides, legal activists and outsourced workers exited not only the firm but also the trade union system, which in their eyes did not represent them anymore, and reconstituted themselves as new political subjects at the margins of these institutions, as well as in productive tension with them. Going back to the episode that begins this chapter, it became clear during my fieldwork that workers' power comes from simple acts of coming together, sharing, pooling and *commoning* knowledge, imagination and material resources. But as the economic crisis increased, these acts of commoning were progressively countered by enclosures – including back sourcing, privatisations, regionalisation and sectionalism – created by CSN, the SMSF and the municipality.

Notes

1. Henceforth called Builders' Union in the text.
2. The CBS-CNS contracted on a 'man per hour' basis.
3. The non-monetary component of their wages was nearly inexistent, which halved the cost of their labour.
4. Braga (2016) links the surge of labour militancy in Brazil since 2003 to the political activism of subcontracted, precarious and informal workers.
5. *The Economist* in the special report 'The Rise of the Superstars' (17 September 2016) argues the big consolidated corporate form of the early twentieth century is making a comeback after it was declared dead by management gurus in the 1980s and 1990s due to their excessive rigidity, labour costs and slowness vis-à-vis the more flexible and lean firms. According to the article, the return of the corporate form is driven by the monopolies in the high-tech sector but also their greater efficiency in bypassing tax regulation and lobbying governments. To this, it must be added the fall in labour standard and wages in developed countries. For instance, Brazilian workers are much more expensive than US or UK ones. The McKinsey Global Institute, the consultancy's research arm, calculates that 10 per cent of the world's public companies generate 80

per cent of all profits. Firms with more than $1 billion in annual revenue account for nearly 60 per cent of total global revenues and 65 per cent of market capitalisation.

6. The minimum wage is currently R$998 per month (as of April 2019).
7. Personal communication, Sabrina Dias, 15 March 2019.
8. I thank Sabrina Dias (2010) for sharing her MA thesis, 'Dentro da usina mas fora da "família"', which this section partly relies on.
9. The FEM was founded in 1960s as a specialised department producing specialised products for the civic construction sector.
10. *Feia* in Portuguese means 'ugly', so *feinha* would translate roughly as 'little ugly'.
11. I am following the classification of workers into a first generation (retired); a second generation (above fifty years) and a third generation (between eighteen and fifty) made by Crovetto (2012: 38–58).
12. Some departments – for instance, the GGFM – have a staggering 60 per cent turnover rate. Surprisingly, the GGFM management argues such high turnover reduces wage levels and the political costs of skilled labour.
13. Brazilian steelmakers continue to export semi-finished industrial products and import high-value machinery (Amann 2004).
14. Comparisons between the salaries of direct workers and those of outsourced workers are difficult to make. For instance, outsourced workers have the right to a minimum salary – higher than the minimum wage – established by law, whereas direct workers do not. Personal communication, Sabrina Dias, 20 March 2019
15. Although the plan of outsourced workers was regional and CSN workers had a better national plan.
16. Which closed down in 2008 because of economic crisis.
17. Including Catholic organisations, neighbourhood associations, the Landless Workers' Movement, black organisations and the Movement for the Ethics in the Politics.
18. Led by Inácio Lula da Silva until the early 1990s.
19. The state gave to VW the following support: (1) an electric station for the sole use of the plant of a value of $4 million, (2) a gas network of the value of $7 million, (3) a $2 million grant to the municipality of Resende for the construction of electricity, water and transport facilities to the plant.
20. Including exemption from corporate taxes and the gift of three million square metres of industrial land.
21. Seventy per cent of them are maximum thirty-four years old and 60 per cent are married and with children; 95 per cent of them has a permanent job contract; 65 per cent gains between three and five *salários mínimos* (the highest wage level in the region) and 69 per cent own a home, 59 per cent of which has already extinguished their mortgages (Ramalho and Santana 2006).
22. Recently, MAN AG was bought back from Volkswagen.
23. It declined from 62 per cent in 2001 to 27 per cent in 2009.
24. The institution is called the Sectoral Chamber of the Automotive Industry.
25. At the national level, from 1991 to 1997, privatisation led to a 30 per cent reduction of waged employment and 10 per cent increase of informal labour mainly, among women and black workers (DIEESE 2000).
26. Thirty per cent of the total new jobs created since 2000 (Pochmann 2012: 232).
27. Of domestic workers, 61.8 per cent are non-white (IBGE, PME 2006).
28. For a detailed account of the 'June movement', see Mollona (2018a); Saad-Filho and Morais (2014); Singer (2014).
29. More recently, the union relocated to a modern building in the city centre.
30. From industrial construction and civic construction workers to concrete-making workers.
31. Resende, Volta Redonda, Barra Mansa, Porto Real, Quatis, Angra dos Reis and Itatiaia.

32. Index-linking the annual adjustments of salaries and wages below the inflation rate while mortgage repayments continued to be index-linked to full inflation.
33. Law 5764 of 16 December 1971.
34. Name of the co-operative, rights and duties, capital threshold, mode of administration and due collection, rules of the assembly, instances of dissolution and instance of reform of the statute.
35. Thirty per cent are from a working-class or rural background, and only 30 per cent of their fathers and 20 per cent of their mothers completed secondary education (Pessanha and Morel 1991: 86).
36. Ives Gandra da Silva Filho, the former president of the TST who passed the law, in an interview with the *Folha de São Paulo* explicitly welcomes the 80 per cent reduction of lawsuits based on supposed moral damages by employers and the weakening of the power of labour judges (Fernandes, 2018).
37. The three *varas* of Volta Redonda process fifteen thousand lawsuits per year.
38. A mandatory salary paid to the workers in November. Its value should be equal to the wage, which is received by workers in twelve months of the year.
39. Argentina's main Peronist trade union.
40. A radical wing of the CGT inspired by Guevarism and Third-Worldism.
41. Elsewhere, I deal extensively on the relationship between slave labour and wagework in the Paraíba Valley in the early twentieth century (Mollona 2018c).
42. According to a recent CUT (2014) report, 80 per cent of slave labour comes from outsourced jobs.
43. I read Mezzadra's (2011) notion of 'postcolonial capitalism' as a step in this direction.

CONCLUSION

Towards an Anthropology of Uneven and Combined Development

In this book, I have discussed the Brazilian model of neo-extractivism by looking at how such capitalist development materialises in everyday practices and in social relations unfolding both at the factory level and in the city. During the 'pink tide' in Latin America, neo-extractivism was endorsed by left-wing 'inclusive' populist governments (Judis 2016), such as those of Kirchner, Lula, Morales and Correa, that combined contradictory measures such as formal wage increases, foreign exchange and price control, universal programmes of poverty reduction, labour deregulation, tight fiscal policy and public debt reduction.

Diverting from the monetarist dogma, these left-wing governments across Latin America renationalised companies, set in motion massive programmes of poverty reduction and urban participation that empowered women, indigenous and black minorities and created a new cross-sectional alliance between the middle class, the working class and informal and rural masses. Leaders like Morales, Lula, Chavez and Correa cast neo-extractivism as the 'southern alternative' to the monetarist economic orthodoxy of the Global North. Yet, they endorsed the 'there is no alternative' (TINA) ideology and formed a 'new extractivist consensus' based on the primarisation of the economy, large-scale enterprises, the enclosure and commodification of nature, a focus on exportation and a tendency for mono-production and monoculture (Svampa 2017).

Examples of this model are the extensive development of strip mining in Chile and Brazil, the expansion of the petroleum and energy frontiers in Ecuador and Bolivia and the generalisation of the agribusiness model (soy and biofuel) in Argentina and Brazil. The TINA ideology underpinning the populism of the Left in Latin America is clearly articulated by Bolivian Vice President Álvaro García Linera (2012: 2) in *Geopolitics of the Amazon*, which argues that without using extractivism in support of social programmes, Bolivia and the Amazon will fall back under the control of right-wing landed elites. But unlike the Andean world (Bolivia, Peru, Ecuador), where extractivism takes the form of intensive commodity extraction and land grabbing led by foreign corporations (Svampa 2017: 70), the extractive operations of capital in Volta Redonda materialise in terms of de-industrialisation, commodity extraction, urban rents, new favelisation, infrastructure development and financialisation of poverty led by the Brazilian state in partnership with the traditional trade unions and powerful national oligarchies.

As well as revealing a profound crisis of the Latin American Left, neo-extractivism sparked new political struggles based on new forms of *commoning* and a new political imaginary – such as *buen vivir*, 'the right of nature', the 'defense of the commons' and 'ethics of care' – and a 'radical deepening of democracy' (Svampa 2017: 77). This communal ethos is particularly strong in Latin America, which, like other peripheral societies, functions as a 'factory of solidarity' (78). For Maristella Svampa, the new self-managed structures of co-operation of popular sectors in the Andean world emerge from the absence of centralised capitalist markets and states. Instead, in urban contexts that are socially atomised and marked by unequal modernisation such as Argentina, Venezuela and Brazil, the 'preexisting communal ethos – if it exists – is weak, and new solidarities need to be generated' (77). While Svampa's analysis is correct, her claim that Latin American commons exist outside or at the margin of capitalism and have a pre-modern and communitarian ethos underestimates the historical entanglement between capitalism in the centre and non-capitalism in the periphery as well as the complex class relations and forms of capital extraction characterising, for instance, the Brazilian context I describe in this book.

Veering away from Svampa's discussion of extractivism as a phenomenon limited to Latin American regions and to the extraction of natural resources in non-urban or rural sectors, I consider extractivism also in relation to finance – particularly the hegemony of rents and the financialisation of poverty – and to various processes of primitive accumulation. I adopt an expanded notion of extractivism (Gago and Mezzadra 2017), intended as generalised form of capital extraction based on the maximisation of rents instead of wages and embedded in specific social relations of production as they unfold in workplaces and in rural and urban spaces. Thus, extractivism, intended as the multiple, heterogeneous and uneven assemblage emerging from the global operations of capital, actualises Trotsky's 'uneven and combined development' framework within the current capitalist context. In its mixture of land enclosures, favelisation, gentrification and housing speculation; financial deterritorialisation, logistical expansion and de-industrialisation; and precarised wagework, Bolsa Família and indentured labour, Volta Redonda reflects the diversity and complexity of the global economic landscape.

The global assemblage of Fordist factories, petty capitalist sweatshops, flexible and modular firms, high-tech conglomerates, self-employment, informal and illegal businesses and urban farming kept together by the superfast circulation of capital shows no convergence towards the standard northern 'variety' of stock market, lean and post-industrial capitalism (Hall and Soskice 2001) and neither towards the brazilianisation (read informalisation) of society predicted by Ulrich Beck (2000). For instance, the rise of 'platform' and 'gig' capitalism in the United States and United Kingdom sparked new collaborative modes of production and sharing economies but also resurrected the old hierarchical and centralised model of monopoly capitalism (Srnicek 2016) and led to the back sourcing of precarious labour.

In fact, the mixture of state capitalism and corporate hierarchy in this book reflects a broader trend in the Global South, where Fordism continues to be the engine of capitalist growth in the form of old state monopolies or partly state-controlled global conglomerates like the CSN. In many neo-developmental regimes across the globe, the economy is controlled either directly or indirectly by the state, unlike in the Global North, where corporations control the state. To be sure, *Brazilian Steel Town* makes clear that the corporatist alliance between states, companies and trade unions; the opaque cross-participations and pyramidal structures through which the state controls the economy; and the despotic rule of industrial oligarchies through which the neo-extractivist model is implemented in Brazil is as exploitative as the dismantling, the dematerialisation and deterritorialisation associated with the extractive operations of capital in the United Kingdom.

Indeed, it is interesting to compare the uneven trajectories of capital extraction in the steel towns of Sheffield and Volta Redonda. Unlike the post-industrial combination of urban rents and immaterial labour in Sheffield, Volta Redonda combines the post-industrial economy with commodity extraction, Fordism and land grabbing – an economic morphology that Francisco Oliveira (2005) describes as 'duckbilled platypus'. These different modes of extractivism reflect different teleologies of economic progress. In Sheffield, economic progress took the linear path of progressive deterritorialisation and dematerialisation of life – from land, to factories to global finance – with capital mutating from land rents, to wages, to urban rents. In Volta Redonda, economic progress followed the classical predatory colonial pattern whereby industry, land and finance constitute three interlacing forces and fixes of capitalism.

Moreover, there are strong similarities between the 'inclusive' extractivism of the Workers' Party (PT) in Brazil and that of the early New Labour in the United Kingdom. Both Lula and Blair combined financial growth and social redistribution, framing the latter as a struggle against poverty and not against inequality; they both continued in their predecessors' (Cardoso and Thatcher) legacies of deindustrialising and privatising the economy and expanding the financial and service sectors, and both 'moralised' the stock exchange. Both Lula and Blair co-opted the trade union in their corporatist and statist business vision, erasing class through inclusive identitarian and multiculturalist narratives – about ethnic communities in the United Kingdom and black and indigenous people and rural *campesinos* (peasants) in Brazil. Lastly, they both cast themselves as charismatic leaders who challenged the existing Conservative and Labour elites proposing a mediatory framework to the advantage of 'the poor', and both depolarised politics by occupying 'the middle ground'.

But there are also striking differences in the two modes of extractivism. In Sheffield, deindustrialisation took the extreme form of mass closures and forced economic reconversions. In Volta Redonda, it led CSN to morph into a decentralised financial-industrial complex and turned the urban landscape into a hybrid between a decaying steel town, a service and financial hub and a housing market for the new affluent working class in the region. Indeed, the (neo)extractivism of Lula internalised finance and imperialism within the state and its industrial apparatus, whereas the New Labour model externalised the imperialist state into the global apparatus of finance. In Brazil, financial capitalism feeds the corporatist alliance between trade unions, the state and corporations, whereas in the United Kingdom, it underpins their dismantling, dematerialisation and deterritorialisation. Moreover, although Brazilian extractivism shows a nationalist resistance to foreign capital, it is nonetheless structurally dependent on it and on foreign technology, and follows a pattern of energy inequality that pits high-pollution and low-value-added industrial processes in the South against low-pollution and high-value-added processes in the North.

The Brazilian working-class experience I have described in this book has three important historical legacies: (1) a mixed economy that blurs private and public spheres, (2) a populist consensus between power elites and industrial masses that produced class emancipation without political emancipation and (3) the articulation of class struggle and anti-colonialist struggle (Mollona 2014). The historical convergence of anti-colonialism and anti-capitalism emerged in cross-sectional collaborations between civic, rural and labour movements that toppled the military rule in the 1980s and later propelled the PT to power. In Brazil, it was the legacy of slavery that brought together anti-capitalist and anticolonial struggles. Fighting for a freedom that encompassed both social demands and labour demands, slaves infused the labour movement of the early nineteenth century with powerful support structures, forms of struggles and civic consciousness.

As José Aricó (2013) notes in *Marx and Latin America*, these anti-colonialist cross-sectional alliances in Latin America negate the classical Marxist[1] hypothesis of the vanguard of the industrial proletariat and reflect a 'southern epistemology of struggle' (Sousa Santos 2014) that combines cultural difference and economic equality or, to paraphrase Nancy Fraser (1995), struggles for 'cultural recognition' and struggles for 'economic redistribution'. If the pauperisation and precarisation of the working class is part of the 'standard' global trajectory of capitalism (and one of the failures of inclusive populism), then the political resurgence of women, subcontractors and marginal labourers tell us of a uniquely Brazilian (or perhaps Latin American) experience of class struggle. Under capitalism, the work of cleaners, maids, garbage and maintenance workers – that is, the reproductive work of looking after and caring for people, machines or discarded objects – is under-valorised by not only capital but also the mainstream trade unions.

In this book, I have shown how human labour is negated and abstracted through a mixture of direct control and violence and through discoursive and practical articulations of the commodity form on the shop floor in the fetish forms of technology, money and labour. In chapter 2, I showed how the systematic dehumanisation of the workers and the misrecognition of their labour takes place first, through technological fetishism, which splits the labour force between capital intensive and labour intensive departments. With machines, money is the other capitalist fetish that obscures the value of the labour of steelworkers. Younger and less skilled or auxiliary workers value labour through quantifiable and monetary measures and support those shop floor reorganisations that introduce the market onto the line and deskill labour. Unlike them, older workers on traditional operative tasks resist intensification of production, the monetisation and commodification of their labour and the externalisation of their knowledge into external mechanical apparatuses.

Lastly, narratives centred on the productivity of the land and the value of nature obscure how deeply entangled industry is with the dynamics of urban rents and the commodification of nature. Since the 1990s, the land ceased to be a terrain of working-class struggle in Volta Redonda and became an investment zone for foreign capitalists, a regional space for cross-sectional political alliances or an item of household management and speculation. Land commodification and rentierism reached the 'affluent' working class in the form of home ownership financed with sub-prime mortgages and households' debts. But subcontracted workers, builders and informal workers challenged this hegemonic productivist culture through practices of commoning, maintenance, care for the machines and solidarity. Women were at the forefront of these struggles in defence of the commons, for the valorisation of reproductive labour (Federici 2012), the rights to the land and an 'ethic of care' (Svampa 2017: 77). Outsourced maintenance workers also challenged working-class productivism through their reproductive labour of caring for, maintaining and looking after the machines – a labour that cannot be automatised, standardised or intensified.

By coming together, marginal workers became aware that their labour was fundamental for the functioning of capitalism – cleaners were necessary for the affluent working class to stay in employment and to live up to its bourgeoise living standards; subcontractors were necessary for CSN to break even; builders were necessary for the municipality to continue to invest, expand and stay afloat. These processes of self-awareness created spaces of valorisation of labour alternative to those of capitalism. Zones of exclusion and 'exilic spaces' *within capitalism* allow for practices of labour commoning, co-operation and non-commodified life to emerge, which challenge the capitalist status quo. The acts of coming together, sharing and connecting experiences *across* the economic spectrum show the cultural and political agency of the working class in the emancipatory Gramscian sense.

Cleaners, outsourced maintenance workers and builders were both outside the internal labour market (and at the edges of formal wagework) and deeply entangled with the corporate world of CSN and global construction firms. Their acts of commoning were effective precisely because they threatened to disrupt the capillary and dispersed infrastructure of state capitalism, which they ultimately supported. In other words, their commoning disrupted capitalism *from within* and not as an *exit* or a space located outside capitalism. Indeed, my uneven development framework shows that relationships between capital and labour fluidly cross the boundaries of economic institutions – and the thresholds between *commons, hierarchies* and *markets,* which are never closed or fixed. For instance, the waves of outsourcing of the 1990s led to increased political militancy and commoning by subcontracted workers, which led CSN to re-internalise work and to shift from 'the mean and lean' Toyotist model to a new regime of precarisation of *internal* labour.

As well as being affected by the global operations of capital, the boundaries between labour and capital are made and unmade by the multi-scalar, geographically dispersed and temporally uneven operations of the state. In fact, I have described CSN, the Metalworkers' Union and the municipality of Volta Redonda as 'effects' of the Brazilian neo-extractivist state. Economically, the UPV is not an industrial firm in the classical sense because its steelmaking activity is economically volatile and often runs at a loss, while the main profit of CSN group comes from iron ore extraction, urban rents, financial speculation and logistics. In fact, it could be argued the UPV primarily covers the 'state functions' of sustaining economically the local municipality through a steady cash flows of municipal taxes, disciplining the working class of Volta Redonda (still a powerful social constituency) through a mixture of employment and repression and facilitating the extractive operations of the state through its logistical network of land, ports and railways.

As the effects of the 'triple' alliance between the state, the industrial trade unions and industrial capitalists reached Volta Redonda, weakening the local trade union movement and dividing the workers, CSN instituted a new despotic regime based on precarious employment, low wages and hazardous working conditions. Such a despotic turn of capital was a reaction to the acts of commoning and sharing – of resources, information and labour – by outsourced workers I describe in chapter 6 as a counter-systemic force *within* capitalism. Because of this triple alliance, the local Metalworkers' Union

(SMSF) morphed from being a political institution into operating like an economic extension of the managerial state through programmes of financial literacy and legal consultancy. Likewise, the municipality shifted its modus operandi from providing political support to anti-capitalist urban movements in the 1980s to being a pro-business regional economic hub from 2000s onwards. The de-politicisation of the municipality and of the trade union and their functioning as economic agents of capitalism are paralleled by the process of politicisation of CSN, which, despite being private, covers the state functions of policing, infrastructural development and employment of a progressively obsolete workforce. During the 2018 presidential election, the CSN president became active in parliamentary politics as the presidential running mate of Ciro Gomes.

The history of CSN also tells the story of the imperialist movements of capital (Fontes 2010) both at the level of the shop floor and at the level of civil society. CSN was set up with US capital and technology at the verge of the Second World War, when Roosevelt coaxed Brazil against the Axis powers. Its early developments, and successive expansions, reorganisations, privatisations and financialisation were moulded on the technological standards, market logic and interests of US capital, as they were translated by the technical experts and financial consultants of McKee, Arthur D. Little and KPMG. The dependent industrial development of today reiterates in different forms, the previous colonial mode of production (Gorender 1978) in the Paraíba Valley, entangling together industrialisation, favelisation and new forms of indentured labour under the violent and fearful rule of the national bourgeoisie (Fernandes 1975; Fontes 2010) – a violence that returned with vengeance with the electoral success of right-wing populist Jair Bolsonaro.[2]

The imperialist movements of capital penetrated national institutions such as the BNDES (whose economists are trained in prestigious US universities), the CUT (whose pension fund capitalism was an import from the UK trade union movement) and the Metalworkers' Union (whose cadres were trained in finance and participatory budget by experts from São Paulo). Even those civic movements that articulated their opposition to capital through forms of ethics struggle or moral indignation – the moralisation of the stock exchange by Union Force or the Movement for Ethics in Politics – reflected the broader neo-liberal shift from class struggle to poverty reduction, and the 'third way' opened between capital and labour by the new European Left and endorsed by Fernando Henrique Cardoso in the 1990s (Fontes 2010).

But the emergence of a militant Left associated with the Socialism and Freedom Party and the Free Fare Movement in support of precarious women, youth and ethnic, migrant and other marginal class constituencies is an important development in Brazil. Another centrepiece of working-class militancy in Brazil is the CLT, although in its contradictory tensions between reinforcing state control and corporative contracts on the one hand and protecting marginal labour constituencies on the other hand. Given how important the CLT is for the Brazilian working class, it is not surprising that former President Temer initiated its dismantling in 2017.

The End of Left Wing Populism?

In an article written in 1991 amidst the monetary crisis, the economist Luiz Bresser-Pereira describes the economic policies of populist parties in Brazil. The populism of the Left is characterised by wage increases, radical programmes of income redistribution and a nostalgic longing for the developmentalism of the 1950s informed by nationalism, economic protectionism and anti-imperialism, hostility towards export-oriented policies and focus on internal markets. Indeed, wage increases appear the most effective redistributive mechanism in a country with nearly the most uneven income redistribution in the word. But, according to Bresser-Pereira, the inflation generated by wage increases ultimately hits the poorer social constituencies. The populism of the Right, on the other hand, is informed by 'subordinated' internationalism, monetarism (especially control of inflation through wage control and high interests rates) and a mild developmentalism based on the co-operation between labour and capital.

Lulismo overcame the economic differences between leftist and rightist populism by mixing together neo-liberalism and developmentalism; high interest rates and minimum wage increases; popular credit and fiscal austerity; artificial valuation of currency and private sector incentives; nationalisation and extreme subcontracting by foreign conglomerates; subsidised credit and household debt; poverty relief and working-class precarisation. The collapse of the commodity boom showed the limits of such a schizophrenic political economy and the failure of Lula's socialist project, which was merely *political* but said nothing about how to democratise the economy, even though a big part of the militant base of the PT originated from the structures of participatory economy it had developed in municipalities across the country in the 1980s. From 2002 to 2014, the PT turned thirty million poor people into the 'new' working class and forty million traditional workers into 'lower middle class' (Singer 2018a: 87). But it did so through top-down policies of income redistribution and employment generation, which, if anything, decreased economic democracy and direct participation at the point of production.

Indeed, Singer (2018a: 99–127) notes that under the PT, 'Class C' brought together the new outsourced and exploited workforce in the service economy and the old precarised industrial working class. Such 'intermediate' dispersed, and alienated social strata representing 55 per cent of the population had no class allegiances to the PT. Elsewhere, I discuss how the wave of mass demonstrations that shook Brazil in June 2013 brought together this 'magmatic' Class C and the traditional middle class in an anti-systemic movement that initiated with the radical Left but was quickly appropriated by the Right and recast as an anti-state movement (Mollona 2018a). Likewise, the Petrobras corruption scandal had a huge symbolic resonance as a morally failed state project and generated a powerful anti-statist front bringing together the precarious Class C, the radical Left and the traditional middle class. This united anti-statist front ultimately led to Rousseff's impeachment. It is also important to highlight the difference between the conservative populism of Vargas and that of Lula. Vargas attempted to tackle the social question without upsetting the political order historically associated with Brazil's dependent development. Unlike him, Lula reached out to the dispossessed rural and urban lumpen and attempted to turn this historically marginalised social constituency into an active political agent within the new state. Indeed, Lula's biggest success was to win over the trust of the traditionally conservative and anti-labour marginal working-class at the expense of the Brazilian Social Democracy Party – the party of the bourgeoisie.

The PT's ascent to power was part of a broader liberation struggle in the Global South, especially in India and Latin America, informed by horizontal and cross-sectional politics and which, based on the experience of slavery and colonial dependency, combined struggles for cultural recognition and struggles for economic redistribution. Unlike the populism of the dominant classes, this new kind of 'socialist populism'[3] sublimated and contaminated working-class ideology with the language and experiences of subaltern classes and brought together labour movement and civil rights activism. These subaltern struggles in the periphery contributed to a phenomenal counterhegemonic political movement in Europe – including feminism, black and ecological activism – much of which was sadly co-opted into the 'new capitalist spirit' of New Labour in the United Kingdom, Podemos in Spain and Syriza in Greece, which turned the critique of Eurocommunism levied by Gramscian public intellectuals such as Hall, Laclau and Mouffe into a new hegemonic discourse.

Socialist populism, even when inspired by a Gramscian framework, cannot be a radical emancipatory project if it remains on a merely discoursive and political level without addressing the economic base of working-class emancipation and how culture articulates itself at the point of production as Gramsci did in his early work.[4] The failure of the inclusive populist neo-extractivist regimes in Latin America is reflected in the collapse of those left-wing populist parties that emerged in Europe after the 2008 crisis in the same way the recent election of right-wing populist and evangelical Christian Jair Bolsonaro (from the Social Liberal Party) to the presidency mirrors the rise of authoritarian neo-liberalism (Saad-Filho 2018) in Europe. In this ethnography, I have shown the cracks of *lulismo* were already evident back in 2008. If the formula of people-led capitalism appeared to work politically, in economic terms it only reinforced Brazil's financial, technological and environmental dependency on foreign capital and the volatile global commodity cycles; the power of financial oligarchies vis-à-vis the industrial bourgeoisie; and the progressive deskilling, precarisation and informalisation of the working-class I describe here.

Indeed, the crisis of the PT in Brazil and that of the Latin American Left speaks more broadly to the crisis of the global Left for the way, once in power, it has 'moralised' the economy and at the same time 'de-economised politics', as if economic equality could be achieved by simple discoursive articulations. According to the economist Wolfgang Streeck (2016), capitalism is currently unable to sustain its 'healthy' oscillation between economy and politics – Polanyi's double movement – and hence is in terminal crisis. Thus, the formula that combines capitalist growth and social redistribution that left-wing populist governments achieved through a mixture of financialisation and state redistribution is also in deep crisis. Consequently, the moralisation of the economy– through Lula's 'people's capitalism' or the indignation of Podemos – is pointless without the articulation of a clear post-capitalist strategy.

The 're-economisation of politics' based on the revival of the working class as a political subject and on an internationalist post-capitalist project is the priority of socialism in the twenty-first century. Capitalism relies on the separation between economics and politics based on the ossification of life into external labour ideologies and economic principles. Thus, post-capitalism calls for their reconnection and for the construction of a society in which forms of livelihood and social relations escape the commodity form and are unique, autonomous and fluid but also relational, open and sustainable – as in Castoriadis's 'magmatic' utopian democracy. What kind of material organisation, which community of practices and what forms of political imagination are associated with post-capitalist life? What kinds of social relations of production and what structures of exchange and circulation can be built outside the commodity form to tackle the material basis of inequality and at the same time increase participatory democracy?

From this Brazilian ethnography, the following directions for transitioning into post-capitalism emerge: (1) the exploitation and commodification of nature and extractivist practices produce economic dispossession, ecological disruption and social inequality, and they are incompatible with any socialist project; (2) new leaderships within the labour and civic movements are forming with women and the youth at their forefront and revolving around practices of commoning; (3) new working-class solidarities and joint forms of activism are emerging along global commodity chains and within corporate conglomerates (Mello 2016; Ramalho and Santos 2018). Hence, a logistical analysis of the extractive operations of capital is necessary to coordinate these joint actions. On the other hand, cross-sectional solidarities and logistical labour activism are hindered by the militant particularism and factory-based activism generated by the current climate of economic protectionism and renationalisation of the economy; (4) in Latin America, questions of spatial justice and rights to the city have been central in the surge of powerful social and labour movements in the 1980s. These movements and issues have been co-opted by left-wing, inclusive populist governments in the 2000s but are undergoing a revitalisation. Thus, it is important to consider work-based activism in its entanglement with issues of spatial justice and the political economy of cities. (5) Under capitalism, the reproductive jobs of looking after objects, people and machines are systematically undervalued and precarised. But a new logic of valorisation of life, revolving around those human actions that contribute to the reproduction of life, is emerging especially among the precarised and informal working class. (6) In Brazil, as in other peripheries of the Global South, the tradition of anti-slavery that animates struggles for civic emancipation and insurgent acts of commoning by its vast, precarised, informal, illegal and unwaged labour force is an important source of working-class power.

It is hoped that, as new and more brutal processes of primitive accumulation hit the Global North, creating new forms of spatial segregation, land grabbing and social expulsions, the experiences of labour and urban activism in this book will be of inspiration to the global Left. The central aim of *Brazilian Steel Town* is to bring the Brazilian experience to bear on the future of the working class worldwide in the belief that its basic needs, agency and capacity for anti-capitalist articulation are universal and cross-cultural and that, as Leo Panitch (2001: 389) argues, 'a new internationalism, able to reinvent solidarity in the era of globalization, is fundamental for the future of labour'. A few days after Bolsonaro's election to the presidency, sociologist André Singer (2018b) wrote an article in the *Folha de São Paulo*, describing his imaginary phone call with Stefan Zweig. Zweig was worried for the popular support gained by Bolsonaro, which reminded him of Hitler's ascent to power in Germany back in the 1930s. Singer tried to reassure him. But Zweig remained staunchly pessimistic. Singer's article seems to suggest the ghosts of Europe continue to haunt Brazil, whose future looks like Europe's past.

But the kind of despotic capitalism that emerged under the Workers' Party – based on the re-emergence of nation-states as despotic capitalist subjects and powerful financial-industrial-trade-union complexes engaged in 'schizophrenic' processes of deindustrialisation and re-industrialisation and in forms of capital governance, which simultaneously formalise and precarise labour – seems to prefigure and pre-enact the kind of economic future that is shaping before our very eyes in the United States and the United Kingdom in the era of Trump and of Brexit. In this sense, paraphrasing Zweig, Brazil is indeed 'the land of the future'.

Notes

1. Despite these Marxists approaches, according to Aricó, Marx did acknowledge the centrality of anticolonial struggles in the peripheries for working-class struggles in the centres in his discussion of the Irish question.
2. Bolsonaro was elected by 67 per cent of Volta Redonda residents.
3. Laclau argues 'socialist populism is the most advanced form of working-class ideology' (2005: 200). Likewise, Judis (2016) distinguishes between authoritarian and left-wing populism.
4. See Gramsci (1994) for his early writing on Factory Council published in the *L'Ordine Nuovo*.

REFERENCES

Abolafia, Mitchell Y. 2004. 'Framing Moves: Interpretive Politics at the Federal Reserve'. *Journal of Public Administration Research and Theory* 4(3): 349–370.

Abreu, Alice R. de P., Huw Beynon and José Ricardo Ramalho, 2000. '"The Dream Factory": VW's Modular Production System in Resende, Brazil'. *Work, Employment and Society* 14(2): 265–282.

Abrucio, Franco L., and Márcia M. Soares, 2001. *Redes federativas no Brasil: Cooperacão intermunicipal no grande ABC*. São Paulo: Fundação Konrad Adenauer.

Agamben, Giorgio. 2005. 'Movement.' *Generation Online*, accessed 18 April 2019. http://www.generation-online.org/p/fpagamben3.htm.

Amann, Edmund. 2004. 'Trade Reform, Market Liberalisation and Industrial Performance in Brazil'. In Amann and Chang 2004: 75–89.

Amann, Edmund, and Ha-Joon Chang, eds. 2004. *Brazil and South Korea: Economic Crisis and Restructuring*. Washington, DC: Brookings Institution Press.

Amann, Edmund, and Fred I. Nixson. 1999. 'Globalization and the Brazilian Steel Industry: 1988–97'. *Journal of Development Studies* 35(6): 59–88.

Amann, Edmund, Germano Mendes de Paula and João Carlos Ferraz. 2004. 'Ownership Structure in the Post-privatized Brazilian Steel Industry: Complexity, Instability and the Lingering Role of the State.' Oxford Centre on Regulation and Competition Working Paper no. 75.

Anderson, Perry. 2016. 'The Heirs of Gramsci'. *New Left Review* 100: 71–98.

Andrade, Oswald de. 1924. 'Manifesto da Poesia Pau-Brasil'. *Correo da Manhã*, 18 March 1924.

Antunes, Ricardo, and Marco Aurélio Santana. 2014. 'The Dilemmas of the New Unionism in Brazil: Breaks and Continuities'. *Latin American Perspectives* 41(5): 10–21.

Arbix, Glauco. 1996. *Uma aposta no futuro: Os primeiros anos da câmar setorial da indústria automobilística*. São Paulo: Scritta.

———. 2000. 'Guerra fiscal e competição intermunicipal por novos investimentos no setor automotivo'. *Dados* 43(1): 5–43.

Arendt, Hannah. 1958. *The Human Condition*. Chicago: University of Chicago Press.

Aricó, José. 2013. *Marx and Latin America*. Trans. David Broder. Leiden: Brill.

Ashwin, Sarah. 1999. *Russian Workers: The Anatomy of Patience*. Manchester: Manchester University Press.

Badaró Mattos, Marcelo. 2008. *Escravizados e livres: Experiências comuns na formação da classe trabalhadora Carioca (1850–1910)*. Rio de Janeiro: Bom Texto.

Baer, Werner, and Paul Beckerman. 1987. 'The Decline and Fall of Brazil's Cruzado'. University of Illinois College of Commerce of Business Administration Faculty Working Paper no. 1393.

Balibar, Etienne, and Immanuel Wallerstein. 1991. *Race, Nation, Class: Ambiguous Identities*. London: Verso.

Banerjee, Abhijit V., and Esther Duflo. 2008. 'What Is Middle Class about the Middle Classes around the World?' *Journal of Economic Perspectives* 22(2): 3–28.

Barkan, Joshua. 2013. *Corporate Sovereignty: Law and Government under Capitalism*. Minneapolis: University of Minnesota Press.

Barros, Maurício. 1999. *Labour Relations and the New Unionism in Contemporary Brazil*. London: Palgrave Macmillan.

Bazerra, Gustavo. 2017. 'Sindicalismo Ajustado ao Neoliberalismo: os metalúrgicos do sul Fluminense. *Caderno CRH* 30(80): 371–387.

Beck, Ulrich, 2000. *Brave New World of Work*. New York: Polity.

Bedê, Waldyr Amaral. 2004. *Volta Redonda na Era Vargas (1941–1964)*. Volta Redonda: Secretaria Municipal de Cultura / Prefeitura Municipal de Volta Redonda.

Bellofiore, Ricardo. 2009. 'A Ghost Turned into a Vampire: The Concept of Capital and Living Labor.' In Bellofiore Riccardo and Roberto Fineshi (eds), *Re-reading Marx: New Perspectives after the Critical Edition*. London: Palgrave Macmillan, 178–194.

Benson, Todd. 2005. 'Brazil's Steel Giant and Its Company Town Are on the Outs'. *New York Times*, 17 May.

Beynon, Huw. 1973. *Working for Ford*. London: Allen Lane.

Beynon, Huw, and Robin Blackburn. 1972. *Perceptions of Work: Variations within a Factory*. Cambridge: Cambridge University Press.

Beynon, Huw, and José Ricardo Ramalho. 2000. 'Democracy and the Organization of Class Struggle in Brazil.' *Socialist Register* 37: 229–237.

Bhattacharya, Adrindam, Thomas Bradtke, Tenbite Ermias, Whitney Haring-Smith, David Lee, Eudardo León, Michael Meyer, et al. 2013. 'Introducing the 2013 BCG Global Challengers'. Boston Consulting Group, 15 January. https://www.bcg.com/publications/2013/globalization-growth-introducing-2013-bcg-global-challengers.aspx.

Birman, Patricia. 1996. 'Cultos de Possessão e Pentecostalismo no Brasil: passagens'. *Religião e Sociedade* 17(2): 90-109.

Boito, Armando, Jr. 1999. *Política neoliberal e sindicalismo no Brasil*, São Paulo: Xamã Editora.

———. 2007. 'Class Relations in Brazil's New Neoliberal Phase'. *Latin American Perspectives* 34(5): 115–131.

Boltanski, Luc, and Eve Chiapello. 2007. *The New Spirit of Capitalism*. London: Verso.

Braga, Ruy. 2016. 'Terra em transe: O fim do lulismo e o retorno da luta de classes'. In Singer and Loureiro 2016: 55–92.

Braga, Ruy, and Sean Purdy. 2018. 'A Precarious Hegemony: Neo-liberalism, Social Struggles, and the End of Lulismo in Brazil'. *Globalization* 16(2): 201–215.

Braverman, Harry. 1974. *Labor and Monopoly Capital: The Degradation of Work in the Twentieth Century*. New York: Monthly Press.

Breman, Jan. 2013. 'The Precariat: A Bogus Concept?' *New Left Review* 84: 130–138.

Brenner, Johanna, and Nancy Fraser. 2017. 'What Is Progressive Neoliberalism? A Debate'. *Dissent* 64(2): 130–140.

Bresser-Pereira, Luiz Carlos. 1986. 'Inflação inercial e Plano Cruzado'. *Revista de Economia Política* 6(3): 9–24.

———. 1991. 'Populism and Economic Policy in Brazil'. *Journal of Interamerican Studies and World Affairs* 33(2): 3–21.

———. 2017. *The Political Construction of Brazil: Society, Economy and State since Independence*. Boulder, CO: Lynne Rienner.

Buarque de Holanda, Sérgio. 1936. *Raízes do Brasil*. São Paulo: Companhia das Letras.

Burawoy, Michael. 1985. *The Politics of Production: Factory Regimes under Capitalism and Socialism*. London: Verso.

Cabral de Souza, Claudia Virginia. 1992. 'Pelo Espaço da Cidade: Aspectos da Vida e do Conflito Urbano em Volta Redonda'. PhD diss., Universidade Federal de Rio de Janeiro.

Camboriú, Balneário. 2013. 'Housing in Brazil: If You Build It'. *The Economist*, 16 February.

Cardoso, Adalberto. 2010. 'Uma Utopia Brasileira: Vargas e a construção do estado de bem-estar numa sociedade estruturalmente desigual'. *Dados* 53(4): 775–819.

Carrier, James, 1992. 'Emerging Alienation in Production: A Maussian History'. *Journal of Royal Anthropological Institute* 27(3): 539–558.
Carrier, James and Don Kalb, eds. 2014. *Anthropologies of Class: Power, Practice, and Inequality*. Cambridge: Cambridge University Press.
Carsten, Janet. 1989. 'Cooking Money: Gender and the Symbolic Transformation of Means of Exchange in a Malay Fishing Community'. In Parry and Bloch 1989: 117–141.
Caruso, Danilo. 2009. 'Reestruturação produtiva e movimento operário em Volta Redonda'. Master's thesis, Universidade Federal Fluminense.
Carvalho, Laura. 2018. *Valsa brasileira: Do boom ao caos econômico*. São Paulo: Todavia.
Castoriadis, Cornelius. 1994. 'The Logic of Magmas and the Question of Autonomy'. *Philosophy and Social Criticism* 20(1–2): 123–154.
Castro Gomes, Angela de. 1987. *A invenção do trabalhismo*. Rio de Janeiro: Editora FGV.
Chakrabarty, Dipesh. 1989. *Rethinking Working Class History: Bengal 1890–1940*. Princeton, NJ: Princeton University Press.
———. 2000. *Provincializing Europe*. Princeton, NJ: Princeton University Press.
Chalcraft, John T. 2005. 'Pluralizing Capital, Challenging Eurocentrism: Toward Post-Marxist Historiography'. *Radical History Review* 91: 13–39.
Chalhoub, Sidney. 2006. 'The Politics of Silence: Race and Citizenship in Nineteenth-Century Brazil'. *Slavery & Abolition* 27(1): 73–87.
———. 2011. 'The Precariousness of Freedom in a Slave Society (Brazil in the Nineteenth Century)'. *International Review of Social History* 56: 405–439.
———. 2012. *A Força da escravidão: Ilegalidade e costume no Brasil oitocentista*. São Paulo: Companha da Letras.
Chandavarkar, Rajnarayan. 1994. *The Origins of Industrial Capitalism in India: Business Strategies and the Working Classes in Bombay, 1900–1940*. Cambridge: Cambridge University Press.
Chen, Mel Y. 2012. *Animacies: Biopolitics, Racial Mattering, and Queer Affect*. Durham, NC: Duke University Press.
Chibber, Vivek. 2013. *Postcolonial Theory and the Specter of Capital*. London: Verso.
Clastres, Pierre. 1974. *Society against the State: Essays in Political Anthropology*. Cambridge, MA: Zone Books. MIT Press.
CMV (Comissão Municipal da Verdade). 2015. *Relatório final*. Volta Redonda: CMV.
Coase, Ronald H. 1937. 'The Nature of the Firm'. *Economica* 4(16): 386–405.
Comaroff, Jean, and John Comaroff. 2000. 'Millennial Capitalism: First Thoughts on a Second Coming'. *Public Culture* 12(2): 291–343.
Coronil, Fernando. 1997. *The Magical State: Nature, Money and Modernity in Venezuela*. Chicago: Chicago University Press.
Costa, Alkindar. 1978. *Volta Redonda ontem e hoje: Visão histórica e estática*. Volta Redonda: Masiero
Crovetto, Gonzalo Díaz, ed. 2012. 'Situando trajetórias na Cidade do Aço: formação, acesso e trabalho entre diferentes gerações de trabalhadores as Companhia Siderúrgica Nacional (CSN) em Volta Redonda'. *MEDEA (Models and Their Effects on Development Path): An Ethnographic and Comparative Approach*.
CSN (Companhia Siderúrgica Nacional). 2010. *Formulário de Referência*. Volta Redonda: CSN.
———. 2015. *Relatório anual*. Volta Redonda: CSN.
CUT (Central Única dos Trabalhadores). 2014. *Terceirização e Desenvolvimento: Uma conta que não fecha – Dossiêr acerca do impacto da Terceirização sobre os trabalhadores e propostas para garantir a igualdade de direitos*. São Paulo: CUT.
Dalton, George. 1965. 'Primitive Money'. *American Anthropologist* 67(1): 44–65.
DaMatta, Roberto. 1983. *Carnavais, malandros e heróis: Para uma sociologia do dilema brasileiro*. Rio de Janeiro: Zahar Editores.

D'Araujo Soares, Maria Celina. 2009. 'Déficits democráticos em lento viés de baixa'. *Conjuntura Economica* 63: 32–35.

Davis, Mike. 2006. *Planet of Slums*. London: Verso.

De Angelis, Massimo. 2018. *Omnia Sunt Communia: On the Commons and the Transformation to Postcapitalism*. London: Zed

Denning, Michael. 2010. 'Wageless Life'. *New Left Review* 66: 79–97.

Devons, Ely. 1961. *Essays in Economics*. London: Allen & Unwin.

Dias, Sabrina de Oliveira Moura. 2010. 'Dentro da usina mas fora da "família": Trabalhadores e terceirização na Companhia Siderúrgica Nacional (CSN)'. Master's thesis, Universidade Federal do Rio de Janeiro.

———. 2015. 'A súmula nº 331 e os limites da terceirização no Brasil'. *Anais do XIV Encontro Nacional da ABET*.

DIEESE (Departamento Intersindical de Estatísticas e Estudos Socioconômicos). 2000. Anuário estatístico. São Bernardo do Campo: Departmanento Intersindical de Estudos Estatísticos e Sócio-econômicos.

Dinius, Oliver Jürgen. 2011. *Brazil's Steel City: Developmentalism, Strategic Power and Industrial Relations in Volta Redonda, 1941–1964*. Stanford, CA: Stanford University Press.

Druck, Maria da Graça. 1999. *Terceirização: (Des) fordizando a fábrica, um estudo do complexo petroquímico*. São Paulo: Boitempo,

Dunn, Elizabeth Cullen. 2004. *Privatizing Poland: Baby Food, Big Business, and the Remaking of Labor*. Ithaca, NY: Cornell University Press.

Dupré, Georges, and Pierre-Philippe Rey. 1973. 'Reflections on the Pertinence of a Theory of the History of Exchange'. *Economy and Society* 2(2): 131–163.

Eccles, Robert G. 1981. 'The Quasifirm in the Construction Industry'. *Journal of Economic Behavior & Organization* 2(4): 335–357.

Eliade, Mircea. 1978. *The Forge and the Crucible: The Origins and Structure of Alchemy: Origins and Structures of Alchemy*. Chicago: Chicago University Press.

Escobar, Arturo. 1999. 'After Nature: Steps to an Anti-essentialist Political Ecology'. *Current Anthropology* 40(1): 3.

Estevez, Alejandra, and Raphael da Costa Lima. 2015. 'A articulação empresarial-militar na cidade de Volta Redonda: violações aos direitos humanos da classe trabalhadora'. In Ozanan Vicente Carrara (ed.), *Direitos Humanos na América Latina*. São Leopoldo: Karywa, 1–23.

Euripides. (484–406 BCE) 2002. *Heracles and Other Plays*. London: Penguin.

Evans, Peter B. 1979. *Dependent Development: The Alliance of Multinational, State, and Local Capital in Brazil*. Princeton, NJ: Princeton University Press.

Federici, Silvia. 2004. *Caliban and the Witch: Women, the Body and Primitive Accumulation*. New York: Autonomedia.

———. 2012. *Revolution at Point Zero: Household, Reproduction and Feminist Struggle*. Oakland, CA: PM Press.

Felício, Marcelo. 2008. 'Os Netos da CSN'. Master's thesis, Universidade Federal Fluminense, Volta Redonda.

Fernandes, Anaïs. 2018. 'Justiça supermãe foi um dos estopins para a reforma trabalhista. Entrevista a Ives Gandra Martins Filho'. *Folha de São Paulo*, 10 November.

Fernandes, Florestan. 1975. *A revolução burguesa no Brasil: Ensaio de interpretação sociológica*. Rio de Janeiro: Zahar.

Fishlow, Albert. 2011. *Starting Over: Brazil since 1985*. Washington, DC: Brookings Institution Press.

Fontes, Virginia. 2010. *O Brasil e o capital-imperialismo: Teoria e história*. Rio de Janeiro: Editora UFRJ.

Fontes, Virginia, and Ana Garcia. 2014. 'Brazil's New Imperial Capitalism'. *Socialist Register* 50: 207–226.

Francisco, Elaine Marlova. 2015. 'Desafios para a organização dos trabalhadores metalúrgicos no Sul Fluminense: uma década de experiências'. In Ramalho and Fortes 2015: 161–180.
Fraser, Nancy. 1995. 'From Redistribution to Recognition? Dilemmas of Justice in a "Post-socialist" Age'. *New Left Review* 212: 68–93.
———. 2013. 'How Feminism Became Capitalism's Handmaiden – How to Reclaim It'. *The Guardian*, 14 October.
———. 2016. 'Contradictions of Capital and Care'. *New Left Review* 100: 99–117.
French, John D. 2004. *Drowning in Laws: Labor Law and Brazilian Political Culture*. Chapel Hill: University of North Carolina Press.
Freyre, Gilberto. 1987. *The Masters and the Slaves (Casa-Grande & Senzala): A Study in the Development of Brazilian Civilization*. Berkeley: University of California Press.
FS (Força Sindical). 1993. *Um projeto para o Brasil: A proposta da Força Sindical*. São Paulo: Geração Editorial.
Fujikawa Nes, Egil. 2016. 'Social Classes in Brazil'. *Brazil Buisness*, 26 January. https://thebrazilbusiness.com/article/social-classes-in-brazil-1453802521.
Gago, Veronica, and Sandro Mezzadra. 2017. 'A Critique of the Expanded Operations of Capital: Towards an Expanded Notion of Extractivism'. *Rethinking Marxism: A Journal of Economics, Culture and Society* 29(4): 574–591.
Garnier, Tony. 1917. *Une cité industrielle*. Princeton, NJ: Princeton Architectural Press.
Giambiagi, Fabio. 2005. 'Rompendo com a Ruptura: O governo Lula (2003–2004)'. In Fabio Giambiagi, André Villela, Lavinia Barros de Castro and Jennifer Hermann, *Economia Brasileira Contemporânea*. Rio de Janeiro: Elsevier, 197–238.
Gibson-Graham, J. K. 1996. *The End of Capitalism (as We Know It): A Feminist Critique of Political Economy*. Oxford: Blackwell.
Godelier, Maurice. 1966. *Rationality and Irrationality in Economics*. Trans. Brian Pearce. London: Verso.
———. 1986. *The Making of Great Men: Male Domination and Power among the New Guinea Baruya*. Trans. Rupert Swyer. Cambridge: Cambridge University Press.
Goldstein, Andrea. 1999. 'Brazilian Privatizations in International Perspective: The Rocky Path from State Capitalism to Regulatory Capitalism'. *Industrial and Corporate Change* 8(4): 673–711.
Goldstein, Andrea, and Ben Ross Schneider. 2004. 'Big Business in Brazil: States and Markets in the Corporate Reorganization of the 1990s'. In Amann and Chang 2004: 48–74.
Goldthorpe, John H., David Lockwood, Frank Bechhofer and Jennifer Platt. 1968. 'The Affluent Worker and the Thesis of Embourgeoisement: Some Preliminary Research Findings'. *Sociology* 1(1): 11–31.
Gorender, Jacob. 1978. *O escravismo colonial*. São Paulo: Editora Atica.
Graciolli, Edilson José. 2007. *Privatização da CSN: Da luta de classes à parceria*. São Paulo: Expressão Popular.
Graeber, David. 2004. *Fragments of an Anarchist Anthropology*. Chicago: Prickly Paradigm Press.
———. 2013. *Debt: The First 5000 Years*. New York: Melville House.
———. 2015. *The Utopia of Rules: On Technology, Stupidity and the Secret Joys of Bureaucracy*. New York: Melville House.
Gramsci, Antonio. 1971. *Selections from the Prison Notebook*. London: Lawrence & Wishart.
———. 1994. *Gramsci: Pre-prison Writings*. Ed. Richard Bellamy. Cambridge: Cambridge University Press.
Guha, Ranajit. 1999. *Elementary Aspects of Peasant Insurgency in Colonial India*. Durham, NC: Duke University Press.

Haggard, Stephan, and Robert R. Kaufman. 1995. *The Political Economy of Democratic Transitions*. Princeton, NJ: Princeton University Press.
Hall, Peter A., and David Soskice 2001. *Varieties of Capitalism: The Institutional Foundations of Comparative Advantage*. Oxford: Oxford University Press.
Hall, Stuart. 1970. 'Black Britons'. *Community* 1: 3–5.
———. 1980b. 'Thatcherism: A New Stage?' *Marxism Today* (February), 26–28.
Hall, Stuart, and Tony Jefferson. 1975. *Resistance through Rituals: Youth Subcultures in Post-war Britain*. Birmingham: Centre for Contemporary Cultural Studies.
Harman, Graham. 2002. *Tool-Being: Heidegger and the Metaphysics of Objects*. Chicago: Open Court Publishing
Hart, Keith. 1973. 'Informal Income Opportunities and Urban Employment in Ghana'. *The Journal of Modern African Studies* 11(1): 61–89.
———. 1986. 'Heads or Tails? Two Sides of the Coin'. *Man* 21(4): 637–656.
Harvey, David. 2012. *Rebel Cities: From the Right to the City to the Urban Revolution*. London: Verso.
Hayek, Friedrich. 1944. *The Road to Serfdom*. London: Routledge.
Hayter, Teresa, and David Harvey. 1993. *The Factory and the City: Story of the Cowley Automobile Workers in Oxford*. London: Thompson Learning.
Heinrich, Michael. 2012. *An Introduction to the Three Volumes of Karl Marx's Capital*. New York: Monthly Press.
Hirschman, Albert O. 1972. *Exit, Voice, and Loyalty Responses to Decline in Firms, Organizations, and States*. Cambridge, MA: Harvard University Press.
———. 1977. *The Passions and the Interests: Political Arguments for Capitalism before Its Triumph*. Princeton, NJ: Princeton University Press.
———. 1981. *Essays in Trespassing: Economics to Politics and Beyond*. Cambridge: Cambridge University Press.
Hobson, John. (1904) 2011. *Imperialism: A Study*. New York: Cosimo.
Hodgson, Geoffrey. 2015. *Conceptualizing Capitalism: Institutions, Evolution, Future*. Chicago: University of Chicago Press.
Holmes, Douglas R. 2013. *Economy of Words: Communicative Imperatives in Central Banks*. Chicago: University of Chicago Press.
———. 2016. 'Central Bank and Magic Talk'. *Anthropology Today* 32(6): 3–7.
Holston, James. 2007. *Insurgent Citizenship: Disjunctions of Democracy and Modernity in Brazil*. Princeton, NJ: Princeton University Press.
Humphrey, John. 1982. *Capitalist Control and Workers' Struggle in the Brazilian Auto Industry*. Princeton, NJ: Princeton University Press.
Huws, Ursula. 2003. *The Making of the Cybertariat: Virtual Work in a Real World*. New York: Monthly Review.
———. 2014. *Labor in the Global Digital Economy: The Cybertariat Comes of Age*. New York: Monthly Review.
Ingold, Tim. 2000. *The Perception of the Environment: Essays in Livelihood, Dwelling and Skill*. London: Routledge.
IBGE (Instituto Brasileiro de Geografia e Estatistica). 2006. 'Perfil dos Trabaladhores Domésticos'. *Pesquisa Mensal de Emprego – (PME)*. (March)
IPEA (Instituto de Pesquisa Economica Applicada). 2012. 'A década inclusiva (2001–2011): Disegualdade, pobreza e políticas da renda'. *Comunicados do IPEA* 155 (25 September).
IPPU-VR (Instituto de Pesquisa e Planejamento de Volta Redonda). 2013. 'Retiro'. *Cadernos de Bairros*. Accessed 29 April 2019. http://www.ippuvr.com.br/nuvem/caderno_bairros/RETIRO.pdf.
Issa, Daniela. 2007. 'Praxis of Empowerment: *Mística* and Mobilization in Brazil's Landless Rural Workers' Movement'. *Latin American Perspectives* 34(2): 124–138.

Jardim, Maria Apparecida Chaves. 2009. 'Domesticacao e/ou "moralizacao do capitalismo" no governo Lula: Inclusão social via mercado e via fundos pensao'. *Dados* 52(1): 123–159.
Jaspers, Karl. 1953. *The Origin and Goal of History*. London: Routledge.
Jessop, Bob, Tom Ling, Simon Bromley and Kevin Bonnett. 1985. 'Thatcherism and the Politics of Hegemony: A Reply to Stuart Hall'. *New Left Review* 153: 87–101.
Judis, John. 2016. *The Populist Explosion: How the Great recession Transformed American and European Politics*. New York: Columbia Global Report.
Kalb, Don. 1998. *Expanding Class: Power and Everyday Politics in Industrial Communities, the Netherlands, 1850–1950*. Durham, NC: Duke University Press.
Kalb, Don, and Massimiliano Mollona, eds. 2018. *Worldwide Mobilizations: Class Struggles and Urban Commoning*. New York: Berghahn Books.
Kalm, Sarah. 2010. 'Liberalizing Movements? The Political Rationality of Global Migration Management'. In Martin Geiger and Antoine Pecaud (eds), *The Politics of International Migration Management*. London: Palgrave Macmillan, 21–44.
Kasmir, Sharryn, and August Carbonella. 2008. 'Dispossession and the Anthropology of Labor'. *Critique of Anthropology* 28(1): 5–25.
Kasmir, Sharryn, and Lesley Gill. 2016. 'History, Politics, Space, Labor: On Unevenness as an Anthropological Concept'. *Current Anthropology* 59(4): 355–377.
Kearney, Michael. 1996. *Reconceptualizing the Peasantry: Anthropology in Global Perspective*. Boulder, CO: Westview Press.
Kropotkin, Peter. (1899) 1919. *Fields, Factories and Workshops*. London: Nelson.
Laclau, Ernesto. 2005. *On Populist Reason*. London: Verso.
Laclau, Ernesto, and Chantal Mouffe, 1985. *Hegemony and Socialist Strategy: Towards a Radical Democratic Politics*. London: Verso.
Lambek, Michael. 2013. 'The Value of (Performative) Acts'. *HAU: Journal of Ethnographic Theory* 3(2): 141–160.
Lazzarato, Maurizio. 1996. 'Immaterial Labour'. In Paolo Virno and Michael Hardt (eds), *Radical Thought in Italy: A Potential Politics*. Mineapolis: University of Minnesota Press, 142–157.
Leahy, Joe. 2012. 'FT Interviews: Dilma Rousseff'. *Financial Times*, 2 October.
———. 2016. 'Brazil Urged to Hold New Pension and Budget Reforms.' *Financial Times*, 22 October.
Lee, Benjamin, and Edward LiPuma. 2002. 'Cultures of Circulation: The Imaginations of Modernity'. *Public Culture* 14(1): 191–213.
Lee, Ching Kwan. 1998. *Gender and the South China Miracle: The Worlds of Factory Women*. Berkeley: University of California Press.
———. 2007. *Against the Law: Labor Protests in China's Rustbelt and Sunbelt*. Berkeley: University of California Press.
Lelis, Francismara de Oliveira. 2008. 'Mulheres Forjadas em Aço e Fogo: Militância feminina em Volta Redonda (1974–1985)'. Master's thesis, Centro Universitário Gerardo di Biase.
Lenin, Vladimir Illich. (1910) 2010. *Imperialism: The Highest Stage of Capitalism*. London: Penguin.
Lévi-Strauss, Claude, Augé Marc and Maurice Godelier. 1976. 'Anthropology, History and Ideology'. *Critique of Anthropology* 2(6): 44–55.
Lima, Raphael J. da Costa. 2010. 'A "Reinvenção" de uma Cidade Industrial: Volta Redonda e o pós-privatização da Companhia Siderúrgica Nacional'. PhD diss., Universidade Federal do Rio de Janeiro.
———. 2014. 'CSN e Volta Redonda: Uma relação histórica de dependência e controle'. *Política & Sociedade (Impresso)* 12(25): 41–64.
———. 2016. 'Delineando o cenário da privatização da Companhia Siderúrgica Nacional (CSN): Uma perspectiva de interpretação do processo a partir da mobilização dos agentes políticos locais'. *Perfiles Económicos* 2: 87–127.

———. 2018. 'Da luta sindical à luta pela cidade: O pos-privatização da Companhia Siderúrgica Nacional (CSN) e o *Coletivo Terras de Volta*'. In Silvia Simonassi and Daniel Diosimo (eds), *Trabajadores y sindicatos en Latinoamérica*. Buenos Aires: Imago Mundi, 37–48.

Linera, Álvaro García. 2013. *Geopolitics of the Amazon: Landlord Hereditary Power and Capitalist Accumulation*. Trans. Richard Fidler. New York: Climate & Capitalism.

LiPuma, Edward, and Benjamin Lee. 2004. *Financial Derivatives and the Globalization of Risk*. Durham, NC: Duke University Press.

Lopes, Alberto da Costa. 1992. 'A aventura da cidade industrial de Tony Garnier em Volta Redonda'. PhD diss., Universidade Federal de Rio De Janeiro.

Lopes, Francisco L. 1986. *O choque heterodoxo: Combate à inflação e reforma monetária*. Rio de Janeiro: Campus.

Lopes, José Sergio Leite. 1988. *A tecelagem dos conflitos de classe na 'Cidade das Chaminés'*. São Paulo: Editora Macro Zero.

———, ed. 2004. *A ambientalicação dos conflitos sociais: Partecipação e controle publico da poluição industrial*. Rio de Janeiro: Relume Dumará.

Lula da Silva, Luiz Inácio. 2008. 'Speech of Luiz Inácio da Silva, President of Brazil, at the Plenary Meeting of the G20 Ministers of Finance'. São Paulo, 8 November. http://www.g20.utoronto.ca/2008/2008lula1108-en.html.

Lupton, Tom. 1963. *On the Shop Floor: Two Studies of Workshop Organization and Output*. London: Pergamon Press.

Luz, Nícia Vilela. 1978. *A Luta pela Industrialização do Brasil*. São Paulo: Alfa Omega.

Mainwaring, Scott, Timothy Power and Farnando Bizzarro. 2018. 'The Uneven Institutionalization of a Party System: Brazil'. In Scott Mainwaring (ed.), *Party Systems in Latin America: Institutionalization, Decay and Collapse*. Cambridge: Cambridge University Press, 164–201.

Mangabeira, Wilma. 1993. *Os dilemas do novo sindicalismo: Democracia e política em Volta Redonda*. Rio de Janeiro: Relume Dumará.

Marazzi, Christian. 2009. *The Violence of Financial Capitalism*. Trans. Kristina Lebedeva. London: MIT Press.

Marinho, Helder. 2010. 'Brazil's Lula Leaves Office with 83% Approval Rating, Folha Says'. *Bloomberg*, 19 December. https://www.bloomberg.com/news/articles/2010-12-19/brazil-s-lula-leaves-office-with-83-approval-rating-folha-says.

Marini, Ruy M. 1977. 'World Capital Accumulation and Sub-imperialism'. *Cuadernos Politicos* 12: 21–39.

Marquese, Rafael, and Dale Tomich. 2015. 'O Vale do Paraíba escravista e a formação do mercado mundial do café no século XIX nos quadros da Segunda Escravidão'. In Mariana Mauze and Ricardo Salles (eds), *O Vale do Paraíba e o Império do Brasil*. Rio de Janeiro: Letras, 21–56.

Martin, Randy. 2015. *Knowledge LTD: Toward a Social Logic of the Derivative*. Philadelphia: Temple University Press.

Martins, Carlos Estevam. 1974. *Tecnocracia e capitalismo: As politicas dos técnicos no Brasil*. São Paulo: Brasiliense.

Marx, Karl. (1852) 1994. *The Eighteenth Brumaire of Louis Bonaparte*. London: International Publisher.

———. (1858) 1973. *Grundrisse: Foundations of the Critique of Political Economy*. London: Penguin.

———. (1867) 1990. *Capital: Critique of Political Economy*. Vol. 1. London: Penguin.

———. (1867) 1992. *Capital: Critique of Political Economy*. Vol. 2. London: Penguin.

Marx, Karl, and Frederick Engels. (1835–1895) 1987. *Collected Works*. Vol. 25. London: Lawrence & Wishart.

———. (1848) 2012. *The Communist Manifesto*. London: Verso.

Massey, Doreen, and Hilary Wainwright. 1985. 'Beyond the Coalfield: The Workers of the Miners' Support Groups'. In Huw Beynon (ed.), *Digging Deeper: Issues in the Miners' Strike*. London: Verso, 148–168.

Mattoso, Katia M. de Queirós. 1986. *To Be a Slave in Brazil*. New Brunswick, NJ: Rutgers University Press.

Maurer, Bill. 2005. *Mutual Life, Limited: Islamic Banking, Alternative Currencies, Lateral Reason*. Princeton, NJ: Princeton University Press.

Mauss, Marcel. (1950) 2002. *The Gift: The Form and Reason for Exchange in Archaic Societies*. London: Routledge.

Meillassoux, Claude. 1964. *Anthropologie economique des gouro de Coite D' Ivoire*. Paris: Mouton.

———. 1972. 'From Reproduction to Production'. *Economy and Society* 1, no. 1: 93–105.

———. 1981. *Maidens, Meal and Money: Capitalism and the Domestic Community*. Cambridge: Cambridge University Press.

———. 1986. *Anthropologie de 'esclavage: Le ventre de fer et d'argent*. Paris: Presses Universitaires de France.

Mello e Silva, Leonardo. 2016. 'Inovações do sindicalismo brasileiro em tempos de globalização e o trabalho sob tensão'. In Singer and Loureiro 2016: 93–122.

Meyer, John W., and Brian Rowan. 1977. 'Institutional Organizations: Formal Structure as Myth and Ceremony'. *American Journal of Sociology*, 83 (2) 340–63.

Mezzadra, Sandro. 2011. 'How Many Histories of Labour? Towards a Theory of Postcolonial Capitalism'. *Postcolonial Studies* 14(2): 151–170.

Mezzadra, Sandro, and Brett Neilson. 2013. *Border as Method, or, the Multiplication of Labor*. Durham, NC: Duke University Press.

MGI (McKinsey Global Institute). 1993. *Global Interpretation of the New Regionalism*. New York: McKinsey & Co.

Miers, Susanne, and Igor Kopytoff. 1977. *Slavery in Africa: Historical and Anthropological Perspectives*. Madison: University of Wisconsin Press.

Mitchell, Timothy. 2011. *Carbon Democracy: Political Power in the Age of Oil*. London: Verso.

Mollona, Massimiliano. 2005. 'Gifts of Labour: Steel Production and Technological Imagination in an Area of Urban Deprivation, Sheffield, UK'. *Critique of Anthropology* 25(2): 177–198.

———. 2009. *Made in Sheffield: An Ethnography of Industrial Work and Politics*. Oxford: Berghahn Books.

———. 2014. 'Working-Class Politics in a Brazilian Steel Town'. In Carrier and Kalb 2014: 149–163.

———. 2018a. 'The Brazilian "June" Revolution: Urban Struggles, Composite Articulations and New Class Analysis'. In Kalb and Mollona 2018: 163–186.

———. 2018b. 'Authoritarian Brazil Redux?' FocaalBlog, 6 October. https://www.focaalblog.com/2018/10/06/massimiliano-mollona-authoritarian-brazil-redux.

———. 2018c. 'Workers are People in the Process of Becoming. Reflections on Wagework and Slavery at the Dawn of Capitalism'. LSE Friday Seminar.

Montero, Alfred P. 1998. 'State Interests and the New Industrial Policy in Brazil: The Privatization of Steel, 1990–1994'. *Journal of Interamerican Studies and World Affairs* 40(3): 27–62.

Morel, Regina Lúcia Moraes. 1989. 'A ferro e fogo: Construção e crise da família siderúrgica o caso de Volta Redonda (1941–1969)'. PhD diss., Universidade de São Paulo.

Morel, Regina Lúcia Moraes, and Elina G. da Fonte Pessanha. 2006. 'Magistrados do trabalho no Brasil: Entre a tradição e a mudança'. *Estudos Históricos* 37: 29–53.

———. 2007. 'A justiça do trabalho'. *Tempo Social* 19(2): 87–109.

Munck, Ronaldo. 2011. 'Globalizing the Working-Class Concept: A Comment'. University of Liverpool Working Paper.

Musacchio, Aldo, and Sergio G. Lazzarini. 2014. *Reinventing State Capitalism: Leviathan in Business, Brazil and Beyond*. Cambridge, MA: Harvard University Press.

Narotzky, Susana, and Niko Besnier. 2014. 'Crisis, Value, and Hope: Rethinking the Economy'. *Current Anthropology* 55(9): 4–16.

Nash, June. 1979. *We Eat the Mines and the Mines Eat Us: Dependency and Exploitation in Bolivian Tin Mines*. New York: Columbia University Press.

Negri, Antonio. 1991. *Marx beyond Marx: Lessons on the Grundrisse*. Trans. Harry Cleaver, Michael Ryan and Maurizio Viano. Ed. Jim Fleming. New York: Autonomedia.

Neiburg, Federico. 2010. 'Sick Currencies and Public Numbers'. *Anthropological Theory* 10(1): 96–102.

Neri, Marcello. 2011. *A nova classe média o lado brilhante da base da pirâmide*. São Paulo: Saraiva.

O'Donnell, Guillermo. 1973. *Modernization and Bureaucratic Authoritarianism: Studies in South American Politics*. Berkeley: University of California Press.

Oliveira, Francisco de. 2005. *Crítica à razão dualista / O ornitorrinco*. São Paulo: Boitempo.

———. 2006. 'O momento Lênin'. *Novo Estudos* 75: 23–47.

Ong, Aiwa. 1988. 'The Production of Possession: Spirits and the Multinational Corporation in Malaysia'. *American Ethnologist* 15(1): 28–42.

Ortiz, Horacio. 2013. 'Financial Value: Economic, Moral, Political, Global'. *HAU: Journal of Ethnographic Theory* 3(1): 64–79.

Ouchi, William G. 1980. 'Markets, Bureaucracies, and Clans'. *Administrative Science Quarterly* 25(1): 129–141.

Overing Kaplan, Joanna. 1975. *The Piaroa: A People of the Orinoco Basin – A Study in Kinship and Marriage*. Oxford: Clarendon Press.

Panitch, Leo. 2001. 'Reflections on Strategies for Labour'. In Leo Panitch and Colin Leys (eds), *Working Classes: Global Realities*. London: Merlin Press, 367–392.

Pargendler, Mariana. 2012. 'State Ownership and Corporate Governance'. *Fordham Law Review* 8(6): 2917–2973.

———. 2013. 'Politics in the Origins: The Making of Corporate Law in Nineteenth-Century Brazil'. *American Journal of Comparative Law* 60(3): 805–849.

Parry, Jonathan. 1986. 'The Gift, the Indian Gift, and the "Indian Gift"'. *Man* 21(3): 453–473.

———. 1989. 'On the Moral Perils of Exchange'. In Parry and Bloch 1989: 64–94.

———. 2018. 'Precarity, Class and the Neoliberal Subject'. In Chris Hann and Jonathan Parry (eds), *Industrial Labor on the Margins of Capitalism: Precarity, Class and the Neoliberal Subject*. Oxford: Berghahn Books, 5–15.

Parry, Jonathan, and Maurice Bloch, eds. 1989. *Money and the Morality of Exchange*. Cambridge: Cambridge University Press.

Pearson, Samantha. 2015. 'Brazil's Middle Class Starts to Lose Ground'. *Financial Times*, 12 May.

Pereira, Sérgio Martins. 2007. 'Sindicalismo e privatização: O caso da Companhia Siderúrgica Nacional'. PhD diss., Universidade Federal do Rio de Janeiro.

Perrot, Michelle. 1988. *Os excluídos da história: Operário, mulheres e prisioneiros*. Trans. Denise Bottmann. Rio de Janeiro: Paz e Terra.

Pessanha, Elina G. da Fonte. 2001. 'Metalúrgicos, sempre operários navais'. In Ramalho and Santana 2001b: 11–23.

Pessanha, Elina G. da Fonte, and Regina Lúcia Moraes Morel. 1991. 'Gerações operárias: Rupturas e continuidades na experiência de metalúrgicos no Rio de Janeiro'. *Revista Brasileira de Ciências Sociais* 6(17): 68–83.

Pinheiro, Armando Castelar. 2002. 'The Brazilian Privatisation Experience: What's Next?' University of Oxford Centre for Brazilian Studies Working Paper no. CBS-30-02.

Pinheiro, Armando Castelar, and Fabio Giambiagi. 1997. 'Lucratividade, dividendos e investimento das empresas estatais: Uma contribuição para o debate sobre a privatização no Brasil'. *Revista Brasileira de Economia* 51(1): 93–131.

Pinney, Christopher. 2000. 'On Living in the Kal(i)yug'. In Jonathan P. Parry, Jan Breman and Karin Kapadia (eds), *The Worlds of Indian Industrial Labour*. London: Sage, 77–106.

Piore, Michael J., and Charles F. Sabel. 1984. *The Second Industrial Divide: Possibilities for Prosperity*. New York: Basic Books.

Piquet, Roberto. 1985. 'Moradia operária em Volta Redonda: De símbolo do populismo a lógica capitalista'. *Espaço e Debates* 5(16): 83–105.

Pitts, Fredrick Harry. 2018. *Critiquing Capitalism Today: New Ways to read Marx*. London: Palgrave Macmillan.

Pochmann, Marcio. 2012. *Nova classe média? O trabalho na base da pirâmide social brasileira*. São Paulo: Boitempo.

Polanyi, Karl. (1944) 2001. *The Great Transformation: The Political and Economic Origins of our Time*. Boston: Beacon Press.

Prahalad, Coimbatore K., 2004. *The Fortune at the Bottom of the Pyramid: Eradicating Poverty through Profits*. Philadelphia: Wharton School of Management.

Ramalho, José Ricardo. 1989. *Estado patrão e luta operária: O caso FNM*. São Paulo: Paz e Terra.

———. 2005. 'Novas conjunturas industriais e participação local em estratégias de desenvolvimento'. *Dados* 48(3): 147–178.

———. 2012. 'Fórum Demissão Zero: Crise e ação coletiva no Sul Fluminense'. In Ramalho and Fortes 2012: 175–200.

Ramalho, José Ricardo, and Alexandre Fortes, eds. 2015. *Desenvolvimento, trabalho e cidadania: Baixada e Sul fluminense*. Rio de Janeiro: Letras.

Ramalho, José Ricardo, and Marco Aurélio Santana. 2001a. 'Flexibilidade ã francesa: Trabalhadores na Peugeot Citroên brasileira'. *Tempo Social* 18(1): 115–132.

———, eds. 2001b. *Trabalho e tradição sindical no Rio de Janeiro: A trejetória dos metalúrgicos*. Rio de Janeiro: DP&A.

———. 2001c. 'Tradição sindical e as mudanças econômicas dos anos 1990: O caso dos metalúrgicos do Rio de Janeiro'. In Ramalho and Santana 2001b: 17–44.

———, eds. 2004. 'The VW's Modular System, Regional Development and Worker's Organisation in Resende, Brazil'. In Paul Stewart and Elsie Charron (eds), *Work and Employment Relations in the Automobile Industry*, vol. 1. London: Palgrave Macmillan, 45–57.

———, eds. 2006. *Trabalho e desenvolvimento regional: Efeitos sociais da indústria automobilística no Rio de Janeiro*. Rio de Janeiro: Mauad.

Ramalho, José Ricardo, Iram Jácome Rodrigues and Jefferson José da Conceição. 2009. 'Reestruturação industrial, sindicato e território Alternativas políticas em momentos de crise na região do ABC em São Paulo, Brasil'. *Revista Crítica de Ciências Sociais* 85: 47–167.

Ramalho, José Ricardo, Rodrigo Salles Pereira dos Santos and Raphael Jonathas da Costa Lima. 2013. 'Estratégias de Desenvolvimento industrial e dinâmicas territoriais de contestação social e confronto politico'. *Sociologia & Antropologia* 3(5): 175–200.

Ramalho, José Ricardo and Rodrigo Salles Pereira dos Santos. 2018. 'Trabalho a Ação Sindical em Redes Globais de Produção. *Tempo Social* 30: 9–29.

Rapoza, Kenneth. 2015. 'Brazil's Crazy Credit Card Rates Are Practically Usury'. *Forbes*, 9 December.

Reddy, William M. 1987. *Money and Liberty in Modern Europe: A Critique of Historical Understanding*. Cambridge: Cambridge University Press.

Reid, Michael. 2014. *Brazil. The Troubled Rise of a Global Power*. New Haven: Yale University Press.

Reis, João José. 2008. *Domingo Sodré, um sacerdote africano: Escravidão, liberdade e candomblé Bahia do século XIX*. São Paulo: Companhia Das Letras.

Resende, André Lara. 1985. 'A moeda indexada: nem mágica nem panaceia'. *Revista de Economia Política* 5(2): 124–130.

Ricci, Rudá. 2010. *Lulismo: Da era dos movimentos sociais à ascensão da nova classe média brasileira*. Rio de Janeiro: Contraponto.

Ritter, Camila D., Gabriel McCrate, R. Henrik Nilsson, Philip M. Fearnside, Ulrika Palme and Alexandre Antonelli. 'Environmental Impact Assessment in Brazilian Amazonia: Challenges and Prospects to Assess Biodiversity'. *Biological Conservation* 206: 161–168.

Roberts, Robert. 1971. *The Classic Slum: Salford Life in the First Quarter of the Century*. London: Penguin.

Rocha, Jan. 'Brazil Prepares to Roll Back Green Laws'. *Climate News Network*, 12 May. https://climatenewsnetwork.net/18257-2.

Rodrigues, Iram Jácome, and Ramalho, José Ricardo. 2010. 'Sindicato, crise econômica e estratégias regionais - novas dimensões da participação política no ABC Paulista'. *Caderno CRH* 23: 339–351.

Rofel, Lisa. 1999. *Other Modernities: Gendered Yearnings in China after Socialism*. Berkeley: University of California Press.

Rose, Sonya O. 1992. *Limited Livelihoods: Gender and Class in Nineteenth-Century England*. Berkeley: University of California Press.

Roseberry, William. 1988. 'Political Economy'. *Annual Review of Anthropology* 17: 161–185.

Ross, Andrews. 2004. *No Collar: The Humane Workplace and Its Hidden Costs*. Philadelphia: Temple University Press.

Rudnyckyj, Daromir. 2011. 'Circulating Tears and Managing Hearts: Governing through Affect in an Indonesian Steel Factory'. *Anthropological Theory* 11(1): 63–87.

Ruggeri, Laura. 2007. '"Palm Springs": Imagineering California in Hong Kong'. In Mike Davis and Daniel Bertrand Monk (eds), *Evil Paradises: Dreamworlds of Neoliberalism*. Berkeley: University of California Press, 106–117.

Saad-Filho, Alfredo, and Lecio Morais. 2014. 'Mass Protest: Brazilian Spring or Brazilian Malaise?' *Socialist Register* 50: 227–246

———. 2018. 'Privilege versus Democracy in Brazil'. *Jacobin*, 27 October.

Sahlins, Marshall. 1972. *Stone Age Economics*. New York: Routledge.

Salerno, Mario Sergio. 1998. 'A indústria automobilística na virada do século'. In Glauco Arbix and Mauro Zilbovicius (eds), *De JK a FHC: A reinvenção dos carros*. São Paulo: Scritta, 503–522.

Sansi-Roca, Roger. 2007. '"Dinheiro Vivo": Money and Religion in Brazil'. *Critique of Anthropology* 27(3): 319–339.

Santana, Marco Aurélio. 2000a. 'Trabalho, trabalhadores e sindicatos em meio ao vendaval contemporâneo'. *Dados* 43 (2). http://dx.doi.org/10.1590/S0011-52582000000200007.

———. 2000b. 'As centrais sindicais brasileiras e a reestruturação produtiva: Análises e propostas'. *Sociologias* 2(4): 186–225.

———. 2006. 'Trabalhadores e política no Sul fluminense: a experiência de Volta Redonda nos anos 1980'. In Ramalho and Santana 2006: 159–174.

———. 2010. 'Ruptura geracional induzida e estratégias de gestão: A experiência nas montadoras do sul fluminense'. *Educação & Sociedade* 31(111): 12–26.

———. 2013. 'Os sindicatos e o golpe de Estado de 1964: A experiência dos metalúrgicos do Rio de Janeiro'. *História, Memória e Política* 10(7): 33–60.

Santana, Marco Aurélio, and Massimiliano Mollona. 2013. 'Trabalho e ação coletiva: memória, espaço e identidades sociais na cidade do aço'. *Horizontes Antropológicos* 19(39): 125–148.

Santos, Milton. 2000. *Por uma outra globalização: Do pensamento único à consciência universal*. Rio de Janeiro: Record.

Santos, Rodrigo. 2006. 'A construção social da região: Desenvolvimento regional e mobilização sócio-política no Sul Fluminense'. Master's thesis, Universidade Federal do Rio de Janeiro.

Sassen, Saskia. 2014. *Expulsions: Brutality and Complexity in the Global Economy*. Cambridge, MA: Harvard University Press.
Schneider, Ben Ross. 1991. *Politics within the State: Elite Bureaucrats and Industrial Policy in Authoritarian Brazil*. Pittsburgh: University of Pittsburgh Press.
———. 1992. 'A privatização no governo Collor: Triunfo do liberalismo ou colapso do Estado desenvolvimentista?' *Revista de Economia Política* 12(1): 5–18.
Schneider, Ben Ross, and David Soskice. 2009. 'Inequality in Developed Countries and Latin America: Coordinated, Liberal and Hierarchical Systems'. *Economy and Society* 38(1): 17–52.
Schwarcz, Lilia M., and Heloisa M. Starling. 2018. *Brazil: A Biography*. London: Farrar Straus and Giroux.
Schwarz, Roberto. 1992. *Misplaced Ideas: Essays on Brazilian Culture*. Ed. John A. Gledson. London: Verso.
Scopes, Kim. 1992. 'Understanding the New Labour Movement in the "Third World": The Emergence of Social Movement Unionism'. *Critical Sociology* 19(2): 81–102.
Seidman, Gay W. 1994. *Manufacturing Militance: Workers' Movements in Brazil and South Africa, 1970–1985*. Berkeley: University of California Press.
Sholette, Gregory. 2011. *Dark Matter: Art and Politics in the Age of Enterprise Culture*. London: Pluto Press.
Sigaud, Lygia. 1980. *Greve nos engenhos*. Rio de Janeiro: Paz e Terra.
Silver, Beverly J. 2003. *Forces of Labor: Workers' Movements and Globalization Since 1870*. Cambridge: Cambridge University Press.
Simmel, Georg. (1918) 2010. *The View of Life: Four Metaphysical Essays with Journal Aphorisms*. Trans. John A. Y. Andrews and Donald N. Levine. Chicago: Chicago University Press.
Simon, Herbert A. 1991. 'Bounded Rationality and Organizational Learning'. *Organization Science* 2(1): 125–134.
Singer, André. 2012. *Os sentidos do Lulismo: Reforma gradual e pacto conservador*. São Paulo: Letras.
———. 2014. 'Rebellion in Brazil: Social and Political Complexion of the June Events'. *New Left Review* 85: 19–37.
———. 2015. 'Cutucando onças com varas curtas: O ensaio desenvolvimentista no primeiro mandato de Dilma Rousseff (2011–2014)'. *Novos Estudos* 102: 43–71.
———. 2018a. *O lulismo em crise: Um quebra-cabeça periodo Dilma (2011–2016)*. São Paulo: Letras.
———. 2018b. 'Jeito fascista'. *Folha de S. Paulo*, 27 October.
Singer, André, and Isabel Loureiro, eds. 2016. *As contradições do lulismo: A que ponto chegamos?* São Paulo: Boitempo.
Smith, Gavin. 1991. *Livelihood and Resistance: Peasants and the Politics of Land in Peru*. Berkeley: University of California Press.
———. 1999. *Confronting the Present: Towards a Politically Engaged Anthropology*. London: Berg.
SMVR (Sindicato dos Metalúrgicos do Sul Fluminense). 2008. '6 horas já!' *Boletim 9 de Novembro*, Edição 281, 16 June.
Sohn-Rethel, Alfred, 1978. *Intellectual and Manual Labour: A Critique of Epistemology*. London: Macmillan Press.
Sola, Lourdes. 1991. 'Heterodox Shock in Brazil: Tecnicos, Politicians, and Democracy'. *Journal of Latin American Studies* 23(1): 163–195.
———. 1993. 'Estado, reformas estruturais e democratização no Brasil'. *Revista USP* 17: 156–175.
Sorj, Bernardo, Sergio Fausto, Timothy Thompson and Fernando Henrique Cardoso. 2011. 'The Sociologist and the Politician: An Interview with Fernando Henrique Cardoso'. *Latin American Perspectives* 38(3): 169–193.

Sousa Santos, Boaventura de. 2014. *Epistemologies of the South: Justice Against Epistemicide*. London: Routledge.
Souza, Celina. 1997. *Constitutional Engineering in Brazil: The Politics of Federalism and Decentralization*. New York: St. Martin's Press.
Souza, Jessé, 2017. *A elite do atraso*. Lisbon: LeYa.
Srnicek, Nick. 2016. *Platform Capitalism (Theory Redux)*. London: Polity.
Standing, Guy. 2011. *The Precariat: The New Dangerous Class*. London: Bloomsbury.
———. 2016. *The Corruption of Capitalism: Why Rentiers Thrive and Work Does Not Pay*. London: Biteback Publishing.
Stein, Stanley J. 1957. *Vassouras: A Brazilian Coffee County, 1850–1900: The Roles of Planter and Slave in a Plantation Society*. Cambridge, MA: Harvard University Press.
Stepan, Alfred. 1988. *Rethinking Military Politics, Brazil and the Southern Cone*. Princeton, NJ: Princeton University Press.
Stolcke, Verena. 1988. *Coffee Planters, Workers and Wives: Class Conflict and Gender Relations on São Paolo Plantations, 1850–1980*. New York: St. Martin's Press.
Streeck, Wolfgang. 2016. *How Will Capitalism End? Essays on a Failing System*. London: Verso.
Svampa, Maristella. 2017. 'Commodities Consensus: Neoextractivism and Enclosure of the Commons in Latin America'. *South Atlantic Quarterly* 114(1): 65–82.
Taschner, Pasternak. 1995. 'Squatter Settlements and Slums in Brazil: Twenty Years of Research and Policy'. In Brian C. Aldrich and Ranvinder S. Sandhu (eds), *Housing the Urban Poor: A Guide to Policy and Practice in the South*. London: Zed Books, 185–222.
Taussig, Michael. 1980. *The Devil and Commodity Fetishism in South America*. Chapel Hill: University of North Carolina Press.
Terray, Emmanuel. 1972. *Marxism and Primitive Societies*. New York: Monthly Review Press.
Tomba, Massimiliano. 2013. 'Marx the Historical Materialist: Re-reading the Eighteenth Brumaire'. *Historical Materialism* 21(2): 21–46.
Tomich, Dale W. 2011. *Pelo prisma da escravidão: Trabalho, capital e economia Mundial*. São Paulo: Editora da Universidade de São Paulo.
Torres, Alberto. 1938. *O problema nacional brasileiro: Introdução a um programa de organização nacional*. São Paulo: Companhia Editora Nacional.
Trebat, Thomas J. 1983. *Brazil's State-Owned Enterprises: A Case Study of the State as Entrepreneur*. Cambridge: Cambridge University Press.
Tversky, Amos, and Daniel Kahneman. 1986. 'Rational Choice and the Framing of Decisions'. *Journal of Business* 59(4): 251–278.
Valenze, Deborah. 1995. *The First Industrial Woman*. Oxford: Oxford University Press.
Van der Linden, Marcel, and Leo Lucassen, eds. 2012. *Working on Labour: Essays in Honour of Jan Lucassen*. Leiden: Brill.
Veblen, Thorstein. (1899) 1994. *The Theory of the Leisure Class: An Economic Study of Institutions*. New York: Macmillan.
———. (1904) 2005. *The Theory of Business Enterprise*. New York. Cosimo.
———. (1914) 2006a. *The Instinct of Workmanship and the State of the Industrial Arts*. New York: Cosimo.
———. (1915) 2006b. *Imperial Germany and the Industrial Revolution*. New York: Cosimo.
Velasco, Licínio, Jr., 1997. 'A economia política das políticas públicas: fatores que favoreceram as privatizações no período 1985/94'. *Texto para Discussão* 54.
———. 1999. 'Privatização: Mitos e falsas percepções'. In Fabio Giambiagi and Moreira Maurício (eds), *A economia brasileira nos anos 90*. Rio de Janeiro: Banco Nacional de Desenvolvimento Econômico e Social, 183–216.
———. 2010. 'Documento histórico: A privatização no sistema BNDES'. *Revista do BNDES* 33: 307–382.

Veltmeyer, Henri, and James Petra. 2014. *The New Extractivism: A Post-neoliberal Development Model or Imperialism of the Twenty-First Century?* London: Zed Books.
Venancio Filho, Alberto. 1968. *A intervencão do Estado no Dominio Econômico*. Rio de Janeiro: Fundação Getulio Vargas.
Vercellone, Carlo. 2013. 'The Becoming Rent of Profit? The New Articulation of Wage, Rent and Profit'. *Knowledge Cultures* 1(2): 194–207.
Verdery, Katherine. 1996. *What Was Socialism, and What Comes Next?* Princeton, NJ: Princeton University Press.
Viana, Francisco José de Oliveira. 1955. *Instituições politicas brasileiras*. Rio de Janeiro: J. Olympio.
Vianna, Luiz Werneck. 1978. *Liberalismo e sindicato no Brasil*. São Paulo: Paz e Terra.
———. 2004. *A revolução passiva: Iberismo e americanismo no Brasil*. Rio de Janeiro: Revan.
———. 2007. 'O Estado Novo do PT'. *Gramsci e o Brasil*, 10 July. http://www.acessa.com/gramsci/?page=visualizar&id=755.
Villela, Annibal Villanova, and Wilson Suzigan. 1973. *Politica do governo e crescimento da economia brasileira 1889–1945*. Rio de Janeiro: Instituto de Pesquisa Economica Applicada.
Wainwright, Hilary, and Sue Branford, eds. 2006. *In The Eye of the Storm: Left-Wing Activists Discuss the Political Crisis in Brazil*. Amsterdam: Transnational Institute.
Wallerstein, Immanuel. 1983. *Historical Capitalism*. London: Verso.
Watkins, Susan. 2018. 'Which Feminism?' *New Left Review* 109: 5–76.
Weeks, Kathi. 2011. *The Problem with Work: Feminism, Marxism, Antiwork Politics, and Postwork Imaginaries*. Durham, NC: Duke University Press.
Weffort, Francisco. 1978. *O populismo na politica Brasileira*. Rio de Janeiro: Paz e Terra.
Weick, Karl E. 1976. 'Educational Organizations as Loosely Coupled Systems'. *Administrative Science Quarterly* 21(1): 1–19.
Weiner, Annette B. 1992. *Inalienable Possessions: The Paradox of Keeping-While Giving*. Berkeley: University of California Press.
Wengrow, David, and David Graeber. 2015. 'Farewell to the 'Childhood of Man': Ritual, Seasonality, and the Origins of Inequality'. *Journal of the Royal Anthropological Institute* 21(3): 597–619.
Westwood, Sallie. 1987. *All Day, Every Day: Factory and Family in the Making of Women's Lives*. London: Pluto Press.
Williams, Raymond. (1961) 1984. *The Long Revolution*. London: Penguin.
Williamson, Oliver E. 1981. 'The Economics of Organizations: The Transaction Cost Approach'. *American Journal of Sociology* 87(3): 548–577.
Willis, Paul. 1978. *Learning to Labour: How Working Class Kids Get Working Class Jobs*. London: Routledge.
Wolf, Eric. 1982. *Europe and the People without History*. Berkeley: University of California Press.
Wright, Eric Olin. 2000. 'Working-Class Power, Capitalist-Class Interests, and Class Compromise'. *American Journal of Sociology* 105(4): 957–1002.
Zibechi, Raúl. 2014. *The New Brazil: Regional Imperialism and the New Democracy*. Trans. Ramor Ryan. London: AK Press.
Zweig, Stefan. 1941. *Brazil: Land of the Future*. New York: Viking.
———. 1943. *The World of Yesterday*. New York: Viking.

Index

Focaal
Journal of Global and Historical Anthropology

Managing and Lead Editor
Luisa Steur, *University of Amsterdam*

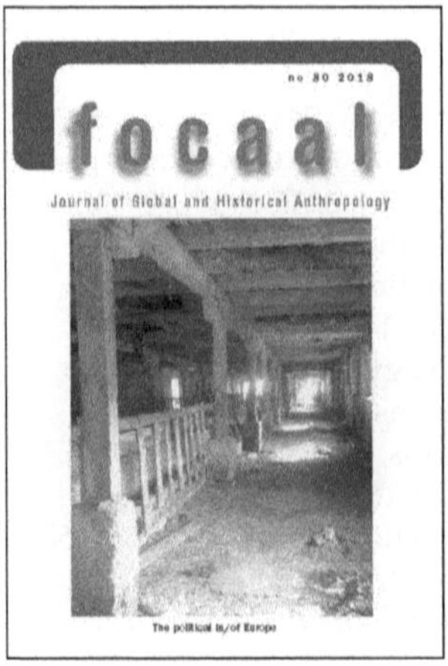

Aims & Scope

Focaal is a peer-reviewed journal advocating an approach that rests in the simultaneity of ethnography, processual analysis, local insights, and global vision. It is at the heart of debates on the ongoing conjunction of anthropology and history, as well as the incorporation of local research settings in the wider spatial networks of coercion, imagination, and exchange that are often glossed as "globalization" or "empire."

Seeking contributions on all world regions, *Focaal* is unique among anthropology journals for consistently rejecting the old separations between "at home" and "abroad," "center" and "periphery." The journal therefore strives for the resurrection of an "anthropology at large" that can accommodate issues of the global south, postsocialism, mobility, metropolitan experience, capitalist power, and popular resistance into integrated perspectives.

Recent Articles

- **Safe Milk and Risky Quinoa: The Lottery and Precarity of Farming in Peru**
  Astrid B. Stensrud
- **The Promise of Education and its Paradox in Rural Flores, East Indonesia**
  Thijs Schut
- **Elite Ethnography in an Insecure Place: The Methodological Implications of "Studying Up" in Pakistan**
  Rosita Armytage
- **Contending with School Reform: Neoliberal Restructuring, Racial Politics, and Resistance in Post-Katrina New Orleans**
  Mathilde Lind Gustavussen

berghahnjournals.com/focaal

berghahn journals
NEW YORK · OXFORD

ISSN 0920-1297 (Print) • ISSN 1558-5263 (Online)
Volume 2019, 3 issues p.a.

www.ingramcontent.com/pod-product-compliance
Ingram Content Group UK Ltd.
Pitfield, Milton Keynes, MK11 3LW, UK
UKHW021836100926
8556IPUK00004B/372

9 781789 204339

Well Integrity for Oil, Gas, and Subsurface Storage Operations

Ahmed Alsubaih · Kamy Sepehrnoori

Well Integrity for Oil, Gas, and Subsurface Storage Operations

Lifecycle Management, Technologies, and Energy Transition Challenges

Springer

Ahmed Alsubaih
The University of Texas at Austin
Austin, TX, USA

Kamy Sepehrnoori
The University of Texas at Austin
Austin, TX, USA

ISBN 978-3-032-15232-9 ISBN 978-3-032-15233-6 (eBook)
https://doi.org/10.1007/978-3-032-15233-6

© The Editor(s) (if applicable) and The Author(s), under exclusive license to Springer Nature Switzerland AG 2026

This work is subject to copyright. All rights are solely and exclusively licensed by the Publisher, whether the whole or part of the material is concerned, specifically the rights of translation, reprinting, reuse of illustrations, recitation, broadcasting, reproduction on microfilms or in any other physical way, and transmission or information storage and retrieval, electronic adaptation, computer software, or by similar or dissimilar methodology now known or hereafter developed.
The use of general descriptive names, registered names, trademarks, service marks, etc. in this publication does not imply, even in the absence of a specific statement, that such names are exempt from the relevant protective laws and regulations and therefore free for general use.
The publisher, the authors and the editors are safe to assume that the advice and information in this book are believed to be true and accurate at the date of publication. Neither the publisher nor the authors or the editors give a warranty, expressed or implied, with respect to the material contained herein or for any errors or omissions that may have been made. The publisher remains neutral with regard to jurisdictional claims in published maps and institutional affiliations.

This Springer imprint is published by the registered company Springer Nature Switzerland AG
The registered company address is: Gewerbestrasse 11, 6330 Cham, Switzerland

If disposing of this product, please recycle the paper.

To my dearest son, Ali,
You are the soft whisper in my soul, the reason behind every heartbeat,
The light that dances in my eyes and the quiet strength that carries me forward.
In your presence, I see a better world—brighter, kinder, full of wonder.
This book, like every dream I've chased, is for you
A token of a father's infinite love,
And a promise that no matter where life takes you,
You will always have my heart beside you.

To my beloved parents
Your love and sacrifices are the foundation of all I am and strive to be.

To my brothers, Ihsan, Awab, and Ibrahim
Thank you for being my steadfast companions and pillars of strength.

To my dear friends, Mr. Mahdi Swadi Dr. Ahmed Alsaedi and Mohammed Rahim
Your friendship and support have enriched this journey.

To Mark Agnew
Thank you for your kind support and encouragement throughout my path.

With deep gratitude,

—Ahmed Alsubaih

With heartfelt gratitude and love,

—Ahmed Alsubaih

To Professor May Wheeler

—Kamy Sepehrnoori

Preface

The extraction of subsurface resources, primarily oil and gas, has powered global development for over a century. However, this process involves creating conduits between the high-pressure, often hostile, downhole environment and the surface. Ensuring these conduits, the wells, maintain their integrity—preventing the uncontrolled release of fluids—is arguably the most critical responsibility of the oil and gas industry. Well integrity is not merely a technical discipline; it is the foundation upon which safe operations, environmental stewardship, economic viability, and public trust are built. Failures, ranging from persistent minor leaks to catastrophic blowouts like the Macondo disaster, serve as stark reminders of the potential consequences when containment is lost. Crucially, well integrity is not a static state achieved at construction but a dynamic condition that evolves over time. Materials degrade, operational stresses accumulate, and environments change.

This book, *Well Integrity for Oil, Gas, and Subsurface Storage Operations: Lifecycle Management, Technologies, and Energy Transition Challenges*, aims to provide a comprehensive, structured understanding of this vital subject. Moving beyond a superficial overview, it delves into the fundamental principles; the practical application of these principles throughout the well lifecycle; the technologies employed for construction, monitoring, and remediation; and the management systems required to ensure integrity is maintained proactively. We explore not only how wells can fail but why, examining root causes from design flaws to operational practices and material degradation, and how these factors change over the life of a well.

The scope encompasses the entire lifecycle, from the initial geological assessment influencing design, through drilling, cementing, completion, decades of operation and potential interventions, to the final, crucial phase of permanent plug and abandonment. It addresses the core components—casing, tubing, cement, packers, seals, wellheads, valves—and the multitude of threats they face, including corrosion, erosion, mechanical stress, high pressure, high temperature, and chemical attack.

This book is intended for a broad audience within the energy sector. Petroleum engineers across disciplines (drilling, completions, production, reservoir, abandonment), geoscientists involved in well planning, service company specialists providing integrity-related equipment and services, technical managers responsible for asset integrity, and regulatory personnel overseeing industry practices will find valuable information herein. It also serves as an advanced text for students pursuing careers in petroleum engineering or related fields, providing a deeper understanding than typically covered in introductory courses.

The book is structured logically into six parts. Part I establishes the fundamental principles, defining well integrity, explaining the critical barrier philosophy, and outlining lifecycle considerations. Part II focuses on well construction and initial integrity assurance, covering design aspects of casing and tubing, the crucial role of primary cementing, completion equipment selection, and initial verification testing. Part III delves into operational well integrity, examining monitoring techniques like annulus pressure management, common failure modes such as corrosion and cement degradation, and the logging and inspection tools used for evaluation. Part V addresses the organizational framework, detailing Well Integrity Management Systems (WIMS), risk assessment methodologies, data management, regulatory context, and the vital role of human factors. Part V tackles specialized topics and advanced concepts, including integrity in harsh environments, unconventional resources, EOR and water injection, subsurface storage, intervention integrity, remediation techniques, permanent abandonment, standards versus technology, and statistical insights. Finally, Part VI looks to the future, discussing advanced monitoring technologies, the impact of digitalization and AI, and the evolving landscape of well integrity.

Ultimately, maintaining well integrity is a continuous journey, demanding vigilance, technical rigor, robust systems, and an unwavering commitment to safety and environmental responsibility. This book aims to equip its readers with the knowledge and understanding necessary to navigate this journey successfully.

Austin, USA

Ahmed Alsubaih
Kamy Sepehrnoori

Competing Interests The authors have no competing interests to declare that are relevant to the content of this manuscript.

Contents

Part I
Fundamentals of Well Integrity

Chapter 1
Introduction to Well Integrity

Abstract Part 1 introduces the foundational principles of well integrity as a dynamic, time-dependent challenge critical to the safe and sustainable operation of subsurface wells throughout their entire lifecycle. Beginning with a precise definition rooted in international standards (NORSOK D-010, ISO 16530-1), this chapter expands the concept to emphasize the evolving nature of integrity threats, including mechanical degradation, thermal cycling, and chemical deterioration of barriers. The chapter explores core pillars of well integrity—containment, control, structural soundness, and longevity—and examines how failures in these elements can manifest as sustained casing pressure, annular flow, or catastrophic blowouts. Through a detailed historical review of well integrity failures, including landmark events like the Ekofisk Bravo, Montara, and Macondo blowouts, the chapter traces the evolution of regulatory standards and industry practices, culminating in modern risk-based and lifecycle-oriented integrity frameworks. A compelling business case is presented, highlighting how integrity management enhances safety, environmental protection, economic performance, and corporate reputation. Furthermore, the chapter addresses how the energy transition reshapes integrity requirements in challenging applications such as carbon capture and storage (CCS), geothermal wells, and underground hydrogen storage (UHS), all of which impose unique thermal, chemical, and cyclic loading stresses on well components. The discussion underscores the necessity of adapting materials, monitoring systems, and lifecycle planning to new energy paradigms. Finally, the chapter introduces the concept of integrity touchpoints across the well lifecycle—from concept selection and design to abandonment—demonstrating how early decisions have long-term implications. Through engineering fundamentals such as stress–strain relationships and safety factor calculations, readers are equipped with the mechanical basis for analyzing barrier performance under complex downhole conditions. This foundational chapter sets the stage for the specialized well integrity challenges and solutions detailed in subsequent sections of the book.

© The Author(s), under exclusive license to Springer Nature Switzerland AG 2026
A. Alsubaih and K. Sepehrnoori, *Well Integrity for Oil, Gas, and Subsurface Storage Operations*, https://doi.org/10.1007/978-3-032-15233-6_1

Offshore drilling rig with smoke plume during the ST 220 incident in the Gulf of Mexico (2015). *Source* U.S. Bureau of Safety and Environmental Enforcement (BSEE, 2015)

Chapter's Objectives

- Provide a clear and comprehensive definition of well integrity, emphasizing its dynamic, time-dependent nature.
- Review the historical context, key incidents, and the evolution of standards that have shaped current well integrity practices.
- Establish the compelling business case for prioritizing well integrity, including safety, environmental, economic, and reputational aspects.
- Discuss the relevance and adaptation of well integrity principles for energy transition applications.
- Introduce the concept of well lifecycle management as it pertains to integrity, stressing the importance of early-phase decisions.

1.1 Defining Well Integrity: A Dynamic, Time-Dependent Challenge

At its core, well integrity is the application of technical, operational, and organizational solutions to reduce the risk of the uncontrolled release of formation fluids and pressures throughout the entire lifecycle of a well (based on NORSOK D-010, ISO 16530-1). However, this definition requires immediate expansion: well integrity is a time-dependent challenge. A well that demonstrates integrity today may develop leaks or lose containment capability in the future due to the cumulative effects of operational stresses and evolving downhole conditions.

While often simplified to "preventing leaks," well integrity encompasses a broader mandate essential for responsible resource extraction and utilization. It implies a multi-faceted state of the well system, requiring:

- **Containment**: Ensuring that fluids (hydrocarbons, formation water, injection fluids, treatment chemicals) and pressures remain confined within their intended pathways—including the primary conduit (tubing or inner casing), monitored secondary containment envelopes (annuli), and the designated reservoir interval. This containment prevents migration not only to the external environment (atmosphere, seawater, groundwater) but also to unintended subsurface formations outside the target reservoir. Effective containment relies on an integrated system of physical barriers, including casing strings, cement sheaths, packers, reservoir seals, and wellhead seals (Dhawan, 2024; Duguid et al., 2019; Fjågesund, 2018).
- **Control**: Maintaining the operational capability to monitor pressures and flows, and to safely shut down or secure the well when necessary. This includes the functionality of safety systems like downhole safety valves (DHSVs) and surface safety valves (SSVs) (Millet et al., 2009).
- **Structural Soundness**: Ensuring the well's physical components—primarily casing strings, tubing, cement sheaths, packers, hangers, the wellhead, and Christmas tree—possess the mechanical strength and durability to withstand all anticipated loads throughout their design life. These loads include internal and external pressures, axial forces (tension, compression), bending moments, torque, vibrations, and crucially, stresses induced by pressure fluctuations, thermal cycling, fluid chemistry changes, and mechanical loading *over time* (Ibukun et al., 2024).
- **Longevity**: Recognizing that integrity must be maintained not just during the productive phase, but *from the moment the well is constructed until it is permanently and safely abandoned*. This necessitates understanding and managing time-dependent degradation mechanisms like corrosion, fatigue, and material property changes (Semwogerere et al., 2025).

Therefore, well integrity is not a static state certified at completion but a dynamic condition that evolves over time. Materials degrade, operational stresses accumulate, and environments change. Key well components, particularly the interfaces between cement and steel or cement and formation, are vulnerable to debonding, cracking, and chemical degradation over years or decades of exposure to downhole conditions (Alberdi-Pagola & Fischer, 2023; Kiran et al., 2017). Common integrity problems resulting from this time-dependent degradation include (Askari et al., 2021; Ibukun et al., 2024; Jaculli et al., 2020; Patel et al., 2019; Solovyeva et al., 2023):

- Sustained casing pressure (SCP) due to gas or fluid migration through failed primary or secondary barriers.
- Wellhead movement or leakage from wellhead/X-tree seals and valves.

- Leakage from casing or tubing connections (threads or seals).
- Corrosion-induced metal loss in casing and completion strings, potentially leading to pinholes or collapse.
- Formation of scale restricting access or damaging equipment.
- Flow behind pipe due to poor cement bonding or subsequent cement degradation (mud channels, voids, fractures).
- Fluid communication between different geological zones or to unintended formations.

These issues highlight that well integrity is intrinsically linked to the quality and durability of its physical barriers. Regulations typically mandate at least two independent and verifiable barriers between hydrocarbon-bearing zones or high-pressure zones and the environment (NORSOK D-010, ISO 16530, ANP Resolution 46). Hydraulic barriers (like cement sheaths) and mechanical barriers (casing, packers, wellhead) work together, with set cement often providing crucial support and protection for the primary mechanical barriers (Kiran et al., 2017; Vo et al., 2008). Failure of these barriers, often due to time-dependent degradation, is a root cause of many oil and gas accidents (Ibukun et al., 2024). Figure 1.1 shows the typical well diagram with developing leakage in the annulus.

The consequences of lost integrity are severe. Beyond the immediate safety risks (blowouts, fires, explosions) and environmental damage (oil spills, gas emissions, groundwater contamination), failures impact operational performance. Field data suggest the actual operational lifespan of many wells is limited to between 7 and 25 years, often due to compromised integrity, preventing them from achieving their full planned production potential (Al-Ghasham et al., 2005). This highlights the critical business case for robust integrity management; preventing failures maximizes asset value. Furthermore, the scale of the challenge is significant; global reviews indicate that tens of thousands of wells have experienced leakage or substantial casing pressure, underscoring the persistent nature of integrity issues across the industry.

The dynamic nature is further complicated by evolving reservoir conditions. Wells initially producing sweet fluids may encounter sour conditions (H_2S) later in life due to reservoir chemistry changes or microbial activity, accelerating corrosion and potentially degrading cement (Ibukun et al., 2024; Jackson et al., 2013). Pressure and temperature changes during production or injection alter the stress field on the wellbore, challenging barrier durability over time. Even conventional wells can show structural impairment years after completion, often stemming from latent cement and casing deficiencies or cumulative stress effects.

This necessitates a lifecycle approach, where decisions made during the design and construction phases have profound and often irreversible impacts on long-term well integrity as shown in Fig. 1.2. Barrier design, material selection, cement slurry formulation, placement practices, initial verification, monitoring, and inspection regimes must all anticipate and manage time-dependent risks from concept to permanent abandonment. Only through comprehensive, adaptive integrity management, backed by robust data and continuous vigilance, can operators ensure safety, environmental protection, and long-term performance.

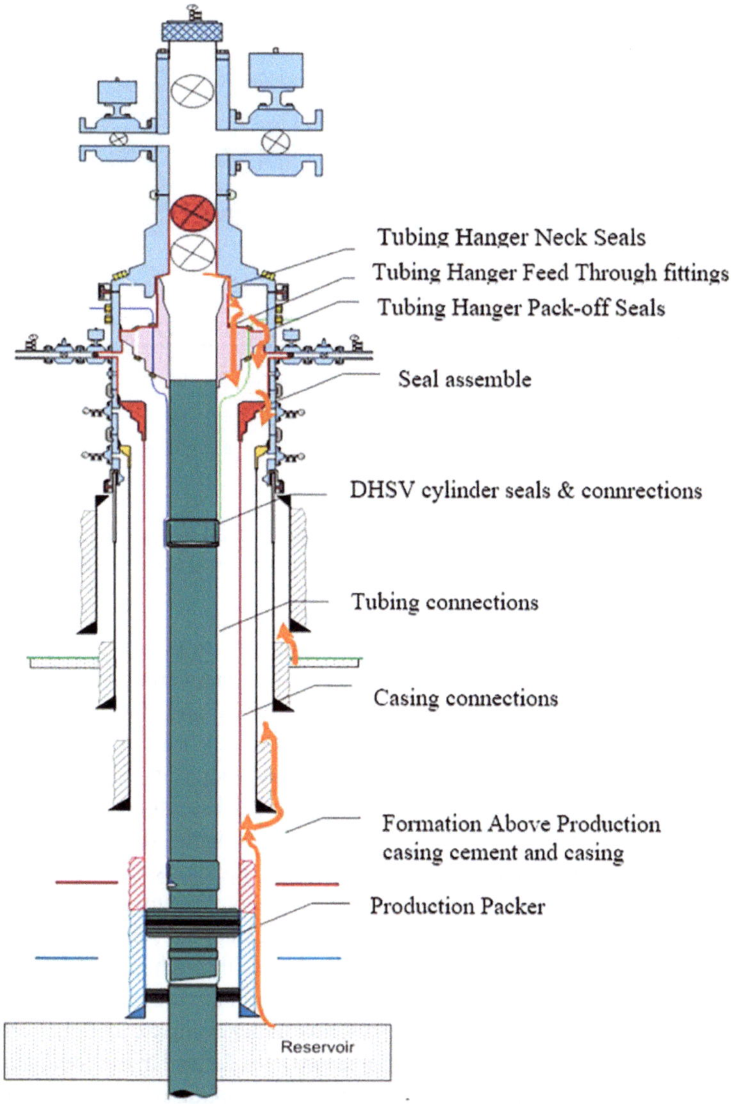

Fig. 1.1 Diagram illustrating a typical wellbore cross section, labeling casing strings, annuli, tubing, cement, wellhead, and Christmas tree components, with arrows showing potential flow paths

It is crucial to distinguish well integrity from, yet understand its relationship with, well control. Well control primarily focuses on preventing blowouts (uncontrolled flow) during active operations where the primary containment system is deliberately opened or potentially compromised (e.g., drilling, intervention). It relies heavily on drilling fluid hydrostatic pressure, blowout preventers (BOPs), and operational

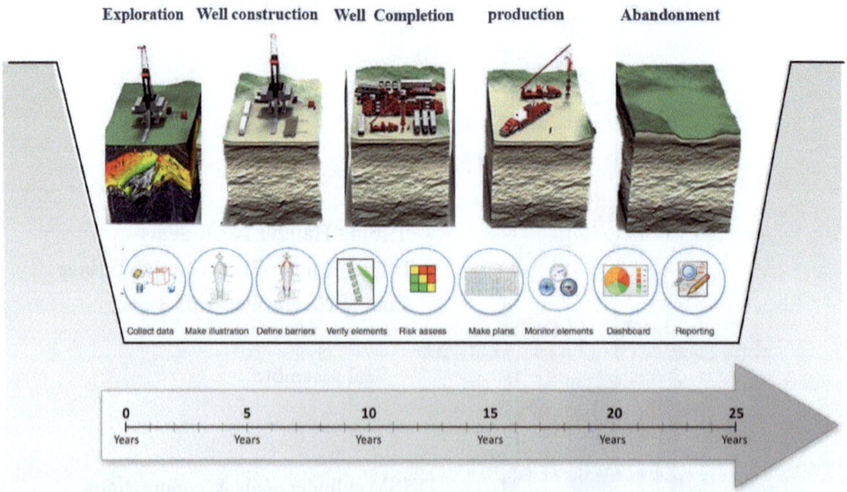

Fig. 1.2 Simplified flowchart showing the key well lifecycle phases (exploration, construction, completion, production, abandonment) with small icons representing key activities in each phase

procedures. Well integrity, on the other hand, focuses on the long-term reliability and sealing capability of the permanent well barrier system components throughout all phases, including static periods (production, shut-in) and after abandonment. A failure in well integrity (e.g., a casing leak) can directly lead to a well-controlled situation. Conversely, poor well control practices or operational errors can damage well integrity components. To support the engineering assessment of well integrity and ensure a robust mechanical design, several fundamental stress and safety relationships are utilized throughout this book. These equations form the foundation for analyzing the structural soundness of well barriers. To support the engineering assessment of well integrity and ensure a robust mechanical design, several fundamental stress and strain relationships are employed throughout this book. These equations are rooted in continuum mechanics and form the foundation for analyzing the structural performance of wellbore systems under complex loading conditions (Karev et al., 2020). Specifically, the strain tensor ε_{ij} is defined in the context of small deformation theory and can be expressed as a combination of elastic and inelastic (plastic) components:

$$\varepsilon_{ij}^{T} = \varepsilon_{ij}^{E} + \varepsilon_{ij}^{p} \tag{1.1}$$

where the total strain ε_{ij}^{T} is the sum of the elastic strain ε_{ij}^{E} and the inelastic (plastic) strain ε_{ij}^{p}. Under the assumption of compatibility and small deformations, the strain tensor is related to the displacement vector u_i via the Cauchy relation:

$$\varepsilon_{ij}^{T} = \frac{1}{2}\left(u_{i,j} + u_{j,i}\right) \tag{1.2}$$

The components of the stress tensor σ_{ij} are governed by the equilibrium equations in the absence of body forces:

$$\sigma_{ij,j} = 0 \tag{1.3}$$

To close the system, the constitutive relationship between stress and strain must be defined. For an anisotropic medium, this relationship is described by Hooke's law using a fourth-rank elasticity tensor Λ_{ijkl}:

$$\sigma_{ij} = \Lambda_{ijkl}\varepsilon_{kl}^{E} \tag{1.4}$$

In the special case of isotropic materials, the stress–strain relationship simplifies using Lamé's constants λ and μ, giving:

$$\sigma_{ij} = \lambda\delta_{ij}\varepsilon_{kk}^{E} + 2\mu\varepsilon_{ij}^{E} \tag{1.5}$$

These tensorial relationships enable precise mechanical evaluation of the well system under various loading paths, including those induced by pressure cycling, thermal gradients, and chemical changes. They are critical for designing casing, tubing, and cement systems capable of withstanding the mechanical and environmental stresses encountered over the entire lifecycle of the well.

$$\text{Safety Factor} = \frac{\text{Failure Stress}}{\text{Allowable Stress}} \tag{1.6}$$

The principles of well integrity are universally applicable, extending beyond conventional oil and gas production to unconventional resources (Chap. 19), EOR projects (Chap. 22), water injection wells (Chap. 23), geothermal wells (Chap. 21), and emerging energy transition applications like Carbon Capture and Storage (CCS) and Underground Hydrogen Storage (UHS) (Chap. 24), each presenting unique, often more severe, integrity challenges to be discussed later in this book. Figure 1.3 provides a comprehensive historical overview of the number of wells drilled annually in the U.S. from 1859 to 2022, including oil, gas, and dry wells, with key events annotated to highlight major technological and geopolitical influences on drilling activity.

1.2 Historical Context, Incidents, and the Evolution of Standards

The modern focus on formalized well integrity management has been significantly shaped by lessons learned from major accidents and the subsequent evolution of industry standards and regulations. While preventing leaks have always been an implicit goal for operational efficiency, catastrophic events have served as powerful

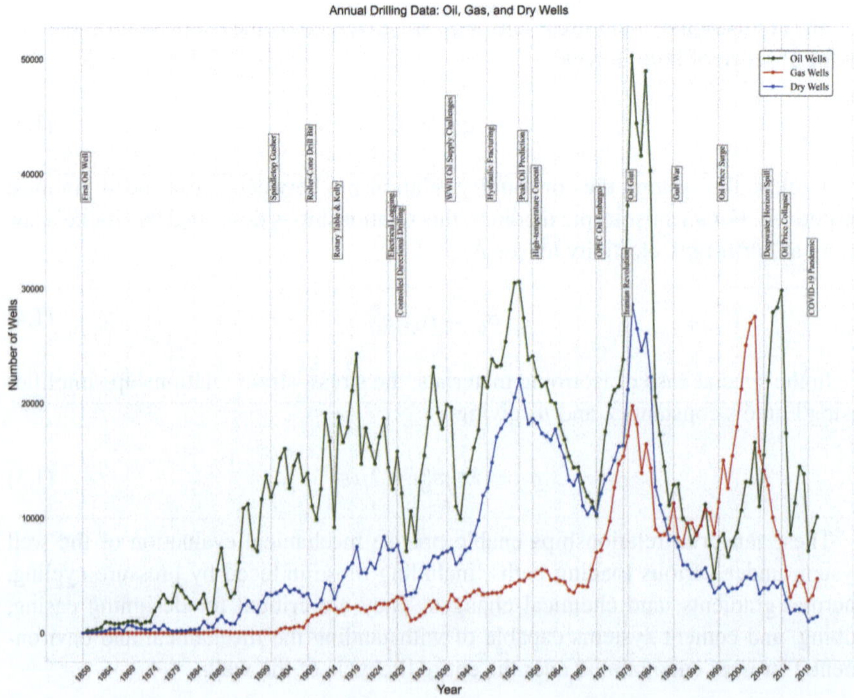

Fig. 1.3 Annual drilling activity in the U.S. (1859–2022), showing the number of oil wells (green), gas wells (red), and dry wells (blue). Major events (e.g., Spindletop, OPEC embargo, Deepwater Horizon) are annotated to show the correlation between historical events and drilling trends

catalysts, exposing systemic weaknesses and driving improvements in practices, technology, and management systems.

- **Early Blowouts and Operational Failures**: Throughout the twentieth century, numerous blowouts occurred, often related to well control issues during drilling or operational errors. While primarily well control incidents, some exposed or led to compromised well integrity components. The 1977 Ekofisk Bravo blowout in Norway, for instance, highlighted deficiencies in operational procedures and response, although directly related to valve reassembly (Law, 2011). The Piper Alpha disaster (1988), though initiated by operational failures on a platform, escalated dramatically due to compromised riser integrity and the inability to remotely shut-in wells feeding the platform, underscoring the need for reliable, safety–critical well barriers and integrated emergency shutdown systems (Tucker, 2016).

- **Recognition of Latent Failures**: Incidents like Saga Petroleum's underground blowout (1989) in Norway and Statoil's blowout on Snorre (2004) further emphasized vulnerabilities in well design, the long-term performance of cement barriers, and the control of annulus pressures, pushing the industry toward a more proactive stance on integrity management beyond active operations (Anunciação Jaculli et al., 2022).
- **Formalization of Standards—NORSOK D-010**: Recognizing the need for a systematic, documented approach to well barriers, the Norwegian petroleum industry, under the direction of the regulator (PSA, now Havtil), initiated the development of the NORSOK D-010 standard. First released in 1996, this highly influential and relatively prescriptive standard formalized the concept of well barrier elements, the two-barrier principle, and detailed requirements for lifecycle integrity management from drilling to abandonment. It has undergone several revisions, incorporating lessons learned and technological advancements, with the latest (Rev. 5) published in 2021 (NORSOK D-010, 2021).
- **SCP Recognition and API RP 90**: The increasing awareness of Sustained Casing Pressure (SCP) as a prevalent indicator of deeper barrier failure prompted focused industry guidance. The American Petroleum Institute (API) released API Recommended Practice 90: Annular Casing Pressure Management for Offshore Wells in 2006 (building on earlier industry efforts), providing a structured framework for monitoring, diagnosis, risk assessment, and management of SCP, which remains a significant integrity challenge today (API RP 90).
- **Deepwater Challenges and API RP 96**: Incidents in complex environments, such as the Montara blowout (Australia, 2009), which involved issues with well design and barrier verification in a remote location, highlighted the need for specific guidance for deepwater operations. API subsequently issued API RP 96: Deepwater Well Design and Construction in 2013, addressing unique integrity considerations for these challenging environments (API RP 96, 2013; Hayes, 2012).
- **Macondo (Deepwater Horizon, 2010)—A Global Catalyst**: This catastrophic blowout in the Gulf of Mexico, resulting in significant loss of life and environmental damage, became a defining moment for well integrity globally (Fig. 1.4). Investigations identified multiple systemic failures spanning technical, operational, and organizational domains, including crucial lapses in cementing practices and evaluation, misinterpretation of a negative pressure test (a vital barrier verification step), deficiencies in the barrier philosophy application, poor management of change, and underlying safety culture issues (Deepwater Horizon Accident Investigation Report Executive Summary, 2010). The incident directly led to sweeping regulatory reforms in the US (e.g., establishment of BSEE with enhanced oversight) and intensified global scrutiny on rigorous well design, cementing, testing, operator Well Integrity Management Systems (WIMS), and organizational safety culture (Anunciação Jaculli et al., 2022). It reinforced the absolute necessity of the multi-barrier principle and rigorous verification at every critical step.

Fig. 1.4 Illustration of Deepwater Horizon oil spill (French-McCay et al., 2021). Premiere Explorer of Venice, La., stands by near a controlled burn of spilled oil from in the Deepwater Horizon/BP oil spill in the Gulf of Mexico June 9. Coast Guard Photo by Petty Officer First Class John Masson

- **International Standardization—ISO 16530**: In the wake of increased global focus, the International Organization for Standardization (ISO) developed the ISO 16530 series on well integrity. ISO/TS 16530-2 (2014) initially focused on the operational phase, followed by the broader ISO 16530-1: Life cycle governance (2017). This standard provides a high-level, less prescriptive management system framework for establishing, implementing, and improving a WIMS based on risk management and lifecycle principles, offering flexibility while outlining essential elements (ISO 16530-1, ISO 16530-2, Jaculli 2022). The development of ISO 16530-3 specifically for P&A integrity is also underway.
- **Regional- and Operator-Specific Frameworks**: Many regulatory bodies (e.g., Brazil's ANP Resolution No. 46, n.d.; UK's OPRED, 2011) and major operating companies (e.g., Equinor, BP, ADNOC) have developed their own detailed well integrity standards, guidelines, and management systems, often building upon international frameworks like ISO and NORSOK, tailored to local conditions and corporate risk appetites (OEUK Guidelines, 2011). The availability of commercial Well Integrity Management Software (WIMS) starting around 2007 provided operators with tools to support these systematic approaches, consolidating data and workflows.

This historical evolution demonstrates a clear trend: major failures drive regulatory change and standard development, leading to increasingly formalized, risk-based, and lifecycle-oriented approaches to managing well integrity. Understanding this history underscores that effective integrity management is not merely about compliance but about learning from the past to proactively manage risks in an inherently hazardous, time-dependent industry. Table 1.1 presents historical blowout accidents due to integrity issues.

Table 1.1 Summary table of major well integrity incidents and their key lessons learned that influenced standard/regulatory development

Incident name/well	Location	Primary suspected cause(s)	Severity/impact	Key lessons/significance	References
Lakeview Gusher #1	Kern County, California, USA Mar. 1910–Sep. 1911	High formation pressure encountered unexpectedly; inadequate casing and early well control equipment (lack of effective BOP)	Estimated 9 million barrels released; largest accidental oil spill in US history; created Lakeview pool	Demonstrated immense reservoir power; spurred need for better pressure control technology	The Lakeview Gusher (2021)
Gassi Touil (GT-16 Well)	Sahara Desert, Algeria Nov. 1961–Apr. 1962	Drilling into unexpectedly high-pressure gas formation; loss of well control during drilling	Massive gas fire ("Devil's Cigarette Lighter"), burned ~550 M cubic ft/day; visible from space	Pioneered large-scale gas well fire fighting; highlighted remote logistics challenges	Oil & Gas: Fire in the Desert I TIME (1961)
Ekofisk Bravo (Platform B)	Norwegian North Sea 4/1/1977	*Well intervention (workover).* Incorrect installation of a downhole safety valve	Significant oil spill (~200 k barrels); first major North Sea blowout; no fatalities	Led to major tightening of Norwegian regulations (NPD/Ptil); emphasized procedural discipline in well interventions	Law (2011)
Ixtoc I	Bay of Campeche, Mexico June 1979–Mar. 1980	Loss of drilling mud circulation, BOP failure during drilling operations	Massive oil spill (~3.3 M barrels); extensive environmental damage in Gulf of Mexico	Highlighted deepwater well control challenges, environmental impact, BOP reliability	Soto et al. (2014)
Piper Alpha	Tucker (2016)	*Production platform disaster.* Gas leak during maintenance; failure of permit-to-work and safety systems; design flaws exacerbated fire spread	Catastrophic fire/explosion; 167 fatalities; total platform destruction	*Fundamentally changed offshore safety culture.* Led to "Safety Case" regime (risk assessment, management systems). Not a drilling blowout, but profoundly influential	

(continued)

Table 1.1 (continued)

Incident name/well	Location	Primary suspected cause(s)	Severity/impact	Key lessons/significance	References
Sidoarjo Mud Flow (Lusi)	East Java, Indonesia May 2006–Ongoing	*Highly Debated.* Linked by many studies to drilling operations (kick/ underground blowout) at Banjar Panji-1. Others cite preceding earthquake	Ongoing mud volcano eruption; >60k displaced; vast areas inundated; long-term environmental/social disaster	Highlights potential links between drilling and geology; catastrophic long-term impact; site assessment importance; remediation difficulty	Tingay et al. (2015)
Montara Wellhead Platform	Timor Sea, Australia Aug.–Nov. 2009	Failure of cemented primary barrier (casing shoe); failure of secondary barrier (corrosion cap); poor well construction/suspension practices	Uncontrolled release (74 days); significant oil spill (~30–40 k barrels); environmental impact concerns	Highlighted operator oversight failures, well integrity management issues, regulatory gaps. Led to reforms (NOPSEMA establishment)	Hayes (2012)
Macondo (Deepwater Horizon)	Gulf of Mexico, USA April–July 2010	Complex cascade: Cement integrity failure, pressure test misinterpretation, kick detection delays, BOP failure (Blind Shear Ram functionality)	11 fatalities; largest marine oil spill (~4.9M barrels); extensive environmental/ economic damage	Major overhaul of US offshore regulations (BSEE formation); focus on deepwater well control, cementing, risk management, BOP reliability/testing	Response to the Montara Wellhead Platform Incident | Australian Maritime Safety Authority (2020)
Elgin PUQ Gas Leak	UK North Sea Mar.–May 2012	*Well integrity failure.* Annular pressure build-up (casing failure/corrosion) in a production well led to surface leak	Major gas leak; prolonged evacuation; large exclusion zone; complex well kill. High ignition potential	Highlighted challenges of aging assets, HPHT wells, annulus pressure management, long-term well integrity monitoring	Gibson (2016)
Pryor Trust Gas Well Blowout	Pittsburg County, Oklahoma, USA1/1/2018	Loss of well control during tripping; potential issues with BOP activation/ effectiveness	Rig fire; 5 fatalities	Reinforced importance of primary control (mud), trip procedures, BOP functionality/testing, crew response, even onshore	CSB Issues Final Report on Fatal Gas Well Blowout - ProQuest (2019)

1.3 The Business Case for Well Integrity

While safety and environmental protection are paramount ethical and societal drivers for well integrity, there is also a compelling business case. Investing in robust well integrity management is not simply a cost center; it is essential for sustainable and profitable operations. Failures incur direct and indirect costs that can far outweigh the investment in prevention. The key business drivers include the following:

- **Safety**: Preventing uncontrolled releases protects personnel working on the well site or platform, as well as the public in surrounding areas, from fires, explosions, and exposure to hazardous substances (like H2S). Accidents lead to injuries, fatalities, emergency response costs, investigations, and potential litigation, representing immense human and financial tolls.
- **Environmental Protection**: Well integrity failures are a primary cause of oil spills and gas leaks, which can devastate local ecosystems, contaminate soil and groundwater, harm wildlife, and contribute to air pollution and greenhouse gas emissions (including potent greenhouse gases like methane). Cleanup costs can be astronomical (e.g., Macondo cleanup and restoration costs exceeded $60 billion), severe regulatory fines, and long-term environmental remediation obligations. Maintaining containment is fundamental to environmental stewardship and meeting societal expectations (Gyo Lee et al., 2018).
- **Economic Performance**
  - **Production Optimization**: Leaks in tubing, casing, or packers can lead to lost production (deferment or ultimate loss of reserves), crossflow between zones, or inefficient injection. Maintaining integrity ensures hydrocarbons flow efficiently from the reservoir to the surface, maximizing recovery. As noted in Sect. 1.1, integrity failures can significantly shorten the operational life of wells, preventing them from achieving planned production.
  - **Reduced Remediation Costs**: Proactive monitoring and preventative maintenance are significantly cheaper than reactive remediation of failed components (e.g., complex cement squeezes, casing patches, major workovers to replace failed packers or tubing). Unplanned interventions due to integrity failures lead to costly downtime (Non-Productive Time—NPT), loss of production, and high intervention expenses. Squeeze cementing, for instance, a common remediation for cement sheath failures, has varying success rates (estimated at 50–80% in some studies depending on the cause and complexity) and can be very expensive, highlighting the value of getting the primary cement job right (Alberdi-Pagola & Fischer, 2023; Ma et al., 2020).
  - **Asset Longevity**: Effective integrity management, particularly corrosion control and managing cyclic loading, helps extend the productive life of wells and associated facilities, maximizing return on the initial investment. Premature well failure or abandonment due to integrity issues represents a significant loss of asset value.

- **Reduced Insurance Costs**: Operators with strong integrity track records and robust management systems may benefit from lower insurance premiums.
- **Avoiding P&A Liabilities**: Delaying necessary P&A can increase costs as rigs and services become more expensive or regulations become stricter. Proactive P&A planning and efficient execution, while costly, address a known future liability. The global P&A liability is estimated to be in the hundreds of billions of dollars, making cost-effective and reliable abandonment essential.

- **Reputation and Social License to Operate**: Major well integrity failures cause significant reputational damage, eroding public trust and potentially jeopardizing an operator's social license to operate in certain regions. Stakeholders (investors, governments, communities, NGOs) increasingly demand demonstrable commitment to safe and environmentally sound practices. A strong integrity record enhances corporate reputation and stakeholder confidence.
- **Regulatory Compliance**: Non-compliance with well integrity regulations can lead to enforced shutdowns, significant fines, and potentially the revocation of operating licenses. Proactive integrity management ensures compliance and avoids these penalties.

In essence, well integrity is integral to operational excellence and responsible resource development. It minimizes risks, protects value, and ensures the long-term sustainability of the business. Viewing integrity solely as a compliance burden misses the significant value it delivers across safety, environmental, operational, and financial performance metrics.

1.4 Well Integrity in the Energy Transition

As the global energy system evolves, the principles and practices of well integrity remain critically important and, in some cases, face new, more challenging requirements. Leveraging decades of oil and gas experience is essential, but adaptation is required for applications like Carbon Capture and Storage (CCS), Geothermal Energy, and Underground Hydrogen Storage (UHS).

- **Carbon Capture and Storage (CCS) (See Chap. 24)**: CCS involves capturing CO_2 emissions and injecting them deep underground for permanent storage. Well integrity is paramount for CCS success.
  - **Long-Term Containment**: Unlike oil/gas wells with finite production lives, CCS wells must ensure CO_2 containment for centuries or millennia. This demands exceptional long-term performance of cement, casing, and plugs (Ibukun et al., 2024; Kiran et al., 2017a).
  - **CO_2 Corrosivity**: Wet CO_2 is corrosive to carbon steel and degrades Portland cement (carbonation). Material selection (CRAs, specialized cements/alternatives), and understanding long-term degradation mechanisms are critical.

  - **Thermal and Pressure Stresses**: Injection of cold CO_2 and subsequent pressure build-up induce stresses.
  - **Monitoring Requirements**: Long-term Monitoring, Measurement, and Verification (MMV) programs are needed to verify containment, potentially post-abandonment. Evaluating legacy wells within the storage complex is also a significant task (Maury Fernandez et al., 2023).

- **Geothermal Energy (See Chap. 21)**: Geothermal wells extract heat from the earth.

  - **High Temperatures and Thermal Cycling**: Geothermal wells often operate at extreme temperatures with severe thermal cycling, pushing the limits of conventional materials (elastomers, cements, electronics) and inducing significant fatigue stresses.
  - **Corrosive Brines and Scaling**: Geothermal fluids can be highly saline and contain corrosive species and dissolved solids, requiring careful material selection and scaling management.

- **Underground Hydrogen Storage (UHS) (See Chap. 24)**: Storing hydrogen underground is explored to buffer renewable energy supplies.

  - **Material Compatibility (Hydrogen Embrittlement)**: Hydrogen can cause embrittlement in certain steels under pressure and stress, affecting casing and tubing integrity. Material selection and qualification testing under hydrogen service conditions are vital (Maury Fernandez et al., 2024).
  - **Small Molecule Leakage**: Hydrogen's small size increases the risk of permeation through materials (seals, cement matrix). Research into low-permeability barriers and seals is crucial.
  - **Cyclic Loading**: Frequent injection and withdrawal cycles associated with energy storage induce fatigue stresses, potentially exacerbated by hydrogen embrittlement.
  - **Biogeochemical Reactions**: Potential for subsurface microbial activity converting hydrogen and producing corrosive byproducts.
  - **Natural Hydrogen Extraction**: Similar integrity challenges exist for potential wells accessing natural hydrogen accumulations (Sun et al., 2025).

While these applications present unique and often more demanding technical challenges, the core well integrity framework developed for oil and gas—barrier philosophy, lifecycle management, risk assessment, material science principles, monitoring techniques, and abandonment strategies—provides the essential foundation for ensuring safety and containment in these new energy frontiers. Adaptation, rigorous qualification, and further research are needed, particularly regarding long-term material performance and sealing verification. Figure 1.5 shows different well integrity mechanisms and causes for various well types.

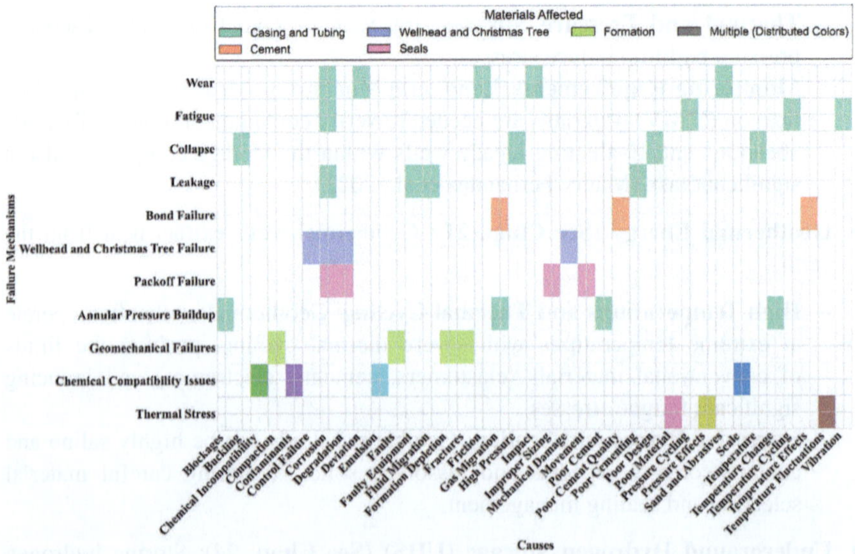

Fig. 1.5 Different well integrity failure mechanisms and causes

1.5 Overview of the Well Lifecycle and Integrity Touchpoints

Managing well integrity effectively requires a holistic perspective that spans the entire life of the well, often conceptualized in distinct phases. Integrity considerations are critical at each stage, and decisions made, and actions taken early in the lifecycle, particularly during design and construction, can have profound, long-lasting, and often irreversible impacts on integrity in subsequent phases. A “cradle-to-grave” (or, more accurately, “concept-to-perpetuity” considering abandonment) perspective is essential. Poor design, inadequate material selection, or compromised construction quality can introduce latent defects that manifest as integrity failures years or decades later (Iriogbe et al., 2024).

The typical phases of a well’s life include the following:

1. **Feasibility/Concept Select**: Initial assessment, not detailed here, but where fundamental concepts influencing long-term design life are considered.
2. **Design and Planning (Phase 1)**: This initial phase lays the groundwork. Key integrity activities include defining the Basis of Design (BoD) that explicitly considers the entire anticipated lifecycle and worst-case credible conditions, performing geological and geomechanical assessments, conducting risk assessments (HAZID) across the lifecycle, selecting appropriate materials suited for future as well as initial conditions, designing casing strings and connections, planning the cementing program for zonal isolation under expected conditions,

establishing the initial barrier strategy (WBS), and planning for future monitoring and abandonment.

3. **Construction (Drilling, Cementing, Completion) (Phase 2)**: Execution quality is paramount here. This involves drilling the well safely while maintaining wellbore stability and managing drilling fluids to support subsequent cementing, running casing correctly, executing the primary cement job effectively to achieve robust and durable zonal isolation, installing completion equipment (packers, DHSV, tubing, wellhead) according to detailed procedures, and performing crucial initial pressure tests to verify barrier integrity. Meticulous documentation of all "as-built" details, events (e.g., losses, tight spots), and test results is vital as baseline data for future management. Failures in mud removal, centralization, or slurry placement during cementing directly compromise long-term isolation.
4. **Operation (Production/Injection) (Phase 3)**: This is typically the longest phase. The focus shifts to managing time-dependent degradation and cumulative stress. Integrity management focuses on vigilant and often continuous monitoring of well parameters (especially annulus pressures—SCP is a key indicator), adhering to safe operating envelopes, conducting routine inspections and functional tests (e.g., valve tests), implementing corrosion mitigation and monitoring programs, managing operational changes (MoC) via a formal process, and performing periodic integrity diagnostics (e.g., logging) to detect degradation early.
5. **Intervention (Phase 4)**: When work needs to be done inside the well (e.g., logging, stimulation, and repair), integrity must be carefully managed. This involves meticulous planning focusing on establishing and verifying temporary barriers while permanent barriers are bypassed or manipulated, ensuring well control, executing the intervention without damaging existing permanent barriers, and rigorously verifying permanent barrier integrity upon completion.
6. **Abandonment (Plug and Abandonment—P&A) (Phase 5)**: The final stage aims to permanently isolate porous formations and seal the wellbore against any future leakage in perpetuity, without reliance on future monitoring or maintenance. This involves designing a P&A program with permanent barriers (plugs) made from durable materials, placing these barriers effectively (often cement, sometimes alternatives), and rigorously verifying their location and sealing integrity. The challenge of demonstrating very long-term containment is significant.

Subsequent chapters in this book will delve into the specific well integrity challenges, technologies, and management practices associated with each of these lifecycle phases and specialized environments in detail. Understanding this lifecycle perspective reinforces that well integrity is not an isolated task but an integrated process requiring continuous attention and diligence from concept to decommissioning, with a clear line of consequence linking early decisions and execution quality to long-term performance.

References

Alberdi-Pagola, P., & Fischer, G. (2023). Review of integrity loss detection and quantification due to cracking in cemented wells. *SPE Journal, 965*. http://onepetro.org/SJ/article-pdf/28/03/965/3120397/spe-214303-pa.pdf

Al-Ghasham, Y. Y., Catte, D. R., & Al-Haji, A. A. (2005). *Integrated analysis of down-hole corrosion logs to investigate casing leaks*. https://doi.org/10.2118/93184-MS

Anunciação Jaculli, M., Choueri, N., da Mata, C. R., Aparecida Silva Leite, A. G., Pelaquim Mendes, J. R., & Colombo, D. (2022). Well safety and integrity evaluation of offshore wells: A review of past, present, and future. *Journal of Petroleum Science and Engineering, 212*. https://doi.org/10.1016/j.petrol.2022.110329

API RP 96. (2013). https://www.apiwebstore.org/standards/96?utm_term=api%20rp%2096&utm_campaign=US+Brand&utm_source=bing&utm_medium=ppc&hsa_acc=4890137417&hsa_cam=11564532636&hsa_grp=1355700341218503&hsa_ad=&hsa_src=o&hsa_tgt=kwd-84732449229502:loc-4126&hsa_kw=api%20rp%2096&hsa_mt=e&hsa_net=adwords&hsa_ver=3&msclkid=b62f7a421336112d1bde124cfdb36c01

Askari, M., Aliofkhazraei, M., Jafari, R., Hamghalam, P., & Hajizadeh, A. (2021). Downhole corrosion inhibitors for oil and gas production—A review. *Applied Surface Science Advances, 6*. https://doi.org/10.1016/j.apsadv.2021.100128

Brazil's ANP Resolution No. 46. (n.d.). Retrieved April 26, 2025, from https://www.legisweb.com.br/legislacao/?id=330646

Bureau of Safety and Environmental Enforcement (BSEE). (2015). ST 220 Panel Investigation Report. U.S. Department of the Interior, 2015. Public domain (U.S. Government work under 17 U.S.C. § 105). Available at: https://www.bsee.gov/sites/bsee.gov/files/panel-investigation/incident-and-investigations/st-220-panel-report9-8-2015.pdf

CSB Issues Final Report on Fatal Gas Well Blowout - ProQuest. (n.d.). Retrieved April 19, 2025, from https://www.proquest.com/openview/78d16680837e9cc9810d62a81bdb570c/1?cbl=47267&pq-origsite=gscholar

Deepwater Horizon Accident Investigation Report Executive Summary. (2010).

Dhawan, S. (2024). Technology focus: Well integrity (January 2024). *Journal of Petroleum Technology, 76*(01), 74–75. https://doi.org/10.2118/0124-0074-JPT

Duguid, A., Zaluski, W., El-Kaseeh, G., Lee, S. Y., & Piercy, M. (2019). Well integrity risk assessment to inform containment risk monitoring for carbon capture, utilization, and storage, applied to the Weyburn-Midale Field, Canada. *International Journal of Greenhouse Gas Control, 86*, 226–238. https://doi.org/10.1016/J.IJGGC.2016.10.018

Fjågesund, T. (2018). Guest editorial: Robust well barrier and well integrity management. *Journal of Petroleum Technology, 70*(12), 14–15. https://doi.org/10.2118/1218-0014-JPT

French-McCay, D. P., Jayko, K., Li, Z., Spaulding, M. L., Crowley, D., Mendelsohn, D., Horn, M., Isaji, T., Kim, Y. H., Fontenault, J., & Rowe, J. J. (2021). Oil fate and mass balance for the Deepwater Horizon oil spill. *Marine Pollution Bulletin, 171*, Article 112681. https://doi.org/10.1016/J.MARPOLBUL.2021.112681

Gibson, M. (2016). HPHT field development experience transfer. In *Society of Petroleum Engineers—IADC/SPE Asia Pacific Drilling Technology Conference*. https://doi.org/10.2118/180660-MS

Gyo Lee, Y., Garza-Gomez, X., & Lee, R. M. (2018). Ultimate costs of the disaster: Seven years after the deepwater horizon oil spill. *Journal of Corporate Accounting and Finance, 29*(1), 69–79. https://doi.org/10.1002/JCAF.22306;PAGEGROUP:STRING:PUBLICATION

Hayes, J. (2012). Operator competence and capacity—Lessons from the Montara blowout. *Safety Science, 50*(3), 563–574. https://doi.org/10.1016/J.SSCI.2011.10.009

Ibukun, M., Elyan, E., Amish, M., Njuguna, J., & Oluyemi, G. F. (2024). A review of well life cycle integrity challenges in the oil and gas industry and its implications for sustained casing pressure (SCP). In *Energies* (Vol. 17, Issue 22). Multidisciplinary Digital Publishing Institute (MDPI). https://doi.org/10.3390/en17225562

Jackson, R. B., Vengosh, A., Darrah, T. H., Warner, N. R., Down, A., Poreda, R. J., Osborn, S. G., Zhao, K., & Karr, J. D. (2013). Increased stray gas abundance in a subset of drinking water wells near Marcellus shale gas extraction. *Proceedings of the National Academy of Sciences of the United States of America, 110*(28), 11250–11255. https://doi.org/10.1073/PNAS.1221635110/SUPPL_FILE/PNAS.201221635SI.PDF

Jaculli, M. A., Mendes, R. P., & Miura, K. (2020). *Methodology for assessing wellhead integrity during its life cycle.* http://asmedigitalcollection.asme.org/OMAE/proceedings-pdf/OMAE2020/84430/V011T11A033/6607289/v011t11a033-omae2020-18468.pdf

Karev, V., Kovalenko, Y., & Ustinov, K. (2020). *Geomechanics of oil and gas wells.* https://doi.org/10.1007/978-3-030-26608-0

Kiran, R., Teodoriu, C., Dadmohammadi, Y., Nygaard, R., Wood, D., Mokhtari, M., & Salehi, S. (2017). Identification and evaluation of well integrity and causes of failure of well integrity barriers (A review). *Journal of Natural Gas Science and Engineering, 45*, 511–526. https://doi.org/10.1016/J.JNGSE.2017.05.009

Law, R. (2011). The Ekofisk Bravo Blowout, 1977. In *Oil spill science and technology* (pp. 1107–1108). https://doi.org/10.1016/B978-1-85617-943-0.10034-6

Ma, L., Zhang, K., Xie, J., Yuan, L., Geng, H., & Ning, K. (2020). Research progress and prospect of well integrity technology. *Journal of Power and Energy Engineering, 08*(07), 45–54. https://doi.org/10.4236/jpee.2020.87004

Maury Fernandez, D., Emadi, H., Hussain, A., & Thiyagarajan, S. R. (2024). A holistic review on wellbore integrity challenges associated with underground hydrogen storage. *International Journal of Hydrogen Energy, 57*, pp. 240–262. Elsevier Ltd. https://doi.org/10.1016/j.ijhydene.2023.12.230

Maury Fernandez, D., Emadi, H., Thiyagarajan, S., Hussain, A., & Ispas, I. (2023). Effects of hydrogen on Class H Well cement's properties under geological storage conditions. In *57th US Rock Mechanics/Geomechanics Symposium.* https://doi.org/10.56952/ARMA-2023-0735

Millet, F., Petit, H., Wallez, G., Lalanne, P., Ducasse, A., & Barzu, E. (2009). Improving well safety with an innovative surface-controlled subsurface safety valve. *Journal of Petroleum Technology, 61*(01), 75–77. https://doi.org/10.2118/0109-0075-JPT

NORSOK D-010. (2021). https://online.standard.no/en/norsok-d-010-2021

OEUK Guidelines. (n.d.). Retrieved April 26, 2025, from https://oeuk.org.uk/product/cumulative-risk-guidelines-issue-1/

Oil & Gas: Fire in the Desert | TIME. (n.d.). Retrieved April 19, 2025, from https://time.com/archive/6623264/oil-gas-fire-in-the-desert/

Oziegbe Iriogbe, H., Ebeh, C. O., & Onita, F. B. (2024). *Well integrity management and optimization: A review of techniques and tools.* https://doi.org/10.56781/ijsrr.2024.5.1.0041

Patel, H., Salehi, S., Ahmed, R., & Teodoriu, C. (2019). Review of elastomer seal assemblies in oil & gas wells: Performance evaluation, failure mechanisms, and gaps in industry standards. *Journal of Petroleum Science and Engineering, 179*, 1046–1062. https://doi.org/10.1016/j.petrol.2019.05.019

Response to the Montara Wellhead Platform incident | Australian Maritime Safety Authority. (n.d.). Retrieved April 19, 2025, from https://www.amsa.gov.au/marine-environment/incidents-and-exercises/response-montara-wellhead-platform-incident

Semwogerere, D., Sangesland, S., Vatn, J., Pavlov, A., & Colombo, D. (2025). Well integrity and late life extension—A current industry state of practice and literature review. *Geoenergy Science and Engineering, 244*, Article 213419. https://doi.org/10.1016/J.GEOEN.2024.213419

Solovyeva, V. A., Almuhammadi, K. H., & Badeghaish, W. O. (2023). Current downhole corrosion control solutions and trends in the oil and gas industry: A review. In *Materials* (Vol. 16, Issue 5). MDPI. https://doi.org/10.3390/ma16051795

Soto, L. A., Botello, A. V., Licea-Durán, S., Lizárraga-Partida, M. L., & Yáñez-Arancibia, A. (2014). The environmental legacy of the Ixtoc-I oil spill in Campeche Sound, southwestern Gulf of Mexico. *Frontiers in Marine Science, 1*(NOV), 116502. https://doi.org/10.3389/FMARS.2014.00057/BIBTEX

Sun, B., Zhang, M., Sun, Q., Zhong, J., & Shao, G. (2025). Review on natural hydrogen wells safety. *Nature Communications, 16*(1), 1–14. https://doi.org/10.1038/s41467-024-55773-y

Tingay, M. R. P., Rudolph, M. L., Manga, M., Davies, R. J., & Wang, C.-Y. (2015). *Initiation of the Lusi mudflow disaster*. https://doi.org/10.1038/ngeo2472

The Lakeview Gusher. (n.d.). Retrieved April 19, 2025, from http://www.sjvgeology.org/history/lakeview.html

Tucker, E. (2016). *Regulating safety, regulating profit: Cost-cutting, injury and death in the British North Sea after Piper Alpha.* 186–210. https://doi.org/10.4324/9781315223322-13

UK's OPRED. (n.d.). Retrieved April 26, 2025, from https://www.gov.uk/government/organisations/offshore-petroleum-regulator-for-environment-and-decommissioning

Vo, D., Ramamurthy, A. S., Qu, J., & Zhao, X. P. (2008). Containment wells to form hydraulic barriers along site boundaries. *Journal of Hazardous Materials, 160*(1), 240–243. https://doi.org/10.1016/J.JHAZMAT.2008.02.124

Chapter 2
The Barrier Philosophy: Principles of Containment

Abstract This chapter introduces the barrier philosophy as the cornerstone of well integrity management, focusing on the principles of containment across the entire well lifecycle. Central to this philosophy is the systematic design, verification, and maintenance of *well barrier elements* (WBEs) and their integration into *well barrier envelopes* to prevent uncontrolled fluid or pressure migration. This chapter provides a rigorous definition of WBEs and barrier envelopes, emphasizing their functional roles, reliability requirements, and the physical and operational conditions they must endure. The chapter elaborates on the **two-barrier principle**, mandating at least two independent, verifiable barriers between any pressure source and the environment to ensure redundancy and reduce the risk of total containment failure. Key engineering concepts—including pressure gradients, hydrostatic calculations, annular pressure build-up, and probabilistic reliability models—are introduced to support the evaluation of barrier performance. Case-specific configurations for production, drilling, intervention, and abandonment phases are illustrated, highlighting the evolving nature of primary and secondary barriers across operational contexts. The importance of *independence* and *verifiability* in barrier design is addressed in depth, detailing how shared failure modes, inadequate testing, or poor documentation can compromise integrity. The chapter also introduces **Well Barrier Schematics (WBS)** as vital visual tools for risk communication, operational planning, and compliance. Standardized WBS enhance understanding of barrier status, verification history, and potential leak paths, facilitating safer decision-making. Finally, the application of the barrier philosophy across design, construction, production, and decommissioning phases is detailed, illustrating its adaptability and relevance in both conventional and evolving energy systems. By grounding the discussion in international standards (e.g., ISO 16530, NORSOK D-010, API RP90) and engineering practice, this chapter establishes the foundational mindset and technical framework necessary for robust containment assurance in all well operations.

© The Author(s), under exclusive license to Springer Nature Switzerland AG 2026
A. Alsubaih and K. Sepehrnoori, *Well Integrity for Oil, Gas, and Subsurface Storage Operations*, https://doi.org/10.1007/978-3-032-15233-6_2

Two drilling rigs located near Naft Shahr #24 well during relief well drilling operations to control a gas well blowout (surface fire visible in the background). *Source* ICONSMAT/Authors: Hamzeh Ghorbani et al. (2021)

Chapter's Objectives

- Explain the core concepts of well barriers: elements and envelopes.
- Detail the two-barrier (or multi-barrier) principle.
- Emphasize the requirements of independence and verifiability.
- Differentiate between primary and secondary barriers.
- Illustrate barrier concepts using diagrams and operational examples.
- Discuss the application and evolution of the barrier philosophy across different operational phases.

2.1 Core Concepts: Barrier Elements and Barrier Envelopes

The foundation of effective well integrity management rests upon the systematic implementation and application of the barrier philosophy (e.g., API RP90, 2021; ISO 16530-1, 2017; NORSOK D-010, 2021). This philosophy, codified and refined in key industry standards such as API RP90, ISO 16530, and NORSOK D-010, provides a structured and logical framework for thinking about and ensuring the containment of wellbore fluids and pressure throughout the well's lifecycle. Its primary objective is to prevent the unintended or uncontrolled flow of fluids and pressure from a source (like a reservoir or pressured annulus) to a potential escape route, which is typically the external environment (surface, seabed) or an unwanted subsurface zone (like another formation or aquifer). This is achieved by identifying, designing, installing, verifying, monitoring, and maintaining physical obstacles—well barriers—that impede flow along potential leak paths. Achieving this goal reduces the risk of uncontrolled releases, thereby safeguarding personnel, protecting the environment, and preventing significant economic losses (Alawad & Mohammad, 2016; Khalifeh & Saasen, 2020a, 2020b; Tayab et al., 2016).

At the heart of this philosophy are two critical, interrelated concepts: the Well Barrier Element (WBE) and the Well Barrier Envelope (often simply called a Well Barrier).

- **Well Barrier Element(WBE)**: This is defined as a single, physical object or component within the well system that is designed, installed, and verified to prevent the unintentional flow of wellbore fluids and pressure along a specific potential leak (Khalifeh & Saasen, 2020a). A WBE may or may not be capable of preventing flow by itself; often, its role is to function in conjunction with other WBEs to form a complete Well Barrier Envelope. A barrier element must be capable of withstanding the maximum anticipated conditions it may encounter during its service life, including pressure differentials, mechanical loads (tension, compression, bending), temperature fluctuations, and the chemical environment posed by wellbore fluids (IADC, 2020). Examples of common barrier elements include (NORSOK D-010, 2021) the following:
  - A length of casing or tubing pipe body with sufficient wall thickness and appropriate material properties.
  - A threaded casing or tubing connection (e.g., premium or API) properly made up.
  - A column of competent, well-bonded cement sheath behind casing or liner, providing annular isolation.
  - An elastomeric or metal-to-metal seal within downhole equipment like packers, bridge plugs, or liner hangers, designed to seal an annulus or the tubing/casing bore.
  - The sealing mechanism (e.g., gate/seat, flapper/seat, ball/seat) of a critical valve, such as a Downhole Safety Valve (DHSV), Surface Safety Valve (SSV), Master Valve, Wing Valve, or Annulus Valve.
  - A seal assembly in the wellhead or Christmas tree (e.g., casing hanger seal, tubing hanger seal, valve stem packing).
  - A column of kill-weight fluid (e.g., drilling mud, brine) correctly placed and monitored within the wellbore, often used as a temporary barrier element during drilling or interventions.
  - A mechanical plug (e.g., bridge plug, packer set in tubing/casing) designed for isolation.
  - In certain situations, a competent, impermeable formation itself (e.g., cap rock, shale, salt) can be considered a WBE.

It is crucial that each barrier element is "fit for purpose" (Davies et al., 2014). This means it must be meticulously designed, manufactured, selected, installed, tested, and maintained to reliably withstand the specific and most severe anticipated pressures, temperatures, fluids (including corrosive components like H_2S or CO_2), and mechanical loads it may be subjected to throughout its intended service life (Ahmed & Salehi, 2021a; Fanailoo et al., 2017). Material selection is critical to ensure resistance to degradation mechanisms like corrosion or elastomer breakdown. The condition, verification status, and performance capability of every WBE designated as part of a barrier envelope must be known, documented, and tracked within the Well Integrity Management System (WIMS) (Corneliussen et al., 2007; Yakoot et al., 2021).

- **Well Barrier Envelope (or Well Barrier)**: This is not a single component but rather a **system** or **envelope** comprising one or more *dependent* well barrier elements working together to provide a complete and continuous seal against flow (Chakravorty, 2021; Choueri et al., 2024). It prevents flow from a specific pressure source (e.g., reservoir pressure, injected fluid pressure, hydrostatic pressure from a fluid column, or pressure in an annulus) to a specific destination (usually the external environment or another subsurface formation). For a barrier envelope to be considered competent and qualified, all its constituent barrier elements must be functional and verified (Ashena, 2025). The elements within an envelope are interdependent; if even one essential element within the envelope fails or is compromised such that the intended seal is lost, the entire barrier envelope is considered to have failed.

Think of a barrier envelope like a chain: it is only as strong as its weakest link. If you are relying on the production tubing string, the production packer seal, and the downhole safety valve (DHSV) sealing mechanism to act together as a primary barrier envelope containing reservoir pressure, the failure of any one of these components (e.g., a tubing leak, packer seal failure, or DHSV leak) breaches that specific barrier envelope. Understanding the distinction between individual elements (the building blocks) and the complete envelope (the system) they form is critical for correctly identifying, verifying, and managing well barriers throughout the well's life. A **Well Barrier Schematic (WBS)** is a vital visual tool used to define, document, and communicate these envelopes and their status (Fjågesund, 2015a, 2015b). To visualize the distinction between Well Barrier Elements (WBEs) and Well Barrier Envelopes, Fig. 2.1 provides a comprehensive schematic of a production well. The primary barrier envelope includes the production tubing, packer seals, tubing hanger seals, and the tree valves controlling reservoir fluids. The secondary barrier envelope, designed to contain pressure in case of a primary failure, consists of the production casing, cement sheath, casing hanger seals, and annular valves. This schematic also highlights typical failure paths—including casing leaks, tubing ruptures, and packer failures—emphasizing the importance of each element's integrity within the overall envelope. To support the implementation of the barrier philosophy and ensure that well barriers can reliably withstand the mechanical stresses and pressures they face throughout the well lifecycle, several fundamental engineering equations are introduced in this chapter. These relationships are essential for evaluating the integrity of barrier elements (such as casing, tubing, and cement) under various loading conditions. The vertical pressure gradient in a fluid column is defined as the change in pressure with respect to height:

$$\nabla P = \frac{\Delta P}{\Delta h} \tag{2.1}$$

The hydrostatic pressure exerted by a static fluid column is determined by multiplying the fluid density, gravitational acceleration, and the vertical height of the fluid column:

$$P = \rho g h \tag{2.2}$$

In this equation, P is the hydrostatic pressure (Pa or psi), ρ is the fluid density (kg/m^3), g is the gravitational acceleration (9.81 m/s^2), and h is the vertical height of the fluid column (m).

To evaluate the overall integrity of a barrier system, the total system reliability can be calculated using the reliability of each independent barrier component:

$$R_{\text{system}} = 1 - \prod_{i=1}^{n} (1 - R_i) \tag{2.3}$$

Here, R_{system} represents the overall reliability of the barrier envelope and R_i is the reliability of the individual i-th barrier element.

Another critical factor is **annular pressure build-up (APB)** due to temperature changes in sealed annuli, common in injection or geothermal wells. Thermal expansion of trapped fluids can lead to casing burst or seal failure if not mitigated:

$$\Delta P = \beta \cdot \Delta T \cdot K \tag{2.4}$$

In this expression, ΔP is the pressure increase (Pa), β is the thermal expansion coefficient of the fluid (1/°C), ΔT is the temperature rise (°C), and K is the bulk modulus (Pa). This relationship emphasizes the importance of fluid compressibility and well design in managing thermal loads.

When evaluating fluid flow through micro-annuli or degraded cement, pressure loss can be estimated using a simplified form of Darcy's law for narrow channels, applicable to gas or fluid migration behind casing:

$$\Delta P = \frac{12 \mu Q L}{k h^3} \tag{2.5}$$

where μ is the dynamic viscosity of the migrating fluid (Pa s), Q is the volumetric flow rate (m^3/s), L is the length of the flow path (m), k is the permeability (m^2), and h is the width of the micro-annular gap (m). This formulation helps quantify leakage risk and informs decisions regarding remediation.

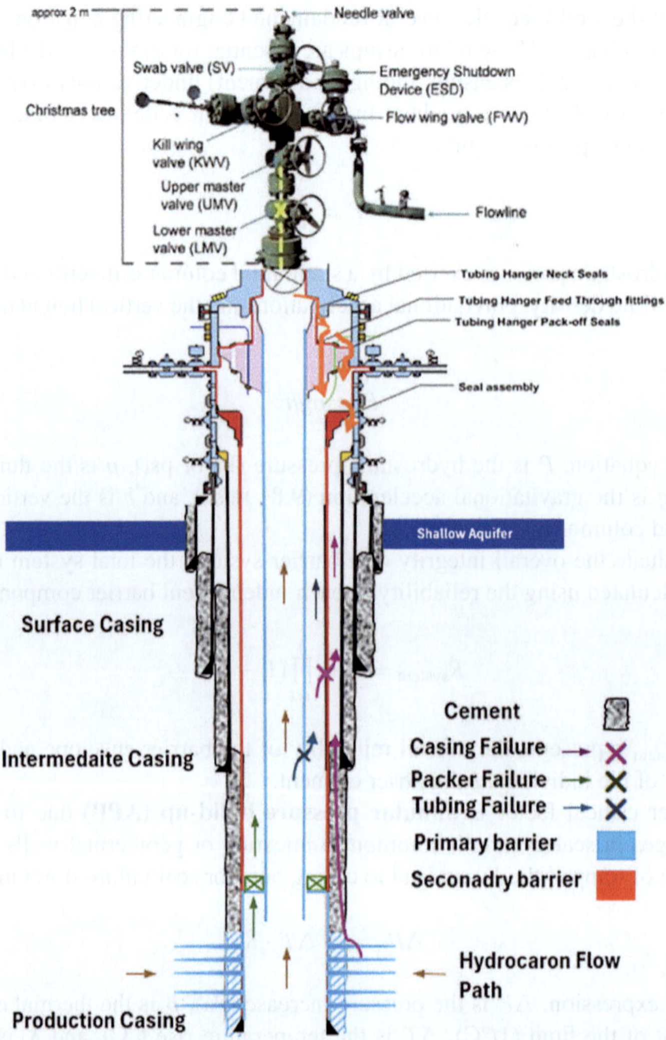

Fig. 2.1 Schematic of well barrier elements and envelopes during the production phase. The primary barrier includes the tubing, packer, tubing hanger, and valves within the Christmas tree. The secondary barrier encompasses the production casing, cement, casing hanger seals, and annulus valves. This diagram also illustrates typical failure paths such as casing, tubing, and packer failures

The reliability of barriers over time can also be modeled using an exponential decay function, assuming a constant failure rate. This is particularly relevant in probabilistic well integrity assessments and aging infrastructure:

$$R(t) = e^{-\lambda t} \tag{2.6}$$

Here, $R(t)$ is the reliability at time t and λ is the constant failure rate (1/year or other time units). As time progresses, the reliability of well components diminishes unless proactive maintenance or barrier replacement is performed.

2.2 The Two-Barrier (or Multi-barrier) Principle

The most widely adopted implementation of the barrier philosophy in the oil and gas industry is the **two-barrier principle** (sometimes referred to as the dual barrier principle (NORSOK D-010, 2021; Torbergsen et al., 2012). This fundamental principle, employed since the early twentieth century and enshrined in leading industry standards like NORSOK D-010 and ISO 16530-2, mandates that:

> During all well activities, at least two independent and tested well barriers shall be available between a potential source of inflow (e.g., reservoir pressure) and the external environment.

This means that for any significant pressure source that could potentially drive fluid flow out of the wellbore (particularly wells capable of natural flow or penetrating overpressured zones) (Davies et al., 2014), there must be two distinct, complete barrier envelopes in place to prevent that flow from reaching the surface, seabed, or other unintended locations (like a shallow aquifer or another formation. As illustrated in Fig. 2.2, during the drilling phase, the primary barrier typically consists of the hydrostatic column of drilling fluid, which provides well control under normal conditions. The secondary barrier is made up of the previously set casing strings, cement sheath, wellhead seals, and, if required, the blowout preventer (BOP). These two independent envelopes provide layered protection, ensuring that if the mud column is lost or compromised, pressure cannot reach the surface or shallow aquifers. Crucially, the failure of one barrier envelope should not compromise the second barrier envelope (IADC, 2020).

While "two barriers" is the common terminology and the minimum standard for containing the primary pressure source during phases like production or shut-in, the underlying principle is about **redundancy** and layers of protection. In some complex situations or specific operational phases (like drilling with multiple openhole sections or managing multiple pressure zones), more than two barriers might be conceptually identified or required to manage different potential flow paths simultaneously. During intervention operations, such as coiled tubing activities, barrier definitions shift to accommodate the active work zone. Figure 2.3 shows an example of this, where the coiled tubing BOP, plug, and hydrostatic fluid column serve as the primary barrier envelope. The secondary barrier envelope includes the production casing, cement sheath, wellhead, and annulus valves. Maintaining two verified, independent barriers during live well interventions is crucial for safe operations and regulatory compliance. However, the minimum standard for containing the primary reservoir pressure during production or shut-in phases is almost universally two independent barriers. Exceptions allowing a single barrier may exist, for instance,

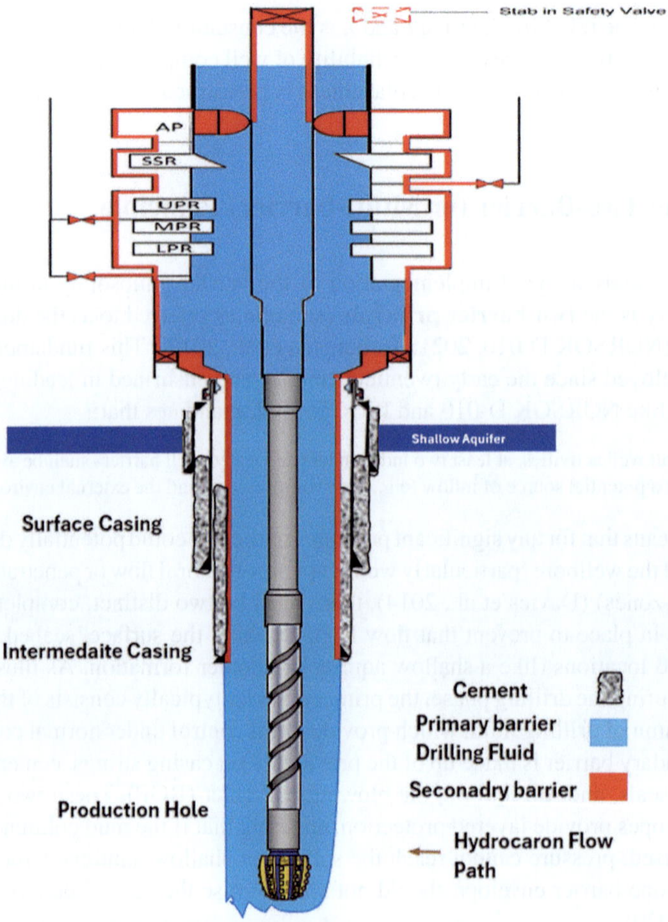

Fig. 2.2 Illustration of well barriers during the drilling phase. The primary barrier is provided by the hydrostatic head of the drilling fluid, while the secondary barrier includes the surface and intermediate casings, cement, and wellhead system. This setup aligns with the two-barrier principle applied in active drilling operations

during abandonment in normally pressured zones, but typically require rigorous risk assessment and justification.

The rationale behind the two-barrier principle is straightforward risk management (Duvsete Tuset et al., 2014; Khelfaoui & Zouyed, 2023; Taleghani & Santo, 2023; Xue et al., 2013):

- Single components (WBEs) and complete barrier envelopes can and do fail, due to design flaws, manufacturing defects, installation errors, operational stresses, degradation over time (as discussed in Chaps. 1, 9, 10 and 11), or human error.

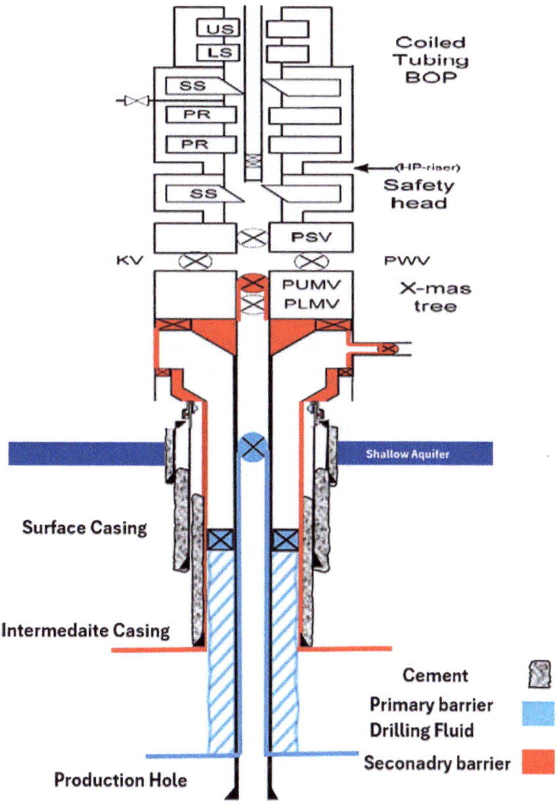

Fig. 2.3 A coiled tubing intervention configuration, where temporary WBEs such as a coiled tubing BOP stack, plugs, and control valves act as the primary barrier. The pre-existing casing, cement, and annulus valves maintain their role as the secondary barrier. This schematic reinforces how intervention-specific envelopes are defined while ensuring compliance with the two-barrier rule

- Relying on a single barrier envelope provides no redundancy. If that barrier fails, containment is immediately lost, potentially leading to an uncontrolled release with severe consequences (Khalifeh & Saasen, 2020b).
- Requiring two independent barriers significantly reduces the probability of simultaneous failure leading to a loss of containment. The presence of a second, independent barrier provides crucial time to detect the failure of the first barrier (e.g., via monitoring) and take corrective action (e.g., shut in the well, perform repairs, activate emergency systems).

This principle applies not just to the main reservoir pressure but to any significant pressure source within the well, such as shallower high-pressure gas zones or aquifers that could potentially flow to the surface via an annulus if not adequately isolated.

2.3 Independence and Verifiability of Barriers

For the two-barrier principle to provide effective protection, the two-barrier envelopes must meet two critical criteria: **independence** and **verifiability** (Skogdalen et al., 2011).

- **Independence**: Two barriers (or barrier envelopes) are considered independent if the failure of one barrier does not cause the failure of the other barrier. Furthermore, they should not rely on the same single, critical barrier element (a common element) for their function against a specific pressure source, nor be susceptible to the same common cause failure mechanism (Eni Norge, 2012; NORSOK D-010, 2021; Vignes & Aadnøy, 2010).
  - *Example of Independence*: During the production phase, the **Primary Barrier Envelope** might consist of the production tubing string, the production packer seals, the downhole safety valve (DHSV) sealing mechanism, the tubing hanger seals, and the Christmas tree production valves. The **Secondary Barrier Envelope** would typically comprise the production casing string, the cement sheath above the packer in the production casing annulus, the casing hanger seals (if applicable), the wellhead seals isolating the production casing annulus, and the Christmas tree annulus valves. These rely on largely different sets of physical components and sealing interfaces. Failure of the packer seal (primary barrier failure) does not automatically cause the production casing, its cement sheath, or the casing hanger seals (secondary barrier elements) to fail.
  - *Example of Lack of Independence*: Relying on two gate valves in series mounted on the same valve block in the Christmas tree as two independent surface barrier envelopes against failure of the block itself would lack independence, as a single crack or failure in the valve block casting could compromise both valve cavities simultaneously. Similarly, relying solely on cement inside casing and cement outside the same casing at the exact same depth might lack independence if a severe geomechanical event (like fault slip or casing deformation) could damage both simultaneously at that location (Casero & Rylance, 2020). Shared or common barrier elements, like a Christmas tree body or Lower Riser Package (LRP) body, represent single points of failure that compromise both envelopes if they fail.
  - Independence requires careful design and analysis to identify and mitigate potential common cause failures or single-point vulnerabilities that could defeat multiple intended. Regulations and standards often emphasize minimizing common failure modes.

- **Verifiability**: A barrier envelope (and its constituent elements) is considered verifiable if its integrity (i.e., its ability to prevent flow and hold pressure) can be demonstrated and confirmed through systematic testing, monitoring, or inspection (King & King, 2013; Skogdalen et al., 2011). This is essential to have confidence that the barriers assumed to be in place are functional and have not failed or degraded below acceptable limits. Verification methods include the following:

- **Pressure Testing**: The most common and direct method (Kiran et al., 2017). This involves applying a differential pressure across the barrier element or envelope, typically in the direction of potential flow (positive test), and observing if it holds pressure within specified limits (e.g., maximum allowable leak rate or pressure drop over time). Alternatively, pressure can be bled off on the downstream side to monitor for flow or pressure build-up (inflow testing or negative testing). Testing against the direction of flow may be used with risk assessment if testing in the direction of flow is not feasible. Examples include casing pressure tests (leak-off test—LOT, formation integrity test—FIT), annulus pressure tests, and valve leak tests (Abimbola et al., 2016; Alberty & McLean, 2014; Wojtanowicz, 2016).
- **Monitoring**: Continuous or periodic monitoring of parameters that indicate barrier status or potential failure. Annulus pressure monitoring is key; the presence of Sustained Casing Pressure (SCP) is a primary indicator of a potential barrier failure. Distinguishing SCP from thermal effects (TIP) or operator-imposed pressure (OIP) is crucial (Schulz et al., 2025). Real-time data from sensors like fiber optics (DTS, DAS) or dedicated annular monitors can also indicate leaks or verify seal integrity (Sharma et al., 2023). Monitoring hydrostatic columns (fluid level, density) is vital when they serve as barriers.
- **Logging**: Running specialized tools into the well to assess the condition or placement of barrier elements. Cement evaluation logs (CBL, VDL, Ultrasonic) are used to infer cement placement quality and bonding, although they are indirect methods subject to interpretation and do not provide a direct measure of hydraulic sealing capacity (Khelfaoui & Zouyed, 2023). Corrosion logs (caliper, electromagnetic thickness) assess casing/tubing metal loss. Leak detection logs (temperature, acoustic, noise) help pinpoint leak locations. Tagging the top of cement or a mechanical plug verifies its position (Al Khamis et al., 2014).
- **Visual Inspection**: Applicable mainly for surface components, such as wellhead seals and valve positions, and for assessing external conditions. Often part of routine maintenance schedules.
- **Functional Testing**: Regularly operating active components like DHSVs or SSVs confirms their mechanical ability to move to the closed/safe position, which is part of their barrier function. This is usually combined with leak rate testing across the closed valve to verify sealing integrity.

Barriers that cannot be reliably tested, monitored, or inspected provide significantly less assurance of containment. The methods, frequency, and acceptance criteria (e.g., maximum allowable leak rates, pressure test holding periods) for verification are key components of a well integrity management system (WIMS) and must be clearly defined. Test results must be documented, deviations from procedures or acceptance criteria must be formally managed (e.g., via Management of Change—MOC), and failures must trigger investigation and corrective actions. A significant challenge remains in reliably verifying the *long-term* integrity of permanent barriers, especially under harsh downhole conditions.

2.4 Primary Versus Secondary Barriers

The two barriers required by the principle are typically categorized based on their role during normal operations:

- **Primary Barrier (or First Barrier)**: This is the barrier envelope designed to contain wellbore pressure and fluids during normal operations (e.g., production, injection, routine shut-in) (ISO/TS 16530-2, 2014; NORSOK D-010, 2021). It is the envelope whose components are normally and continuously in contact with the primary source of pressure (e.g., reservoir fluids, injection fluids). Its components are actively involved in controlling and containing the well flow under standard operating conditions.
  - *Typical Primary Barrier Elements (Production Phase)*: Production tubing string body and connections, production packer seals, downhole safety valve (DHSV/SCSSV) sealing mechanism, tubing hanger seals/pickoff, Christmas tree valves in the primary production flow path (e.g., swab valve, upper and lower master valves, production wing valve).
- **Secondary Barrier (or Second Barrier)**: This barrier envelope is designed to contain wellbore pressure and fluids *if the primary barrier fails*. It acts as a backup or contingency system. During normal operations, it is typically not directly exposed to the full differential pressure from the primary source (or is exposed to it only via leakage through the primary barrier) but must be capable of containing the pressure if the primary barrier is breached. The presence of pressure in an annulus contained by the secondary barrier (e.g., SCP in the A-annulus) is often a primary indicator that the primary barrier has failed.
  - *Typical Secondary Barrier Elements (Production Phase)*: Production casing string body and connections, cement sheath above the packer in the production casing annulus (A-annulus), casing hanger seals (if applicable), wellhead seals isolating the production casing annulus (A-annulus), and Christmas tree valves providing access to the production casing annulus (e.g., annulus wing valve, annulus master valve).

The specific components comprising the primary and secondary barriers **change depending on the operational phase of the well** and the specific activity being performed. For example:

- **Drilling Phase**: The primary barrier is typically the hydrostatic pressure provided by the drilling fluid column, supplemented by the Blowout Preventer (BOP) stack when needed (e.g., during an influx or kick). The secondary barrier is the previously set casing string and its verified cement sheath, along with the wellhead system and the closed BOP (Tangstad, 2014).
- **Completion/Workover (Live Well)**: Barrier definitions depend heavily on the specific operation. Often involves establishing temporary barriers (e.g., mechanical plugs set in tubing/casing, kill-weight fluid columns, wireline/coiled tubing

BOPs) as the primary barrier controlling the wellbore below the work area, while the permanent casing, cement, and wellhead/Xmas tree act as the secondary barrier above the work area. Requires careful planning and execution (often using specific WBS for each step) to ensure two independent barriers are always maintained when manipulating tubing or downhole equipment.

- **Shut-in (Temporary)**: Similar barrier envelopes to the production/injection phase, but with key valves (e.g., DHSV, Master Valves) closed and verified. Verification often involves monitoring annulus pressures for signs of primary barrier leakage (SCP build-up).
- **Abandonment (Permanent)**: The philosophy shifts to establishing multiple, permanent barrier plugs (typically cement, sometimes alternatives like bismuth alloys) set at strategic locations across potential flow paths (in casing/tubing and across annuli). These are designed to provide long-term, often eternal, primary and secondary (and sometimes tertiary) sealing against fluid migration to the surface or crossflow between formations. This approach is visualized in Fig. 2.4, which presents a well barrier schematic during permanent abandonment. The primary barrier consists of one or more cement plugs across critical flow paths (e.g., perforated intervals, casing shoes, formation tops). The open-hole to surface well barrier—typically a long cement column between the casing and formation—acts as a secondary barrier, providing long-term annular isolation. These barriers are designed for eternal containment, eliminating the need for future surveillance. The goal is perpetual containment with no reliance on monitoring or maintenance. Verification of long-term integrity is a key challenge.

Defining, documenting (via WBS), and verifying the status of both primary and secondary barriers for each specific operational phase and activity is a fundamental task in well integrity management and a key output of operational planning and risk assessment.

It's important to distinguish between **barrier failure** and **well integrity failure**. Barrier failure means one or more elements within an envelope are compromised, leading to the failure of that envelope (e.g., primary barrier failure leading to SCP). This indicates a loss of redundancy and increased risk, often requiring intervention. **Well integrity failure** implies a complete loss of containment, where *all* barriers needed along a potential leak path have failed, creating an actual leak to the environment or another formation. True well integrity failures resulting in leaks are statistically much rarer than individual barrier failures.

2.5 Barrier Diagrams and Schematics (WBS)

Visual tools are invaluable for understanding, defining, documenting, and communicating well barrier status (Ahmed & Salehi, 2021b). **Well Barrier Schematics (WBS)** or Barrier Diagrams, introduced in the early 1990s, are simplified drawings of the wellbore, subsurface components, and surface equipment (wellhead/Xmas tree)

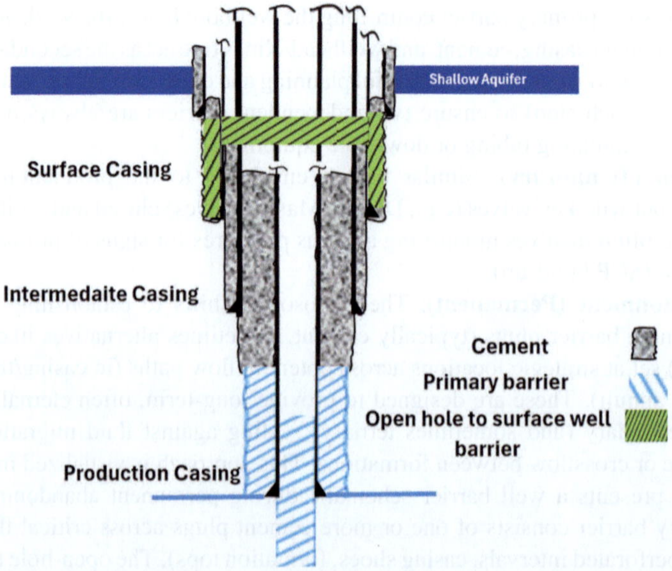

Fig. 2.4 Final barrier configuration during permanent well abandonment. The figure depicts the primary cement plug barrier across the production casing and the open-hole to surface well barrier. These permanent seals are designed for long-term containment without reliance on future monitoring

that clearly illustrate the well's containment system for a specific operational status or phase (Fjågesund, 2015a, 2015b). They provide an immediate, clear snapshot, facilitating a common understanding among all stakeholders and reducing ambiguity. Standardization in their creation and interpretation is crucial.

A comprehensive and effective WBS should contain specific essential information to accurately reflect the well's barrier status:

- **Visual Representation**: A clear schematic drawing of the wellbore, surface equipment (wellhead/Xmas tree), and relevant downhole components (casings, tubing, packers, plugs, valves, cement tops, perforations, etc.).
- **Barrier Envelope Depiction**: The primary barrier envelope is clearly marked (conventionally in blue), and the secondary barrier envelope is clearly marked (conventionally in red).
- **WBE Identification**: All individual WBEs forming part of the defined primary and secondary envelopes must be explicitly identified on the schematic or in an accompanying table/legend.
- **Formation Integrity**: If the formation itself is considered part of a barrier envelope (e.g., cap rock, competent shale/salt for annular seal), its extent and assumed integrity (e.g., verified by FIT/LOT) should be indicated.
- **Pressure Sources and Paths**: Reservoirs, potential sources of inflow (including pressures), and potential leak paths should be illustrated or clearly understood from the schematic. Aligning the WBS with formation pressure profiles

(pore pressure and fracture gradient) can enhance understanding of containment margins.
- **Verification and Monitoring Status**: A table or notes listing the WBEs should detail the requirements or status of their initial verification (e.g., pressure test results, log confirmation, date tested) and any ongoing monitoring requirements (e.g., annulus pressure monitoring frequency, function testing schedule).
- **Well Geometry**: All casing strings, liners, and tubing should be shown with their sizes, weights, grades, and setting depths (both measured depth, MD and true vertical depth, TVD), particularly if they function as WBEs. Cement tops (verified versus assumed) and plug locations/lengths must be accurately depicted.
- **Component Positioning**: All components should be shown in their correct relative vertical and spatial positions.
- **Well Identification and Status**: Essential header information including field/facility name, well name/number, well type (producer, injector, etc.), current well status (producing, shut-in, abandoned, under intervention), current well pressure conditions, WBS revision number and date, and traceability (prepared by, checked by, approved by).
- **Status Indication**: A clear distinction between a WBS representing the *planned* configuration versus the *executed* or *as-built* configuration is necessary.
- **Impairments**: Any known failed, degraded, uncertain, or non-verified WBE must be clearly and unambiguously highlighted on the schematic (e.g., using specific symbols or colors).
- **Notes**: A dedicated field for recording important integrity-related information, anomalies, deviations from standard, operational constraints, or references to supporting documents.

These diagrams are typically standardized within a company or region and may be color-coded or use standardized symbols to represent different barrier elements, their location, and their current status. A WBS should be created and maintained for each distinct operational phase of the well (e.g., drilling phase WBS, production phase WBS, workover phase WBS, and abandonment phase WBS). WBS are essential tools for:

- **Well Design and Planning**: Ensuring the design incorporates the required barriers for construction, completion, intervention, and abandonment.
- **Risk Assessment**: Serving as the foundation for well integrity risk assessment (e.g., HAZID, HAZOP) by helping teams visualize failure paths, controls, identify vulnerabilities, evaluate barrier independence, assess the impact of failures, and provide input for reliability modeling (e.g., defining states for Markov models).
- **Operational Procedure Development and Review**: Ensuring procedures correctly account for barrier status and requirements.
- **Communication**: Providing standardized visual language for engineers, operations personnel, service providers, management, and regulators, ensuring everyone understands how containment is achieved.
- **Operational Safety**: Reviewed before operations to ensure personnel understand the current barrier status and requirements.

- **Compliance**: Demonstrating adherence to the two-barrier policy to regulators and auditors.
- **Incident investigation**: Helping to understand which barriers failed and why.
- **Training**: Educating personnel on barrier principles and specific well configurations.

Crucially, the WBS must be treated as a **dynamic risk management tool**, not merely a static record. Its value depends on its accuracy and timeliness. The WBS must be formally updated whenever the well configuration changes during specific operations, when a new component assumes a barrier function, or if the status of an existing WBE changes (e.g., failure detected, repair performed, verification completed). Maintaining an accurate, up-to-date WBS is fundamental to the entire WIM process.

2.6 Application and Evolution of the Barrier Philosophy Across Operational Phases

The rigorous application of the barrier philosophy—demanding independence and verifiability and clearly defining primary and secondary envelopes for every operational phase—is the bedrock upon which robust well integrity management is built. As the well progresses through its lifecycle, the specific configuration, focus, and challenges related to the barriers change:

- **Design Phase**: The barrier philosophy guides the well design, ensuring casing programs, cementing plans, material selection, and completion designs incorporate sufficient, robust, independent, and verifiable barrier elements to control potential pressure sources identified during geological assessment and risk analysis. Potential future operational phases (interventions) and abandonment requirements are considered.
- **Construction Phase (Drilling and Completion)**: During drilling, the focus is on managing dynamic barriers—such as drilling fluid and BOPs—while relying on previously installed static barriers, including casing, cement, and the wellhead. Testing (FIT/LOT), cement evaluation (logging), and functional testing of equipment (e.g., BOPs) are critical. As each casing string is run and cemented, new static barrier elements are established and verified. During the completion installation, the permanent production/injection barriers (tubing, packer, DHSV, wellhead, and Xmas tree) are installed and rigorously pressure tested and verified.
- **Operational Phase (Production and Injection)**: The focus shifts to monitoring and maintaining the integrity of the installed permanent primary and secondary barriers. This involves surveillance (e.g., SCP monitoring, visual inspections), testing (e.g., valve leak tests, annulus tests), preventative maintenance (e.g., wellhead servicing, valve greasing), and managing changes (MoC). Degradation of barrier elements over time due to mechanisms like corrosion, erosion, or material

fatigue becomes a primary concern. Maximum Allowable Annulus Surface Pressure (MAASP) limits, based on the verified strength of annular barrier elements, must be established and respected.

- **InterventionPhase (Workovers, Well Servicing)**: This phase requires temporary barriers (e.g., fluids, plugs, intervention BOPs/lubricators) to be explicitly defined, installed, and verified whenever permanent barriers are intentionally breached or manipulated. Well control equipment acts as a critical temporary barrier system. Maintaining two independent barriers throughout the intervention is paramount and requires careful planning, documented via a specific WBS for different operational steps.

The evolution of the barrier philosophy, driven by learning from incidents and technological advancements, continues to refine the definition of "independent" and "verifiable" in increasingly complex well designs and operating environments. Understanding and correctly applying this philosophy is fundamental for anyone involved in ensuring well integrity.

References

Abimbola, M., Khan, F., & Khakzad, N. (2016). Risk-based safety analysis of well integrity operations. *Safety Science, 84*, 149–160. https://doi.org/10.1016/J.SSCI.2015.12.009

Ahmed, S., & Salehi, S. (2021a). Failure mechanisms of the wellbore mechanical barrier systems: Implications for well integrity. *Journal of Energy Resources Technology, Transactions of the ASME, 143*(7). https://doi.org/10.1115/1.4050694/1105947

Ahmed, S., & Salehi, S. (2021b). Failure mechanisms of the wellbore mechanical barrier systems: Implications for well integrity. *Journal of Energy Resources Technology, Transactions of the ASME, 143*(7). https://doi.org/10.1115/1.4050694

Al Khamis, M. N., Al Khalewi, F., Al Hanabi, M., Al Yateem, K., Al Qatari, A., & Al Muailu, H. (2014). A comprehensive approach of well integrity surveillance. In *Society of petroleum engineers—International petroleum technology conference 2014, IPTC 2014: Unlocking energy through innovation, technology and capability* (Vol. 3, pp. 2233–2243). https://doi.org/10.3997/2214-4609-PDB.395.IPTC-17465-MS/CITE/REFWORKS

Alawad, M. N. J., & Mohammad, H. A. (2016). Engineering management and inspection schedule of petroleum well integrity. *Journal of Emerging Trends in Engineering and Applied Sciences, 7*(3), 109–117.

Alberty, M. W., & McLean, M. R. (2014). The use of modeling to enhance the analysis of formation-pressure integrity tests. *SPE Drilling & Completion, 29*(04), 431–437. https://doi.org/10.2118/167945-PA

API RP90. (2021). https://www.api.org/products-and-services/standards/important-standards-announcements/90-1

Ashena, R. (2025). Barriers. *The practical handbook of well control* (pp. 20–31). https://doi.org/10.1201/9781003473770-2/BARRIERS-RAHMAN-ASHENA

Casero, A., & Rylance, M. (2020). The unconventional unconventionals: Tectonically influenced regions, stress states and casing failures. In *Society of petroleum engineers—SPE hydraulic fracturing technology conference and exhibition 2020, HFTC 2020* (pp. 4–6). https://doi.org/10.2118/199710-MS

Chakravorty, A. (2021). *Use of machine learning algorithms to predict well barrier elements and envelopes in drilling operations.* https://uis.brage.unit.no/uis-xmlui/handle/11250/2829441

Choueri, N., Jaculli, M. A., Leite, A. G. A. S., & Mendes, J. R. P. (2024). Use of graphs to assess well safety in drilling projects and during operations by identification of available barrier elements and consolidation of barrier envelopes. *Petroleum Research, 9*(3), 418–431. https://doi.org/10.1016/J.PTLRS.2024.02.002

Corneliussen, K., Sørli, F., Haga, H. B., Tenold, E., Menezes, C., Grimbert, B., & Owren, K. (2007). Well integrity management system (WIMS)—A systematic way of describing the actual and historic integrity status of operational wells. In *Proceedings—SPE annual technical conference and exhibition* (Vol. 4, pp. 2576–2583). https://doi.org/10.2118/110347-MS

Davies, R. J., Almond, S., Ward, R. S., Jackson, R. B., Adams, C., Worrall, F., Herringshaw, L. G., Gluyas, J. G., & Whitehead, M. A. (2014). Oil and gas wells and their integrity: Implications for shale and unconventional resource exploitation. *Marine and Petroleum Geology, 56*, 239–254. https://doi.org/10.1016/J.MARPETGEO.2014.03.001

Duvsete Tuset, G., Saeternes, S., & Norske Veritas, D. (2014). *Operational barrier elements in critical drilling operations: Understanding failure mechanisms in the primary well barrier* (Vol. 124). https://ntnuopen.ntnu.no/ntnu-xmlui/handle/11250/238897

Eni Norge. (2012). *An introduction to well integrity.*

Fanailoo, P., Buchmiller, D., Ouyang, S., & Allen, E. (2017). Risk based approach to well plugging and abandoning—Reducing costs while verifying risk. In *Proceedings of the annual offshore technology conference* (Vol. 4, pp. 2728–2744). https://doi.org/10.4043/27921-MS

Fjågesund, T. (2015a). *Using schematics for managing well barriers.* https://jpt.spe.org/using-schematics-managing-well-barriers

Fjågesund, T. (2015b). *Using schematics for managing well barriers.* JPT. https://jpt.spe.org/using-schematics-managing-well-barriers

Ghorbani H., Abdali M. R., Mohamadian N., Wood D. A. (2021). Petroleum well blowouts as a threat to drilling operation and wellbore sustainability: Causes, Prevention, Safety and Emergency response. *Journal of Construction Materials by ICONSMAT*. Licensed under CC BY-NC 4.0. https://iconsmat.com.au/wp-content/uploads/2021/02/si1.1r.pdf

IADC. (2020). *Well control barrier philosophy.*

ISO 16530-1. (2017). https://www.iso.org/standard/63192.html

ISO/TS 16530-2. (2014). https://www.iso.org/standard/57056.html

Khalifeh, M., & Saasen, A. (2020a). *General principles of well barriers* (pp. 11–69). https://doi.org/10.1007/978-3-030-39970-2_2

Khalifeh, M., & Saasen, A. (2020b, January). *Introduction to permanent plug and abandonment of wells* (p. 273). https://doi.org/10.1007/978-3-030-39970-2

Khelfaoui, O., & Zouyed, M. (2023). *Applications of advanced well integrity evaluation technologies for critical decision making.* http://dspace.univ-ouargla.dz/jspui/handle/123456789/34574

King, G. E., & King, D. E. (2013). Environmental risk arising from well-construction failure—Differences between barrier and well failure, and estimates of failure frequency across common well types, locations, and well age. *SPE Production & Operations, 28*(04), 323–344. https://doi.org/10.2118/166142-PA

Kiran, R., Teodoriu, C., Dadmohammadi, Y., Nygaard, R., Wood, D., Mokhtari, M., & Salehi, S. (2017). Identification and evaluation of well integrity and causes of failure of well integrity barriers (a review). *Journal of Natural Gas Science and Engineering, 45*, 511–526. https://doi.org/10.1016/J.JNGSE.2017.05.009

NORSOK D-010. (2021). https://online.standard.no/en/norsok-d-010-2021

Schulz, M., Talukdar, M., Pfander, I., Lackey, G., Röckel, L., Wojtanowicz, A., Schilling, F., & Blum, P. (2025). Global regulations and practices for well integrity management: a review. *Journal of Petroleum Exploration and Production Technology, 15*(5), 1–18. https://doi.org/10.1007/S13202-025-01997-7

Sharma, J., Santos, O., Ogunsanwo, O., Ekechukwu, G. K., Cuny, T., Almeida, M., & Chen, Y. (2023). Fiber-optic DAS and DTS for monitoring riser gas migration. *Journal of Petroleum Science and Engineering, 220*, 111157. https://doi.org/10.1016/J.PETROL.2022.111157

Skogdalen, J. E., Utne, I. B., & Vinnem, J. E. (2011). Developing safety indicators for preventing offshore oil and gas deepwater drilling blowouts. *Safety Science, 49*(8–9), 1187–1199. https://doi.org/10.1016/J.SSCI.2011.03.012

Taleghani, A. D., & Santo, L. (2023). *Wellbore integrity.* https://doi.org/10.1007/978-3-031-19024-7

Tangstad, J. R. (2014). *Assessment of dynamic barriers in oil/gas well operations: Vurdering av dynamiske barrierer i olje/gass brønner* (Vol. 104). https://ntnuopen.ntnu.no/ntnu-xmlui/handle/11250/241058

Tayab, M. R., Abu Aishah, A. N. M., Valappil, S., & Draz, A. M. A. (2016). Assessment and strengthening of well control barriers to prevent well blowout. In *Society of petroleum engineers—Abu Dhabi international petroleum exhibition and conference, 2016-January.* https://doi.org/10.2118/182899-MS

Torbergsen, H.-E. B., Haga, H. B., Aadnøy, B. S., Sæby, J., Johnsen, S., Rausand, M., & Lundeteigen, M. A. (2012). *An introduction to well integrity.*

Vignes, B., & Aadnøy, B. S. (2010). Well-integrity issues offshore Norway. *SPE Production & Operations, 25*(02), 145–150. https://doi.org/10.2118/112535-PA

Wojtanowicz, A. K. (2016). Environmental control of well integrity. *Environmental Technology in the Oil Industry*, 61–100. https://doi.org/10.1007/978-3-319-24334-4_3/FIG.S/22

Xue, L., Fan, J., Rausand, M., & Zhang, L. (2013). A safety barrier-based accident model for offshore drilling blowouts. *Journal of Loss Prevention in the Process Industries, 26*(1), 164–171. https://doi.org/10.1016/J.JLP.2012.10.008

Yakoot, M. S., Elgibaly, A. A., Ragab, A. M. S., & Mahmoud, O. (2021). Well integrity management in mature fields: A state-of-the-art review on the system structure and maturity. *Journal of Petroleum Exploration and Production, 11*(4), 1833–1853. https://doi.org/10.1007/S13202-021-01154-W/FIG.S/21

Chapter 3
Well Lifecycle Management and Integrity Requirements

Abstract This chapter offers a comprehensive analysis of well integrity challenges throughout the complete lifecycle of a well, from initial design and construction to intervention and final abandonment. The chapter begins by defining lifecycle-based well integrity and identifying key touchpoints—critical phases where decisions, failures, or stressors impact long-term containment and structural performance. Using field examples and case histories, this chapter explores how mechanical, chemical, and operational loads evolve with time, affecting the performance of well barrier elements such as casing, cement, packers, and valves. Each lifecycle stage is examined in detail as follows:

- **Design and Construction**: Emphasis is placed on formation characterization, barrier material selection, centralization strategies, and the implications of well trajectory on zonal isolation. Poor design decisions at this stage are shown to have cascading effects on long-term integrity.
- **Drilling Phase**: This section covers common failures including wellbore instability, poor cement placement, and casing deformation due to unexpected formation pressure, lost circulation, or thermal shock.
- **Completion and Production**: Challenges include sustained casing pressure (SCP), micro-annulus formation, thermal cycling, corrosion, scale build-up, and gas migration—each evaluated through failure case studies and supported by quantitative analysis of stress and flow behavior.
- **Well Intervention**: The integrity risks posed by operations such as stimulation, artificial lift installation, and sidetracking are discussed, with attention to integrity testing, barrier revalidation, and mechanical damage.
- **Abandonment and Post-Abandonment**: The long-term risks associated with legacy wells are explored, including orphaned well reactivation, plug degradation, and CO_2/methane leakage pathways, particularly under evolving subsurface conditions.

The chapter integrates stress-path modeling, fluid–structure interaction analysis, and case-based diagnostics to interpret degradation mechanisms across time. It emphasizes the critical need for **designing for the full lifecycle** rather than merely for initial production. Additionally, it introduces **Digital Well Integrity Twins**, a

© The Author(s), under exclusive license to Springer Nature Switzerland AG 2026

A. Alsubaih and K. Sepehrnoori, *Well Integrity for Oil, Gas, and Subsurface Storage Operations*, https://doi.org/10.1007/978-3-032-15233-6_3

modern solution for integrating time-lapse data, simulation, and AI into lifecycle-based risk prediction and integrity monitoring. By demonstrating how decisions made at each phase can affect future integrity, this chapter sets the stage for advanced mitigation strategies, including adaptive cement designs and predictive leakage modeling, discussed in later chapters.

Pump jack unit in operation with a drilling rig in the background, illustrating the transition from drilling to production phase. *Source* The Authors

Chapter's Objectives

- Detail the specific well integrity objectives and activities within each phase of the well lifecycle.
- Highlight the interdependencies between lifecycle phases regarding integrity, stressing the importance of early-phase decisions.
- Emphasize the importance of decisions made during design and construction on long-term integrity.
- Introduce key integrity considerations during operation, intervention, and abandonment.

3.1 Introduction to Lifecycle Management

A well is not a static object but a dynamic system that evolves through distinct phases, from initial conception to final decommissioning. Effective well integrity management requires a holistic approach that considers and integrates integrity requirements across this entire well lifecycle. Decisions, actions, or omissions in one phase can have significant, often irreversible, consequences for integrity in subsequent phases. A "cradle-to-grave" (or more accurately, "concept-to-perpetuity" considering abandonment) perspective is essential for proactively managing risks and ensuring long-term containment.

The typical phases of a well's life include the following:

1. Feasibility/Concept Select: Initial assessment, not detailed here, but where the need for the well and its overall objectives are defined, influencing long-term design life considerations.
2. Design and Planning: Defining the well's objectives and how it will be built and operated.
3. Construction: Physically drilling and completing the well.
4. Operation: Producing from or injecting into the well.
5. Handover: Transferring responsibility between phases (e.g., construction to operations).
6. Intervention: Performing work inside the well after initial completion.
7. Suspension (Temporary): Putting a well in a safe, shut-in state for a period.
8. Abandonment (Permanent): Permanently sealing and decommissioning the well.

This chapter examines the key integrity objectives, activities, and challenges associated with each major phase after initial feasibility, stressing how actions in early phases dictate potential performance and issues later. To support a structured and consistent approach, well integrity management must align with each phase of the well's lifecycle, from initial design through to permanent abandonment. Each phase introduces unique integrity challenges and decision points that, if mismanaged, can compromise the entire system. A visual overview of the integrity responsibilities associated with each phase is presented in Fig. 3.1, which highlights the interconnected nature of lifecycle decisions and the critical role of feedback through lessons learned. This holistic framework emphasizes that well integrity is not achieved in isolation, but through a continuous loop of design, execution, monitoring, and adaptation. A set of fundamental equations that quantify key aspects of well integrity across design, construction, operation, and abandonment phases. These relationships help engineers and integrity managers calculate the mechanical loads that barrier elements must withstand and assess the performance of annular pressure containment. In field applications, the Maximum Allowable Annular Surface Pressure (MAASP) is often calculated using mud weight values rather than absolute pressures, especially during well design and kick tolerance analysis. This formulation is based on the pressure margin between the formation's fracture resistance at the casing shoe and the current mud weight in the well. The MAASP is given by

$$\text{MAASP} = (\text{EMW} - \text{MW}_{\text{mud}}) \times 0.052 \times \text{TVD}_{\text{shoe}} \quad (3.1)$$

In this equation, EMW refers to the equivalent mud weight at which the formation at the casing shoe would fracture (typically expressed in pounds per gallon, ppg), MW_{mud} is the current mud weight in use (ppg), and TVD_{shoe} is the true vertical depth of the casing shoe (ft). The constant 0.052 is a unit conversion factor that relates mud weight (ppg) and depth (ft) to pressure (psi).

This formulation provides a practical limit for surface pressure during operations such as well control or pressure testing, ensuring that pressure at the shoe remains below the fracture gradient to prevent lost circulation or formation breakdown. In

Fig. 3.1 Summary of the key integrity responsibilities and actions associated with each major phase of the well lifecycle: basis of design, construction, operation, intervention, suspension, and abandonment. Each segment highlights core tasks such as pressure control, barrier verification, material selection, and regulatory compliance. At the center, "Lessons Learned" symbolizes the importance of feedback and knowledge transfer between phases, ensuring continuous improvement through a well integrity management system (WIMS)

addition to the Maximum Allowable Annular Surface Pressure (MAASP), two key parameters must be considered for pressure integrity design at the casing shoe: **fracture pressure** and **formation breakdown pressure**. These define the upper limits for safe pressure containment and help avoid formation damage during well control, cementing, or injection operations.

The **fracture pressure** represents the minimum pressure required to initiate a tensile fracture in the formation adjacent to the wellbore. It is governed by the least principal in-situ stress—typically the minimum horizontal stress—and the rock's tensile strength. The fracture pressure is calculated as follows:

$$P_{\text{fracture}} = \sigma_h + T_0 \tag{3.2}$$

Here, σ_h is the minimum horizontal stress in the formation (psi) and T_0 is the tensile strength of the rock (psi). This expression is widely used in well integrity modeling and fracturing design, particularly for vertical wells.

The formation breakdown pressure is typically higher than the fracture pressure and represents the pressure at which a fracture not only initiates but also propagates. In basic vertical well cases, this value incorporates stress anisotropy and pore pressure (Aadnøy & Looyeh, 2019). However, in more complex wellbore conditions—such as inclined or horizontal wells, or where significant temperature variation is expected (e.g., in geothermal, HPHT, or CO_2 injection wells)—the breakdown pressure must additionally account for thermal stresses and the orientation of in-situ stresses.

The extended expression for breakdown pressure in these thermally and directionally influenced scenarios is given by

$$P_b = 3\sigma_{\min} - \sigma_{\max} - p + \sigma_T + T_0 \tag{3.3}$$

In this equation, $\sigma_{\min}$ and $\sigma_{\max}$ represent the minimum and maximum horizontal in-situ stresses, respectively, expressed in pounds per square inch (psi). The term p refers to the pore pressure within the formation (psi), while T_0 denotes the tensile strength of the rock (psi), indicating the point at which tensile failure initiates. The variable σ_T accounts for the additional thermal stress (psi) generated because of temperature changes in the wellbore environment, which may arise from fluid injection, production operations, or geothermal gradients.

This formulation provides a more accurate prediction of the pressure at which fractures initiate and propagate, especially in anisotropic or thermally affected formations. When designing for well integrity under thermal cycling conditions—such as CO_2 injection, high-rate production, or steam injection—incorporating the thermal stress component σ_T is essential. It ensures the prediction of safe operating limits and prevents premature barrier failure, loss of containment, or wellbore instability.

Together, Eqs. (3.1) through (3.3) define the safe operational pressure window and form a critical part of the casing and cement design framework for long-term wellbore integrity.

3.2 Phase 1: Design and Planning—Laying the Foundation

This is arguably the most critical phase for ensuring long-term well integrity, as fundamental design choices made here dictate the well's inherent robustness and resilience against anticipated future conditions and degradation mechanisms (Moreira et al., 2024). Errors, inadequate consideration of future loads or environments, or cost-driven compromises at this stage can introduce latent integrity issues that are extremely difficult, costly, or sometimes impossible to rectify later. Key integrity-focused activities include the following:

- **Defining the Basis of Design (BoD)**: This foundational document captures all the assumptions and requirements the well must meet (Daghmouni et al., 2010). From an integrity perspective, the BoD must accurately define the full range of anticipated conditions throughout the entire well lifecycle, including the most severe credible scenarios (ISO 16530-1, 2017; NORSOK D-010, 2021). This includes the following:
  - Anticipated pressures (initial, flowing, shut-in, maximum stimulation/injection pressures, reservoir depletion effects, potential future re-pressurization, or EOR pressures).
  - Anticipated temperatures (static geothermal gradient, flowing temperatures, maximum/minimum temperatures under all operational modes, thermal cycling amplitudes and frequencies, potential future thermal EOR temperatures).
  - Fluid properties (composition, initial and anticipated future H_2S, CO_2 content, salinity, chlorides, solids potential, density, viscosity, corrosivity, potential for scale deposition).
  - Geological environment (formation types, strengths, potential hazards like shallow gas, reactive shales, mobile salt zones, fractured zones, aquifers).
  - Geomechanical stresses (in-situ stress state, potential for compaction, subsidence, fault reactivation, creep).
  - Planned operational loads (drawdown, injection rates, stimulation treatments, artificial lift methods).
  - Design life of the well for production/injection.
  - Anticipated types and frequency of future interventions.
  - Applicable regulatory requirements and standards.
  - The planned final abandonment strategy and required long-term barrier performance.
- **Risk Assessment**: Performing systematic hazard identification (HAZID) and lifecycle-based risk assessments (e.g., preliminary HAZOP, FMECA on critical components) to identify potential well integrity threats at every stage of the well's life, from drilling to abandonment (Brandt & Gl, 2016). This proactive identification informs design choices and mitigation strategies (see Chap. 14). Examples include assessing risks of shallow gas blowouts, poor cement jobs, casing failure modes under cyclic loading, SCP occurrences due to time-dependent degradation,

or long-term P&A leakage due to material degradation (Jackson, 2014; Patel & Salehi, 2021; Yousuf et al., 2021).

- **Material Selection**: Choosing appropriate materials for casing, tubing, cement, elastomers (seals), and wellhead components is crucial (Bellarby, 2009; Mohammed et al., 2022; Supervisor & Astor Brechan, 2019). This selection must be forward-looking, considering the most severe anticipated future conditions identified in the BoD, not just initial conditions. Selection criteria include the following:
  - Strength requirements based on all anticipated load calculations (burst, collapse, tension, compression, triaxial, fatigue).
  - Corrosion resistance (considering initial and potential future H_2S, CO_2, temperature, chlorides—often involving NACE MR0175/ISO 15156 for sour service, and other standards for CO_2/chloride). Choosing Corrosion-Resistant Alloys (CRAs) upfront is often necessary if future corrosive conditions are anticipated, even if initial fluids are sweet (Askari et al., 2021).
  - Temperature limitations (especially for elastomers—requiring high-temperature FKM/FFKM or MTM seals for HPHT/Geothermal).
  - **Tubulars** must be compatible with fluids (e.g., acid gases, high-salinity brines) and capable of resisting corrosion and fatigue over time.
  - **Elastomers and seals** must be rated for expected thermal cycles and chemical exposures.
  - **Cement systems** must be engineered to resist degradation due to pressure, temperature, and fluid–rock interactions, and to provide both zonal isolation and mechanical support.
  - Chemical compatibility (elastomers versus wellbore fluids, completion fluids, inhibitors, injected EOR/storage fluids).
  - Resistance to hydrogen embrittlement for potential UHS applications (Maury Fernandez et al., 2024).
  - Cost versus reliability and longevity trade-offs.

- **Design Phase: Engineering for Integrity from the Start**

The foundation of well integrity is established during the well planning and design phase. This includes the following:

Casing Point Selection: Identifying formation transitions and fracture gradients to determine the safest and most technically appropriate depths for casing setting.

Tubular Selection: Choosing casing and tubing materials and grades that meet API or ISO standards, capable of withstanding anticipated mechanical loads, chemical exposures (CO_2, H_2S), and temperature cycles.

Well Trajectory Design: Avoiding doglegs or abrupt inclination changes that may induce casing fatigue or compromise cement sheath bonding. Safety margins: Incorporating sufficient design factors for burst, collapse, and axial loads to account for uncertainties during drilling and production.

- **Casing Design**: Designing the number of casing strings, their setting depths, weights, grades, and connections to withstand all anticipated loads (including thermal stresses, fatigue from cycling, and geomechanical loads like compaction or salt creep) with appropriate safety factors for the entire design life (Rahman & Chilingar, 1995). Design must ensure sufficient formation strength (fracture gradient, confirmed by LOT/FIT) exists at each casing shoe to safely drill the next section and control potential influxes (Chap. 4).
- **Cementing Program Design**: Planning the primary cement jobs for each casing string to achieve effective and durable zonal isolation, preventing inter-zonal crossflow and migration of fluids behind casing to surface. This involves selecting appropriate cement slurries and additives designed for long-term stability and resistance to anticipated downhole conditions (temperature, pressure, CO_2, H_2S, sulfates) and operational stresses (thermal/pressure cycling), designing effective spacer trains for mud removal, calculating volumes, and planning placement procedures to maximize the chance of achieving a uniform, well-bonded sheath (Chap. 5) (Arbad et al., 2022; De Andrade & Sangesland, 2016; Schreppers, 2015).
- **Completion Design**: Selecting and designing the completion string components (tubing, packer, DHSV, expansion joints, etc.) and the wellhead/Xmas tree to safely contain fluids, control flow, allow monitoring, and meet barrier requirements. This includes selecting packers and seals rated for anticipated P/T/fluid conditions and cyclic loading, and DHSVs meeting safety standards (Chap. 6).
- **Initial Barrier Strategy**: Explicitly defining the primary and secondary barrier envelopes for the planned construction and initial operational phases using Well Barrier Schematics (WBS). This is a living document that evolves with the well.
- **Contingency Planning**: Developing preliminary plans for addressing potential integrity issues identified during risk assessment or those inherent to the design (e.g., strategies for SCP management, potential remediation techniques, planning for intervention access).

Thoroughness and diligence in the design phase, based on accurate data and sound engineering principles that consider the *entire lifecycle*, significantly enhance the probability of achieving and maintaining well integrity throughout the well's life. Failures that are "designed in" due to inadequate foresight are the most difficult and costly to correct. To illustrate the long-term implications of inadequate design, material selection, or lifecycle planning, Fig. 3.2 presents a visual summary of actual casing and tubing failure cases encountered in field operations. These failures highlight the types of damage that can result from corrosion, poor cementing, mechanical overload, and unaddressed annular pressure build-up—many of which can be prevented or mitigated through diligent design-phase decisions. Well integrity must be applied systematically as a procedure at every operational step. Table 3.1 is a structured integrity framework.

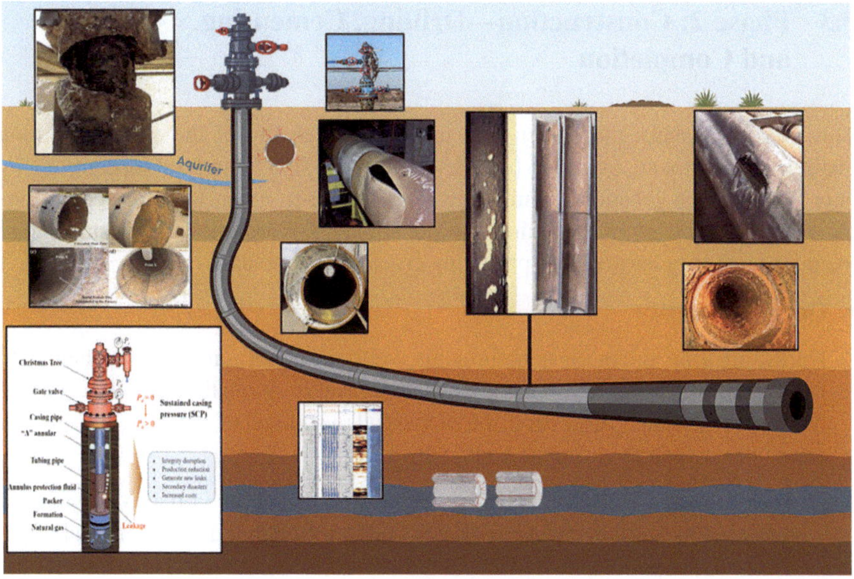

Fig. 3.2 Visual documentation of common casing and tubing failures encountered in oil and gas wells. This figure illustrates 11 distinct failure mechanisms using both field photographs and schematic representations. (1) Shows internal corrosion severely affecting casing integrity. (2) Highlights circumferential wear due to erosion and chemical attack. (3) Presents a split seam failure caused by mechanical stress. (4) Captures deformation in the pipe body resulting from collapse or buckling. (5) Provides evidence of casing damage identified through imaging logs. (6) Displays extensive corrosion and wall thinning. (7) Depicts a hole in the casing, likely caused by localized corrosion or abrasive wear. (8) Reveals longitudinal cracking and pitting corrosion. (9) Shows a surface wellhead (Christmas tree) configuration prone to pressure retention and leaks. (10) Illustrates an annular pressure leak path that compromises zonal isolation. Finally, (11) represents inadequate cement bonding and debonding from casing surfaces. Together, these examples underscore the need for robust well design, ongoing surveillance, and timely interventions to ensure long-term wellbore integrity

Table 3.1 Summary of well integrity actions required at each key phase of the well construction process. Each phase involves specific practices to preserve barrier function and ensure long-term well performance

Well phase	Integrity action
Planning	Apply best practices in well trajectory, casing design, and formation evaluation
Material prep	Inspect, protect, and validate casing, tubing, and cement materials
Drilling	Control ROP, maintain proper mud weight, and minimize ECD fluctuations
Casing running	Use thread protectors, control running speed, apply correct torque, monitor fill
Cementing	Design tailored slurries, ensure effective displacement, and monitor in real time
Post-cementing	Respect WOC time, drill carefully through accessories, and evaluate cement bond logs

3.3 Phase 2: Construction—Drilling, Cementing, and Completion

This phase translates the design into the physical reality of the well. Execution quality is paramount; even the best design can be significantly compromised by poor construction practices, human error, or failure to adapt to actual conditions encountered. Meticulous execution and documentation are critical for establishing the initial integrity barriers and providing crucial baseline data.

- **Drilling Integrity**

  During drilling operations, optimizing wellbore conditions is fundamental to long-term well integrity. The goal is to preserve formation stability, ensure barrier performance, and enable the successful placement of subsequent structural components such as casing and cement.

  – **Maintaining Wellbore Stability**: Preventing borehole collapse is essential for enabling smooth casing operations and minimizing the risk of stuck pipe, washouts, or poor cement placement. Proper wellbore support also reduces the likelihood of stress concentrations that may later propagate as micro-annuli or casing deformation (Alsubaih & Nygaard, 2016).
  – **Managing Drilling Fluid and Hydraulics**: The drilling fluid serves as the primary pressure barrier during the open-hole phase. Selecting the appropriate mud weight and rheological properties is critical to achieving well control, avoiding kicks, minimizing formation damage, and preventing fluid losses that can impair cement bonding. A mud weight that is too low can lead to wellbore collapse or influx; too high, and it may fracture the formation, initiating lost circulation or formation breakdown (Alsubaih et al., 2018).
  – **Drilling Parameter Optimization**: Drilling dynamics such as Rate of Penetration (ROP), Weight on Bit (WOB), Rotations Per Minute (RPM), and surge/swab pressures must be carefully managed. Excessively fast ROP can result in poor hole cleaning, surge pressures, or excessive Equivalent Circulating Density (ECD), all of which may reduce cement sheath quality and lead to long-term integrity failure.
  – **Bit and Hydraulic System Selection**: The bit type and bottomhole assembly (BHA) should be selected to ensure clean hole cutting and efficient cuttings transport, without exceeding fracture gradients. Hydraulic parameters must be modeled to maintain effective annular velocity and prevent cuttings from settling.
  – **Running Casing Considerations**: When running casing, it is essential to minimize wear on previously installed casing strings and reduce borehole disturbances that could impair the cement bond. Surge/swab effects should be modeled to avoid overpressuring the formation. Filling the casing during running operations and applying appropriate torque at connections ensure long-term pressure integrity.

- **Formation Integrity Testing(FIT/LOT)**: After drilling out each casing shoe, it is critical to perform a Formation Integrity Test (FIT) or Leak-Off Test (LOT) to validate that the exposed formation and cement sheath can withstand the pressures required to drill the next hole section safely. These tests confirm the strength of the secondary barrier at the shoe and validate the ECD limits for the next section (Liegenfeld, 2019). Refer to Chap. 7 for detailed procedures.
- **Shallow Hazard Mitigation**: In regions prone to shallow gas, overpressured sands, or unconsolidated formations, implementing mitigation measures—such as pilot hole drilling, conductor casing pre-setting, and controlled mud weight windows—is critical to isolate hazards before they compromise the wellbore.
- **Meticulous Documentation and Monitoring**: Throughout the drilling process, maintaining comprehensive records is vital. Logs should capture real-time and static data on:
- Drilling parameters (ROP, WOB, torque, RPM).
- Mud properties and hydraulics.
- Pressures (standpipe, annular, ECD).
- Gains and losses.
- Indicators of instability (tight spots, overpull, fill on bottom).

These records form the foundation for post-drill integrity assessments and enable effective root-cause analysis in the event of barrier failure.

- **Cementing Operations Integrity**

  Cementing is one of the most critical steps in establishing permanent well barriers:

  - Execution Quality: Achieving effective zonal isolation relies heavily on operational execution. Poor mud removal and inadequate casing centralization are primary causes of cementing failure (Jung & Frigaard, 2022; Sauer, 1987), leading to mud channels and leak paths that undermine long-term integrity. Proper hole conditioning, effective mud removal using turbulent flow or optimized spacers, correct slurry mixing and pumping rates, use of centralizers as per design (based on standoff calculations for the actual wellpath), and pipe movement (rotation/reciprocation) if applicable are crucial (Al-Dhafeeri et al., 2019; Kiran et al., 2017). Deviations from the planned procedure must be managed and documented.
  - Cement slurry design must be tailored to withstand thermal cycles, chemical exposure, and formation pressures.
  - Spacer and preflush systems should ensure effective mud removal, thereby enhancing the quality of the cement bond.
  - ECD management during displacement is essential to avoid losses. The circulating density should always remain below the fracture gradient.
  - Use of pipe movement (reciprocation and rotation) improves displacement efficiency and cement bonding.
  - Monitoring slurry properties (density, viscosity, thickening time) during pumping ensures quality control in real time.

  - Wait-on-cement (WOC) time must be respected to allow full compressive strength development before drilling ahead.
  - Slurry Quality Control: Verifying that the cement slurry properties (density, rheology, fluid loss, thickening time) meet the design specifications *at the rig site* before pumping is critical.
  - Verification: Running cement evaluation logs (CBL/VDL, Ultrasonic) after the cement has set provides crucial (though not infallible) verification of cement placement and bonding quality (Khelfaoui & Zouyed, 2023). These logs provide essential baseline data for assessing cement degradation over time. Remedial cementing (squeeze jobs) may be required if results are unacceptable and compromise barrier requirements.
  - Meticulous Documentation: Precise records of the *actual* cementing operation parameters (pump rates, pressures, volumes, densities, returns, plug bump pressure, any deviations, WOC time) and the cement evaluation log reports are fundamental "as-built" integrity data.
  - After casing and cement are in place:

- **Avoid using aggressive bits or drilling parameters** when re-entering through casing shoes or float collars to prevent casing damage or wear.
- **Completion Installation Integrity**
  - Equipment Handling: Avoiding damage to sensitive equipment like packers, seals, and valve sealing surfaces during handling and installation. Damage during installation can lead to immediate or premature failure of critical barrier elements (Ahmed & Salehi, 2021).
  - Correct Procedures: Following manufacturer and operator procedures meticulously for making up connections (correct torque), setting packers, installing hangers, and function testing components like DHSVs. Deviations or errors in these steps directly compromise initial barrier integrity.
  - Initial Pressure Testing: Performing comprehensive pressure tests on the completion string, packer seals, annuli, wellhead seals, and Xmas tree components to verify initial integrity against defined pressure requirements before handover to operations (Chap. 7) (API Specification 6A, 2019). Documenting these baseline tests (pressure, duration, decline, test medium) is vital for future monitoring and diagnosis.
  - Meticulous Documentation: Detailed records of completion component serial numbers, installation depths, setting procedures, test results, and wellhead stack-up details constitute the critical "as-built" data for the completion system.

Quality control and assurance (QA/QC) processes, clear operational procedures, competent personnel, diligent supervision, and meticulous documentation are essential throughout the construction phase to ensure the well is built according to design specifications and that initial barrier integrity is established and verified as per the design intent. Latent defects introduced during construction are a common cause of integrity failures years later, highlighting the lasting impact of execution quality.

3.4 Phase 3: Operation—Production/Injection

This is typically the longest phase, where the well generates revenue or serves its injection purpose. The focus shifts from construction to managing time-dependent degradation, cumulative stresses, and operational risks. Proactive monitoring, maintenance, and adherence to established procedures are crucial for maintaining integrity over the well's planned life, which may range from years to decades.

- **Surveillance and Monitoring**: This is the primary activity for detecting potential integrity issues as they develop over time.
  - Annulus Pressure Monitoring: Regular (often continuous for critical wells) monitoring of all accessible annuli pressures is fundamental for detecting leaks in primary or secondary barriers (manifesting as SCP—sustained casing pressure). SCP is a primary indicator of barrier failure and requires systematic management (Chap. 8).
  - Production/Injection Parameters: Monitoring flow rates, pressures, temperatures, and fluid composition (water cut, gas–liquid ratio, H_2S, CO_2, salinity) can provide indirect indicators of integrity issues (e.g., unexpected changes suggesting communication, pressure changes indicating restrictions or leaks) (Lackey & Dilmore, 2021).
  - Corrosion Monitoring: Implementing planned corrosion inhibition programs, monitoring corrosion rates (e.g., coupons, probes, sampling for iron counts), and potentially performing periodic corrosion logging to assess cumulative metal loss over time (Chaps. 9 and 12).
  - Scaling Monitoring: Monitoring for injectivity decline (in injectors) or pressure drops/restrictions (in producers), and sampling fluids to assess scaling potential.
- **Maintaining Operating Envelopes**: Ensuring the well is operated within its defined safe limits for pressure, temperature, flow rates, and fluid corrosivity. Excursions outside these limits, or failure to consider cumulative stress effects, can overstress components or accelerate degradation (fatigue, corrosion).
- **Routine Testing and Inspection**: Performing periodic tests and inspections to directly verify barrier function and condition.
  - Valve Functional Testing: Regularly testing the operation of safety-critical valves (DHSV, SSVs, annulus valves) to ensure they function mechanically.
  - Valve Leak Testing: Periodically testing the sealing integrity of closed safety valves against differential pressure.
  - Annulus Integrity Testing: Periodic pressure testing of accessible annuli (where safe and feasible) to verify the integrity of the outer casing strings, cement, and wellhead seals.
  - Surface Equipment Inspection: Regular visual inspection of the wellhead and Xmas tree for leaks, corrosion, or damage.

  - Downhole Integrity Logging: Performing periodic logs (e.g., cement evaluation, caliper, EM thickness) in wells with known degradation risks or after specific events to assess the condition of downhole barriers and measure cumulative metal loss or cement deterioration (Chap. 12).

- **Maintenance Procedures**: Implementing planned preventative maintenance for integrity-critical equipment (e.g., valve servicing, wellhead maintenance, calibration of pressure gauges/sensors).
- **Management of Change (MoC)**: A formal process to evaluate and control any changes (operational, equipment, procedural) that could impact well integrity. This is critical as even seemingly minor changes can introduce new risks or affect existing barriers if not properly assessed (Chap. 17).
- **Data Management and Analysis**: Systematically recording all monitoring data, test results, inspection/logging data, maintenance activities, and operational parameters. Analyzing trends over time (e.g., SCP build-up rates, corrosion rates, injectivity changes) is crucial for identifying potential emerging issues proactively and planning interventions before failure occurs (Chap. 15).

The goal during the operational phase is to manage the time-dependent degradation of the barriers established during construction, detect any degradation early through vigilant monitoring, manage risks within the WIMS framework, and plan interventions or maintenance as needed to ensure continued safe and reliable operation throughout the well's planned life.

3.5 Phase 4: Intervention

Wells often require interventions after initial completion for activities like logging, stimulation (acidizing, fracturing), scale removal, changing lift methods, or repairing failed components (e.g., replacing a DHSV, squeezing off perforations). Maintaining well integrity during these operations is critical, as the well's permanent barriers are often deliberately manipulated or bypassed, and potential pressure sources are being accessed or disturbed. Interventions carry inherent risks and must be meticulously planned to ensure well control and integrity are maintained.

- **Planning for Barriers**: Intervention programs must explicitly define the primary and secondary barrier envelopes during the intervention activity itself. This often involves the planned use of temporary barriers.

  - Temporary Barrier Elements: These can include mechanical plugs (bridge plugs, packers set in tubing/casing), a sufficient column of kill-weight fluid (mud, brine), shear-seal BOP rams, or valves within the intervention equipment stack used at surface (e.g., lubricator valves, coiled tubing BOPs).
  - Verification: Crucially, temporary barriers must also be verifiable (e.g., pressure tested after setting) to confirm their sealing integrity before relying on them.

  – Well Barrier Schematics: Updating and using WBS diagrams specifically for the intervention phases is essential to visualize and communicate the active barriers at each step of the operation.

- **Well Control**: Intervention operations, especially on live wells, require robust well control equipment (e.g., intervention BOP stack, lubricator, snubbing unit) and rigorous procedures to manage pressure if temporary downhole barriers fail or if an influx occurs. The surface pressure control equipment (PCE) acts as a vital temporary barrier (Chap. 25).
- **Avoiding Damage**: Intervention activities (e.g., running tools, pumping fluids, perforating, milling) must be performed carefully to avoid inadvertently damaging existing permanent barrier elements like casing, cement sheath, permanent packers, or the wellhead. Such damage could introduce new integrity issues that manifest later.
- **Restoring Permanent Barriers**: After the intervention is complete and temporary barriers are removed (if applicable), it is crucial to correctly reinstate and rigorously verify the integrity of the well's permanent primary and secondary barrier envelopes through appropriate pressure tests or other means before returning the well to its normal operational state or shut-in status. Documenting this post-intervention verification is vital.

Interventions inherently carry risks, and meticulous planning emphasizing the temporary barrier philosophy, robust well control, competent execution, and thorough post-intervention verification are essential to ensure well integrity is maintained throughout the operation and successfully restored afterward.

3.6 Phase 5: Suspension (Temporary)

Temporary suspension involves shutting in a well for a period (days, months, or even years) to potentially return it to service later. Integrity requirements are crucial to ensure safe, static containment during the idle period and readiness for future operations (Al Hosani et al., 2025; Stein, 2018).

- **Barrier Requirements**: Two independent, verified barriers are still required between potential pressure sources and the environment. The specific barriers may be the normal operational barriers (if verified) or may involve setting temporary plugs or changing annular fluid densities to achieve the required barrier standard.
- **Monitoring**: While potentially less frequent than for actively producing wells, monitoring (especially annulus pressures) is still typically required to detect potential leaks that develop during the idle period.
- **Maintenance**: Equipment might still require some maintenance (e.g., valve lubrication) during suspension.

- **Return to Service Verification**: Before returning a suspended well to operation, a comprehensive integrity verification program (pressure tests, function tests) is mandatory to confirm that the barriers remain intact and functional after the idle period.

3.7 Phase 6: Abandonment (Plug and Abandonment—P&A)

When a well reaches the end of its useful life, or if exploration proves unsuccessful, it must be permanently abandoned. The objective of P&A is fundamentally different from operational integrity: it is to achieve permanent isolation of porous/permeable formations and seal the wellbore against any future migration of fluids to the surface or crossflow between zones, effectively restoring or exceeding the geological seal disrupted by drilling. This must be achieved without reliance on future monitoring or maintenance, in perpetuity (King & Valencia, 2014; Van Der Kuip et al., 2011).

- **Permanent Barrier Philosophy**: P&A typically involves setting multiple permanent barriers (plugs) made from durable materials at strategic locations within the wellbore. These barriers must seal both inside the casing and across all annuli (achieving a seal across the full wellbore cross section from formation to formation or competent caprock). The location, length, and material of these plugs are typically specified by regulations and standards based on risk assessment and the need to isolate specific zones (Chap. 27).
- **Barrier Materials**: Cement (specifically formulated for P&A) is the most common permanent barrier material. However, research and application of alternative materials (bismuth alloys, geopolymers, resins) are increasing, driven by the need for proven long-term durability and potentially better sealing properties in certain situations or over geological timescales (Chap. 27).
- **Placement and Verification**: Ensuring plugs are placed at the correct depth and achieve a robust hydraulic seal across the entire wellbore cross section is critical. This often involves wellbore preparation, potentially cutting and pulling casing to set plugs in open hole or across annuli ("rock-to-rock" barriers), and rigorously verifying their location (e.g., tagging with tubing/wireline) and sealing integrity (e.g., pressure testing above the plug, potential inflow testing below, specialized logging) (Chap. 27).
- **Long-Term Perspective**: The major challenge in P&A is demonstrating that the barriers will remain effective for geological timescales (hundreds or thousands of years). This involves understanding the long-term degradation mechanisms of barrier materials in the downhole environment (chemical attack, geomechanical stress, thermal effects) and designing/qualifying plugs accordingly (Chap. 27).
- **Regulatory Requirements**: P&A operations are heavily regulated, with specific requirements regarding the number, location, length, material, and verification methods for permanent barriers. Compliance demonstration is critical.

Effective P&A is crucial for protecting the environment (groundwater, surface emissions) and avoiding significant future liabilities associated with leaking abandoned wells. It represents the final, critical step in responsible lifecycle management, demanding meticulous planning, execution, and verification, focusing on *perpetual* containment. A key technique in plug placement is the balanced plug method, which seeks to equalize the hydrostatic pressures in the annulus and the drill pipe at the base of the plug (Khalifeh & Saasen, 2020). This ensures a stable, precise placement without fluid displacement or plug contamination. The pressure balance can be expressed as

$$\Delta P_{CD} + \Delta P_{WD} + \Delta P_{MD} = \Delta P_{CA} + \Delta P_{WA} + \Delta P_{MA} \tag{3.4}$$

which simplifies to

$$P_D = P_A \tag{3.5}$$

In this equation, ΔP_{CD} is the hydrostatic pressure is exerted by cement inside the workstring, ΔP_{WD} by spacer fluid, and ΔP_{MD} by mud in the workstring. Similarly, ΔP_{CA}, ΔP_{WA}, and ΔP_{MA} are the hydrostatic pressures of cement, spacer, and mud in the annulus, respectively. Achieving a pressure balance ensures that the plug remains stationary during placement and prevents channeling or fallback.

3.8 Interdependencies Across the Lifecycle

It is vital to recognize the strong links and dependencies between these lifecycle phases:

- **Design to Construction**: Inadequate design parameters (e.g., underestimating future pressure/temperature, selecting inappropriate materials) cannot be fully overcome by excellent construction execution. Conversely, poor execution (e.g., bad cement job, damaged casing) can undermine even a perfect design.
- **Construction to Operation**: Latent defects introduced during drilling (e.g., borehole instability affecting cementing) or completion installation (e.g., damaged seals) often manifest as integrity issues (e.g., SCP, leaks) during the operational phase, potentially years later (Jackson 2014). Poor baseline data acquisition during construction hinders effective monitoring and diagnostics later.
- **Operation to Intervention**: Integrity issues detected during operation necessitate intervention. The operational history (e.g., corrosion rates, scaling) informs intervention planning and risk assessment.
- **Interventionto Operation**: Damage caused during intervention can compromise long-term integrity during subsequent operation or abandonment. Verification post-intervention is critical.

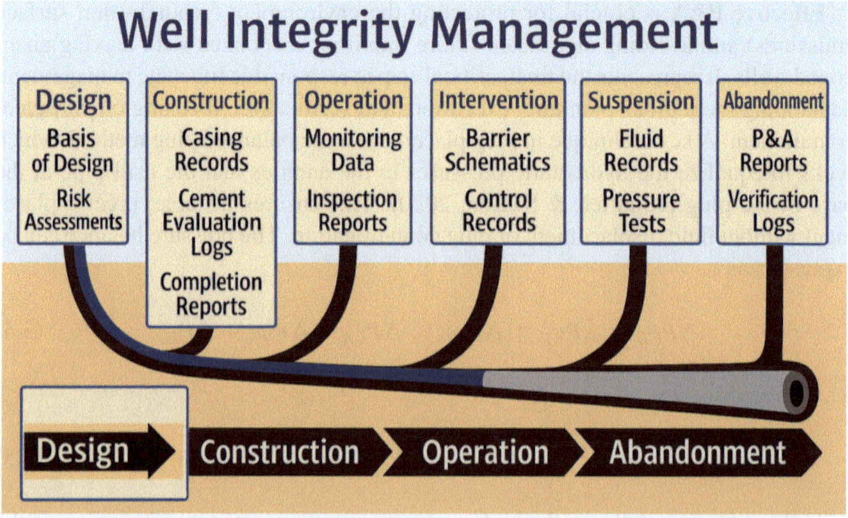

Fig. 3.3 Conceptual graphic showing the types of data and documentation required for integrity management at different lifecycle phase transitions. This diagram emphasizes the importance of systematically generating, maintaining, and transferring well integrity documentation throughout the entire well lifecycle. The design phase establishes the foundation through Basis of Design and risk assessments. Construction produces critical records such as casing data, cement evaluation logs, and completion reports. During operation, integrity depends on monitoring data and inspection reports. Interventions require validated barrier schematics and control documentation. Suspension activities demand fluid and pressure test records, while final abandonment relies on P&A reports and verification logs to ensure regulatory closure and environmental safety. This flow underscores that lifecycle integrity is built upon consistent, traceable information management

- **All Phases to Abandonment**: The well's construction history (casing program, actual cement quality, materials used) and operational history (e.g., SCP history, material degradation observed) heavily influence the complexity, cost, and strategy required for effective permanent abandonment. Designing for decommissioning (considering P&A requirements and potential future conditions during the initial design phase) can significantly simplify and reduce the cost/risk of abandonment.

A successful well integrity program requires seamless integration of knowledge, data, processes, and responsibilities across all these lifecycle phases, managed within a robust Well Integrity Management System (WIMS), which will be discussed in Part 4. These interdependencies are made actionable through structured documentation that carries design assumptions, operational data, and verification records from one phase to the next. Figure 3.3 illustrates how well integrity management relies on phase-specific data being available, validated, and communicated across the entire lifecycle. Recognizing and actively managing interdependence and the long-term consequences of early-phase decisions is fundamental to achieving sustainable well integrity.

References

Aadnøy, B. S., & Looyeh, R. (2019). *Petroleum rock mechanics: Drilling operations and well design.*

Ahmed, S., & Salehi, S. (2021). Failure mechanisms of the wellbore mechanical barrier systems: Implications for well integrity. *Journal of Energy Resources Technology, Transactions of the ASME, 143*(7). https://doi.org/10.1115/1.4050694

Al Hosani, E. A., Hakro, I. A., Ahmed, A. H., Ali, M. S., Daba, A. M., Offshore, A., Dhabi, A., Jaffrey, A., & Gordon, N. (2025). Enhancing subsea asset integrity: Holistic risk management and monitoring strategy for mud line suspended wells. *Offshore Technology Conference.* https://doi.org/10.4043/35594-MS

Al-Dhafeeri, A. M., Aramco, S., Fouad, S., Zkery, A., Ahmed, S., Sdeiri, A., & Operations, K. J. (2019). Lessons learned from effective well integrity management in offshore field. In *Proceedings—SPE annual technical conference and exhibition, 2019-September* (p. 195915). https://doi.org/10.2118/195915-MS

Alsubaih, A. A., & Nygaard, R. (2016). *Shale instability of deviated wellbores in Southern Iraqi fields.* OnePetro. https://dx-doi-org.ezproxy.lib.utexas.edu/

Alsubaih, A., Albadran, F., & Alkanaani, N. (2018). Mechanical specific energy and statistical techniques to maximizing the drilling rates for production section of Mishrif wells in southern Iraq fields. In *Proceedings of the SPE/IADC Middle East drilling technology conference and exhibition, 2018-January.* https://doi.org/10.2118/189354-ms

API Specification 6A. (2019). *API | API Specification 6A* (21st ed.). https://www.api.org/products-and-services/standards/important-standards-announcements/spec-6a

Arbad, N., Emadi, H., & Watson, M. (2022). A comprehensive review of geothermal cementing from well integrity perspective. *Journal of Petroleum Science and Engineering, 217*, 110869. https://doi.org/10.1016/J.PETROL.2022.110869

Askari, M., Aliofkhazraei, M., Jafari, R., Hamghalam, P., & Hajizadeh, A. (2021). Downhole corrosion inhibitors for oil and gas production—A review. *Applied Surface Science Advances, 6.* https://doi.org/10.1016/j.apsadv.2021.100128

Bellarby, J. (2009). Chapter 8 material selection. *Developments in Petroleum Science, 56*(C), 433–472. https://doi.org/10.1016/S0376-7361(08)00208-2

Brandt, H., & Gl, D. (2016). Well integrity management—Application of risk assessment to optimize your decisions. In *Offshore technology conference Asia 2016, OTCA 2016* (pp. 4615–4623). https://doi.org/10.4043/26817-MS

Daghmouni, H., El Manci, M., Rady, S. A. A., & Shattawa, Y. (2010). Well integrity management system WIMS development. In *Society of petroleum engineers—14th Abu Dhabi international petroleum exhibition and conference 2010, ADIPEC 2010* (Vol. 2, pp. 1421–1427). https://doi.org/10.2118/137966-MS

De Andrade, J., & Sangesland, S. (2016). Cement sheath failure mechanisms: Numerical estimates to design for long-term well integrity. *Journal of Petroleum Science and Engineering, 147*, 682–698. https://doi.org/10.1016/j.petrol.2016.08.032

ISO 16530-1. (2017). https://www.iso.org/standard/63192.html

Jackson, R. B. (2014). The integrity of oil and gas wells. *Proceedings of the National Academy of Sciences of the United States of America, 111*(30), 10902. https://doi.org/10.1073/PNAS.1410786111

Jung, H., & Frigaard, I. A. (2022). Evaluation of common cementing practices affecting primary cementing quality. *Journal of Petroleum Science and Engineering, 208*, 109622. https://doi.org/10.1016/J.PETROL.2021.109622

Khalifeh, M., & Saasen, A. (2020). *Introduction to permanent plug and abandonment of wells* (Vol. 12). https://doi.org/10.1007/978-3-030-39970-2

Khelfaoui, O., & Zouyed, M. (2023). *Applications of advanced well integrity evaluation technologies for critical decision making.* http://dspace.univ-ouargla.dz/jspui/handle/123456789/34574

King, G. E., & Valencia, R. L. (2014). Environmental risk and well integrity of plugged and abandoned wells. In *Proceedings—SPE annual technical conference and exhibition* (Vol. 6, pp. 4852–4868). https://doi.org/10.2118/170949-MS

Kiran, R., Teodoriu, C., Dadmohammadi, Y., Nygaard, R., Wood, D., Mokhtari, M., & Salehi, S. (2017). Identification and evaluation of well integrity and causes of failure of well integrity barriers (a review). *Journal of Natural Gas Science and Engineering, 45*, 511–526. https://doi.org/10.1016/J.JNGSE.2017.05.009

Lackey, G., & Dilmore, R. M. (2021). *NETL well integrity workshop: Identifying well integrity research needs for subsurface energy infrastructure*. https://doi.org/10.2172/1828877

Liegenfeld, A. T. (2019). *Scoring well construction process and impact on well integrity*. https://doi.org/10.34901/MUL.PUB.2024.197

Maury Fernandez, D., Emadi, H., Hussain, A., & Thiyagarajan, S. R. (2024). A holistic review on wellbore integrity challenges associated with underground hydrogen storage. *International Journal of Hydrogen Energy, 57*, 240–262. https://doi.org/10.1016/j.ijhydene.2023.12.230

Mohammed, A. I., Bartlett, M., Oyeneyin, B., Kayvantash, K., & Njuguna, J. (2022). Multi-criteria material selection for casing pipe in shale gas wells application. *Journal of Petroleum Exploration and Production Technology, 12*(12), 3183–3199. https://doi.org/10.1007/S13202-022-01506-0/TABLES/5

Moreira, N., Lazaro, A. F., Minucci, F. R., Aranha, P. E., Weidlich, M. C., Fernandes, F. B., Nardi, F. B., Radespiel, E. S., Okama, C. O., Palmieri, I., Santos, C. M. D., Terra, F. S., & Brasileiro, P. (2024). *Well integrity management based on new operating limits—A new method for extending the well lifespan and changes in operating conditions*. https://doi.org/10.4043/35440-MS

NORSOK D-010. (2021). https://online.standard.no/en/norsok-d-010-2021

Patel, H., & Salehi, S. (2021). Structural integrity of liner cement in oil and gas wells: Parametric study, sensitivity analysis, and risk assessment. *Engineering Failure Analysis, 122*, 105203. https://doi.org/10.1016/J.ENGFAILANAL.2020.105203

Rahman, S. S., & Chilingar, G. V. (1995). *Casing design: Theory and practice* (p. 373).

Sauer, C. W. (1987). Mud displacement during cementing state of the art. *Journal of Petroleum Technology, 39*(09), 1091–1101. https://doi.org/10.2118/14197-PA

Schreppers, G. (2015). *A framework for wellbore cement integrity analysis*. OnePetro. https://dx-doi-org.ezproxy.lib.utexas.edu/

Stein, A. (2018). Verification of barriers in a suspension or abandonment phase. In *Offshore technology conference Asia 2018, OTCA 2018*. https://doi.org/10.2118/191301-MS

Supervisor, T., & Astor Brechan, B. (2019). *Advanced tubing design*. https://ntnuopen.ntnu.no/ntnu-xmlui/handle/11250/2625928

Van Der Kuip, M. D. C., Benedictus, T., Wildgust, N., & Aiken, T. (2011). High-level integrity assessment of abandoned wells. *Energy Procedia, 4*, 5320–5326. https://doi.org/10.1016/J.EGYPRO.2011.02.513

Yousuf, N., Olayiwola, O., Guo, B., & Liu, N. (2021). A comprehensive review on the loss of wellbore integrity due to cement failure and available remedial methods. *Journal of Petroleum Science and Engineering, 207*, 109123. https://doi.org/10.1016/J.PETROL.2021.109123

Part II
Well Construction and Initial Integrity Assurance

Part I established the fundamental principles of well integrity, the critical barrier philosophy, and the importance of a lifecycle perspective, highlighting how early-phase decisions and execution quality have lasting impacts. Part II delves specifically into the crucial construction phase, where the well's primary structural and sealing components—the casing strings, cement sheaths, completion equipment, and wellhead—are designed and installed. The engineering considerations and operational practices employed during tubular design, cementing, completion equipment installation, and initial verification testing form the bedrock upon which long-term well integrity is built. Getting these steps right is essential for establishing robust and verifiable well barriers from the outset and minimizing the potential for latent defects that could lead to failures years later.

Chapter 4
Casing and Tubing Design for Integrity

Abstract This chapter addresses the foundational role of casing and tubing systems in maintaining long-term well integrity, emphasizing their function as both structural and pressure-containment barriers. It presents a comprehensive framework for the mechanical and environmental design of well tubulars, incorporating industry standards (e.g., API 5CT) and accounting for lifecycle load scenarios including burst, collapse, tension, compression, and triaxial stresses. The chapter begins by introducing the mud window as a guiding constraint for casing depth selection and progresses to detailed analytical models for stress evaluation under downhole conditions. Material selection strategies are discussed in the context of temperature effects and sour service environments (e.g., H_2S), with particular focus on grade-specific performance and degradation. Additionally, the integrity of tubular connections is rigorously examined, comparing API and premium designs with respect to sealability, mechanical robustness, and fatigue resistance. Design factors and safety margins are highlighted to address uncertainties, manufacturing tolerances, and operational variability. Finally, casing wear mechanisms and predictive models are introduced to safeguard against progressive damage in deviated wells. Through this integrated approach, the chapter equips engineers with the critical design principles and calculations needed to ensure tubular reliability across drilling, production, and abandonment phases.

© The Author(s), under exclusive license to Springer Nature Switzerland AG 2026

A. Alsubaih and K. Sepehrnoori, *Well Integrity for Oil, Gas, and Subsurface Storage Operations*, https://doi.org/10.1007/978-3-032-15233-6_4

Drilling rig with casing joints prepared on racks prior to running into the wellbore. *Source* The Authors

Chapter's Objectives

- Introduce relevant API specifications and material grades for casing and tubing.
- Explain the primary load cases considered in tubular design, including lifecycle loads.
- Discuss the importance of connection selection and performance, especially under fatigue.

- Outline the concept of design factors and safety margins.
- Address the critical impact of temperature and sour service (H_2S) environments on material selection.
- Briefly cover wear considerations during drilling and operations.

4.1 Introduction to Well Tubulars

Casing and tubing strings are fundamental components of almost every oil and gas well, forming the primary structural elements and critical barrier elements.

- **Casing**: Large-diameter pipe strings set sequentially inside the drilled hole and typically cemented in place. They provide structural support to the wellbore, prevent formation collapse, isolate problematic zones (e.g., lost circulation zones, unstable shales, aquifers), and form the primary conduits for drilling subsequent, deeper sections. Importantly, casing strings and their associated cement sheaths form key secondary (and sometimes primary) barrier envelopes against external formation pressures and fluid migration. The long-term integrity of the casing is essential for supporting the well and providing containment throughout its operational life and into abandonment.
- **Tubing**: Smaller-diameter pipe string installed inside the final casing string (production casing), typically after the well is drilled to total depth. It serves as the primary conduit for conveying produced fluids to the surface or injection fluids to the reservoir, protecting the production casing from direct exposure to corrosive fluids, high pressures, and wear from flow. The tubing string is a critical element of the *primary barrierenvelope* during production/injection operations (NORSOK D-010, 2021).

Ensuring the mechanical integrity of these tubular strings throughout the well's life is paramount. This requires careful design considering the materials used, the loads anticipated *over the entire lifecycle*, and the connections joining individual pipe segments (Aadnøy et al., 2003a). The selection of casing and tubing depths is critical for maintaining well integrity, and it is largely determined by the mud window—an essential concept that defines the pressure range within which drilling fluids must operate. As illustrated in Fig. 4.1, the mud window must ensure that the drilling fluid density stays between the pore pressure and the fracture pressure (Alsubaih et al., 2018, 2019, 2020, 2024, 2024b, 2025) . The casing depths are strategically designed to prevent formation collapse or fluid loss during drilling operations. Figure 4.1 demonstrates how the design of the mud window influences casing selection at each stage of drilling. By understanding the mud window, engineers can select appropriate casing depths to optimize the integrity of the wellbore throughout its lifecycle. Once casing and tubing depths are selected based on mud window constraints, the mechanical design must account for the most critical loading scenarios these tubulars may experience over their lifetime. Among the most important are collapse pressure loading and burst pressure loading. Collapse loading occurs when external

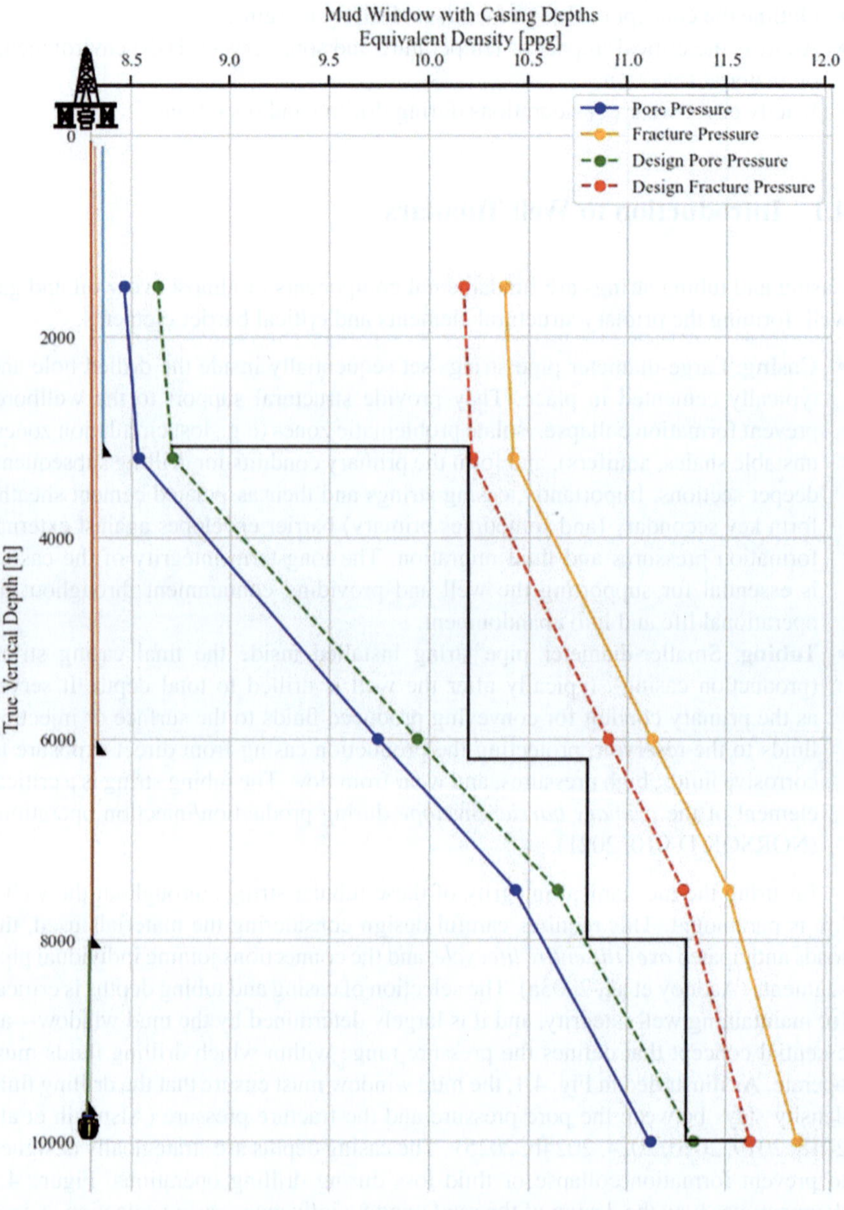

Fig. 4.1 Illustration of the concept of selecting casing depths based on the mud window, where the drilling fluid density should remain between the pore pressure (blue curve) and fracture gradient (orange curve). The static and dynamic densities are shown within the mud window, highlighting the necessary balance for well stability. The casing depths (surface casing, intermediate casing, production casing, and liner) are marked with labels to represent the typical choices for each stage of drilling, ensuring safe operations while maintaining well integrity

pressure exceeds the internal pressure inside the casing, potentially causing inward failure of the pipe wall. Conversely, burst loading occurs when high internal pressure from reservoir fluids or trapped gases exceeds the external support pressure, risking outward rupture of the casing. These failure modes are evaluated using simplified but conservative analytical models to ensure structural integrity under worst-case assumptions.

Collapse pressure loading for production casing is expressed as

$$\Delta p = \left[p_o(0) + \gamma_o x\right] - \left[p_i(0) - \gamma_i x\right] \tag{4.1}$$

Assuming the casing is fully evacuated and the external pressure results solely from mud hydrostatics, this simplifies to

$$p(x) = 0.052\gamma_o x \tag{4.2}$$

where γ_o is the mud density (in ppg) and x is the depth (ft). This condition is commonly used for collapse design because it is conservative and protects against external pressure overloads during cementing or well evacuation.

For burst pressure loading, the internal pressure distribution caused by reservoir gas trapped in the casing is given by

$$p_i(x) = p_f - 0.052\gamma_g(L - x) \tag{4.3}$$

where p_f is the reservoir pressure, γ_g is the gas density (ppg), and L is the total casing length. This profile represents the worst-case gas expansion scenario and is essential for verifying the burst resistance of the casing.

4.2 API Specifications and Material Grades

The American Petroleum Institute (API) publishes globally recognized standards governing the specification, manufacturing, and testing of casing and tubing. The primary standard is **API Specification 5CT: Specification for Casing and Tubing** (API | API Specification 5CT, 10th Edition, 2019). This standard defines.

- **Steel Grades**: These denote the minimum yield strength of the pipe material (e.g., J55, K55, N80, L80, C90, T95, P110, Q125). The number indicates the minimum yield strength in thousands of psi (ksi). Letters often denote specific properties or service conditions (e.g., L80 is a controlled yield strength grade often used for sour service applications, C90/T95 are specifically for sour service and have hardness limitations per NACE standards). Higher grades offer greater strength but may have different susceptibility to certain failure mechanisms like sulfide stress cracking (SSC) in sour environments.

- **Dimensions and Weights**: Standardized outside diameters (OD), wall thicknesses, and nominal weights per foot/meter.
- **Manufacturing Processes**: Requirements for seamless or electric-welded (ERW) pipe production.
- **Chemical Composition**: Limits on various elements in the steel to control mechanical properties and susceptibility to certain failure mechanisms.
- **Mechanical Properties**: Minimum requirements for yield strength, tensile strength, and sometimes toughness or hardness, depending on the grade and service conditions.
- **Inspectionand Testing**: Mandated non-destructive testing (NDT) and hydrostatic pressure testing requirements during manufacturing to ensure quality.

Selecting the appropriate API grade is a critical design decision that depends on the anticipated stresses (burst, collapse, tension, compression, fatigue, triaxial), temperatures, and, crucially, the corrosivity of the wellbore environment (particularly the presence of H2S and CO_2). Material selection must consider the most severe conditions anticipated throughout the well's planned operating life (as defined in the BoD), not just initial conditions. Figure 4.2 highlights two critical aspects of casing design: the frequency of use for various API grades across different well applications (Plot A), and the distribution of failure mechanisms associated with each grade (Plot B). The left plot reveals that P110 and N-80 are the most widely used grades, particularly in high-stress environments such as HPHT/geothermal and shale gas wells. This preference is likely driven by their high yield strength and availability. However, the second plot demonstrates that these high-strength grades are also more prone to a wider range of failure mechanisms, including buckling, fatigue, wear/erosion, and shear failure. In contrast, lower-strength grades like J55, X56, and L80 show fewer failure modes, suggesting that their limited use in extreme environments correlates with lower exposure to high stress and aggressive fluids. The "Several" line further emphasizes that multi-mode failures are concentrated in higher-grade materials, reinforcing the importance of considering not only strength but also toughness, environmental compatibility, and connection reliability when selecting casing materials. Together, these plots reinforce a central design principle: higher strength alone does not guarantee integrity. Without careful consideration of service conditions, cyclic loads, and connection sealing reliability, the risk of failure can increase—especially in complex or sour environments. This underscores the need for a balanced approach to casing grade selection, where mechanical capacity, corrosion resistance, and fatigue life are optimized in tandem.

4.3 Load Cases in Tubular Design

Understanding the mechanical stresses acting on casing and tubing is critical to ensuring long-term well integrity (King & Valencia, 2016). These tubulars are exposed to a combination of pressure, temperature, and axial loads over the course

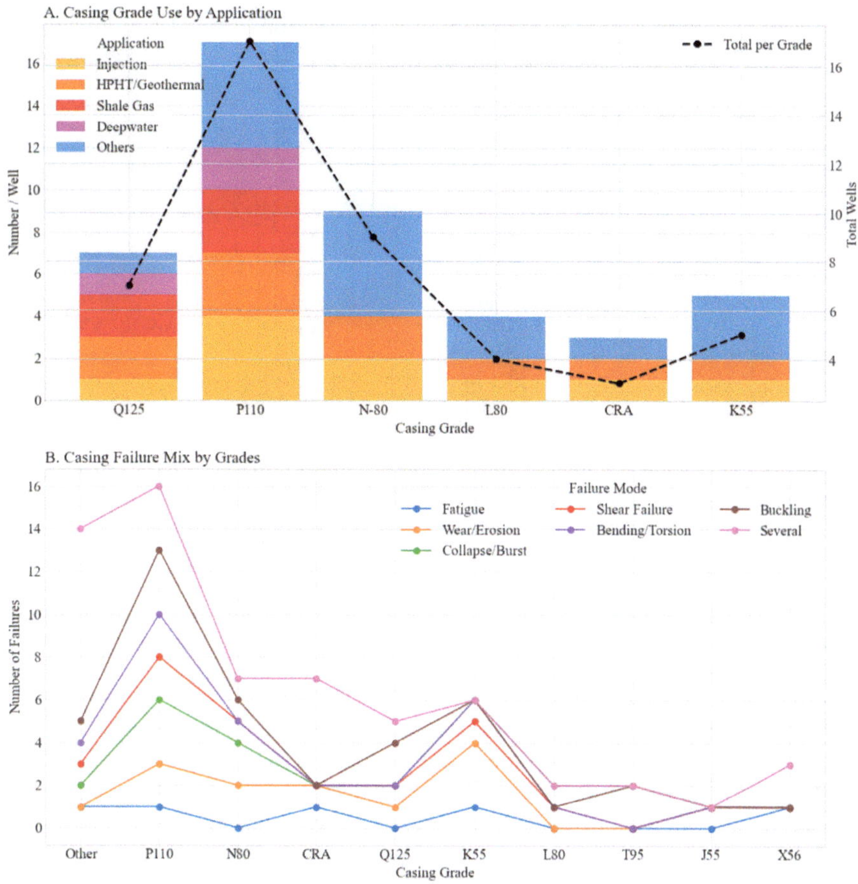

Fig. 4.2 Casing-grade utilization and associated failure trends (Mohammed et al., 2019) **a** Stacked bar chart showing the frequency of casing grade use across different applications such as injection, HPHT/geothermal, shale gas, deepwater, and others. The dashed black line represents the cumulative number of wells using each grade. Grades like P110 and N-80 are widely applied across demanding environments and **b** Line plot illustrating the distribution of casing failure mechanisms (e.g., fatigue, buckling, wear, collapse) by grade. Notably, high-strength grades such as P110 exhibit a broader range of failure modes, highlighting the need for tailored design practices and robust connection selection

of the well's life. The primary load cases considered in design—burst, collapse, tension, and compression/buckling—represent the fundamental modes of mechanical failure that tubulars must be engineered to resist. Figure 4.3 provides a visual summary of these four primary load cases. It illustrates how internal and external pressures, as well as axial tensile and compressive forces, affect a pipe segment within a realistic subsurface geological structure. Each case is shown in the context of a wellbore traversing multiple geological formations, emphasizing the downhole conditions under which failure risks are evaluated. Casing and tubing strings are

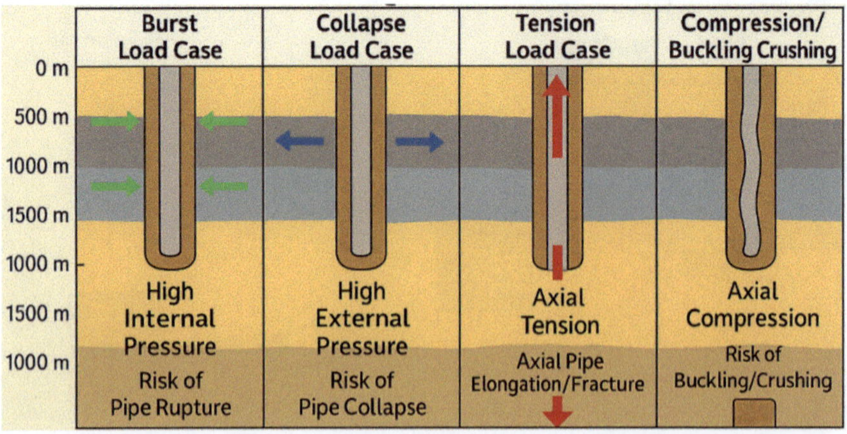

Fig. 4.3 Diagrams illustrating the four primary load cases (Burst, Collapse, Tension, Compression/Buckling) on a pipe segment in a subsurface oil well. Each case highlights the mechanical stress applied to the casing or tubing, as well as the associated risk to structural integrity

subjected to complex loading conditions throughout their life, from installation to abandonment. Designs must ensure the pipe can withstand the worst-case credible scenarios for several primary load types and their combinations:

- **Burst**: Caused by internal pressure exceeding the external pressure leading to tangential tensile stress (hoop stress) in the pipe wall (Khandka, 2007; Schneider, 1982). Critical scenarios often involve high surface pressure during operations (production, injection, stimulation) or pressure testing, potentially combined with light fluid or gas columns inside the pipe (e.g., gas influx during drilling or depletion causing hydrostatic pressure loss (Dall'Acqua et al., 2013). The pipe's resistance depends on its yield strength, wall thickness, and OD.
- **Collapse**: Caused by external pressure exceeding the internal pressure, causing the pipe wall to experience compressive stress (Lin et al., 2020). Critical scenarios include cementing operations (high external fluid density), full evacuation of the pipe (e.g., during swabbing, packer setting with an empty annulus), or exposure to high external pressures from mobile formations like salt or swelling shales (Hørte et al., 2024; Huang & Gao, 2015). Collapse resistance is highly dependent on the pipe's OD/thickness ratio (D/t), yield strength, ovality, and axial loading.
- **Tension**: Axial tensile load attempting to pull the pipe apart. Primarily caused by the hanging weight of the string itself, but can be significantly increased by pressure effects (e.g., piston forces acting upward under internal pressure), temperature changes (cooling causing contraction), or overpull during operations (Klementich & Jellison, 1986; Zhang et al., 2019). Tensile strength is governed by the pipe body's cross-sectional area and the steel's tensile strength, but often the connection strength is the limiting factor. Fatigue from cyclic tension loading (e.g., due to thermal cycling, wave motion on offshore risers) can also lead to failure over time. As part of the design process, it is essential to understand the behavior

of casing materials under different loading conditions and temperature environments. Figure 4.2 illustrates the stress–strain behavior of different casing grades at varying temperatures, which is a critical factor in ensuring that the selected materials can withstand the pressure and temperature extremes encountered during drilling and production operations. The ability to predict the material response to strain and temperature variations directly informs the design choices for casing selection, ensuring long-term wellbore integrity and optimal performance in harsh conditions.

- **Compression**: Axial compressive load attempting to crush the pipe lengthwise. Can be caused by the string's own weight if supported from below (e.g., landed on bottom), temperature increases causing expansion (especially if constrained by cement or packers), or pressure-induced buckling (Jellison & Brock, 2000). Buckling itself, where the string contorts into a helical or sinusoidal shape under compression, induces high bending stresses that must be considered. Geomechanical loads from formation movement (e.g., compaction, subsidence, fault movement) can also induce significant compressive or shear loads on the casing that must be designed for, particularly in challenging environments (Discussion in Chap. 18).
- **Triaxial Stress**: In reality, pipes are often subjected to combinations of burst/collapse pressure, axial tension/compression, and bending simultaneously (Johnson et al., 1987). The material's yield strength under this combined stress state (evaluated using criteria like the von Mises yield criterion) determines its resistance. Triaxial stress analysis is essential for accurately assessing integrity under complex loading, particularly in HPHT wells (where thermal and pressure loads combine severely) or wells with significant bending/buckling or geomechanical stress (Ref. Discussion in Chap. 18). Figure 4.5 illustrates the stress behavior of L80 casing, which is commonly used in moderately harsh well conditions. The figure highlights key loading cases like burst, collapse, and tension, alongside the corresponding axial stress and tubing pressure differentials. These stress states are critical when assessing casing failure mechanisms, helping engineers optimize casing design and material selection for preventing failure under extreme operational conditions. Understanding these interactions ensures that the selected casing will perform reliably over the well's lifespan, providing the necessary strength and flexibility.

Design calculations involve determining the maximum anticipated loads for each credible scenario *throughout the well's life* (from installation through abandonment) and ensuring the selected pipe has sufficient resistance, incorporating appropriate design factors. This requires a comprehensive understanding of the operating envelope and potential future events. A thorough understanding of mechanical load cases—burst, collapse, tension, and compression/buckling—is fundamental for reliable tubular design. The burst pressure (Eq. 4.1), collapse pressure (Eq. 4.2), and axial load capacity (Eq. 4.3) equations form the backbone of this evaluation, guiding safe tubular selection.

The burst pressure is estimated using Barlow's formula, accounting for internal pressure exceeding external formation pressure:

$$P_{\text{burst}} = 0.875\frac{2S_{\text{YS}} \cdot t}{D_o} \quad (4.4)$$

where P_{burst} is the burst pressure, S_{YS} is the yield strength of the steel, t is the wall thickness, and D_o is the outside diameter. The factor 0.875 accounts for the API pipe wall thickness tolerance of 12.5% less than the nominal wall thickness.

For collapse pressure, the API collapse formulas guide the elastic collapse mode, incorporating material elasticity and geometry:

$$P_{\text{collapse, elastic}} = \frac{2E}{1-\nu^2}\left(\frac{t}{D_o}\right)^3 \quad (4.5)$$

where E is Young's modulus and ν is Poisson's ratio.

The axial load capacity essential for tension and compression loading, particularly relevant to hanging string weight and thermal expansion stresses, is determined by

$$F_{\text{axial}} = A \cdot S_{\text{YS}} \quad (4.6)$$

where A is the pipe's cross-sectional area and S_{YS} is the steel's yield strength.

In real-world well conditions, casing is not only subjected to external and internal pressures but also to axial tensile loads. Axial tensile stress can significantly reduce a casing's resistance to collapse, particularly under deep or deviated well conditions where tensile forces are more pronounced. To account for this, the casing's yield strength must be adjusted using a triaxial correction factor that reflects the combined effect of axial and radial stresses on the pipe body. The derated yield strength under axial stress is calculated using the following relationship:

$$Y_{pa} = \left[\sqrt{1 - 0.75\left(\frac{\sigma_a}{Y_p}\right)^2} - \frac{\sigma_a}{2Y_p}\right] Y_p \quad (4.7)$$

where Y_{pa} is the adjusted (or derated) yield strength under axial load (psi), σ_a is the axial tensile stress (psi), and Y_p is the original yield strength of the pipe material (psi). This equation is essential for evaluating casing performance in environments where significant tension coexists with external pressure loading.

To define the complete strength envelope of a casing under triaxial stress conditions, **triaxial collapse equations** must be used. These equations incorporate the effects of axial stress, hoop stress, internal pressure, and external pressure in a unified form. The **triaxial collapse pressure** P_o can be evaluated using the following expression derived from a complex stress state (Maurer, 1993):

$$P_o = \frac{-\sigma_a + 2CP_i - P_i \pm \sqrt{-3\sigma_a^2 - 6\sigma_a P_i - 3P_i^2 + 4\sigma_v^2}}{2C} \quad (4.8)$$

or

$$P_i = \frac{C\sigma_a - 2\sigma_a + 2C^2P_o - CP_o \pm \sqrt{3C^2\sigma_a - 6C^2\sigma_a P_o - 3C^2P_o^2 + 4(C^2 - C + 1)\sigma_v^2}}{2(C^2 - C + 1)} \quad (4.9)$$

where σ_a is the axial stress (psi), σ_v is the hoop stress (psi), P_i and P_o are the internal and external pressures (psi), respectively, and C is a geometry-dependent constant based on casing dimensions. These triaxial models are vital for advanced casing design, especially in deepwater, HPHT, or extended-reach wells, where uniaxial assumptions no longer provide sufficient accuracy for collapse resistance predictions.

To evaluate the onset of plastic deformation under combined axial, radial, and tangential stresses in casing and tubing, the von Mises Equivalent (VME) stress criterion is widely used. Based on the maximum distortion energy theory, this approach provides a robust metric for assessing whether a material has yielded under multi-axial loading. The VME stress combines the three principal stresses (axial, tangential, radial) into a single scalar value for comparison against the material yield strength. To compute the von Mises equivalent stress accurately, it is essential to understand the three principal stresses acting on the casing: axial, radial, and tangential (hoop) stresses. These are illustrated in Fig. 4.2, showing their orientation on a tubular cross section.

Casing strings are subjected to a combination of axial, radial, and circumferential (hoop) stresses that together define the overall loading state experienced by the pipe. To assess the combined effect of these stresses, the von Mises equivalent stress criterion is commonly used. This criterion provides a single scalar value that accounts for the interaction between the principal stresses and is used to predict yielding in ductile materials. The von Mises equivalent stress is given by the expression:

$$\sigma_{\text{VME}} = \frac{1}{\sqrt{2}}\left[(\sigma_a - \sigma_r)^2 + (\sigma_t - \sigma_r)^2 + (\sigma_r - \sigma_a)^2\right]^{0.5} \quad (4.10)$$

In this equation, σ_{VME} is the equivalent stress (psi), σ_a is the axial stress, σ_r is the radial stress, and σ_t is the tangential or hoop stress, all expressed in psi. This relationship is essential in well design to ensure that casing strings remain within elastic limits when subjected to multi-axial stress states.

The total axial load acting on the casing string, denoted as F_a, arises from several sources that collectively define the axial force profile along the length of the casing. These include the weight of the casing, buoyancy effects, pressure differentials, thermal expansion or contraction, and additional mechanical loads induced by wellbore curvature or handling operations. The total axial load can be described by the equation:

$$F_a = F_w + F_b + F_p + F_t + F_m \quad (4.11)$$

where F_w is the axial load due to the weight of the casing in air, F_b accounts for the reduction in effective weight due to buoyancy in drilling or completion fluids, F_p is the net axial force resulting from internal and external pressure differences such as ballooning and piston effects, F_t is the thermally induced axial force from temperature fluctuations along the wellbore, and F_m includes mechanical effects such as bending in doglegs and forces applied by surface handling equipment.

Among these, the casing weight is typically the most significant contributor to axial load, particularly in vertical and moderately deviated wells. The load resulting from casing weight is calculated as the product of the nominal casing weight per unit length and the total string length:

$$F_w = W_n \cdot L \tag{4.12}$$

The nominal casing weight W_n is dependent on the casing geometry and steel density. It is defined as

$$W_n = \rho_s \cdot A_s \tag{4.13}$$

where ρ_s is the density of steel, typically around 490 lbm/ft^3, and A_s is the cross-sectional area of the steel. The area A_s is calculated using the outer diameter D and wall thickness t of the casing through the following expression:

$$A_s = \frac{\pi}{4}\left(D^2 - (D - 2t)^2\right) \tag{4.14}$$

As compressive forces increase—particularly during thermal recovery operations, deep completions, or inclined wells—there is an increased risk of structural instability due to buckling. The critical load at which elastic buckling occurs can be estimated using Euler's column stability theory:

$$F_{\text{cr}} = \frac{\pi^2 EI}{L_e^2} \tag{4.15}$$

In this formula, F_{cr} is the critical buckling load, E is Young's modulus for steel, approximately 30 million psi, I is the moment of inertia of the casing cross section, and L_e is the effective unsupported length of the casing, which is determined by the boundary conditions (such as pinned–pinned or fixed–fixed support). The moment of inertia I for a hollow circular casing cross section is given by

$$I = \frac{\pi}{64}\left(D^4 - (D - 2t)^4\right) \tag{4.16}$$

These equations together form the foundation of axial loading and buckling design for casing strings, allowing engineers to ensure that tubulars retain their structural integrity under the complex mechanical and thermal loading conditions expected throughout the well lifecycle (Figs. 4.4 and 4.6).

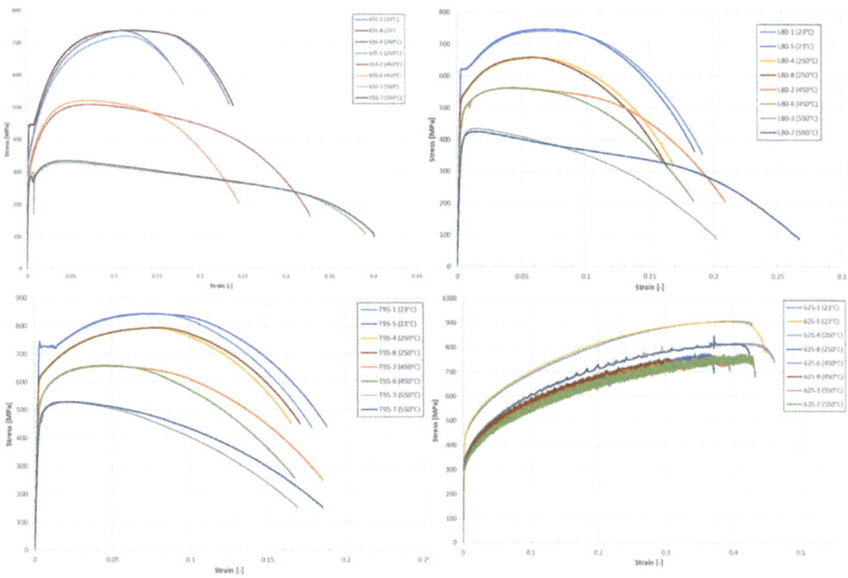

Fig. 4.4 Stress-strain curves for various casing grades (L80, T95, K55, and Inconel 625) at different temperatures (23°C, 250°C, 350°C, 450°C, and 550°C). The curves represent the material behavior under increasing strain at elevated temperatures, showing variations in tensile strength and ductility (Dillingh et al., 2021)

4.4 Connection Selection and Performance

Individual lengths of casing or tubing (typically ~40 ft/12 m) are joined together by connections (couplings or integral joints) to form a continuous string. The integrity of these connections is as critical as the pipe body itself. Connections must be provided:

- **Pressure Containment(Sealability)**: Prevent leakage of fluids/pressure from inside the string to the outside (internal pressure containment) or vice versa (external pressure containment) under a range of operational conditions, including internal and external pressure, tension, compression, bending, and temperature cycling (Hamilton & David Pattillo, 2019a; Li et al., 2019; Well Casing Connection, 1987). Sealing mechanisms vary:
  - **Thread Seal**: Achieved primarily by interference fit and thread compound in simpler API connections (e.g., LTC—long thread coupling, STC—short thread coupling, BTC—buttress thread coupling). Often insufficient for high pressure or gas service due to inherent leak paths along the threads.
  - **Metal-to-MetalSeal (MTM)**: A dedicated sealing surface within the connection where high contact stress between precisely machined metal surfaces creates a reliable seal. Common in "premium" connections designed for critical

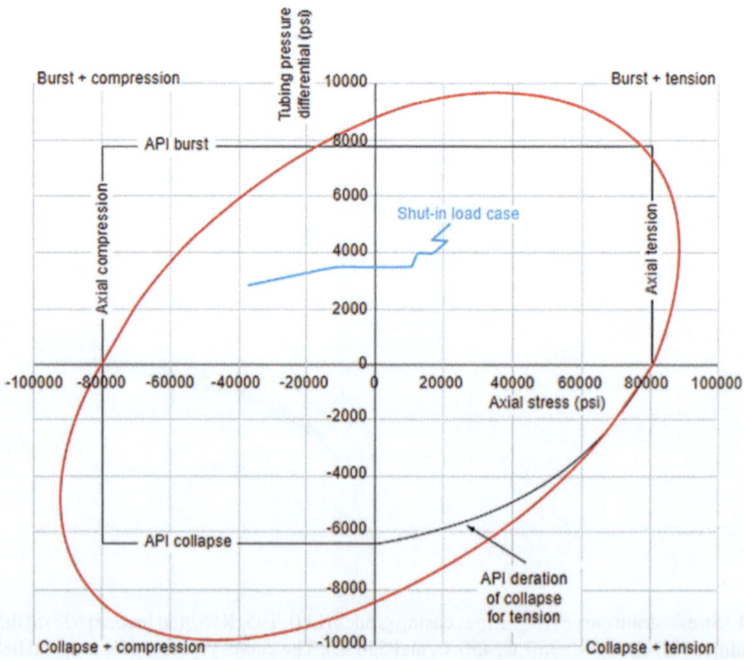

Fig. 4.5 Stress–strain plot for L80 casing grade under varying conditions. The diagram represents the relationship between axial stress and tubing pressure differential (burst, compression, collapse, and tension) (Bellarby, 2009)

service (HPHT, gas wells, sour service). Requires precise machining, careful handling, and controlled makeup to energize the seal correctly. Less susceptible to chemical degradation than elastomers.

- **Elastomer Seal**: Some connections incorporate O-rings or other elastomeric seals, often as a secondary sealing mechanism. Performance is limited by elastomer temperature and chemical compatibility (as discussed in Chap. 11).

- **Mechanical Strength**: Transmit axial loads (tension, compression) and resist bending and torque without failure. Connection tensile and compressive strength should ideally match or exceed the pipe body strength, although this is not always the case, especially for simpler API connections. Resistance to fatigue from cyclic axial or bending loads is a critical requirement, particularly in gas storage wells (frequent cycling) or wells subjected to significant thermal cycling or structural movement (Ref Discussion in Chaps. 24 and 21).
- Operational Robustness: Connections should be relatively easy and reliable to make up correctly on the rig floor under operational conditions (e.g., cold weather, limited space), resistant to damage during handling and running, and resistant to galling (seizing) during makeup or breakout. Torque shoulders in premium connections provide a positive indication of correct makeup.

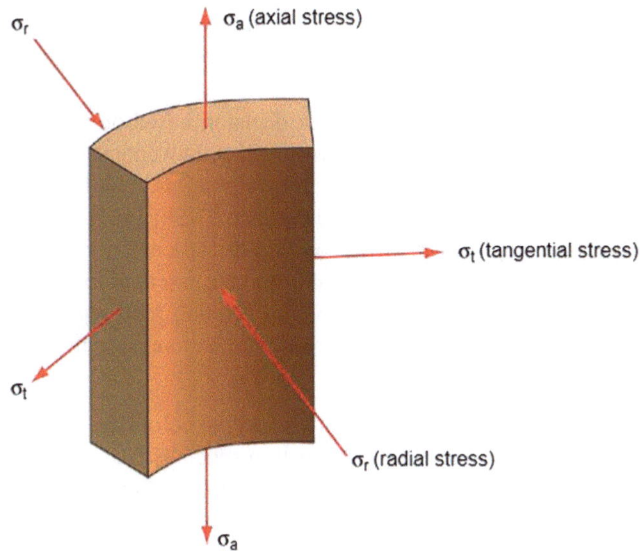

Fig. 4.6 Schematic representation of the principal stresses acting on casing: axial stress (σ_a), tangential or hoop stress (σ_t), and radial stress (σ_r). These stresses form the basis for von Mises yield criterion evaluation in well integrity analysis

API Connections: Generally lower cost, widely available, suitable for less demanding applications (lower pressure, primarily oil service, less severe temperature cycling). Sealability relies primarily on thread interference and compound, making them less reliable for gas or high pressure/thermal cycling service (Gao et al., 2013).

Premium Connections: Proprietary designs offered by various manufacturers, engineered for specific demanding applications (HPHT, sour service, gas wells, thermal wells, fatigue resistance, high bending loads) (Bradley et al., 2005). Typically feature robust thread forms for high strength and galling resistance, metal-to-metal seals for superior pressure containment (especially gas-tightness), and often torque shoulders for precise makeup control. They are more expensive but offer significantly higher integrity assurance and are often mandatory for critical strings like production tubing in HPHT gas wells or liners in thermal operations. In **premium connections**, the final **makeup torque** T_D is a key parameter to ensure connection reliability (Eq. 4.5). It is composed of four main components: the leading torque T_A, the thread interference torque T_{ti}, the additional makeup torque from the sealing interface T_{se}, and the shoulder torque T_{sh}. This relationship is captured by

$$T_D = T_A + T_{ti} + T_{se} + T_{sh} \tag{4.17}$$

This formulation (Eq. 4.4) provides a quantitative framework for understanding and controlling the makeup process to ensure a reliable seal and mechanical performance of the connection.

Connection selection is a critical design decision based on the well's pressure rating, fluid type (gas presence is a key driver for premium connections), temperature range and cycling frequency, potential for bending/fatigue, axial loads, and overall criticality of the barrier (Hamilton & David Pattillo, 2019b; Tao & Xie, 2013). Proper handling, inspection, application of appropriate thread compound, and meticulous makeup procedures (controlling torque and turns) are essential for ensuring connection integrity during installation. A leaking connection represents a direct breach of the tubular barrier element. Several considerations to ensure the integrity of the casing should be implemented as follows

- The casing thread should be cleaned and inspected before the casing running job by the casing running team and the inspection specialist.
- The thread protector should always be in place up to running the casing.
- Makeup torque for each casing connection must be within specified tolerances—under-torque may cause leaks, while over-torque risks yielding or connection failure. The progression of makeup torque during casing connection installation is illustrated in Fig. 4.7, which highlights key engagement stages—from initial thread contact to final shoulder torque—essential for achieving proper sealing and structural integrity. Improper handling, installation errors, and environmental exposure can lead to serious thread damage, compromising the pressure integrity and mechanical performance of casing connections, as shown in Fig. 4.8.

:

4.5 Design Factors and Safety Margins

Tubular design does not simply aim to match the pipe's theoretical resistance to the maximum predicted load. Design Factors (DF) or Safety Factors (SF) are applied to account for uncertainties and potential variations in predicted loads, actual material properties (which can vary within API specifications), manufacturing tolerances, installation quality, and potential degradation over time (e.g., minor corrosion) (Aadnøy et al., 2003b; Rahman & Chilingarian, 1995). A design factor is typically defined as the ratio of the pipe's rated resistance (e.g., API or VME rated burst pressure, collapse resistance, tensile strength) to the maximum anticipated load for a specific load case.

Commonly used minimum design factors vary depending on operator standards, regulations, specific application, and the criticality of the well string. Illustrative values might be [illustrative values, actual values depend on specific standards and policies]:

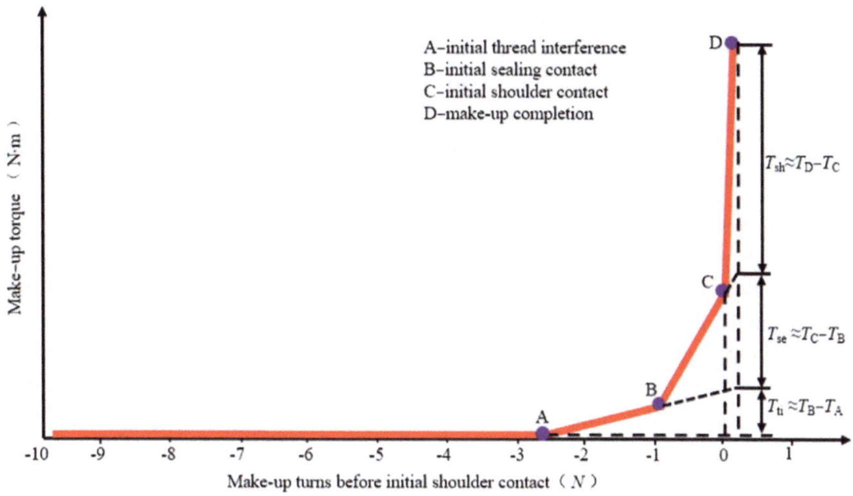

Fig. 4.7 Makeup torque versus makeup turns illustrating the engagement process of a threaded connection. Point A marks initial thread interference, B indicates the start of sealing contact, C represents initial shoulder contact, and D signifies final makeup completion. The torque components—interference torque (T_{ti}), sealing torque (T_{se}), and shoulder torque (T_{sh})—are derived from the incremental torque increases at each stage of engagement. Proper control of these phases is crucial to achieving reliable hydraulic and mechanical sealing without over-torque or thread damage (Yang et al., 2023)

Fig. 4.8 Examples of damaged casing threads. The left image shows a "jumped thread" resulting from cross threading during makeup, while the right image illustrates a corroded and weakened connection due to prolonged exposure to harsh conditions. Both conditions pose significant risks to selling effectiveness and long-term well integrity

- Burst: Typically ranging from 1.1 to 1.25 or higher. A lower factor might be accepted for casing exposed only to hydrostatic pressure, while higher factors are used for production tubing potentially exposed to high surface pressures or stimulation pressures.
- Collapse: Typically ranging from 1.0 to 1.15. Collapse loads during cementing or evacuation can be critical.

- Tension: Often ranging from 1.6 to 2.0 based on yield strength for hanging weight scenarios. Factors may be lower when considering tension associated with pressure effects or thermal contraction if these loads are managed.
- Triaxial: Typically assessed by ensuring the combined stress remains below a certain percentage (e.g., 85–95%) of the material's derated yield strength under triaxial conditions. The exact percentage is effectively a design factor.

Higher design factors may be used for more critical strings (e.g., production casing, high-pressure liners), in uncertain environments, where data quality is lower, or where the consequences of failure are severe (e.g., wells in highly populated areas or environmentally sensitive zones, HPHT wells). The selection of appropriate design factors is a key aspect of operator policy, risk management, and regulatory compliance. It represents a safety margin to account for uncertainties and potential deviations from ideal conditions.

4.6 Impact of Temperature and Sour Service (H_2S)

Operating conditions significantly influence material performance and must be factored into design, often requiring material property derating and careful selection based on environmental compatibility:

- **Temperature**
  - **Derating**: Steel yield strength and tensile strength decrease at elevated temperatures. Design calculations *must* use the strength derated for the maximum anticipated operating temperature expected over the well's life (API 5CT). API standards guide temperature derating curves for different steel grades.
  - **Thermal Stresses**: Temperature changes during startup, shutdown, production, injection, or stimulation cause the tubulars to expand (on heating) or contract (on cooling). If this movement is constrained (e.g., by cement, packers, or wellhead hangers), significant axial stresses (compressive on heating, tensile on cooling) are induced (He et al., 2025; Laaouar et al., 2022; Phi et al., 2019). These thermal stresses must be accurately calculated and added to other loads (pressure, weight) in the design analysis (often requiring triaxial analysis). Severe thermal cycling, common in geothermal or thermal EOR wells, induces fatigue loads on connections and can degrade cement bonds (see Chaps. 21 and 22).
  - **Elastomer Limits**: High temperatures are a primary limitation for elastomeric seals used in some connections, packers, and completion equipment. Elastomers undergo irreversible degradation (hardening, compression set, loss of elasticity) above their temperature rating (see Chap. 11). Selecting elastomers qualified for the maximum anticipated temperature is critical (ISO 23,936–1, 2022).

- **Sour Service (H2S)**: Hydrogen sulfide (H2S) is highly dangerous and extremely corrosive. Its presence poses a significant threat through Sulfide Stress Cracking (SSC) and various forms of sour corrosion (see Chap. 9).
  - **Sulfide Stress Cracking (SSC)**: A brittle fracture mechanism occurring in susceptible steels under tensile stress (applied or residual) in the presence of H2S and water. Preventing SSC requires strict adherence to material selection guidelines defined by **NACE MR0175 / ISO15,156** (NACE MR0175/ISO 15156). This standard provides limits on the use of carbon steels, low-alloy steels, and corrosion-resistant alloys (CRAs) based on H2S partial pressure, pH, temperature, and chloride content. It often restricts the use of high-strength steel grades or imposes specific heat treatment and hardness requirements (e.g., L80, C90, T95 grades are designed for sour resistance and hardness limits). Selecting materials compliant with NACE MR0175/ISO 15156 during initial design is essential if there is any credible risk of encountering sour conditions at any point in the well's life (including potential future reservoir souring).
  - **Sour Corrosion**: H2S contributes to general and localized corrosion (pitting). Material selection (CRAs) or chemical inhibition may be necessary depending on the severity (see Chap. 9).

Designing for HPHT (high pressure/high temperature) wells or wells in highly corrosive (sour/high CO_2) environments combines these challenges, demanding careful triaxial stress analysis, use of temperature-derated properties, selection of high strength, potentially CRA materials resistant to the specific temperature and fluid chemistry (including H2S), selection of appropriately rated elastomers/seals, and use of high-performance premium connections (see Chap. 18). Temperature derating of yield strength is critical for ensuring casing performance in high-temperature wells. Equation 4.6 captures this adjustment:

$$S_{\mathrm{YS,T}} = S_{\mathrm{YS,\,room}} \cdot R_T \tag{4.18}$$

where R_T is the temperature derating factor provided in the API standards. This ensures the selected casing and tubing grades can withstand anticipated elevated temperatures during well operations.

4.7 Wear Considerations

During drilling, the rotation of the drill string inside previously set casing, particularly in deviated or horizontal sections with significant doglegs, can cause abrasive wear, reducing the casing wall thickness (Schoenmakers, 1987; Williamson & Driico, 1981; Zhang et al., 2016). Similarly, operations involving running tools, wireline, or coiled tubing (especially with mechanical tools or tractors) during the well's life can cause wear on production tubing or casing.

- **Prediction**: Wear can be predicted using empirical models based on factors like contact force between the rotating/moving string and the casing (influenced by tension, compression, bending due to doglegs), rotation hours/wireline distance, mud properties (lubricity, solids content), and tool joint/wireline characteristics.
- **Mitigation**: Wear can be mitigated by using non-rotating drill pipe protectors, controlling dogleg severity during directional drilling, optimizing drilling fluid properties, and minimizing unnecessary wireline or coiled tubing runs, or using appropriate rollers/centralizers on tools.
- **Design Allowance**: In some cases, particularly for long drilling sections inside critical casing (like the production casing), an allowance for anticipated wear (sacrificial wall thickness) may be incorporated into the initial casing design, ensuring sufficient remaining wall thickness and structural integrity after predicted wear occurs. Alternatively, wear can be monitored periodically using specialized logging tools (MFC, electromagnetic thickness) later in the well's life (see Chap. 12).

Casing wear is a significant concern during drilling and intervention operations, particularly in highly deviated wells where the drill string contacts the casing wall. Prolonged or repeated tool joint contact can lead to thinning of the casing wall, ultimately compromising well integrity. Accurate wear prediction is thus critical for maintaining safe operations and planning well workovers.

Casing wear is a critical concern in deviated and horizontal wells, where repeated contact between the rotating drill string and the casing wall can lead to material loss, localized thinning, and, ultimately, mechanical failure of the casing (Alsaghir et al., 2023). To quantify this wear behavior, the **casing wear factor** K is used. This factor represents the volume of casing material removed per unit side load and sliding distance, offering a normalized index for evaluating casing damage due to tool joint interaction. The casing wear factor is expressed as follows:

$$K = \frac{V_w}{P \cdot L} \tag{4.19}$$

In this equation, K is the wear factor with typical units of $mm^3/(N \cdot mm)$ or $MPa-^1$, V_w is the wear volume (mm^3), P is the side load acting between the tool joint and casing wall (N), and L is the cumulative sliding distance (mm). This formulation enables the comparison of wear severity across different drilling conditions and casing materials and is widely used in wear testing protocols and simulation models.

The **sliding distance** L, which represents the total distance traversed by the tool joint along the casing surface during drilling, is calculated by combining the effects of rotational motion and axial drilling penetration. It is given by

$$L = \pi DNt + (\mathrm{ROP} \cdot t) \tag{4.20}$$

Here, D is the tool joint diameter (mm), N is the rotational speed (rpm), t is the duration of drilling or wear testing (min), and ROP is the rate of penetration (mm/min). The first term in the equation accounts for the circumferential movement of the tool joint, while the second term captures the linear advancement of the bit. Together, they describe the full path over which casing contact occurs.

These equations are vital for assessing casing wear risk, particularly in high-angle wells, extended-reach drilling, and operations involving frequent drilling fluid circulation or high ROP. Predicting casing wear accurately enables engineers to select appropriate casing grades, implement tool joint hardbanding, or adopt casing-friendly drilling practices to extend well life and maintain barrier integrity.

In conclusion, robust casing and tubing design is a complex process requiring careful consideration of anticipated loads *across the entire lifecycle*, environmental conditions (P, T, H_2S, CO_2, chlorides), material properties and their degradation, connection performance (including fatigue), and operational factors like wear. Decisions made in the design phase regarding materials and connection types are fundamental to achieving long-term tubular integrity and reliability.

References

Aadnøy, B. S., Aasen, J. A., & Pollard, M. (2003a). *Casing design-review of design methodology.*

Aadnøy, B. S., Aasen, J. A., & Pollard, M. (2003b). *Casing design-review of design methodology.*

Alsubaih, A. A., Hayfaa, L. S., Nuhad, A., Erfan, M. A. lawe, Ahmed, I. A., Swadi, M., & Tariq, K. K. (2019). *Investigation of the rock strength anisotropy on the wellbore stability analysis.* OnePetro. https://dx-doi-org.ezproxy.lib.utexas.edu/

Alsaghir, S., Osman, O., Merah, N., Al-Ibrahim, A., Baig, M., Al-Shaarawi, A., & Alshalan, M. (2023). Casing wear volume and factor under rotation and sliding motion. *Procedia Structural Integrity, 47*, 437–447. https://doi.org/10.1016/J.PROSTR.2023.07.081

Alsubaih, A., Albadran, F., Abbood, N., & Alkanaani, N. (2018). *Optimization of the tripping parameters to prevent the surge related wellbore instability.* OnePetro. https://dx-doi-org.ezproxy.lib.utexas.edu/

Alsubaih, A. A., Hayfaa, S., Alboshawi, H. S., Alaa, A., Al, lawe, Erfan, M., Sawdi, M., Al, A., & Lubna, H. (2020). *Reservoir geomechanics simulation to optimize reservoir management in Southern Iraq fields.* OnePetro. https://dx.doi.org/

Alsubaih, A., Sepehrnoori, K., & Delshad, M. (2024b). Environmental impacts of orphaned and abandoned wells: Methane emissions, and implications for carbon storage. *Applied Sciences, 14*(24), 11518. https://doi.org/10.3390/APP142411518

Alsubaih, A., Sepehrnoori, K., Lopez Manriquez, A., Espinoza, D. N., Allawi, H. R., & Hayfaa, S. (2024). *A statistical approach to estimate the drilling mud window based on field data and ten failure criteria in depleted reservoirs-new modified failure criterion.* https://doi.org/10.56952/ARMA-2024-0799

Alsubaih, A., Sepehrnoori, K., & Delshad, M. (2025). Development and optimization of self-healing cement for CO_2 injection and storage wells: Enhancing long-term wellbore integrity in extreme subsurface conditions. *Applied Sciences 15*(10), 5428. https://doi.org/10.3390/APP15105428

API I API Specification 5CT, 10th Edition. (2019). https://www.api.org/products-and-services/standards/important-standards-announcements/standard-5ct

Bellarby, J. (2009). Chapter 9 tubing stress analysis. *Developments in Petroleum Science, 56*(C), 473–556. https://doi.org/10.1016/S0376-7361(08)00209-4

Bradley, A. B., Nagasaku, S., & Verger, E. (2005). *Premium connection design, testing, and installation for HPHT sour wells.* https://doi.org/10.2118/97585-MS

Dall'Acqua, D., Hodder, M., & Kaiser, T. M. V. (2013). Burst and collapse responses of production casing in thermal applications. *SPE Drilling & Completion, 28*(01), 93–104. https://doi.org/10.2118/151810-PA

Dillingh, B., Skúlason Kaldal, G., Thorbjornsson, I., Wollenweber, J., & Vercauteren, F. (2021). *Tensile testing of casing material at elevated temperatures up to 550°C.*

Gao, L. X., Sun, K. Z., & Zhang, Y. (2013). Development of premium threaded connections for casing and tubing. *Advanced Materials Research, 744*, 53–57. https://doi.org/10.4028/WWW.SCIENTIFIC.NET/AMR.744.53

Gehong, D. H., Lingling, L., Kang, L., Dong, Z., & Zhang, G. Y. (n.d.). Buckling analysis of non-cementing casing string in borehole. *Journal of Failure Analysis and Prevention, 18.* https://doi.org/10.1007/s11668-017-0332-1

Hamilton, K., & David Pattillo, P. (2019a). Developing an evaluation method for casing connections used in hydraulically fractured wells. *Society of Petroleum Engineers-SPE Hydraulic Fracturing Technology Conference and Exhibition 2019, HFTC 2019.* https://doi.org/10.2118/194369-MS

Hamilton, K., & David Pattillo, P. (2019b). Developing an evaluation method for casing connections used in hydraulically fractured wells. *Society of Petroleum Engineers-SPE Hydraulic Fracturing Technology Conference and Exhibition 2019, HFTC 2019.* https://doi.org/10.2118/194369-MS

He, Y., Song, Y., Hu, S., Liu, H., & Ge, X. (2025). Mechanical evaluation of casing in multiple thermal recovery cycles for offshore heavy oil wells. *Journal of Marine Science and Engineering 13*(3), 597. https://doi.org/10.3390/JMSE13030597

Hørte, T., Liu, T., Bjørset, A., & Kjørholt, H. (2024). *Collapse capacities of casings with internal wear based on a structural reliability approach.* https://doi.org/10.1115/OMAE2024-127244

Huang, W., & Gao, D. (2015). A theoretical study of the critical external pressure for casing collapse. *Journal of Natural Gas Science and Engineering, 27*, 290–297. https://doi.org/10.1016/J.JNGSE.2015.08.063

Jellison, M. J., & Brock, J. N. (2000). The impact of compression forces on casing-string designs and connectors. *SPE Drilling & Completion, 15*(04), 241–248. https://doi.org/10.2118/67608-PA

Johnson, R., Jellison, M. J., & Klementich, E. F. (1987). Triaxial-load-capacity diagrams provide a new approach to casing and tubing design analysis. *SPE Drilling Engineering, 2*(03), 268–274. https://doi.org/10.2118/13434-PA

Khandka, R. K. (2007). Leakage behind casing. *tpg 4920 drilling specialization thesis work.* https://ntnuopen.ntnu.no/ntnu-xmlui/handle/11250/239275

King, G. E., & Valencia, R. L. (2016). Well integrity for fracturing and re-fracturing: What is needed and why? *Society of Petroleum Engineers-SPE Hydraulic Fracturing Technology Conference, HFTC 2016*, 9–11. https://doi.org/10.2118/179120-MS

Klementich, E. F., & Jellison, M. J. (1986). A service-life model for casing strings. *Spe Drilling Engineering, 1*(02), 141–152. https://doi.org/10.2118/12361-PA

Laaouar, Z. L., Kalbaza, Y. R., & Merzougui, I. (2022). *Well integrity management: Annulus pressure, causes and solutions.* http://dspace.univ-ouargla.dz/jspui/handle/123456789/32165

Li, Y., Cao, Y., Dou, Y., Yu, Y., Zhang, J., & Zhang, L. (2019). Simulation of sealing ability for premium connection based on ISO XE "ISO" 13679 CAL IV tests. *Procedia Structural Integrity, 22*, 43–50. https://doi.org/10.1016/J.PROSTR.2020.01.006

Lin, B., Coe, D., Harris, R., & Thomas, T. (2020). On some factors affecting casing collapse resistance under external pressure. *Proceedings-SPE Annual Technical Conference and Exhibition.* https://doi.org/10.2118/201352-MS

Maurer, B. (1993). *Triaxial stress analysis model (TRIAXl) Theory and User's Manual.*

Mohammed, A. I., Oyeneyin, B., Atchison, B., & Njuguna, J. (2019). Casing structural integrity and failure modes in a range of well types-a review. In *Journal of Natural Gas Science and Engineering* (Vol. 68). Elsevier B.V. https://doi.org/10.1016/j.jngse.2019.05.011

NORSOK D-010. (2021). https://online.standard.no/en/norsok-d-010-2021

Phi, T., Elgaddafi, R., Al Ramadan, M., Fahd, K., Ahmed, R., & Teodoriu, C. (2019). Well integrity issues: Extreme high-pressure high-temperature wells and geothermal wells a review. *Society*

of Petroleum Engineers-SPE Thermal Well Integrity and Design Symposium 2019, TWID 2019, 19–21. https://doi.org/10.2118/198687-MS
Rahman, S. S., & Chilingarian, G. V. (1995). Casing design-theory and practice. In *Developments in Petroleum Science* (Vol. 42, Issue C). https://doi.org/10.1016/S0376-7361(06)80014-2
Schneider, W. P. (1982). Casing and tubing connection stresses. *Journal of Petroleum Technology, 34*(08), 1851–1862. https://doi.org/10.2118/10861-PA
Schoenmakers, J. M. (1987). Casing wear during drilling: Simulation, prediction, and control. *SPE Drilling Engineering, 2*(04), 375–381. https://doi.org/10.2118/14761-PA
Skúlason Kaldal, G., Jónsson, M. Þ., Pálsson, H., Karlsdóttir, S. N., & Þorbjörnsson, I. Ö. (n.d.). load history and buckling of the production casing in a high temperature geothermal well. *PROCEEDINGS, Thirty-Sixth Workshop on Geothermal Reservoir Engineering.*
Tao, G., & Xie, J. (2013). Analytical evaluation of casing connections for thermal well applications. *Society of Petroleum Engineers-SPE Heavy Oil Conference Canada, 2013*(2), 1379–1393. https://doi.org/10.2118/165493-MS
Well casing connection. (1987). https://scispace.com/papers/well-casing-connection-4j5ja16pfg
Williamson, J. S., & Driico, S. (1981). Casing wear: The effect of contact pressure. *Journal of Petroleum Technology, 33*(12), 2382–2388. https://doi.org/10.2118/10236-PA
ISO 23936–1. (2022). *ISO 23936–1:2022-Oil and gas industries including lower carbon energy—non-metallic materials in contact with media related to oil and gas production—Part 1: Thermoplastics.* https://www.iso.org/standard/81995.html
Yang, B., Xu, H., Xiang, S., Zhang, Z., Su, K., & Yang, Y. (2023). Effects of make-up torque on the sealability of sphere-type premium connection for tubing and casing strings. *Processes 11*(1), 256. https://doi.org/10.3390/PR11010256
Zhang, Q., Lian, Z., Lin, T., Deng, Z., Xu, D., & Gan, Q. (2016). Casing wear analysis helps verify the feasibility of gas drilling in directional wells. *Journal of Natural Gas Science and Engineering, 35*, 291–298. https://doi.org/10.1016/J.JNGSE.2016.08.066
Zhang, B., Guan, Z., Lu, N., Hasan, A. R., Wang, Q., & Xu, B. (2019). Trapped annular pressure caused by thermal expansion in oil and gas wells: A review of prediction approaches, risk assessment and mitigation strategies. *Journal of Petroleum Science and Engineering, 172*, 70–82. https://doi.org/10.1016/J.PETROL.2018.09.041

Chapter 5
Primary Cementing for Zonal Isolation

Abstract This chapter presents a comprehensive overview of primary cementing as a foundational operation in well construction, emphasizing its critical role in establishing long-term zonal isolation and structural well integrity. The discussion begins with the core functions of well cement—including hydraulic sealing, mechanical support, corrosion protection, and well control—and proceeds to explore the chemistry of Portland cement, hydration reactions, and the extensive use of additives to tailor slurry performance for downhole environments. It details key design principles for slurry formulation, covering essential parameters such as density, rheology, thickening time, fluid loss, strength development, shrinkage/expansion, and gas migration resistance. The chapter emphasizes the importance of proper mud removal, the use of spacers and centralizers, and dynamic cementing techniques to ensure effective bonding between the cement, casing, and formation. It outlines hydraulic and rheological modeling frameworks, including pressure drop equations, gel strength development, and displacement efficiency metrics. Practical cement placement strategies are analyzed in the context of preventing gas migration, ensuring proper plug sequencing, and maintaining hydrostatic control. Finally, the chapter addresses the cement setting and curing process, WOC timing, bond strength development, and the implications of shrinkage and degradation. Through detailed equations, field practices, and case-referenced insights, the chapter equips engineers with the critical knowledge needed to design and execute successful primary cementing operations that prevent sustained casing pressure, formation crossflow, and long-term integrity failures.

© The Author(s), under exclusive license to Springer Nature Switzerland AG 2026
A. Alsubaih and K. Sepehrnoori, *Well Integrity for Oil, Gas, and Subsurface Storage Operations*, https://doi.org/10.1007/978-3-032-15233-6_5

Cementing operation site showing bulk cement silos, cementing units, and surface equipment layout. *Source* The Authors

Chapter's Objectives

- Explain the critical functions of well cement, with a focus on zonal isolation durability.
- Describe basic cement chemistry and the role of common additives in achieving desired properties.
- Outline key slurry design principles for achieving integrity under downhole conditions.
- Detail the absolute importance of mud removal, spacers, and centralization for bond quality.
- Discuss cement placement techniques and strategies for gas migration prevention.
- Explain the cement setting process and the concept of Wait-on-Cement (WOC) time.

5.1 The Critical Functions of Well Cement

Primary cementing involves pumping a cement slurry down the casing and up into the annular space between the casing and the open hole (or previous casing string). After the cement sets and hardens, it performs several crucial functions vital for well integrity and performance throughout its life (Fig. 5.1):

- **Zonal Isolation**: This is arguably the most important integrity function. The set cement sheath forms a hydraulic seal in the annulus, preventing the migration of fluids (oil, gas, water) between different geological formations (inter-zonal flow) and preventing flow from subsurface formations up the annulus to the surface (annular migration) (Dabzhu, 2023; Kopp et al., 2000; Ladva et al., 2005). Effective and durable zonal isolation is essential for:
  - Preventing blowouts or Sustained Casing Pressure (SCP) caused by formation gas migrating to the surface or shallow zones.

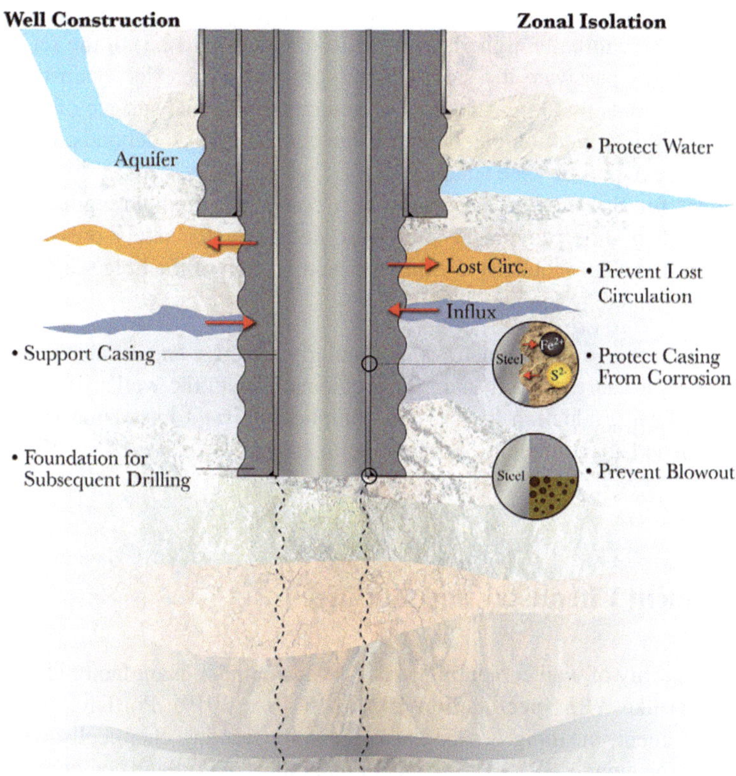

Fig. 5.1 Diagram illustrating the functions of well cement in the annulus: zonal isolation, structural support, corrosion protection (Liu, 2021)

  - Protecting freshwater aquifers from contamination by saline water or hydrocarbons.
  - Ensuring efficient hydrocarbon production only from the intended reservoir zones.
  - Allowing effective reservoir management (e.g., preventing unwanted water influx into oil zones, controlling injected fluids in EOR/Storage wells).
  - Providing isolation barriers required for subsequent operations like hydraulic fracturing (confining fracturing growth to the target zone) or safe abandonment.
  - Maintaining pressure differentials necessary for well control and operational procedures.
- **Structural Support**: The cement sheath provides mechanical support to the casing string, helping it resist external collapse pressures (especially in mobile formations like salt) and distributes loads imposed by the formation (e.g., geomechanical stress) or subsequent operations (e.g., drilling vibrations, perforating shock). It prevents casing buckling and centralizes the pipe within the borehole, which in turn facilitates the achievement of a uniform cement sheath.
- **Casing Corrosion Protection**: By filling the annulus and bonding effectively to the casing, the initially high alkaline environment (~pH 12.5) of the set Portland cement helps passivate the steel surface, providing a barrier against corrosive formation fluids that might otherwise contact the external surface of the casing. However, this protection can be lost if the cement cracks, debonds, or degrades chemically over time (See Chap. 10).
- **Support for Subsequent Operations**: A competent, high-strength cement job at the casing shoe provides a solid base necessary for safely pressure testing the casing and subsequently drilling the next section of the hole (supporting the required fracture gradient).

Failure to achieve these functions, particularly effective and durable zonal isolation, is a major cause of well integrity problems later in the well's life, leading to SCP, crossflow, production losses, and complicating (and increasing the cost of) remediation and abandonment efforts (Alberdi-Pagola & Fischer, 2023; Kiran et al., 2017).

5.2 Cement Chemistry and Additives

The vast majority of well cementing uses Portland cement, manufactured according to standards like **API Specification 10A** (API 10A, 2019). Portland cement is a hydraulic cement, meaning it sets and hardens by reacting chemically with water (hydration) (Naaman, 2001). The primary compounds in Portland cement clinker are calcium silicates (C_3S, C_2S), calcium aluminate (C_3A), and calcium aluminoferrite (C_4AF) (Herfort et al., 2010). When mixed with water, these compounds hydrate to form a complex, interlocking matrix of calcium silicate hydrates (C-S-H gel) and calcium hydroxide (CH), providing strength and low permeability.

API defines several classes of Portland cement (e.g., Class A, B, C, G, H, J) suitable for different well depths, temperatures, pressures, and sulfate resistance requirements. Classes G and H are the most common base cements used in well cementing globally due to their availability, predictable performance with additives, and suitability for a wide range of temperature/pressure conditions.

Pure Portland cement mixed only with water is rarely suitable for the diverse and often challenging downhole conditions. A wide range of **additives** are blended with the base cement and mix water (along with other components like weighting materials) to modify the slurry and set cement properties for successful placement and durable integrity:

- **Accelerators**: Speed up the hydration reaction, reducing the waiting-on-cement (WOC) time before subsequent operations can resume, especially at lower temperatures (e.g., calcium chloride, sodium chloride, triethanolamine). Must be used carefully to avoid flash setting or compromising long-term properties.
- **Retarders**: Slow down the hydration reaction, increasing the slurry thickening time (the time it remains pumpable) to allow placement at higher temperatures and greater depths before it sets (e.g., lignosulfonates, organic acids, phosphonates). Crucial for preventing premature setting in the drill pipe or annulus during long pump times or in hot wells.
- **Fluid Loss Control Agents**: Reduce the rate at which water filters out of the slurry into permeable formations under differential pressure. Excessive fluid loss can cause premature dehydration and bridging of the slurry in the annulus, leading to voids and poor placement. The filtrate can also damage sensitive formations. (e.g., cellulose derivatives, synthetic polymers, latex).
- **Dispersants (Friction Reducers)**: Improve slurry flow properties by reducing viscosity and yield stress, allowing easier pumping at lower pressures and often enabling turbulent flow (desirable for mud removal) at feasible pump rates (e.g., polynaphthalene sulfonates—PNS, polycarboxylates).
- **Extenders (Lightweight Additives)**: Reduce slurry density for use in formations with low fracture gradients to prevent lost circulation. They also typically increase slurry yield volume per sack of cement. (e.g., bentonite, pozzolans like fly ash or silica fume, diatomaceous earth, hollow glass or ceramic microspheres). It can sometimes reduce strength and durability if not properly designed.
- **Weighting Agents**: Increase slurry density to control high-pressure formations. (e.g., barite, hematite, manganese oxide). Proper dispersion is needed to prevent the settling of solids.
- **Gas Migration Control Additives**: Help prevent formation gas from channeling through the cement slurry during the critical transition period before it develops sufficient gel strength to resist gas entry. This period occurs after the hydrostatic pressure of the slurry column diminishes (due to hydration shrinkage or fluid loss) but before the cement develops adequate static gel strength. (e.g., latex polymers, specific fluid loss agents creating impermeable filter cakes, expanding

additives, and thixotropic agents for rapid gel strength development). Preventing gas migration is critical for avoiding annular SCP (Ref Discussion in Chap. 8).

- **Other Additives**: Anti-foam agents (to prevent air entrainment during mixing), silica flour (mandatory at high temperatures > ~110 °C/230 °F to prevent strength retrogression), fibers (to improve toughness and resistance to cracking), expanding agents (to counteract setting shrinkage and potentially induce beneficial expansion), lost circulation materials (LCM) (to seal off permeable zones during placement), dispersible synthetic polymers or elastomers (to improve flexibility and bond durability).

The selection and concentration of additives are critical parts of the slurry design process, tailored to the specific well conditions (temperature and pressure profile, formation types, fluid types, wellbore geometry), operational constraints (pump rates, mixing equipment), and the desired long-term performance properties (strength, durability, low permeability, resistance to degradation).The density of the cement slurry, ρ_{slurry}, is critical for hydrostatic control:

$$\rho_{slurry} = \frac{W_{cement} + W_{additives}}{V_{slurry}} \tag{5.1}$$

The slurry density, denoted as ρ_{slurry}, , is calculated as the total weight of cement and additives ($W_{cement} + W_{additives}$) divided by the total slurry volume (V_{slurry}), and is typically expressed in kilograms per cubic meter (kg/m^3) or pounds per gallon (lb/gal). This parameter is critical for ensuring that the hydrostatic pressure exerted by the cement column is sufficient to control formation pressures during placement.

5.3 Slurry Design Principles for Integrity

The goal of slurry design is to formulate a cement mixture that can be successfully pumped into place under downhole conditions and will set to provide durable, long-term zonal isolation and structural support. This involves balancing numerous, sometimes conflicting, requirements. Key design parameters and principles include:

- **Density**: Must be high enough to provide sufficient hydrostatic pressure to control formation pressures and prevent influx during placement, but low enough to avoid exceeding the formation fracture gradient at the casing shoe or other weak points (Ahmed et al., 2019; Alsubaih et al., 2014, 2018, 2020, 2024, 2025), which would cause lost circulation and potentially a failed cement job (Mahmoud et al., 2024). Density is controlled by the water-to-cement ratio and the use of extenders or weighting agents.
- **Rheology (Flow Properties)**: Determines how easily the slurry can be pumped and its effectiveness in displacing drilling mud from the annulus. Measured by parameters like plastic viscosity (PV) and yield point (YP). Lower viscosity generally allows for turbulent flow at lower pump rates, which is highly desirable for

effective mud removal (Foroushan et al., 2021). Slurry rheology must also be stable over temperature and pressure changes during pumping. Accurately modeling cement slurry rheology is essential for predicting flow behavior during placement and ensuring effective mud displacement. Depending on slurry composition and downhole conditions, slurries can exhibit Bingham Plastic, shear-thinning, or yield-pseudoplastic behavior. Table 5.1 summarizes a range of rheological models applied in cementing and drilling applications, each offering different levels of complexity and accuracy for field and simulation use.

- **Thickening Time**: The time the slurry remains fluid enough to be pumped under the downhole temperature and pressure conditions it will experience (Abdelaal & Elkatatny, 2023). Must be long enough to allow placement plus a sufficient safety margin for potential operational delays, but not excessively long, which could increase risks like particle settling (sedimentation) or gas migration during the waiting period. Determined by rigorous laboratory testing in a pressurized consistometer simulating downhole temperature and pressure (HPHT consistometer for high temperatures) (API RP 10B-2). Cement thickening time is evaluated using a consistometer, which simulates wellbore temperature and pressure conditions while tracking slurry consistency. Figure 5.2 presents a typical consistometer test, where consistency rises with time until a rapid increase indicates the onset of setting. The thickening time (often marked at 70 Bc) defines the operational limit for safe placement before the slurry begins to gel or set. Ensuring this time is long enough for placement—but not excessively long—is critical for a successful cement job.
- **Fluid Loss**: The rate of water loss from the slurry into permeable formations under differential pressure. Must be controlled to prevent premature slurry dehydration and bridging of the slurry in the annulus, ensure sufficient water remains for hydration, and minimize filtrate invasion into the formation (Deccod & Baret, 1994). Measured using a static or dynamic fluid loss cell (Dynamic fluid loss better simulates flow conditions in the annulus). API standards suggest maximum fluid loss values for different applications.
- **Compressive Strength Development**: The rate at which the cement gains strength after placement. It needs to develop sufficient strength quickly enough to support the casing and allow subsequent operations (e.g., drilling out the shoe) within a reasonable Waiting-on-Cement (WOC) time, but excessive early strength can sometimes lead to brittleness and increased susceptibility to cracking under stress (Barati Harooni et al., 2025). Compressive strength is typically measured by crushing cured cement cubes or using non-destructive ultrasonic cement analyzers (UCA) simulating downhole conditions. Standards define required minimum compressive strengths before specific operations can proceed.
- **Permeability and Porosity**: Low permeability and porosity of the set cement are essential for effective hydraulic sealing and preventing fluid migration through the cement matrix itself (Renpu, 2011). These properties are controlled by the water-to-cement ratio, hydration extent, and certain additives.

Table 5.1 Rheological models used to describe non-Newtonian behavior of cement slurries and drilling fluids. These models characterize the relationship between shear stress (τ) and shear rate ($\dot{\gamma}$), and include classical, empirical, and advanced constitutive formulations applied in wellbore hydraulics and cement placement (Tao et al., 2020)

Model (Author(s))	Model
(Bingham, 1922)	$\tau = \tau_y + \eta_p \dot{\gamma}$
Bingham (tensor form)	$T = \left[\frac{\tau_y}{\left(\frac{1}{2} D{:}D\right)^{1/2}} + \eta_p \right] D$
Herschel–Bulkley (1926) (Velayati et al., 2015)	$\tau = \tau_y + K\dot{\gamma}^n$
Power-Law (1929) (Li et al., 2016a)	$\tau = K\dot{\gamma}^n$
Williamson (1929) (Tao et al., 2019)	$\tau = \eta_0 \dot{\gamma} + \tau_y \frac{\dot{\gamma}}{\gamma + \dot{\gamma}}$
Eyring (1936) (Taylor, 1997)	$\tau = \sum_{i=1}^{n} a_i \sinh^{-1}(\gamma b_i)$
Sisko (1958) (Brothers et al., n.d.)	$\tau = a\dot{\gamma} + b\dot{\gamma}^c$
Casson (1959) (Scherer et al., 2010)	$\sqrt{\tau} = \sqrt{\tau_y} + \sqrt{\eta_p \dot{\gamma}}$
Shangraw–Grim–Mattocks (RP 10B, 1997)	$\tau = \tau_y + \eta_0 \dot{\gamma} + \alpha_1 [1 - \exp(-\alpha_2 \gamma)]$
Ellis (1965) (Kumar, 2002)	$\dot{\gamma} = a\tau + b\tau^c$
Robertson and Stiff (1976) (Saasen et al., 1991)	$\tau = a(\dot{\gamma} + b)^c$
Robertson and Stiff (tensor) (Mindess, 2019)	$\dot{\gamma}_{ij} = \frac{2A^{1/n} \Pi_1^{1/n}}{\Pi_2^{1-2/n}} e_{ij}$
Vom Berg (1979) (Vlachou & Piau, 1999)	$\tau = \tau_y + a\sinh^{-1}(\gamma / b)$
Lapasin (1979) (Vlachou & Piau, 1999)	$\tau = \tau_\infty + (\tau_0 - \tau_\infty)\exp(-bt_s)$
Quemada (1984) (Scholarship@western & Shahriar, 2011)	$\tau = \left(\frac{1+\sqrt{a\dot{\gamma}}}{b+c\sqrt{a\dot{\gamma}}} \right)^2 \dot{\gamma}$
Modified Casson (1985) (Rice et al., 2012)	$\dot{\gamma} = a + b\tau^{1/2} + c\tau$
Atzeni (1985) (Guan et al., 2014)	$\tau = \tau_y + a\dot{\gamma} + \sinh^{-1}(\dot{\gamma}/c)$
Lapasin (1983) (Guan et al., 2015)	$\tau = \tau_y + \left[A_0 + (A_1 - A_0)\exp\left(-\frac{\dot{\gamma}}{a+b\dot{\gamma}} \right) \right] \sinh^{-1} \frac{\dot{\gamma}}{c}$
De Kee (1994) (Chindaprasirt et al., 2005)	$\tau = \tau_y + \eta_p \dot{\gamma} \exp(-a\dot{\gamma})$
Modified Bingham (Schowalter, 1978)	$\tau = \tau_y + \eta_p \dot{\gamma} + c\dot{\gamma}^2$
Yahia and Khayat (CW Macosko, 1994)	$\tau = \tau_y + 2\left(\sqrt{\tau_y \eta_p}\right)\sqrt{\dot{\gamma}}\exp(-a\dot{\gamma})$
Vipulanandan (Bird et al., 1987)	$\tau = \tau_y + \frac{a}{\dot{\gamma} + b\dot{\gamma}^2}$
Yuan (2015) (Carreau & De Kee, 2021)	$\tau_\infty + (\tau_0 + ct^m - \tau_\infty)\exp(-bt_s)$

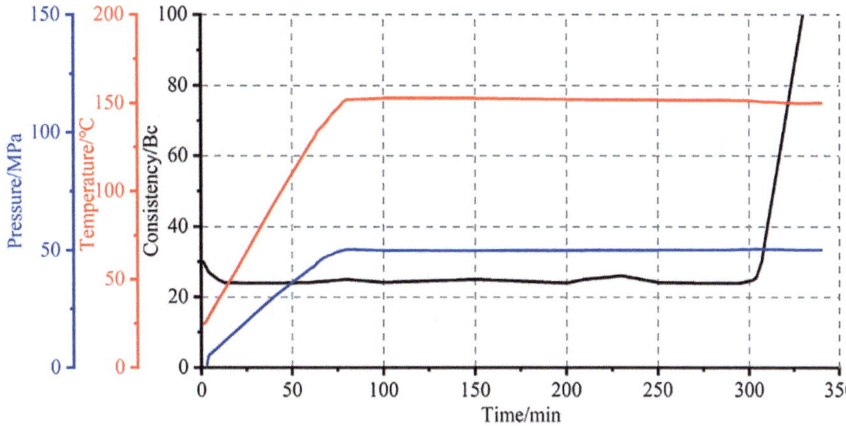

Fig. 5.2 Consistometer test results showing cement slurry consistency (Bc), applied pressure, and temperature over time. The test simulates downhole conditions to determine the slurry's thickening time, defined as the point when consistency exceeds a critical value (typically 70 Bc). This helps define the safe operational window for placement before setting begins

- **Shrinkage/Expansion**: Ideally, the cement should exhibit zero or slight beneficial expansion upon setting to ensure a tight seal against the casing and formation interfaces, counteracting setting shrinkage. Excessive shrinkage can lead to micro-annuli (small gaps) at the casing-cement interface, creating critical leak paths (Kiran et al., 2017). Additives (e.g., expansive agents, flexible additives) can be used to control shrinkage or induce controlled expansion (Lian et al., 2024).
- **Gas Migration Resistance**: As discussed in Sect. 5.2, specific properties (e.g., rapid static gel strength development, low fluid loss, low YP, use of gas-blocking additives) are critical for designing slurries that can effectively resist formation gas entry and channeling before the cement develops sufficient strength to hold back the gas pressure (See Sect. 5.5).
- **Durability and Degradation Resistance**: For long-term integrity, especially in challenging environments (HPHT, sour, high CO_2, geothermal), the set cement must resist long-term chemical attack (carbonation, sulfate attack, leaching) and mechanical degradation under cyclic operational stresses (Chap. 10). This requires careful selection of base cement and additives (e.g., silica flour for high temp, pozzolans for CO_2/sulfate resistance, flexible additives for cyclic loads).

Slurry design is an engineered laboratory process, rigorously testing different formulations under simulated downhole conditions (temperature, pressure, sometimes fluid exposure) until all performance criteria for pumpability, placement, early strength, and long-term integrity are met for the specific well application and anticipated lifecycle conditions. Slurry design cannot be done in isolation; it must consider the actual conditions encountered during drilling, including potential lost circulation zones or challenges to effective mud removal. The fluid loss of the cement slurry to the formation is often calculated as:

$$V_{\text{loss}} = C_{\text{LF}} \cdot \sqrt{\Delta P \cdot t} \tag{5.2}$$

The fluid loss volume, denoted as V_{loss}, is typically measured in milliliters (mL) or cubic centimeters (cm^3) and reflects the amount of filtrate expelled from the cement slurry under differential pressure during a standardized test period. It is influenced by the fluid loss coefficient C_{LF} (expressed in $\text{mL/min}^{1/2}$), the differential pressure ΔP (in pascals or psi), and the test duration t (in minutes). Monitoring this parameter is essential to assess the slurry's ability to minimize filtrate invasion into permeable formations and maintain proper placement.

Additionally, the development of compressive strength in cement over time can be described by an exponential growth function:

$$S_c(t) = S_{c,\max} \cdot \left(1 - e^{-kt}\right) \tag{5.3}$$

where $S_c(t)$ is the compressive strength at time t, $S_{c,\max}$ is the maximum strength achievable, and k is the strength development constant. Time t may be expressed in hours or days, depending on the testing protocol. This equation is widely used to predict the mechanical performance of cement during early setting and curing periods.

The success of primary cementing depends heavily on accurately calculating the required **surface pumping pressure**, which ensures that the cement slurry is delivered to the desired depth without exceeding equipment limitations or compromising wellbore integrity. Proper hydraulic design minimizes the risk of **formation fracturing, lost circulation**, or **poor zonal isolation** due to under displacement or channeling. Pumping pressure must overcome both the hydrostatic pressure of the slurry column and the frictional losses along the casing and annulus.

During primary cementing operations, the frictional pressure loss encountered as cement slurry travels through the casing must be accurately estimated to ensure that surface pumping pressures remain within operational limits. This frictional loss, denoted as ΔP_f, is calculated using the equation:

$$\Delta P_f = \frac{2 \cdot f \cdot \rho_{\text{slurry}} \cdot v^2 \cdot L}{D_i \cdot 144} \tag{5.4}$$

where f is the dimensionless friction factor (dependent on the Reynolds number), ρ_{slurry} is the slurry density in lbm/ft^3, v is the average flow velocity derived from the pump rate Q and internal cross-sectional area A_i, D_i is the internal diameter of the casing in inches, and L is the pipe length in feet. The factor 144 ensures unit consistency for pressure in psi. The velocity v is computed as Q/A_i, with the pump rate Q typically converted from barrels per minute using the conversion bbl/min$\div$0.1781, and the cross-sectional area A_i calculated as $\frac{\pi}{4}\left(\frac{D_i}{12}\right)^2$ in ft^2.

The total surface pumping pressure required to successfully place the cement slurry, P_{pump}, is the sum of the hydrostatic pressure of the cement column P_h, the calculated frictional pressure loss ΔP_f, and any surface-related equipment or blowout preventer (BOP) resistance pressure, denoted as P_{surface}. This relationship is expressed as

$$P_{\text{pump}} = P_h + \Delta P_f + P_{\text{surface}} \tag{5.5}$$

Understanding and correctly applying these calculations are essential for designing reliable pumping schedules and maintaining wellbore integrity during cement placement.

In addition to pressure and flow considerations, accurately modeling the rheological behavior of cement slurries is essential for predicting placement efficiency, pressure losses, and flow regimes in the wellbore. Table 5.1 summarizes a range of rheological models that describe non-Newtonian fluid behavior, capturing the complex relationship between shear stress and shear rate under downhole conditions.

Another key factor influencing slurry performance is the development of static gel strength, which contributes to a reduction in bottomhole pressure as the cement sets. As the cement slurry develops yield strength after placement, it can support itself in the annulus and reduce the hydrostatic pressure acting on the formation. This effect becomes particularly relevant in long vertical or inclined cement columns and is represented by the equation

$$P_{BHP} = P_{BHP,0} - \frac{2\tau_Y L}{R_w - R_c} \tag{5.6}$$

where P_{BHP} is the bottomhole pressure during gelation, $P_{BHP,0}$ is the initial hydrostatic pressure at the base of the cement column, τ_Y is the yield strength of the cement (in pressure units), L is the length of the cement column, R_w is the wellbore radius, and R_c is the outer radius of the casing. This pressure drop, proportional to cement yield stress and column height, can be substantial in deep wells.

Mechanically, the axial load that a cement bond can sustain is a function of the cement's compressive strength and the geometry of the casing. The maximum load the bond can resist before failure is described by

$$F = 0.969 \cdot S \cdot d \cdot h \tag{5.7}$$

where F is the bond load (lbf), S is the compressive strength of the set cement (psi), d is the casing's outer diameter (in), and h is the height of the cement sheath (ft). This relationship is frequently used in post-cement job evaluations to determine the structural support capacity of the cemented casing.

Cement shrinkage, especially during hydration, may compromise long-term bonding and sealing performance. The absolute volumetric shrinkage due to the hydration of Portland cement phases can be calculated using

$$\Delta V_{ab} = K_{hyd1}[C_3S] + K_{hyd2}[C_2S] + K_{hyd3}[C_3A] + K_{hyd4}[C_4AF] \tag{5.8}$$

where ΔV_{ab} is the absolute shrinkage, $[C_3S]$, $[C_2S]$, $[C_3A]$, $[C_4AF]$ are the respective mass fractions of cement phases (alite, belite, aluminate, and ferrite), and K_{hyd1} through K_{hyd4} are empirically derived shrinkage coefficients associated with each phase. This equation highlights the impact of cement composition on dimensional stability.

Lastly, to evaluate the effectiveness of the cement as a barrier, the Index of Zonal Isolation (IZI) compares flow through the cement sheath to that through the reservoir (Nelson & Guillot, 2006). It is defined by

$$I_{ZI} = \frac{q_{res}}{q_{leak}} = \frac{k_{res}h_{res}(\Delta I)}{k\pi\left[(r_{wb})^2 - (r_{csg})^2\right]\left(\ln\left(\frac{r_{res}}{r_{wb}}\right) + s\right)} \tag{5.9}$$

where q_{res} is the intended flow rate from the reservoir, q_{leak} is the leakage flow through the cement, k_{res} is the permeability of the reservoir rock, h_{res} is the height of the producing zone, ΔI is the pressure differential, r_{wb} and r_{csg} are the wellbore and casing radii, respectively, r_{res} is the radial distance to the reservoir boundary, s is the skin factor, and k is the cement permeability. A high IZI reflects effective zonal isolation, essential for long-term production and environmental safety.

5.4 Mud Removal, Spacers, and Centralization—Critical for Bonding

Achieving a good bond between the set cement and *both* the outside of the casing and the inside of the borehole wall is absolutely critical for creating a competent hydraulic seal and ensuring long-term zonal isolation. This bond can only be achieved if the drilling mud coating the casing and the borehole wall is effectively removed *before* the cement slurry arrives in the annulus. Poor mud removal is a primary cause of cementing failure, leading to mud channels, voids, and poor bonding that act as persistent leak paths for SCP and crossflow (Guo et al., 2015; Pelipenko & Frigaard, 2004; Shadravan et al., 2015). When the drilling mud and filter cake are not effectively displaced, they act as barriers that prevent the cement from bonding directly to the formation. This results in the formation of micro-annuli or flow channels that can allow gas or fluid migration, even if the cement itself is properly set. Figure 5.3 shows an example of this failure mechanism, where poor mud removal has led to filtrate invasion and channeling, compromising the long-term well integrity. Key factors influencing mud removal efficiency include:

- **Hole Condition**: The borehole should be relatively in-gauge (not excessively washed out) and free of thick, immobile filter cakes (solids deposited from the drilling mud onto permeable formation surfaces) (Pelipenko & Frigaard,

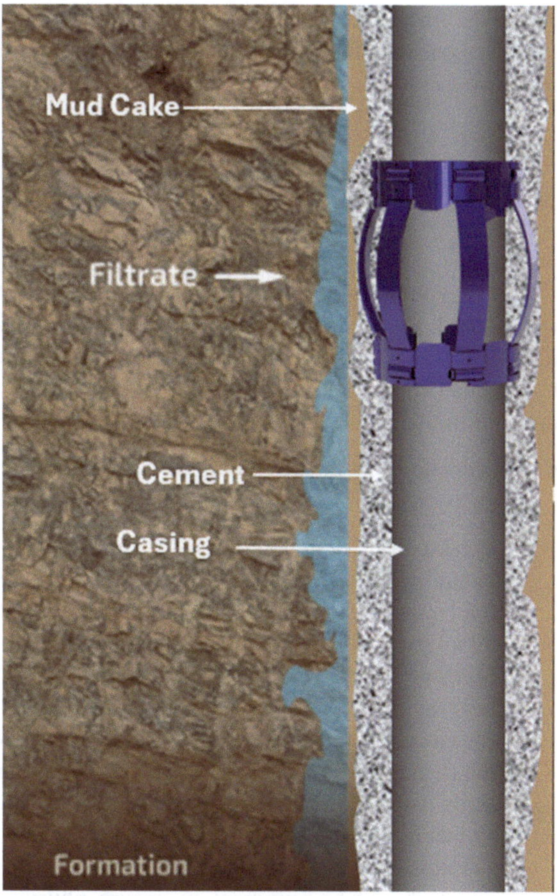

Fig. 5.3 Illustration of poor mud removal leading to incomplete cement bonding. Residual mud cake and filtrate prevent intimate contact between the cement and formation, creating fluid channels that can compromise zonal isolation and lead to sustained casing pressure (SCP)

2004). Proper drilling fluid design (fluid loss control) and hole cleaning (circulation, potentially running scrapers) just before running casing are important prerequisites.

- **Mud Properties**: The drilling mud remaining in the hole just before cementing should be mobile and easily displaced by the fluids pumped ahead of and behind the cement. Thick, viscous, or highly gelled mud, or settled solids beds are extremely difficult to remove. Conditioning the mud (circulating and thinning it) just before running casing and cementing is the standard practice (Beirute et al., 1991).

In addition to fluid loss, thickening time, and rheological behavior, other performance parameters are essential to characterize the interaction of cement with the formation and casing during and after placement. One important factor is the thickness of the cement filter cake, particularly in the absence of a pre-existing mud cake. The filter cake thickness can be estimated using a filtration-based relationship that considers fluid loss volume over time, exposed area, permeability, pressure differential, and the viscosity of the filtrate. This relationship is expressed as

$$\left(h_{fc}\right)_{cem} = K_{dep} \cdot \frac{V(t)}{A} = \sqrt{\frac{2\left(k_f\right)_{cem} \cdot K_{dep} \cdot \Delta p \cdot t}{\mu_{fil}}} \tag{5.10}$$

where $\left(h_{fc}\right)_{cem}$ is the cement filter cake thickness, K_{dep} is the deposition coefficient, $V(t)$ is the volume of filtrate collected over time t, A is the area over which the filtrate is measured, $\left(k_f\right)_{cem}$ is the permeability of the filtercake, Δp is the applied pressure differential, and μ_{fil} is the filtrate viscosity. This formulation enables estimation of how much cementitious material will adhere to the wellbore wall during pressure-driven filtration.

- **Spacers and Pre-flushes**: Pumping carefully designed fluids ahead of the cement slurry is crucial for effective mud displacement (Newhall, 2006).
  - **Pre-flushes**: Thin, low-viscosity fluids (e.g., water, chemical washes like mutual solvents) designed to break up gelled mud, thin mobile mud, and initiate fluid movement in the annulus (Guillot et al., 2007). Chemical washes can help dissolve or disperse filter cake components.
  - **Spacers**: Fluids are designed to be compatible with *both* the preceding drilling mud and the following cement slurry, preventing them from mixing (which can cause detrimental reactions or thickening). Spacers typically have tailored rheological properties (often high viscosity, sometimes high density) to physically displace the mud piston-like ahead of the cement. Spacers often contain surfactants or wetting agents to leave surfaces water-wet, which improves subsequent cement bonding (Adesanya et al., 2016). The volume of the spacer pump should be sufficient to ensure physical separation along the entire annulus.

Displacement efficiency is also central to achieving zonal isolation and preventing mud channeling. A common metric for this is the volumetric displacement efficiency, defined as

$$\eta_{vol} = \frac{V_i}{V_a} \tag{5.11}$$

where η_{vol} is the ratio of the volume V_i of the displacing fluid that replaces the mud to the total annular volume V_a. This provides a direct measure of how effectively the cement or spacer displaces drilling mud from the annular space, a key indicator of bonding potential and zonal integrity.

In deviated wells, the frictional pressure gradient generated by the displacing fluid must be high enough to overcome the gelled mud and displace it fully (Lavrov & Torsæter, 2016a). This required gradient is governed by the gel strength of the mud, well geometry, and fluid density contrast, and is given by

$$\left(\frac{dP_{fr}}{dz}\right)^{(sp)} = \frac{2\tau_{gel}^{(mud)}}{(R_o - R_i)s} - \left[\rho^{(sp)} - \rho^{(mud)}\right]g\cos i \tag{5.12}$$

where $\left(\frac{dP_{fr}}{dz}\right)^{(sp)}$ is the frictional pressure gradient of the spacer fluid, $\tau_{gel}^{(mud)}$ is the gel strength of the mud to be displaced, R_o and R_i are the outer and inner radii of the annulus, s is the eccentricity correction factor, $\rho^{(sp)}$ and $\rho^{(mud)}$ are the densities of the spacer and mud fluids, respectively, g is gravitational acceleration, and i is the inclination angle of the wellbore. This equation reflects the importance of density contrast and flow regime in effective mud removal.

- **Flow Regimes**: Pumping the cement slurry and spacers at sufficient rates to achieve turbulent flow in the annulus is highly effective at scouring mud from the casing and formation surfaces due to increased shear stress. If turbulent flow cannot be achieved due to pressure limitations (e.g., risk of fracturing the formation), achieving effective laminar flow (high flow rate, specific rheology profile with a flat velocity profile) is the next best option. Plug flow (low rate, high viscosity) is generally least effective for mud removal. Flow regime is determined by fluid properties (density, viscosity, yield point), flow rate, and annular geometry.
- **Pipe Movement**: or called dynamic cementation—Rotating and/or reciprocating (moving up and down) the casing string while circulating mud and pumping cement/spacers significantly improves mud displacement efficiency (Guzman et al., 2018; Holt & Lahoti, 2012; Holt et al., 2013). This physical movement helps break up stagnant mud pockets, disrupts filter cake, and distributes the cement more evenly around the annulus, preventing channeling (Kiran et al., 2017). This is often standard practice, especially for critical casing strings like production liners, where feasible.
- **Centralization**: Ensuring the casing is centered within the borehole and previous casing string (achieving adequate standoff) is vital for obtaining a uniform cement sheath around the entire circumference. The standoff defined in Eq. (5.13) If the casing lies against the borehole wall (poor standoff), the narrow annular gap on the low side makes mud removal extremely difficult, leading to trapped mud and creating a high-permeability channel when the mud gels or dehydrates(Guzman et al., 2018; Holt et al., 2013). Centralizers (bow-spring or solid-body types) are attached to the casing exterior at planned intervals to maintain adequate standoff (ideally > 67–70%, or even 80–90% in critical applications). Calculations are performed during the design phase to determine the number, type, and placement of centralizers needed based on hole deviation, casing properties, well path curvature (dogleg severity), and centralizer restoring force against borehole side forces. Poor centralization is a major, preventable cause of cement failure and SCP. Figure 5.4 shows how the casing is centered throughout the interval. Sections with low standoff (below 60–70%) indicate points where the casing lies close to the borehole wall, which impedes cement placement and mud removal on the low side. This can result in eccentric cement sheaths and unbonded channels that

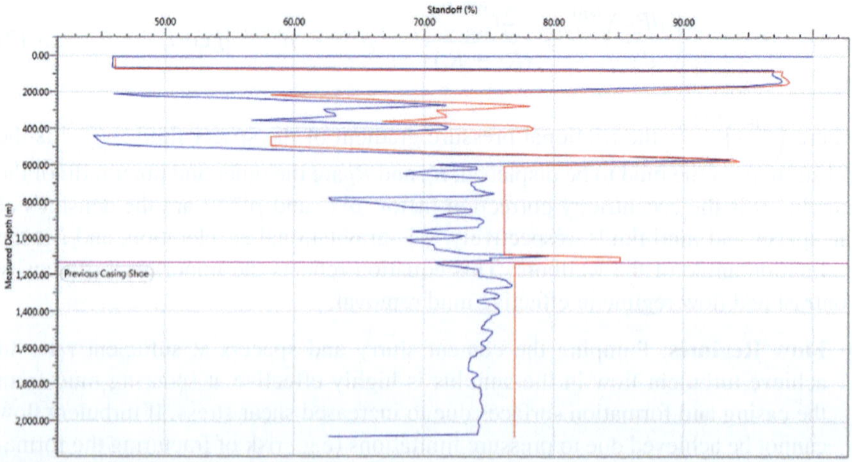

Fig. 5.4 Casing centralization log showing standoff percentage versus depth. Adequate standoff (>70%) is essential to ensure uniform cement distribution around the casing. Low standoff values indicate casing contact with the borehole wall, increasing the risk of poor mud displacement and cement channeling

jeopardize long-term zonal isolation. Proper centralizer selection and placement design, based on simulations and standoff analysis, are essential for avoiding such issues.

Effective centralization ensures standoff:

$$\text{Standoff (\%)} = \frac{W_{\min}}{R_w - R_c} \times 100 \tag{5.13}$$

where $W_{\min}$ is the minimum gap width between the casing and the wellbore wall (in meters), R_w is the wellbore radius (in meters), and R_c is the outer radius of the casing (in meters).

Effective mud removal and proper centralization require careful planning (including rheology compatibility, volumes, standoff calculations, and centralizer placement) and vigilant execution of the entire displacement process, considering all these factors in combination. Failure in this step directly translates to failure in zonal isolation and creates latent integrity issues.

Accurate centralizer spacing is essential to ensure adequate casing standoff and effective cement placement, particularly in deviated or directional wells where gravitational and geometric effects can compromise uniform cement coverage. The force acting on each centralizer, which reflects both mechanical tension and weight distribution along the wellbore, can be quantified using the expression:

$$F = 2T \cdot \sin\left(\frac{\text{DLS} \cdot \text{CS}}{2}\right) + (WF)_b \cdot \text{CS} \cdot \sin\phi \tag{5.14}$$

Here, F denotes the total force on each centralizer in pounds-force (lbf), T is the axial tension in the pipe (lbf), DLS is the dogleg severity expressed in degrees per 100 feet, CS is the spacing between centralizers (ft), $(WF)_b$ represents the buoyancy-corrected pipe weight per foot (lbf/ft), and ϕ is the inclination angle of the well in degrees. To determine the optimal spacing of the centralizers, Eq. (5.14) is applied:

$$\mathrm{CS} = \frac{F}{0.0175 \cdot T \cdot \mathrm{DLS} + (WF)_b \cdot \sin\phi} \tag{5.15}$$

This equation accounts for the interaction between mechanical loading and trajectory-induced deviation forces. All parameters—including axial tension, inclination, buoyancy-adjusted pipe weight, and curvature severity—are crucial to achieving a mechanically balanced and hydraulically effective cementing operation.

5.5 Cement Placement Techniques and Gas Migration Prevention

Cement placement in the well involves a carefully orchestrated sequence of fluid displacements to ensure proper slurry delivery and effective mud removal. Figure 5.5 provides a visual overview of the typical primary cementing setup, illustrating the key components involved—cementing unit, plugs, spacers, centralizers, float collar, and guide shoe. Understanding the role of each element in this sequence is essential to prevent fluid contamination, ensure full annular fill, and reduce the risk of gas migration. Cement placement involves pumping the designed slurry volume down the casing string, typically displacing drilling mud or other fluids already inside the casing. A **bottom plug** is usually launched ahead of the cement slurry to separate it from the mud inside the casing and wipe the casing wall. A **top plug** is launched after the slurry volume to separate it from the displacement fluid (e.g., brine, mud, water) pumped behind it. The bottom plug lands on a float collar near the casing shoe and ruptures, allowing cement to flow out into the annulus. The top plug lands on the bottom plug (or a landing collar), creating a seal that allows pressure to build up ("bumping the plug"), indicating that the planned volume of cement has been displaced into the annulus (Fig. 5.6).

- **Pump Rates**: Flow rates are carefully controlled during cement placement. High rates are desirable to achieve turbulent flow for mud displacement (Sect. 5.4) but must be below the rate that would cause excessive surge pressure downhole/ Equivalent circulation density or exceed the formation fracture pressure. Rates may be varied during the job (e.g., slower rate through the shoe track or critical weak zones) (Isgenderov et al., 2014; Khaldy et al., 2023).

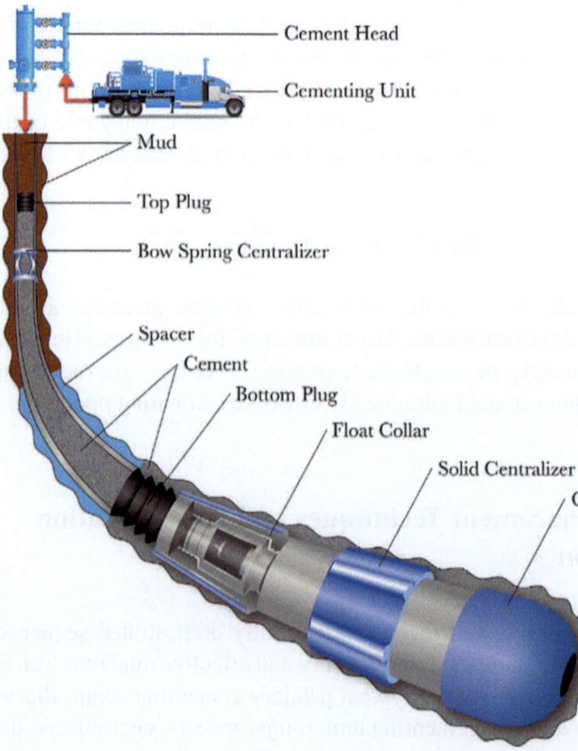

Fig. 5.5 Schematic of primary cementing operations in a wellbore. The cementing unit pumps the slurry through the cement head, displacing drilling mud with a bottom plug, spacer, cement, and top plug. Centralizers (bow spring and solid) help ensure proper casing standoff, while the float collar and guide shoe assist in directing flow and preventing backflow (Liu, 2021)

- **Pressure Monitoring**: Surface pressures are continuously monitored during placement to detect potential issues like lost circulation (sudden pressure drop), bridging or premature setting (sudden pressure increase), or successful plug bumping (pressure increase confirming end of displacement).
- **Cement Volume**: Pumping the correct volume of cement slurry is critical to ensure adequate fill height in the annulus as designed to cover all zones requiring isolation. Excess volume can lead to cement above the planned height (cost/cleanup issue), while insufficient volume leaves zones uncemented. Volume calculations must account for caliper logs if available, hole washout, and shrinkage.

Gas Migration Prevention: As discussed in Sects. 5.1 and 5.2, preventing formation gas influx and channeling through the cement slurry before it develops sufficient gel strength to hold back the gas pressure is critical for avoiding annular SCP, particularly when cementing across gas-bearing formations. Strategies include:

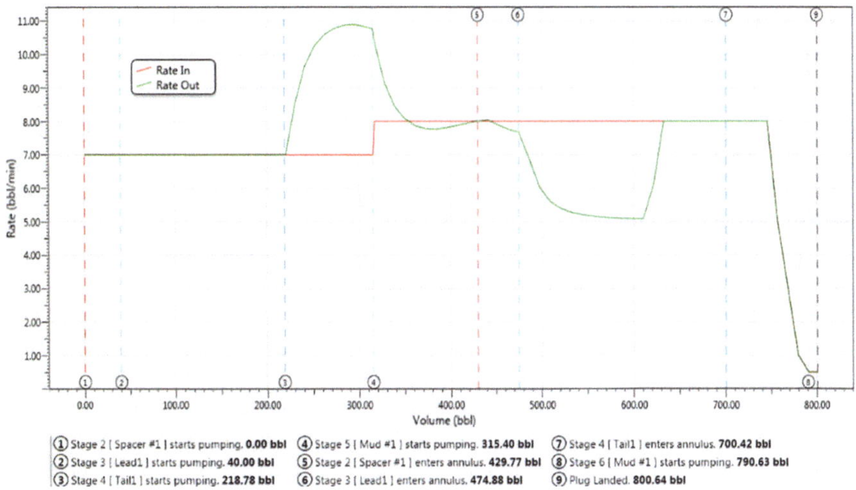

Fig. 5.6 Primary cementing pump schedules

- **Minimizing Transition Time**: Designing slurries with a short transition time from fluid to semi-solid (measurable static gel strength). This reduces the vulnerable period when gas pressure can exceed the diminishing hydrostatic pressure of the slurry column (due to fluid loss, slurry settling, and hydration shrinkage) (Prabhakar et al., 2019).
- **Optimized Slurry Properties**: Using slurries with low fluid loss, specific rheology (low YP, thixotropy), and potentially gas-blocking additives (like latex, expanding agents) that create a more impermeable barrier or develop static gel strength rapidly (Ref Discussion in Sect. 5.2).
- **Maintaining Hydrostatic Pressure**: Designing the fluid column densities (mud, spacer, cement, displacement fluid) to maintain sufficient hydrostatic pressure overbalance against the gas zone for as long as possible during placement and WOC. Sometimes applying surface pressure to the annulus after placement can help maintain pressure on the slurry column while it develops strength.
- **Specialized Cement Systems**: Utilizing systems designed to generate internal pressure upon setting (expanding cements) or to have extremely short transition times.

Successful cement placement relies on meticulous planning based on accurate well data and slurry testing, careful execution according to the plan by competent personnel, real-time monitoring of pressures and returns, and effective response to any deviations. The quality of the primary cement job directly impacts the well's ability to provide zonal isolation and support for decades, emphasizing the long-term consequences of execution.

5.6 Setting and Hydration

Once placed in the annulus, the cement slurry undergoes hydration, a chemical reaction with water that transitions it from a fluid to a solid, load-bearing material (Lavrov & Torsæter, 2016b). This process generates heat (heat of hydration) and develops compressive strength, tensile strength, and low permeability over time. The rate of strength development is primarily influenced by downhole temperature, pressure, and the cement/additive formulation (Cao et al., 2024; Pang et al., 2022).

- **Waiting-on-Cement (WOC) Time**: This is the minimum period required for the cement to develop adequate compressive strength before proceeding with subsequent operations (e.g., nippling down the BOP, drilling out the shoe track, or perforating). WOC time depends on the slurry design, wellbore temperature, and the operational strength requirement—such as supporting casing weight, ensuring zonal isolation, or enabling pressure testing. (API RP 10B-2, 2024) provides standardized procedures for testing and reporting the compressive strength of well cements, but does not prescribe a specific minimum threshold. In practice, thresholds such as 50 psi (0.34 MPa) for basic mechanical support or 500 psi (3.45 MPa) for formation integrity tests and leak-off tests are commonly used based on operational or regulatory requirements.
- **Strength Development**: Cement strength continues to increase beyond the minimum WOC time for days or weeks. Final strength and durability depend on full hydration and resistance to degradation over time.
- **Bond Strength**: The strength of the hydraulic and mechanical bond between the set cement and the casing and formation is critical for zonal isolation and structural support. Bond strength develops as the cement sets and is heavily influenced by the effectiveness of mud removal, centralization, and shrinkage control during placement.

Achieving a successful primary cement job, characterized by complete annular fill across zones of interest, good bonding to both casing and formation, adequate compressive strength, low permeability, and freedom from leak paths, is fundamental to establishing durable well integrity from the start. Failures in primary cementing are latent defects that often lead to integrity problems like SCP years after the well is put on production and are typically difficult and expensive to remediate later in the well's life, highlighting the importance of getting it right during construction. The vulnerability of the cement to gas migration is most pronounced during the transition time, when the slurry changes from a pumpable fluid to a semi-solid with measurable gel strength. This period is marked by a decline in hydrostatic pressure as hydration progresses, potentially allowing gas to enter the annulus before the cement develops sufficient resistance. Figure 5.7 illustrates this critical window, showing the relationship between hydrostatic pressure, static gel strength, and compressive strength development over time. It underscores the importance of minimizing transition time and designing slurries that rapidly build gel strength to resist gas intrusion during this vulnerable period.

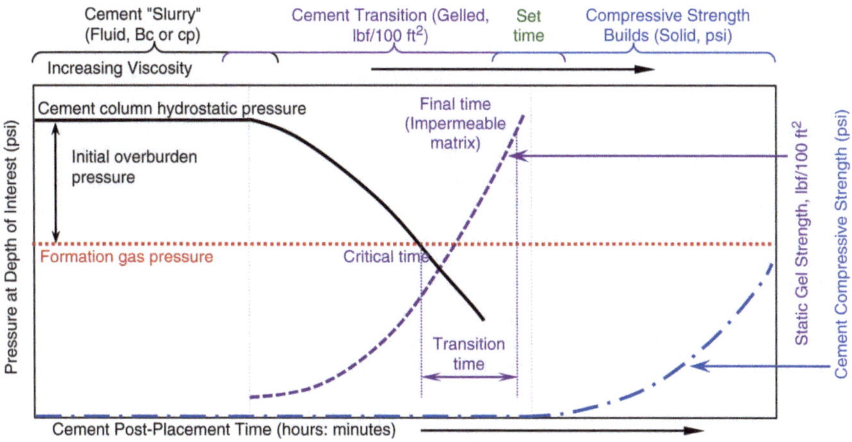

Fig. 5.7 Evolution of cement properties post-placement showing the increase in gel strength and compressive strength. The graph highlights the "transition time" and "critical time" where the cement is most vulnerable to gas migration due to loss of hydrostatic pressure before sufficient gel strength and set strength develop (Li et al., 2016a, 2016b)

References

Abdelaal, A., & Elkatatny, S. (2023). Synergy of retarders and superplasticizers for thickening time enhancement of hematite based fly ash geopolymers. *Geoenergy Science and Engineering, 224*, Article 211641. https://doi.org/10.1016/J.GEOEN.2023.211641

Adesanya, O., Mofunlewi, S. S., Ighodaro, O., Dickson, T., Esegbue, A., & Ilesanmi, O. (2016) Improving cementing efficiency by optimizing the spacer train: Case histories. In *Society of Petroleum Engineers—SPE Nigeria Annual International Conference and Exhibition.* https://doi.org/10.2118/184237-MS

Alberdi-Pagola, P., & Fischer, G. (2023). Review of integrity loss detection and quantification due to cracking in cemented wells. *SPE Journal, 965.* http://onepetro.org/SJ/article-pdf/28/03/965/3120397/spe-214303-pa.pdf

Ahmed, A. A., Hayfaa, L. S., Nuhad, A., Erfan, M. A. lawe, Ahmed, I. A., Swadi, M., & Tariq, K. K. (2019). Investigation of the rock strength anisotropy on the wellbore stability analysis. OnePetro. https://dx-doi-org.ezproxy.lib.utexas.edu/

Alsubaih, A. A., Hayfaa, S., Alboshawi, H. S., Alaa, A., Al, lawe, Erfan, M., Sawdi, M., Al, A., & Lubna, H. (2020). Reservoir geomechanics simulation to optimize reservoir management in southern Iraq fields. OnePetro. https://dx.doi.org/

Alsubaih, A., Albadran, F., Abbood, N., & Alkanaani, N. (2018). Optimization of the tripping parameters to prevent the surge related wellbore instability. OnePetro. https://dx-doi-org.ezproxy.lib.utexas.edu/

Alsubaih, A., Sepehrnoori, K., & Delshad, M. (2024). Environmental impacts of orphaned and abandoned wells: Methane Emissions, and Implications for carbon storage. *Applied Sciences, 14*(24), 11518. https://doi.org/10.3390/APP142411518

Alsubaih, A., Sepehrnoori, K., & Delshad, M. (2025). Development and optimization of self-healing cement for CO_2 injection and storage wells: enhancing long-term wellbore integrity in extreme subsurface conditions. *Applied Sciences, 15*(10), 5428. https://doi.org/10.3390/APP15105428

Alsubaih, A., Sepehrnoori, K., Lopez Manriquez, A., Espinoza, D. N., Allawi, H. R., & Hayfaa, S. (2024). A statistical approach to estimate the drilling mud window based on field data and ten failure criteria in depleted reservoirs—New modified failure criterion. https://doi.org/10.56952/ARMA-2024-0799

API 10A. (2019). *API SPEC 10A: Cements and materials for well cement*. https://www.apiwebstore.org/standards/10A?utm_term=api%2010a&utm_campaign=Webstore+-+US&utm_source=adwords&utm_medium=ppc&hsa_acc=4890137417&hsa_cam=19885414257&hsa_grp=148201952835&hsa_ad=652489415992&hsa_src=g&hsa_tgt=aud-1016965943913:kwd-2033893061006&hsa_kw=api%2010a&hsa_mt=e&hsa_net=adwords&hsa_ver=3&gad_source=1&gclid=Cj0KCQjwoNzABhDbARIsALfY8VMSyYUv8o8qefbNGedP3NvSwAMwlLXkeTljrgqNtG4A5ePGPmJAw7UaAto7EALw_wcB

API RP 10B-2. (2024). *Recommended practice for testing well cements*. www.api.org

Applications, C. M.-M., & 1994, undefined. (n.d.). Rheology principles. *Cir.Nii.Ac.Jp*. Retrieved July 1, 2025, from https://cir.nii.ac.jp/crid/1573387449190639744

Barati Harooni, A., Maroof, H., Abdollahi Khoshmardan, M., & Pourali, O. (2025). A novel predictive model for the compressive strength development in class G cement slurries at different temperatures, retarder, and salt concentrations. *Case Studies in Construction Materials, 22*, Article e04360. https://doi.org/10.1016/J.CSCM.2025.E04360

Beirute, R. M., Sabins, F. L., & Ravi, K. V. (1991). Large-scale experiments show proper hole conditioning: A critical requirement for successful cementing operations. In *Proceedings—SPE Annual Technical Conference and Exhibition, Pi*(pt 1) (pp. 135–150). https://doi.org/10.2118/22774-MS

Bingham, E. (1922). *Fluidity and plasticity*. https://books.google.com/books?hl=en&lr=&id=h8MKAAAAIAAJ&oi=fnd&pg=PA1&ots=_10VLxIQeo&sig=sfjXAOk9wkPo-T-tKMtbERwThM0

Bird, R. B., Armstrong, R. C., & Hassager, O. (1987). *Dynamics of polymeric liquids* vol. 1, Fluid mechanics (2nd ed.) https://inis.iaea.org/records/v53wc-st124

Brothers, L., AV Palmer—US Patent 6, 244,343, & 2001, undefined. (n.d.). Cementing in deep water offshore wells. *Google Patents*. Retrieved July 1, 2025, from https://patents.google.com/patent/US6244343B1/en

Cao, Z., Liu, J., Xiao, J., Zhang, J., & Xu, Y. (2024). Strength characteristics and energy evolution of cement stone under true-triaxial loading conditions. *Construction and Building Materials, 432*, Article 136690. https://doi.org/10.1016/J.CONBUILDMAT.2024.136690

Carreau, P. J., & De Kee, D. (2021). Carreau: Rheology of polymeric systems: Principles—Google Scholar. https://scholar.google.com/scholar_lookup?title=Rheology+of+Polymeric+Systems:+Principles+and+Applications&author=Carreau,+P.J.&author=De+Kee,+D.C.R.&author=Chhabra,+R.P.&publication_year=1997

Chindaprasirt, P., Jaturapitakkul, C., & Sinsiri, T. (2005). Effect of fly ash fineness on compressive strength and pore size of blended cement paste. *Cement and Concrete Composites, 27*(4), 425–428. https://doi.org/10.1016/J.CEMCONCOMP.2004.07.003

Dabzhu. (2023). *Geological analysis and cement composition for improving zonal isolation—ProQuest*. https://www.proquest.com/openview/18981afc00e02d78dd72378b80e4393d/1?cbl=18750&diss=y&pq-origsite=gscholar

Deccod, G., & Baret, J. F. (1994). How fluid loss influences primary cementing: Literature review and methodology. *SPE Drilling & Completion, 9*(02), 133–138. https://doi.org/10.2118/25150-PA

Foroushan, H. K., Lund, B., Ytrehus, J. D., & Saasen, A. (2021). Cement placement: An overview of fluid displacement techniques and modelling. *Energies 14*(3), 573. https://doi.org/10.3390/EN14030573

Guan, S., Rice, J. A., Li, C., & Gu, C. (2014). Automated DC offset calibration strategy for structural health monitoring XE "monitoring" based on portable CW radar sensor. *IEEE Transactions on Instrumentation and Measurement, 63*(12), 3111–3118. https://doi.org/10.1109/TIM.2014.2317298

Guan, S., Rice, J. A., Li, C., Li, Y., & Wang, G. (2015). Dynamic and static structural displacement measurement using backscattering DC coupled radar. *Smart Structures and Systems, 16*(3), 521–535. https://doi.org/10.12989/SSS.2015.16.3.521

Guillot, D. J., Desroches, J., & Frigaard, I. (2007). *Are Preflushes Really Contributing to Mud Displacement During Primary Cementing?* https://doi.org/10.2118/105903-MS

Guo, Y., Qi, J., Zheng, Y. S., Taoutaou, S., Wang, R., An, Y., & Guo, H. (2015). Cementing optimization through an enhanced ultrasonic imaging tool. In *Society of Petroleum Engineers—SPE/IATMI Asia Pacific Oil and Gas Conference and Exhibition, APOGCE 2015*. https://doi.org/10.2118/176069-MS

Guzman, J., Mavares, F., Monasterios, E., & Massirrubi, L. (2018). Casing centralization and pipe movement in cementing operations for improved displacement efficiency. In *Society of Petroleum Engineers—SPE Trinidad and Tobago Section Energy Resources Conference 2018*. https://doi.org/10.2118/191255-MS

Herfort, D., Moir, G. K., Johansen, V., Sorrentino, F., & Bolio Arceo, H. (2010). The chemistry of Portland XE "Portland" cement clinker. *Advances in Cement Research, 22*(4), 187–194. https://doi.org/10.1680/ADCR.2010.22.4.187

Holt, C., & Lahoti, N. (2012). Dynamic cementation improves wellbore construction and reduces the hazards of groundwater contamination in shale plays. In *Society of Petroleum Engineers—SPE Canadian Unconventional Resources Conference 2012, CURC 2012* (vol. 2, pp. 1099–1106). https://doi.org/10.2118/162739-MS

Holt, C., Lahoti, N., & Fortier, V. (2013). Dynamic cementation: A solution to well integrity problems. In *SPE/IADC Drilling Conference, Proceedings* (vol. 1, pp. 634–641). https://doi.org/10.2118/163459-MS

Isgenderov, I., Bogaerts, M., Osayande, V., Kurawle, I., Kanahuati, A. I., & Aghaguluyev, J. (2014). ECD management solves lost circulation issues in cementing narrow pressure window: A case study. In *Society of Petroleum Engineers—IADC/SPE Asia Pacific Drilling Technology Conference 2014: Driving Sustainable Growth Through Technology and Innovation* (pp. 292–303). https://doi.org/10.2118/170491-MS

Khaldy, A., Hassan, J., Gupta, S. K., Baloushi, Z. A., Alqattan, F. A., Datta, A., Monterio, K., Kunchur, B., Hussain, M. K., Ellis, C., Singh, R., & Kalgutkar, N. (2023). Unconventional cementing technique to achieve good zonal isolation in primary casing cementation jobs for weaker formations or loss zones. In *Society of Petroleum Engineers—ADIPEC, ADIP*. https://doi.org/10.2118/216828-MS

Kiran, R., Teodoriu, C., Dadmohammadi, Y., Nygaard, R., Wood, D., Mokhtari, M., & Salehi, S. (2017). Identification and evaluation of well integrity and causes of failure of well integrity barriers (a review). *Journal of Natural Gas Science and Engineering, 45*, 511–526. https://doi.org/10.1016/J.JNGSE.2017.05.009

Kopp, K., Reed, S., Foreman, J., Carty, B., & Griffith, J. (2000). Foamed cement vs. conventional cement for zonal isolation—Case histories. *SPE Reservoir Engineering (Society of Petroleum Engineers), A*, 75–84. https://doi.org/10.2118/62895-MS

Kumar, S., Singh, C. J., Singh, R. P., & Kachari, J. (2002). Microfine cement: Special superfine Portland cement—Google Scholar. https://scholar.google.com/scholar_lookup?title=Microfine+cement:+Special+superfine+Portland+cement&conference=Proceedings+of+the+7th+NCB+International+Seminar&author=Kumar,+S.&author=Singh,+C.J.&author=Singh,+R.P.&author=Kachari,+J.&publication_year=2002

Ladva, H. K. J., Craster, B., Jones, T. G. J., Goldsmith, G., & Scott, D. (2005). The cement-to-formation interface in zonal isolation. *SPE Drilling & Completion, 20*(03), 186–197. https://doi.org/10.2118/88016-PA

Lavrov, A., & Torsæter, M. (2016a). *Physics and Mechanics of Primary Well Cementing*. https://doi.org/10.1007/978-3-319-43165-9

Lavrov, A., & Torsæter, M. (2016b). Properties of well cement. *SpringerBriefs in Petroleum Geoscience and Engineering, 9–23*. https://doi.org/10.1007/978-3-319-43165-9_2/TABLES/1

Li, Z., Vandenbossche, J. M., Iannacchione, A. T., Brigham, J. C., & Kutchko, B. G. (2016a). Theory-based review of limitations with static gel strength in cement/matrix characterization. *SPE Drilling and Completion, 31*(2), 145–158. https://doi.org/10.2118/178923-PA

Li, Z., Vandenbossche, J. M., Iannacchione, A. T., Brigham, J. C., & Kutchko, B. G. (2016b). Theory-based review of limitations with static gel strength in cement/matrix characterization. *SPE Drilling & Completion, 31*(02), 145–158. https://doi.org/10.2118/178923-PA

Lian, J., Feng, Y., Yang, Z., Wang, Y., Lu, X., Zhang, H., Xiao, W., Xu, K., & Liu, W. (2024). Experimental investigation of shrinkage/expansion of oil-well cement. In *International Geomechanics Conference*. https://doi.org/10.56952/IGS-2024-0835

Liu, G. (2021). Introduction to cementing engineering. *Applied Well Cementing Engineering, 1–15*. https://doi.org/10.1016/B978-0-12-821956-0.00004-3

Mahmoud, A. A., Abdelaal, A., Adjei, S., & Elkatatny, S. (2024). Foamed cement applications in oil industry based on field experience: A comprehensive review. *ACS Omega, 9*(9), 9961. https://doi.org/10.1021/ACSOMEGA.3C07580

Macosko, C. W. (1994). Rheology: Principles, Measurements, and Applications

Mindess, S. (2019). *Developments in the formulation and reinforcement of concrete*. https://books.google.com/books?hl=en&lr=&id=YGGdDwAAQBAJ&oi=fnd&pg=PP1&ots=1e_vIAg8tA&sig=0v3DfLeSpxp0Yb8TcwVwmVgTk7o

Naaman, A. E. (2001). Reinforced concrete. *Encyclopedia of Materials: Science and Technology, 8095–8109*. https://doi.org/10.1016/B0-08-043152-6/01454-6

Nelson, E. B., & Guillot, D. (2006). *Well cementing* (ed.). Schlumberger.

Newhall, C. (2006). Improving cement bond in the Appalachian basin with adjustments to Preflush and spacer design. *SPE Eastern Regional Meeting, 2006*, 270–279. https://doi.org/10.2118/104576-MS

Pang, X., Sun, L., Chen, M., Xian, M., Cheng, G., Liu, Y., & Qin, J. (2022). Influence of curing temperature on the hydration and strength development of Class G Portland XE "Portland" cement. *Cement and Concrete Research, 156*, Article 106776. https://doi.org/10.1016/J.CEMCONRES.2022.106776

Pelipenko, S., & Frigaard, I. A. (2004). Mud removal and cement placement during primary cementing of an oil well. *Journal of Engineering Mathematics, 48*(1), 1–26. https://doi.org/10.1023/B:ENGI.0000009499.63859.F0/METRICS

Prabhakar, A., Lee, N., Ong, K. C. G., Zhang, M., Moon, J., Cheng, A., & Kong, K. H. (2019). Enhancing the gel transition time and right-angle-set property of oil well cement slurries by incorporating CSA cement and gypsum. In *International Petroleum Technology Conference 2019, IPTC 2019*. https://doi.org/10.2523/IPTC-19224-MS

Renpu, W. (2011). Production casing and cementing. *Advanced Well Completion Engineering, 221–294*. https://doi.org/10.1016/B978-0-12-385868-9.00009-9

Rice, J., Gu, C., Li, C., & Guan, S. (2012). A radar-based sensor network for bridge displacement measurements. In *Sensors and smart structures technologies for civil, mechanical, and aerospace systems*. https://doi.org/10.1117/12.914796.SHORT

RP 10B. (1997). American Petroleum Institute: RP 10B: Recommended—Google Scholar. https://scholar.google.com/scholar_lookup?title=10B+Recommended+Practice+for+Testing+Oil+Well+Cements+and+Cement+Additives&author=American+Petroleum+Institute+(API)&publication_year=1997

Saasen, A., Rafoss, E., & Behzadi, A. (1991). Experimental investigation of rheology and thickening time of class G oil well cement slurries containing glycerin. *Cement and Concrete Research, 21*(5), 911–916. https://doi.org/10.1016/0008-8846(91)90186-L

Scherer, G. W., Funkhouser, G. P., & Peethamparan, S. (2010). Effect of pressure on early hydration of class H and white cement. *Cement and Concrete Research, 40*(6), 845–850. https://doi.org/10.1016/J.CEMCONRES.2010.01.013

Scholarship@western, S., & Shahriar, A. (2011). *Investigation on rheology of oil well cement slurries*. https://search.proquest.com/openview/04edd2eb66b210b68fb4082c33d78126/1?pq-origsite=gscholar&cbl=18750&diss=y

Schowalter, W. (1978). *Mechanics of non-Newtonian fluids*. https://library.wur.nl/WebQuery/titel/134280

Shadravan, A., Narvaez, G., Alegria, A., Carman, P., Perez, C., & Erger, R. (2015). Engineering the mud-spacer-cement rheological hierarchy improves wellbore integrity. In *Society of Petroleum Engineers—SPE E and P Health, Safety, Security and Environmental Conference—Americas* (pp. 374–387). https://doi.org/10.2118/173534-MS

Tao, C., Kutchko, B. G., Rosenbaum, E., & Massoudi, M. (2020). A review of rheological modeling of cement slurry in oil well applications. *Energies, 13*(3), 570. https://doi.org/10.3390/EN13030570

Tao, C., Kutchko, B. G., Rosenbaum, E., Wu, W. T., & Massoudi, M. (2019). Steady flow of a cement slurry. *Energies, 12*(13). https://doi.org/10.3390/EN12132604

Taylor, H. F. W. (1997). *Cement chemistry*. Thomas Telford Publishing. https://doi.org/10.1680/CC.25929.FM

Velayati, A., Tokhmechi, B., Soltanian, H., & Kazemzadeh, E. (2015). Cement slurry optimization and assessment of additives according to a proposed plan. *Journal of Natural Gas Science and Engineering, 23*, 165–170. https://doi.org/10.1016/J.JNGSE.2015.01.037

Vlachou, P. V., & Piau, J. M. (1999). Physicochemical study of the hydration process of an oil well cement slurry before setting. *Cement and Concrete Research, 29*(1), 27–36. https://doi.org/10.1016/S0008-8846(98)00182-3

Chapter 6
Completion Equipment and Wellhead Integrity

Abstract This chapter provides an in-depth examination of verification practices critical for confirming the mechanical and hydraulic integrity of well barriers immediately after construction. It begins by establishing the regulatory and operational rationale for initial integrity testing, followed by a breakdown of industry-standard procedures including pressure tests (e.g., casing pressure tests, shoe tests, and leak-off tests), inflow tests, and negative pressure tests. Emphasis is placed on test design, interpretation of pressure-volume-time (PVT) behavior, and diagnostic criteria to identify potential barrier weaknesses. The chapter also introduces logging technologies essential for evaluating the quality of cement placement and casing-to-formation bonding. Cement evaluation logs (such as CBL, VDL, and ultrasonic imaging) are reviewed in terms of operational principles, resolution capabilities, and interpretation workflows. Case studies are presented to highlight common diagnostic scenarios, such as channeling, micro-annuli, and uncemented zones. Further, the role of temperature and noise logs in dynamic cement integrity assessment is discussed. The chapter also addresses the uncertainty and limitations of various logs, offering guidelines for data integration and risk-based decisions in well acceptance. Finally, best practices for documentation, regulatory compliance, and long-term digital recordkeeping are outlined. By combining rigorous testing protocols with advanced logging interpretation, this chapter ensures that wells are placed into service with verified and reliable barriers capable of withstanding subsequent operational loads.

© The Author(s), under exclusive license to Springer Nature Switzerland AG 2026
A. Alsubaih and K. Sepehrnoori, *Well Integrity for Oil, Gas, and Subsurface Storage Operations*, https://doi.org/10.1007/978-3-032-15233-6_6

Wellhead (Bohrlochkopf) high-pressure valve assembly. *Source* Hartmann Valves (2008)

Chapter's Objectives

- Describe the function and types of downhole packers and seals, emphasizing seal integrity.
- Explain the operation and importance of Downhole Safety Valves (DHSVs) as critical primary barrier elements.
- Detail tubing hangers and their sealing mechanisms within the wellhead.
- Outline the components and integrity aspects of wellhead systems (API 6A).
- Describe Christmas Tree equipment and its role in surface control and barrier provision.
- Emphasize the importance of correct installation procedures for establishing initial integrity.

6.1 Introduction to Completion Equipment

Once the final casing string (production casing or liner) is cemented and initial integrity confirmed, the well is "completed" by installing specialized equipment inside the casing and at the surface. This completion equipment forms the primary conduit for production or injection and incorporates critical barrier elements. Key components include the production tubing string, downhole packer(s), downhole safety valve(s), tubing hanger, wellhead, and Christmas tree (Xmas tree). Ensuring the mechanical and sealing integrity of each component, and their proper installation and interface with other wellbore elements (casing, cement), is vital for safe, efficient, and reliable operation throughout the well's life. These components form the primary operational barriers, and their seals interact with the casing and wellhead, which are part of the secondary barrier envelope.

6.2 Packer Systems

A packer is a downhole tool designed to provide a pressure seal between the outside of the production tubing string and the inside of the casing (or sometimes open hole) below or within the production/injection zone (Gao et al., 2019; Gavioli & Vicario, 2012; Li et al., 2021; Sadana et al., 2020). Its primary functions are:

- To isolate the production/injection interval below or across the packer from the annular space above it (the tubing-casing annulus, or A-annulus).
- To force produced fluids to flow up the tubing, or injection fluids down the tubing, thereby protecting the production casing from direct exposure to reservoir pressure, corrosive fluids, high flow velocities, or operational pressure fluctuations (e.g., during stimulation).
- To allow pressure to be applied down the annulus (e.g., for gas lift, chemical injection) without communicating with the reservoir.
- To allow the separation of multiple production zones in a single wellbore (dual or multiple completions).
- To provide an anchor point for the tubing string to manage axial movement.

Packers are critical barrier elements, forming part of both the primary barrier envelope (by sealing the tubing to casing annulus) and the secondary barrier envelope (as the lower boundary for the sealed A-annulus). The integrity of the packer seal is paramount; failure constitutes a breach of the primary barrier, potentially leading to SCP in the A-annulus or unwanted crossflow (Hans-Emil Bensnes Torbergsen, 2012).

Types of Packers:

- **Permanent Packers**: Designed to be set once in the casing and remain in place for the life of the well, or until milled out during abandonment or a major workover (Hadzihafizovic, 2024). They generally offer higher performance ratings (pressure, temperature) and reliability for long-term applications due to their robust construction and fewer potential leak paths compared to retrievable types. Setting mechanisms can be electric wireline, hydraulic (pressure applied down the tubing), or hydro-static (using wellbore fluid pressure acting on a piston).
- **Retrievable Packers**: Designed to be unset and retrieved from the well using tubing manipulation (rotation, tension, compression) or specialized retrieving tools run on wireline or coiled tubing. They offer operational flexibility (e.g.,

for easy recompletion or repair) but may have lower performance ratings or potentially more complex sealing mechanisms with more leak paths than permanent packers. Setting mechanisms include mechanical (tubing manipulation), hydraulic, or hydrostatic.

Figure 6.1 shows common types of downhole isolation tools used to achieve zonal isolation and form pressure barriers during completion. The selection between mechanical, hydraulic, and retrievable packers—or bridge plugs—depends on operational objectives, anticipated well conditions, and whether future access or removal is required. Understanding their construction and activation mechanisms is essential for designing a robust barrier strategy. Setting Mechanisms: Determine how the packer's sealing elements and anchoring slips are activated to grip the casing and create the seal. Common types include mechanical (weight-set, tension-set, rotation-set), hydraulic (differential pressure applied), and hydro-static (using wellbore pressure acting on a piston).

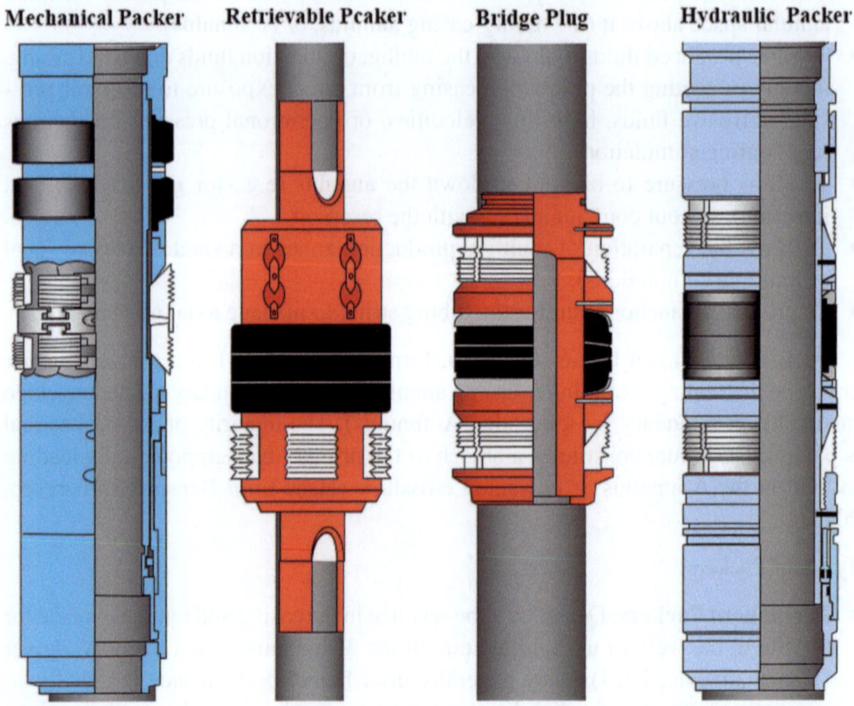

Fig. 6.1 Shows common types of downhole isolation tools used to achieve zonal isolation and form pressure barriers during completion. The selection between mechanical, hydraulic, and retrievable packers—or bridge plugs—depends on operational objectives, anticipated well conditions, and whether future access or removal is required. Understanding their construction and activation mechanisms is essential for designing a robust barrier strategy

Seal Elements: The heart of the packer's sealing capability. Typically comprise one or more elastomeric elements that are compressed and extruded outwards against the casing wall, often combined with non-elastomeric backup elements (e.g., made of plastic like PEEK, or metal) to prevent extrusion of the elastomer into the clearance gap under high pressure and temperature (Chen et al., 2022).

Elastomer Selection: Must be compatible with expected wellbore fluids (oil, gas, H_2S, CO_2, brine, water, inhibitors, completion fluids, injected EOR/Storage fluids), temperatures, and pressures over the well's life. Common elastomers include Nitrile (NBR), Hydrogenated Nitrile (HNBR), Fluoroelastomers (FKM/Viton®), and Perfluoroelastomers (FFKM/Kalrez®) for increasingly harsh conditions (See Chap. 11) (Kwatia et al., 2017).Elastomer compatibility is a critical factor in packer seal performance and longevity, especially in challenging wellbore environments involving brine, sour hydrocarbons, or high temperatures. Table 6.1 summarizes the performance characteristics of commonly used elastomers across key operational parameters. Selection must balance chemical resistance, mechanical properties, and cost depending on the application. ISO 23936 Part 2 provides guidance on elastomer selection and qualification for oil and gas environments (ISO 23936-2). Qualification standards like ISO 14310 (API 11D1) define rigorous testing protocols (including pressure and temperature cycling) and validation grades (e.g., V3, V0 - V0 indicating gas-tight sealing) for downhole seal assemblies to demonstrate performance under simulated downhole conditions (ISO 14310/API 11D1).

Seal Design: Can be single elements or multiple stacked elements with anti-extrusion features designed to handle high pressures and differential temperatures without failure (see Chap. 11).

Selection Criteria: Choosing the right packer is crucial for well integrity and operational success (Ceccarelli et al., 2009a, 2009b; Dethlefs & Chastain, 2012; Javora et al., 2006). Key factors include:

- Pressure Rating: Must exceed the maximum anticipated differential pressure across the seal (internal or external) during all operational modes, including stimulation or potential intervention pressures.
- Temperature Rating: Must be suitable for the maximum expected bottom-hole or flowing temperature, considering the long-term temperature limits and degradation characteristics of the seal elements.
- Fluid Compatibility: Elastomers and metallic components must resist degradation from wellbore fluids (H_2S, CO_2, inhibitors, EOR/Storage fluids, etc.) over time. Compatibility testing may be required.
- Casing Size and Weight: Must be designed for the specific casing dimensions and tolerances.
- Tubing Size and Weight: Must be compatible with the production tubing string.
- Operational Requirements: Permanent versus retrievable, desired setting mechanism, need for feed-through lines (e.g., for chemical injection, electrical submersible pump cables, fiber optics, control lines).

Table 6.1 Comparison of elastomer properties and chemical compatibility for downhole seal applications. Elastomer performance varies significantly based on fluid exposure, temperature, and well environment. Kalrez® and Aflas® offer superior resistance in harsh service conditions (steam, brine, sour crude), while NBR and HNBR are more cost-effective for moderate environments (Kwatia et al., 2017)

Property	NBR	HNBR	Viton® (FKM)	Aflas® (FEPM)	Kalrez® (FFKM)
Max. temperature (°F)	250	300	400	400	620
Tensile strength (psi)	200–3500	1500–3500	500–2000	1900	2000
Steam compatibility	Poor	Fair to good	Poor	Excellent	Excellent
Brine high density (Na/CaBr)	Poor	Fair to good	Excellent	Excellent	Excellent
Brine low density (Ca/NaCl)	Excellent	Excellent	Excellent	Excellent	Excellent
Crude oil, sour (<2000 ppm H_2S)	Poor	Excellent	Fair to good	Excellent	Excellent
Drilling mud, diesel based	Fair to good	Excellent	Fair to good	Fair to good	Excellent
Hydraulic fluid, oil/water (HFA)	Fair to good	Excellent	Poor	Excellent	Excellent

- Reliability and Validation: Preference often given to packers with proven field history or validated performance according to rigorous industry standards (e.g., ISO 14310/API 11D1 V0 rating for gas wells).

Failure of a packer seal element is a common cause of primary barrier failure, potentially leading to SCP or unwanted fluid communication between zones. Rigorous selection, installation, and testing (as part of initial completion testing—Chap. 7) are essential to minimize this risk.

Hydraulic set packers are actuated by applying differential pressure between the tubing and annulus, often using a plug or a dropped ball to isolate pressure below the packer. The differential pressure creates a piston force that drives the slips to bite into the casing and set the packer. This force must be calculated to ensure a reliable setting and prevent unintentional movement due to thermal or ballooning effects. For a single-diameter tubing completion, the additional axial force locked into the system during setting can be estimated by the following equation:

$$F_{\text{set}} = P_{\text{set}} A_i (1 - 2\mu) \tag{6.1}$$

The variable F_{set} represents the force applied at the set pressure, expressed in pounds-force (lbf), while P_{set} denotes the set pressure in pounds per square inch (psi). The term A_i refers to the internal cross-sectional area of the tubing, measured in square inches (in^2), and μ stands for Poisson's ratio, which typically has a value of 0.3 for steel.

6.3 Downhole Safety Valves (DHSVs)

The Downhole Safety Valve (DHSV), also known as a Subsurface Safety Valve (SSSV), is a critical safety barrier element installed in the production tubing string, typically several hundred feet below the surface or seabed within the well's pressure envelope. Its primary function is to automatically shut off flow up the tubing in the event of an emergency or failure of surface equipment (e.g., damage to the Xmas tree, riser failure offshore), preventing an uncontrolled release (blowout) from the tubing bore.

Functionality:

- Most DHSVs used in oil and gas wells are **surface controlled (SCSSV)**. They are held in the open position by hydraulic pressure applied from the surface through a dedicated small-bore hydraulic control line strapped to the outside of the production tubing. Figure 6.2 illustrates the internal mechanism of a surface-controlled subsurface safety valve (SCSSV). In the open position, hydraulic pressure applied via a control line overcomes the spring force and holds the flapper open. If hydraulic pressure is lost—intentionally during an emergency shutdown or unintentionally due to failure—the fail-safe spring automatically closes the flapper, providing a reliable primary barrier to isolate flow from the reservoir.
- They are designed to be fail-safe: if the hydraulic control pressure is lost (either intentionally by the control system during an emergency shut-down (ESD), or unintentionally due to control line damage, surface hydraulic leak, or power/control system failure), an internal mechanism (typically a spring and/or wellbore pressure acting on a piston) forces the valve closed, shutting off flow from the tubing bore. This fail-safe closure is a critical safety feature.
- The closure mechanism is usually a hinged flapper or a ball that rotates to seal against a polished seat ring.

Types:

- **Tubing-Retrievable SCSSV (TR-SCSSV)**: The valve body is an integral part of the production tubing string. Retrieving or repairing the valve requires pulling the entire tubing string to the surface (a major workover). These generally offer a larger bore for flow, potentially higher performance ratings, and potentially higher reliability due to fewer leak paths compared to wireline retrievable types. They are the preferred type in many critical applications.

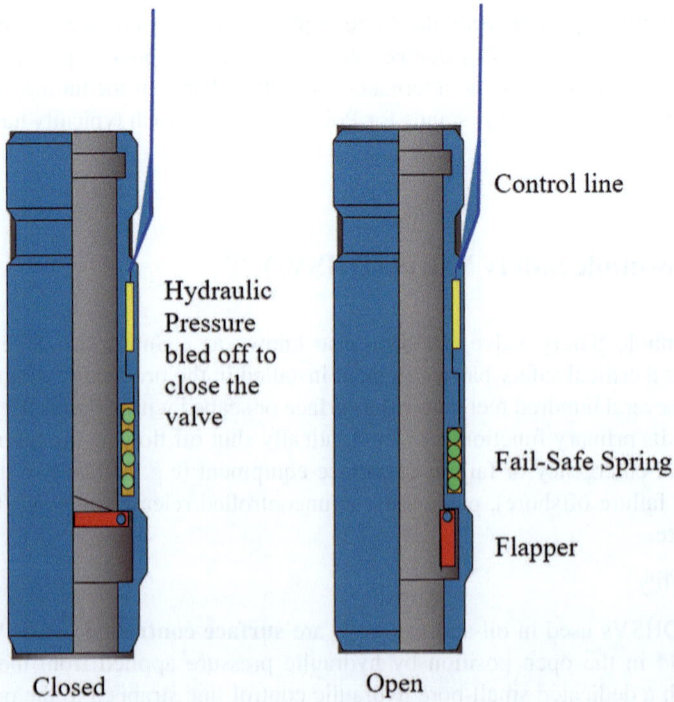

Fig. 6.2 Schematic of a Surface-Controlled Subsurface Safety Valve (SCSSV) in both closed (left) and open (right) positions. The valve is held open by hydraulic pressure from the surface via a control line. In the event of hydraulic pressure loss, a fail-safe spring automatically closes the flapper, isolating the tubing bore and preventing uncontrolled flow

- **Wireline-Retrievable SCSSV (WR-SCSSV)**: The valve mechanism is housed in a cartridge that can be installed and retrieved using wireline operations inside a dedicated landing nipple incorporated into the tubing string at the desired setting depth. This allows for easier repair/replacement without a full workover but introduces additional seal points (at the interface between the cartridge and the landing nipple) and typically has a smaller internal diameter (bore) than a TR-SCSSV, which can restrict flow or intervention access.

Surface Safety Valves (SSVs) are critical wellhead components designed to shut in the well in the event of emergency or control line failure. One important aspect of SSV design is determining the maximum depth at which the valve can be reliably set to fail-close. This is based on maintaining enough hydrostatic pressure in the control line to overcome spring tension and ensure valve closure when needed. If the valve is positioned too deep, the hydrostatic column can keep the valve open unintentionally. Therefore, the maximum fail-close depth must be calculated using the following relationship:

$$D_{\max} = \frac{P_{\text{vc}} - P_{\text{mc}}}{\rho_f} \tag{6.2}$$

In evaluating depth-related closure conditions, $D_{\max}$ indicates the maximum fail-close setting depth in feet (ft). The recorded valve closing pressure is represented by P_{vc} and is measured in pounds per square inch absolute (psia), while P_{mc} corresponds to the manufacturer's recommended closing safety margin in psi. Lastly, ρ_f is the density of the control line fluid, expressed in psi per foot (psi/ft).

Integrity Aspects:

- **Sealing Mechanism**: The flapper/ball and seat sealing surfaces must provide a tight shut-off seal against full well pressure when closed (sealing from below). Leakage past the closure mechanism (through-valve leak) constitutes a failure of this critical primary barrier element. **Regular leak testing (e.g., via inflow testing or pressure tests from above per API RP 14B) is required by regulations and standards** (Ref API RP 14B/ISO 10417).
- **Control System Integrity**: The hydraulic control line (typically stainless-steel tubing), surface control panel, and associated fittings must maintain pressure integrity to keep the valve open. A leak in the control line can prevent the valve from being held open or cause it to close inadvertently ("nuisance trip"). Control line material selection and connection integrity are important to prevent corrosion or leaks.
- **Fail-Safe Operation**: The closing mechanism (spring, piston, seals) must function reliably and close the valve within a specified time (e.g., typically < 30–60 s) when hydraulic pressure is removed. **Regular functional testing verifies this capability** (API RP 14B/ISO 10417).
- **Opening Capability**: The valve must reliably reopen when hydraulic pressure is reapplied. Obstructions (scale, debris, hydrates) in the tubing bore or component failure can prevent opening.
- **Body and Connection Integrity**: The valve body and its connections to the tubing string must withstand wellbore pressure, temperature, and loads.
- **Material Compatibility**: Valve components (body, seals, internal parts) must be compatible with wellbore fluids (H_2S, CO_2, etc.) and operating conditions.

DHSVs are mandatory safety devices in most offshore wells and many high-risk onshore wells, particularly those near populated areas or environmentally sensitive locations. Their reliable function, verified through regular testing and maintenance as part of the WIMS, is a cornerstone of well integrity and safety management. Failure to close or leakage when closed represents a significant safety risk.

6.4 Tubing Hangers and Annular Sealing

The tubing hanger is a crucial component that sits inside the wellhead (specifically, within the tubing head spool or tubing bonnet) and serves multiple vital functions related to well integrity and operation (Bellarby, 2009; Renpu, 2011). It lands the production tubing string within the wellhead.

Functions:

- **Support Tubing Weight**: It supports the hanging weight of the entire production tubing string, transferring the tensile load to the wellhead body, which is bolted to the surface casing.
- **Seal Tubing Annulus (A-Annulus)**: It provides seals (typically elastomeric or metal-to-metal) to prevent pressure and fluids from leaking up the annular space between the production tubing string and the production casing (the A-annulus) past the hanger into the atmosphere. This is a critical secondary barrier element at the surface, containing pressure that might build up in the A-annulus due to primary barrier failure (e.g., packer or tubing leak) or deliberate pressurization.
- **Provide Penetrations**: Often incorporates ports or conduits for the DHSV hydraulic control line, chemical injection lines, electrical cables (for Electrical Submersible Pumps—ESPs), or fiber optic lines to pass through from the tubing-casing annulus (A-annulus) to the surface control system or chemical injection skids. Reliable seals are required around these penetrations to prevent leakage.
- **Allow Access**: Provides controlled access to the production tubing bore (via the top profile) for installing the Xmas tree and performing wireline or coiled tubing interventions (after removing the tree or using a lubricator).

Sealing Mechanisms: The sealing interface between the tubing hanger and the internal bore of the tubing head spool is critical.

- **Elastomeric Seals**: O-rings or moulded seals are common, often with anti-extrusion backup rings. Material selection (HNBR, FKM, FFKM) is critical based on anticipated temperature, pressure, and fluid compatibility (especially H_2S, CO_2, EOR/Storage fluids) (See Chap. 11) (Khasanova et al., 2025; Slay & Ferrell, 2008; Xu et al., 2016).
- **Metal-to-Metal (MTM) Seals**: Increasingly used for HPHT or critical sour service wells where elastomer performance or longevity is uncertain or insufficient (van Dort, 2009). These require precise machining of the hanger and tubing head sealing surfaces and high actuation forces (usually via lockdown screws in the tubing head spool) to energize the seal by creating high contact stress (see Chap. 11). They offer higher reliability and longevity under severe conditions but are less forgiving of surface imperfections, damage, or debris.

The integrity of the tubing hanger seals is vital for maintaining the secondary barrier in the A-annulus at the surface. Leaks past the tubing hanger can pressurize the surface annulus (within the wellhead) and potentially lead to external leaks if the outer wellhead seals also fail. Verification typically involves pressure testing the A-annulus above the hanger seals after installation (Chap. 7).

6.5 Wellhead Systems (API 6A)

The wellhead is the pressure-containing equipment installed at the surface (or seabed for subsea wells) that provides the interface between the downhole casing strings and the surface facilities (via the Christmas tree). Figure 6.3 provides a cross-sectional and modular breakdown of a conventional surface wellhead assembly. It highlights the structural and functional relationships among the tubing head, casing spools, casing hangers, and wellhead body components. Understanding these elements is essential for ensuring both mechanical support and hydraulic integrity at the surface. It is a critical component of the secondary barrier envelope and supports the entire well structure from the surface.

Key Components:

- **Casing Head Housing (or Starter Head)**: The lowest part, typically attached to the surface casing string, often via a threaded connection or welding. It supports the subsequent casing string(s) via casing hangers landed within its profile.

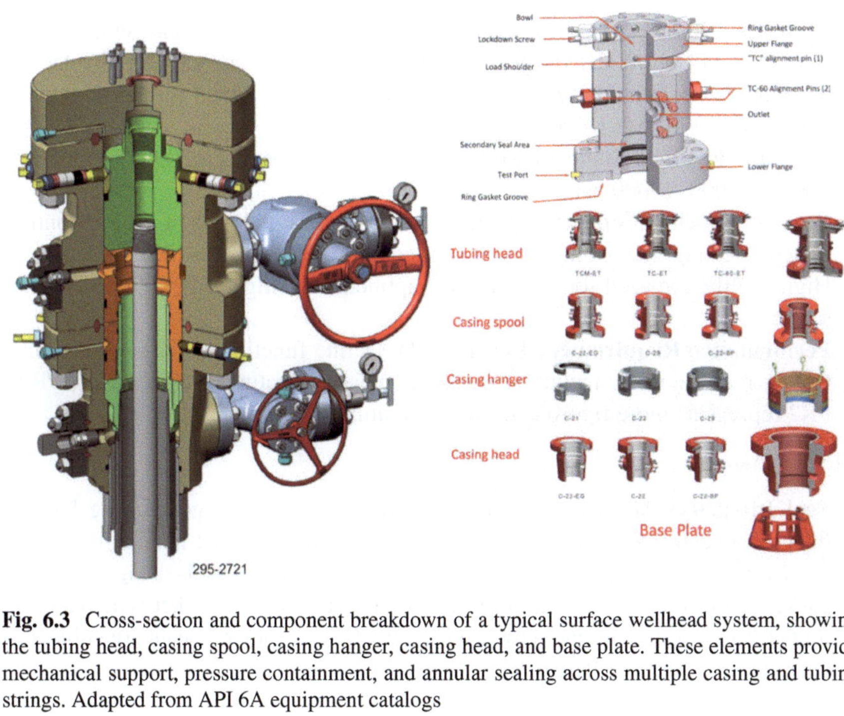

Fig. 6.3 Cross-section and component breakdown of a typical surface wellhead system, showing the tubing head, casing spool, casing hanger, casing head, and base plate. These elements provide mechanical support, pressure containment, and annular sealing across multiple casing and tubing strings. Adapted from API 6A equipment catalogs

- **Casing Spools**: Installed on top of the casing head housing or a previous spool. Each spool houses a casing hanger for the next casing string and provides side outlets (flanged or studded connections) for annulus access/monitoring (e.g., B-annulus outlet, C-annulus outlet). These outlets are fitted with valves.
- **Casing Hangers**: Sit inside the casing head or spools, supporting the hanging weight of a casing string and providing seals (primary and often secondary, often elastomeric or MTM) to isolate the annulus below from the annulus above. Mandrel-type hangers are common (landed on a shoulder); slip-type hangers grip the casing using wedges. Sealing relies on compression of elastomers or energization of MTM surfaces, often activated mechanically or by the casing weight.
- **Tubing Head Spool (or Tubing Bonnet)**: The uppermost spool (or an adapter on compact wellheads) that houses the tubing hanger. Provides side outlets for the production casing annulus (A-annulus) access, fitted with valves.
- **Connectors**: Flanges (most common, utilizing ring gaskets) or specialized clamp connectors are used to join the different housings and spools. Sealing relies on ring gaskets (typically metal API ring joints like R, RX, BX types for flanged connections, specified by API 6A) or specialized seals for clamped connections. The integrity of these connection seals is vital.

API Specification 6A: Specification for Wellhead and Christmas Tree Equipment (API 6A) is the governing standard for the design, manufacturing, and testing of surface wellhead and Christmas tree equipment. It defines requirements based on:

- **Material Classes**: Specifies materials based on temperature and retained fluid (e.g., AA, BB up to HH for general/sour service/high temp). Material selection depends on operating temperature, pressure, and H2S content (often referencing NACE MR0175/ISO 15156 for sour service).
- **Pressure Ratings**: Standard working pressure ratings (e.g., 2000, 3000, 5000, 10000, 15000, 20000 psi).
- **Product Specification Levels (PSL)**: Define increasing levels of quality control, material testing, traceability, and documentation requirements (PSL 1, 2, 3G, 4). Higher PSLs are used for more critical applications (higher pressure/temp, sour service).
- **Performance Requirement Levels (PR)**: Define functional and environmental testing requirements, including pressure and temperature cycling (PR1, PR2). PR2 represents more rigorous testing to simulate operational stresses.

Integrity Aspects:

- **Seal Integrity**: All seals within the wellhead (casing hanger, tubing hanger, flange/connector seals, valve stem packings on outlets) must maintain integrity against pressure, temperature, and wellbore fluids. Flange seals using metal ring gaskets require correct installation (clean grooves, correct gasket type/material, proper bolt torque sequence) to seal effectively. Elastomeric seals are subject to degradation over time (Chap. 11).

- **Structural Integrity**: The wellhead body and connectors must withstand the combined loads from hanging tubulars (axial tension), internal pressure, and external environmental loads (e.g., thermal expansion forces from tubulars, bending moments from risers on offshore platforms, wind loads). Advanced modeling (e.g., FEA) is sometimes employed, particularly for complex offshore wellhead systems, to assess structural stress and deformation under dynamic and static loads, ensuring adequate design margins for long-term reliability (Jaculli et al., 2020).
- **Annulus Access Valve Integrity**: Valves on the side outlets (A, B, C annuli) must seal correctly to allow safe monitoring (SCP), pressure testing, or intervention in the annuli. These are part of the secondary barrier envelope when the annulus is pressurized.

Regular inspection, maintenance (e.g., greasing valves, checking bolt torques, external corrosion control), and periodic pressure testing (where feasible and safe) are necessary to ensure ongoing wellhead integrity throughout the well's life.

6.6 Christmas Tree Equipment

The Christmas Tree (Xmas Tree or "tree") is the assembly of valves, spools, chokes, and fittings installed on top of the wellhead. Its primary purpose is to control the flow of fluids from the well (production) or into the well (injection), provide access for well interventions, and act as a crucial surface barrier system (Crumpton, 2018). Figure 6.4 shows a schematic of a surface Christmas Tree, highlighting the arrangement of critical valves and flow paths. Each valve plays a specific role in flow control, safety, and well intervention. Proper functioning of these components is essential for surface barrier integrity and operational flexibility throughout the life of the well. It is the interface between the well and the flowline/surface facilities.

Key Components (for a typical production tree):

- **Tree Connector**: Connects the base of the tree (often a Lower Master Valve or a Tree Connector Spool) to the top flange of the wellhead. Utilizes high-pressure flange or clamp connectors with metal ring gaskets or specialized seals.
- **Master Valves (Lower and Upper)**: Two valves in series on the main vertical bore (production path). The lower master valve is typically left open during normal operations and closed only for maintenance, major interventions, or emergencies as a safety barrier. The upper master valve is used more frequently for routine shut ins. Having two in series provides redundancy for this critical primary barrier function at surface.
- **Production Wing Valve(s)**: Valves on the side arms (wings) directing flow from the vertical bore to the production flowline. Control the flow rate during production and provide isolation.

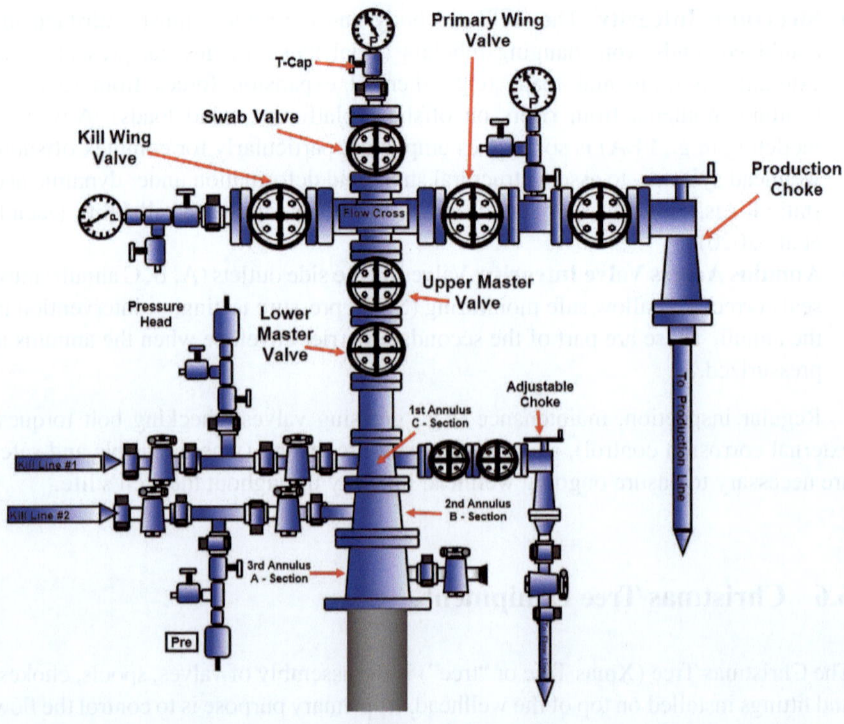

Fig. 6.4 Surface Christmas Tree schematic showing major components and valve locations. The tree includes master valves (upper and lower), wing valves (production and kill), swab valve, adjustable and production chokes, and annulus access points. These components collectively control flow, enable interventions, and serve as primary and secondary barrier systems at the surface

- **Swab Valve**: The valve at the very top of the tree, providing vertical access into the tubing bore for interventions (wireline, coiled tubing) when the master valves are closed. Acts as the upper boundary of the primary barrier envelope during these operations.
- **Choke**: A device, either fixed or adjustable (controlled manually or remotely), used to control the flow rate and reduce the pressure of fluids coming from the well. Can be part of the production wing or a separate flow loop. Subject to severe erosion.
- **Annulus Valves (e.g., Annulus Wing Valve, Annulus Master Valve, Annulus Swab Valve)**: Provide controlled access to the production casing annulus (A-annulus) via the tubing head outlet. These are part of the **secondary barrier envelope** when the A-annulus is pressurized. Similar valve configurations (master/wing) might be used on the annulus outlets for redundancy.
- **Sensors/Instrumentation**: Ports for pressure gauges, temperature sensors, chemical injection points, sampling points, etc.

Types: Can be composite (multiple stacked valve blocks and spools) or solid block (valves machined into a single block of steel, reducing the number of potential leak paths). Designs vary based on pressure rating, temperature, flow rates, complexity (e.g., dual completion trees), and location (surface vs. subsea). Subsea trees are significantly more complex, incorporating remote actuation, ROV interfaces, and sophisticated control systems (See Chap. 20).

Integrity Aspects:

- **Valve Sealing**: All valves (master, wing, swab, annulus) must provide reliable sealing when closed. The gate and seat sealing surfaces in gate valves (or ball/seat in ball valves) are susceptible to erosion (from high velocity flow or solids), corrosion pitting, or mechanical damage (e.g., closing on debris) over time, which can prevent a tight shut-off and lead to through-valve leakage–a failure of a critical barrier element (see Chap. 11). Regular testing (e.g., inflow tests, double block and bleed tests where possible) verifies sealing integrity.
- **Stem Packing**: Dynamic seals around the moving valve stems (gate or ball stem, actuator stem) are prone to wear and relaxation over time, leading to external leaks. Regular maintenance (e.g., injecting sealant/packing, tightening packing glands where possible) may be required, but eventually, replacement might be necessary (see Chap. 11). Elastomer degradation also affects stem packing longevity.
- **Body/Flange/Connector Integrity**: Similar to the wellhead, the tree body, flanges, and connectors must maintain structural and sealing integrity under pressure and environmental loads. API 6A specifications (Material Class, PSL, PR) apply to these components.
- **Actuation**: Surface safety valves (SSVs), often the upper master, production wing, or annulus valves, may be equipped with actuators (hydraulic, pneumatic, or electric) linked to the platform or facility Emergency Shut Down (ESD) safety system to provide automatic shut-in capability. The reliable function of the actuator and its control system is part of the overall safety barrier system.
- **Choke Integrity**: Subject to severe erosion from high velocity flow and solids, particularly for production trees. Material selection (hardened materials) and design (e.g., angle bean chokes, external sleeve chokes) are critical to maintain flow control and prevent erosion through the choke body.
- **Vibration**: High flow rates can induce vibrations in tree components and pipework, potentially contributing to fatigue damage over time.

Leaks in the wellhead or Christmas tree can range from minor drips to significant releases, posing safety and environmental risks. Maintaining the integrity of these surface components through regular inspection, testing, and maintenance, and prompt remediation of detected leaks, is crucial for upholding well integrity.

6.7 Installation Best Practices

Ensuring the integrity of completion equipment and the wellhead starts with proper installation. Even high-specification equipment can be compromised by poor handling, incorrect procedures, or insufficient attention to detail during assembly on the rig floor. Common issues arise from poor practices during the construction phase that introduce latent defects:

- **Handling Damage**: Nicking seal surfaces, damaging threads, bending components, or dropping equipment during transport, lifting, or rig floor handling. Protection of precision-machined sealing areas is paramount.
- **Cleanliness**: Debris (pipe dope, dirt, sand, metal shavings, welding slag) in seal areas, ring grooves, or threads can prevent proper sealing, cause galling during makeup, or damage surfaces. Thorough cleaning of all critical sealing interfaces before makeup or assembly is essential.
- **Correct Procedures**: Using the right thread compounds (correct type, amount, and application method), applying correct torque values and turns for connection makeup (using calibrated torque wrenches or makeup machines), following proper sequences for setting packers and hangers (e.g., applying correct set force), and adhering to procedures for installing ring gaskets and tightening flange bolts (correct bolt tensioning or torque sequence) are crucial. Manufacturer recommendations and operator specific procedures must be followed meticulously and documented. Failure to adhere to these procedures is a common root cause of initial and latent integrity issues.
- **Alignment and Stabbing**: Ensuring components are properly aligned before stabbing (e.g., stabbing the tubing hanger into the tubing head) to avoid damaging seals or threads. Use of guide tools or running guidelines is often required.
- **Testing**: Performing function tests (e.g., cycling valves, setting packers) and pressure tests at appropriate stages during installation verifies components are working correctly and seals are holding pressure *before* proceeding to the next step or handing the well over to operations (Chap. 7). Documenting these tests is vital.
- **Competency and Supervision**: Ensuring personnel performing critical installation tasks are competent, trained, and supervised adequately.

Competent personnel, adequate supervision, use of correct and calibrated tools, and adherence to detailed procedures and QA/QC processes are fundamental for establishing the initial integrity of the completion and wellhead system. Latent defects introduced during installation due to poor practices are notoriously difficult and expensive to identify and rectify later in the well's life.

References

Bellarby, J. (2009). Well completion design. *Developments in Petroleum Science, 56*, 726.

Ceccarelli, T., Albino, E., Watson, G., & Deffieux, D. (2009a). Deepwater completion designs: A review of current best practices. In *Society of Petroleum Engineers—SPE/IATMI Asia Pacific Oil and Gas Conference and Exhibition 2009, APOGCE 09* (vol. 1, pp. 392–413). https://doi.org/10.2118/122518-MS

Ceccarelli, T., Albino, E., Watson, G., & Deffieux, D. (2009b). Deepwater completion designs: A review of current best practices. In *Society of Petroleum Engineers—SPE/IATMI Asia Pacific Oil and Gas Conference and Exhibition 2009, APOGCE 09* (vol. 1, pp. 392–413). https://doi.org/10.2118/122518-MS

Chen, Y., Liu, X., & Li, C. (2022). Development of rubber packing element for 105 MPa/215 °C deep-well test packer. *Materials, 15*(6). https://doi.org/10.3390/MA15062024

Crumpton, H. (2018). Well control for completions and interventions. *Well Control for Completions and Interventions, 1–814*. https://doi.org/10.1016/C2014-0-00602-8

Dethlefs, J. C., & Chastain, B. (2012). Assessing well-integrity risk: A qualitative model. *SPE Drilling & Completion, 27*(02), 294–302. https://doi.org/10.2118/142854-PA

Gao, B., Wang, L., & Hu, T. (2019). Testing string dynamics when Jarring Stuck Packer. OnePetro. https://dx.doi.org/

Gavioli, P., & Vicario, R. (2012). The evolution of the role of openhole packers in advanced horizontal completions: From novel technology to a critical key to success. *SPE Drilling & Completion, 27*(01), 75–93. https://doi.org/10.2118/132846-PA

Hadzihafizovic, D. (2024). *The packer and recommended tourque*. https://papers.ssrn.com/abstract=5032187

Hans-Emil Bensnes Torbergsen, H. B. H. S. S. A. S. J. R. M. A. L. (2012). *An introduction to well integrity*.

Hartmann Valves. (2008). "Wellhead Bohrlochkopf." Wikimedia Commons, 30 June 2008. Licensed under CC BY 3.0. https://commons.wikimedia.org/wiki/File:Wellhead_Bohrlochkopf.JPG

Jaculli, M. A., Mendes, R. P., & Miura, K. (2020). Methodology for assessing wellhead integrity during its life cycle. http://asmedigitalcollection.asme.org/OMAE/proceedings-pdf/OMAE2020/84430/V011T11A033/6607289/v011t11a033-omae2020-18468.pdf

Javora, P. M., Stevens, R., Devine, C., Jeu, S., Simmons, M., Firmin, G., Poole, G., Franklin, B., Qu, Q. (2006). Deepwater completion challenges redefine best practices for completion and packer fluid selection. In *Proceedings—SPE Annual Technical Conference and Exhibition* (vol. 7, pp. 4415–4424). https://doi.org/10.2118/103209-MS

Khasanova, L., Koishymanova, S., Neelalohith, S., Fornitz, K. H., Kashefi, K., Tea, K., Rogowska, M., Mathiesen, T., Rasmussen, L., Yusoff, M. K., & Ismail, I. (2025). Material feasibility & survivability testing for carbon capture and storage well construction. In *SPE—International Association of Drilling Contractors Drilling Conference Proceedings*. https://doi.org/10.2118/223736-MS

Kwatia, G., Research Assistant Chinedum Ezeakacha, G., Research Assistant Saeed Salehi, G., & Investigator By, P. (2017). *Literature report of elastomer sealing materials and cement systems submission date: Project title: Studying "fitness for service" of the sealing assemblies and cement system in shallow well designs by conducting scaled laboratory testing, leakage modelling and risk assessment AUTHORS*.

Li, C., Guan, Z., Zhang, B., Wang, Q., Xie, H., Yan, Y., & Han, C. (2021). Failure and mitigation study of packer in the deepwater HTHP gas well considering the temperature-pressure effect during well completion test. *Case Studies in Thermal Engineering, 26*. https://doi.org/10.1016/J.CSITE.2021.101021

Renpu, W. (2011). *Advanced well completion engineering* (p. 736).

Sadana, A., Kovalchuk, A., & Cook, C. (2020). delayed oil swell packer for openhole zonal isolation of long laterals wells. In *International Petroleum Technology Conference 2020, IPTC 2020*. https://doi.org/10.2523/IPTC-19967-MS

Slay, J. B., & Ferrell, K. (2008). Proper performance testing to maintain seal integrity in deepwater completions. https://doi.org/10.4043/19626-MS

van Dort, R. (2009). Metal-to-metal seals meet downhole hazard demands. *Journal of Petroleum Technology, 61*(01), 24–26. https://doi.org/10.2118/0109-0024-JPT

Xu, Z., Zhao, L., & Prieto, C. A. (2016). Ultra-high temperature rated elastic carbon composite packer for reliable downhole zone isolation. In *Proceedings—SPE Annual Technical Conference and Exhibition.* https://doi.org/10.2118/181614-MS

Chapter 7
Initial Integrity Verification and Testing

Abstract This chapter explores the critical role of completion equipment and wellhead systems in establishing and maintaining well integrity throughout the life of the well. It begins with an overview of completion objectives and system classification (e.g., open hole vs. cased hole, single vs. multizone, and intelligent completions), highlighting how equipment selection directly influences barrier reliability and intervention capability. The chapter details the design, functionality, and installation practices of key completion components including packers, tubing, subsurface safety valves (SSSVs), flow control devices, and production liners. Emphasis is placed on the mechanical integrity, pressure ratings, corrosion resistance, and seal effectiveness of each component under expected downhole conditions. Wellhead systems and Christmas trees are also thoroughly examined in terms of structural design, load-bearing capacity, pressure containment, and barrier redundancy. API standards (e.g., API 6A, 17D) and testing requirements are integrated to ensure compliance and operational safety. The chapter includes failure case analyses related to completion leakage, packer slippage, tubing collapse, and wellhead seal erosion, supported by root-cause diagnostic techniques and integrity management workflows. Additional sections address the importance of elastomer compatibility, thermal cycling effects, and wellhead fatigue under cyclic loads. By consolidating engineering principles with practical field insights, the chapter equips engineers with a robust foundation for selecting, installing, and maintaining completion and wellhead systems that uphold barrier integrity under dynamic production, injection, and stimulation conditions.

© The Author(s), under exclusive license to Springer Nature Switzerland AG 2026

A. Alsubaih and K. Sepehrnoori, *Well Integrity for Oil, Gas, and Subsurface Storage Operations*, https://doi.org/10.1007/978-3-032-15233-6_7

Pressure gauge and filtration system used in surface well monitoring and control operations. *Source* The Authors

Chapter's Objectives

- Explain the principles and importance of pressure testing for barrier verification.
- Describe standard tests performed on casing strings after cementing.
- Detail Leak-Off Tests (LOT) and Formation Integrity Tests (FIT) and their significance.
- Outline integrity tests performed on the completion string, packer, DHSV, and wellhead.

- Emphasize the critical role of acquiring accurate baseline integrity data during construction.

7.1 The Principle of Pressure Testing

Pressure testing is the primary method used to verify the initial integrity (i.e., the sealing capability and structural soundness) of well barrier elements and envelopes after they are installed or modified during the construction phase (Henry et al., 2024). The basic principle involves applying a differential pressure across the component or system being tested and observing its ability to hold that pressure without significant leakage or failure (e.g., pressure drop, visible leak, deformation).

Key Aspects of Pressure Testing for Integrity Verification:

- **Test Medium**: Can be liquid (typically water, brine, drilling mud, or completion fluid) or gas (nitrogen or natural gas). Liquid tests are generally preferred for safety reasons (liquids are far less compressible, storing less hazardous energy than pressurized gas) and ease of leak detection (visible fluid or pressure drop) (Simpson, 2020). Gas tests are sometimes necessary to simulate operating conditions (gas wells) or test for gas-tight sealing required by standards (e.g., V0 packer rating) but require stricter safety precautions due to the higher stored energy and less visible nature of leaks (Zhang et al., 2020). API RP 7SC provides guidance on the safe use of gas for pressure testing (API RP 7SC).
- **Test Pressure**: The target pressure applied during the test. This is typically specified in the well program based on regulatory requirements, operator standards, and the maximum anticipated pressure the component or barrier might credibly experience during its life, often with a safety margin applied (e.g., testing to 80% of component yield strength or casing burst rating, or 1.2 times maximum anticipated operating pressure) (API R ECOMMENDED P RACTICE 100-1, 2015). Test pressures should not exceed the rated working pressure or design limits of the components being tested, considering potential weak points in the string.
- **Test Duration**: The length of time the test pressure is held. Durations vary depending on the component being tested, test type, wellbore volume, fluid compressibility, and regulatory requirements, typically ranging from 10 min to several hours (Islam et al., 2016; Shan et al., 2020). Longer durations provide more confidence, especially when looking for small leaks, as minor pressure changes become more apparent. Pressure should stabilize within a defined acceptable drop over the duration.
- **Success Criteria**: Clear, predefined criteria that define what constitutes a successful test. These usually involve maintaining the applied pressure within a specified tolerance (e.g., less than 5–10% pressure drop relative to the initial or stabilized pressure) over the defined duration, with no visible leaks (for surface equipment) or audible leak noise (EPA, 2006). The acceptance criteria must be

clearly documented and understood by all personnel performing the test. Deviations from acceptance criteria require investigation and corrective action; a failed test means the barrier is NOT verified as competent.

- **Safety**: Pressure testing involves significant stored energy and potential hazards. Operations must be conducted according to strict safety procedures, using properly calibrated gauges and pressure recording equipment, establishing appropriate exclusion zones (especially for gas testing), and having appropriate pressure relief systems in place to prevent over-pressurization. Personnel must be competent and aware of the risks.

Pressure testing provides a direct, quantitative measure of a barrier's ability to contain pressure at a specific point in time. It is a fundamental, mandatory verification step throughout the construction and initial completion phase, confirming that installed components meet the required sealing and structural performance standards before the well is put into operation.

7.2 Casing Pressure Tests

After a casing string is run into the hole and cemented (and the cement has reached sufficient compressive strength—WOC time elapsed), the casing string itself is pressure tested before drilling out the shoe track and the cement inside the casing shoe. This crucial test verifies the pressure integrity of the newly installed casing string and its components.

- **Purpose**: To confirm:
  - The pressure integrity of the casing pipe body along its entire length.
  - The pressure integrity of all the casing connections joining the individual pipe segments.
  - The integrity of the float collar and float shoe valves at the bottom of the string (by holding pressure from above and preventing flow downwards).
- **Procedure (Typical)**:
  1. Run the drill string or a dedicated test tool into the casing, often positioning it just above the float collar or cement plug inside the shoe track.
  2. Close the annular preventer or seal element of the Blowout Preventer (BOP) stack around the drill string or test tool. This isolates the casing annulus from the surface.
  3. Pump fluid (typically drilling mud, brine, or kill-weight fluid) into the casing string to increase the internal pressure up to the specified test pressure. This pressure is typically based on the Maximum Anticipated Surface Pressure (MASP) that the casing string might see during drilling the next section or during future operations, incorporating a design factor.
  4. Once the test pressure is reached, shut in the pressure source (pumps).

5. Hold the pressure for the required duration (e.g., 10, 30 min, or longer, depending on regulations/standards).
6. Continuously monitor and record the pressure on surface using calibrated gauges and preferably a pressure chart recorder or electronic data acquisition system.
7. Compare the pressure decline over the test duration to the predetermined acceptance criteria (Sect. 7.1).

A successful test confirms that the casing string can hold the required internal pressure without significant leakage, providing confidence in its integrity before drilling ahead into potentially higher-pressure formations or exposing the string to operational pressures. Failure to hold pressure indicates a leak path (most commonly a connection leak, or rarely, a pipe body defect) that must be investigated and potentially remediated before proceeding. This test verifies the pressure containment capability of a key secondary barrier element.

7.3 Leak-Off Tests (LOT) and Formation Integrity Tests (FIT)

After successfully pressure testing the casing string, drilling out the casing shoe, the cement plug inside the shoe track, and a short section of new formation below the shoe (typically 5–20 ft/1.5–6 m of open hole), a Formation Integrity Test (FIT) or a Leak-Off Test (LOT) is performed (Al Ramadan, 2024). These tests are critical for determining the maximum pressure that the formation *directly below the casing shoe* and the cement seal around the shoe can withstand without fracturing or leaking excessively into the formation. To better understand the interpretation of formation breakdown and fracture propagation behavior during pressure tests, Fig. 7.1 illustrates a representative pressure–volume plot from a combined FIT, LOT, and XLOT operation. The labeled points on the curve demonstrate critical values used in assessing formation strength and fracture behavior, including leak-off pressure (LOP/FIP), fracture propagation pressure (FPP), and instantaneous shut-in pressure (ISIP). This information is essential for safely drilling the next section of the hole and managing equivalent circulating densities (ECD).

- **Formation Integrity Test (FIT)**: A test where pressure is applied in the wellbore up to a predetermined limit (based on estimates of formation strength or the required mud weight/ECD for drilling the next section) and held for a short period (SPE, 2025). The primary goal is to *confirm* that the formation at the shoe and the cement seal can withstand this required pressure without breaking down or exhibiting significant leak-off. It does *not* intentionally fracture the formation. The test pressure is typically equivalent to the planned mud weight for the next section plus a safety margin, or a regulatory-defined minimum equivalent mud weight (Alsubaih et al., 2018, 2019, 2020, 2024a, 2024b, 2025).

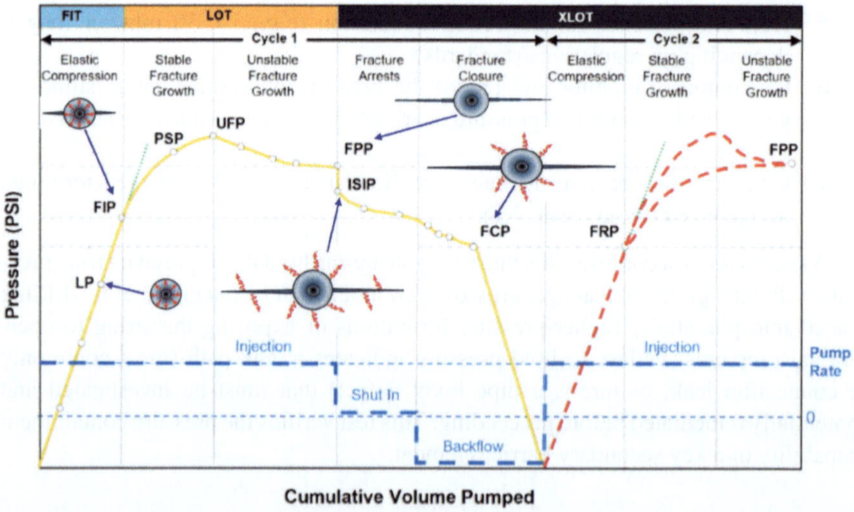

Fig. 7.1 Representative pressure versus cumulative volume curve showing the stages of Formation Integrity Test (FIT), Leak-Off Test (LOT), and Extended Leak-Off Test (XLOT). Key pressure points include leak-off pressure (LOP), fracture initiation pressure (FIP), fracture closure pressure (FCP), instantaneous shut-in pressure (ISIP), and fracture propagation pressure (FPP). The plot also distinguishes between elastic compression, stable and unstable fracture growth, fracture arrest, and closure, providing insight into subsurface mechanical responses

- **Leak-Off Test (LOT)**: A more extensive test where pressure is increased incrementally in the wellbore until the formation begins to accept fluid at a significantly higher rate than expected from compressibility, indicating either fluid permeating into the formation (leak-off) or the initiation of a hydraulic fracture. The pressure at which this deviation occurs is defined as the **Leak-Off Pressure (LOP)** or **Fracture Initiation Pressure (FIP)**. This pressure defines the upper limit of the wellbore pressure that the weakest point of the tested interval (the formation at the shoe or the shoe cement seal) can tolerate before breaking down (Addis et al., 1998). Pumping may continue slightly further to determine the Fracture Propagation Pressure (FPP).
- **Procedure (Typical)**:
    1. Ensure the casing shoe track is drilled out and a small, specified section of new formation is exposed below the shoe. The wellbore fluid is typically drilling mud or brine.
    2. Run the drill string into the wellbore and position the bit or a test tool just above the casing shoe.
    3. Close the Blowout Preventers (BOPs)—typically the pipe rams—around the drill pipe, sealing the annulus at the surface.
    4. Slowly pump drilling fluid (or other wellbore fluid) into the wellbore *down the drill pipe* at a constant, low rate, accurately monitoring the pump pressure

at the surface. The closed BOPs prevent fluid from circulating up the annulus, so pressure builds up below the rams.

5. Plot the surface pressure increases against the volume of fluid pumped. Initially, the pressure increases linearly as the incompressible fluid in the sealed volume below the BOPs is compressed.
6. For a **FIT**, stop pumping when the predetermined target test pressure is reached and hold this pressure for the required duration (e.g., 10–15 min), monitoring for pressure decline. A successful FIT shows minimal pressure drops.
7. For a **LOT**, continue pumping slowly beyond the linear pressure increase. The pressure plot will deviate from the linear trend as the fluid begins to leak off into the formation or a fracture initiates. The pressure at this deviation point is the LOP/FIP. Pumping may continue slightly further.
8. Carefully bleed off the pressure from the wellbore in a controlled manner.

- **Importance of Integrity**:
  - The LOT/FIT result provides a critical data point for safe drilling operations. It determines the Maximum Allowable Equivalent Circulating Density (MAECD) or Maximum Allowable Mud Weight (MAMW) that can be used while drilling the next hole section *without risking formation fracture at the shoe*. Exceeding this limit with mud weight or ECD can lead to lost circulation, a well control issue if a permeable zone is exposed higher up with insufficient hydrostatic pressure, or even an underground blowout.
  - It provides verification of the cement seal integrity around the casing shoe. A very low LOP/FIP result might indicate a poor cement job or channel around the shoe allowing premature leakage into the formation at low pressures, which could require remedial action.

LOT/FIT results are critical inputs for safe drilling procedures and represent a direct verification of the combined barrier provided by the formation strength and the cement seal at the casing shoe–a key element of the secondary barrier system during the drilling phase. Before drilling the next section of the well, the **maximum allowable mud weight** that can be safely used is calculated based on the formation strength determined by the LOT or FIT. This weight is expressed as an equivalent mud weight (EMW), a useful parameter for planning drilling operations. The EMW is calculated by converting the measured leak-off or fracture initiation pressure to an equivalent mud density.

$$\text{EMW} = \frac{P_{\text{LOP}}}{0.052 \cdot \text{True Vertical Depth}} \tag{7.1}$$

where, EMW represents the Equivalent Mud Weight, expressed in pounds per gallon (ppg), and P_{lop} refers to the Leak-Off Pressure or Fracture Initiation Pressure, measured in pounds per square inch (psi). The True Vertical Depth corresponds to the measured vertical depth of the well in feet (ft). The constant 0.052 is a unit conversion factor that relates psi to ppg·ft and is used to align pressure and mud weight units.

This formula provides a direct estimate of the equivalent mud weight that can be used during subsequent drilling operations without fracturing the formation at the casing shoe.

7.4 Completion Integrity Tests

After the completion equipment (production tubing, packer, DHSV, etc.) is installed inside the production casing, a series of pressure tests are performed to verify the integrity of the primary barrier envelope components and associated seals before initiating production or injection (Setiawan et al., 2016; Varma et al., 2024). These tests confirm that the system designed to contain and control the flow path is sealed as intended.

- **Tubing String Test**: The production tubing string itself is pressure tested to confirm the integrity of the pipe body and, crucially, all the tubing connections (which are subjected to high internal pressures and potential fatigue during operations). This is typically done by setting a temporary test plug or isolation device near the bottom of the tubing string (e.g., in a profile nipple, or above the packer) and applying pressure from the surface down the tubing (Setiawan et al., 2016; Yugay et al., 2023a, 2023b). Alternatively, if the DHSV is landed, the test may be performed against the closed DHSV (provided the DHSV's test rating allows).
  - **Purpose**: To verify that the primary conduit for production/injection is pressure tight.
  - **Procedure**: Set the plug or close the valve. Apply internal pressure up to the specified test pressure (e.g., based on MAASP or component ratings). Hold pressure for the duration. Monitor pressure drops against acceptance criteria.
- **Packer Seal Test**: The integrity of the seal provided by the production packer between the tubing and the production casing is verified (Yugay et al., 2023a, 2023b). This is a critical test of the primary barrier envelope's lower component. Figure 7.2 illustrates a typical pressure test diagram during a packer seal verification test, showing the pressure buildup, hold period, and the stability of pressure over time that indicates a successful test according to defined acceptance criteria.
  - **Purpose**: To confirm the packer is effectively isolating the annulus above from the zone below.
  - **Procedure**: Typically involves applying pressure to the A-annulus (between tubing and production casing) via the annulus outlet valves on the wellhead, up to a specified test pressure (often based on the packer's rating or potential future annulus pressure build-up—APB). The pressure is held, and the annulus volume/pressure is monitored for leaks (e.g., pressure drop, or monitoring for flow into the tubing below the packer, if possible, although direct inflow testing

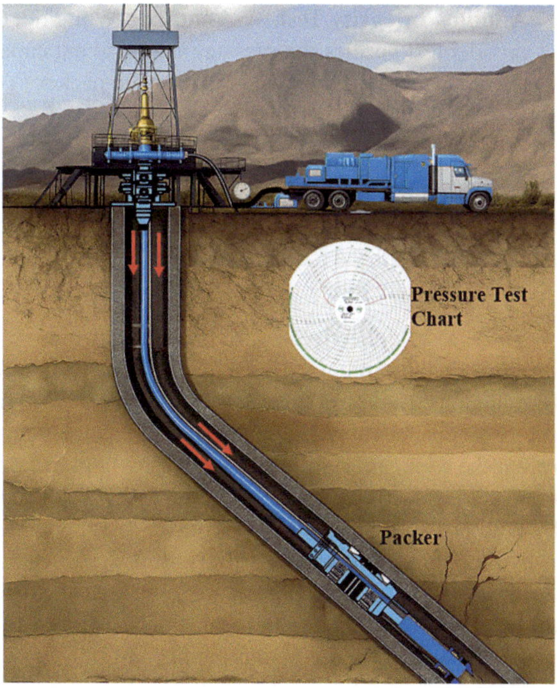

Fig. 7.2 Example pressure test plot for a packer seal integrity test. The chart displays the applied pressure profile over time, with a stable hold period indicating no significant leak or pressure drop, thereby confirming the effectiveness of the packer seal as part of the primary well barrier

is less common for packer seals). Alternatively, in some cases, pressure might be applied down to the tubing below the packer (if access is possible, e.g., via a sliding sleeve or bullheading) and the annulus monitored for pressure increase.

- **Downhole Safety Valve (DHSV) Test**: DHSVs are safety–critical devices requiring specific testing after installation.
  - **Functional Test**: Verifying the fail-safe operation. The hydraulic control pressure is bled off at the surface, and the valve is observed (e.g., via pressure monitoring, flow check, or audible indication) to close fully and within the specified time. Hydraulic pressure is then reapplied, and the valve is verified to reopen (Eni Norge, 2012).
  - **Leak Test**: Verifying the sealing integrity of the closed valve against pressure from below (reservoir side). This is a direct test of its barrier function. Typically performed by applying pressure from the surface down the tubing against the closed DHSV and monitoring for leakage through the valve (e.g., monitoring for pressure build-up below the valve if accessible, or monitoring for flow at the surface after bleeding pressure above the valve). Acceptance criteria often specify a maximum allowable leak rate (e.g., per API RP 14B) (API RP 14B/ ISO 10,417).

- **Annulus Integrity Test (A-Annulus)**: While related to the packer seal test, this broader test verifies the pressure containment capability of the entire A-annulus envelope from the wellhead down to the packer (Yugay et al., 2023a, 2023b).
  - **Purpose**: To verify the secondary barrier between the production tubing and production casing from the packer up to the wellhead.
  - **Procedure**: As with the packer test, pressure is applied to the A-annulus via the wellhead outlet up to a specified test pressure (often based on the casing burst rating, wellhead rating, or anticipated APB) and held for a duration, monitoring for pressure drop.

These tests collectively verify the initial integrity of the primary production/injection conduit (tubing, packer, DHSV) and its interface with the secondary barrier (production casing, wellhead seals). In cases where pressure tests fail or show unexpected trends, advanced diagnostics such as temperature and acoustic logging can reveal the root cause. Figure 7.3 presents field examples of integrity failures identified during testing, including a packer leak and tubing string leakage confirmed by both pressure behavior and acoustic-log anomalies. During packer seal testing, it is important to assess the **differential pressure** acting across the packer's seal elements. This differential pressure represents the driving force that could compromise the seal integrity, and it forms the basis for evaluating whether the packer can maintain isolation under operational conditions.

$$P_{\text{packer}} = P_{\text{A-annulus}} - P_{\text{tubing}} \tag{7.2}$$

where, ΔP_{packer} denotes the differential pressure acting across the packer seal in pounds per square inch (psi). This differential arises between the **A-annulus pressure**, represented by $\mathbf{P_{a\text{-}annulus}}$, and the internal pressure in the production tubing, P_{tubing}, both also measured in psi.

A successful packer seal test confirms that ΔP_{packer} is safely contained within the packer's rated pressure differential capacity.

During initial completion and mechanical barrier verification, tools such as packers and mechanical anchors apply compressive axial forces that are transferred to the casing through slips. These slips engage with the casing wall using a tapered cone mechanism, creating radial outward forces that can contribute to localized casing expansion or burst-like loading. This effect must be evaluated, particularly in HPHT wells or wells with aging casings. The following equations describe the induced radial force and its contribution to total burst pressure at the slip location.

Radial force applied to casing due to slips:

$$F_r = \frac{F_a(1 - \mu \tan\alpha)}{(\mu + \tan\alpha)} \tag{7.3}$$

Resulting burst pressure on the casing:

$$p_{\text{burst}} = \Delta p_{\text{casing}} + \frac{F_r}{\text{Slip Area}} \tag{7.4}$$

(a)

(b)

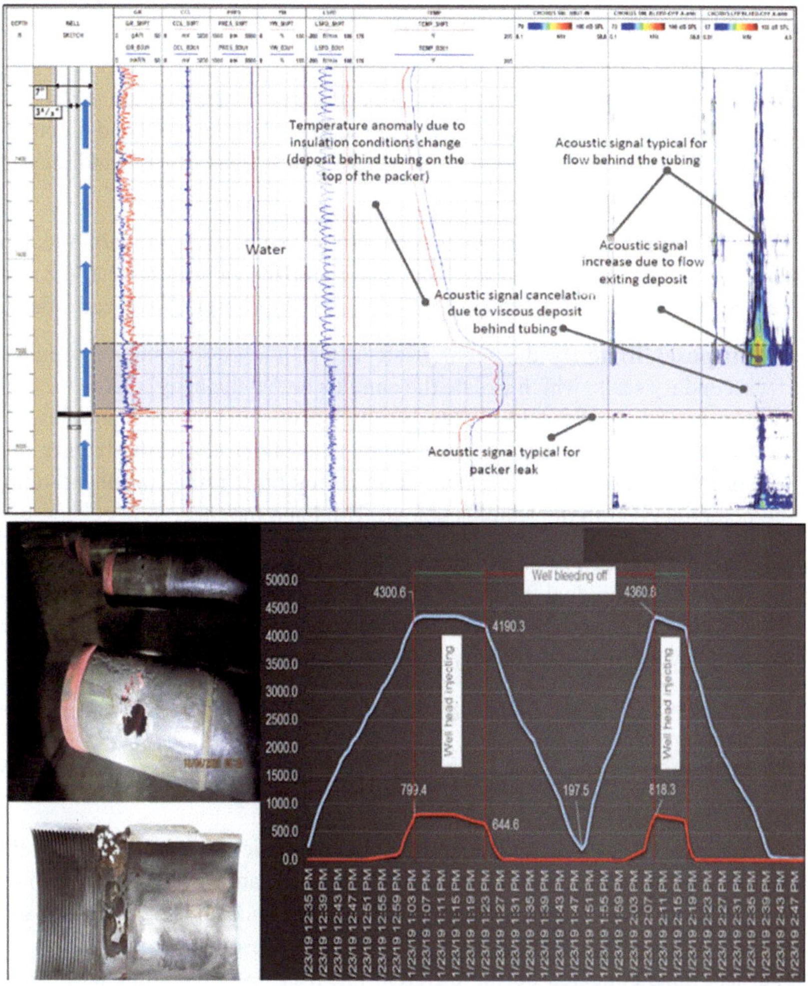

Fig. 7.3 **a** Acoustic and temperature log interpretation indicating a packer leak and flow behind tubing in a water-injected well. Notable signals include acoustic suppression due to viscous deposits and a temperature anomaly above the packer. **b** Evidence of tubing leakage identified through repeated pressure testing and wellhead bleed-off response, with visual confirmation of localized corrosion on the tubing body. These observations emphasize the importance of integrating pressure testing with diagnostic tools for early detection of completion failures (Yugay et al., 2023a, 2023b)

where the **radial force**, denoted as $\mathbf{F_r}$, and the **axial load**, $\mathbf{F_a}$, are both expressed in pounds-force (lbf). The term $\boldsymbol{\mu}$ represents the coefficient of friction between contacting surfaces, and $\boldsymbol{\alpha}$ is the cone angle of the slip in degrees. The pressure differential across the casing, **Δp_casing**, is expressed in psi, and the **Slip Area** refers to the contact surface area of the slips, given in square inches (in^2).

7.5 Wellhead and Annulus Seal Tests

Beyond the A-annulus, the sealing integrity of the intermediate and surface casing annuli (B-annulus, C-annulus, etc.) and the seals within the wellhead assembly are also typically verified during the construction phase (Das & El-Mandrawy, 2014; Yin & Gao, 2014; D. Zhu et al., 2021).

- **Casing Hanger Seal Tests**: After each casing string is run, landed in its hanger, and the hanger seal is set (mechanically energized or weight-set), the annulus *above* the hanger seal is pressure tested via the appropriate wellhead side outlet.
  - **Purpose**: To verify the sealing integrity of the casing hanger primary and secondary seals, which isolate the annulus below the hanger from the annulus above. This verifies a key secondary barrier element for each annulus.
  - **Procedure**: Apply pressure to the annulus outlet above the hanger up to a specified test pressure (e.g., based on casing burst rating, wellhead rating, or potential APB in the annulus below). Hold and monitor.
- **Wellhead Connection Tests**: During the installation (stack-up) of the wellhead components, individual flange or connector seals (using ring gaskets or other seal types) may be tested (often via dedicated test ports) as they are assembled to ensure they are sealing correctly before proceeding to the next step.
  - **Purpose**: To verify the integrity of the seals between wellhead spools and housings.
  - **Procedure**: Apply pressure to the test port on the flange or connector and monitor for leakage.

These tests verify the initial integrity of the static seals within the wellhead structure itself and the isolation between successive annuli. A variety of pressure test records are used to validate the sealing performance and structural capacity of key well barrier components. Figure 7.4 presents representative pressure charts and digital test plots collected during different integrity tests, including surface equipment, completion string, annuli, packer, and casing. When testing the **annulus seal integrity** (such as in casing hanger or wellhead connection tests), the applied **test pressure** must be carefully selected based on expected operational pressures and safety factors. This ensures the test is rigorous enough to verify integrity but not excessive enough to damage the component.

$$P_{\text{test}} = P_{\text{op}} \cdot \text{Test Factor} \tag{7.5}$$

where the variable $\mathbf{P_{test}}$ corresponds to the applied test pressure in pounds per square inch (psi), while $\mathbf{P_{op}}$ indicates the maximum anticipated operating pressure under standard conditions. The **Test Factor** is a dimensionless safety multiplier, typically ranging between 1.1 and 1.5, used to ensure the tested system can safely withstand pressures beyond expected operational loads. This relationship defines a conservative and safe approach to verifying the wellhead and annulus seals.

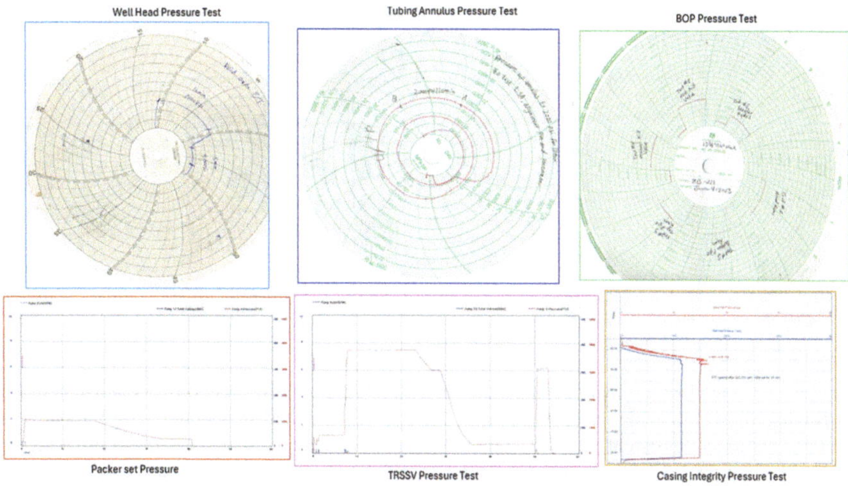

Fig. 7.4 Collection of pressure test records and charts. Top row (analog chart recorders): Wellhead pressure test, Tubing Annulus pressure test, and Blowout Preventer (BOP) pressure test. Bottom row (digital plots): Packer set pressure test, Tubing Retrievable Subsurface Safety Valve (TRSSV) test, and casing integrity pressure test. These charts are used to evaluate whether each component meets defined pressure-holding requirements during installation and prior to well operation

7.6 Importance of Baseline Data Acquisition

Beyond simply passing or failing a test, the initial integrity verification phase during construction is absolutely crucial for acquiring comprehensive and accurate baseline data. This data represents the 'as-built' integrity state of the well and is invaluable for future integrity management, monitoring, diagnostics, and remediation efforts throughout the well's life. Key baseline data includes:

- **Accurate Test Records**: Documenting the exact test pressures applied, durations, acceptable and actual pressure declines, test fluids used, environmental conditions, and results for *all* pressure tests performed (casing tests, LOT/FIT, completion tests, annulus tests, valve tests). These records should be signed off and archived in the well file and WIMS database.
- **Cement Evaluation Logs**: Obtaining good-quality CBL/VDL and/or ultrasonic logs provides a baseline map of the cement quality, bond strength (casing-cement), and presence of potential channels or voids behind the casing immediately after the cement has set. This serves as the reference point for assessing cement degradation later (Chap. 12).
- **Caliper and EM Thickness Logs**: Running baseline caliper logs (MFC) or electromagnetic thickness logs inside critical casing strings document the initial internal diameter, ovality, and wall thickness distribution. This provides a crucial reference for measuring cumulative wear or internal corrosion metal loss during subsequent inspection logs (Chap. 12). Figure 7.5 shows a combination of

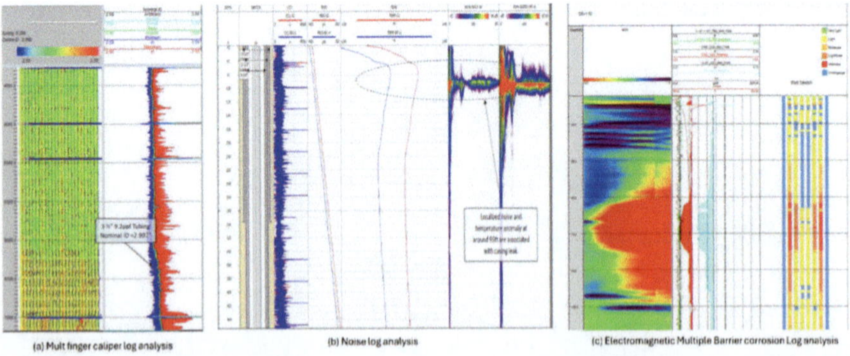

Fig. 7.5 Advanced well integrity diagnostics using multiple logging technologies. **a** Multi-finger caliper log showing casing deformation and ovality; **b** Noise and temperature log indicating a localized anomaly consistent with a casing leak; **c** Electromagnetic multiple barrier corrosion log identifying internal and external casing corrosion and metal loss. Together, these tools provide a comprehensive view of barrier degradation and help pinpoint leak locations and severity (Varma et al., 2024)

advanced log interpretations used to identify casing deformation, leaks, and corrosion across multiple barriers.

- **Initial Annulus Pressures**: Recording any stabilized pressures observed in all accessible annuli after completion, but before operations commence. This establishes the initial condition of the sealed annuli.
- **'As-Built' Documentation**: Ensuring all design drawings, construction records (drilling reports, cementing reports, tubular running reports, completion installation reports), equipment serial numbers, material certificates, and test certificates are complete, accurate, and readily accessible within the well file and WIMS database. This documentation provides the context needed to understand the well's construction history and interpret future integrity data.

This baseline data is invaluable for long-term integrity management. Comparing later test results, log data, or monitoring trends against the baseline allows engineers to definitively identify changes or degradation over time (e.g., increasing pressure decline during tests, bond deterioration on cement logs, increasing internal diameter/metal loss on corrosion logs, development of SCP). Without accurate baseline data, interpreting subsequent integrity assessments becomes significantly more difficult, potentially leading to misdiagnosis or delayed action. Establishing this comprehensive initial integrity fingerprint is a critical investment in the long-term, time-dependent management of the well's integrity.

References

Addis, M. A., Hanssen, T. H., Yassir, N., Willoughby, D. R., & Enever, J. (1998). A comparison of leak-off test and extended leak-off test data for stress estimation. *SPE/ISRM Rock Mechanics in Petroleum Engineering*. https://doi.org/10.2118/47235-MS

Al Ramadan, M. (2024). Overview of leak-off test and formation integrity test: Test interpretation and pitfalls. https://doi.org/10.1021/acsomega.4c01385

Ahmed, A. A., Hayfaa, L. S., Nuhad, A., Erfan, M. A. lawe, Ahmed, I. A., Swadi, M., & Tariq, K. K. (2019). Investigation of the rock strength anisotropy on the wellbore stability analysis. OnePetro. https://dx-doi-org.ezproxy.lib.utexas.edu/

Alsubaih, A. A., Hayfaa, S., Alboshawi, H. S., Alaa, A., Al, lawe, Erfan, M., Sawdi, M., Al, A., & Lubna, H. (2020). Reservoir geomechanics simulation to optimize reservoir management in southern Iraq fields. OnePetro. https://dx.doi.org/

Alsubaih, A., Albadran, F., Abbood, N., & Alkanaani, N. (2018). Optimization of the tripping parameters to prevent the surge related wellbore instability. OnePetro. https://dx-doi-org.ezproxy.lib.utexas.edu/

Alsubaih, A., Sepehrnoori, K., & Delshad, M. (2024a). Environmental impacts of orphaned and abandoned wells: Methane emissions, and implications for carbon storage. *Applied Sciences, 14*(24), 11518. https://doi.org/10.3390/APP142411518

Alsubaih, A., Sepehrnoori, K., Lopez Manriquez, A., Espinoza, D. N., Allawi, H. R., & Hayfaa, S. (2024b). A statistical approach to estimate the drilling mud window based on field data and ten failure criteria in depleted reservoirs—New modified failure criterion. https://doi.org/10.56952/ARMA-2024-0799

Alsubaih, A., Sepehrnoori, K., & Delshad, M. (2025). Development and optimization of self-healing cement for CO_2 injection and storage wells: enhancing long-term wellbore integrity in extreme subsurface conditions. *Applied Sciences, 15*(10), 5428. https://doi.org/10.3390/APP15105428

API R ECOMMENDED P RACTICE 100-1. (2015). *Hydraulic fracturing-well integrity and fracture containment.*

Das, P. K., & El-Mandrawy, W. M. (2014). A safe and efficient method of surface interception to access outermost sealed annulus charged with abnormal gas pressure—Relieves well integrity concerns. In *Society of Petroleum Engineers—International Petroleum Technology Conference 2014, IPTC 2014—Innovation and Collaboration: Keys to Affordable Energy* (vol. 4, pp. 3494–3501). https://doi.org/10.2523/IPTC-18122-MS

Eni Norge. (2012). *An introduction to well integrity.*

EPA. (2006). *Ground water section guidance no. 39.* https://www.epa.gov/sites/default/files/2017-03/documents/r8_guidance_39_-_miti.pdf

Henry, O. I., Chukwuemeka, O. E., & Femi, B. O. (2024). Well integrity management and optimization: A review of techniques and tools. *International Journal of Scholarly Research and Reviews, 5*(1), 079–087. https://doi.org/10.56781/ijsrr.2024.5.1.0041

Islam, M. A., Helstrup, O. A., Moi, S., & Carlsen, L. A. (2016). Automated pressure integrity testing APIT-A step change approach. In *Society of Petroleum Engineers—Abu Dhabi International Petroleum Exhibition and Conference 2016*. https://doi.org/10.2118/183273-MS

Setiawan, T., Ghazali, R. B., Granados, L. P., Chandrakalatharan, J., Zubbir, A. U., Hanafi, M. M. M., & Sepulveda, W. (2016). Samarang downhole integrity diagnostic techniques. In *Society of Petroleum Engineers—IADC/SPE Asia Pacific Drilling Technology Conference*. https://doi.org/10.2118/180680-MS

Shan, Y., Gao, W., Wang, X., Xu, M., Cao, C., Zhang, C., Du, H., Jin, X., Su, X., & Wang, W. (2020). Well integrity-sealant pressure barrier for high delta P, V0 sealing and long durability. In *Society of Petroleum Engineers—SPE Asia Pacific Oil and Gas Conference and Exhibition 2020, APOG 2020*. https://doi.org/10.2118/202371-MS

Simpson, D. A. (2020). Comparative risks of hydrostatic and pneumatic pipeline testing. *Journal of Pressure Vessel Technology, Transactions of the ASME, 142*(5). https://doi.org/10.1115/1.4046981

SPE. (2025). *Formation integrity test*. https://doi.org/10.2118/PW0386

Varma, E., Piccone, G., Alhammadi, A., Salah, M., Emam, A., Azher, M., Al Hammadi, M. M., Alshehhi, H. M. A. J., Mohamad, A. A., Marwat, Q., Gul, N., & Kumar, R. (2024). Rigless well intervention using tubing patch technology helps restore inactive wells. In *Society of Petroleum Engineers—ADIPEC*. https://doi.org/10.2118/222604-MS

Yin, F., & Gao, D. (2014). Improved calculation of multiple annuli pressure buildup in subsea HPHT wells. In *Society of Petroleum Engineers—IADC/SPE Asia Pacific Drilling Technology Conference 2014: Driving Sustainable Growth Through Technology and Innovation* (pp. 820–828). https://doi.org/10.2118/170553-MS

Yugay, A., Alkarbi, M. A. M. A., Berdiyev, Y., Utama, I., Al Hamadi, M. T., Hassan, K., Al Marzooqui, A., Alshehhi, H. M. A. J., Rincon, G., Al Menhali, S., Ramadan, T., Al Baloushi, M. S., Elhanbouley, S. H., Metlyaev, E., Respati, I., Eliatar, R., & Setioko, E. (2023a). Rigless repair of packer leak via annular cement plug. In *SPE Middle East Oil and Gas Show and Conference, MEOS, Proceedings*. https://doi.org/10.2118/213733-MS

Yugay, A., Zhukin, A., Vorozhikhin, S., Girgis, F., & Alkarbi, M. A. (2023b). Optimization of the well diagnostics investigation. In *Society of Petroleum Engineers—ADIPEC, ADIP*. https://doi.org/10.2118/216484-MS

Zhang, Q., Sun, S., & Yang, W. (2020). *Simulated comprehensive test system for gas sealing performance (2020) | Zhang Qiang*. https://scispace.com/papers/simulated-comprehensive-test-system-for-gas-sealing-42fh285t18

Zhu, D., Fan, Y., Zhang, H., Li, Y., Zhang, L., Wang, C., Wang, X., Chen, H., Lu, L., & Duan, Y. (2021). Integrity evaluation and management in high temperature high pressure and high production rate wells in Southwest China. In *Society of Petroleum Engineers—SPE 2021 Symposium Compilation. SYM*. https://doi.org/10.2118/208443-MS

Chapter 8
Annulus Pressure Monitoring and Management

Abstract This chapter investigates the most prevalent well integrity failures originating from the construction phase, emphasizing how design, material, and execution flaws can lead to long-term barrier degradation. Drawing from field statistics, forensic analyses, and post-abandonment studies, the chapter categorizes failures into mechanical, hydraulic, and chemical types—including poor cement bonding, casing collapse, micro-annulus formation, channeling, annular pressure buildup, and wellhead seal erosion. Each failure mode is traced to its root cause, such as improper centralization, inadequate mud removal, poor cement slurry design, casing wear, poor torque and drag modeling, and ineffective pressure testing. The chapter uses failure trees and fault-mode-effect analyses (FMEA) to illustrate how minor construction oversights can propagate into significant integrity breaches. Case studies demonstrate how substandard cement placement in high-deviation wells, under-designed casing strings in HPHT conditions, or failure to identify gas-cut cement can result in costly interventions, sustained casing pressure (SCP), and regulatory non-compliance. Attention is also given to failures associated with premature degradation from CO_2, H_2S, and brine exposure in early service life. Corrective recommendations are provided, linking failures to preventive measures such as risk-based cement design, load path modeling, and enhanced diagnostics during well construction. By uncovering the causality between early-phase errors and later-life failures, this chapter establishes a feedback loop for continuous improvement in well design and execution.

© The Author(s), under exclusive license to Springer Nature Switzerland AG 2026
A. Alsubaih and K. Sepehrnoori, *Well Integrity for Oil, Gas, and Subsurface Storage Operations*, https://doi.org/10.1007/978-3-032-15233-6_8

Surface pressure gauges during integrity monitoring and pressure recording at the wellsite. *Source* The Authors

Chapter's Objectives

- Define Sustained Casing Pressure (SCP) and explain its significance as a key integrity indicator.
- Identify potential sources of annular pressure, distinguishing between thermal and leakage sources.
- Describe methods for monitoring annulus pressures.
- Explain diagnostic tests used to characterize SCP (bleed-off/build-up).
- Discuss the interpretation of annulus pressure behavior and its link to barrier failure.
- Outline regulatory and industry guidance (API RP 90) for SCP management.
- Introduce risk assessment approaches for evaluating SCP.

8.1 The Significance of Annulus Pressure

Most wells are constructed with multiple concentric casing strings, creating annular spaces between them (e.g., A-annulus between tubing and production casing, B-annulus between production casing and intermediate/surface casing, C-annulus between intermediate casing and surface casing/conductor, etc.). These annuli are typically sealed at the top by the wellhead system and at the bottom by cement sheaths, packers, or landing collars.

During normal operations, these annuli are generally expected to be at relatively low pressure, often open to the atmosphere at surface wellheads or sometimes deliberately filled with an inert fluid and sealed (especially offshore or in sensitive areas). The presence of unexpected or abnormally high pressure in an annulus is a primary indicator that a well barrier element designed to isolate that annulus has failed, allowing fluid (gas, oil, or water) from a higher-pressure source to leak into that annular space. Monitoring these pressures is one of the most fundamental surveillance activities for operational well integrity. Monitoring annulus pressure provides crucial, often the earliest, evidence of barrier degradation or failure that has occurred over time since initial construction. Figure 8.1 illustrates a typical multi-casing well structure with labeled annuli (A, B, C), showing how gas migration or fluid leakage into these annular spaces can signal barrier failure. The diagram highlights the placement of annulus monitors and the flow behavior of potential leakage paths. The variability in sustained casing pressure (SCP) magnitude and its distribution across casing types and sizes underscores the importance of tailored pressure monitoring strategies. Figure 8.2 presents a comprehensive visual summary of sustained casing pressure (SCP) data, including the cumulative frequency of SCP magnitudes by casing type, SCP occurrences by casing size, and comparative SCP percentages from self-approved and combined well databases.

8.2 Sustained Casing Pressure (SCP)

The key term associated with problematic annulus pressure is **Sustained Casing Pressure (SCP)**. SCP is not merely the presence of pressure but specifically pressure that rebuilds after being bled down, indicating a persistent leak (Al-Ansari et al., 2015; Bourgoyne et al., 1999; Rocha-Valadez et al., 2014). Industry standards, like API Recommended Practice 90 (API RP 90): Annular Casing Pressure Management for Offshore Wells, provide widely accepted definitions (API RP 90):

> Any measurable casing pressure in an annulus that rebuilds after being bled down, indicating a leak path from a higher pressure source.

This definition highlights two crucial aspects that differentiate SCP from transient pressure fluctuations:

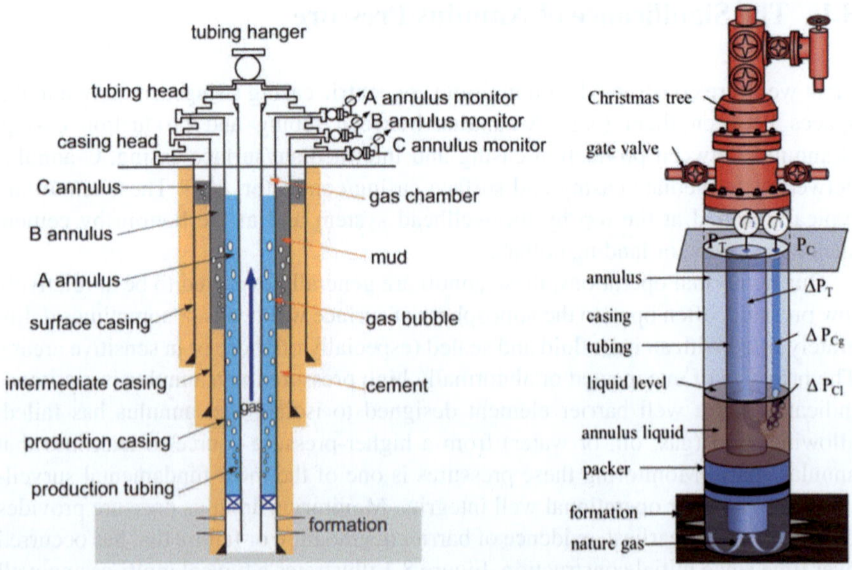

Fig. 8.1 Combined schematic showing the structure and pressure dynamics in a typical oil or gas well. **Left**: Vertical cross-section of a well with multiple concentric casing strings and annuli (A, B, C), showing annulus monitors, potential gas migration, and fluid interfaces (Zhu et al., 2012a, 2012b). **Right**: Pressure distribution along the wellbore, with annotations for differential pressures across tubing (ΔP_T), casing (ΔP_Cg), and annulus liquid column (ΔP_Cl), as well as highlighting the roles of packers, formation gas, and wellhead components such as the Christmas tree and gate valves. It highlights critical zones for pressure monitoring and fluid communication, relevant to sustained casing pressure (SCP) interpretation (Wu et al., 2018).

1. **Rebuilds After Bleed-Down**: Pressure caused by temporary effects like heating of trapped annular fluid during startup (thermal expansion—Annular Pressure Buildup or APB) will typically dissipate or stabilize at a lower level once bled down and temperature stabilizes (Al-Ansari et al., 2015). SCP, however, will consistently reappear and rebuild towards its original (or a new) equilibrium pressure because there is an ongoing leak from a higher-pressure source feeding fluid/pressure into the annulus.
2. **Indicates a Leak Path**: Confirmed SCP is direct evidence that a barrier designed to isolate that annulus from a higher-pressure source has been compromised (Rocha-Valadez et al., 2014; Roth et al., 2008; Wu et al., 2018). This means either the primary barrier has failed (leaking into the A-annulus from the production conduit) or a secondary barrier has failed (e.g., casing/cement leak allowing deeper formation pressure into the B, C, or outer annuli).

The presence of confirmed SCP signifies a potential well integrity issue that requires investigation, assessment, and management. It is a symptom of underlying barrier failure that has developed over time, potentially due to corrosion, cement degradation, seal failure, or connection leaks. Statistical reviews across mature basins

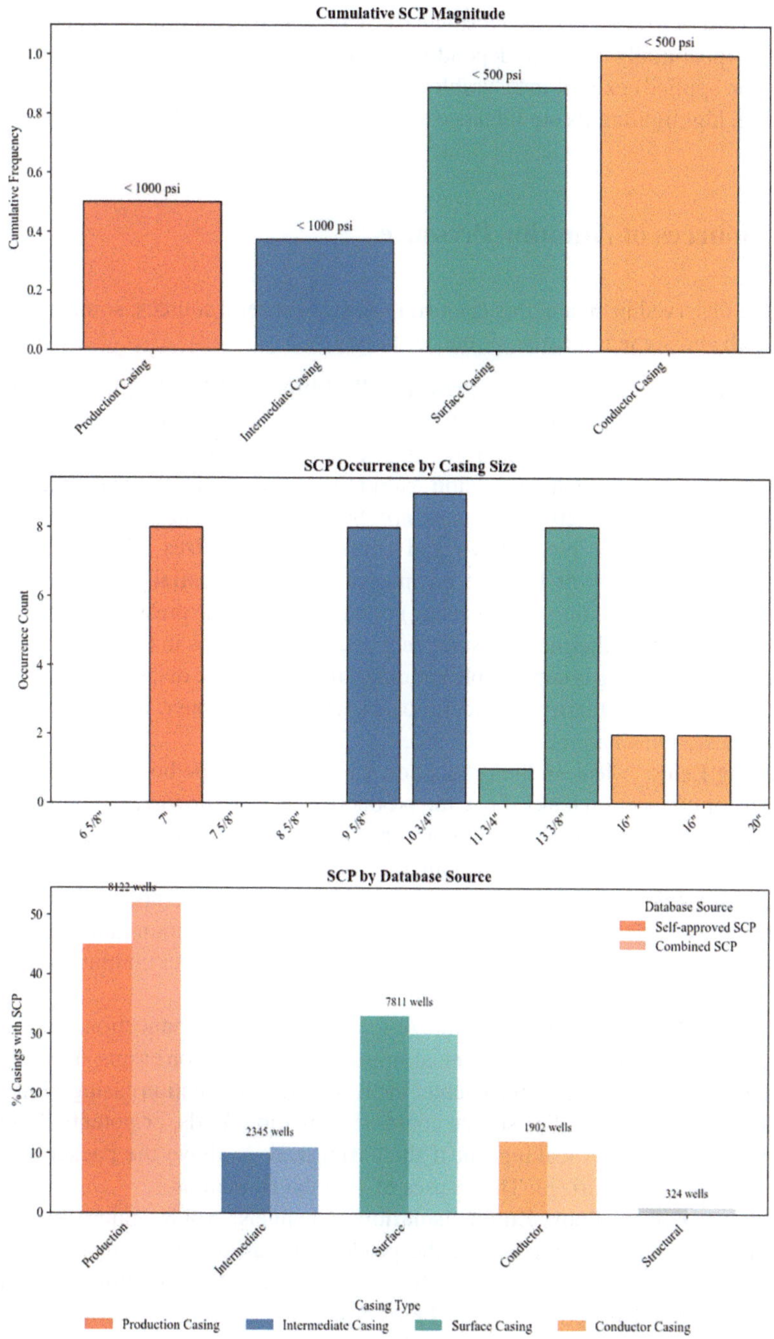

Fig. 8.2 Summary of sustained casing pressure (SCP) metrics: (top) cumulative SCP magnitude categorized by casing type; (middle) SCP occurrence count by casing size; (bottom) percentage of casings with SCP in both self-approved and combined datasets, including total number of wells analyzed per casing type. Data adapted from Bourgoyne et al. (2004a), Wojtanowicz et al. (2001a, 2001b), Xu and Wojtanowicz (2017)

like the Gulf of Mexico and the North Sea have indicated that a substantial percentage of wells (potentially 10–45% depending on region, age of wells, and specific SCP definition applied) exhibit measurable SCP in at least one annulus at some point in their life, highlighting its significance as a common integrity challenge (Chap. 29).

8.3 Sources of Annular Pressure

Pressure observed in an annulus can originate from several sources, some benign (if managed) and some indicative of barrier failure:

- **Thermal Effects (Annular Pressure Buildup—APB)**: Heating of trapped, incompressible fluids (typically liquids) in a sealed annulus during well startup, changes in operating temperature, or circulation can cause significant pressure increases due to thermal expansion within the constrained volume (Ibukun et al., 2024; Kiran et al., 2017). This is *not* SCP if the pressure does not consistently rebuild after bleed-down and temperature stabilizes. However, uncontrolled APB can itself become an integrity threat, potentially over-pressuring and damaging (bursting or collapsing) casing strings if the pressure exceeds their rating. Managing thermal pressures (e.g., via nitrogen caps in the annulus, bleed/feed systems, using compressible annular fluids, or rupture discs in some designs) is an important integrity consideration in HPHT, deepwater, and thermal wells (Discussion in Chap. 18, API RP 90).
- **Tubing Leak**: A leak in the production tubing string (pipe body corrosion hole, crack, or connection leak) above the production packer allows reservoir pressure/fluids to enter the A-annulus (tubing x production casing) (Dou et al., 2023; Zhang et al., 2021). This is a common source of SCP in the A-annulus.
- **Packer Leak**: Failure of the production packer seal elements or packer body allows reservoir pressure/fluids from below the packer to communicate directly into the A-annulus (Zhu et al., 2012a, 2012b). This is another common source of A-annulus SCP.
- **Production Casing Leak**: A failure (e.g., corrosion hole, connection leak, crack, or shear failure due to geomechanical stress) in the production casing string allows fluids from outside the casing to enter the B-annulus (production casing x intermediate/surface casing). The source could be formation fluids, or potentially fluids from the A-annulus leaking out if the casing leak is above the packer and the A-annulus is pressurized. This causes SCP in the B-annulus.
- **Cement Failure/Poor Zonal Isolation**: Channels, voids, micro-annuli, or degraded cement sheath behind the production casing or intermediate casing strings can allow formation fluids (gas, oil, or water) from shallower or deeper zones (isolated by other cement barriers or casing strings) to migrate up the annulus, pressurizing the B-annulus, C-annulus, etc. Poor primary cementing and subsequent cement degradation are common causes of SCP in outer annuli (B, C, etc.).

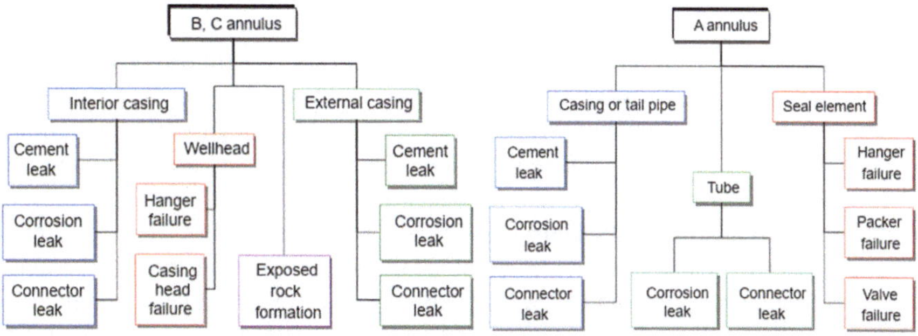

Fig. 8.3 Failure tree diagram illustrating potential leakage pathways and mechanical failure modes in the A-annulus and B/C-annuli. The A-annulus typically experiences failures related to tubing integrity, seal elements (e.g., packer, hanger, valve), or tail pipe connections. In contrast, the B and C annuli are more susceptible to issues originating from casing corrosion, cement degradation, and wellhead sealing failures. This categorization provides a logical framework for identifying and addressing the source of annular pressure anomalies during well integrity assessments (Kinik & Wojtanowicz, 2011a)

- **Wellhead/Hanger Seal Leak**: Failure of the casing hanger or tubing hanger seals in the wellhead can allow pressure communication between adjacent annuli (e.g., A-annulus pressure leaking into the B-annulus if the production casing hanger seal fails, or C-annulus pressure leaking to surface if the surface casing hanger seal fails).
- **Operator Induced Pressure**: Pressure deliberately applied during operations like annulus pressure testing or circulation activities. This is expected and temporary; it is not SCP unless it fails to bleed off correctly, indicating a potential blockage or issue.

Identifying the specific source of the pressure is the first critical step in diagnosing and managing SCP (Kinik & Wojtanowicz, 2011a). This requires analyzing pressure behavior, fluid type, well history, and potentially conducting diagnostic tests. A systematic understanding of leakage mechanisms is essential for diagnosing sustained casing pressure (SCP) and managing annular integrity. Figure 8.3 presents a failure tree outlining the primary leakage paths and failure modes associated with pressure buildup in the A-annulus and B/C-annuli.

8.4 Monitoring Techniques

Effective SCP management starts with reliable monitoring of annulus pressures throughout the well's operational life. This vigilance is crucial for detecting the development of SCP early, allowing for timely assessment and management (Bourgoyne et al., 2004b).

- **Pressure Gauges**: Simple, calibrated mechanical Bourdon tube or digital pressure gauges installed on the annulus valve outlets at the wellhead provide direct pressure readings. Regular visual inspection and recording of these pressures (e.g., daily, weekly, or monthly, depending on well criticality and operational philosophy) is the minimum requirement. Ensuring gauges are calibrated and functioning correctly is important for data accuracy.
- **Pressure Transmitters/Sensors**: Electronic pressure sensors provide continuous pressure readings that can be transmitted to local data logging systems, control rooms, or integrated SCADA (Supervisory Control and Data Acquisition) systems. This allows for real-time or near-real-time monitoring, automated data logging, trend analysis, and automated alarming if pressures exceed predefined limits (e.g., alert level, action level). Increasingly common, especially for critical wells, offshore facilities, and subsea wells where direct access is limited (See Chap. 20).
- **Monitoring Frequency**: The required frequency of annulus pressure monitoring varies depending on well criticality, history, potential risk, regulatory requirements, and operator policy. It can range from continuous (e.g., critical HPHT or subsea wells) to daily, weekly, or monthly manual readings for lower-risk wells. Wells with known SCP often require more frequent monitoring to track build-up behavior.
- **Data Logging and Archiving**: Systematically recording all monitored pressures over time (whether manual readings or electronic data) is essential for identifying trends (stable pressure, increasing pressure/build-up rate, decreasing pressure), correlating pressure changes with well operations (startup, shutdown, stimulation), and providing the historical data necessary for performing diagnostic analyses and long-term trend assessment. This data should be stored securely within the well integrity database (Chap. 15).
- **Wellhead Inspection**: Routine visual inspection of annulus valve outlets and surrounding wellhead components can sometimes reveal external leakage (e.g., fluid stains, audible hissing of gas) associated with high annular pressure, although external leaks are typically less common than contained SCP.

Monitoring should cover all accessible annuli (A, B, C, etc.) to provide a comprehensive picture of the well's annular integrity status.

8.5 Diagnostic Testing for SCP Characterization

When SCP is detected, diagnostic tests are performed to understand its source, severity, potential risks, and to estimate the pressure of the feeding source. The most common and fundamental diagnostic procedure is the **bleed-down/build-up test** (Wojtanowicz et al., 2001a, 2001b):

1. **Initial Shut-in**: Ensure the annulus is shut-in at the surface valve and allow the pressure to stabilize. Record the stabilized shut-in pressure (the initial SCP value) and the time of shut-in.
2. **Bleed-Down**: Carefully and in a controlled manner, open the annulus valve and bleed the trapped pressure/fluid volume down to near zero (or atmospheric pressure), while carefully measuring the volume of fluid (gas or liquid) produced and noting the time taken. Safety precautions are essential, especially if flammable gas, H2S, or high pressures are expected. Fluids may be routed to a safe collection system (e.g., flare, tank, venting with gas detection). Note any changes in fluid type during the bleed-down.
3. **Shut-in and Build-up Monitoring**: Immediately after bleeding down, close the annulus valve and begin monitoring the pressure build-up over time. Record pressure readings frequently initially (e.g., every minute for the first few minutes), then at increasing intervals (e.g., 5 min, 15 min, 1 h, 6 h, 24 h, etc.) until the pressure stabilizes or the build-up rate slows significantly.
4. **Repeat (Optional)**: The bleed/build-up cycle may be repeated (if safe and operationally feasible) to confirm the behavior and potentially refine the diagnosis.

Interpretation of Bleed/Build-Up Data (See Conceptual Plot–Fig. 8.4):

- **Confirmation of SCP**: Pressure rebuilding after bleeding down confirms that a leak path exists from a higher-pressure source, differentiating it from temporary thermal pressure (APB).
- **Build-up Rate**: A rapid pressure build-up generally indicates a larger leak path or higher differential pressure from the source compared to a slow build-up. A rapidly increasing build-up rate might suggest two-phase flow or complex leak geometry.
- **Stabilized Pressure (Estimated Source Pressure)**: The final pressure the annulus builds back up to (if it stabilizes) often provides an estimate of the pressure of the source feeding the leak. Comparing this pressure to known formation or wellbore pressures helps identify the potential source zone.
- **Fluid Type/Volume**: The type (gas, oil, water, mud) and volume of fluid bled off provide crucial clues about the leak source. Gas bleed-off often indicates communication with a gas cap or gas-bearing zone, or potentially a tubing/packer leak if producing gas. Water might indicate communication with an aquifer, re-injected water (in injectors), or a casing leak below the water table. Traces of mud might indicate residual drilling fluid from construction.
- **Build-up Signature**: Plotting pressure versus time (or log-time) can sometimes reveal diagnostic signatures characteristic of specific leak types (e.g., linear flow, radial flow), although interpretation can be complex and often requires specialized expertise.

To effectively manage sustained casing pressure (SCP), it is critical to understand the underlying pressure behavior during diagnostic testing. The most applied method involves conducting bleed-off and build-up tests, which provide valuable

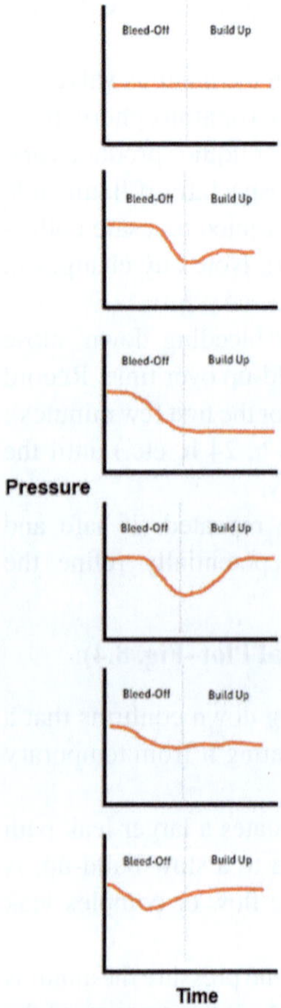

Test Behavior	Bleed-Off Response	Build-Up Response	Interpretation
No pressure	No pressure to bleed off	No build-up after valve closure	No SCP
Thermal pressure	Pressure bleeds to zero and stays zero	Pressure rebuilds slowly and stabilizes	Thermal expansion (e.g., heating fluids in annulus)
Trapped pressure	Pressure bleeds to zero and stays zero	No rebuild after long monitoring	Trapped pressure – no active source
Recharge pressure	Pressure bleeds to zero	Pressure builds back to original or increasing	Likely formation influx / barrier failure
Constant pressure	Pressure does not bleed down fully	Remains constant or rises slightly	Active recharge / external communication
High rate recharge	Pressure quickly rebuilds after bleed	Rapid pressure buildup	Significant leakage – failure in barrier system

Fig. 8.4 Pressure diagnostic chart for sustained casing pressure (SCP), illustrating typical bleed-off and build-up pressure responses over time. Each test behavior represents a distinct scenario ranging from benign (e.g., thermal pressure) to critical (e.g., recharge or high-rate recharge pressures indicating possible barrier failure). The vertical dashed line separates the bleed-off period from the build-up phase, enabling interpretation of pressure sources and annular integrity conditions

insight into whether annular pressure is caused by thermal effects, trapped fluids, or a more serious issue such as formation influx or barrier failure. Figure 8.4 presents a diagnostic chart summarizing the typical pressure–time responses for various SCP scenarios. By observing these trends and their associated interpretations, engineers can determine the root cause of annular pressure anomalies and make informed decisions regarding remediation strategies, in accordance with API RP 90 guidelines.

Other Diagnostic Techniques:

- **Annulus Fluid Sampling**: Collecting samples of the fluid bled from the annulus for chemical analysis (e.g., salinity, hydrocarbon composition, gas chromatography for gas composition, tracers) can help pinpoint the origin of the leaking fluid by comparing its 'fingerprint' to known reservoir or formation fluids, or injected fluids. This is particularly useful for confirming water sources or differentiating gas sources (Birkle & AlRamadan, 2023).
- **Pressure Steps/Interference Testing**: Systematically changing the pressure in adjacent conduits (e.g., tubing pressure, pressure in adjacent annulus, reservoir pressure via nearby wells) while monitoring the SCP annulus pressure can sometimes reveal pressure communication paths and quantify the degree of connectivity.
- **Logging**: In some cases, specialized logging tools (e.g., noise logs, temperature logs, fluid identification logs, pulsed neutron logs) that run inside the wellbore (if operational constraints allow access) might help locate the depth of the leak entry point into the annulus or diagnose flow behind pipe (Vasilyev et al., 2019) (see Chap. 12).

Rigorous diagnostic testing and analysis are essential for accurately characterizing the SCP and identifying the failed barrier and pressure source, which is the basis for risk assessment and deciding on the appropriate management strategy. To further illustrate the behavior of pressure communication and diagnostics during a tubing-to-annulus leak, Fig. 8.5 presents a progressive sequence of pressure evolution scenarios. The diagrams incorporate geological layers to contextualize the physical leakage environment, offering a clearer understanding of pressure redistribution across depths.

These scenarios support diagnosis of sustained casing pressure by identifying pressure gradients and equilibrium states across depths (z_t and z_l) (Wu et al., 2018).

8.6 Interpretation, Risk Assessment, and Management Strategies

Interpreting SCP data is a key skill that involves synthesizing information from monitoring trends, bleed/build-up tests, fluid analysis, diagnostic logs, and detailed knowledge of the well's construction ('as-built' data) and operational history. The key questions to answer are:

- What is the source of the leak? (Tubing, packer, casing, cement failure, wellhead seal?)
- What is the fluid type and composition? (Gas—with/without H_2S/CO_2, oil, water—type?)
- How severe is the leak? (Based on build-up rate and bled volume? Is it stable, increasing, or decreasing over time?)
- What is the estimated pressure of the source feeding the leak?

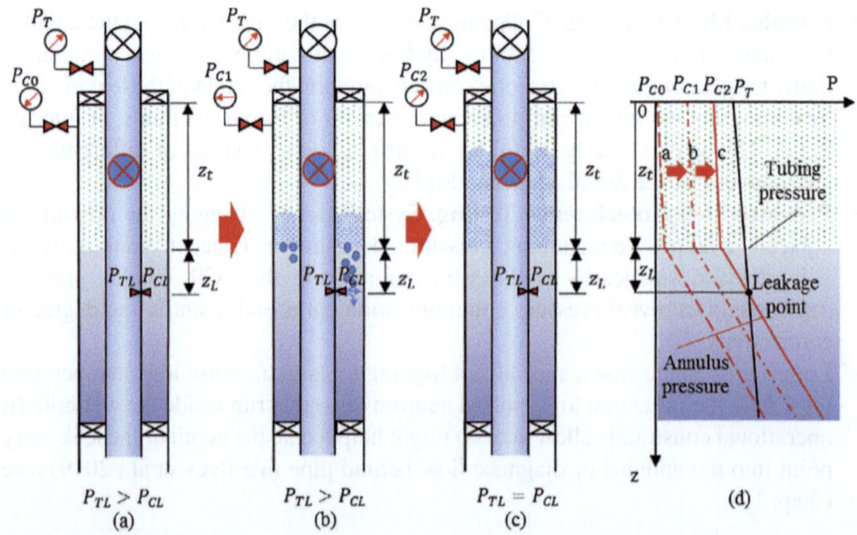

Fig. 8.5 Process of pressure behavior changes between tubing and annulus during various leakage scenarios across subsurface formations
(a) Perfect condition—no leakage; pressure in the tubing remains higher than in the annulus.
(b) Initial leakage state—small bubbles migrate across the tubing breach into the annulus, increasing annulus pressure.
(c) Advanced leakage—tubing and annulus pressures equalize, indicating loss of isolation.
(d) Pressure profile corresponding to the tubing-annulus system, showing the vertical distribution of tubing pressure, annulus pressure, and the leakage point in relation to geological layers.

- What is the maximum potential pressure the annulus could credibly reach if shut in (Estimated MAASP—Maximum Allowable Annulus Shut-in Pressure)?
- Can the outer casing string(s) and wellhead safely contain this potential pressure? (Comparing Estimated MAASP/MAWOP against component ratings).
- Is there a credible path for the leak to escalate and reach the surface, seabed, or shallow aquifers? (Considering the integrity of outer barriers).

To evaluate the severity of sustained casing pressure (SCP) in oil and gas wells, pressure trends were analyzed across various casing types. Based on dynamic threshold classification and well-specific maximum pressure behavior, each well was grouped into low and high SCP categories to assess potential risks to well integrity. Figure 8.6 illustrates the pressure-time profiles for each casing type, highlighting the distinction between wells exhibiting low SCP and those classified as high SCP based on casing-specific thresholds (200 psi for production and intermediate casings, 150 psi for surface and conductor casings). This visualization enables rapid identification of anomalous wells and supports diagnostic interpretation for annulus pressure buildup behavior

Risk Assessment: Once the SCP is characterized, a risk assessment is performed to determine the potential consequences of the SCP and the associated risk level.

Fig. 8.6 Classification of SCP behavior across wells for different casing types. The pressure profiles are separated into Low SCP and High SCP groups based on threshold criteria specific to each casing type (Production: 200 psi, Intermediate: 200 psi, Surface: 150 psi, Conductor: 150 psi). The plots reveal varying magnitudes and temporal behavior of annulus pressure buildup, highlighting wells at greater risk of barrier failure and sustained pressure communication in the data obtained from Wojtanowicz et al. (2001a, 2001b)

This informs the management strategy and often utilizes a risk matrix (Chap. 14) (D'Alesio et al., 2010, 2019; Kinik & Wojtanowicz, 2011b). Key factors considered in the risk assessment include:

- **Pressure Level versus Casing/Wellhead Rating:** Comparing the estimated MAASP against the pressure rating of the specific casing string and wellhead components forming the outer barrier (considering design factors, age, wear, and potential temperature derating). Is there a risk of burst failure of the outer casing or wellhead? API RP 90 provides methodologies for calculating MAASP and Maximum Allowable Wellhead Operating Pressure (MAWOP) for annuli (API RP 90).
- **Fluid Type**: Gas leaks generally pose a significantly higher risk than liquid leaks due to flammability, potential for gas migration to the surface (even through minor defects), higher stored energy (compressibility), and potential presence of H2S. H2S presence significantly increases safety risks.
- **Build-up Rate**: Faster build-up rates are often associated with higher risk due to a larger leak path and potentially higher potential flow rates if containment is lost.
- **Leak Location and Potential Escalation Path**: A leak high up in the wellbore near the surface or seabed might be considered higher risk than a deep leak contained by multiple intact outer strings and competent cement. The integrity of the outer barriers (casing, cement, wellhead seals) determines the potential for uncontrolled release if the inner containment fails. Is there evidence of flow behind multiple casing strings?
- **Well Location and Environment**: SCP in wells on platforms with occupied facilities, near populated areas onshore, in environmentally sensitive zones (shallow water, near coral reefs, over aquifers), or in deepwater (higher environmental consequences) is typically assessed as higher risk.
- **Regulatory Requirements**: Specific regulations or operator policies may mandate certain actions based on SCP level or build-up rate (e.g., mandatory bleed-off frequency, required intervention within a timeframe).

Management Strategies: Based on the risk assessment, a management strategy is developed and implemented within the WIMS. Options range from:

- **Continued Monitoring**: For very low, stable pressures with low risk and no credible escalation path, the strategy may be continued vigilant monitoring, potentially with increased frequency or additional surveillance (e.g., regular bleed-down/build-up tests to track rate changes). This only manages the symptom, not the cause.
- **Bleeding (Vent or Bleed-to-Flare)**: Periodically bleeding the pressure to keep it below acceptable limits (e.g., below a regulatory threshold or a percentage of the MAWOP). This manages pressure but does not fix the underlying barrier failure. Bleeding gas to the atmosphere is generally prohibited due to environmental regulations (methane emissions) and safety risks; flaring or capturing/routing to facilities is preferred where possible.

- **Investigation and Repair (Remediation)**: Performing further diagnostic logging (See Chap. 12) or interventions to pinpoint the leak source and implement a permanent repair (e.g., cement squeeze to seal cement/casing leak, casing patch, packer replacement, tubing replacement) (See Chap. 26). This is often the preferred long-term solution for higher-risk SCP but can be complex and expensive.
- **Well Shut-in or Restriction**: If the risk is deemed unacceptable (e.g., pressure exceeding safe limits, high-risk fluid type, inability to confirm containment) or repairs are not feasible/successful, the well may need to be shut-in, restricted (rate/pressure), or permanently abandoned to eliminate the pressure source or contain the risk.

API RP 90 provides a widely adopted framework specifically for managing SCP in offshore wells, but its principles are often applied onshore as well. It outlines classifications of SCP, monitoring requirements, diagnostic testing, methods for calculating MAASP/MAWOP, risk assessment considerations, guidelines for corrective actions based on risk level, and documentation requirements. Operators typically have their internal standards based on or exceeding API RP 90 guidelines, setting specific pressure limits, build-up rate thresholds, and required actions for different annuli and risk levels. Managing SCP effectively through proactive monitoring, accurate diagnosis, robust risk assessment, and appropriate management strategies is a critical aspect of ensuring continued well integrity during the operational phase. Ignoring or inadequately managing SCP has been a contributing factor in the past incidents.

8.7 Calculation of Sustained Annulus Pressure

The presence of sustained casing pressure (SCP) in an annulus arises from various mechanisms, including thermal expansion of trapped fluids, changes in the annulus volume due to mechanical deformation, and potential influx or leakage of fluids from adjacent formations (Yin & Gao, 2015). Understanding and quantifying these pressure contributions is essential for accurate assessment of well integrity and the risk associated with SCP. The following section introduces a differential expression for SCP and details how to evaluate the impact of thermal expansion, compressibility, and fluid influx on the sustained annulus pressure.

8.7.1 General Expression for Sustained Pressure Change

Sustained casing pressure can be modeled as a function of temperature (T), annular volume (V_{ann}), and fluid mass (m), as given by the general form:

$$p = p(T, V_{ann}, m) \tag{8.1}$$

The change in annulus pressure (Δp) due to variations in these parameters is described by its total differential:

$$\Delta p = \left(\frac{\partial p}{\partial T}\right)\Delta T + \left(\frac{\partial p}{\partial V_{\text{ann}}}\right)\Delta V_{\text{ann}} + \left(\frac{\partial p}{\partial m}\right)\Delta m \tag{8.2}$$

These partial derivatives represent the sensitivity of annulus pressure to changes in temperature, annular volume, and fluid mass, respectively. This formulation forms the foundation for diagnosing and interpreting the physical drivers of SCP in wellbores.

8.7.2 Coefficient of Thermal Expansion

To quantify the impact of temperature changes on annulus pressure, the coefficient of thermal expansion (α_T) is introduced. It represents the change in unit volume induced by a unit change in temperature:

$$\alpha_T = \frac{\Delta V}{V \cdot \Delta T} \tag{8.3}$$

Here, ΔV is the change in fluid volume, V is the initial volume, and ΔT is the change in temperature. This coefficient is critical in thermal wells and HPHT (High-Pressure, High-Temperature) environments, where thermal effects can significantly contribute to SCP.

8.7.3 Coefficient of Compressibility

In addition to temperature-driven changes, the compressibility of the annulus fluid itself also plays a major role in SCP behavior. The coefficient of compressibility (κ_T) is defined as:

$$\kappa_T = \frac{\Delta V}{V \cdot \Delta p} \tag{8.4}$$

This expression relates the volume change to the applied pressure change, helping to capture how compressible the trapped fluid is under varying conditions. In gas-filled annuli, κ_T can be orders of magnitude larger than in liquid-filled annuli, making it an essential parameter in SCP assessments.

8.7.4 Final Expression for Sustained Annulus Pressure

By substituting the coefficients of thermal expansion and compressibility (Eqs. 8.3 and 8.4) into the general differential expression (Eq. 8.2), a comprehensive formula for the sustained casing pressure change can be obtained:

$$\Delta p = \frac{\alpha_T}{\kappa_T} \cdot \Delta T - \frac{1}{\kappa_T \cdot V_{\text{ann}}} \cdot \Delta V_{\text{ann}} + \frac{1}{\kappa_T \cdot V_l} \cdot \Delta V_l \tag{8.5}$$

where $\mathbf{\Delta T}$ denotes the temperature change, typically measured in degrees Fahrenheit or Celsius. The term $\mathbf{\Delta V_{ann}}$ represents the change in annular volume, which may result from casing deformation, thermal expansion, or other mechanical effects. Meanwhile, $\mathbf{\Delta V_l}$ refers to the change in fluid volume, which can be attributed to either influx into the annulus or leakage losses from the system.

This equation highlights the three principal contributors to SCP:

1. Thermal expansion of the fluid,
2. Mechanical changes in the annulus volume,
3. Fluid influx or leakage.

References

Al-Ansari, A. A., Al-Refai, I. M., Al-Beshri, M. H., Pino, R. M., Leon, G. A., Knudsen, K., & Sanabria, A. E. (2015). Thermal activated resin to avoid pressure build-up in casing-casing annulus (CCA). In *Society of Petroleum Engineers—SPE Offshore Europe Conference and Exhibition, OE 2015* (pp. 8–11). https://doi.org/10.2118/175425-MS

Birkle, P., & Al Ramadan, H. A. (2023). Geochemical techniques to identify fluid sources in highly pressured casing-casing annuli (CCA). *Geochemistry, 83*(2), Article 125955. https://doi.org/10.1016/J.CHEMER.2023.125955

Bourgoyne, A. T., Scott, S. L., & Regg, J. B. (1999). Sustained casing pressure in offshore producing wells. In *Proceedings of the Annual Offshore Technology Conference* (vol. 3, pp. 625–637). https://doi.org/10.4043/11029-MS

Bourgoyne, A. T., Stuart Scott, L. L., & Wojciech Manowski, L. (2004a). *A review of sustained casing pressure occurring on the OCS.*

Bourgoyne, A. T., Stuart Scott, L. L., & Wojciech Manowski, L. (2004b). *A review of sustained casing pressure occurring on the OCS.*

D'Alesio, P., Caramanico, L., Angelucci, E., & Scalzitti, L. (2019). Wells life extension through a risk-based well ageing study. In *Society of Petroleum Engineers—SPE Kuwait Oil and Gas Show and Conference 2019, KOGS 2019* (vol. 15). https://doi.org/10.2118/198162-MS

D'Alesio, P., Poloni, R., Valente, P., & Magarini, P. A. (2010). Well integrity assessment and assurance: The operational approach for three CO_2 storage fields in Italy. In *Proceedings—SPE Annual Technical Conference and Exhibition* (vol. 1, pp. 494–505). https://doi.org/10.2118/133056-MS

Dou, N., Hu, Z., Wang, Z., He, H., & Yang, S. (2023). Risk assessment of annular pressure caused by tubing leakage in offshore gas wells with high CO_2. *Geofluids, 2023*(1), 9072217. https://doi.org/10.1155/2023/9072217

Ibukun, M., Elyan, E., Amish, M., Njuguna, J., & Oluyemi, G. F. (2024). A review of well life cycle integrity challenges in the oil and gas industry and its implications for sustained casing pressure (SCP). In *Energies* (vol. 17, no. 22). Multidisciplinary Digital Publishing Institute (MDPI). https://doi.org/10.3390/en17225562

Kinik, K., & Wojtanowicz, A. K. (2011a). Identifying environmental risk of sustained casing pressure. In *Society of Petroleum Engineers—SPE Americas E and P Health, Safety, Security, and Environmental Conference 2011* (pp. 457–468). https://doi.org/10.2118/143713-MS

Kinik, K., & Wojtanowicz, A. K. (2011b). Identifying environmental risk of sustained casing pressure. In *Society of Petroleum Engineers—SPE Americas E and P Health, Safety, Security, and Environmental Conference 2011* (pp. 457–468). https://doi.org/10.2118/143713-MS

Kiran, R., Teodoriu, C., Dadmohammadi, Y., Nygaard, R., Wood, D., Mokhtari, M., & Salehi, S. (2017). Identification and evaluation of well integrity and causes of failure of well integrity barriers (A review). *Journal of Natural Gas Science and Engineering, 45*, 511–526. https://doi.org/10.1016/J.JNGSE.2017.05.009

Rocha-Valadez, T., Hasan, A. R., Mannan, S., & Kabir, C. S. (2014). Assessing wellbore integrity in sustained-casing-pressure annulus. *SPE Drilling & Completion, 29*(01), 131–138. https://doi.org/10.2118/169814-PA

Roth, J., Reeves, C., Johnson, C. R., De Bruijn, G., Bellabarba, M., Le Roy-Delage, S., & Bulte-Loyer, H. (2008). Innovative hydraulic isolation material preserves well integrity. In *SPE/IADC Drilling Conference, Proceedings* (vol. 2, pp. 1033–1046). https://doi.org/10.2118/112715-MS

Vasilyev, S., Boltunov, N., Shkaran, A., Kirpichikova, E., Minakhmetova, R., Pimantyev, V., Nagimov, V., & Zaripova, A. (2019). Determination the source of sustained annulus pressure in high GOR oil wells by using acoustic logging. In *Society of Petroleum Engineers—SPE Annual Caspian Technical Conference 2019, CTC 2019* (pp. 16–18). https://doi.org/10.2118/198385-MS

Wojtanowicz, A., … S. N.-F. Report. L. (2001). *Diagnosis and remediation of sustained casing pressure in wells*. Wellintegrity.Net. Final Report. Louisiana State University, Virginia. Submitted to: US. http://wellintegrity.net/Documents/MMS%20-%20Diagnosis%20of%20SCP%20-%202001-07-31.pdf

Wojtanowicz, A. K., Nishikawa, S., & Rong, X. (2001). Diagnosis and remediation of sustained casing pressure in wells.

Wu, S., Zhang, L., Fan, J., Zhang, X., Liu, D., & Wang, D. (2018). A leakage diagnosis testing model for gas wells with sustained casing pressure from offshore platform. *Journal of Natural Gas Science and Engineering, 55*, 276–287. https://doi.org/10.1016/J.JNGSE.2018.05.006

Xu, R., & Wojtanowicz, A. K. (2017). Pressure buildup test analysis in wells with sustained casing pressure. *Journal of Natural Gas Science and Engineering, 38*, 608–620. https://doi.org/10.1016/J.JNGSE.2016.12.033

Yin, F., & Gao, D. (2015). Prediction of sustained production casing pressure and casing design for shale gas horizontal wells. *Journal of Natural Gas Science and Engineering, 25*, 159–165. https://doi.org/10.1016/J.JNGSE.2015.04.038

Zhang, B., Xu, Z., Lu, N., Liu, H., Liu, J., Hu, Z., Xie, J., & Wang, K. (2021). Characteristics of sustained annular pressure and fluid distribution in high pressure and high temperature gas wells considering multiple leakage of tubing string. *Journal of Petroleum Science and Engineering, 196*, Article 108083. https://doi.org/10.1016/J.PETROL.2020.108083

Zhu, H., Lin, Y., Zeng, D., Zhang, D., Chen, H., & Wang, W. (2012a). Mechanism and prediction analysis of sustained casing pressure in "A" annulus of CO_2 injection well. *Journal of Petroleum Science and Engineering, 92–93*, 1–10. https://doi.org/10.1016/J.PETROL.2012.06.013

Zhu, H., Lin, Y., Zeng, D., Zhang, D., & Wang, F. (2012b). Calculation analysis of sustained casing pressure in gas wells. *Petroleum Science, 9*(1), 66–74. https://doi.org/10.1007/S12182-012-0184-Y/METRICS

Chapter 9
Corrosion and Erosion Mechanisms and Control

Abstract This chapter provides a thorough examination of corrosion and erosion phenomena that compromise well integrity throughout the lifecycle of oil and gas wells. It begins with the fundamental principles of electrochemical corrosion, emphasizing the roles of anodic and cathodic reactions, electrolytes, and metallic pathways in downhole environments. The chapter then explores internal corrosion mechanisms driven by corrosive species such as CO_2, H_2S, O_2, microbiological agents, and organic acids, detailing their electrochemical pathways, scale formation and breakdown, and specific degradation forms like pitting, hydrogen-induced cracking, and sulfide stress cracking. External corrosion processes affecting the outer surfaces of casing strings due to formation fluids, cement pore water, and environmental exposure are also examined. The synergistic impact of erosion and erosion-corrosion is discussed, highlighting how high-velocity flow and entrained solids exacerbate metal loss beyond the sum of their individual effects. Recognizing these threats, the chapter presents a suite of corrosion control strategies including material selection, use of corrosion-resistant alloys (CRAs), chemical inhibition, protective coatings and linings, operational controls, and cathodic protection. It also evaluates the strengths and limitations of various corrosion monitoring techniques such as corrosion coupons, electrical resistance probes, and inline inspection tools. Given the challenges of predicting localized corrosion, ensuring inhibitor effectiveness in multiphase flow, and accessing downhole regions, this chapter emphasizes an integrated, risk-based approach to corrosion management. By consolidating industry standards (API, ISO, NORSOK) and case-specific mitigation practices, it equips engineers and asset managers with the tools to proactively address time-dependent metal degradation and safeguard wellbore structural integrity in increasingly aggressive operational environments.

© The Author(s), under exclusive license to Springer Nature Switzerland AG 2026

A. Alsubaih and K. Sepehrnoori, *Well Integrity for Oil, Gas, and Subsurface Storage Operations*, https://doi.org/10.1007/978-3-032-15233-6_9

Rusted casing pipes illustrating external corrosion—a common integrity threat in oil and gas wells. *Source* The Authors

Chapter's Objectives

- Explain the fundamental principles of electrochemical corrosion as it applies to well environments.
- Describe common forms of internal corrosion in wells (CO_2, H_2S, O_2, MIC, organic acids).
- Discuss external corrosion mechanisms affecting casing strings in annuli.
- Outline erosion and erosion-corrosion phenomena and their synergistic effects.
- Review primary corrosion control strategies: material selection, chemical inhibition, coatings, and operational controls.
- Briefly introduce cathodic protection for external corrosion.
- Discuss challenges in corrosion monitoring and prediction in downhole environments.

9.1 Introduction: The Threat of Metal Loss

Casing and tubing strings, as well as wellhead and completion components, are predominantly made of carbon or low-alloy steel. While strong and relatively inexpensive, steel is susceptible to corrosion—the gradual degradation of metal through electrochemical or chemical reactions with its environment (Kain, 2011). Over time, corrosion causes metal loss, reducing the wall thickness of tubulars and potentially leading to leaks, loss of tensile/compressive/burst/collapse strength, or catastrophic failure. Erosion, the physical removal of material by flowing fluids containing solids, can also cause metal loss, often working synergistically with corrosion (erosion-corrosion) (Abdul-Majid et al., 2017; Jain et al., 2016; Klever, 1992). Corrosion is a time-dependent degradation mechanism, with damage accumulating over the operational life of the well. It is a leading cause of integrity failures, particularly in aging wells or those operating in aggressive environments. Understanding, predicting, monitoring, and managing corrosion and erosion are therefore fundamental aspects of maintaining well integrity throughout the operational life. To illustrate how corrosion manifests under diverse operational scenarios, Fig. 9.1 highlights typical failure locations and associated corrosion mechanisms observed in oil and gas pipelines, particularly under conditions such as CO_2/H_2S exposure, high flow velocity, acidic environments, and supercritical CO_2 injection.

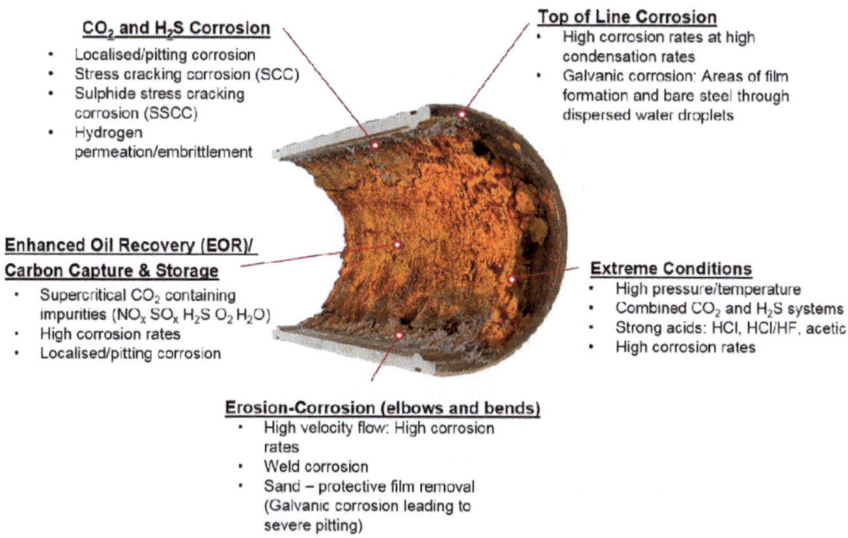

Fig. 9.1 Representative internal corrosion mechanisms affecting pipelines in oil and gas operations. The annotated cross-section identifies damage caused by CO_2 and H_2S corrosion, top-of-line condensation corrosion, erosion-corrosion at elbows and welds, extreme pressure/temperature environments, and enhanced oil recovery (EOR)/carbon capture and storage (CCS) conditions. Each region of corrosion is associated with localized pitting, galvanic interaction, or film degradation, leading to accelerated metal loss (*Corrosion and Damage Mechanisms | Mat Integrity*, 2022)

9.2 Fundamentals of Electrochemical Corrosion

Most corrosion of steel in downhole and surface well environments is an electrochemical process occurring at the interface between the metal and an electrolyte (an ionically conductive fluid, typically water containing dissolved salts) (Davis et al., 2016). This process involves the formation of microscopic or macroscopic electrochemical cells with four key components:

1. **Anode**: The site on the metal surface where oxidation occurs–the metal (Iron in steel) loses electrons and dissolves into the electrolyte as positive ions (e.g., $Fe \rightarrow Fe^{2+} + 2e^{-}$) This is where metal loss occurs.
2. **Cathode**: The site on the metal surface where reduction occurs—a chemical species in the electrolyte gains electrons provided by the anodic reaction. Common cathodic reactions in well environments include:
   - Reduction of hydrogen ions (in acidic solutions): $2H^{+} + 2e^{-}\ H_2$ (hydrogen gas)
   - Reduction of dissolved oxygen (in neutral or alkaline water): $O_2 + 2H_2O + 4e^{-} \rightarrow 4OH^{-}$ (hydroxide ions)—Highly corrosive!
   - Reduction of water (in de-aerated water, less common): $2H_2O + 2e^{-} \rightarrow H_2 + 2OH^{-}$
3. **Metallic Path**: The metal itself provides a pathway for electrons generated at the anode to flow to the cathode.
4. **Electrolyte**: An ionically conductive medium in contact with both the anode and cathode, allowing ions to migrate and complete the electrical circuit, facilitating the flow of charge. Formation water, completion brines, injected water, condensed water in gas streams, and water associated with oil production act as electrolytes. Pure hydrocarbons are generally not corrosive unless water is present.

Corrosion occurs because different areas on the steel surface (due to variations in composition, stress, surface finish, or presence of deposits) or different metals in electrical contact (galvanic corrosion) can have slightly different electrochemical potentials, causing one area to act as an anode and another as a cathode (Comizzoli et al., 1986; Song et al., 2012). The rate of corrosion depends on numerous factors, including temperature, pressure, fluid composition (presence of corrosive agents like CO_2, H_2S, O_2, pH, chloride concentration), fluid velocity (influencing mass transfer and shear stress), the formation of protective or non-protective corrosion products (scales), and the metallurgical state of the steel. To better visualize the interaction between these components, Fig. 9.2 illustrates a simplified corrosion cell, highlighting the flow of electrons through the metallic path and the movement of ions through the electrolyte, which together drive the electrochemical corrosion process. For electrochemical corrosion processes, the **corrosion current density** (i_{corr}) can be used to estimate the corrosion rate via Faraday's law:

$$\text{Corrosion Rate (mm/year)} = \frac{0.00327 \times i_{corr} \times \text{EW}}{\rho} \tag{9.1}$$

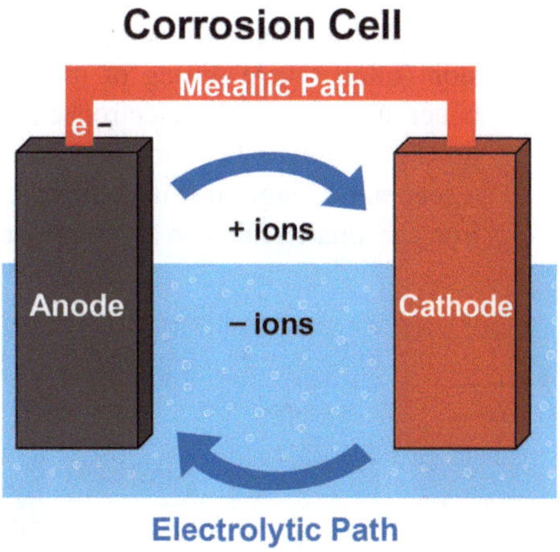

Fig. 9.2 Schematic of a basic corrosion cell showing the key components: the anode where metal dissolution occurs, the cathode where reduction reactions take place, the metallic path for electron flow, and the electrolytic path for ion movement. This electrochemical system underpins most forms of corrosion observed in oilfield environments

where i_{corr} is the corrosion current density in microamperes per square centimeter ($\mu A/cm^2$), **EW** is the equivalent weight of the metal in grams per equivalent, and **ρ** is the density of the metal in grams per cubic centimeter (g/cm^3). This approach is more common in laboratory electrochemical testing or with field probes (LPR or electrochemical noise probes).

9.3 Internal Corrosion Mechanisms

Corrosion occurring on the inside surfaces of tubing and casing strings, and on the internal components of packers, valves, and wellheads (Al-Janabi, 2020; Franco, 2020), is driven by the composition and conditions of the produced or injected fluids flowing within the wellbore. Corrosion in oil and gas wells manifests in multiple forms, each with distinct visual and structural characteristics. To enhance understanding, Fig. 9.3 presents a side-by-side comparison between schematic illustrations and actual field photographs of various corrosion types. This dual representation bridges theoretical descriptions with real-world degradation patterns observed in pipelines, casings, and equipment surfaces.

- **CO_2 Corrosion (Sweet Corrosion)**: Carbon dioxide (CO_2) dissolves in water to form carbonic acid (H_2CO_3), a weak acid that is highly corrosive to carbon steel (Porcayo-Calderon et al., 2022).
  - **Mechanism**: Carbonic acid dissociates in water, providing hydrogen ions (H^+) for the cathodic reaction. The anodic reaction is the dissolution of iron (Fe). Protective iron carbonate ($FeCO_3$) scales can form on the steel surface, which can significantly reduce corrosion rates if they are dense, adherent, and continuous. However, their stability depends heavily on temperature, pH, CO_2 partial pressure, and flow conditions. If the scale is porous, non-adherent, removed

by high flow rates or pigging, or disrupted by pitting, rapid localized corrosion (often severe pitting or mesa attack) can occur beneath the scale. To better understand the electrochemical processes and environmental conditions governing CO_2 corrosion and $FeCO_3$ scale behavior, Fig. 9.4 presents a schematic diagram illustrating the formation and breakdown of protective iron carbonate scales on steel surfaces.

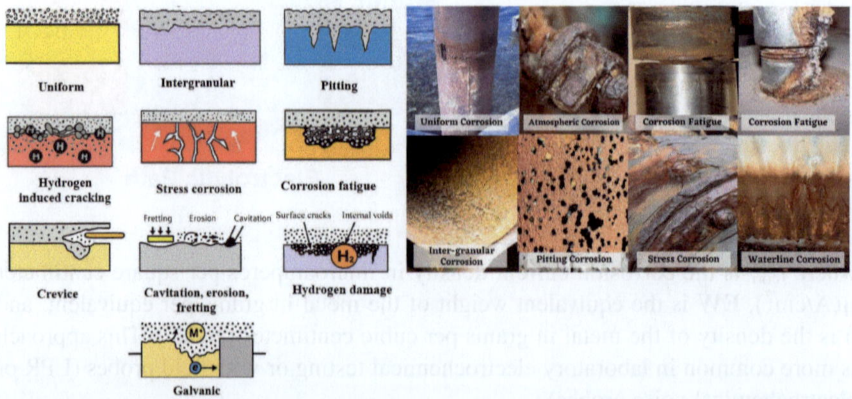

Fig. 9.3 Comparison of corrosion types through schematic diagrams (left) and real-world field photographs (right). The illustrations show classic corrosion patterns such as uniform, intergranular, galvanic, pitting, crevice, stress corrosion cracking, hydrogen-induced damage, corrosion fatigue, and combined fretting, erosion, and cavitation. These forms are commonly encountered in oilfield operations and contribute significantly to metal loss and failure

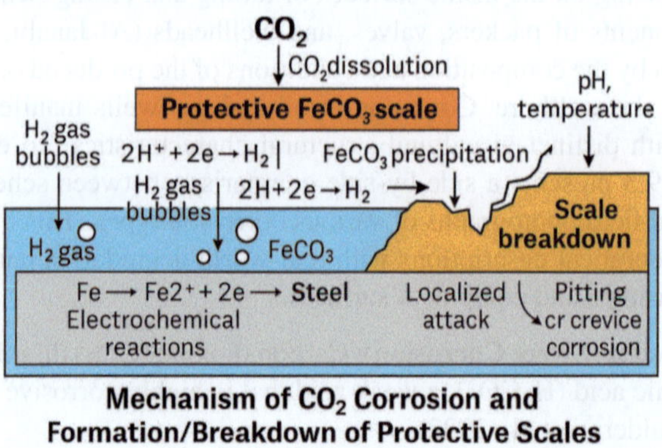

Fig. 9.4 Mechanism of CO_2 corrosion and formation/breakdown of protective $FeCO_3$ scales. CO_2 dissolves in water forming carbonic acid, which dissociates and participates in electrochemical reactions. Iron dissolves anodically, while hydrogen gas evolves at the cathode. $FeCO_3$ precipitates to form a protective scale, but this scale may break down under the influence of environmental factors such as pH, temperature, and flow-induced mechanical stress, leading to localized corrosion such as pitting and crevice attack

- **Factors**: Corrosion rates increase significantly with CO_2 partial pressure (higher partial pressure means more dissolved CO_2 and lower pH), temperature (up to a certain point, then potentially decreasing at very high temperatures if stable, protective scales form), and flow velocity (inhibiting scale formation or causing erosion of existing scale) (Obot et al., 2023). The presence of a free water phase wetting the steel surface is essential for electrochemical corrosion to occur.
- **Prediction**: Numerous empirical and mechanistic models exist to predict CO_2 corrosion rates based on environmental parameters (temperature, pressure, water chemistry, CO_2 partial pressure, flow velocity) [e.g., NORSOK M-506, models by de Waard-Milliams, ECE, etc.] (de Waard et al., 1991; Olsen, 2003; Olsen et al., 2005). These models inform material selection and inhibition strategies.

- **H_2S Corrosion (Sour Corrosion)**: Hydrogen sulfide (H_2S) is extremely dangerous (toxic, flammable) and highly corrosive (Sotoodeh, 2021).
  - **Mechanism**: H_2S dissolves in water and acts as a weak acid. It reacts with steel to form iron sulfide (FeS) scales. FeS scales can sometimes be protective, similar to $FeCO_3$, but they are often porous, brittle, non-adherent, or conductive, leading to localized corrosion (pitting, blistering) due to galvanic effects between different FeS phases or between FeS and the underlying steel.
  - **Sulfide Stress Cracking (SSC)**: As discussed in Sect. 4.6, the hydrogen atoms generated during the cathodic reaction in sour environments (especially at low pH) can be absorbed into the steel lattice. In susceptible materials (primarily high-strength steels with certain microstructures or hard weld zones) under tensile stress (applied or residual), these hydrogen atoms can cause embrittlement and sudden, brittle fracture at stresses well below the material's yield strength (Muthukumar, 2014). This is SSC, a major, potentially catastrophic, integrity threat. Preventing SSC requires strict adherence to material selection and heat treatment guidelines defined by NACE MR0175/ ISO15,156, which specifies limits on steel hardness and structure based on H_2S partial pressure, temperature, pH, and chloride concentration (NACE MR0175/ISO 15156). Sulfide Stress Cracking (SSC) is a critical form of environmentally assisted cracking that occurs when high-strength steels are exposed to H_2S-containing environments under tensile stress. The mechanism involves hydrogen generation, absorption into the steel matrix, and accumulation at crack tips and grain boundaries, eventually leading to embrittlement and brittle fracture. Figure 9.5 illustrates the SSC process and hydrogen transport within the steel.
  - **Hydrogen Induced Cracking (HIC)/Stepwise Cracking (SWC)**: Hydrogen atoms can also recombine at internal defects or inclusions within the steel (e.g., manganese sulfide inclusions) to form molecular hydrogen (H_2), creating high internal pressures that cause blisters or stepwise internal cracks, particularly in lower-strength steels (Muthukumar, 2014). This is less common in typical casing/tubing steels manufactured to API standards but is a known threat in certain plate steels used for facilities.

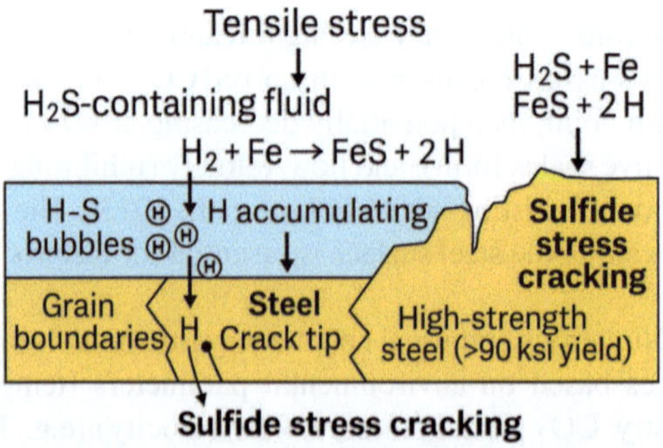

Fig. 9.5 Diagram illustrating the mechanism of Sulfide Stress Cracking (SSC) in high-strength steels exposed to H_2S-containing environments. Atomic hydrogen generated during the corrosion reaction diffuses into the steel, accumulates at crack tips and grain boundaries, and contributes to embrittlement and crack propagation under tensile stress

- **Oxygen Corrosion**: Dissolved oxygen (O_2) is a very aggressive corrodent, acting as an efficient cathodic reactant that drives rapid iron dissolution. Fortunately, produced fluids from most reservoirs are typically anaerobic (oxygen-free) (Yan et al., 2019). However, oxygen ingress can occur through:
  - Injection fluids (water, polymers, stimulation fluids) if not properly deoxygenated (See Chap. 23).
  - Leaks in surface equipment or packer/wellhead seals allowing air contact (e.g., in annulus fluids open to atmosphere).
  - Workover or completion fluids are exposed to air. Oxygen corrosion often results in severe, highly localized pitting. Maintaining anaerobic conditions in systems where oxygen is not expected is critical; even small amounts of oxygen can initiate significant corrosion.
- **Microbiologically Influenced Corrosion (MIC)**: Certain types of microorganisms, particularly Sulfate-Reducing Bacteria (SRB) and Acid-Producing Bacteria (APB), can thrive in downhole environments (especially in stagnant zones, under deposits, or in water injection systems if water quality is not controlled) (Telegdi et al., 2017).
  - **Mechanism**: SRB consume sulfate ions and produce H_2S as a metabolic byproduct, leading to localized sour corrosion and pitting, often occurring beneath protective biofilms or deposits where oxygen levels are low (anaerobic). APB produces organic acids that can lower local pH and increase corrosion. MIC often causes very rapid, highly localized pitting, which is difficult to predict or monitor with conventional methods (Discussion in Sect. 24.4—MIC relevance to UHS).To visualize the complex interplay between microbial activity and electrochemical reactions in corrosion, Fig. 9.6 illustrates the mechanism of Microbiologically Influenced Corrosion (MIC) under a biofilm,

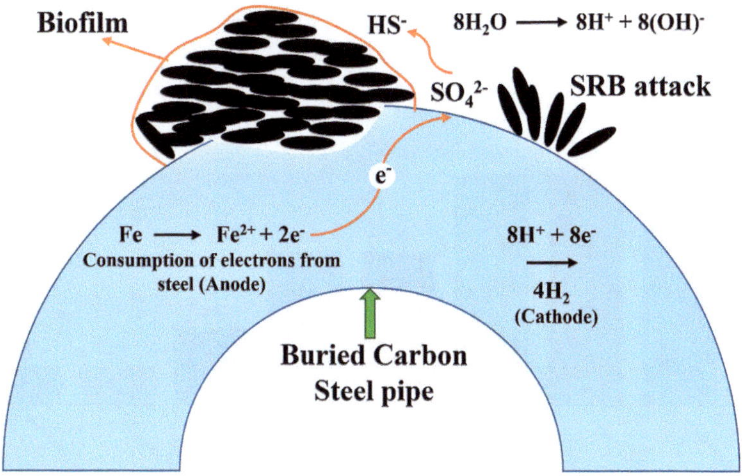

Fig. 9.6 Diagram illustrating the mechanism of Microbiologically Influenced Corrosion (MIC) under a biofilm, where sulfate-reducing bacteria (SRBs) induce localized corrosion by consuming electrons from the steel surface and producing corrosive hydrogen sulfide species (Khouzani et al., 2019)

highlighting the role of sulfate-reducing bacteria (SRB) in accelerating steel degradation beneath deposits.

– **Detection/Mitigation**: Difficult to predict and diagnose. Requires specialized sampling and analysis techniques (molecular methods like quantitative PCR are improving detection). Control involves using biocides (batch or continuous), regular cleaning (pigging) to remove deposits and biofilms, and maintaining flow to prevent stagnation (Kannan et al., 2018).

- **Organic Acid Corrosion**: Naturally occurring organic acids (e.g., acetic, propionic) present in some reservoir fluids can contribute to corrosion, particularly at higher temperatures and pressures, although typically less aggressive than CO_2 or H_2S. Their presence can also influence the pH and stability of protective scales (Scribner, 2001).

Understanding the relative occurrence of each internal corrosion type provides valuable insight into where mitigation and monitoring efforts should be prioritized. Figure 9.7 presents a breakdown of corrosion-related failures by type, highlighting CO_2 and H_2S corrosion as the most dominant mechanisms observed in oil and gas operations. This frequency distribution underscores the importance of selecting materials and control strategies tailored to these environments.

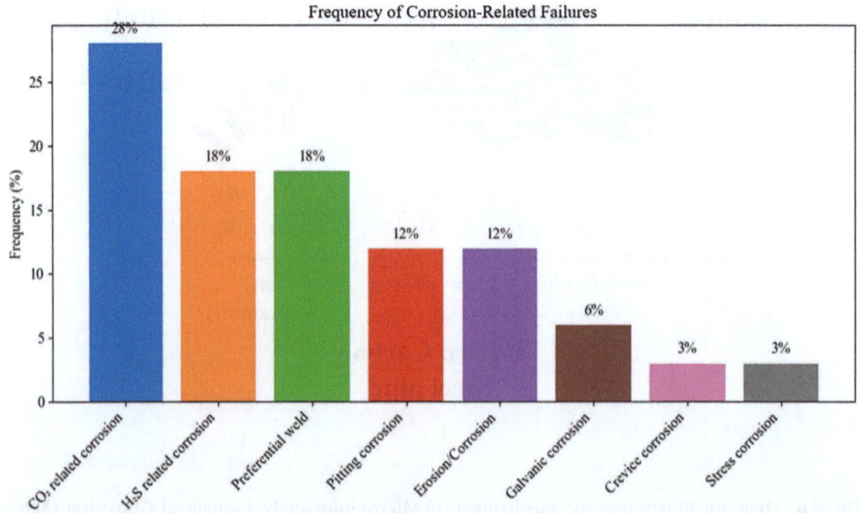

Fig. 9.7 Relative frequency of different corrosion-related failure mechanisms observed in oil and gas well systems. CO_2 and H_2S corrosion dominate internal degradation cases, while pitting, erosion-corrosion, and galvanic corrosion also pose significant localized threats (Kadhim & Ali, 2021)

9.4 External Corrosion Mechanisms

The exterior surfaces of casing strings, and potentially subsea structures, are exposed to formation fluids, muds, or cement pore water in the annuli, or seawater/soil externally.

- **Corrosive Formation Fluids**: If zonal isolation by cement is poor, corrosive brines or fluids containing CO_2/H_2S migrating from adjacent formations can contact the casing exterior in the annulus, causing corrosion similar to internal mechanisms (Wasim & Djukic, 2022).
- **Oxygen Corrosion (Shallow Depths)**: Near the surface, oxygenated groundwater (onshore) or seawater (in offshore wells, especially in the splash zone or if annulus cement does not reach surface) can contact the outer casing strings or conductor pipe (Ismail & Adan, 2014; Schremp et al., 1961). If protective coatings or cathodic protection are absent or compromised, this leads to aggressive external corrosion and metal loss. Splash zones on offshore platforms are particularly vulnerable due to repeated wet-dry cycling and high oxygen levels.
- **Galvanic Corrosion**: Occurs when two dissimilar metals are electrically connected in an electrolyte. The more electrochemically active metal becomes the anode and corrodes preferentially (e.g., connection between a carbon steel casing and a more noble CRA wellhead component, or differences in metallurgical state along a single casing string). Can also occur due to differences in the

environment along the casing string (e.g., different soil types with varying resistivity, aerated vs. non-aerated zones in an annulus) (O'Dea et al., 2023; Schremp et al., 1961).

- **Stray Current Corrosion**: Electrical currents from external sources (e.g., interference from cathodic protection systems for other nearby structures, high-voltage power lines, welding currents) can sometimes flow through the well casing, causing localized anodic sites and rapid corrosion where the current leaves the pipe (Hanson, 2003).
- **Cement Pore Water**: Initially, the high pH (~12.5) of set Portland cement pore water passivates the steel surface, protecting against general corrosion (Sumra et al., 2020). However, over long periods or if the cement is chemically altered (e.g., by CO_2 carbonation reducing pH, or leaching of calcium hydroxide by flowing water), this passive layer can break down, allowing external corrosion to initiate, particularly in the presence of aggressive species like chlorides.

Managing external corrosion is crucial for maintaining casing integrity and preventing leaks to the environment or loss of structural support.

9.5 Erosion and Erosion-Corrosion

- **Erosion**: Erosion is the physical removal of material from a surface by the mechanical action of flowing fluids, especially those containing solid particles (e.g., sand produced from the reservoir, proppant flowback after fracturing, solids in drilling mud or injection water) (Arefi et al., 2004; Laronge & Ward Thomas, 1999; Venkatesh, 1986).
  - **Mechanism**: High-velocity fluid impingement or abrasive action of particles wears away the metal surface. Typically occurs at points of high turbulence, flow direction change (bends, elbows, tees), restrictions (chokes, valves, perforation tunnels), or where flow impinges on a surface.
  - **Factors**: Rate increases significantly with flow velocity (often exponentially), particle concentration, particle size/hardness/shape, and impingement angle. High flow rates in production tubing or near perforations are common areas for erosion.
- **Erosion-Corrosion**: This is a highly aggressive synergy where erosion and corrosion act together, resulting in metal loss rates far exceeding the sum of erosion and corrosion acting alone (Rao & Mulky, 2023).
  - **Mechanism**: The mechanical action of erosion (or even flow shear stress alone, sometimes called flow-accelerated corrosion) continuously removes any protective corrosion product layers (scales like $FeCO_3$, FeS, or passive films) from the metal surface. This prevents the formation or reformation of these layers, constantly exposing fresh, active metal underneath to the corrosive environment, dramatically accelerating the overall metal loss rate.

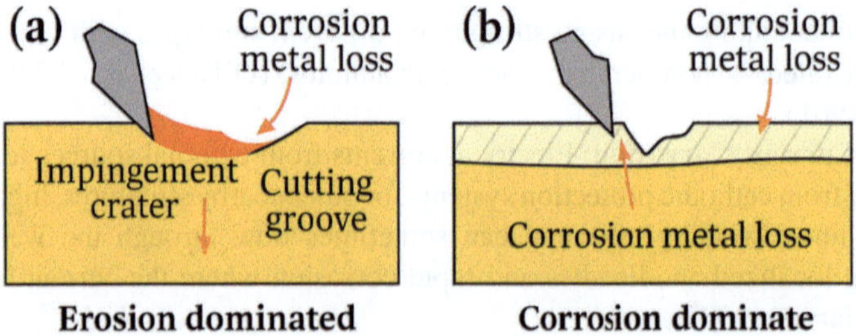

Fig. 9.8 Diagrams comparing metal loss mechanisms: **a** Erosion-dominated loss characterized by impingement craters and cutting grooves due to high-velocity particle impact, and **b** corrosion-dominated loss resulting from gradual dissolution of the passive layer and material underneath

- **Occurrence**: Common in high-velocity production streams containing both solid particles (sand) and corrosive species (CO_2, H_2S, organic acids). Particularly aggressive at points of high turbulence or impact (chokes, bends, valves, production nipples, near perforations).

Managing erosion-corrosion often involves controlling flow velocities (e.g., using larger diameter tubing, choking back the well to flow below critical velocity), managing solids production (sand control completions, limiting flow rates when sand is produced), using erosion-resistant materials or coatings in critical areas (chokes, valves, nipples), and ensuring that corrosion inhibition is effective even under high flow/shear conditions. Figure 9.8 illustrates the distinction between erosion-dominated and corrosion-dominated mechanisms of metal loss, which are critical for understanding material degradation under high-velocity flow and chemically aggressive environments.

9.6 Corrosion Control Strategies

Maintaining well integrity requires a multi-faceted, proactive approach to corrosion control throughout the well lifecycle, integrated into the WIMS. Strategies are selected based on the predicted corrosion mechanisms, severity, required design life, and cost-effectiveness.

- **Materials Selection (Often the Primary Strategy)**: Choosing materials inherently resistant to the anticipated environment is often the most reliable and long-term approach, especially for critical components or in severe service.
  - **Carbon Steel Limits**: Using carbon/low-alloy steels only within their acceptable operating limits for temperature, pressure, H_2S partial pressure, CO_2 partial pressure, pH, and chloride content, as defined by standards like NACE MR0175/ISO 15,156 (for sour service) and operator guidelines based on corrosion prediction models.

- **Corrosion Resistant Alloys (CRAs)**: Employing stainless steels (e.g., 13Cr, Super 13Cr), duplex stainless steels, super duplex stainless steels, or nickel-based alloys (e.g., Inconel) for tubing, casing, wellhead components, valves, and packers in highly corrosive environments (high H_2S, CO_2, chlorides, high temperatures) (Askari et al., 2021). CRAs are significantly more expensive than carbon steel, driving up well construction costs. Selection depends on a detailed assessment of the environment and required service life, balancing cost and performance. CRA-lined pipe or clad components offer potential cost savings but introduce integrity considerations for the weldment and liner integrity.

- **Corrosion Inhibition**: Injecting chemical inhibitors into the produced or injected fluids to reduce the corrosion rate on metal surfaces.
  - **Mechanism**: Inhibitors are chemicals that adsorb onto the metal surface, forming a protective film that interferes with the anodic or cathodic reactions, or both. Different inhibitor chemistries (e.g., imidazolines, quaternary ammonium salts, film-forming amines, phosphonates) are effective against different corrosion types (CO_2, H_2S, acid).
  - **Application**: Can be applied via continuous injection (via capillary string strapped to tubing, or injection mandrel), periodic batch treatments (pumping down tubing/annulus), or squeeze treatments (forcing inhibitor into the formation near the wellbore).
  - **Effectiveness**: Requires careful selection based on fluid compatibility (oil, water, gas, other chemicals), temperature stability, effectiveness at required concentrations, and environmental regulations (e.g., biodegradability, toxicity) (Askari et al., 2021; Solovyeva et al., 2023). Maintaining continuous presence of the inhibitor film is critical, requiring consistent injection or regular batching. Monitoring for inhibitor residual and effectiveness is crucial. Advanced delivery systems, like encapsulated inhibitors or stimuli-responsive polymers, are emerging areas of research (Solovyeva et al., 2023).

- **Coatings and Linings**: Applying a physical barrier layer between the corrosive fluid and the steel surface.
  - **Internal Coatings**: Thin film coatings (e.g., fusion-bonded epoxy—FBE, phenolic, polyurethanes, polyketones) applied internally to tubing or casing provide a barrier but are susceptible to damage during handling, installation (particularly connection makeup), wireline operations, or abrasive flow. Their effectiveness relies on maintaining coating integrity. Advanced organic and inorganic coatings, including nanomaterial-enhanced composites and fluoropolymers, offer improved resistance to harsh conditions and potentially self-healing properties (Solovyeva et al., 2023). Qualification standards (ASTM, NACE, ISO) are used to assess coating performance regarding adhesion, abrasion resistance, chemical resistance, and cathodic disbonding (Solovyeva, 2023).

  - **Linings**: Thicker linings, sometimes metallic (e.g., CRA-lined pipe) or non-metallic (e.g., GRE—glass reinforced epoxy liners inserted into existing casing for rehabilitation), can offer robust protection but add complexity and cost.
  - **External Coatings**: Coatings (FBE, coal tar epoxy, polyurethanes) are commonly applied to surface casing or conductor pipe, particularly in the splash zone or near-surface sections exposed to soil or seawater, often used in conjunction with cathodic protection.

- **Operational Controls**:

  - **Dehydration**: Removing water from gas streams significantly reduces internal corrosion rates, as water is required for the electrochemical process.
  - **Flow Rate Control**: Limiting flow velocities can reduce erosion-corrosion and support protective scale formation, although too low a velocity might allow water/solids to drop out and cause issues like under-deposit corrosion or plugging.
  - **Solids Management**: Implementing sand control measures (e.g., gravel packs, screens, frac packs) minimizes sand production and thus erosion.
  - **Water Quality Control (Injectors)**: Treating injected water (deaeration, filtration, biocides, scale/corrosion inhibitors) is critical for preventing internal corrosion, scaling, and MIC in injection wells and associated facilities (see Chap. 23).
  - **pH Stabilization**: Adjusting fluid pH (where feasible and compatible) can sometimes reduce corrosion rates by promoting protective scale formation or reducing acidity.

- **Cathodic Protection (CP)**: Used primarily to control external corrosion of buried pipelines or the submerged portions of offshore structures and outer well casings (Lawson et al., 2024). It works by making the entire steel structure the cathode of an electrochemical cell, thereby preventing iron dissolution (the anodic reaction). This is achieved by connecting it to a more active metal (sacrificial anode, e.g., zinc, aluminum) or impressing a current from an external power source (impressed current CP–ICCP) using inert anodes. CP is highly effective for external corrosion control when properly designed, installed, and maintained, and often works in conjunction with external coatings.
- **Corrosion Monitoring**: Regularly monitoring corrosion rates is essential to verify the effectiveness of control strategies, assess the rate of degradation, and detect potential problems early. Techniques include:

  - **Corrosion Coupons**: Weighted strips of steel (or CRA) exposed to the process stream, removed periodically, and weight loss measured to calculate average corrosion rate over the exposure period. Provide average, not localized, rates.
  - **Electrical Resistance (ER) Probes**: Measure the change in electrical resistance of a wire element or thin film exposed to the fluid, which correlates with cumulative metal loss. Can provide real-time or near-real-time data.

- **Linear Polarization Resistance (LPR) Probes**: Electrochemical technique providing instantaneous corrosion rate measurements but requires a conductive fluid path (water). Sensitive to changes but less useful for long-term cumulative loss.
- **Fluid Analysis**: Monitoring iron counts, manganese counts, pH, H_2S/CO_2 levels, organic acid content, chloride levels, bacterial counts, and inhibitor residuals in produced fluids. Increasing iron counts can indicate active corrosion.
- **Inspection Logging**: Periodic logging using multi-finger calipers (MFC) or electromagnetic thickness (EM) tools to directly measure internal and/or external metal loss and assess remaining wall thickness (See Chap. 12). Comparing logs run at different times quantifies the metal loss rate over that period.

To mitigate the detrimental effects of corrosion in oil and gas systems, a range of corrosion control strategies are employed depending on the operational environment, materials in use, and economic feasibility. These strategies encompass both passive and active protection approaches, as summarized in Fig. 9.9. One of the most direct methods to determine corrosion rate involves measuring the weight loss of a metallic coupon or component exposed to the corrosive environment:

$$\text{Corrosion Rate (mm/year)} = \frac{87.6 \times W}{\rho \times A \times t} \tag{9.2}$$

where **W** is the total mass loss due to corrosion in milligrams (mg), **ρ** is again the density of the metal (g/cm^3), **A** is the surface area exposed to the corrosive medium in square centimeters (cm^2), and **t** is the duration of exposure in hours. This formula is typically used when laboratory coupon testing or in-situ coupon retrieval is conducted (e.g., as discussed in Sect. 9.6 on corrosion monitoring).

9.7 Challenges in Corrosion Monitoring and Prediction

Despite advances, predicting and monitoring downhole corrosion accurately remains challenging:

- **Complex Environments**: Downhole conditions (P/T gradients, multiphase flow, unpredictable water wetting, transient conditions during startups/shutdowns) are difficult to replicate accurately in lab tests or models.
- **Localized Corrosion**: Pitting and MIC are highly localized and difficult to predict or monitor with average rate measurements (coupons, ER probes). Inspection logging is often needed to find these.
- **Inhibitor Effectiveness**: Ensuring inhibitors reach the target surface at effective concentrations in complex multiphase flow regimes is difficult. Monitoring residuals or water chemistry might not reflect actual protection level at the steel surface.

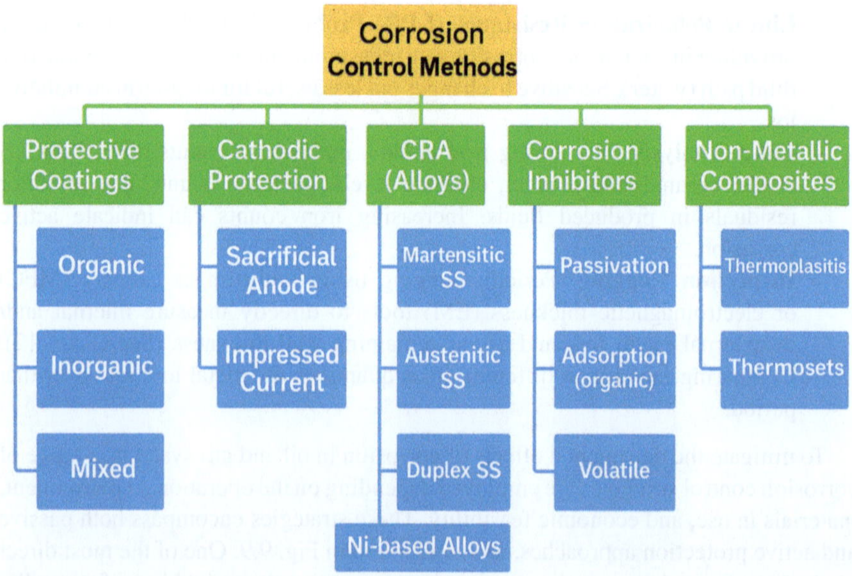

Fig. 9.9 Classification of corrosion control methods used in oil and gas systems, including protective coatings, cathodic protection, corrosion-resistant alloys (CRA), corrosion inhibitors, and non-metallic composite materials (Solovyeva et al., 2023)

- **Downhole Access**: High temperatures limit the use of some monitoring tools. Access for logging or retrieving downhole coupons/probes can be restricted by wellbore geometry or scale.
- **Data Integration**: Integrating data from various monitoring techniques, operational parameters, and prediction models is needed for a holistic view.

Effective corrosion and erosion management requires an integrated strategy combining robust design and material selection based on accurate prediction models, supported by appropriate inhibition/operational controls, and verified through diligent monitoring and periodic inspection using suitable tools. Table 9.1 summarizes the most widely adopted standards from organizations such as API, ISO, and NORSOK, providing a reference framework for engineers and asset managers tasked with ensuring long-term well integrity. Failure to manage these time-dependent threats is a leading cause of loss of well integrity during the operational phase, particularly in mature assets. Alternatively, the corrosion rate can be estimated based on the measured penetration depth (e.g., from caliper or ultrasonic logs):

$$\text{Corrosion Rate (mm/year)} = \frac{\text{Penetration Depth (mm)}}{\text{Exposure Time (years)}} \tag{9.3}$$

Table 9.1 Summary of key industry standards governing casing, tubing, and material selection in petroleum applications

Standard (organization and number)	Full title (or common name)	Description	Latest known edition/revision
API standards			
API Spec 5CT	Specification for Casing and Tubing	Specifies technical requirements for steel casing and tubing pipes used in oil wells, including dimensions, grades, and connections	10th Ed., Mar 2018 (incl. Add. 3, Nov 2023)
API Spec 5B	Threading, Gauging, and Thread Inspection of Casing, Tubing, and Line Pipe Threads	Covers threading, gauging, and inspection of casing, tubing, and line pipe threads	16th Ed., Apr 2017 (incl. Add. 1, May 2023)
API TR 5C3	Technical Report on Calculating Performance Properties of Pipe Used as Casing or Tubing	Provides formulas and calculations for determining performance properties of casing and tubing	7th Edition (March 2022)
ISO standards			
ISO 11,960	Petroleum and Natural Gas Industries—Steel Pipes for Use as Casing or Tubing for Wells	Specifies technical delivery conditions for steel pipes (casing, tubing, and pup joints), coupling stock, coupling material, and accessory materials	2020 Edition
ISO 13,680	Petroleum and Natural Gas Industries—Corrosion-Resistant Alloy Seamless Tubes for Use as Casing, Tubing, and Coupling Stock—Technical Delivery Conditions	Specifies requirements for corrosion-resistant alloy seamless tubes used as casing, tubing, and coupling stock	2020 Edition
ISO 13,679	Petroleum and Natural Gas Industries—Procedures for Testing Casing and Tubing Connections	Establishes procedures for testing casing and tubing connections for the oil and gas industries	2019 Edition
ISO 15,463	Petroleum and Natural Gas Industries— Field Inspection of New Casing, Tubing, and Plain-End Drill Pipe	Specifies requirements and provides recommendations for field inspection of new casing, tubing, and plain-end drill pipe	2003 Edition (Confirmed 2021)

(continued)

Table 9.1 (continued)

Standard (organization and number)	Full title (or common name)	Description	Latest known edition/revision
NORSOK standards			
NORSOK M-001	Materials Selection	Provides guidance and requirements for material selection and corrosion protection for hydrocarbon production and processing facilities	Rev. 5 (September 2014)
NORSOK M-630	Material Data Sheets for Piping	Specifies material data sheets (MDS) and requirements for piping materials, including tubing	Rev. 6 (October 2013)

References

Abdul-Majid, I., Ayotunde, A., Shihab, B., Abdullah, S., & Blessing, O. (2017). Unique cost saving solution applied in tight gas field in sultanate of Oman to mitigate corrosion and well integrity damage. In *Society of Petroleum Engineers—SPE Abu Dhabi International Petroleum Exhibition and Conference 2017*. https://doi.org/10.2118/188692-MS

Al-Janabi, Y. T. (2020). An overview of corrosion in oil and gas industry: Upstream, midstream, and downstream sectors. *Corrosion Inhibitors in the Oil and Gas Industry, 3–39*. https://doi.org/10.1002/9783527822140.CH1;CSUBTYPE:STRING:EDITED

Arefi, B., Settari, A., & Angman, P. (2004). Simulation of erosion in drilling tools for oil and gas industry. In *Canadian International Petroleum Conference 2004, CIPC 2004*. https://doi.org/10.2118/2004-261

Askari, M., Aliofkhazraei, M., Jafari, R., Hamghalam, P., & Hajizadeh, A. (2021). Downhole corrosion inhibitors for oil and gas production—A review. *Applied Surface Science Advances, 6*. https://doi.org/10.1016/j.apsadv.2021.100128

Corrosion and Damage Mechanisms | Mat Integrity. (2022). https://matintegrity.wixsite.com/inspection-corrosion/group/corrosion-and-damage-mechanisms/about

Comizzoli, R. B., Frankenthal, R. P., Milner, P. C., & Sinclair, J. D. (1986). Corrosion of electronic materials and devices. *Science, 234*(4774), 340–345. https://doi.org/10.1126/SCIENCE.234.4774.340;WEBSITE:WEBSITE:AAAS-SITE;JOURNAL:JOURNAL:SCIENCE;WGROUP:STRING:PUBLICATION

Davis, R., Naguib, M., & Snider, B. (2016). *Understanding and mitigating downhole corrosion and wear failures*.

de Waard, C., Lotz, U., & Milliams, D. E. (1991). Predictive model for CO_2 corrosion engineering in wet natural gas pipelines. *Corrosion, 1991*, 1–16. https://doi.org/10.5006/C1991-91577

Franco, R. J. (2020). Internal corrosive XE "corrosive" environment control. *Corrosion Control in Petroleum Production, 371–418*. https://doi.org/10.5006/37646-CH09

Hanson, H. R. (2003). Stray-current corrosion. *Corrosion: Fundamentals, Testing, and Protection*, 214–215. https://doi.org/10.31399/ASM.HB.V13A.A0003608

Ismail, A., & Adan, N. H. (2014). Effect of oxygen concentration on corrosion rate of carbon steel in seawater. *American Journal of Engineering Research (AJER), 03*, 64–67. www.ajer.org

Jain, S., Al Hamadi, M. A., Alghasra, A. M., Saada, M., & Amin, A. H. (2016). Well integrity management: Analysis of multi-barrier corrosion in mature wells using advanced magnetic impulse technology. In *Society of Petroleum Engineers—Abu Dhabi International Petroleum Exhibition and Conference 2016*. https://doi.org/10.2118/182968-MS

Kadhim, M. G., & Ali, Dr. M. T. (2021). A critical review on corrosion and its prevention in the oilfield equipment. *Journal of Petroleum Research and Studies*, *7*(2), 162–189. https://doi.org/10.52716/JPRS.V7I2.195

Kain, V. (2011). Corrosion-resistant materials. *Functional Materials: Preparation, Processing and Applications, 507–547*. https://doi.org/10.1016/B978-0-12-385142-0.00012-X

Kannan, P., Su, S. S., Mannan, M. S., Castaneda, H., & Vaddiraju, S. (2018). A review of characterization and quantification tools for microbiologically influenced corrosion in the oil and gas industry: Current and future trends. *Industrial and Engineering Chemistry Research, 57*(42), 13895–13922. https://doi.org/10.1021/ACS.IECR.8B02211/SUPPL_FILE/IE8B02211_SI_001.PDF

Khouzani, M. K., Bahrami, A., Hosseini-Abari, A., Khandouzi, M., & Taheri, P. (2019). Microbiologically influenced corrosion of a pipeline in a petrochemical plant. *Metals, 9*(4), 459. https://doi.org/10.3390/MET9040459

Klever, F. J. (1992). Burst strength of corroded pipe: Flow stress revisited. In *Proceedings of the Annual Offshore Technology Conference* (pp. 417–431). https://doi.org/10.4043/7029-MS

Laronge, T. M., & Ward Thomas M, M. A. (1999). The basics and not so basics of water corrosion processes altered by flow changes. OnePetro. https://dx-doi-org.ezproxy.lib.utexas.edu/

Lawson, K., Land, J. C., & Kowalski, A. R. (2024). Vintage pipeline enhanced corrosion management analysis. OnePetro. https://dx-doi-org.ezproxy.lib.utexas.edu/

Muthukumar, N. (2014). Petroleum products transporting pipeline corrosion—A review. *The Role of Colloidal Systems in Environmental Protection, 527–571*. https://doi.org/10.1016/B978-0-444-63283-8.00021-1

Obot, I. B., Sorour, A. A., Verma, C., Al-Khaldi, T. A., & Rushaid, A. S. (2023). Key parameters affecting sweet and sour corrosion: Impact on corrosion risk assessment and inhibition. *Engineering Failure Analysis, 145*, Article 107008. https://doi.org/10.1016/J.ENGFAILANAL.2022.107008

O'Dea, V., Wenger, D., & Cheshire, B. (2023). Galvanic corrosion in water and wastewater structures: Coupling stainless steel and carbon steel metals leads to accelerated corrosion. OnePetro. https://dx-doi-org.ezproxy.lib.utexas.edu/

Olsen, S. (2003). CO_2 corrosion prediction by use of the NORSOK M-506 model—Guidelines and limitations. *Corrosion, 2003*, 1–12. https://doi.org/10.5006/C2003-03623

Olsen, S., Halvorsen, A. M., Lunde, P. G., & Nyborg, R. (2005). CO_2 corrosion prediction model—Basic principles. OnePetro. https://dx.doi.org/

Porcayo-Calderon, J., Canto, J., Martinez-de-la-Escalera, L. M., & Neri, A. (2022). Sweet corrosion XE "sweet corrosion" inhibition by CO_2 capture. *Molecules, 27*(16), 5209. https://doi.org/10.3390/MOLECULES27165209

Rao, P., & Mulky, L. (2023). Erosion XE "Erosion"—C++6orrosion of materials in industrial equipment: A review. *ChemElectroChem, 10*(16), Article e202300152. https://doi.org/10.1002/CELC.202300152;REQUESTEDJOURNAL:JOURNAL:21960216;WEBSITE:WEBSITE:CHEMISTRY-EUROPE;WGROUP:STRING:PUBLICATION

Schremp, F. W., Chittum, J. F., & Arczynski, T. S. (1961). Use of oxygen scavengers to control external corrosion of oil-string casing. *Journal of Petroleum Technology, 13*(07), 703–711. https://doi.org/10.2118/1606-G-PA

Scribner, L. A. (2001). Corrosion by organic acids. OnePetro. https://dx.doi.org/

Solovyeva, V. A., Almuhammadi, K. H., & Badeghaish, W. O. (2023). Current downhole corrosion control solutions and trends in the oil and gas industry: A review. *Materials, 16*(5), 1795. https://doi.org/10.3390/MA16051795

Song, J., Wang, L., Zibart, A., & Koch, C. (2012). Corrosion protection of electrically conductive surfaces. *Metals, 2*(4), 450–477. https://doi.org/10.3390/MET2040450

Sotoodeh, K. (2021). Material selection and corrosion. *Subsea Valves and Actuators for the Oil and Gas Industry, 421–457*. https://doi.org/10.1016/B978-0-323-90605-0.00004-9

Sumra, Y., Payam, S., & Zainah, I. (2020). The pH of cement-based materials: A review. *Journal of Wuhan University of Technology-Materials Science and Engineering, 35*(5), 908–924. https://doi.org/10.1007/S11595-020-2337-Y

Telegdi, J., Shaban, A., & Trif, L. (2017). Microbiologically influenced corrosion (MIC XE "MIC"). *Trends in Oil and Gas Corrosion Research and Technologies: Production and Transmission, 191–214*. https://doi.org/10.1016/B978-0-08-101105-8.00008-5

Venkatesh, E. S. (1986). *Erosion damage in oil and gas wells*. https://doi.org/10.2118/15183-MS

Wasim, M., & Djukic, M. B. (2022). External corrosion of oil and gas pipelines: A review of failure mechanisms and predictive preventions. *Journal of Natural Gas Science and Engineering, 100*, Article 104467. https://doi.org/10.1016/J.JNGSE.2022.104467

Yan, X., Wang, Y., Du, Q., Jiang, W., & Li, R. (2019). Research progress on factors affecting oxygen corrosion and countermeasures in oilfield development. https://doi.org/10.1051/e3sconf/201913101031

Chapter 10
Cement Sheath Failure Mechanisms and Degradation

Abstract This chapter presents a comprehensive analysis of the mechanisms responsible for the degradation and failure of cement sheaths in oil, gas, and carbon storage wells. Serving as a primary barrier for zonal isolation and casing support, the cement sheath is critical for maintaining long-term well integrity. However, it is subject to a variety of mechanical, thermal, chemical, and geomechanical stresses throughout the well's lifecycle, all of which can compromise its performance. This chapter begins by examining mechanical failure modes such as debonding, radial and shear cracking, and micro-annulus formation—conditions often exacerbated by poor mud removal, casing expansion, thermal cycling, or insufficient cement properties. The chapter also explores the cumulative effects of operational stresses like pressure and temperature fluctuations, which drive fatigue damage in the cement matrix and casing interfaces over time. In addition to mechanical degradation, the chapter delves into chemical attack mechanisms including carbonation, sulfate attack, acid leaching, and magnesium infiltration—each of which alters the microstructure of the cement and increases permeability. These reactions are particularly relevant in CO_2-rich and aggressive saline environments typical of carbon storage and geothermal wells. Advanced topics include long-term performance strategies such as the incorporation of flexible or fiber-reinforced cement, elastomeric and pozzolanic additives, and self-healing formulations. Emphasis is placed on understanding time-dependent degradation under coupled thermal, mechanical, and chemical (TMC) conditions, with modeling and laboratory simulations guiding cement system design for extended durability. The influence of geomechanical stresses—due to formation movement, reservoir compaction, salt creep, and fault reactivation—is also explored, highlighting the necessity of stress-informed cement design. Finally, the chapter outlines mitigation and remediation approaches, including cement squeezes, polymer gels, and section milling, while also referencing key industry standards (API, ISO, NORSOK) for cement performance verification. Through a multidisciplinary lens, this chapter underscores the importance of robust cement design, predictive modeling, and proactive remediation in preventing barrier failure and ensuring long-term subsurface containment in a variety of well applications, including permanent abandonment (P&A), CO_2 storage, and thermal recovery operations.

© The Author(s), under exclusive license to Springer Nature Switzerland AG 2026

A. Alsubaih and K. Sepehrnoori, *Well Integrity for Oil, Gas, and Subsurface Storage Operations*, https://doi.org/10.1007/978-3-032-15233-6_10

Adapted figure from "Towards the design of an enriched concrete with enhanced dissipation performances. *Source* Elsevier (Scerrato et al. 2016)

Chapter's Objectives

- Describe mechanical failure modes of the cement sheath (debonding, cracking) and their causes.
- Explain the impact of operational stresses (pressure, temperature cycling) on cement integrity and durability over time.
- Detail common chemical degradation mechanisms (carbonation, sulfate attack, leaching, acid attack).
- Discuss considerations for long-term cement performance and durability in various downhole environments.
- Outline the influence of geomechanical stresses on cement sheath integrity.
- Introduce common mitigation and remediation strategies for cement sheath failures.

10.1 Introduction: The Importance of the Cement Barrier

As established in Chap. 5, the hardened cement sheath placed in the annular space between casing and formation serves as a critical barrier element within the wellbore system. Its primary functions are to ensure zonal isolation by preventing fluid migration between geological formations, to provide mechanical support for the casing string, and to offer initial corrosion protection by buffering steel components from aggressive subsurface fluids (Burdylo & Birch, 1990; Gu et al., 2024).

However, like all engineered materials deployed in harsh environments, cement is not immune to long-term degradation. Over the operational life of the well, the cement sheath is exposed to a variety of mechanical, thermal, and chemical stresses (De Andrade & Sangesland, 2016b; Li et al., 2023). These stresses may originate from well construction and stimulation operations, routine production and injection cycles, or long-term geomechanical changes in the surrounding formation. In combination, these loads can initiate damage mechanisms such as debonding, cracking, micro-annulus formation, and chemical alteration of the cement matrix. Leakage in wells typically arises from micro-cracks that develop and propagate due to factors such as cement job design, downhole stresses, chemical interactions with subsurface fluids, and temperature fluctuations (De Andrade & Sangesland, 2016a; De Souza & Sanchez, 2023; Ma et al., 2018).

If the cement barrier fails to maintain its hydraulic and mechanical integrity, the consequences can be severe. Loss of zonal isolation may lead to the development of sustained casing pressure (SCP), formation crossflow, and the unintended migration of formation fluids—including hydrocarbons, brine, or CO_2—into non-target zones. Such migration pathways can result in environmental contamination, including aquifer pollution and surface seeps, while also compromising well performance, reducing injectivity or productivity, and increasing the risk of casing corrosion and deformation. In severe cases, the structural role of the cement may be lost altogether, resulting in loss of casing centralization or mechanical instability in the well.

Importantly, cement degradation is not always immediate. It is often progressive and time-dependent, influenced by cumulative loading cycles, exposure duration, and evolving downhole conditions. For example, pressure and temperature cycling can induce fatigue at the casing–cement interface, while chemical degradation—such as carbonation, sulfate attack, acid leaching, and magnesium infiltration—can slowly alter the microstructure of the cement, reduce strength, and increase permeability.

Understanding the full spectrum of cement sheath failure mechanisms—including their causes, progression, and interdependence—is therefore vital for ensuring well integrity. This knowledge supports the design of durable cement systems; the interpretation of annular pressure anomalies; and the planning of monitoring, maintenance, and remediation strategies throughout the well's life.

In the following sections, this chapter systematically explores the mechanical, thermal, chemical, and geomechanical factors that contribute to cement sheath degradation. Emphasis is placed on how these mechanisms evolve over time, how they interact, and what operational and design practices can mitigate their effects.

10.2 Mechanical Failure Modes

These failures relate to the physical breakdown of the cement sheath or its bond to the casing/formation, often initiated or exacerbated by mechanical stresses imposed during well construction or operation. Quantitative evaluation of cement sheath cracking reveals typical crack widths ranging from 1 to 500 μm, leading to effective permeability increases (10^{-17} to 10^{-12} m^2) and potential leak rates of 1–10,000 mL/min for gas or 1–1,000 mL/min for oil, depending on crack geometry and pressure differential (Alberdi-Pagola & Fischer, 2023). Completely undamaged cement sheaths are rare in practice; therefore, design strategies often aim to minimize crack size and maintain overall seal effectiveness rather than prevent cracking entirely (Alberdi-Pagola & Fischer, 2023). Figure 10.1 complements this context by summarizing a range of micro-annulus thicknesses reported in the literature, highlighting the variability in annular separation widths observed across different fields and laboratory studies. These measurements directly influence the hydraulic conductivity of compromised wellbore sections and inform cement design targets for long-term integrity. Figure 10.2 provides a comprehensive schematic summarizing key causes and leakage pathways associated with mechanical failure of the cement sheath in wellbores. These failure modes—often initiated by suboptimal cement placement, material deficiencies, or severe operational conditions—result in the development of micro-annuli, fractures, and debonded interfaces that compromise zonal isolation. Understanding the spatial relationship between these causes and their resultant leakage geometries is fundamental to diagnosing, preventing, and remediating well integrity failures.

- **Debonding (Micro-annulus Formation)**: This is the loss of intimate contact and hydraulic seal at the interfaces between the set cement and the casing (casing–cement interface) or the cement and the formation wall (cement–formation interface) (Orlic et al., 2018a). Figure 10.3 illustrates the interfacial transition zone (ITZ) between the cement sheath and the casing or formation. The ITZ represents a mechanically and chemically distinct zone where bonding is often weakest, making it the most likely location for debonding and micro-annulus formation. Diagram (A) shows a side view of the wellbore, with formation fluids migrating through the interfacial region. Diagram (B) presents a top-down cross section, highlighting the concentric annular nature of the ITZ. These interfaces are especially vulnerable to stress-induced separation during pressure or thermal cycling, poor mud removal, or cement shrinkage, and often represent the initial site of

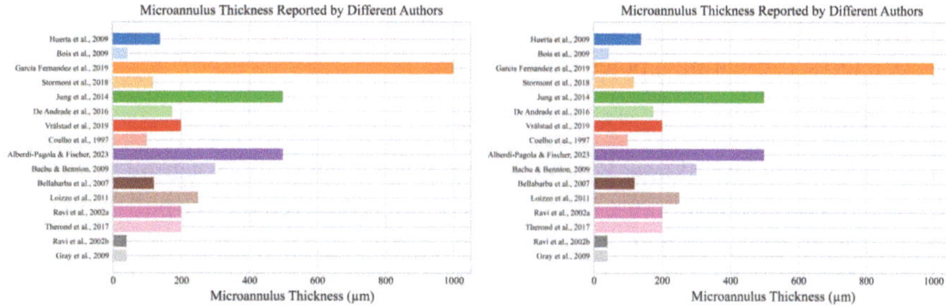

Fig. 10.1 Cement micro-annulus thickness measurements reported by different authors (comparison view). This figure compares two identical datasets of reported micro-annulus thicknesses (in micrometers, μm) from various studies, visualized side-by-side to emphasize reproducibility and consistency in literature data. Each bar represents a unique source, reflecting variability in micro-annulus formation due to different measurement techniques, well conditions, and interpretations. Micro-annulus gaps as small as 50 μm and as large as 1000 μm have been documented, underscoring their critical role as potential leakage pathways affecting annular seal integrity

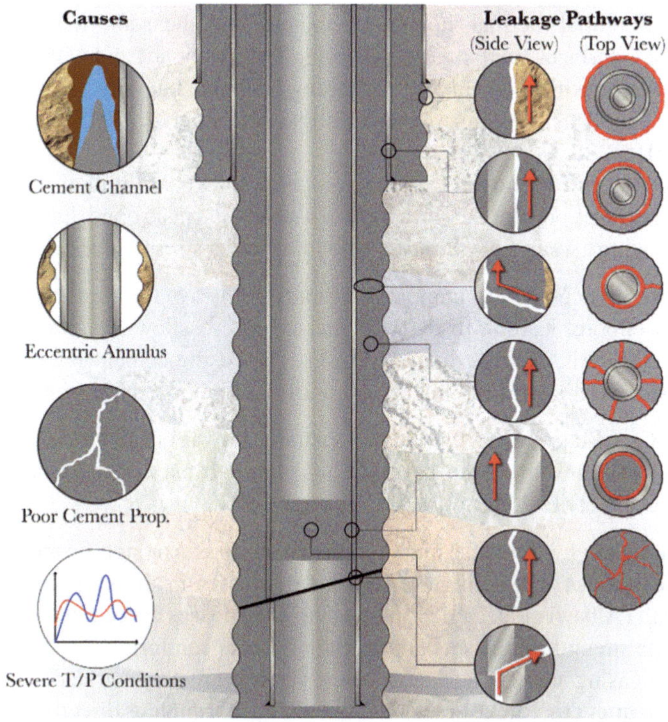

Fig. 10.2 Cross-sectional schematic illustrating common mechanical failure causes and resulting leakage pathways in cement sheaths. Causes include cement channels, eccentric annuli, poor cement properties, and severe temperature/pressure cycling. Leakage pathways are visualized from both sides and top views, including micro-annuli, radial cracks, longitudinal fractures, and interface debonding, all of which can compromise hydraulic isolation (Liu, 2021)

integrity loss in the cement barrier. A small gap, or micro-annulus, forms along the interface, creating a potential leak path.

- **Causes**

  Cement Shrinkage: Volumetric shrinkage during cement setting and hydration (especially before sufficient bond strength develops) can pull the cement away from the interfaces (Gu et al., 2022).
  Pressure Changes: Applying internal casing pressure (e.g., during pressure testing—Chap. 7, or operational pressure cycles) causes the casing to expand radially outward, potentially breaking the bond or initiating a micro-annulus at the casing–cement interface. Subsequent pressure reduction may not fully restore contact due to hysteresis. External pressure changes can also cause casing deflection.
  Temperature Changes: Heating the casing (e.g., during production startup, steam injection) causes it to expand radially and axially more than the surrounding cement and formation. Cooling causes contraction. These differential thermal movements induce high shear and tensile stresses at the interfaces (particularly the casing–cement bond) and within the cement sheath, which are highly damaging to bond integrity. Thermal cycling exacerbates this effect through fatigue (Kiran et al., 2017; Torsæter et al., 2015).
  Poor Mud Removal: Residual mud film, filter cake, or mud channels at the interfaces prevent the formation of a strong intrinsic bond during hydration, leaving a weak zone prone to debonding under stress (Jafarzade et al., 2014).

- **Consequences**: Micro-annuli are primary pathways for fluid migration along the wellbore, leading to SCP or inter-zonal crossflow, compromising zonal isolation (Alkhamis & Imqam, 2021). The casing-cement interface is often considered the weakest link and the most common location for micro-annuli.

• **Radial Cracking**: Cracks propagating radially outward from the casing (or sometimes radially inward from the formation) through the thickness of the cement sheath(Dong et al., 2022).

  - **Causes**: High internal casing pressure (burst loads) or significant temperature increases can induce high tensile hoop stresses in the surrounding cement sheath (Albawi et al., 2013). If these tensile stresses exceed the relatively low tensile strength of the set cement, radial cracks can form. These often initiate at the casing–cement interface, where stresses are concentrated.
  - **Consequences**: Radial cracks can link the casing annulus directly to the formation, bypassing zonal isolation, or interconnect with longitudinal cracks or debonded interfaces to create complex, high-permeability leak paths for fluid migration.

• **Axial/Shear Cracking**: Cracks running parallel or obliquely to the wellbore axis within the cement sheath (Orlic et al., 2018b).

- **Causes**: Can be induced by shear stresses during drilling subsequent sections (vibrations, eccentric rotation), perforating operations (shock waves), hydraulic fracturing treatments (high pressure and stress changes near perforations), or significant geomechanical loads from formation movement (e.g., fault slip, compaction, salt creep) causing casing deformation (ovality, shear). Casing deformation (e.g., ovalization due to collapse loads) can also induce complex cracking patterns (Dusseault et al., 2001; Kim et al., 2016a; Li et al., 2023).
- **Consequences**: Can compromise the structural integrity of the sheath and, if interconnected, create pathways for fluid migration parallel to the wellbore axis, compromising zonal isolation.

- **Shattering/Pulverization**: Complete mechanical breakdown of the cement sheath into fragments or powder, often localized to a specific interval.
  - **Causes**: Can occur due to high impact loads (e.g., perforating directly through brittle cement with high-charge density guns), extreme stress concentrations, or severe crushing loads in combination with other factors, particularly in brittle cement formulations.
  - **Consequences**: Results in a complete loss of both hydraulic sealing and structural support in the affected zone.

The mechanical properties of the set cement—its compressive strength, tensile strength, Young's modulus (stiffness), and Poisson's ratio—influence its susceptibility to these failure modes. Traditional oil well cements are often strong in compression but relatively weak and brittle in tension. Modern cement system designs sometimes incorporate additives (e.g., flexible polymers, fibers, and elastomers) to enhance flexibility, toughness, and resistance to cracking and debonding under stress and cyclic loading (Al-Yami et al., 2018a; Segre et al., 2006; Wray et al., 2009). To provide a quantitative framework for understanding debonding and micro-annulus formation at cement interfaces, the Cohesive Zone Model (CZM) is used. It simulates the traction-separation behavior of the interface and predicts failure initiation and propagation. The normal and shear tractions (t_n, t_s, t_t) are related to displacement jumps (d_n, d_s, d_t) through the **stiffness matrix** K_e:

$$\begin{bmatrix} t_n \\ t_s \\ t_t \end{bmatrix} = \begin{bmatrix} K_{nn} & 0 & 0 \\ 0 & K_{ss} & 0 \\ 0 & 0 & K_{tt} \end{bmatrix} \begin{bmatrix} d_n \\ d_s \\ d_t \end{bmatrix} \tag{10.1}$$

The parameters K_{nn}, K_{ss}, and K_{tt} represent the normal and shear stiffness components (in GPa/m), while d_n, d_s, and d_t denote the corresponding normal and shear separation displacements.

The quadratic nominal stress criterion for damage initiation:

$$\left(\frac{t_n}{t_{n0}}\right)^2 + \left(\frac{t_s}{t_{s0}}\right)^2 + \left(\frac{t_t}{t_{t0}}\right)^2 = 1 \tag{10.2}$$

where t_{n0}, t_{s0}, t_{t0} are the interface strengths (MPa) for each mode.

The energy-based criterion for fracture propagation:

$$G_c = G_n + G_s + G_t \tag{10.3}$$

The terms G_n, G_s, and G_t refer to the fracture energies (in J/m^2) associated with Mode I (opening), Mode II (sliding), and Mode III (tearing), respectively, and G_c denotes the critical fracture energy threshold beyond which complete material separation occurs.

Finally, the effective displacement at failure is obtained from:

$$D = \sqrt{(d_n)^2 + (d_s)^2 + (d_t)^2} \tag{10.4}$$

The variables d_n, d_s, and d_t correspond to displacements in the opening, sliding, and tearing directions, respectively, and are used to evaluate the progressive damage process leading to full decohesion of the material interface.

To quantitatively assess cement sheath failure under mechanical loading, various failure criteria originally developed for brittle materials and rocks can be applied to the cement matrix. These include Mohr–Coulomb, Mogi–Coulomb, Drucker–Prager, and others. They help predict the onset of cracking and failure in the cement, considering different stress paths (triaxial, octahedral shear, etc.). Table 10.1 summarizes these failure envelopes.

10.3 Impact of Operational Stresses

Well operations inevitably impose mechanical and thermal stresses on the casing–cement–formation system. While initial stresses are applied during construction (casing running, cementing), these operational stresses accumulate over time and can initiate or exacerbate mechanical damage (cracking, debonding) that degrades integrity (Zhou et al., 2024). Figure 10.3 illustrates the progressive damage induced by pressure and thermal cycling, showing how casing expansion/contraction stresses lead to debonding at the interfaces and cracking of the cement sheath over time.

- **Pressure Cycling**: Repeated cycles of high and low pressures inside the casing (e.g., during production fluctuations, batch injection, cyclic steam stimulation—CSS, pressure testing, stimulation treatments, gas storage injection/withdrawal cycles) cause the casing to repeatedly expand and contract radially and axially. This cyclic loading fatigues the bond at the casing–cement interface and can

Table 10.1 Summary of key failure criteria for cement sheath integrity assessment (Alsubaih et al., 2024)

No.	Criterion	Equation(s)
1	Mohr–Coulomb	$\sigma_1 = UCS + \sigma_3 q$, where $q = \frac{1+\sin\Phi}{1-\sin\Phi}$
2	Mogi–Coulomb	$\tau_{oct} = \alpha + b\sigma_m$ $\alpha = \frac{C_o 2\sqrt{2}}{3(q+1)}$ $b = \frac{(q-1)2\sqrt{2}}{(q+1)3}$
3	Von Mises	$\sqrt{J_2} = \sqrt{\frac{1}{6}\left[(\sigma_1-\sigma_2)^2 + (\sigma_1-\sigma_3)^2 + (\sigma_2-\sigma_3)^2\right]}$or $\sqrt{J_2} = \frac{\sigma_1}{\sqrt{3}}$
4	Drucker–Prager	$\sqrt{J_2} = k + \alpha J_1, J_1 = \frac{1}{3}(\sigma_1+\sigma_2+\sigma_3)$ $\alpha = \frac{3\sin\Phi}{\sqrt{9+3\sin^2\Phi}}, k = \frac{3C_o\sin\Phi}{2\sqrt{2}\sqrt{9+3\sin^2\Phi}}$ or $\alpha = \frac{\sqrt{3}(q-1)}{2+q}, k = \frac{\sqrt{3}C_o}{2+q}$
5	Modified Wiebols–Cook	$\sqrt{J_2} = A + BJ_1 + CJ_1^2$
6	Hoek–Brown	$\sigma_1 = \sigma_3 + \sqrt{mC_o\sigma_3 + sC_o^2}$
7	Modified Lade	$\frac{I_1^3}{I_3^3} = 27 + \eta$
8	Modified Griffith	$\sigma_1\left[\sqrt{u^2+1} - u\right] - \sigma_3\left[\sqrt{u^2+1} - u\right] = 4T_o$
9	Murrell	$(\sigma_1-\sigma_2)^2 + (\sigma_1-\sigma_3)^2 + (\sigma_2-\sigma_3)^2 =$ $24T_o(\sigma_1+\sigma_2+\sigma_3)$
10	Stassi D'Alia	$(\sigma_1-\sigma_2)^2 + (\sigma_1-\sigma_3)^2 + (\sigma_2-\sigma_3)^2 =$ $2(C_o - T_o)(\sigma_1+\sigma_2+\sigma_3) + 2C_oT_o$
11	Alsubaih et al.	$\frac{1}{3}\sqrt{(\sigma_1-\sigma_2)^2 + (\sigma_1-\sigma_3)^2 + (\sigma_2-\sigma_3)^2} =$ $A + \frac{\sqrt{3}(q-1)}{2+q}J_1 + CJ_1^2$ where $A = \frac{C_o}{\sqrt{3}} - \frac{C_o}{3}B - \frac{C_o^2}{9}C$ $B = \frac{\sqrt{3}(q-1)}{q+2} - \frac{C}{3}\left[2C_o + (q+2)\sigma_3\right]$ $C_1 = (1+0.6u_i)C_o$ $C = \frac{\sqrt{27}}{2C_1+(q-1)\sigma_3-C_o}\left(\frac{C_1+(q-1)\sigma_3-C_o}{2C_1+(2q-1)\sigma_3-C_o} - \frac{q-1}{q+2}\right)$

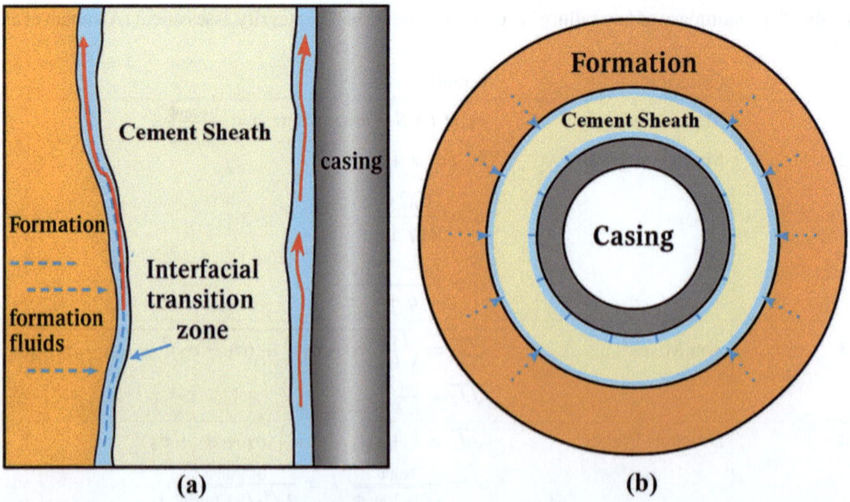

Fig. 10.3 Cross-sectional diagrams illustrating the interfacial transition zone (ITZ) between the cement sheath and casing/formation. **a** Side view showing the flow of formation fluids through micro-annuli along the ITZ. **b** Top-down view of the annular geometry, highlighting the cement sheath bounded by two ITZs where bond failure and leakage are most likely to initiate

initiate or propagate micro-annuli and cracks within the cement sheath, leading to progressive degradation of the hydraulic seal over time (Alberdi-Pagola & Fischer, 2024; Kiran et al., 2017). Cement fatigue under cyclic loading can cause microfractures that compromise the hydraulic seal.

Thermal Cycling: Similar to pressure cycling, repeated changes in temperature (e.g., production startups/shutdowns, steam injection cycles in thermal recovery wells, hot fluid circulation during cleanouts, injection of cold fluids like CO_2 or gas) cause significant differential expansion and contraction between the casing (which typically has a higher thermal expansion coefficient than cement and formation rock) (De Andrade et al., 2015). This induces cyclic shear and tensile stresses at the interfaces and within the cement, which are highly damaging to bond integrity and can cause fatigue cracking. HPHT wells, geothermal wells, thermal EOR wells (steam injection), and gas storage wells (due to Joule–Thomson cooling/heating during cycles) are particularly prone to severe thermal cycling effects and associated cement damage [(Han et al., 2023), Discussion in Chapters 18, 21, 22, 24]. Thermal stresses caused by temperature changes (thermal cycling) can be estimated by

$$\sigma_{\text{thermal}} = E \cdot \alpha_{\text{cement}} \cdot \Delta T \tag{10.5}$$

where E is Young's modulus of the cement, α_{cement} is the coefficient of thermal expansion, and ΔT represents the temperature change, all of which contribute to thermally induced strain in the cement matrix.

- **Hydraulic Fracturing**: High pressures and transient stress changes associated with hydraulic fracturing operations, especially multi-stage fracturing in unconventional wells, can impose significant stresses on the cement sheath near the perforated interval (Kim et al., 2016b; Wang et al., 2021). If the cement integrity (bond and mechanical properties) is already marginal or if pressures are excessive, this can cause cracking or debonding. Ensuring good zonal isolation by competent cement around the interval to be fractured is critical to contain fracture growth within the target zone and prevent undesirable propagation outside the target formation via channels or cracks behind pipe.
- **Vibrations**: Drilling vibrations (e.g., from PDC bit whirl), high flow rate turbulence, or artificial lift operations (e.g., ESP or PCP vibrations) can induce vibrations in the tubulars that may transmit loads to the cement sheath and contribute to long-term fatigue damage of the bond or the cement matrix.
- **Axial Loading Changes**: Changes in axial loads on the casing (e.g., due to thermal effects, ballooning/piston effects from internal/external pressure, or weight transfer) can induce stresses in the cement sheath or affect bond integrity.

Designing cement systems and casing programs that can withstand these operational stresses *over the entire well life* is a key challenge. This often involves balancing strength requirements with the need for some flexibility and toughness to accommodate cyclic loading and deformation without brittle failure (e.g., using lower Young's modulus cements, incorporating elastomers/fibers), and performing detailed stress analysis during the design phase.

10.4 Chemical Degradation Mechanisms

In addition to mechanical stresses, the cement sheath can be chemically attacked and degraded by certain components of formation fluids it encounters, particularly if zonal isolation is breached (e.g., via micro-annuli, cracks, or channels) or if corrosive fluids permeate through the cement matrix itself over long periods. Chemical degradation is a time-dependent process, with the extent of damage accumulating over decades of exposure.

- **Carbonation (CO_2 Attack)**: Dissolved CO_2 in formation water, injected CO_2, or produced fluids containing CO_2 can react with the alkaline hydration products of Portland cement, primarily calcium hydroxide ($Ca(OH)_2$) and calcium silicate hydrates (C-S-H gel) (Alsubaih et al., 2025; Kutchko et al., 2007; Omosebi et al., 2016).
  - **Mechanism**: CO_2 reacts with CH to form calcium carbonate ($CaCO_3$). This reaction initially reduces permeability in the carbonated zone but also consumes $Ca(OH)_2$, lowering the pH. At lower pH, the primary strength-giving C-S-H gel becomes unstable and decalcifies (calcium is leached out), losing its binding capacity and converting to silica gel. The overall result is a significant

loss of strength, increased porosity, and increased permeability in the degraded zone over time.

  - **Occurrence**: A major concern for long-term integrity in CO_2-rich reservoirs (sour/sweet gas), gas storage wells (especially when cushion gas mixes with formation water), and *particularly CCS wells* where the cement is exposed to high concentrations of wet CO_2 for extended periods. The rate of degradation is influenced by CO_2 partial pressure, temperature, pressure, water saturation, and flow rate (Gaurina-Međimurec, 2010; Omosebi et al., 2015, 2017).

- **Sulfate Attack**: Sulfates (SO_4^{2-}) present in formation water or injection water can react with the calcium aluminate hydrate phases (formed from C_3A hydration) and calcium hydroxide in Portland cement (Al-Dulaijan et al., 2003).
  - **Mechanism**: Reactions form expansive minerals like ettringite and gypsum. These expansive minerals generate internal stresses within the cement matrix, leading to expansion, cracking, softening, loss of strength, and loss of bonding (Pouya et al., 2022).
  - **Occurrence**: Can be an issue in formation with high sulfate concentrations or when using high-sulfate injection water (see Chap. 23). Using sulfate-resistant Portland cements (API Type II or V, with low C_3A content) or non-Portland-based cements designed for sulfate resistance can mitigate this risk.
- **Acid Attack (H_2S, Organic Acids)**: Acidic fluids (low pH) can directly dissolve the alkaline hydration products ($Ca(OH)_2$ and C-S-H) of Portland cement, leading to increased porosity, permeability, and loss of structural integrity (Mackay, 2019). H_2S dissolved in water forms a weak acid, and its presence can accelerate cement degradation. Organic acids present in some reservoir fluids can also be aggressive, particularly at high temperatures.
- **Leaching**: Flowing formation water (even seemingly non-aggressive water) can slowly dissolve and carry away soluble components of the cement matrix, particularly calcium hydroxide ($Ca(OH)_2$), over very long periods. This process increases porosity and permeability, potentially creating or enlarging leak paths, especially in fractured or already damaged cement. This is a long-term process that can be accelerated by high flow rates or low pH fluids (Duchesne & Bertron, 2013).
- **Magnesium Attack**: Brines rich in magnesium chloride or sulfate can also degrade cement through complex reactions that replace calcium phases with weaker magnesium phases like brucite ($Mg(OH)_2$) and magnesium silicate hydrates, leading to strength loss and increased permeability (Poole et al., 1994).

The rate and severity of chemical degradation depend on the cement formulation (e.g., water-to-cement ratio, use of pozzolans like fly ash or slag which consume CH and refine pore structure can improve resistance to some forms of attack); the concentration and type of aggressive species; and temperature, pressure, and fluid flow rates through or along the cement sheath. For chemical degradation, carbonation front depth can be estimated with:

$$d = k\sqrt{t} \tag{10.6}$$

The carbonation front depth d (in mm) can be estimated as a function of the carbonation rate coefficient k (in mm/$\sqrt{}$year) and the exposure time t (in years), following a square-root-of-time relationship commonly used in durability assessments.

10.5 Long-Term Cement Performance and Durability

Ensuring the cement sheath maintains its integrity over the entire well life, potentially decades for production wells and centuries for P&A barriers in CCS wells, requires careful consideration of long-term performance and durability under the specific downhole environment. In recent years, various strategies have been proposed to enhance well integrity, ranging from proactive cement design improvements to the incorporation of advanced cement formulations. These include the use of specialized additives such as calcium carbonate precursors, polymers, and silica gels to improve sealing performance. Traditionally, remedial actions like top jobs or cement squeeze operations have been employed to address integrity issues after they arise. However, these methods are often costly, time-consuming, and disruptive, with a reported success rate of only around 50%, and they typically require well shutdowns, resulting in significant operational expenses (Ma et al., 2018).

An alternative to reactive remediation is the enhancement of cement's mechanical properties to improve its durability and resilience against downhole stresses, particularly those induced by thermal and mechanical cycling. One effective approach involves the incorporation of elastomeric particles into the cement slurry. These particles reduce Young's modulus of the hardened cement, increasing its elasticity and enabling the cement sheath to better absorb and distribute stresses encountered in the wellbore environment (Al-Yami et al., 2018b). While the addition of elastomers can lead to a modest reduction in compressive strength, they play a critical role in improving toughness by absorbing stress and arresting crack propagation (Segre et al., 2006). Nonetheless, there are practical limitations to elastomer loading: as concentration increases, the slurry viscosity also rises, leading to higher equivalent circulating density (ECD) and requiring lower pumping rates to avoid inadvertently fracturing the formation (Wray et al., 2009).

Beyond mechanical enhancement, elastomeric additives also offer functional sealing capabilities through their ability to swell in the presence of hydrocarbons. Based on the Flory–Huggins theory of polymer–solvent interaction, elastomer chains relax and expand volumetrically when exposed to organic solvents (Billmeyer, 1984). In the context of oil and gas production, hydrocarbon fluids act as solvents, initiating the swelling of elastomer particles embedded in the cement. As these fluids migrate through cracks or micro-annuli within the cement sheath, the swelling particles progressively fill the void space. Once the particles expand to a size comparable to the crack width, they can effectively seal the defect and obstruct further fluid flow—providing a self-sealing mechanism that enhances the long-term integrity of the cement barrier.

- **Material Selection**: Choosing cement formulations (or alternative materials for P&A—Chap. 27) specifically designed for long-term durability in the anticipated downhole environment is paramount. This involves considering temperature (requiring silica flour at high temperature), pressure, and potential exposure to aggressive fluids (CO_2, H_2S, sulfates, high salinity). This may involve using specialized Portland cements, blended cements (with high volumes of fly ash, slag, silica fume, metakaolin), or non-Portland-based systems like geopolymers or calcium aluminate cements in particularly aggressive environments where Portland cement durability is questionable. Innovative approaches have been developed to address limitations in cement integrity. Fiber-reinforced cement enhances mechanical properties and durability, while nanoparticle additives improve density and strength, reducing chemical degradation. Pozzolanic-Portland cement lowers permeability and chemical attacks (Benge, 2009), whereas gypsum cement, despite its rapid setting and high early strength, is unsuitable for CO_2 service due to high acid solubility. Microfine cements are effective for repairing small fractures. ThermaLock™ cement resists CO_2 and acids, and latex cement improves elasticity and bonding strength. Resin and plastic cements are ideal for plugging open holes in acidic environments (Peng et al., 2007). EverCRETE™ CO_2 is marketed for CO_2 resistance in carbon capture, storage, and enhanced oil recovery projects (EverCRETE CO2-Resistant Cement-Solar Impulse Efficient Solution, 2022). Recent advancements in self-repairing polymer-cement formulations have been explored, including Calcium Aluminate Phosphate (CaP) cement and Sodium Silicate-Activated Slag (SSAS) cement (Sugama, 2007). These materials offer cost-effective solutions, as they incorporate industrial byproducts from coal combustion and steel manufacturing. To enhance cement resistance in acidic environments, Achang and Radonjic recommended incorporating 5% BWOC olivine microparticles, which act as sacrificial materials available for carbonation.
- **Understanding Degradation Rates**: Research into quantifying the long-term rates of chemical degradation and mechanical damage (e.g., fatigue from cycling) under realistic downhole conditions (coupled chemical–mechanical–thermal processes) is ongoing and crucial for predicting the service life and reliability of cement barriers. Long-term laboratory testing under simulated conditions is often necessary, especially for new formulations. A fundamental aspect of ensuring long-term cement sheath performance is the ability to quantify degradation rates under conditions that realistically simulate the wellbore environment. Cement degradation in the subsurface is a complex, time-dependent process influenced by a combination of mechanical fatigue, chemical dissolution, and thermal cycling—often acting simultaneously and interactively. These coupled chemical–mechanical–thermal (CMT) phenomena are especially critical in high-stress wells, such as those subjected to cyclic injection and production, geothermal energy extraction, or CO_2 storage operations. To accurately predict the service life and failure thresholds of cement barriers, accelerated aging experiments are conducted using laboratory setups that simulate downhole pressure, temperature, fluid chemistry, and

stress conditions (Wu et al., 2020). These include long-duration soaking tests in CO_2-rich brine, thermal cycling chambers, and mechanical loading rigs designed to replicate casing expansion–contraction or formation movement. Advanced analytical tools—such as X-ray computed tomography (CT), scanning electron microscopy (SEM), nuclear magnetic resonance (NMR), and mechanical property testing—are employed to track microstructural evolution, crack development, and degradation kinetics over time.

- **Self-healing Cements**: To enhance long-term sealing performance and reduce the risk of time-dependent damage, self-healing cement systems have been developed that incorporate specialized additives capable of autonomously or naturally repairing cracks and micro-annuli (Alsubaih et al., 2025). These systems include a wide range of materials such as expansive minerals, microcapsules loaded with healing agents, bacteria capable of mineral precipitation, superabsorbent polymers, and flexible or elastomeric inclusions. Self-healing mechanisms can be broadly categorized into autogenous (natural) and autonomous (engineered) approaches (De Souza & Sanchez, 2023; Zhang et al., 2020). Autogenous healing primarily relies on unhydrated cement particles and carbonation reactions that precipitate calcium carbonate ($CaCO_3$) within cracks. Other contributing factors include the swelling of calcium silicate hydrate (C–S–H) gels at crack faces, rehydration of loose cementitious particles, and the physical blocking of cracks by waterborne impurities. While promising, autogenous healing is generally limited to sealing narrow cracks ($<150\ \mu m$) and is most effective under static conditions with abundant moisture or CO_2 exposure.

  By contrast, autonomous healing involves the deliberate incorporation of healing agents not inherently present in the cement matrix. These agents include bacterial spores (e.g., ureolytic or calcite-precipitating strains), organic sealants contained in microcapsules, shape-memory alloys, and nanomaterials that activate in response to mechanical damage or environmental triggers (Arce et al., 2016). In microbial systems, dormant bacteria are activated upon crack ingress of water, initiating biochemical reactions that lead to $CaCO_3$ precipitation and effective sealing (Bhaskar et al., 2017; Ruan et al., 2019). Similarly, microencapsulation techniques encase liquid or solid healing compounds in thin, protective shells. When mechanical damage ruptures the shell, the core material is released to react and seal the crack (Ghosh, 2009; Youngblood & Sottos, 2008). Geopolymer-based cements—formed by the reaction of aluminosilicate precursors (e.g., fly ash, GGBFS) with alkaline activators—also offer enhanced durability through their dense, amorphous Si–O–Al networks and chemical resistance, though their long-term performance under downhole conditions remains under investigation (Fernandez et al., 2021). Additionally, pozzolanic cements, while often favored for cost efficiency and CO_2 reactivity control, exhibit higher porosity and diffusion rates, which may accelerate CO_2 penetration but do not necessarily compromise structural integrity (Kutchko et al., 2009). These advanced self-healing systems represent a frontier in cement technology, providing robust barriers that adapt and recover under cyclic loading, thermal gradients, and chemical exposure—crucial for the evolving demands of long-term well integrity. Development of cement systems containing additives (e.g., expansive minerals, microcapsules with healing agents that release sealant upon cracking, specific bacteria that precipitate minerals) that can potentially seal cracks or micro-annuli that form during the well's life is an active area of research aimed at improving long-term sealing reliability and potentially mitigating time-dependent damage (Azenha et al., 2009; Kiran et al., 2017). However, the practical effectiveness of these systems in oil and gas applications depends on their ability to seal cracks at least as wide as those observed in field cement failures. To contextualize performance thresholds, Fig. 10.4 compares the maximum healed crack widths reported for a range of chemical-based self-healing materials against the benchmark failure crack width observed in oil and gas cements (up to 500 μm, Alberdi-Pagola & Fischer, 2024). This visual comparison highlights the importance of matching self-healing capabilities with realistic field degradation profiles in cement sheath system.

- **P&A Considerations**: Permanent well abandonment (P&A) represents the final and most critical phase of a well's lifecycle, where long-term durability of the cement sheath is paramount. Once a well is abandoned, direct monitoring and maintenance are no longer feasible, meaning that the barriers must remain functional for centuries to millennia without human intervention. As such, P&A plug design must be based on rigorous performance standards and a deep understanding of long-term degradation mechanisms, geomechanical stability, and fluid barrier reliability under evolving subsurface conditions (see Chap. 27).

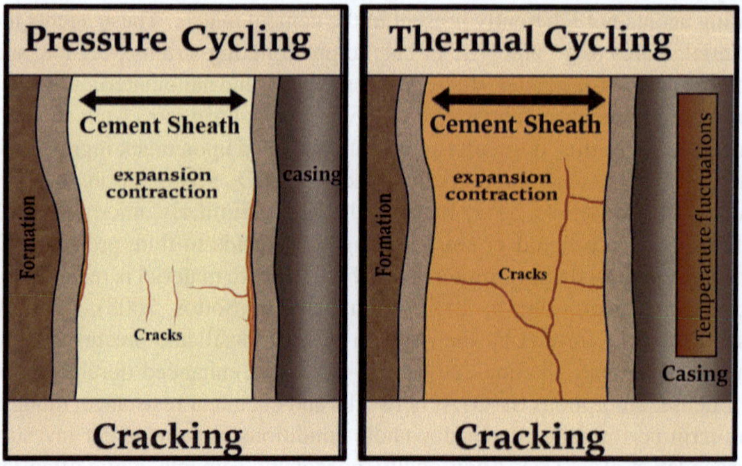

Fig. 10.4 Pressure and thermal cycling effects on cement sheath integrity. Left: Pressure cycling causes the casing to expand and contract radially, inducing tensile and shear stresses at the casing–cement interface, leading to debonding and progressive cracking. Right: Thermal cycling induces differential expansion due to contrasting thermal coefficients between the casing, cement, and formation, contributing to debonding and crack propagation through the cement sheath

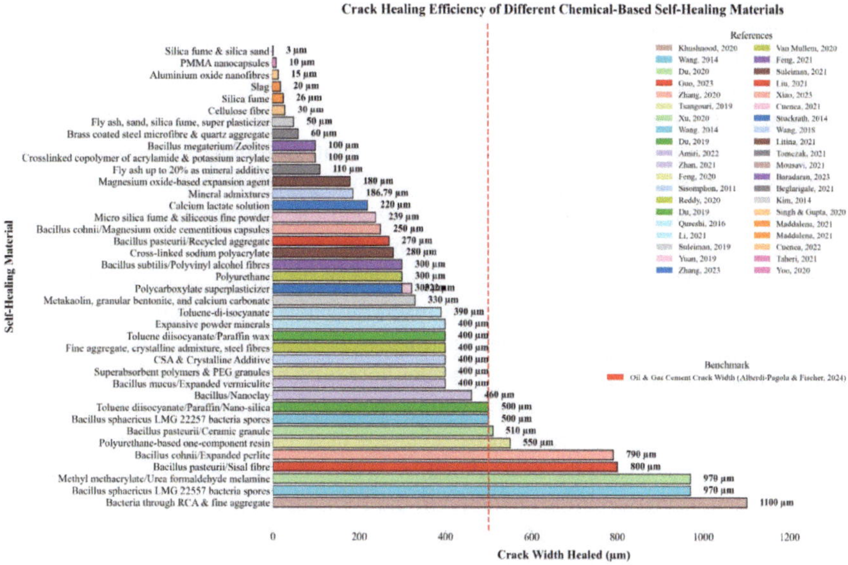

Fig. 10.5 Crack healing efficiency of various chemical-based self-healing materials, showing the maximum crack width, each material can heal. The red dashed line (500 µm) represents the benchmark crack width in oil and gas cement applications, with some materials exceeding this threshold, offering potential for enhanced cement integrity (Stuckrath et al., 2014; Singh & Gupta, 2020; Du et al., 2020; Xu et al., 2020; Liu et al., 2021; Guo et al., 2023; Feng et al., 2021; Baradaran & Sadeghpour, 2023; Amiri et al., 2022; Zhan et al., 2021; Litina et al., 2021; Khushnood et al., 2020; Taheri & Clark, 2021; Tsangouri et al., 2019)

In P&A applications, failure of the cement plug is not tolerable, as it can lead to delayed environmental leakage—such as migration of hydrocarbons, brines, or injected CO_2—into underground sources of drinking water (USDWs) or back to the surface, posing significant environmental, regulatory, and reputational risks. Therefore, material selection must be optimized not only for initial placement performance but also for extreme longevity in chemically and thermally aggressive downhole environments. The use of neat Portland cement alone is often insufficient, particularly in CO_2-rich or sulfate-bearing formations, where chemical degradation can severely compromise integrity over time (Fig. 10.5).

10.6 Influence of Geomechanical Stresses

The cement sheath exists within a complex stress field determined by the surrounding rock formations and the pressures within the wellbore. Changes in these stresses during drilling, production, injection, or due to natural geological processes can impose significant loads on the casing and cement, potentially leading to integrity failures (Meng et al., 2021; Yu et al., 2023; Zhao, 2019).

- **In-situ Stresses**: The natural stresses in the rock (vertical, minimum horizontal, maximum horizontal) influence the stress state within the cement sheath. Differential stresses can exist between formation layers.
- **Formation Movement**: Compaction of reservoir rock due to fluid withdrawal (leading to subsidence at the surface), movement of mobile formations like salt (salt creep), or slip along geological faults can impose significant shear, compressive, or tensile stresses on the well casing and the surrounding cement sheath. This can lead to deformation (ovality, buckling), cracking, or bond failure of the cement. The most severe loads often occur at the interfaces between different formations (e.g., between compacting reservoir and non-compacting overburden).
- **Near-Wellbore Stress Changes**: Drilling itself alters the stress state around the wellbore. Production or injection changes pore pressures and temperatures, which also modifies the effective stresses acting on the cement and formation. Hydraulic fracturing induces significant stress changes near the wellbore. Failure mechanisms can occur within the cement matrix, at the interface between the casing and the cement, or between the cement and the formation. These failures are mostly attributed to hoop stress, radial stress, axial stress, or the mechanical properties of the cement and the rock (Lavrov & Torsæter, 2016).
- **Interaction with Casing Loads**: Geomechanical stresses interact with operational loads on the casing (pressure, tension, compression), influencing the overall triaxial stress state on the casing and the stresses transmitted to the cement sheath (Abraham et al., 2024; Wang et al., 2013).

Accounting for geomechanical loads in casing and cement design, particularly in challenging environments like subsalt wells (salt creep), compacting reservoirs, or tectonically active areas (faulting), is essential. This often involves using advanced geomechanical modeling techniques (e.g., Finite Element Analysis—FEA) to predict the complex stress interactions between the wellbore, formation, casing, and cement, and designing cement systems with appropriate mechanical properties (strength, stiffness, toughness) to withstand the anticipated loads without failure. To further illustrate the role of mechanical strength parameters in cement failure evolution, numerical models have been employed by (Dusseault et al., 2001; Kim et al., 2016; Li et al., 2023) to evaluate shear failure behavior under varying cohesion conditions. As shown in Fig. 10.6, lower cohesion values lead to extensive failure zones within the cement sheath, while higher cohesion significantly limits the failed region. This highlights the importance of selecting cement systems with appropriate mechanical strength—particularly cohesion—for wells exposed to high geomechanical loads or long-term stress evolution. Finite element simulations such as this provide valuable insight into how design parameters influence wellbore integrity across a range of geological scenarios.

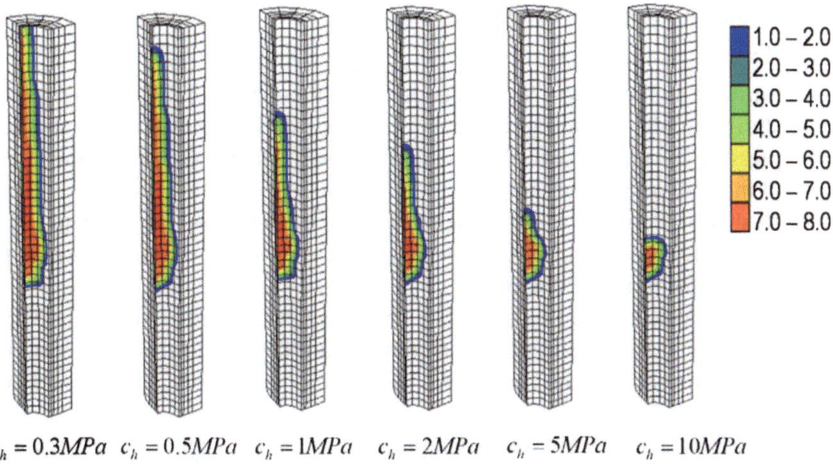

Fig. 10.6 Shear failure distribution in cement sheaths under varying cohesion values (ch = 0.3 MPa to 10 MPa), simulated using a finite element model. The color map indicates the number of failed Gauss points within each element, representing progressive damage intensity. As cohesion increases, failure extent significantly decreases, demonstrating that low-cohesion cements are more susceptible to widespread shear failure, especially under complex loading conditions imposed by geomechanical forces and casing deformation. This underscores the critical role of cement mechanical design in preventing zonal isolation loss due to structural breakdown (Kim et al., 2016).

10.7 Mitigation and Remediation Strategies for Cement Sheath Failures

Managing cement sheath integrity involves both preventative design measures and remedial actions when failures occur.

- **Preventative Strategies**: Focus on getting the primary cement job right and designing for durability:
  - Optimized slurry design for placement and long-term durability (Sects. 5.2, 5.3, 10.4 and 10.5).
  - Ensuring effective mud removal and centralization (Sect. 5.4).
  - Proper placement techniques and gas migration control (Sect. 5.5).
  - Using flexible, low-shrinkage, or expansive cements, potentially incorporating elastomers or fibers, to improve resistance to cracking and debonding under stress and cycling (Al-Yami et al., 2018b; Wray et al., 2009).
  - Considering geomechanical stresses in design (Sect. 10.6).
  - Research and development of self-healing cement systems (Alsubaih et al., 2025).

- **Remedial Actions**: Aim to seal existing leak paths in the cement sheath (see Chap. 26).
  - **Squeeze Cementing**: The most common technique, involving forcing cement slurry into the leak path (via perforations or existing channels) and applying pressure to create a seal. Success rates are variable and depend on the nature of the leak path (Ma et al., 2018).
  - **Resins or Polymer Gels**: Sometimes used as alternatives or supplements to cement for sealing small leaks or in specific conditions, offering potential for better penetration or flexibility.
  - **Section Milling and Plugging**: For severe annular integrity issues, milling out a section of casing and placing a cement plug directly against the formation ("rock-to-rock" barrier) provides a more robust seal but is a costly workover operation (Chap. 27).
  - **Perforate and Circulate**: Attempting to circulate cement up a suspected channel behind casing after perforating, but achieving a reliable seal this way can be difficult.

In summary, the cement sheath, while intended to be a durable barrier, is subject to a variety of mechanical and chemical failure mechanisms over time. Ensuring long-term well integrity requires designing robust cement systems for both initial placement and long-term durability, executing placement effectively to achieve good bonding, understanding how operational and geological stresses can impact the sheath, and having strategies for monitoring and potentially remediating failures. Table 10.2 presents the API, ISO, and NORSOK standards for well cementing operations.

Table 10.2 API, ISO, and NORSOK standards for well cementing operations

Standard (organization and number)	Full title (or common name)	Description	Latest known edition/revision (as per your input and general knowledge)
API standards			
API Spec 10A	Cements and materials for well cementing	Specifies requirements for eight classes of well cements (A–H), including chemical and physical properties for various well conditions	2022 (25th ed.)
API RP 10B-2	Recommended practice for testing of well cements	Outlines procedures for testing well cement properties under simulated well conditions to ensure performance, with enhanced testing protocols in the latest edition	2013 (2nd ed.)
API RP 10B-3	Recommended practice on testing of deepwater well cement formulations	Provides methods for evaluating cement formulations in deepwater environments, addressing low temperatures and high pressures	2016 (1st ed.)
API RP 10B-4	Recommended practice on preparation and testing of foamed cement slurries at atmospheric pressure	Details procedures for preparing and testing foamed cement slurries to reduce density and improve zonal isolation	2015 (2nd ed.)
API RP 10B-5	Recommended practice on setermination of shrinkage and expansion of well cement formulations at atmospheric pressure	Specifies methods to measure dimensional changes (shrinkage/expansion) of cement slurries during curing	2020 (1st ed.)
API RP 10B-6	Recommended practice for determining the static gel strength of cement formulations	Provides procedures to assess static gel strength development in cement slurries to prevent gas migration	2019 (1st ed.)

(continued)

Table 10.2 (continued)

Standard (organization and number)	Full title (or common name)	Description	Latest known edition/ revision (as per your input and general knowledge)
API RP 65-2	Isolating potential flow zones during well construction	Offers best practices for cementing to isolate flow zones, including job evaluation and updated guidance on cement placement	2020 (2nd ed.)
API RP 65-3	Wellbore plugging and abandonment	Guides design, placement, and verification of cement plugs for temporary or permanent well closure, emphasizing environmental compliance	2021 (1st ed.)
ISO standards			
ISO 10426-1	Petroleum and natural gas industries—cements and materials for well cementing—part 1: specification	Aligns with API Spec 10A, specifying requirements for six main classes of well cements. Harmonized with API	2019
ISO 10426-2	Petroleum and natural gas industries—cements and materials for well cementing—part 2: Testing of well cements	Provides testing procedures for well cements under simulated conditions, with refined testing methods in the latest edition	2003
ISO 10426-3	Petroleum and natural gas industries—cements and materials for well cementing—part 3: Testing of deepwater well cement formulations	Focuses on testing cement for deepwater applications, addressing low-temperature challenges	2019

(continued)

Table 10.2 (continued)

Standard (organization and number)	Full title (or common name)	Description	Latest known edition/ revision (as per your input and general knowledge)
ISO 10426-4	Petroleum and natural gas industries—cements and materials for well cementing—part 4: Preparation and testing of foamed cement slurries	Details methods for preparing and testing foamed cement slurries, with updated atmospheric testing protocols	2019
ISO 10426-5	Petroleum and natural gas industries—cements and materials for well cementing—part 5: Determination of shrinkage and expansion of well cements	Specifies procedures to measure cement slurry dimensional changes, with refined measurement techniques	2024
ISO 10426-6	Petroleum and natural gas industries—cements and materials for well cementing—part 6: Methods for determining the static gel strength of cement formulations	Provides methods to determine static gel strength, with enhanced accuracy in gel strength measurement	2020
NORSOK standards			
NORSOK D-010	Well integrity in drilling and well operations	Defines minimum requirements for well design and operations to ensure integrity throughout the well's lifecycle, including updated cementing requirements	Rev. 5 (2021)

References

Abraham, J. J., Carvero, A., Devers, C., Teodoriu, C., & Amani, M. (2024). Stress failure assessment and potential well integrity issues with different oilwell cement classes and formulations in complex wells. In *Society of petroleum engineers—GOTECH Conference 2024*. https://doi.org/10.2118/219117-MS

Albawi, A., Torsaeter, M., & Opedal, N. (2013). Influence of thermal cycling on cement sheath integrity. *78*. https://ntnuopen.ntnu.no/ntnu-xmlui/handle/11250/239986

Alberdi-Pagola, P., & Fischer, G. (2023). Review of integrity loss detection and quantification due to cracking in cemented wells. In *SPE Journal* (Vol. 965). http://onepetro.org/SJ/article-pdf/28/03/965/3120397/spe-214303-pa.pdf

Alberdi-Pagola, P., & Fischer, G. (2024). Quantification of casing expansion-induced cracking in oilwell cement sheaths. *SPE Journal, 29*(02), 714–724. https://doi.org/10.2118/218003-PA

Al-Dulaijan, S. U., Maslehuddin, M., Al-Zahrani, M. M., Sharif, A. M., Shameem, M., & Ibrahim, M. (2003). Sulfate resistance of plain and blended cements exposed to varying concentrations of sodium sulfate. *Cement and Concrete Composites, 25*(4-5 SPEC), 429–437. https://doi.org/10.1016/S0958-9465(02)00083-5

Alkhamis, M., & Imqam, A. (2021). A simple classification of wellbore integrity problems related to fluids migration. *Arabian Journal for Science and Engineering, 46*(7), 6131–6141. https://doi.org/10.1007/S13369-021-05359-3/METRICS

Alsubaih, A., Sepehrnoori, K., Lopez Manriquez, A., Espinoza, D. N., Allawi, H. R., & Hayfaa, S. (2024). *A statistical approach to estimate the drilling mud window based on field data and ten failure criteria in depleted reservoirs—new modified failure criterion.* https://doi.org/10.56952/ARMA-2024-0799

Alsubaih, A., Sepehrnoori, K., & Delshad, M. (2025). Development and optimization of self-healing cement for CO_2 injection and storage wells: Enhancing long-term wellbore integrity in extreme subsurface conditions. *Applied Sciences, 15*(10), 5428. https://doi.org/10.3390/APP15105428

Al-Yami, A. S., Alqam, M. H., Riefky, A., & Shafqat, A. U. (2018a). Self healing durable cement; development, lab testing, and field execution. In *Proceedings of the SPE/IADC Middle East drilling technology conference and exhibition, 2018-January*. https://doi.org/10.2118/189397-MS

Al-Yami, A. S., Alqam, M. H., Riefky, A., & Shafqat, A. U. (2018b). Self healing durable cement; development, lab testing, and field execution. In *Proceedings of the SPE/IADC Middle East drilling technology conference and exhibition, 2018-January*. https://doi.org/10.2118/189397-MS

Amiri, Y., Hassaninasab, S., Chehri, K., & Zahedi, M. (2022). Investigating the effect of adding bacillus bacteria and nano-clay on cement mortar properties. *Case Studies in Construction Materials, 17*, Article e01167. https://doi.org/10.1016/J.CSCM.2022.E01167

Arce, G. A., Hassan, M. M., Mohammad, L. N., & Rupnow, T. (2016). Characterization of self-healing processes induced by calcium nitrate microcapsules in cement mortar. *Journal of Materials in Civil Engineering, 29*(1), 04016189. https://doi.org/10.1061/(ASCE)MT.1943-5533.0001717

Azenha, M., Faria, R., & Ferreira, D. (2009). Identification of early-age concrete temperatures and strains: Monitoring and numerical simulation. *Cement and Concrete Composites, 31*(6), 369–378. https://doi.org/10.1016/J.CEMCONCOMP.2009.03.004

Baradaran, M., & Sadeghpour, M. (2023). Effect of bacteria on the self-healing ability of concrete containing zeolite. *Innovative Infrastructure Solutions, 8*(10), 1–15. https://doi.org/10.1007/S41062-023-01222-6/METRICS

Benge, G. (2009). *SPE/IADC 119267 improving wellbore seal integrity in co 2 injection wells.*

Bhaskar, S., Anwar Hossain, K. M., Lachemi, M., Wolfaardt, G., & Otini Kroukamp, M. (2017). Effect of self-healing on strength and durability of zeolite-immobilized bacterial cementitious mortar composites. *Cement and Concrete Composites, 82*, 23–33. https://doi.org/10.1016/J.CEMCONCOMP.2017.05.013

Billmeyer, F. W. (1984). *A textbook of polymer science (chemistry and technology of polymer) : (Processing and Applications).* https://books.google.com/books/about/Textbook_of_Polymer_Science.html?id=kGnyDwAAQBAJ

Burdylo, L., & Birch, G. (1990). 12 primary cementing techniques. *Developments in Petroleum Science, 28*(C), 12–1. https://doi.org/10.1016/S0376-7361(09)70310-3

De Andrade, J., & Sangesland, S. (2016a). Cement sheath failure mechanisms: numerical estimates to design for long-term well integrity. *Journal of Petroleum Science and Engineering, 147*, 682–698. https://doi.org/10.1016/J.PETROL.2016.08.032

De Andrade, J., Sangesland, S., Todorovic, J., & Vrålstad, T. (2015). Cement sheath integrity during thermal cycling: A novel approach for experimental tests of cement systems. *Society of Petroleum Engineers—SPE Bergen One Day Seminar, 2015*, 377–392. https://doi.org/10.2118/173871-MS

De Souza, D. J., & Sanchez, L. F. M. (2023). Understanding the efficiency of autogenous and autonomous self-healing of conventional concrete mixtures through mechanical and microscopical analysis. *Cement and Concrete Research, 172*, Article 107219. https://doi.org/10.1016/J.CEMCONRES.2023.107219

Dong, X., Duan, Z., Dou, H., Ma, Y., & Gao, D. (2022). A parametric study for radial cracking in cement under different loading events based on the stress intensity factor. *Journal of Energy Resources Technology, Transactions of the ASME, 144*(5). https://doi.org/10.1115/1.4051742/1114465

Duchesne, J., & Bertron, A. (2013). Leaching of cementitious materials by pure water and strong acids (HCl and HNO_3). *RILEM State-of-the-Art Reports, 10*, 91–112. https://doi.org/10.1007/978-94-007-5413-3_4

Dusseault, M. B., Bruno, M. S., & Barrera, J. (2001). Casing shear: Causes, cases, cures. *SPE Drilling & Completion, 16*(02), 98–107. https://doi.org/10.2118/72060-PA

Feng, J., Chen, B., Sun, W., & Wang, Y. (2021). Microbial induced calcium carbonate precipitation study using Bacillus subtilis with application to self-healing concrete preparation and characterization. *Construction and Building Materials, 280*, Article 122460. https://doi.org/10.1016/J.CONBUILDMAT.2021.122460

Fernandez, C. A., Correa, M., Nguyen, M. T., Rod, K. A., Dai, G. L., Cosimbescu, L., Rousseau, R., & Glezakou, V. A. (2021). Progress and challenges in self-healing cementitious materials. *Journal of Materials Science, 56*(1), 201–230. https://doi.org/10.1007/S10853-020-05164-7/FIG.S/13

Gaurina-Međimurec, N. (2010). The influence of Co_2 on well cement. *Rudarsko-Geološko-Naftni Zbornik, 22*(1), 19–25.

Ghosh, S. K. (2009). *Self-healing materials: Fundamentals, design strategies, and applications* (pp. 1–28). https://doi.org/10.1002/9783527625376.CH1

Gu, C., Feng, Y., & Li, X. (2024). Cement sheath integrity in oil and gas wells. *Advances in Oil and Gas Well Engineering*. https://doi.org/10.5772/INTECHOPEN.113052

Gu, C., Li, X., Feng, Y., Deng, J., & Gray, K. (2022). Numerical investigation of cement interface debonding in deviated shale gas wells considering casing eccentricity and residual drilling fluid. *International Journal of Rock Mechanics and Mining Sciences, 158*, Article 105197. https://doi.org/10.1016/J.IJRMMS.2022.105197

Guo, Y., Xiang, K., Wang, H., Liu, X., Ye, Q., & Wang, X. (2023). Experimental study on self-healing and mechanical properties of sisal fiber-loaded microbial concrete. *Materials Research Express, 10*(4), Article 045701. https://doi.org/10.1088/2053-1591/ACC718

Han, X., Feng, F., & Zhang, J. (2023). Study on the whole life cycle integrity of cement interface in heavy oil thermal recovery well under circulating high temperature condition. *Energy, 278*, Article 127873. https://doi.org/10.1016/J.ENERGY.2023.127873

Jafarzade, G., Opedal, N., & Torsaeter, M. (2014). Leakage of CO_2 along annular well cement (Vol. 111). https://ntnuopen.ntnu.no/ntnu-xmlui/handle/11250/240491

Khushnood, R. A., Qureshi, Z. A., Shaheen, N., & Ali, S. (2020). Bio-mineralized self-healing recycled aggregate concrete for sustainable infrastructure. *Science of the Total Environment, 703*, Article 135007. https://doi.org/10.1016/J.SCITOTENV.2019.135007

Kim, D. J., Kang, S. H., & Ahn, T. H. (2014). Mechanical characterization of high-performance steel-fiber reinforced cement composites with self-healing effect. *Materials, 7*(1), 508–526. https://doi.org/10.3390/MA7010508

Kim, J., Moridis, G. J., & Martinez, E. R. (2016a). Investigation of possible wellbore cement failures during hydraulic fracturing operations. *Journal of Petroleum Science and Engineering, 139*, 254–263. https://doi.org/10.1016/j.petrol.2016.01.035

Kiran, R., Teodoriu, C., Dadmohammadi, Y., Nygaard, R., Wood, D., Mokhtari, M., & Salehi, S. (2017). Identification and evaluation of well integrity and causes of failure of well integrity barriers (A review). *Journal of Natural Gas Science and Engineering, 45*, 511–526. https://doi.org/10.1016/J.JNGSE.2017.05.009

Kutchko, B. G., Strazisar, B. R., Dzombak, D. A., Lowry, G. V., & Thauiow, N. (2007). Degradation of well cement by CO_2 under geologic sequestration conditions. *Environmental Science and Technology, 41*(13), 4787–4792. https://doi.org/10.1021/ES062828C/SUPPL_FILE/ES062828CSI20070411_025348.PDF

Kutchko, B. G., Strazisar, B. R., Huerta, N., Lowry, G. V., Dzombak, D. A., & Thaulow, N. (2009). CO_2 reaction with hydrated class H well cement under geologic sequestration conditions: Effects of flyash admixtures. *Environmental Science and Technology, 43*(10), 3947–3952. https://doi.org/10.1021/ES803007E/SUPPL_FILE/ES803007E_SI_001.PDF

Lavrov, A., & Torsæter, M. (2016). Formation stresses, casing pressure, and annular cement. *Springer Briefs in Petroleum Geoscience and Engineering*, 75–91. https://doi.org/10.1007/978-3-319-43165-9_5

Li, X. R., Gu, C. W., Ding, Z. C., & Feng, Y. C. (2023). THM coupled analysis of cement sheath integrity considering well loading history. *Petroleum Science, 20*(1), 447–459. https://doi.org/10.1016/J.PETSCI.2022.09.001

Litina, C., Bumanis, G., Anglani, G., Dudek, M., Maddalena, R., Amenta, M., Papaioannou, S., Pérez, G., Calvo, J. L. G., Asensio, E., Cobos, R. B., Pinto, F. T., Augonis, A., Davies, R., Guerrero, A., Moreno, M. S., Stryszewska, T., Karatasios, I., Tulliani, J. M., & Al-tabbaa, A. (2021). Evaluation of methodologies for assessing self-healing performance of concrete with mineral expansive agents: An interlaboratory study. *Materials, 14*(8), 2024. https://doi.org/10.3390/MA14082024

Ma, C., Deng, J., Wu, R., Yu, B., Zhou, Y., & Lin, S. (2018). A novel self-healing spacer fluid for sustained casing pressure mitigation. *Journal of Petroleum Science and Engineering, 163*, 722–730. https://doi.org/10.1016/J.PETROL.2017.11.027

Mackay, E. (2019). Effect of very low Ph on degradation of class G cement: Implications for Co_2 geological storage. *SSRN Electronic Journal*. https://www.academia.edu/94611071/Effect_of_Very_Low_Ph_on_Degradation_of_Class_G_Cement_Implications_for_Co2_Geological_Storage

Meng, M., Frash, L., William Carey, J., Niu, Z., Zhang, W., Lei, Z., Li, W., & Welch, N. (2021). Predicting cement-sheath integrity with consideration of initial State of stress and thermoporoelastic effects. *SPE Journal, 26*(6), 3505–3528. https://doi.org/10.2118/205344-PA

Olivine—Portland cement as a solution to wellbore integrity issues in CO_2-rich geothermal reservoirs|GRC 2020 Annual Meeting. (n.d.). https://grc2020.mygeoenergynow.org/olivine-portland-cement-solution-wellbore-integrity-issues-co2-rich-geothermal-reservoirs

Omosebi, O. A., Ahmed, R. M., & Shah, S. N. (2017). Mechanisms of cement degradation in HPHT carbonic acid environment. In *Proceedings—SPE international symposium on oilfield chemistry* (pp. 448–468). https://doi.org/10.2118/184567-MS

Omosebi, O., Maheshwari, H., Ahmed, R., Shah, S., Osisanya, S., Hassani, S., DeBruijn, G., Cornell, W., & Simon, D. (2016). Degradation of well cement in HPHT acidic environment: Effects of CO_2 concentration and pressure. *Cement and Concrete Composites, 74*, 54–70. https://doi.org/10.1016/J.CEMCONCOMP.2016.09.006

Omosebi, O., Maheshwari, H., Ahmed, R., Shah, S., Osisanya, S., Santra, A., & Saasen, A. (2015). Investigating temperature effect on degradation of well cement in HPHT carbonic acid environment. *Journal of Natural Gas Science and Engineering, 26*, 1344–1362. https://doi.org/10.1016/J.JNGSE.2015.08.018

Orlic, B., Chitu, A., Brunner, L., Koenen, M., Wollenweber, J., & Schreppers, G.-J. (2018b). *Numerical investigations of cement interface debonding for assessing well integrity risks*. OnePetro. https://dx-doi-org.ezproxy.lib.utexas.edu/

Orlic, B., Chitu, A., Brunner, L., Koenen, M., Wollenweber, J., & Schreppers, G.-J. (2018a). *Numerical investigations of cement interface debonding for assessing well integrity risks*. OnePetro. https://dx-doi-org.ezproxy.lib.utexas.edu/

Peng, S., Fu, J., & Zhang, J. (2007). Borehole casing failure analysis in unconsolidated formations: A case study. *Journal of Petroleum Science and Engineering, 59*(3–4), 226–238. https://doi.org/10.1016/J.PETROL.2007.04.010

Poole, T. S., Wakeley, L., & Young Structures Laboratory, C. (1994). *Individual and combined effects of chloride, sulfate, and magnesium ions on hydrated portland-cement paste*. https://doi.org/10.2172/10147904

Pouya, J., Neji, M., De Windt, L., Péralès, F., Socié, A., & Corvisier, J. (2022). Mineralogical evolution and expansion of cement pastes in a sulfate-confined environment. *Minerals, 13*(1), 1–1. https://doi.org/10.3390/MIN13010001

Pyatina, T., & Sugama, T. (n.d.). *Aluminum-rich cements for high temperature geothermal wells.*

Ruan, S., Qiu, J., Weng, Y., Yang, Y., Yang, E. H., Chu, J., & Unluer, C. (2019). The use of microbial induced carbonate precipitation in healing cracks within reactive magnesia cement-based blends. *Cement and Concrete Research, 115*, 176–188. https://doi.org/10.1016/J.CEMCONRES.2018.10.018

Segre, N., Galves, A. D., Rodrigues, J. A., & Joekes, I. (2006). Tire rubber-cement composites: effect of slag on properties. *Cerâmica, 52*(324), 283–292. https://doi.org/10.1590/S0366-69132006000400012

Scerrato, D., Giorgio, I., Della Corte, A., Madeo, A., Dowling, N. E., & Darve, F. (2016). Towards the design of an enriched concrete with enhanced dissipation performances. *Cement and Concrete Research, 84*, June 1. Licensed from Elsevier under License No. 6127400579508, obtained on Oct 13, 2025

Self-healing of surface cracks in mortars with expansive additive and crystalline additive|TU Delft Repository. (n.d.). https://repository.tudelft.nl/record/uuid:5ca76586-4431-4dcc-865f-d15c2f80224a

Singh, H., & Gupta, R. (2020). Cellulose fiber as bacteria-carrier in mortar: Self-healing quantification using UPV. *Journal of Building Engineering, 28*, Article 101090. https://doi.org/10.1016/J.JOBE.2019.101090

Stuckrath, C., Serpell, R., Valenzuela, L. M., & Lopez, M. (2014). Quantification of chemical and biological calcium carbonate precipitation: Performance of self-healing in reinforced mortar containing chemical admixtures. *Cement and Concrete Composites, 50*, 10–15. https://doi.org/10.1016/J.CEMCONCOMP.2014.02.005

Sugama, T. (2007). *Advanced Cements for Geothermal Wells*. https://doi.org/10.2172/909955

Taheri, S., & Clark, S. M. (2021). Preparation of self-healing additives for concrete via miniemulsion polymerization: formulation and production challenges. *International Journal of Concrete Structures and Materials, 15*(1), 1–15. https://doi.org/10.1186/S40069-020-00449-2/FIG.S/9

Torsæter, M., Todorovic, J., & Lavrov, A. (2015). Structure and debonding at cement–steel and cement–rock interfaces: Effect of geometry and materials. *Construction and Building Materials, 96*, 164–171. https://doi.org/10.1016/J.CONBUILDMAT.2015.08.005

Tsangouri, E., Lelon, J., Minnebo, P., Asaue, H., Shiotani, T., Van Tittelboom, K., De Belie, N., Aggelis, D. G., & Van Hemelrijck, D. (2019). Feasibility study on real-scale, self-healing concrete slab by developing a smart capsules network and assessed by a plethora of advanced monitoring techniques. *Construction and Building Materials, 228*, Article 116780. https://doi.org/10.1016/J.CONBUILDMAT.2019.116780

Wang, X., Qu, Z., & Dou, Y. (2013). Influence of geometrical and physical parameters of cement on loads of casing and cement. *Advanced Materials Research, 634–638*(1), 3591–3594. https://doi.org/10.4028/WWW.SCIENTIFIC.NET/AMR.634-638.3591

Wang, Y., Liu, K., & Gao, D. (2021). Investigation of the interface cracks on the cement sheath stress in shale gas wells during hydraulic fracturing. *Journal of Petroleum Science and Engineering, 205*, Article 108981. https://doi.org/10.1016/J.PETROL.2021.108981

Wray, B., Bedford, D., Leotaud, L., & Hunter, B. (2009). The application of high-density elastic cements to solve HPHT challenges in South Texas: The success story. *Proceedings—SPE Annual Technical Conference and Exhibition, 1*, 135–151. https://doi.org/10.2118/122762-MS

Wu, Y., Patel, H., & Salehi, S. (2020). *Thermal considerations of cement integrity in geothermal wells.*

Youngblood, J. P., & Sottos, N. R. (2008). Bioinspired materials for self-cleaning and self-healing. *MRS Bulletin, 33*(8), 732–738. https://doi.org/10.1557/MRS2008.158/METRICS

Yu, X., Cen, X., Kan, C., Hu, Y., Yang, Y., Tao, S., Chen, X., Chen, X., & Hu, Z. (2023). Numerical simulation analysis of wellbore integrity and casing damage in high-temperature injection and production of shale oil. *Processes, 11*(11). https://doi.org/10.3390/PR11113053

Zhang, W., Zheng, Q., Ashour, A., & Han, B. (2020). Self-healing cement concrete composites for resilient infrastructures: A review. *Composites Part b: Engineering, 189*, Article 107892. https://doi.org/10.1016/J.COMPOSITESB.2020.107892

Zhao, X. (2019). Analysis of integrity of wellbore structure in soft mudstone during drilling and cementing. *Journal of Engineering Science and Technology Review, 12*(2), 27–33. https://doi.org/10.25103/JESTR.122.04

Zhou, N., Deng, K., Lin, Y., Yan, K., Li, C., Li, H., & Xie, P. (2024). A prediction method of wellhead movement with consideration of interface debonding. *Geoenergy Science and Engineering, 239*, Article 212919. https://doi.org/10.1016/J.GEOEN.2024.212919

Part III
Operational Well Integrity: Monitoring and Failure Modes

Chapter 11
Failure Modes of Seals, Packers, and Valves

Abstract This chapter investigates the critical challenges associated with achieving long-term zonal isolation in wells undergoing permanent plug and abandonment (P&A). As global regulations evolve to mandate safe and environmentally secure well closures, ensuring the integrity of P&A operations has become a top priority for both industry and regulators. This chapter presents a detailed assessment of failure pathways, material degradation, and design limitations that compromise permanent isolation over decades to centuries. The chapter begins by reviewing regulatory requirements and industry standards (e.g., API, NORSOK, ISO) governing P&A practices, including barrier placement, verification, and abandonment design life expectations. It then explores the physical and chemical mechanisms that can lead to loss of isolation, including micro-annulus formation, casing corrosion, cement shrinkage, and long-term geochemical reactions in the wellbore environment. Special attention is given to the unique challenges posed by legacy wells, orphaned wells, and deep high-pressure/high-temperature (HPHT) wells, where uncertain construction histories and degraded materials complicate effective isolation. Case studies highlight instances of post-abandonment gas migration and fluid leakage, emphasizing the consequences of incomplete barrier coverage or improper verification. To address these challenges, the chapter evaluates advanced abandonment materials, such as self-healing cements, expandable materials, bismuth alloys, and polymeric gels, comparing their sealing capabilities and long-term stability. The role of formation behavior (e.g., subsidence, creep, fault movement) is also addressed through geomechanical models, underscoring the importance of site-specific design. Barrier placement and verification techniques are critically assessed, including ultrasonic logging, cement bond logging, temperature and noise tools, and pressure testing. The limitations of these tools in detecting subtle failure mechanisms or verifying zonal isolation beyond casing strings are discussed. The chapter concludes with forward-looking recommendations for smart abandonment strategies, including digital monitoring of plugged wells, machine learning-based failure prediction, and the integration of sustainability and lifecycle risk into abandonment planning. Together, these insights promote safer and more effective P&A operations that protect groundwater, surface ecosystems, and public safety over geological timeframes.

© The Author(s), under exclusive license to Springer Nature Switzerland AG 2026
A. Alsubaih and K. Sepehrnoori, *Well Integrity for Oil, Gas, and Subsurface Storage Operations*, https://doi.org/10.1007/978-3-032-15233-6_11

Corroded wellhead flange showing advanced seal degradation, corrosion, and bolt damage—illustrating a well integrity failure point. *Source* The Authors

Chapter's Objectives

- Explain the role of static and dynamic seals as critical barrier elements.
- Describe common failure modes associated with elastomeric seals (chemical, thermal, ED, extrusion, compression set).
- Discuss performance limitations and failure modes of metal-to-metal seals.
- Detail typical failure mechanisms specific to downhole packers.
- Explain common causes of wellhead and Christmas tree seal and valve leaks, including through-valve leakage.
- Outline failure modes specific to Downhole Safety Valves (DHSVs).

11.1 Introduction: The Role of Static and Dynamic Seals

Well integrity is fundamentally dependent on the ability of sealing elements to prevent fluid migration across different pressure regimes in the wellbore and to maintain zonal isolation throughout the life of the well (Ashena, 2025; Duijm, 2009; Saharuddin et al., 2024). Seals serve as essential primary and secondary barriers that contain formation fluids, maintain pressure containment, and enable safe operation and intervention. Failure of these seals—whether in the form of gradual degradation, mechanical failure, or improper installation—can result in sustained casing pressure (SCP), well control events, or costly remedial operations.

Seals are typically made from elastomers (rubbers), thermoplastics (plastics), or metal. Seals are classified broadly into two categories: static and dynamic (Flitney, 2014). Static seals are used between stationary components and are expected to retain their sealing capacity under prolonged exposure to pressure, temperature, and corrosive fluids without relative motion once installed. These are commonly found in packers, casing hangers, tubing hangers, wellhead gaskets, and valve bonnets. In contrast, dynamic seals are subjected to relative movement during operation or intervention, such as valve stem packings, sliding sleeves, or stripper packers in coiled tubing operations. These seals must maintain integrity while accommodating mechanical motion and differential pressure across the sealing interface.

Materials used for sealing elements range from elastomer and thermoplastics to precision-machined metal-to-metal (MTM) interfaces, depending on the expected loading conditions, chemical exposure, thermal profile, and design life (Garfield & Mackenzie, 2007a, 2007b). Elastomeric seals, although flexible and highly conformable, are prone to degradation from CO_2, H_2S, hydrocarbons, amine-based chemicals, and elevated temperature, which can lead to swelling, hardening, blistering, or extrusion. MTM seals, while more chemically robust and durable under HPHT conditions, require high contact stress, precise machining, and proper energization to function effectively.

Seal failure remains one of the most frequent and critical failure modes compromising well integrity. It can occur due to chemical degradation, thermal breakdown, explosive decompression, loss of compression resilience, improper energization, or mechanical wear. In surface components such as wellheads and valves, seal failure can lead to external leaks or loss of control at the surface, while in downhole components like packers or safety valves, it can cause loss of isolation between tubing and annuli, or complete barrier failure.

This chapter explores the performance and failure mechanisms of sealing systems in petroleum wells, with a focus on material behavior, operational failure modes, and failure consequences across a range of critical components. A clear understanding of these mechanisms is essential for optimizing material selection, improving design, enhancing qualification protocols, and establishing effective monitoring and maintenance strategies to uphold well integrity throughout the well's operational life. Failures in sealing systems and structural barriers occur at every stage of the well's lifecycle—from initial drilling and cementing, through workover operations, to long-term production. The distribution and nature of these failures vary depending on operational phase and equipment stressors. Understanding where and why failures occur provides critical context for evaluating the performance of seals, valves, wellheads, and cement systems. Figure 11.1 summarizes statistically significant failure sources during drilling, workover, and production phases, reinforcing the need for reliable barrier components and improved diagnostics in well design and maintenance.

Seal degradation and failure are common time-dependent issues, accumulating damage from chemical exposure, heat, and stress cycling over the operational life of the well.

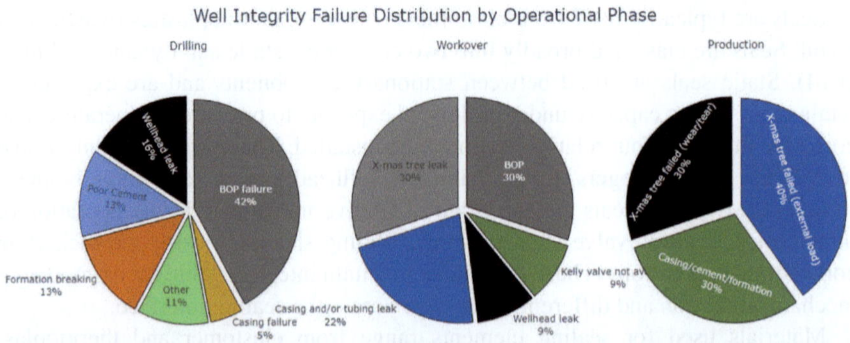

Fig. 11.1 Common failure causes across the drilling, workover, and production phases. During drilling, the most prevalent failures involve BOPs (42%), wellhead leaks (16%), poor cement, and formation breakdown. In workovers, leaks through the Xmas tree and tubing/casing dominate. Production failures are heavily associated with mechanical wear and external loading of the Christmas tree, as well as integrity loss in the casing–cement–formation system. These categories highlight the critical need for robust sealing systems and structural integrity at all stages of well operation (Patel et al., 2019a, 2019b)

11.2 Failure Modes of Elastomeric Seals

Elastomers (rubbers) are widely used for seals in packers, hangers, valves, and other downhole equipment due to their flexibility, resilience, and ability to conform to sealing surfaces and tolerate some surface imperfections (Mody et al., 2013). However, their organic nature makes them susceptible to degradation and distinct failure modes, particularly under the harsh, high-pressure, high-temperature, and chemically aggressive conditions encountered downhole (Patel et al., 2019a, 2019b). Elastomeric seals are critical secondary barriers in various well components (wellheads, packers, liner hangers, BOPs), but are susceptible to multiple failure mechanisms under harsh downhole conditions (Patel et al., 2019a, 2019b). Elastomeric seals are highly vulnerable to a variety of time-dependent failure mechanisms when exposed to the harsh downhole environment. These include explosive decompression (caused by rapid depressurization of gas-saturated elastomers), extrusion into clearance gaps, thermal oxidation, and chemical attack from CO_2, H_2S, and hydrocarbons. Over time, many seals also suffer from compression set, losing their ability to rebound and maintain sealing force. These mechanisms are not mutually exclusive and often act in combination, accelerating degradation. Figure 11.2 visually summarizes several of these typical elastomer failure modes observed in the field and laboratory, reinforcing the need for careful material selection and qualification in well design.

These failures typically result from exposure to high-pressure gas, thermal cycling, incompatible fluids, and mechanical loading (Patel et al., 2019a, 2019b).

Key material degradation and failure modes for elastomeric seals include the following:

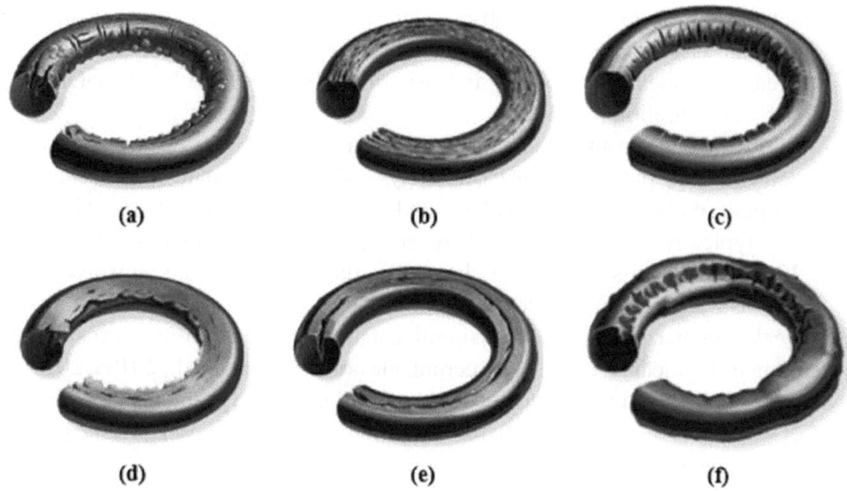

Fig. 11.2 Common failure modes in elastomeric O-rings used in oil and gas well applications: **a** Explosive decompression, **b** Abrasion/friction damage, **c** Thermal degradation (heat aging), **d** Extrusion, **e** Compression set, **f** Chemical degradation

- **Chemical Degradation** is a prevalent failure mechanism in oil and gas well environments. During service, elastomeric components are frequently exposed to chemically aggressive substances such as drilling fluids, completion fluids, fracturing fluids, formation brines, and production fluids containing solvents, acids, caustics, or other corrosive agents. These substances can leach into the elastomer, initiating chemical interactions that weaken the polymer structure. Such degradation may involve breaking polymer chains or alteration of cross-linking, which can significantly impair the elastomer's performance. The effects include excessive swelling, shrinkage, hardening, softening, cracking, and ultimately a loss of mechanical integrity and sealing capability over time.
  - H_2S Attack: Hydrogen sulfide (H_2S) exposure can lead to significant deterioration in elastomer performance, often manifesting as hardening, embrittlement, or specific forms of cracking. H_2S may also chemically interact with additives such as carbon black fillers. Cong et al. (2013) performed aging experiments on hydrogenated nitrile butadiene rubber (HNBR) in aqueous H_2S environments at a pressure of 1000 ± 100 psi and a temperature of 212 °F. Their findings revealed a notable decline in tensile strength, hardness, and elongation capacity. The degradation mechanisms were attributed to homolytic or heterolytic cleavage reactions, which generate macromolecular radicals. These radicals subsequently react with mercapto radicals, initiating a chain reaction that disrupts double and triple bonds in the polymer matrix, leading to the formation of saturated C–S–C linkages. Such transformations significantly alter the mechanical properties of the elastomer. Similarly, Fernández and Castaño (2016) investigated the influence of H_2S concentration on nitrile

butadiene rubber (NBR) and observed the formation of brittle fracture surfaces with increasing H_2S levels. Their study also reported a reduction in tensile strength, elongation at break, and overall resilience of the NBR material as H_2S concentration increased.

- CO_2 Attack: Exposure to supercritical or dense-phase CO_2 can lead to substantial swelling, plasticization (softening), and a reduction in the mechanical strength of elastomers, particularly under high-pressure conditions. Although CO_2 typically does not chemically react with elastomeric materials (Patel et al., 2019a, 2019b), its interaction changes in the presence of an aqueous phase such as water or brine. Under these conditions, CO_2 can form carbonic acid—a weak acid that, when present in significant amounts, can become corrosive and initiate chemical reactions with certain elastomers (Patel et al., 2019a, 2019b). For instance, in nitrile butadiene rubber (NBR), the acid promotes dissociation of weak C–N bonds, leading to the formation of amine groups. Ethylene propylene diene monomer (EPDM) elastomers, with their more stable $C = C$ double bonds, exhibit relatively higher resistance to such degradation. Fluorocarbon elastomers (FKM), characterized by their highly stable C–F bonds with high dissociation energy, show even greater resistance to CO_2-induced deterioration. Fernández and Castaño (2016) found that increasing CO_2 concentration leads to greater volumetric swelling and irreversible deformation in NBR elastomers, with the deformation eventually plateauing at high CO_2 levels. Scanning Electron Microscope (SEM) analysis also revealed a reduction in brittle fracture surfaces as CO_2 concentration increased. Zhu et al. (2017) examined the aging behavior of NBR and HNBR under mechanical compression in both liquid and gaseous CO_2 environments. They observed weight gain in the elastomers after exposure, along with a reduction in hardness, which was more pronounced in the gaseous phase. These findings suggest that mechanical loading exacerbates CO_2-induced degradation.
- Amine/Corrosion Inhibitor/EOR Chemical Attack: Some chemicals used downhole (e.g., certain corrosion inhibitors, scale inhibitors, completion fluid additives, EOR chemicals like alkaline agents, or surfactants) can be aggressive toward certain elastomer types (Askari et al., 2021). Compatibility testing with the specific chemicals is essential.
- Hydrocarbon Swelling: Absorption of hydrocarbons (oil, condensate) can cause excessive swelling, potentially leading to extrusion of the seal into the clearance gap or loss of mechanical strength.
- Material Selection: The most common elastomer materials used in the oil and gas industry fall into six categories: NBR (nitrile butadiene rubber), EPDM (ethylene propylene diene monomer), HNBR (hydrogenated NBR), FKM (fluoroelastomers), FEPM (perfluoroelastomers), and FFKM (perfluorocarbon elastomers). Their chemical and mechanical properties depend on the type of monomer, molecular weight, and degree of cross-linking. As a general rule, FKM, FEPM, and FFKM exhibit superior chemical resistance and withstand higher operating temperatures but are more expensive and less effective at low temperatures (Apple Rubber Products Inc., 2009; Elhard, 2017). In

contrast, more affordable elastomers such as NBR offer better mechanical performance but have limited resistance to heat and aggressive fluids. HNBR provides enhanced thermal and chemical durability over NBR and EPDM, though at a higher cost. Choosing the right elastomer type (e.g., NBR, HNBR, FKM, FFKM, Aflas®) based on its proven performance under expected chemical, pressure, and temperature conditions is essential. ISO 23,936–2 provides detailed guidance on elastomer selection and qualification for oilfield environments. Table 11.1 summarizes industry standards from API, ISO, and NORSOK relevant to elastomer seal assemblies in wellhead and liner hanger systems. Extrusion and nibbling (Fig. 11.6d) arise from cyclic movement or pressure pulsation across sealing interfaces, leading to material loss and impaired retrieval of tools. Compression set (Fig. 11.6e), the permanent deformation of seals, is often due to poor dimensional control and can lead to premature seal failure (Elhard, 2017). Understanding these mechanisms is critical to ensure robust seal performance throughout the well's lifecycle.

- **Thermal Degradation**: Exposing elastomers to temperatures above their maximum continuous operating limit causes irreversible damage to the polymer structure (thermal decomposition, oxidation), leading to hardening, cracking, loss of elasticity (compression set), and eventual seal failure. Each elastomer type has a specific temperature rating that should not be exceeded. Heat aging effects are cumulative over time and accelerate significantly at higher temperatures. The lifespan of an elastomer seal decreases rapidly as its temperature limit is approached or exceeded. Thermal degradation may cause radial cracks or softening when operating temperatures exceed 250–300 °F (Mackenzie & Garfield, 2007).
- **Explosive Decompression(ED)/Rapid Gas Decompression (RGD)**: When elastomers are exposed to high-pressure gas (e.g., methane, CO_2, H_2, N_2) for extended periods, gas molecules can permeate and dissolve into the polymer matrix (Mackenzie & Garfield, 2007). If the external pressure is then released rapidly (e.g., during blowdown, pressure testing, tripping out of the hole), the trapped gas inside the elastomer expands violently, causing internal cracks, blisters, or complete rupture of the seal structure. This is a major concern in high-pressure gas wells, gas storage wells, or during pressure testing with gas. ED-resistant elastomer grades and specific seal designs incorporating reinforcement or less permeable materials are available and necessary in gas service.
- **Extrusion**: If the differential pressure across the seal is too high relative to the seal material's strength, or the clearance gap between the metal components it seals against (the extrusion gap) is too large, the elastomer can be physically forced (extruded) into the gap, leading to permanent damage and leakage. Harder elastomers and the use of anti-extrusion backup rings (made of harder plastic like PEEK, or metal) help prevent this, especially at high pressures and temperatures where elastomers soften (Garfield & Mackenzie, 2007a).
- **Compression Set**: Over time, especially at elevated temperatures, elastomers held under continuous compression (as they are in many static seals like O-rings) can

Table 11.1 Qualification standards for elastomeric materials and sealing components in oil and gas applications

Standard	Organization	Title/description	Primary focus/notes
API 6A	API	Specification for Wellhead and christmas tree equipment	Covers equipment design and materials including elastomers used in sealing components
API 11D1	API	Packers and bridge plugs	Specifies requirements and test procedures for mechanical packers and elastomeric sealing elements
API Spec 17D	API	Subsea wellhead and tree equipment	Includes testing and qualification of sealing systems in subsea environments
API 17TR8	API	High-pressure High-temperature (HPHT) design guidelines	Provides guidance for equipment and seal design under HPHT conditions
ISO 10,423	ISO	Specification for Wellhead and christmas tree equipment	International equivalent to API 6A; includes elastomer and non-metallic sealing requirements
ISO 13,533	ISO	Drilling and production equipment—Drill through equipment	Covers blowout preventers and drill-through equipment including elastomeric sealing materials
ISO 13,628–4	ISO	Subsea production systems—subsea wellhead and tree equipment	Specifies requirements for subsea sealing systems and non-metallic components
ISO 14,310	ISO	Downhole equipment—packers and bridge plugs	Defines performance requirements and testing of elastomeric packers in downhole conditions
ISO 23,936–2	ISO	Non-metallic materials in contact with oil and gas media—part 2: Elastomers	Focuses on chemical compatibility, thermal aging, and qualification of elastomers in oil and gas environments
NORSOK M-710	NORSOK	Qualification of non-metallic sealing materials and manufacturers—polymers	Industry benchmark for elastomer RGD testing, thermal/chemical aging, and sour service exposure
NORSOK D-010	NORSOK	Well integrity in drilling and well operations	Includes integrity requirements for barrier elements such as elastomeric seals
ASTM D471	ASTM	Rubber property—effect of liquids	Measures swelling, hardness, and volume change of elastomers in contact with fluids

(continued)

Table 11.1 (continued)

Standard	Organization	Title/description	Primary focus/notes
ASTM D2000	ASTM	Standard classification system for rubber products in automotive applications	General specification, not oilfield-specific, but used for material reference
NACE TM0187	NACE	Elastomeric materials for rapid gas decompression (RGD) testing	Standardized test method for explosive decompression resistance in gas service

lose their ability to rebound to their original shape when the compressive load is removed. This permanent deformation reduces the sealing force (the resilience needed to conform to surfaces and maintain contact) and can eventually lead to leaks, particularly under subsequent pressure or temperature cycling. Low compression set materials are preferred for long-term sealing applications. The permanent deformation of seals is often due to poor dimensional control and can lead to premature seal failure (Elhard, 2017). Understanding these mechanisms is critical to ensure robust seal performance throughout the well's lifecycle.
- **Mechanical Damage**: Seals can be damaged during installation (e.g., cut or pinched while running through restrictions, assembled incorrectly, or exposed to sharp edges), during operation (e.g., abrasion from flowing solids, wear from moving components in dynamic seals), or during retrieval of equipment. Careful handling, proper tools, and adherence to installation procedures are essential.
- **Improper Installation/Energization**: Incorrect sizing, insufficient squeeze (compression), or failure to properly energize the seal during installation can result in an immediate or early leak path that may worsen over time.

Existing qualification standards (API, ISO, NORSOK, ASTM, NACE) for elastomers often focus on material properties or single-condition testing. However, they frequently lack robust, standardized protocols for testing seal *assemblies* under combined HPHT conditions, accounting for long-term aging effects, chemical compatibility, storage impacts, and the complex loading cycles experienced downhole. This represents a significant gap in ensuring the reliability and predicting the lifespan of elastomer seals for critical well integrity applications, highlighting the need for more rigorous, system-level, and performance-based qualification testing.

11.3 Performance and Failure of Metal-to-Metal (MTM) Seals

Metal-to-metal (MTM) seals are increasingly deployed in critical well environments such as high-pressure/high-temperature (HPHT) reservoirs, corrosive fluid service, gas injection wells, and subsea completions—settings in which elastomeric sealing systems often lack sufficient reliability, longevity, and resistance to degradation mechanisms such as explosive decompression and chemical attack (van Dort, 2009). Unlike elastomeric seals, which rely on elastic deformation and compressibility to achieve conformity, MTM seals operate by bringing two precision-machined metallic surfaces into intimate contact under high applied stress—typically via bolt preload, mechanical compression, or hydraulic actuation. This localized plastic deformation of surface asperities creates a robust, gas-tight barrier that resists leakage even under extreme wellbore conditions.

Traditional rubber, polymer, and plastic-based seals have well-established limitations when exposed to high differential pressures, elevated temperatures, chemically aggressive fluids, or high-pressure gas environments. MTM sealing systems were developed to overcome these limitations and have now evolved into highly engineered solutions capable of eliminating common elastomeric failure modes, including explosive decompression (gasification), thermal degradation, extrusion across clearance gaps, dynamic fatigue from cyclic loading, catastrophic failure under excessive compression loads, and chemical degradation. Unlike elastomers, which can begin to fail at temperatures between 150 and 300°F, MTM seals have demonstrated reliable performance at temperatures exceeding 700°F without loss of sealing capability (Garfield & Mackenzie, 2007b).

One of the most critical differentiators of MTM technology is its immunity to gas permeation and explosive decompression. Elastomeric seals, due to their porous structure and voids formed during manufacturing, can absorb gas under pressure and suffer from blistering, cracking, or rupture when that pressure is released—either suddenly or gradually. MTM seals, being non-porous and metallic, are entirely unaffected by gas saturation or decompression effects, maintaining sealing integrity across decompression cycles and long-term gas exposure.

MTM systems are also inherently resistant to thermal degradation. Elevated temperatures cause elastomers to soften, lose resilience, and enter compression set, ultimately resulting in seal failure. In contrast, MTM seals are constructed from alloys capable of withstanding sustained high temperatures without material breakdown or dimensional instability. Similarly, extrusion failures are commonly observed in elastomeric seals at high-pressure differentials, especially when unsupported by proper anti-extrusion rings. Under extreme loading, rubber elements deform and extrude through small gaps in the sealing interface, leading to leak paths and tool retrieval issues. This risk increases with seal expansion and reduced mechanical strength. Conversely, MTM seals, due to their all-metal construction, withstand high-pressure differentials without extruding or deforming, providing reliable sealing even across tight tolerance gaps.

In addition to their resistance to thermal and chemical breakdown, modern MTM seals now feature significant mechanical adaptability. Since 2000, design advancements have enabled expansion ratios of up to 160% of the original run-in diameter, enabling these seals to conform to varying borehole geometries or pass through restrictions before setting (Garfield & Mackenzie, 2007a). This expansion capability makes MTM seals viable in many applications that previously required inflatable elastomeric devices, especially in high-expansion zones where conformity and anchoring are critical.

The advantages of MTM sealing systems can be clearly defined in four key performance domains (Garfield & Mackenzie, 2007a):

(1) superior expansion capability,
(2) high thermal resistance,
(3) extreme pressure tolerance, and
(4) outstanding chemical compatibility.

These characteristics enable MTM technology to bridge the gap between conventional high-expansion elastomeric seals and the stringent performance demands of today's advanced well environments. As such, MTM sealing represents not just an incremental improvement, but a fundamental shift in how downhole sealing integrity is achieved—paving the way for more durable, efficient, and intervention-free well operations.

Common MTM seal failure modes include the following:

- **Insufficient Seal Stress**: The primary cause of MTM seal leakage. If the contact stress between the sealing surfaces is too low (due to improper design, incorrect energization/makeup during installation, relaxation over time from creep or cyclic loading, or insufficient bolt load in flanges), it may not be enough to overcome microscopic surface imperfections and achieve a tight seal, particularly against small molecules like gas (Stewart, 2019).
- **Surface Damage**: Scratches, dents, corrosion pits, or trapped debris on the precision-machined sealing surfaces can create leak paths. Protection of sealing surfaces during handling, meticulous cleanliness during installation, and careful assembly procedures are critical.
- **Galling**: Seizing or tearing of the metal surfaces during makeup (e.g., when tightening premium connections or energizing wellhead seals) due to high friction and localized cold welding. Galling damages the seal surfaces and prevents proper energization. Use of appropriate, qualified thread compounds/lubricants and controlled makeup procedures (e.g., torque-turn) are necessary to prevent galling.
- **Corrosion**: While often made from CRAs, MTM seals can still be susceptible to localized corrosion mechanisms (e.g., crevice corrosion, galvanic corrosion if dissimilar metals are used) in aggressive environments over time, which can compromise the sealing interface (Garfield & Mackenzie, 2007a).
- **Loss of Seal Stress (Relaxation/Creep)**: Over time, especially under sustained load and elevated temperatures (creep) or repeated pressure/temperature cycling (fatigue), the initial contact stress in the seal can relax due to material deformation or component movement, potentially opening a leak path (Yuan et al., 2024).
- **Thermal Expansion/Contraction**: Differential thermal expansion/contraction between mating metal components can induce movement that challenges the seal's ability to maintain contact stress, particularly in designs that do not accommodate significant movement (Vacculex, 2022).

MTM seals generally require higher precision in manufacturing, more careful handling, cleaner assembly conditions, and often higher actuation forces (e.g., higher

Fig. 11.3 Comparison of retrieved bridge plug seal elements: the elastomeric seal (left) exhibits visible degradation, including missing chunks and tearing, indicating exposure damage or mechanical wear; in contrast, the metal-to-metal (MTM) seal (right) remains intact and undamaged, demonstrating superior resilience under identical downhole conditions (van Dort, 2009)

bolt torque, hydraulic setting pressure) compared to elastomeric seals. They are less forgiving of imperfections but offer superior performance in demanding conditions when properly designed, installed, and qualified. They represent a promising alternative or necessary technology where elastomers are insufficient, but their own limitations require careful consideration. Figure 11.3 presents a direct visual comparison between the post-retrieval conditions of elastomeric and MTM seal elements used in bridge plugs.

11.4 Packer Failure Mechanisms

Packers, as introduced in Sect. 6.2, are critical downhole tools used to establish zonal isolation by sealing the annular space between the tubing and the casing. This isolation is achieved through the interaction of elastomeric sealing elements and mechanical anchoring devices such as slips. A failure of the packer to maintain this seal constitutes a breach of the primary well barrier, which can result in sustained casing pressure (SCP) in the A-annulus or unintended crossflow between formations behind the casing.

As illustrated in Fig. 11.4, the packer comprises several essential components—slips, cone, packing element, and mandrel—each serving a specific function in the overall operation of the tool. The slips act as the anchoring mechanism, gripping the casing wall to resist axial movement. These are driven outward by the cone, which acts as a tapered wedge to force the slips into engagement during the setting process. Simultaneously, the elastomeric packing element is compressed axially between rigid shoulders, causing it to expand radially and form a pressure-tight seal against the casing wall. The mandrel, which runs longitudinally through the center of the tool, provides structural support and serves as the conduit for hydraulic activation.

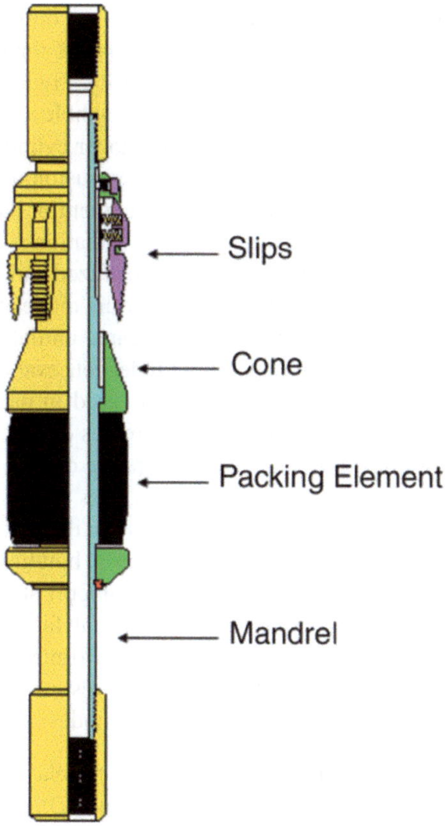

Fig. 11.4 Cross-sectional schematic of a typical packer assembly showing key components: slips for anchoring to the casing wall, cone to drive the slips outward, packing element to form a pressure-tight seal, and mandrel serving as the structural core for load transmission and hydraulic activation

The packer is typically set using hydraulic pressure applied from the tubing. This pressure is transmitted through the central bore and acts on a control piston. When the applied pressure exceeds a pre-set threshold, shear pins or connectors break, allowing the piston to move and trigger the setting sequence. The cone is driven downward, expanding the slips and compressing the packing element to secure the packer in place. Once set, the packer maintains isolation even under varying downhole conditions (Liu et al., 2012).

To unset the packer, the tubing string is lifted. While the slips and cone remain stationary relative to the casing, the mandrel moves upward, releasing the shear ring. This disengages the locking mechanism and allows the outer sleeve to retract. As a result, the slips collapse inward and the packing element relaxes, restoring the tool to its unset condition and allowing for retrieval or repositioning.

However, prolonged exposure to mechanical loads, temperature fluctuations, and differential pressure cycling can degrade the elastomeric packing element. Over time, this can lead to extrusion, fatigue, or irreversible deformation. Additionally, improper packer design—such as sharp geometrical transitions, excessive void spaces, or poorly constrained elastomer volumes—can exacerbate stress concentrations within the sealing system. These issues not only compromise the sealing performance of the packer but also increase the risk of zonal communication, necessitating remedial operations to restore well integrity. Figure 11.5 presents photographic evidence of packer element failure after bottomhole exposure, illustrating the deformation of the support ring, crushing of the rubber cylinder, and characteristic signs of extrusion-related failure. In addition to extrusion and surface crushing, elastomeric packer elements may experience detachment or partial shedding, particularly in geometrically transitional areas. These "mutation zones" often lack proper mechanical retention, making them prone to localized failure under thermal and pressure cycling. Figure 11.6 illustrates such a failure mode observed in an anti-outburst packer, where the rubber sealing element became partially dislodged. These failures compromise sealing integrity and may hinder retrieval operations or sealing reliability in future runs. Rubber sealing elements used in packers are subject to significant mechanical, thermal, and chemical stresses during downhole operations. After exposure to sustained pressure and temperature cycles, their structural integrity may degrade—resulting in material extrusion, loss of elasticity, and eventual mechanical rupture. Figure 11.7 deformation diagram illustrates the post-pressure condition of three rubber cylinders retrieved from a bottomhole environment. Visible shear marks, burst failure, and extrusion of support structures indicate the cumulative effect of internal gas expansion, mechanical fatigue, and reduced elastic resilience. Such damage compromises sealing performance, especially in wells containing corrosive media or elevated temperatures, ultimately increasing the risk of leakage and barrier failure. Common failure modes include the following:

- **Seal Element Failure**: Any of the elastomer failure modes described in Sect. 11.2 (chemical/thermal degradation, ED/RGD, extrusion, compression set, mechanical damage during installation/operation) is the most common cause of packer leakage. For packers utilizing MTM seals, failure modes described in Sect. 11.3 apply.

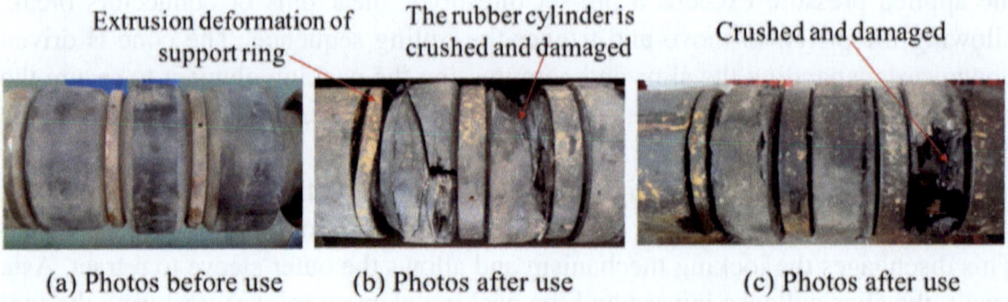

Fig. 11.5 Photographic sequence showing progressive damage to a flat-bottom rubber cylinder structure in a packer seal. **a** Pre-deployment condition; **b** Post-use extrusion deformation and crushing of the rubber seal; **c** Additional evidence of mechanical breakdown. Damage includes loss of elasticity, radial extrusion beyond the support ring, and tearing—indicating high stress and permanent deformation from long-term bottomhole sealing (Liu et al., 2022)

Fig. 11.6 Localized detachment of the rubber seal element in an anti-outburst packer. The failure occurred in a geometric transition zone ("mutation area"), leading to partial shedding and potential sealing loss (Liu et al., 2022)

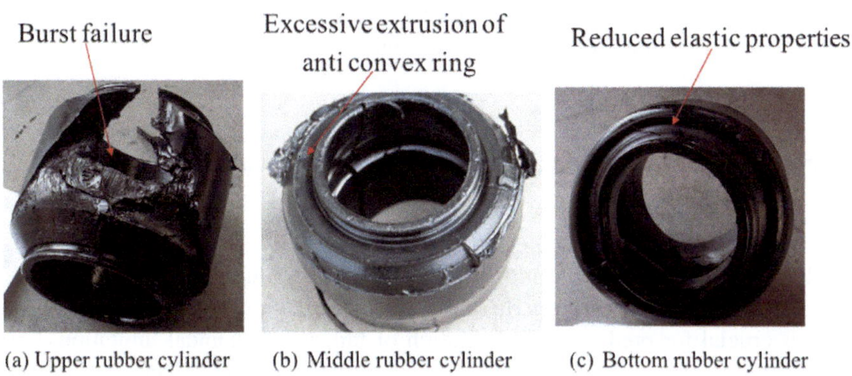

Fig. 11.7 Post-pressure deformation and failure modes of rubber cylinders used in packer seal assemblies. **a** Upper rubber cylinder exhibiting burst failure and shear damage; **b** middle cylinder showing excessive extrusion of the anti-convex support ring; **c** bottom cylinder displaying permanent deformation and reduced elastic recovery. These conditions reflect mechanical fatigue, thermal softening, and gas-induced expansion effects commonly encountered during downhole operations (Liu et al., 2022)

- **Leak Past Slips/Anchoring Mechanism**: In some packer designs, particularly those with complex mechanical setting systems, if the primary seal elements fail significantly, pressure could potentially bypass the intended seal and leak past the anchoring slips or through gaps in the packer body. Slips are primarily designed for mechanical anchoring, not sealing.
- **Failure to Set Correctly**: Improper setting procedures (e.g., insufficient set force), debris on the casing wall preventing slip engagement or full seal element compression, or tool malfunction during the setting sequence can lead to an ineffective initial seal or inability to anchor the packer securely.

- **Failure to Unset (Retrievable Packers)**: Accumulation of scale or deposits (mineral scale, paraffin, asphaltenes) on the packer body, corrosion, or component failure within the unsetting mechanism can prevent retrievable packers from being unset and retrieved from the well, requiring costly milling operations during intervention or abandonment.
- **Leak Through Packer Body/Mandrel**: Cracks, corrosion holes, or manufacturing defects in the main packer body or mandrel can create a leak path, though this is generally less common than seal failure in modern, qualified packers.
- **Failure of Feed-Through Seals**: Packers with penetrations for control lines, chemical injection lines, or cables rely on seals around these lines passing through the packer bore or body. Failure of these specialized feed-through seals creates a leak path for wellbore fluids to migrate up the line or into the annulus above the packer.

Packer failure directly results in loss of isolation between the tubing and the A-annulus from below, breaching the primary barrier and typically leading to SCP. Remediation often requires expensive well intervention (see Chap. 26).

11.5 Mathematical Evaluation of Rubber Seal Deformation and Failure

In addition to qualitative descriptions of packer element failure mechanisms, a quantitative understanding of elastomeric rubber seal deformation under downhole conditions is crucial for predicting the initiation of failure, mechanical limitations, and safe operating envelopes. The three-stage deformation model (Zhang et al., 2018) provides an analytical framework for evaluating the contact pressure, shear stress, and sealing performance of rubber elements in packers under pressure loading, temperature variation, and casing constraints.

11.5.1 Free Deformation Stage

In the free deformation stage, the rubber element expands radially without constraint. The radial displacement of the rubber can be described by

$$\frac{d^2u_r}{dr^2} + \frac{1}{r}\frac{du_r}{dr} - \frac{u_r}{r^2} = 0 \tag{11.1}$$

Boundary Conditions

$$u_r\big|_{r=R_0} = 0, \ u_r\big|_{r=R_{os}} = R_{os} - R_1 \tag{11.2}$$

where u_r is the radial displacement (in meters) and r is the radial position (in meters). R_0 is the initial inner radius (in meters), R_{os} is the outer radius at the start of deformation (in meters), and R_1 is the final inner radius (in meters).

The solution for radial displacement becomes

$$u_r = -\frac{(R_{os}-R_1)R_{os}}{R_{os}^2-R_0^2}r + \frac{(R_{os}-R_1)R_0^2R_{os}}{(R_{os}^2-R_0^2)r} \tag{11.3}$$

The initial sealing force—a key parameter for evaluating the rubber's ability to establish an initial seal—is given by

$$F_t = -2\pi\frac{ER_{os}}{1+\nu}\frac{R_{os}^2-R_1^2}{R_{os}^2-R_0^2}(R_{os}-R_1) \tag{11.4}$$

where E denotes Young's modulus of the rubber (in Pascals) and ν is Poisson's ratio of the rubber material.

This expression highlights how the **elastic modulus** and **radial expansion gap** affect the sealing force.

11.5.2 Constrained Deformation Stage

As the rubber encounters the **casing wall**, further radial deformation is constrained. In this stage, the **radial equilibrium equation** adjusts to incorporate the gradient of contact pressure p:

$$\frac{\partial^2 u}{\partial r^2} + \frac{1}{r}\frac{\partial u}{\partial r} = \frac{1}{G}\frac{\partial p}{\partial z} \tag{11.5}$$

where u represents the radial displacement (in meters), G is the shear modulus of the rubber (in Pascals), p is the contact pressure (in Pascals), and z is the axial position along the packer length (in meters).

The associated **shear stress distribution** within the rubber is

$$\tau(r,z) = -G\frac{\partial u}{\partial r} \tag{11.6}$$

This shear stress distribution is critical for understanding extrusion and material failure risks.

11.5.3 Stable Deformation Stage

In the final stage, a stable sealing contact pressure is established:

$$p_2(z) = \frac{\Delta p}{\cosh\hat{\lambda}}\cosh\hat{\lambda}\left(\frac{z}{H_1} - 1\right) \tag{11.7}$$

The parameter $\hat{\lambda}$, reflecting rubber stiffness and geometric constraints, is given by

$$\hat{\lambda} = H_1\sqrt{\frac{8G(\ln R_{os} - \ln R_0)}{K[(R_{os}^2 + R_0^2)(\ln R_{os} - \ln R_0) - (R_{os}^2 - R_0^2)]}} \tag{11.8}$$

where H_1 is the height of the rubber seal (m) and K is bulk modulus of the rubber (Pa).

These formulas quantify how geometry (casing ID, packer ID, seal height) and rubber stiffness govern the final contact pressure distribution along the seal.

11.5.4 Sealing Performance Criterion

The total maximum contact pressure that ensures safe sealing is

$$p_{max}(z) = p_1(z) + p_2(z) \geq C\Delta p \tag{11.9}$$

where $p_1(z)$ is the pressure from free expansion, Δp is the applied pressure differential, and C is a reliability coefficient (greater than 1) used for safety margin considerations.

This **criterion** ensures the contact pressure exceeds the **pressure differential** by a safe margin.

11.5.5 Shear Stress Failure Limit

To ensure shear stability and avoid extrusion failure, the maximum shear stress is compared to the allowable shear limit:

$$\tau_{max} = \max\{\tau(R_0, 0), \tau(R_{os}, 0)\} \leq [\tau] \tag{11.10}$$

where $[\tau]$ is the allowable shear stress of rubber (Pa), determined from lab tests.

This condition prevents overloading and shear-induced ruptures of the packer element.

11.6 Wellhead and Christmas Tree Seal and Valve Leaks

The wellhead and Xmas tree (as described in Chap. 6) contain numerous static seals (hanger seals, flange gaskets, valve bonnet seals) and dynamic seals (valve stem packings) that are critical for both primary and secondary barrier integrities at the surface. Failures here pose direct safety and environmental risks. A common and often underdiagnosed cause of metal-to-metal (MTM) seal failure in tubing hanger assemblies is pitting corrosion along the critical sealing surface. Unlike gross mechanical damage, pitting can occur microscopically over time due to aggressive wellbore fluids, stagnant zones, or poor surface protection during manufacturing or installation. These pits compromise the surface contact stress required for the MTM seal to energize effectively, leading to leakage despite correct installation. Figure 11.8 illustrates such a case, where visible pitting damage was observed on the tubing hanger sealing surface. The schematic highlights the location of the seal and its interface, while the photographs show the actual corroded sealing surface after disassembly (Tyagi et al., 2022). This emphasizes the importance of proper material selection, surface finish quality, corrosion mitigation strategies, and post-installation inspection to preserve the long-term integrity of wellhead sealing systems.

- **Flange/Connector Seal Leaks**: Leaks past the seals joining different wellhead spools, housings, or the tree to the wellhead.
  - Ring Gasket Issues: The most common cause is improper installation of metal API ring gaskets (R, RX, BX types). This includes using a dirty or damaged ring groove, incorrect gasket type or material, inadequate or uneven bolt tightening (insufficient bolt torque or improper sequence), or damage to the ring gasket

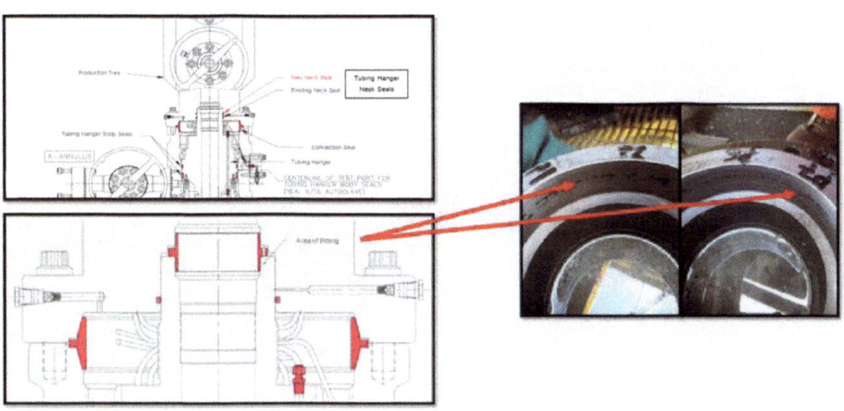

Fig. 11.8 Schematic and photographic evidence of metal-to-metal (MTM) seal failure due to pitting corrosion in a tubing hanger assembly. The left-side diagrams identify the sealing zones, while the photos on the right reveal localized surface pitting that compromised sealing integrity, leading to leakage (Tyagi et al., 2022)

itself. BX gaskets, used in higher pressure ratings, are particularly sensitive to correct installation and pressure energization. Corrosion of ring grooves or gaskets over time can also cause leaks.

– Clamp Seal Issues: Seals in clamp connectors (used as an alternative to flanges) can fail due to elastomer degradation, damage, or improper clamp makeup.
– Corrosion/Erosion Damage: The sealing surfaces themselves can be damaged over time by corrosion or erosion if exposed to flowing fluids, preventing a tight seal even with a new gasket. In addition to seal degradation, structural cracking of the flange body itself presents a serious failure mode for wellhead and Christmas tree assemblies (Zhou et al., 2021). These cracks often initiate at stress concentrations such as bolt holes, seal grooves, or weld interfaces and can propagate over time due to thermal cycling, high-pressure operation, or poor material properties. Figure 11.9 illustrates a case of catastrophic cracking in a four-way flange operating at 243 K. Both circumferential and longitudinal cracks were observed along the upper flange and external surfaces. Internal inspection further revealed through-wall propagation. This highlights the importance of material selection, flange design, and periodic non-destructive testing (NDT) to prevent sudden failure of these critical pressure-containing components.

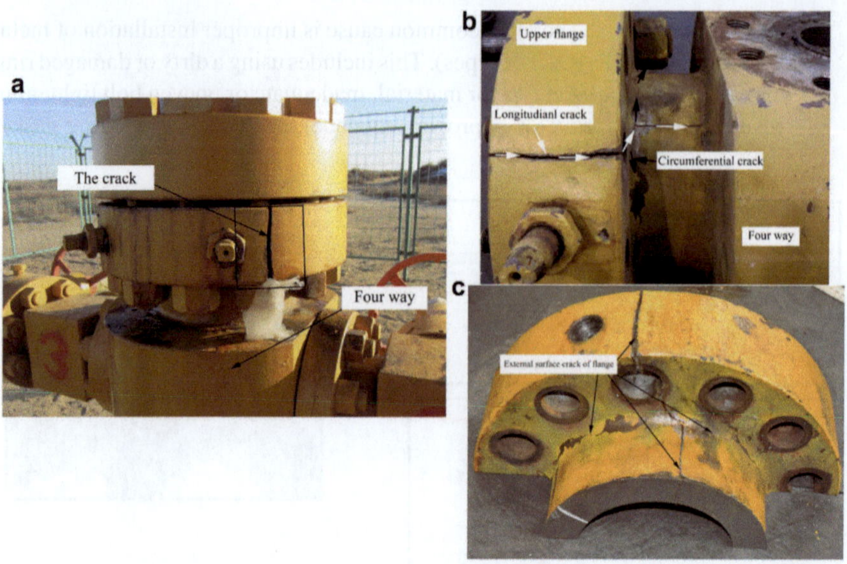

Fig. 11.9 Macro-scale cracking morphology in a four-way Christmas tree flange. **a** Field photograph showing visible longitudinal and circumferential cracks along the flange assembly. **b** Local magnification indicating crack propagation at the upper flange connection. **c** Sectioned flange showing internal and external surface crack development. Operating temperature was 243 K (Zhang et al., 2019)

- **Hanger Seal Leaks (Casing or Tubing Hanger)**: Leaks past the seals isolating the annuli from each other or from the surface
  - Elastomer Failure: Subject to the failure modes in Sect. 11.2 (chemical/thermal degradation, ED, extrusion, compression set, damage). Access for replacement is often difficult, potentially requiring removal of the tree or spools.
  - MTM Seal Failure: Insufficient energization during installation, surface damage, or corrosion (see Sect. 11.3).
  - Penetration Seal Leaks: Failure of seals around control line or cable feed-throughs in the tubing hanger (see Sect. 6.4).
- **Valve Bonnet/Body Seal Leaks**: Static seals between the valve body and bonnet (the cover containing the stem and operating mechanism) can fail due to elastomer degradation, damage, or improper assembly.
- **Valve Stem Packing Leaks**: Dynamic seals around the moving valve stem (allowing the stem to move in/out for gate valves or rotate for ball/plug valves) are prone to wear, degradation (elastomer part of packing), and relaxation over time, leading to external leaks. Regular maintenance (e.g., injecting sealant/packing compound into injection ports, tightening packing glands where provided) may temporarily mitigate minor leaks, but eventual replacement of the packing set might be necessary during intervention (see Chap. 26).
- **Valve Leak Through (Failure to Seal)**: This is when a valve fails to provide a tight shut-off when in the closed position, allowing fluid/pressure to pass through the valve bore (Hadzihafizovic, 2024). This represents a failure of a critical barrier element (e.g., a master valve failing to contain reservoir pressure).
  - Gate/Seat Damage: Erosion from high-velocity flow/solids, corrosion pitting, or mechanical damage (e.g., closing on debris, scale, or sand) to the sealing surfaces of the gate (or ball/plug) and seat ring prevents a tight shut-off. This is a very common failure mode, especially for production master and wing valves that experience erosive or corrosive flow.
  - Obstruction: Debris, scale build-up, paraffin, or hydrates preventing the gate/ball from fully closing onto the seat.
  - Actuator Failure (for actuated valves): Failure of the actuator (hydraulic, pneumatic, electric) to fully stroke the valve to the closed position or apply sufficient force to energize the seal.
- **Valve Failure to Operate**: Valves can seize due to corrosion, scale build-up, lack of lubrication/maintenance, or mechanical failure of internal components, preventing them from being opened or closed when required (Zhou et al., 2021). Regular valve lubrication and functional testing help prevent this.

Leaks in the wellhead or Xmas tree can range from minor drips to significant releases, posing safety and environmental risks and potentially escalating if left unaddressed. Maintaining the integrity of these surface components through regular inspection, testing, and maintenance is crucial for upholding well integrity. Leakage rate assessments are critical for maintaining well integrity, particularly when evaluating potential leaks through the Christmas tree valves or into the annulus. Industry standards such as ISO 10,418, API 14C, and API 14H provide reference limits for acceptable leakage rates, while the draft ISO/TS 16,530–2 introduces best practice thresholds, including a liquid leakage limit of 2 cc/min per inch of valve size and a gas leakage limit of 0.35 scf/min per inch (Raj, 2014). Figure 11.10 (left) compares these limits to actual leakage assessments observed in Christmas tree valves, offering a benchmark for evaluating valve performance. The right-hand side of Fig. 11.10 illustrates leakage rate thresholds for flow into the annulus, distinguishing between leak sources located above and below the surface safety valve (SSSV). These thresholds are categorized into alarm, action, and maximum allowable levels to guide operational decisions and risk management.

To quantitatively assess the severity of gas leaks across sealing elements—such as wellhead flange seals, casing hanger seals, and valve bonnet seals—API14B/ISO10417 standards provide a method to calculate the gas leakage rate using the ideal gas law. The leakage rate q (in scf/hr) is calculated as follows:

$$q = 2122\left(\left(\frac{p_f}{Z_f} - \frac{p_i}{Z_i}\right)\left(\frac{1}{t}\right)\left(\frac{V}{T}\right)\right) \tag{11.11}$$

where q is leakage (flow) rate in standard cubic feet per hour (scf/hr) and p_f, p_i are final and initial pressures (psi), Z_f, Z_i are final and initial compressibility factors (dimensionless), t is test duration (minutes), V is isolated observed volume (ft^3), and T is the absolute temperature of the gas in Rankine (°F + 460).

This formula converts the observed pressure rise in a sealed volume into an equivalent volumetric flow rate, allowing operators to compare measured leakage against regulatory standards and integrity criteria.

11.7 Downhole Safety Valve (DHSV) Failure Modes

Surface and subsurface valve systems work in tandem to maintain well control and isolate pressure during normal operations and emergency scenarios. Among these, Downhole Safety Valves (DHSVs)—as detailed in Sect. 6.3—serve as critical primary barrier elements. They are designed to automatically shut in the well in the event of surface control failure or emergency conditions. Proper DHSV functionality is essential to prevent blowouts, especially in high-pressure or offshore environments. The failure of a DHSV to close, open, or seal properly can result in serious safety risks and loss of well control (Yuan et al., 2024). Figure 11.11 presents a visual overview of the DHSV's role within the well safety barrier envelope and introduces a risk classification framework that incorporates DHSV status alongside other critical elements, such as surface valves and sustained casing pressure. This classification supports risk-informed prioritization for well integrity interventions.

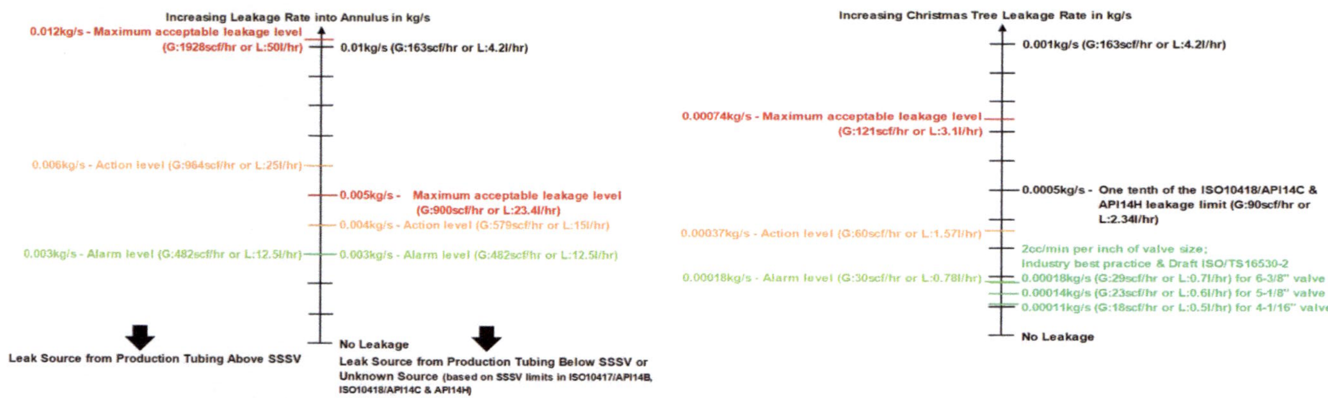

Fig. 11.10 Leakage rate thresholds for well integrity evaluation. Left: Comparison of assessed leakage limits for Christmas tree valves with ISO 10,418/API 14C/API 14H maximum allowable leakage levels and the draft ISO/TS 16,530–2 best practice criteria (2 cc/min per inch of valve size for liquids or 0.35 scf/min per inch for gas). Right: Assessment of leakage rates into the annulus, distinguishing between leak sources above the surface safety valve (SSSV) and those below it or of unknown origin, with alarm, action, and maximum acceptable levels indicated for both gas and liquid flows (Raj, 2014)

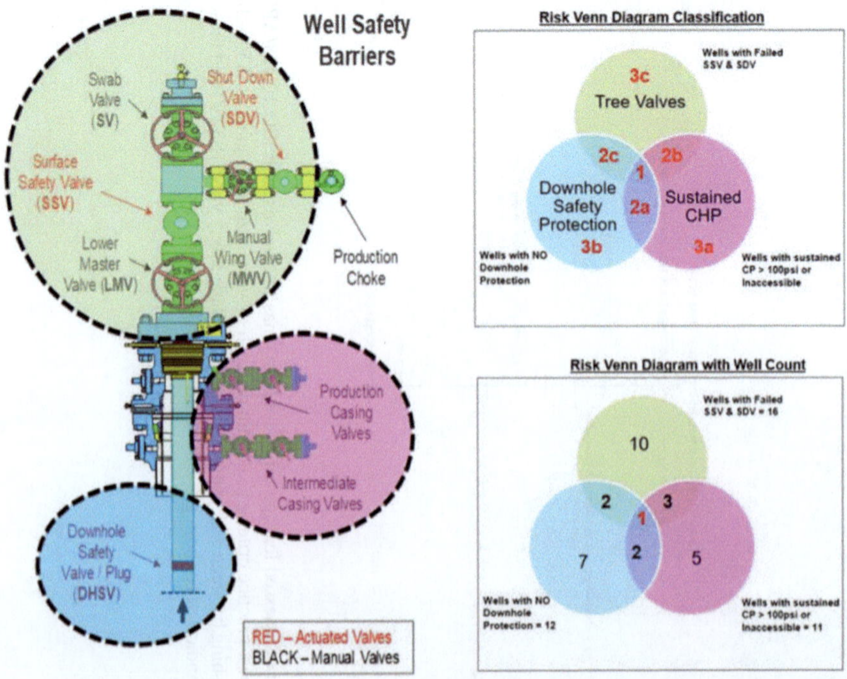

Fig. 11.11 Schematic representation of well safety barriers, including actuated and manual valves across the wellhead, casing, and downhole systems. The Downhole Safety Valve (DHSV) is shown as a critical subsurface barrier. The right-hand side includes Venn diagrams for risk classification and well count, highlighting intersections of failed surface valves (SSV/SDV), absence of downhole safety protection, and sustained casing head pressure (CHP > 100 psi or inaccessible). Color-coded zones illustrate wells at highest risk (e.g., failure of both surface and downhole barriers combined with sustained pressure) (Adeyinka et al., 2020)

- **Failure to Close**: The most critical safety failure mode. If the valve fails to close on demand (loss of control pressure), its primary safety function is defeated (Soerensen et al., 2024). Can be caused by
  - **Control System Failure:** Blockage, leak, or collapse of the hydraulic control line preventing pressure bleed-off; failure of the surface hydraulic power unit or control system; failure of umbilical components (for subsea wells—Chap. 20).
  - **Mechanical Obstruction**: Debris, scale, paraffin, or hydrates settling in the tubing bore or within the valve mechanism physically preventing the flapper/ball from swinging or rotating into the closed position or landing fully on the seat.
  - **Component Failure**: Broken spring, seized piston, damaged linkage, or other mechanical failure in the closing mechanism preventing it from stroking the valve closed.

- **Leakage When Closed (Through-Valve Leak)**: Failure of the flapper/ball or seat sealing surfaces to provide a tight seal against well pressure from below when the valve is commanded closed. Similar causes to other valve leak-throughs (erosion, corrosion damage, obstruction) (Marsh, 2008). Requires periodic leak testing (per API RP 14B/ISO 10,417) to detect and quantify (API RP 14B/ISO 10417). An allowable leak rate (e.g., gas/liquid specific) is usually defined by standards/regulations; exceeding this rate means the valve fails its integrity test.
- **Failure to Open**: Inability to reopen the valve after it has closed. Can be caused by mechanical component failure, scale/debris preventing movement, torsion spring, or control system issues (unable to build sufficient hydraulic pressure to overcome well pressure and spring force) (Soerensen et al., 2024). This results in the well being shut in below the valve.
- **Control Line Leak**: Leakage from the hydraulic control line itself (due to corrosion, connection failure, or damage during installation or operation). A control line leak leads to loss of hydraulic holding pressure, causing the fail-safe valve to close (often resulting in a "nuisance trip" if not an emergency). It also prevents the valve from being reopened.

Regular functional testing and leak testing according to standards (e.g., API RP 14B/ISO 10,417) and regulatory requirements are essential to ensure DHSVs remain reliable primary barriers and safety devices. Failure requires prompt diagnosis and remediation, often involving wireline intervention to install an insert valve (which lands inside the existing valve profile) or a full workover to replace the TR-SCSSV (see Chap. 26).

Understanding these potential failure modes for seals, packers, and valves allows for better material selection and design choices (Chap. 6), targeted monitoring strategies (Chap. 8), effective testing programs (Chap. 7), and more effective preventative maintenance programs to uphold well integrity throughout the operational life.

11.8 Tubing Movement and Packer to Tubing Forces

Understanding the interplay between tubing movement and packer to tubing forces is essential for predicting and preventing packer failures, as well as for ensuring long-term well integrity. Changes in pressure, temperature, and mechanical loading throughout well operations can induce significant axial and radial forces that compromise packer performance (Lubinski et al., 1962). This section presents the key mathematical formulations used to evaluate these effects.

Mathematical Formulations

1. **Packer Setting Force**

The force applied to the packer during setting depends on the pressure difference across the tubing plug and the cross-sectional area of the tubing ends.

$$F_{set} = P_{set} A_i (1 - 2\mu) \tag{11.12}$$

where F_{set} is setting force on the packer, P_{set} is pressure difference, A_i = internal tubing cross-sectional area, and μ is Poisson's ratio.

2. **Hydrostatic Packer Setting Force**

For hydrostatic packers, the locked force depends on the hydrostatic tubing pressure and the difference in external and internal cross-sectional areas.

$$F_{set} = P_{set} (A_o - A_i)(2\mu - 1) \tag{11.13}$$

where A_o is outer tubing cross-sectional area.

3. **Piston Effect Force**

The axial force due to the piston effect is

$$F_a = P_i (A_p - A_i) - P_o (A_p - A_o) \tag{11.14}$$

where P_i is internal tubing pressure, P_o is external pressure, and A_p is packer bore area.

4. **Length Change Due to Piston Effect**

$$\Delta L_1 = \frac{F_a \times L}{(A_o - A_i)E} \tag{11.15}$$

where L is tubing length and E is Young's modulus.

5. **Length Change Due to Ballooning**

$$\Delta L_2 = \frac{2\mu L (A_i \Delta P_i - A_o \Delta P_o)}{(A_o - A_i)E} \tag{11.16}$$

6. **Length Change Due to Buckling**

$$\Delta L_3 = \frac{r^2 A_p}{2} \frac{(\Delta P_i - \Delta P_o)^2}{8EI(w_s + w_i - w_o)} \tag{11.17}$$

where r is radius of tubing, I is moment of inertia, and w_s, w_i, w_o are the weights of tubing, liquid inside tubing, and liquid in casing per unit length, respectively.

7. **Length Change Due to Temperature**

$$\Delta L_4 = L\alpha\Delta T \tag{11.18}$$

where α is thermal expansion coefficient and ΔT is temperature change.

8. **Length Change Due to Packer Setting**

$$\Delta L_5 = \frac{F \times L}{(A_o - A_i)E} \tag{11.19}$$

Total Tubing Movement

$$\Delta L_{\text{total}} = \Delta L_1 + \Delta L_2 + \Delta L_3 + \Delta L_4 + \Delta L_5 \tag{11.20}$$

Packer to Tubing Force

The force to restore tubing to its original position depends on the sign of ΔL_{total}. If $\Delta L_{\text{total}} < 0$, the force is tensile; if $\Delta L_{\text{total}} > 0$, it is compressive.

References

Adeyinka, A., Tsakporhore, A., Arije, O., Adegoke, O., Aliyev, S., Iwueze, C., & Hulett, R. (2020). Rethinking well integrity for sustainability: Adopting a risk based approach. In *Society of petroleum engineers—SPE Nigeria annual international conference and exhibition 2020, NAIC 2020*. https://doi.org/10.2118/203672-MS

Apple Rubber Products Inc. (2009). *Seal design guide welcome to apple rubber*.

Ashena, R. (2025). *Barriers. The practical handbook of well control* (pp. 20–31). https://doi.org/10.1201/9781003473770-2

Askari, M., Aliofkhazraei, M., Jafari, R., Hamghalam, P., & Hajizadeh, A. (2021). Downhole corrosion inhibitors for oil and gas production–a review. *Applied Surface Science Advances*, *6*. https://doi.org/10.1016/j.apsadv.2021.100128

Cong, C. B., Cui, C. C., Meng, X. Y., Lu, S. J., & Zhou, Q (2013). Degradation of hydrogenated nitrile-butadiene rubber in aqueous solutions of H2S or HCl. *Chemical Research in Chinese Universities*, *29*(4), 806–810. https://doi.org/10.1007/S40242-013-2401-7/METRICS

Duijm, N. J. (2009). Safety-barrier diagrams as a safety management tool. *Reliability Engineering and System Safety, 94*(2), 332–341. https://doi.org/10.1016/J.RESS.2008.03.031

Elastomers for fluid containment in offshore oil and gas production: Guidelines and review. (2005). https://search.worldcat.org/title/316540766

Elhard, J.D., & Duguid, A. (2017). *TAP-767AA|Bureau of safety and environmental enforcement*. https://www.bsee.gov/tap-technical-assessment-program/research-on-safety-technology-verification-for-materials-and

Elhard, A. D. (2017). *TAP-767AA|Bureau of safety and environmental enforcement*. https://www.bsee.gov/tap-technical-assessment-program/research-on-safety-technology-verification-for-materials-and

Fernández, C., & Castaño, P. (2016). *Compatibility behavior of elastomers for PCP applications*. OnePetro. https://dx-doi-org.ezproxy.lib.utexas.edu/

Flitney, R. (2014). Chapter five—materials. *Seals and Sealing Handbook*, 369–435. https://doi.org/10.1016/B978-0-08-099416-1.00005-X

Garfield, G., & Mackenzie, G. (2007a). Metal-to-metal sealing technology: bridging the gap between conventional and high-expansion zonal isolation applications. In *Society of petroleum engineers coiled tubing and well intervention conference and exhibition 2007* (pp. 351–356).

Garfield, G., & Mackenzie, G. (2007b). *Zonal isolation applications utilizing new metal-to-metal sealing technology demonstrates potential for offshore and deepwater environments.* OnePetro. https://dx-doi-org.ezproxy.lib.utexas.edu/

Hadzihafizovic, D. (2024). *Xmas tree and oil wellhead.* https://papers.ssrn.com/abstract=4766593

Liu, Y., Guo, F., & Xu, J. (2012). Nalyzing packer deformation of tubular for unsetting process in HTHP wells under variable (T, P) fields. *The Open Petroleum Engineering Journal, 5*, 109–117.

Liu, Y., Qian, L., Zou, J., Xia, C., & Lian, Z. (2022). Study on failure mechanism and sealing performance optimization of compression packer. *Engineering Failure Analysis, 136*, Article 106176. https://doi.org/10.1016/J.ENGFAILANAL.2022.106176

Mackenzie, G., & Garfield, G. (2007). Wellbore isolation intervention devices utilizing a metal-to-metal rather than an elastomeric sealing methodology. *Proceedings—SPE Annual Technical Conference and Exhibition, 2*, 995–999. https://doi.org/10.2118/109791-MS

Marsh, J. (2008). Downhole cra materials performance—failure analysis of a sub surface safety valve. In *Texas At Austin user on* (Vol. 19). OnePetro. https://dx-doi-org.ezproxy.lib.utexas.edu/

Mody, R., Gerrard, D., & Goodson, J. (2013). Elastomers in the oil field. *Rubber Chemistry and Technology, 86*(3), 449–469. https://doi.org/10.5254/RCT.13.86999

Patel, H., Salehi, S., Ahmed, R., & Teodoriu, C. (2019a). Review of elastomer seal assemblies in oil & gas wells: Performance evaluation, failure mechanisms, and gaps in industry standards. *Journal of Petroleum Science and Engineering, 179*, 1046–1062. https://doi.org/10.1016/j.petrol.2019.05.019

Patel, H., Salehi, S., & Teodoriu, C. (2019b). Assessing mechanical integrity of expanding cement. In *Society of Petroleum Engineers—SPE Oklahoma city oil and gas symposium 2019, OKOG 2019*. https://doi.org/10.2118/195225-MS

Raj, M. (2014). *Well integrity—Christmas tree acceptable leakage rate and sustained casing pressure.*

Saharuddin, M. K., Fadzin, K. A., Aziz, M. F., Ghaffar, M. F. A., Kasim, M. H., Mursalim, M. H., Xintee, C., Khalid, M. S., Chua, H. L., Choi, Y. X., & Mokhtar, M. K. M. (2024). Unconventional solution for well operation and maintenance: A successful case history of surface and downhole leak repair using robust chemical sealant. In *Society of petroleum engineers—SPE symposium—well integrity management 2024*. https://doi.org/10.2118/220572-MS

Soerensen, D. S. H., Paterson, J., Lobban, G., Agayev, F., Savelli, P., Holt, S., Craig, R., & Muralidharan, M. (2024). Electromechanical wireline tool for DHSV investigation and remediation. *Texas at Austin User on.* https://doi.org/10.2118/220661-MS

Stewart, M. (2019). Centrifugal pumps. *Surface Production Operations*, 61–309. https://doi.org/10.1016/B978-0-12-809895-0.00003-X

Tyagi, V., Bates, N., & Gurmin, I. (2022). Safety recovery around failed tree seals in an offshore gas well. In *Society of Petroleum Engineers—SPE Asia Pacific oil and gas conference and exhibition 2022, APOG 2022*. https://doi.org/10.2118/210725-MS

Uai, L., Lubinski, A., Althouse, A. W. S., & Logan, A. J. L. (1962). Helical buckling of tubing sealed in packers. *Journal of Petroleum Technology, 14*(06), 655–670. https://doi.org/10.2118/178-PA

Vacculex. (2022). *Analysis of failure causes of mechanical seals from failure modes—vacculex.* https://www.vacculex.com/knowledge/failure-causes-of-mechanical-seals/

van Dort, R. (2009). Metal-to-metal seals meet downhole hazard demands. *Journal of Petroleum Technology, 61*(01), 24–26. https://doi.org/10.2118/0109-0024-JPT

Yan, H., Yin, J., Liu, S., Chen, Z., Jiang, T., Yan, H., Yin, J, Liu, S., Chen, Z., & Jiang, T. (n.d.). *Failure model and multi-parameter leakage rate analysis of metal rubber seals* (Vol. 155). https://doi.org/10.1007/978-981-97-0922-9_83

Yuan, G., Wang, Y., Yang, X., Fang, Y., Ma, R., Ning, K., Guan, M., & Tang, Y. (2024). Study on quantitative evaluation method for failure risk factors of the high-temperature and high-pressure downhole safety valve. *Sustainability, 16*(5), 1896. https://doi.org/10.3390/SU16051896

Zhang, J. M., Wang, X., Zhu, Y., & Du, P. (2019). *Failure analysis of a four-way flange cracking in a KQ65 wellhead christmas tree.* https://doi.org/10.1007/s11668-019-00612-2

Zhang, F., Jiang, X., Wang, H., Song, N., Chen, J., & Duan, J. (2018). Mechanical analysis of sealing performance for compression packer rubber tube. *Mechanics & Industry, 19*(3), 309. https://doi.org/10.1051/MECA/2017045

Zhou, Z., Gong, D., Zeng, Z., & Lin, T. (2021). Failure analysis of a four-way flange erosioning in a KQ52 wellhead Christmas tree. *Engineering Failure Analysis, 129*, Article 105641. https://doi.org/10.1016/J.ENGFAILANAL.2021.105641

Zhu, D., Lin, Y., Ma, H., Zhang, H., Li, Y., Zhang, L., & Deng, K. (2017). Experimental studies on CO_2 corrosion of rubber materials for packer under compressive stress in gas wells. *Engineering Failure Analysis, 80*, 11–23. https://doi.org/10.1016/J.ENGFAILANAL.2017.01.012

Chapter 12
Logging and Inspection Techniques for Integrity Evaluation

Abstract This chapter provides a detailed overview of logging and inspection techniques crucial for evaluating and maintaining well integrity across all phases of a well's life. It begins with cement evaluation tools—such as Cement Bond Logs (CBL), Variable Density Logs (VDL), and advanced ultrasonic imaging—that assess the quality of cement bonding and detect zonal isolation failures. The chapter then covers corrosion logging methods, including electromagnetic thickness tools and mechanical calipers, which identify and quantify internal and external metal loss in casing and tubing. Leak detection technologies are also explored, including temperature logs, acoustic/noise logs, fluid identification tools, and chemical tracers. These techniques are essential for pinpointing the depth and nature of leaks behind casing or within annular spaces. The chapter further addresses routine wellhead inspection and maintenance practices that ensure surface barrier components remain functional. Finally, the chapter introduces Risk-Based Inspection (RBI), a strategic approach that prioritizes monitoring and diagnostics based on the likelihood and consequence of failure. This framework enables efficient resource allocation and enhances long-term well integrity management. Collectively, these tools and strategies support safe operations, timely intervention, and regulatory compliance.

© The Author(s), under exclusive license to Springer Nature Switzerland AG 2026

A. Alsubaih and K. Sepehrnoori, *Well Integrity for Oil, Gas, and Subsurface Storage Operations*, https://doi.org/10.1007/978-3-032-15233-6_12

Logging truck unit and field crew during nighttime operations, illustrating wireline logging and well surveillance activities. *Source* The Authors

Chapter's Objectives

- Explain the principles and applications of cement evaluation logs (CBL/VDL, Ultrasonic) for assessing cement sheath integrity.
- Describe the use of corrosion logging tools (electromagnetic thickness, mechanical calipers) for assessing tubular metal loss.
- Detail logging methods used for leak detection (temperature, acoustic/noise, fluid identification, tracers).
- Discuss common wellhead inspection and maintenance techniques.
- Introduce the concept of Risk-Based Inspection (RBI) strategies for prioritizing surveillance activities.

12.1 Introduction: Diagnosing the Downhole Condition

The structural and functional integrity of a well depends on the continuous performance of multiple barrier elements—tubulars, cement, packers, valves, and seals—throughout the well's lifecycle. While surface-based monitoring techniques, such as annular pressure surveillance (see Chap. 8), can serve as early warning indicators, they often provide only indirect evidence of subsurface degradation or failure. Accurate, location-specific diagnosis of well integrity issues—such as metal loss, cement failure, leak paths, or valve malfunction—requires direct downhole investigation. Logging and inspection techniques are indispensable for this purpose. These tools are deployed into the wellbore on wireline, slickline, or coiled tubing to collect detailed measurements from within the tubing, casing, or annular spaces. Wireline logging involves lowering specialized instruments into the wellbore to acquire real-time or memory-recorded measurements of subsurface conditions. These tools are typically conveyed on a steel-armored cable from a logging truck at the surface and are powered and controlled from the surface in real time. The sensors within the toolstring can measure properties such as acoustic response, resistivity, gamma ray emissions, and borehole geometry. Figure 12.1 illustrates the conceptual setup of a wireline logging operation, highlighting the surface unit, cable deployment, logging tool position within the borehole, and the correlation between tool readings and depth-indexed log data. This visualization helps contextualize the diverse data types acquired during a typical integrity evaluation run, including resistivity logs and formation imaging. Modern logging technology enables engineers to assess the condition of primary and secondary barrier elements, quantify corrosion, detect and locate leaks, evaluate cement bonding, and identify deformations or mechanical damage. In parallel, surface inspection methods focus on critical infrastructure at the wellhead and Christmas tree—where leaks, corrosion, or valve failure can represent a direct loss of containment. The integration of these data—whether collected routinely for baseline monitoring or triggered in response to anomalies—supports informed decision-making for integrity management. It enables operators to evaluate degradation over time, compare with historical baselines, and determine whether barriers continue to perform their intended function or require repair, remediation, or abandonment. Increasingly, these diagnostics are aligned with risk-based inspection (RBI) frameworks that help prioritize efforts and reduce unnecessary interventions, while ensuring the most critical components are closely monitored.

This chapter provides a comprehensive overview of key logging and inspection technologies applied in the field of well integrity. It begins with cement evaluation logs, which are essential for verifying zonal isolation post-cementing and over the well's lifetime. This is followed by corrosion logging tools that assess casing/tubing wall loss due to internal or external mechanisms. Next, leak detection technologies are introduced, including thermal, acoustic, fluid identity, and tracer-based techniques. The chapter then transitions to wellhead inspection methods, which focus on surface-accessible barriers and interfaces, and concludes with a discussion on Risk-Based Inspection (RBI) strategies, which optimize inspection programs based on risk levels, operational conditions, and historical data. Ultimately, logging and inspection are not merely technical routines—they are proactive integrity assurance tools. By choosing the right tool, at the right time, and for the right barrier, operators can uphold containment, reduce downtime, comply with regulatory requirements, and ensure safe, sustainable field operations.

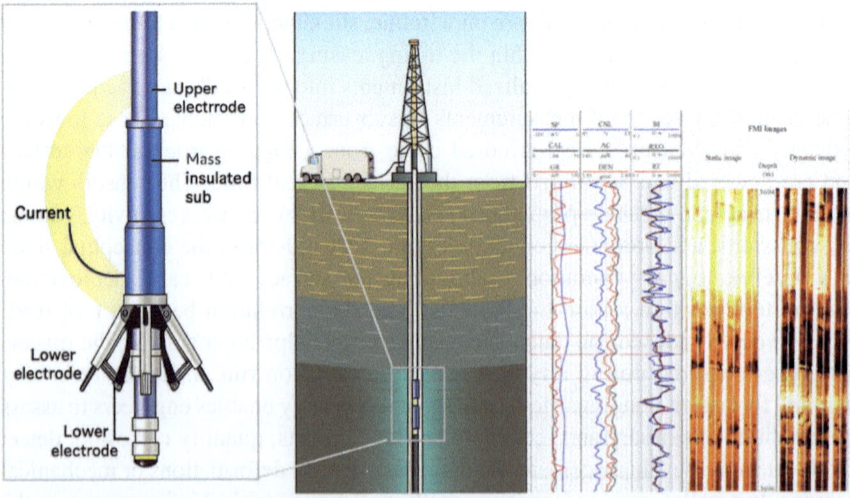

Fig. 12.1 Conceptual illustration of a wireline logging operation. The diagram shows the deployment of a wireline logging tool from a surface truck into the wellbore, where downhole electrodes and sensors measure formation and casing conditions. On the right, example log outputs—including gamma ray, resistivity, density, and borehole imaging (FMI). The integration of log data and visual images enables a detailed assessment of subsurface integrity and formation properties

12.2 Cement Evaluation Logging

Verifying the quality of the primary cement job immediately after placement (as discussed in Chap. 5) and assessing its long-term condition and bond degradation over time are crucial for assessing zonal isolation integrity. Several logging techniques are used:

- **Cement Bond Log (CBL)/Variable Density Log (VDL)**: These are acoustic tools that have been the industry standard for decades for evaluating the cement bond at the casing–cement interface.
  - **Principle**: An acoustic transmitter in the logging tool emits a sound pulse that travels through the wellbore fluid. The pulse interacts with the casing wall, and subsequent signals are received by receivers spaced at different distances from the transmitter (e.g., 3 ft and 5 ft or 8 ft). If the casing is free (not bonded to cement), the sound energy resonates strongly within the casing wall, and a high-amplitude signal travels relatively unimpeded along the casing to the receivers. If the casing is well bonded to a solid material (set cement), energy is coupled *out* of the casing into the cement and potentially the formation, causing the signal traveling *along* the casing to be attenuated (reduced in amplitude) (Bateman, 2015).

– **CBL Output**: The primary measurement is the amplitude of the first arriving signal at a specific receiver (typically the 3 ft receiver). Low amplitude indicates good coupling to the annulus (good bond), while high amplitude indicates poor coupling (poor bond or free pipe). The distribution of cement around the casing is reported by the **Bond Index (BI)**, which correlates the amplitude of the casing arrival time to the relative degree of cement bond to the casing. It quantifies the percentage of the pipe circumference that is cemented:

$$BI = \frac{A_m}{A_c} \tag{12.1}$$

where A_m is the attenuation in the zone of interest (dB/ft) and, A_c is the attenuation in a 100% cemented zone (dB/ft).

The attenuation values A_m and A_c are calculated using:

$$A_m = \frac{-20}{d}\log\left(\frac{\text{Amplitude measured in zone of interest}}{\text{Amplitude in "free pipe"}}\right) \tag{12.2}$$

$$A_c = \frac{-20}{d}\log\left(\frac{\text{Amplitude measured in 100\% scemented sheath}}{\text{Amplitude in "free pipe"}}\right) \tag{12.3}$$

where d is the tool spacing or receiver distance (ft), amplitude measured in zone of interest: actual log measurement amplitude

Amplitude in “free pipe”: reference amplitude for unbonded casing.

These formulas allow quantitative evaluation of cement bond quality in the casing–cement–formation system, providing an essential integrity diagnostic in well construction and lifecycle monitoring.

– **VDL Output**: Displays the full acoustic waveform received by a further receiver (typically the 5 ft or 8 ft receiver) as intensity variations over time (a “wiggly line” display) (Catala et al., 1984). Characteristic patterns help differentiate between:

Free pipe: Strong, clear signals corresponding to acoustic energy traveling only along the casing wall (“casing arrivals”). No or weak signals corresponding to energy traveling slower through the formation.

Well-bonded cement: Weak casing arrivals (energy coupled out) and potentially strong signals corresponding to energy traveling through the cement and formation (“formation arrivals”).

Partially bonded/channeled: Intermediate patterns, often with some casing energy remaining and weak or absent formation arrivals.

– **Applications**: Primarily assesses the bond quality at the casing–cement interface. By analyzing the VDL, it can also provide qualitative indications of bonding to the formation if formation signals are clearly received. Used as a baseline immediately after cementing and potentially later in life to assess bond degradation due to thermal/pressure cycling or chemical attack.

CBL and VDL are also used to determine the isolation between reservoir units by evaluating the cement quality behind casing. CBL measures the loss of acoustic energy, which is proportional to the fraction of the casing perimeter covered by cement. The ratio of CBL/VDL gives an average volumetric assessment of cement in the casing-to-formation annular space. Ultrasonic logs, in comparison, provide a high-resolution scan in real time about the condition of the casing-to-cement bond and casing quality (van Kuijk et al., 2005). These logs can identify the presence of gas and liquid channels in cement, as well as tubular damage, offering insights into the zonal isolation status. These inferences are made based on the difference in acoustic impedance thresholds between various materials.

However, ultrasonic recordings have limitations, particularly when the cement is contaminated with gas. The acoustic impedance of gas-contaminated cement varies depending on the volume and type of gas entrained, and ultrasonic logs often cannot distinguish between gas-contaminated cement and free gas or fluid in the annulus (Coelho de Souza Padilha & da Silva Araujo, 1997; Johns et al., 2011). Furthermore, ultrasonic logs provide post-operational assessments but do not directly measure cement strength. Table 12.1 shows CBL qualitative interpretation of the cement quality.

Additional factors that affect the reliability of both CBL and ultrasonic logs include the following:

- Micro-annuli, especially when filled with liquid due to thermal cycling or mechanical stress, may still appear as bonded zones.
- Mud cake, pipe coatings, or irregular tool movement can create misleading results.
- Eccentric casing and poor centralization can lead to signal bias.
- Lightweight cement may result in low acoustic contrast between unset slurry and hardened cement.
- Cement setting time, affected by temperature, contamination, and mix design, can distort interpretation if logging is performed prematurely (Contraires et al., 2009).
- **Limitations**: Can be affected by tool centralization (eccentered tool gives misleading high amplitude on the side closest to the casing wall), heavy drilling mud properties, fast formations (formation signal arrives before casing signal), thin casing, and micro-annuli (may show good bond even with a small gap if liquid-filled, although gas-filled micro-annuli might be detectable). Provides limited azimuthal information (the signal averages around the borehole circumference). Interpretation requires experience and knowledge of the specific well conditions and tool physics.

- **Ultrasonic Cement Evaluation Tools (e.g., USIT, Isolation Scanner, CAST-V, FMI-CT)**: Ultrasonic Cement Evaluation Tools—such as the Ultrasonic Imaging Tool (USIT), Isolation Scanner, CAST-V, and FMI-CT—represent a class of advanced logging instruments developed to provide high-resolution, azimuthal, and radial information about cement sheath quality, casing integrity, and borehole fluid conditions. These tools operate on the ultrasonic pulse-echo principle but

Table 12.1 CBL qualitative interpretation

Condition	Transit time	CBL amplitude	VDL	Interpretation notes
Free pipe	Normal (~180–220 μs)	High (>60 mV)	– Strong casing arrivals—No formation arrivals	Casing is unbonded; indicates free pipe
Good bond to casing and formation	Normal to high/may be noisy	Low (<30 mV)	– No casing arrivals—Clear formation arrivals	Excellent cement placement and coupling to both casing and formation
Good bond to casing, NOT formation	Normal to high/may be noisy	Low	– No casing arrivals—No formation arrivals	Cement is bonded to casing but debonded from formation (e.g., poor mud removal)
Poor bond to casing	Normal	Medium to high	– Strong casing arrivals—No formation arrivals	Poor cement placement—likely debonded or only partially covered casing
Micro-annulus (liquid-filled)	Normal	Medium to high	– Casing arrivals—Weak or no formation arrivals	Cement present but hydraulic seal may be compromised; micro-annulus filled with liquid
Micro-annulus (gas-filled)	Normal or slightly low	Medium to high	– Casing arrivals—Irregular/absent formation arrivals	More likely to show up as poor bond; gas severely affects acoustic impedance
Channeling	Normal	Medium to high	– Casing arrivals—Patchy or weak formation arrivals	Cement channel or poor centralization; partial or circumferential bond
Eccentered tool	Low or erratic	Low (on near side)	– Varies; may show casing arrivals depending on offset	Results highly affected by tool position; may falsely indicate poor bonding on one side

differ in sensor design, resolution, and signal processing capabilities. The USIT, one of the earliest tools of its kind, generates 360-degree acoustic impedance logs using basic amplitude and travel-time measurements, offering effective detection of bonded and unbonded zones, though with limited radial resolution (Narender Pal, 2006). The isolation scanner enhances this capability through improved signal processing and dynamic range, allowing for more accurate identification of solid, liquid, or gas in the annular space, It identifies corrosion- or drilling-induced wear and is particularly useful in wells with lightweight or gas-contaminated cements (SLP, 2025). The CAST-V (cased-hole acoustic scanning tool—Version V) further advances cement imaging by employing multiple receivers to produce high-fidelity azimuthal images and casing wall thickness measurements, making it ideal for identifying eccentric casing, corrosion, and localized bonding issues (Halliburton, 2025). The FMI-CT, adapted from the Formation Micro-Imager for cased-hole applications, delivers extremely high-resolution ultrasonic images of the internal casing and cement interface, making it especially valuable in high-risk or high-resolution environments such as CO_2 storage wells or HPHT completions. Collectively, these tools offer significantly greater diagnostic precision than conventional CBL/VDL logs, especially where detailed spatial imaging of cement integrity is essential for verifying wellbore isolation and long-term integrity.

- **Principle**: Use a rotating ultrasonic transducer that emits high-frequency sound pulses toward the casing wall and measures the amplitude and travel time of the reflected echo from the internal casing surface. The resonance (or acoustic impedance) of the casing wall is affected by the material it is acoustically coupled to on the outside (solid set cement, liquid mud/water, or gas).
- **Outputs**: Generate high-resolution azimuthal maps (images) showing:

  Acoustic Impedance: Calculated from the reflected echo properties. High impedance indicates solid material on the outside (good cement coupling); low impedance indicates liquid (mud, water) or gas. Allows clear differentiation between solid cement and liquid/gas in the annulus around the circumference (Morris et al., 2007a, 2007b).

  Bond Quality/Status: Often presented as a color map showing areas of good bond, partial bond, liquid-filled areas (channels), or gas-filled areas (Ko Kyi, 2018).

  Casing Thickness/Internal Radius: Most ultrasonic tools simultaneously measure the thickness of the casing wall and the internal casing radius based on the travel time of the ultrasonic pulse. This allows detection of internal and external casing corrosion or wear (Loo & Maric, 2024).

• **Applications**: Provide detailed assessment of cement quality and distribution around the entire circumference of the casing. Can better identify channels (especially gas- or liquid-filled) compared to CBL/VDL. By analyzing later arrivals of the acoustic signal that have traveled through the cement, some tools can provide information about the cement–formation bond, although this is more challenging.

Often preferred for critical cement jobs (HPHT, deepwater, storage wells) or diagnosing complex isolation problems. Also useful for assessing casing corrosion in conjunction with cement quality.

The **attenuation rate** quantifies the energy loss of acoustic waves, helping to assess cement bonding radially and azimuthally:

$$\alpha = \frac{20 \cdot \log_{10}\left(\frac{A_1}{A_2}\right)}{L} \tag{12.4}$$

where α is attenuation rate (dB/ft), A_1, A_2 are amplitudes at two receiver points (mV), and L is distance between receivers (ft).

High α indicates good cement bonding (energy absorbed), low α suggests poor bonding.

Compared to CBL/VDL tools, ultrasonic logs offer significantly improved vertical and radial resolution, enabling a higher fidelity image of the cement sheath and its spatial distribution. Additionally, ultrasonic interpretation is less dependent on the cement's acoustic properties, relying instead on sufficient contrast in acoustic impedance (typically $\geq$0.5 MRayl) between cement and surrounding fluids to distinguish material boundaries (Morris et al., 2007a, 2007b). However, unlike CBL/VDL tools, ultrasonic logs do not directly provide qualitative information on cement-to-formation bonding. Conversely, CBL tools are more tolerant of heavy mud environments, where ultrasonic tools may not perform reliably—making CBL/VDL advantageously in certain high-density fluid conditions.

– **Limitations**: More sensitive to tool centralization and borehole conditions than CBL/VDL. Interpretation can still be complex, especially with lightweight or foam cements, or in very irregular boreholes. Higher cost than standard CBL/VDL. Requires specialized interpretation expertise.

Cement evaluation logs are essential for verifying initial zonal isolation immediately after construction (Chap. 5) and for investigating suspected cement failures later in life (e.g., as a root cause of SCP—Chap. 8, or assessing suitability for intervention/abandonment). Comparing logs run at different times can reveal degradation of the cement bond or matrix over time. Figure 12.3 showed the severity of cement damage for 40 wells using Cement Bond Log (CBL) evaluation. Each vertical column represents a well. Color intensity reflects poor cement index (%), with red indicating low-quality (poor bonding) zones and blue indicating good bonding. Black vertical lines represent casing intervals, and diamonds indicate casing shoe depths. (a) shows 7-inch casing data and (b) shows 9.625-inch casing data. Table 12.1 shows CBL qualitative interpretation summary (Fig. 12.2).

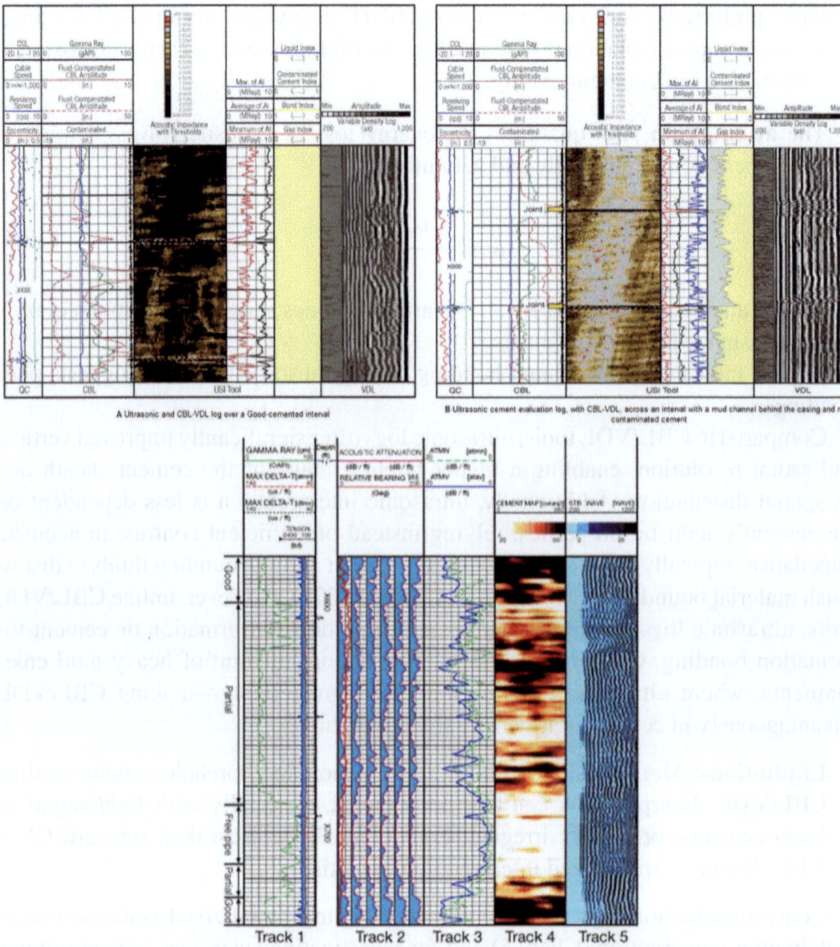

Fig. 12.2 (Top Left, **a**): Ultrasonic and CBL-VDL log over a good-cemented interval showing strong casing–cement bond with high acoustic impedance and clear formation arrivals. (Top Right, **b**): Ultrasonic and CBL-VDL log over a poorly cemented interval indicating a mud channel behind casing and mud-contaminated cement, evidenced by acoustic impedance anomalies and weak VDL formation signals. (Bottom): Track-by-track layout illustrating (Track 1) gamma ray and casing collar log, (Track 2) acoustic attenuation and impedance attributes, (Track 3) relative cement bond index curves, (Track 4) azimuthal ultrasonic cement map revealing bonding variations around the casing, and (Track 5) VDL acoustic waveform for cement quality confirmation. This layout aids in diagnosing cement integrity and zonal isolation (Petro Shine, 2023)

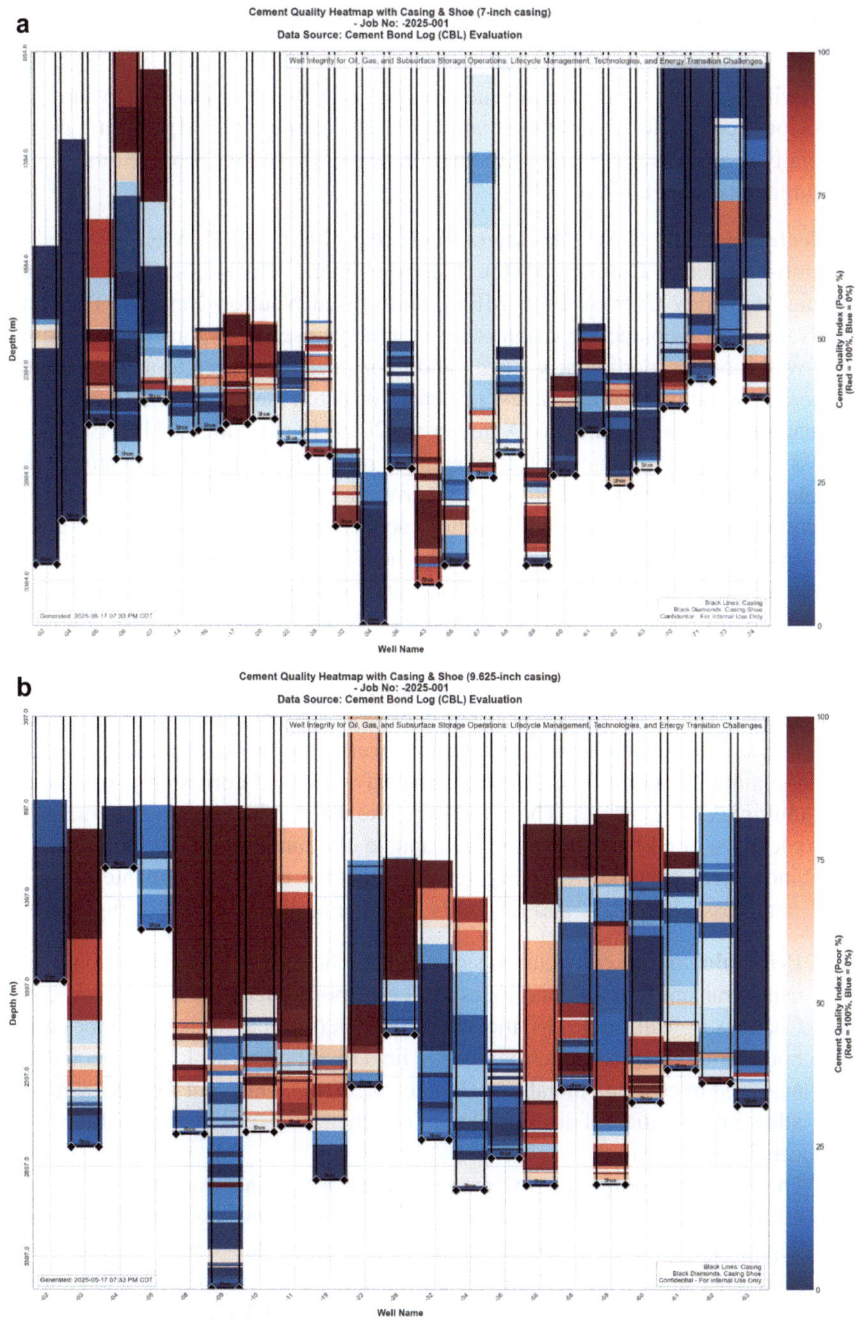

Fig. 12.3 Cement quality heatmaps for 40 wells using Cement Bond Log (CBL) evaluation. Each vertical column represents a well. Color intensity reflects poor cement index (%), with red indicating low-quality (poor bonding) zones and blue indicating good bonding. Black vertical lines represent casing intervals, and diamonds indicate casing shoe depths. **a** shows 7-inch casing data and **b** shows 9.625-inch casing data

12.3 Corrosion Logging Tools

Monitoring the condition of casing and tubing strings for metal loss due to corrosion or erosion is critical for assessing remaining structural integrity and preventing leaks, especially in aging wells or corrosive environments. These logs provide quantitative measurements of metal loss.

- **Electromagnetic (EM) Thickness Tools (e.g., MFC, ETC, MET, PNN)**: The Multi-Frequency Caliper (MFC), Electromagnetic Thickness Caliper (ETC), Magnetic Eddy Current Tool (MET), and Pulsed Neutron Neutron (PNN) are advanced, non-contacting diagnostic instruments used to evaluate the condition of well tubulars, particularly casing and liner integrity. These tools measure the average remaining wall thickness or detect localized metal loss due to corrosion, wear, or mechanical damage (Ajgou et al., 2019; Imrie et al., 2021a). EM thickness tools utilize a range of electromagnetic principles—including magnetic flux leakage, eddy current, and pulse eddy current methods—combined with varying frequencies to assess material conditions across different pipe sizes. Their ability to function without direct contact allows them to operate in fluid-filled wells and evaluate both single and multiple concentric casing strings. This makes them especially effective in mature or complex wells where physical access is limited. EM tools are widely applied in corrosion monitoring, life-of-well analysis, and risk-based integrity evaluations, providing operators with essential data to guide remedial actions, workovers, or permanent well abandonment planning. Electromagnetic thickness tools play a vital role in well integrity diagnostics by enabling non-contact detection of metal loss, internal wear, and corrosion across single or multiple casing strings. These tools operate using principles such as magnetic flux leakage or eddy currents and are capable of quantifying wall thickness even behind concentric pipe layers. Figure 12.4 illustrates a comparative interpretation of metal loss using MFC and EM tools for inner and outer casing strings.
  - **Principle**: Induce eddy currents or magnetic fields into the casing/tubing wall using transmitter coils and measure the response detected by receiver coils. Changes in wall thickness, metal loss defects (internal or external), or casing boundaries affect the electromagnetic field distribution, which is detected and analyzed to estimate metal loss (Brill et al., 2011). Different frequency ranges allow investigation of different depths of penetration (e.g., distinguishing inner versus outer casing).
  - **Outputs**: Provide quantitative measurements of remaining wall thickness (Brill et al., 2011). Often presented as a curve showing average thickness versus depth, statistical distributions (min, max, average thickness), or color maps indicating areas of metal loss (e.g., percentage of original wall thickness). Some advanced tools can identify and size localized defects like pits or holes, and sometimes differentiate between internal and external metal loss, though differentiation can be challenging with some tool types or in complex multi-string environments.

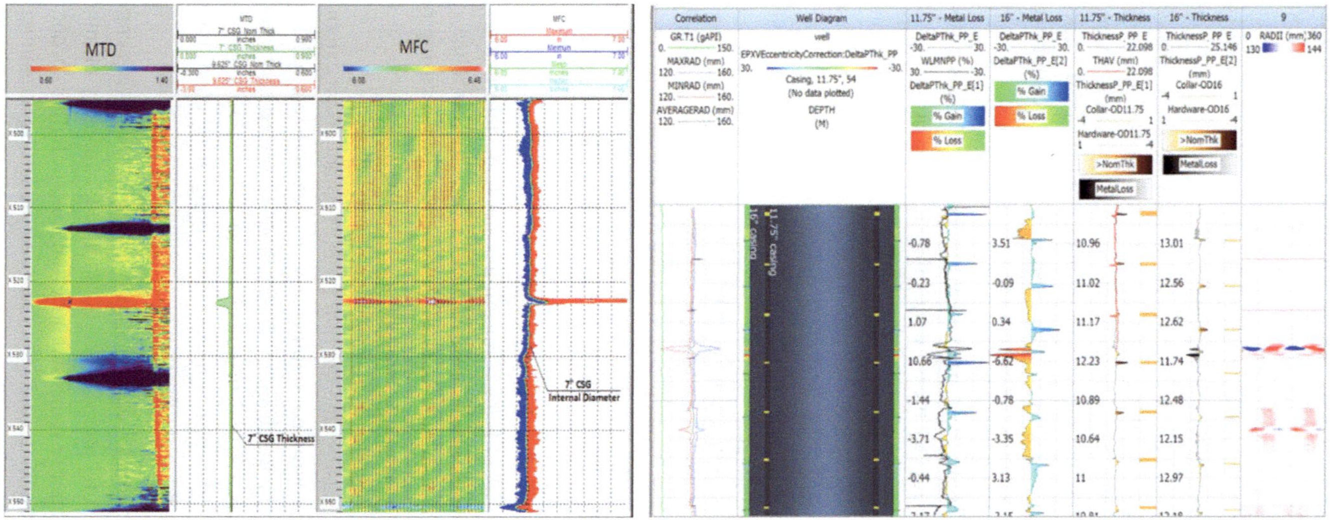

Fig. 12.4 Comparative corrosion log results from Magnetic Flux Leakage Caliper (MFC) and Electromagnetic (EM) thickness tools. The left panels show wall thickness and internal diameter variations detected by MFC across the 7-inch casing, while the right panels illustrate metal loss in the outer 16-inch casing based on EM-derived thickness differentials. MFC provides detailed internal pipe geometry and localized metal loss (Track 3), while EM tools offer non-contact metal loss quantification even across nested tubulars. This combination improves multi-layer corrosion profiling in complex well completions (Ajgou et al., 2019; Imrie et al., 2021a)

- **Applications**: Widely used for routine corrosion monitoring (especially in corrosive wells or mature assets), assessing remaining burst/collapse strength based on remaining wall thickness, and identifying areas requiring remediation (e.g., casing patches). Can often measure thickness of multiple concentric strings (e.g., production casing and intermediate casing simultaneously), which is valuable for assessing integrity in annuli.
- **Limitations**: Accuracy depends on tool type, frequency used, tool centralization, logging speed, calibration, and wellbore environment (e.g., salinity of fluid in the annulus). May have difficulty distinguishing closely spaced defects or accurately characterizing complex pit geometry or long axial cracks. Differentiating internal versus external metal loss is not always definitive with all tools. Interpretation requires understanding tool response.

- **Mechanical Caliper Tools (e.g., MFC—Multi-Finger Caliper)**: Tools with mechanical fingers that directly measure the internal radius of the pipe.
  - **Principle**: Consists of multiple (typically 24 to 60 or more) mechanical arms or fingers that are spring-loaded or hydraulically deployed to contact and track the inner wall of the pipe as the tool is moved along the wellbore (Kotlar, 1988). High-precision sensors (e.g., potentiometers, encoders) measure the radial position of each finger.
  - **Outputs**: Generate a high-resolution 3D map (image) of the internal pipe surface, showing the internal diameter profile versus depth and circumference. Clearly identifies internal metal loss features like pitting, erosion channels, wear grooves, scale build-up, or mechanical deformation (ovalization, buckling, and collapse). Provides accurate measurements of defect depth, diameter, and geometry.
  - **Applications**: Excellent for detailed assessment of internal damage, quantifying pitting severity, identifying connection issues (e.g., deformation, internal scale near threads), measuring internal wear (e.g., from wireline or coiled tubing), and characterizing mechanical deformation of the casing or tubing (Kotlar, 1988). Often run after initial indications from EM tools, as part of a detailed inspection program, or as a baseline log (Chap. 7).
  - **Limitations**: Only measures the *internal* profile (cannot detect external corrosion unless it has progressed significantly and caused internal bulging or failure). Fingers can be damaged by obstructions or severe restrictions in the wellbore. Can be sensitive to debris, scale, or paraffin deposits if not properly cleaned before logging, which can mask underlying metal condition or cause inaccurate readings. Requires open tubing/casing access.

Corrosion logs provide direct measurements of metal loss, allowing engineers to assess the remaining structural integrity and containment capability of tubulars and plan preventative maintenance or remediation before leaks or failures occur. Regular logging over time allows tracking corrosion rates and predicting remaining life.

To calculate the **percentage metal loss** (sometimes reported by EM/MFC tools):

$$\text{Wall Loss (\%)} = \left(1 - \frac{\text{Measured Thickness}}{\text{Original Thickness}}\right) \times 100 \tag{12.5}$$

where **Measured Thickness**: thickness from log and **Original Thickness**: nominal thickness of pipe.

12.3.1 Leak Detection Logging

When SCP (Chap. 8), production anomalies, or other indicators suggest a leak in a barrier element (tubing, casing, packer, cement), specialized logs can help pinpoint the location (depth) of the leak entry or exit point in the wellbore.

- **Temperature Logs (e.g., Differential Temperature, DTS):** Highly sensitive thermometers run along the wellbore.
  - **Principle**: Fluid flowing through a leak path (either into or out of the conduit being logged, or up/down an annulus) often has a different temperature than the surrounding static fluid column or the natural geothermal gradient. This fluid movement creates a thermal anomaly (cooling or heating) at the leak depth. Gas expansion through a leak path causes significant cooling due to the Joule–Thomson effect, making temperature logs very effective for detecting gas leaks. Liquid leaks can cause heating (due to friction) or cooling (due to injection of colder fluids or expansion) (Sharma et al., 2021).
  - **Applications**: Effective for locating leaks, especially gas leaks or cross-flow between zones with different temperatures. Can be run in tubing or casing. Often run under different well conditions (shut-in versus flowing/injecting, before/after pressure changes, during bleed-down) to amplify the thermal anomaly signal. Distributed Temperature Sensing (DTS) using permanently installed fiber optics provides continuous real-time temperature profiles, enabling constant leak surveillance (see Chap. 30).
  - **Limitations**: Require a sufficient temperature difference between the leaking fluid/flow and the static wellbore environment to create a detectable anomaly. Interpretation can be ambiguous if multiple thermal events (e.g., multiple leaks, fluid entries/exits) are present, or if wellbore fluid is circulating. Can be affected by operational activities that alter temperature profiles.
- **Acoustic/Noise Logs (e.g., PLT Noise, DAS)**: Listen for the characteristic sound generated by turbulent fluid flow through a leak path.
  - **Principle**: Fluid moving through a restriction (like a leak orifice, channel, or crack) at velocity generates high-frequency acoustic noise due to turbulence. Sensitive microphones or hydrophones in the logging tool detect this noise (Aslanyan et al., 2015).
  - **Outputs**: Identify depths where anomalous noise levels occur, often presented as spectral maps showing noise intensity at different frequencies versus depth.

Frequency analysis of the noise signature can sometimes help differentiate between different types of flow (e.g., gas versus liquid, flow through different types of restrictions).

  - **Applications**: Useful for locating small leaks or gas migration paths that may not have a strong thermal signature. Can detect flow behind casing (e.g., gas channeling up an annulus) if the tool is run inside the casing. Distributed Acoustic Sensing (DAS) using permanently installed fiber optics provides continuous acoustic monitoring along the wellbore, enabling constant leak surveillance and monitoring of well operations (see Chap. 30).
  - **Limitations**: Highly susceptible to background noise from surface activities (pumps, rig noise), wellbore flow (if run while flowing), or gas lift operations. Requires a relatively quiet wellbore for best results. Interpretation requires distinguishing leak noise from other sources of acoustic energy (e.g., tool noise, natural formation noise, and scale cracking).

Temperature and noise logging techniques are essential tools for detecting subsurface leak paths and behind-casing flow, especially when annular pressure monitoring alone is insufficient. These logs are typically acquired under shut-in (SI) conditions to isolate fluid movement effects. Temperature logs reveal localized anomalies—such as cooling due to gas expansion or heating from fluid friction—that deviate from the expected geothermal gradient. Noise logs, meanwhile, record acoustic energy emitted by turbulent flow through micro-annuli, fractures, or compromised cement. When interpreted together, they enhance diagnostic accuracy for identifying leak zones. Figure 12.5 illustrates these combined logs, where distinct thermal gradients and correlated acoustic signals in two zones confirm fluid migration behind casing, indicating well integrity compromise. This integrated interpretation provides a robust method for evaluating barrier performance and guiding remediation efforts.

- **Fluid Identification Logs (e.g., Fluid Density, Capacitance/Resistivity, Spectrometry, Sample Takers)**: Fluid identification logs play a crucial role in diagnosing the source and nature of wellbore leaks, particularly when pressure and temperature anomalies alone do not provide conclusive evidence of fluid movement. These logs are designed to characterize the composition and behavior of fluids within the wellbore or annular space, using advanced sensing technologies to detect subtle variations in fluid properties. In the context of sustained casing pressure (SCP), annular leaks, or integrity breaches behind casing, the ability to identify the invading fluid—whether it is gas, oil, water, or treatment chemicals—can significantly enhance the diagnosis and remediation planning process. Unlike pressure or temperature logs that detect physical changes, fluid identification logs provide compositional insights, which are often necessary to distinguish between formation influx, crossflow from other zones, or mechanical failure of internal components. These measurements are especially valuable in low-rate flow conditions or shut-in wells, where leaks may not generate strong thermal or pressure signatures but still introduce changes in fluid type that can be measured and confirmed.

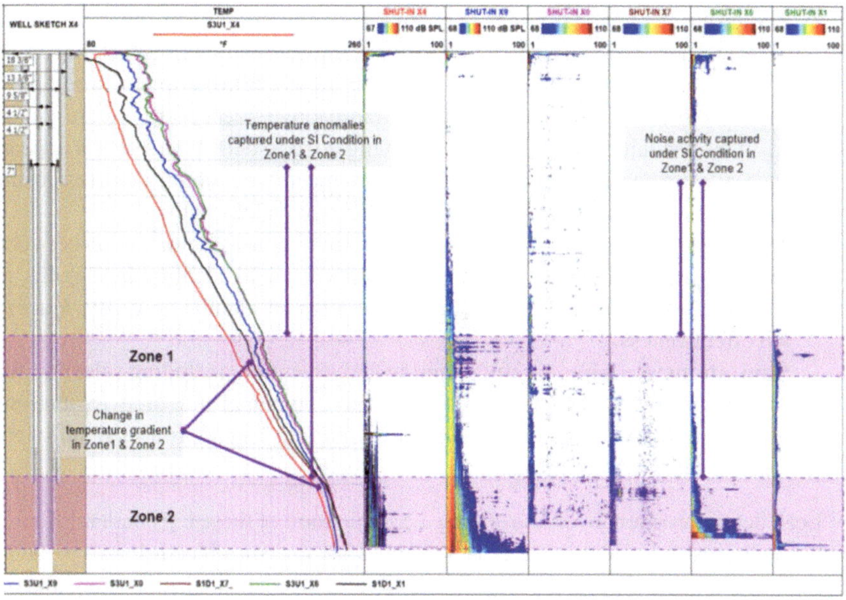

Fig. 12.5 Correlated temperature and noise logging under shut-in conditions, highlighting leak-related anomalies. The temperature log (left tracks) shows two distinct zones (Zone 1 and Zone 2) with deviation from the normal geothermal gradient, suggesting fluid movement. The noise panels (right tracks) confirm elevated acoustic activity in the same intervals, indicating turbulent flow behind the pipe or within annular micro-channels. The well sketch to the far left illustrates casing configuration, assisting in identifying the affected casing string. The combined dataset supports the diagnosis of behind-casing communication and possible barrier failure (Sedrati et al., 2025)

  - **Principle**: Changes in fluid type entering the wellbore from a leak path (e.g., water entering a gas well, oil entering a water-filled annulus) can be detected by sensors measuring fluid density, capacitance (indicating water cut), resistivity, or by collecting fluid samples for surface analysis.
  - **Applications**: Useful for identifying zones of fluid entry, particularly in annuli or shut-in wells where flow rates might be low but fluid composition changes are detectable. Fluid sample takers can provide definitive proof of fluid type at a specific depth. Pulsed Neutron Logs can identify fluid types (hydrocarbon/water saturation) behind casing (Discussion in 12.4).
  - **Limitations**: The leaking fluid to mix with the wellbore fluid and cause a detectable change in property. Sample takers only provide data at discrete points.

- **Chemical/Radioactive Tracers:** Involves injecting a detectable substance into a suspected leaking conduit and monitoring its appearance at an unexpected location (McKinley, 1994).

- **Principle**: A chemical tracer (e.g., unique salts, fluorescent dyes, dissolved gases) or a short-lived radioactive tracer is injected into the suspected leaking conduit (e.g., tubing, an annulus). Logging tools (fluid samplers, specialized chemical sensors, gamma ray tools for radioactive tracers) are then run in an adjacent conduit or the suspected receiving zone to detect the tracer's appearance at unexpected locations (e.g., tracer injected in tubing detected in the A-annulus fluid sample), indicating flow through a leak path.
- **Applications**: Can provide definitive confirmation of communication between specific points or zones in the well. Often used to identify leak paths in complex situations (e.g., communication between annuli) or verify squeeze job effectiveness.
- **Limitations**: Requires careful planning and execution, including selecting the appropriate tracer and detection method. Environmental and safety regulations apply to radioactive tracers. May require multiple logging runs and sufficient time for the tracer to migrate.

Leak detection often involves running a combination of these logs under different well conditions (e.g., flowing versus shut-in, after pressure bleed-down) to build confidence in pinpointing the location and characteristics of the leak source. Table 12.2 summarizes the primary applications and inherent limitations of common wellbore logging methods used to assess cement integrity, casing condition, and barrier performance in cased-hole environments. These tools provide complementary insights and are often interpreted collectively to ensure a comprehensive evaluation of well integrity.

12.4 Wellhead Inspection and Maintenance Techniques

The wellhead and Christmas tree are the most accessible components of the well barrier system. Maintaining their integrity relies on routine surface inspection and planned maintenance activities (Ashena, 2025c).

- **Visual Inspection**: Regularly checking the wellhead, Xmas tree, valves, flanges, pipework, and associated instrumentation (pressure gauges) for visible signs of integrity issues.
  - Leaks: Oil stains, fluid drips, ice formation (Joule–Thomson effect from gas leaks), audible hissing (gas leaks), or visual detection using gas sniffers or leak detection sprays.
  - Corrosion: External rust, pitting, coating damage, or signs of corrosion product build-up.
  - Physical Damage: Dents, cracks, bent components (e.g., impact damage), loose bolts, vibration.

Table 12.2 Updated overview of well logging tools for cement and casing integrity assessment

Logging method	Primary diagnostic use	Key limitations
CBL/VDL	Evaluates cement bond at casing interface and detects free pipe versus bonded zones	Cannot detect vertical cracks, mud channels, gas chimneys, or radial heterogeneity. Affected by micro-annuli and tool bias
Ultrasonic imaging log	Provides high-resolution images of cement quality, casing corrosion, and fluid channels	Limited by lightweight/gas-contaminated cement. Poor performance in heavy muds or eccentric casing
Isolation scanner	Detects gas chimneys, radial bonding variation, casing wear, and channels with azimuthal clarity	Inaccurate in identifying debonding at dry casing or thin vertical defects
Radioactive tracer survey	Confirms leak location and communication between well zones or annuli	Cannot assess cement quality or bonding. Requires tracer injection and is operationally complex
Temperature log/acoustic log	Identifies fluid movement or leaks by thermal anomalies or acoustic energy signatures	Provides no direct information on cement or casing condition. Temperature logs may be inconclusive in equilibrium wells
Corrosion log (EM/MFC tools)	Quantifies metal loss in casing and tubulars—identifies internal/external corrosion zones	No insight into cement bonding or annular isolation. Cannot detect non-metallic degradation
SAPT/VIT	Evaluates annular flow properties and hydraulic communication	Cannot assess cement quality or material integrity directly

  - Valve Positions: Ensuring valves are in the correct operational or shut-in position. Pressure Gauge Readings: Checking and recording readings, comparing against expected values and previous readings.
  - Sealant Levels: Checking levels in valve stem sealant reservoirs or annulus vent lines.

- **Valve Lubrication and Function Testing**: Periodically lubricating valve stems and bodies (injecting grease/sealant through specified injection ports) helps prevent seizing, reduces wear on seals, and facilitates smooth operation. Operating valves through their full range (functional testing—open/close) confirms their mechanical operability. This should be done routinely as part of preventative maintenance.
- **Leak Testing**: Periodically testing valve sealing integrity (e.g., double block and bleed test for gate valves where applicable, inflow tests on master valves) and potentially flange connections (using dedicated test ports or specialized test clamps). These tests verify the barrier function of surface valves and seals.

- **Bolt Torque/Tensioning**: Checking and potentially retightening flange bolts as required, especially after initial installation, significant pressure/temperature cycles, or maintenance. Proper bolt tension is critical for energizing ring gaskets and flange seals.
- **Coating Maintenance**: Inspecting external coatings on the wellhead and tree and repairing damaged areas to prevent external corrosion, particularly in corrosive atmospheres (e.g., offshore, coastal).
- **Annulus Pressure Monitoring**: As discussed in Chap. 8, this is a key surveillance technique performed at the wellhead outlets, involving monitoring pressure gauges or transmitters.

These routine activities, typically performed by trained production operators or maintenance personnel, help identify potential surface integrity issues before they escalate, enabling timely intervention (see Chap. 26).

12.5 Risk-Based Inspection (RBI)

Instead of inspecting all components on a fixed, time-based schedule, Risk-Based Inspection (RBI) methodologies use risk assessment principles to prioritize inspection efforts and optimize schedules.

- **Principle**: RBI identifies potential damage mechanisms (corrosion, erosion, fatigue, cracking, degradation, etc.) relevant to specific equipment items (e.g., specific casing strings, types of packers, sections of tubing, wellhead components). It then assesses the likelihood of failure occurring due to these mechanisms based on operating conditions, material properties, age, inspection history, and analytical models (Akinsanya et al., 2025). Simultaneously, it evaluates the potential consequences of failure (safety, environmental, economic, operational).
- **Application**: Components with a higher calculated risk (higher likelihood *and/or* higher consequence of failure) are targeted for more frequent or more detailed inspection using appropriate techniques (e.g., requiring annual caliper logs in a known corrosive tubing section versus a log every 5 years in a benign environment; prioritizing visual inspection of wellheads in the splash zone; and increasing annulus monitoring frequency on wells with known SCP). Lower-risk components may be inspected less frequently, or different inspection methods may be chosen.
- **Benefits**: Focuses resources (inspection personnel, logging costs, intervention budget) where they are most needed to manage the highest risks, potentially reducing overall inspection costs and minimizing unnecessary interventions while maintaining or improving safety and integrity by concentrating on high-risk areas. Supports data-driven decision-making for surveillance planning.
- **Requirements**: Requires robust data management (Chap. 15) to track historical data, a clear understanding of potential degradation mechanisms relevant to the specific asset and environment, established risk assessment methodologies

(Chap. 14), and defined acceptance criteria for assessing inspection results and triggering actions.

RBI principles can be applied to wellheads, Xmas trees, pipelines, pressure vessels, and are increasingly being adapted and integrated into WIMS to prioritize downhole integrity logging and monitoring efforts based on well-specific risk factors, operational history, and monitoring data (e.g., using AI/ML to predict degradation and inform RBI—Chap. 31). Figure 12.6 illustrates Risk-Based Inspection (RBI) workflow for well integrity. The diagram illustrates the sequential phases of an RBI program tailored to oil and gas wells, starting with data collection, followed by risk assessment, probability and consequence analysis, risk prioritization, and culminating in inspection planning. Each phase is continuously informed by monitoring and feedback to ensure inspection profiles and schedules remain aligned with the evolving risk landscape and well conditions. This proactive, adaptive process enhances decision-making and ensures the long-term integrity of well barrier elements.

In risk-based inspection (often adapted from FMEA/FMECA principles), a general equation for prioritizing inspection:

$$\text{RPN} = \text{Probability of Failure (P)} \times \text{Consequence of Failure (C)} \times \text{Detectability (D)} \tag{12.6}$$

where P is likelihood/probability (scale 1–10), C is consequence/severity (scale 1–10), and D is Detectability (scale 1–10, inverse of ease of detection).

Equation (12.3) helps prioritize wellhead inspections and downhole diagnostics based on operational risk.

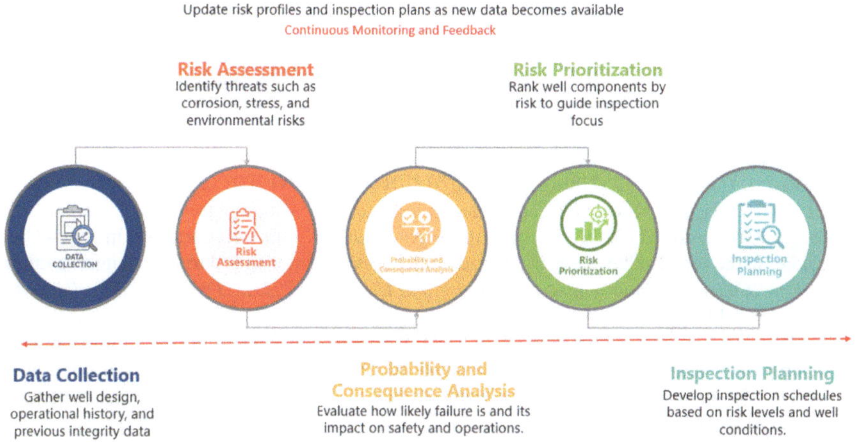

Fig. 12.6 Flowchart illustrating the process of Risk-Based Inspection (RBI) for well integrity. The workflow begins with data collection, followed by risk assessment, probability and consequence analysis, and risk prioritization. Based on this, an inspection plan is developed and continuously refined through monitoring and feedback to ensure well integrity is maintained

Effective operational integrity management relies on using the right monitoring, inspection, and logging tools at the right time, guided by monitoring data, operational history, risk assessment, and RBI principles, to accurately diagnose the condition of well barriers and enable timely interventions before loss of containment occurs.

References

Ajgou, N., Graba, B., Sayah, L., Ismail, M., Yakupov, A., Saada, M., Rourke, M., & Abdelmoula, M. (2019). Effective solutions to well integrity management using multi finger caliper and electromagnetic tool. In *Society of petroleum engineers—SPE gas and oil technology showcase and conference 2019, GOTS 2019*. https://doi.org/10.2118/198570-MS

Akinsanya, A. O., Qin, J., Guan, Y., Sørensen, J. D., & Faber, M. H. (2025). Framework for reliability assessment and risk-based integrity management XE "Integrity management" of sub-surface well subject to temporal and spatial degradation variability. *International Journal of Pressure Vessels and Piping, 217*, Article 105537. https://doi.org/10.1016/J.IJPVP.2025.105537

Ashena, R. (2025). Barriers. *The Practical Handbook of Well Control*, 20–31. https://doi.org/10.1201/9781003473770-2

Aslanyan, A., Aslanyan, I., Karantharath, R., Minakhmetova, R., Kohzadi, H., & Ghanavati, M. (2015). Spectral noise logging integrated with high-precision temperature logging for a multi-well leak detection survey in South Alberta. *Society of Petroleum Engineers - SPE Offshore Europe Conference and Exhibition, OE, 2015*, 8–11. https://doi.org/10.2118/175450-MS

Bateman, R. M. (2015). Cement bond logging. *Cased-Hole Log Analysis and Reservoir Performance Monitoring*, 195–226. https://doi.org/10.1007/978-1-4939-2068-6_12

Brill, T. M., Demichel, C., Nichols, E. A., & Zapata Bermudez, F. (2011). Electromagnetic casing inspection XE "Inspection" tool for corrosion evaluation. *International Petroleum Technology Conference*. https://doi.org/10.2523/14865-MS

Catala, G. N., Stowe, I. D., & Henry, D. J. (1984). A combination of acoustic measurements to evaluate cementations. *Software—Practice and Experience, 1984-September*. https://doi.org/10.2118/13139-MS

de Souza, C., Padilha, S. T., & da Silva Araujo, R. G. (1997). New Approach on cement evaluation XE "cement evaluation" for oil and gas reservoirs using ultrasonic images. *Latin American and Caribbean Petroleum Engineering Conference*. https://doi.org/10.2118/38981-MS

Contraires, S., Loizzo, M., Lecampion, B., & Sharma, S. (2009). Long-term well bore integrity in Otway: Integrating ultrasonic logs, cement petrophysics, and mechanical analysis. *Energy Procedia, 1*(1), 3545–3552. https://doi.org/10.1016/J.EGYPRO.2009.02.148

HALLINURTON. (2025). *CAST-VTM circumferential acoustic scanning tool-visualization.*

Imrie, A., Kozlowski, M., Torky, O., & Wijaya, A. A. (2021). Full well corrosion insight—Case studies in the added value of electromagnetic thickness measurements during well interventions. In *SPE/ICoTA Well Intervention Conference and Exhibition 2021, CTWI 2021*. https://doi.org/10.2118/204431-MS

Johns, J. E., Aloisio, F., & Mayfield, D. R. (2011). Well Integrity analysis in Gulf of Mexico wells using passive ultrasonic leak detection method. In *SPE automotive and composites divisions—coiled tubing and well intervention conference and exhibition 2011* (pp. 90–104). https://doi.org/10.2118/142075-MS

Ko Kyi, K. (2018). *Issues with cement bond and cement evaluation logs-case studies*. https://doi.org/10.2523/IPTC-18538-MS

Kotlar, N. (1988). *NK: Multifinger caliper interpretation and applications-© KAPPA 1988–2020 Multifinger caliper interpretation and applications.*

Loo, S. H., & Maric, C. (2024). Best practices in wells workover planning using four-in-one well integrity suite evaluation. In *APOGCE 2024—SPE Asia Pacific oil and gas conference and exhibition* (pp. 15–17). https://doi.org/10.2118/221291-MS

Morris, C., Sabbagh, L., Wydrinski, R., Hupp, J., Van Kuijk, R., & Froelich, B. (2007a). Application of enhanced ultrasonic measurements for cement and casing evaluation. *SPE/IADC Drilling Conference, Proceedings, 2*, 739–753. https://doi.org/10.2118/105648-MS

Morris, C., Vaeth, J., van Kuijk, R., & Froelich, B. (2007). *AADE-07-NTCE-14 enhanced ultrasonic measurements for cement and casing evaluation.*

Narender Pal, M. M. A. (2006). Value addition using cement evaluation by ultra sonic imaging tool in upper Assam oil fields. In *6th international conference & exposition on petroleum geophysics.* https://spgindia.org/conference/6thconf_kolkata06/97.pdf

Petro Shine. (2023). *Cement evaluation|cement evaluation well logging: assessing cement integrity in oil and gas wells|petro shine.* https://petroshine.com/cement-evaluation/2/

McKinley, J. T. T. (1994). *Temperature, radioactive tracer, and noise logging.* https://www.epa.gov/sites/default/files/2020-02/documents/temperature_radioactive_tracer_and_noise_logging_for_well_integrity-july1994.pdf

Sedrati, K., Almulhim, B. K., Hassan, T. M., & Aramco, S. (2025). Cross-flow behind the casing examination using spectral noise and high precision temperature tool in partially cemented long string gas wells. *GOTECH*. https://doi.org/10.2118/224600-MS

Sharma, J., Adnyana, G., Barnes, D., Mims, D., & Behrens, R. (2021). Temperature logging guidelines and factors that affect measurement accuracy in steamfloods. *Journal of Petroleum Science and Engineering, 196*, Article 107727. https://doi.org/10.1016/J.PETROL.2020.107727

SLP. (2025). *Cement evaluation service.*

van Kuijk, R., Zeroug, S., Froelich, B., Allouche, M., Bose, S., Miller, D., Le Calvez, J. L., Schoepf, V., & Pagnm, A. (2005). A novel ultrasonic cased-hole imager for enhanced cement evaluation XE "cement evaluation." *International Petroleum Technology Conference Proceedings, 2005*, 855–868. https://doi.org/10.2523/IPTC-10546-MS

Part IV
Well Integrity Management Systems (WIMS)

Parts I–III explored the technical foundations of well integrity: the principles, the construction elements, the operational threats and time-dependent degradation mechanisms, and the monitoring/diagnostic tools. However, technology, engineering knowledge, and individual expertise alone are insufficient to consistently guarantee integrity throughout a well's complex and lengthy lifecycle, especially across a portfolio of diverse wells. A systematic, organized, and integrated approach is required to ensure that integrity principles are consistently applied, risks are proactively managed, performance is monitored, and continuous improvement occurs. Part IV focuses on this crucial organizational framework: the Well Integrity Management System (WIMS). This part details the essential components of a WIMS, dives into risk assessment methodologies (including how data-driven approaches are evolving), discusses the critical role of data management and performance indicators, outlines the regulatory and standard landscape that governs practices, and emphasizes the indispensable contribution of human factors and a strong safety culture.

Chapter 13
Framework for Well Integrity Management

Abstract This chapter introduces the foundational framework of a Well Integrity Management System (WIMS), emphasizing the need for a structured, lifecycle-based approach to proactively manage integrity across diverse well portfolios. Unlike isolated technical fixes, WIMS integrates policies, risk assessment, competency, procedures, monitoring, data management, and continuous improvement into a cohesive system. It establishes clear roles, ensures compliance with international standards such as ISO 16530 and NORSOK D-010, and aligns integrity practices with broader Safety Management Systems (SMS). Key components include leadership commitment, hazard identification, operational controls, intervention procedures, and performance monitoring through key performance indicators and audits. The chapter also highlights the importance of robust data systems, cross-functional coordination, and human factors in sustaining WIMS effectiveness. By addressing implementation challenges and leveraging international standards, WIMS transforms well integrity from reactive maintenance into a proactive, risk-informed discipline essential for safe and sustainable operations.

© The Author(s), under exclusive license to Springer Nature Switzerland AG 2026
A. Alsubaih and K. Sepehrnoori, *Well Integrity for Oil, Gas, and Subsurface Storage Operations*, https://doi.org/10.1007/978-3-032-15233-6_13

A pump jack in the Rawlins Field Office (2015). *Source* mypubliclands/U.S. Bureau of Land Management (2015)

Chapter's Objectives

1. Explain the necessity of a systematic, integrated management approach for well integrity.
2. Describe the key elements typically found in a comprehensive WIMS.
3. Discuss the alignment of WIMS with international standards like ISO 16,530 and NORSOK D-010.
4. Highlight the crucial integration of WIMS within broader Safety Management Systems (SMS).
5. Discuss common challenges in implementing and maturing a WIMS.

13.1 The Need for a Systematic Approach

Managing well integrity across a diverse portfolio of wells, each with its unique design, construction history ("as-built" state), operating conditions, degradation profile, and lifecycle stage, is a complex and dynamic undertaking. Relying solely on the expertise of individual engineers, ad hoc procedures, or reactive responses to problems is insufficient to consistently manage the associated risks and ensure long-term containment in a cost-effective and compliant manner (Brekka & Vlay, 1993;

Ghafoor, 2023; Maylor et al., 2006; Miller & Lessard, 2001). Major incident investigations frequently reveal failures not just in specific technical components, but fundamentally in organizational processes, communication, risk assessment execution, management oversight, and learning from experience.

A systematic approach, formalized through a Well Integrity Management System (WIMS), is therefore essential. A WIMS provides a structured framework that integrates policies, procedures, competencies, resources, data management systems, and controls to proactively manage well integrity throughout the entire lifecycle, from concept to permanent abandonment. Its primary goals are as follows:

- To establish clear accountability and responsibilities for well integrity across different departments and roles.
- To ensure compliance with regulatory requirements and applicable industry standards.
- To systematically identify, analyze, and assess well integrity hazards and risks throughout the lifecycle.
- To define, implement, and maintain controls (barriers, procedures, monitoring, inspection, maintenance) to mitigate these risks to an acceptable level (often defined as As Low As Reasonably Practicable—ALARP).
- To ensure well integrity issues and anomalies are detected, reported, evaluated, and addressed in a timely and systematic manner.
- To ensure personnel performing integrity-critical tasks are competent and adequately trained.
- To capture, manage, and analyze data and measure performance to enable learning and drive continuous improvement.
- To provide assurance to management, regulators, and other stakeholders that well integrity is being effectively managed in accordance with policies and requirements.

Effective WIMS provides a structured framework integrating accountability, procedures, technology, and data analysis throughout the well lifecycle (Oziegbe Iriogbe et al., 2024; Yakoot et al., 2021a). Implementing a robust WIMS moves well integrity management from a reactive (fixing problems after they occur) or purely compliance-driven activity toward a proactive, risk-based, predictive (where possible), and continuously improving process embedded within the organization's operational culture. Its importance is amplified in mature fields with aging infrastructure and diverse well stock, where degradation risks are higher (Yakoot et al., 2021a).

13.2 Key Elements of a Well Integrity Management System (WIMS)

While the specific structure and terminology of a Well Integrity Management System (WIMS) may vary across organizations, a comprehensive WIMS typically aligns with the core principles of widely adopted management system frameworks such as ISO 9001 for quality, ISO 14001 for environmental management, and ISO 45001 for occupational health and safety. These standards emphasize governance, risk-based decision-making, and continuous improvement—principles that are equally critical in managing well integrity throughout the lifecycle of a well. To ensure that WIMS is both effective and compliant, several key industry references provide technical and organizational guidance. These include NORSOK Standard D-010 for well integrity in drilling and well operations; OP095 from Oil & Gas UK, which outlines Well Life Cycle Integrity Guidelines; ISO/TS 16,530–2, which covers well integrity during the operational phase; and OLF 117, the Norwegian Oil and Gas Recommended Guidelines for Well Integrity. Additionally, relevant API documents and applicable regulatory requirements are integral to ensuring that WIMS frameworks are both globally informed and locally compliant. Together, these references support the development of a structured, risk-managed, and auditable system to safeguard personnel, the environment, and asset value across all phases of well operation (Wilson, 2015). Key elements often include (based on principles in ISO 16530–1, NORSOK D-010, and general management system structures) the following:

- **Leadership, Policy, and Objectives**
  - **Management Commitment**: Clear and demonstrable commitment from senior leadership emphasizing the absolute priority of well integrity, often through formal statements, resource allocation, and active participation in reviews (Daghmouni et al., 2010).
  - **Well Integrity Policy**: A formal, high-level statement outlining the organization's intentions, principles, and overarching goals regarding well integrity management throughout the lifecycle.
  - **Strategic Objectives**: Defining clear, measurable, and time-bound objectives for well integrity performance and the WIMS itself (e.g., reduce the number of wells with uncontrolled SCP, improve compliance with mandatory testing, and reduce integrity-related NPT).
- **Organization, Roles, Responsibilities, and Competency**
  - **Organizational Structure**: Defining the structure for managing well integrity, including dedicated roles or functions (e.g., well integrity focal points, departmental responsibilities for design, operations, maintenance).
  - **Roles and Responsibilities**: Clearly documenting and communicating who is accountable and responsible for specific well integrity tasks and decisions across the lifecycle (e.g., responsibility for design reviews, approving materials, witnessing critical tests, analyzing monitoring data, performing risk

assessments, approving remediation plans, managing P&A execution) (Yakoot et al., 2021a).

- **Competency Assurance**: Establishing processes for defining the necessary skills, knowledge, training, qualifications, and experience (competencies) for personnel performing integrity-critical tasks. Implementing systems for assessing, assuring, maintaining, and tracking individual competency and providing necessary training and development.

- **Hazard Identification and Risk Assessment/Management**
  - **Systematic Identification**: Processes for identifying potential well integrity hazards and failure modes throughout the lifecycle (e.g., using HAZID, HAZOP, FMECA, lessons learned from incidents, predictive analytics). This includes identifying threats unique to specific well types or environments (e.g., HPHT, sour service, EOR, storage).
  - **Risk Assessment**: Evaluating the likelihood (frequency or probability) of identified failures occurring and analyzing the potential consequences (safety, environmental, asset damage, production loss, reputation) if they do occur (Heydari et al., 2020). This determines the level of risk associated with each failure mode (using qualitative risk matrices or more quantitative methods) (see Chap. 14).
  - **Risk Mitigation/Control**: Defining, implementing, and maintaining measures (technical barriers, operational procedures, monitoring strategies, maintenance plans) to prevent the causes of failure or mitigate the consequences, thereby reducing risks to acceptable levels (ALARP or as defined by policy/regulation).
  - **Risk Register**: Maintaining a dynamic record of identified risks, their assessment, implemented controls, and residual risk levels.

- **Well Design and Construction Standards/Practices**
  - **Design Basis and Standards**: Clear documentation of the Basis of Design (BoD), design parameters, assumptions (including lifecycle considerations), and the technical standards (internal and industry) used for design calculations (casing, cement, completion, wellhead) and material selection.
  - **Construction Procedures**: Detailed, clear, and user-friendly procedures for critical construction activities known to impact integrity (e.g., cementing operations, tubular running and makeup, packer setting, wellhead assembly, initial pressure testing). Incorporating QA/QC checks.
  - **Quality Assurance/Quality Control (QA/QC)**: Processes to ensure materials and equipment meet specifications (e.g., traceability, testing certificates) and that construction activities are performed correctly according to procedures and standards (e.g., witnessed critical steps, documentation signoffs).

- **Operational Procedures and Controls**

- **Operating Envelopes**: Defining safe operational limits (pressure, temperature, rates, fluid composition limits) for each well based on design analysis. Procedures for operating within these limits.
- **Monitoring and Surveillance Procedures**: Defined procedures for conducting annulus pressure monitoring, collecting data, analyzing trends, and responding to anomalies (e.g., SCP bleed/build-up procedures) (Ramadhani et al., 2021). Procedures for utilizing data from advanced monitoring systems (e.g., DTS/DAS).
- **Testing and Inspection Procedures**: Standardized procedures for routine functional tests (e.g., valve cycling), leak tests (e.g., DHSV leak tests, annulus tests), integrity logging runs (e.g., corrosion logs, cement logs), and visual inspections, including defining frequency and acceptance criteria (often based on RBI principles).
- **Maintenance Procedures:** Planned preventative and corrective maintenance procedures for integrity-critical equipment (e.g., valve servicing and greasing, wellhead maintenance, control system checks).
- **Management of Change (MoC)**: A formal, documented process to evaluate and control any temporary or permanent changes (to equipment, procedures, operating conditions, materials, personnel) that could impact well integrity and safety (Daghmouni et al., 2010). This ensures risks associated with change are assessed and mitigated (see Chap. 17).
- **Handover Procedures:** Formal processes for effectively transferring comprehensive well information, integrity status documentation ("as-built" data, anomaly history), risks, and responsibilities between different phases or teams (e.g., from drilling/completions to operations, from operations to abandonment teams).

- **Well InterventionManagement**
  - Specific procedures and risk assessment requirements for planning and executing well intervention activities, ensuring the two-barrier principle is maintained throughout the operation (e.g., temporary barrier requirements and verification), managing well control, and verifying permanent barrier integrity post-intervention (see Chap. 25).
- **Data Management and Documentation**
  - **Well Integrity Data Requirements**: Defining the critical integrity-related data to be collected, recorded, and managed throughout the lifecycle (design details, construction reports, test records, monitoring data, logs, inspection findings, anomaly/failure history, intervention reports, and P&A records).
  - **Data Management System**: Establishing robust, accessible, and ideally integrated systems (databases, software platforms) for storing, validating, retrieving, analyzing, and visualizing well integrity data (Khan et al., 2023). Ensuring data quality and consistency.

  - **Documentation Control**: Ensuring all procedures, standards, design records, test certificates, reports, and data are current, version-controlled, accessible to relevant personnel, and securely archived (see Chap. 15).

- **Performance Monitoring and Improvement**

  - **Key Performance Indicators (KPIs)**: Defining and tracking measurable metrics to monitor the performance of well integrity and the effectiveness of the WIMS itself (see Chap. 15).
  - **Incident Reporting, Investigation, and Learning**: Fostering an open reporting culture for all well integrity anomalies, near misses, and failures. Implementing systematic procedures for investigating incidents to determine root causes (technical, human, organizational). Establishing mechanisms for capturing, disseminating, and implementing lessons learned across the organization to prevent recurrence (see Chap. 17).
  - **Assurance, Audit, and Review**: Periodic internal and potentially external audits to verify compliance with WIMS procedures, regulatory requirements, and standards. Regular management reviews to assess overall WIMS performance against objectives, review KPI trends, audit findings, lessons learned, resource adequacy, and identify strategic opportunities for improvement (see Chap. 16).

These elements are interconnected and should function as an integrated system, driven by a Plan-Do-Check-Act (PDCA) cycle for continuous improvement, ultimately aiming to manage well integrity risks throughout the asset's life. Figure 13.1 presents the key functional components of a typical WIMS, demonstrating how leadership, risk management, operational execution, data control, and performance evaluation interact within a unified system. These elements are structured around a continuous Plan–Do–Check–Act (PDCA) cycle, ensuring well integrity risks are proactively identified, managed, and monitored throughout the life of the well. The Plan–Do–Check–Act (PDCA) cycle is widely recognized in international standards (e.g., ISO 9001, ISO 14001, ISO 45001) as the core engine for managing system performance, ensuring adaptability, and embedding lessons learned. In the context of well integrity, the PDCA cycle reinforces a structured approach to developing policies, assigning organizational roles, conducting risk assessments, executing operational controls, auditing performance, and driving system enhancements. Each stage of the cycle aligns with specific WIMS components, thereby ensuring that integrity objectives are systematically planned, implemented, evaluated, and improved over time. Figure 13.2 illustrates how PDCA elements integrate with WIMS pillars to support safe and sustainable well operations.

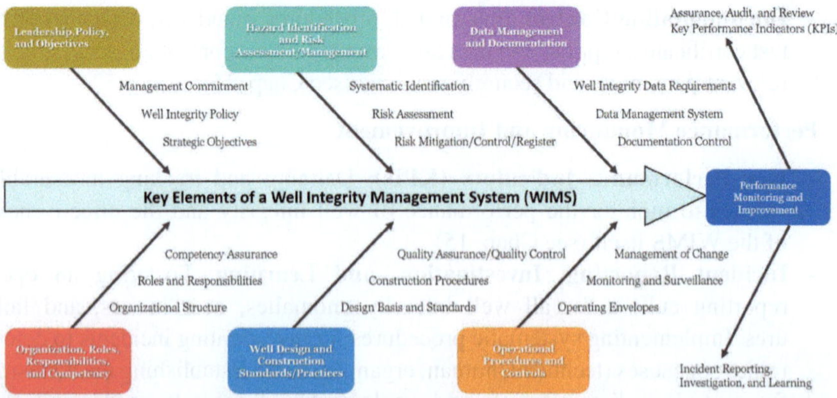

Fig. 13.1 Key elements of a Well Integrity Management System (WIMS), showing the interaction between leadership, risk assessment, operational controls, data management, and performance monitoring. These components function as part of a PDCA-based continuous improvement loop to manage well integrity risks effectively across the well lifecycle

13.3 Alignment with International Standards

Several key international standards provide guidance on the structure and content of a WIMS, promoting consistency and best practices across the industry. Aligning an organization's WIMS with these standards is often a regulatory expectation, a contractual requirement, or a demonstration of due diligence.

- **ISO 16,530: Petroleum and Natural Gas Industries—Well Integrity**: This is a primary reference standard for WIMS frameworks.
  - **ISO 16,530–1: Life Cycle Governance (2017)**: Provides a high-level framework for establishing, implementing, maintaining, and improving a WIMS covering the *entire* well lifecycle. It outlines key principles and management system elements (Leadership, Planning, Support, Operation, Performance Evaluation, Improvement) analogous to ISO 9001 or 45,001 but tailored to well integrity. It emphasizes risk management, the barrier philosophy, and data/documentation control. It is less prescriptive on specific technical details but provides the *structure* for the management system.
  - **ISO 16,530–2: Well Integrity for the Operational Phase (2014)**: Focuses specifically on the requirements and practices for managing well integrity during the operational phase, covering monitoring, testing, maintenance, anomaly management (like SCP), MoC, and documentation relevant to operations. Provides more detail for this specific lifecycle stage within the overall framework of Part 1.
  - **ISO 16,530–3 on P&A Integrity Is Under Development/Consideration**: Acknowledges the critical need for lifecycle integrity to extend through permanent abandonment.

Fig. 13.2 The Plan–Do–Check–Act (PDCA) cycle applied within a Well Integrity Management System. Each phase connects with key WIMS functions—planning policies and roles, implementing organizational and risk controls, checking through audits and monitoring, and acting through continuous improvement—to ensure the integrity of wells over their entire lifecycle

- **NORSOK D-010: Well Integrity in Drilling and Well Operations**: Developed by the Norwegian petroleum industry, this standard is highly influential globally, particularly for offshore operations.
  - **Scope**: Comprehensive, covering the entire lifecycle from drilling to abandonment.
  - **Approach**: Relatively prescriptive, providing detailed technical and operational *requirements* for well barrier elements (defining acceptable components for different service types), barrier schematics (specifying requirements for defining and documenting barriers), testing frequencies (specifying test methods, pressures, durations, and acceptance criteria for various barriers), monitoring (e.g., SCP management details), documentation, and specific operational procedures.

  - **Key Features**: Strong, detailed emphasis on the two-barrier principle, specific requirements for verification methods and acceptance criteria, detailed guidance on SCP management, and specific requirements for P&A barrier placement and verification.
  - **Influence**: Often used as a benchmark for best practice, even where not strictly required by regulation, particularly for defining technical requirements for barrier elements and verification.

- **API Standards and Recommended Practices (RPs)**: The American Petroleum Institute publishes numerous documents that provide technical specifications, recommended practices, and procedures relevant to specific aspects of well integrity. These are widely used globally, either mandated by regulations or adopted by operators as their internal standards. Key examples include (but are not limited to) the following:

  - API RP 90: Annular Casing Pressure Management (as discussed in Chap. 8).
  - API RP 65–2: Isolating Potential Flow Zones During Well Construction (cementing focus).
  - API Std 65–3: Wellbore Plugging and Abandonment.
  - API Spec 5CT: Casing and Tubing Specification (materials, testing).
  - API Spec 10A/RP 10B Series: Well Cement Specifications and Testing Procedures.
  - API Spec 6A: Wellhead and Christmas Tree Equipment Specification.
  - API RP 11D1 (equivalent to ISO 14,310): Downhole Packer and Bridge Plug standards.
  - API RP 14B (equivalent to ISO 10,417): Subsurface Safety Valve Systems—Design, Installation, Operation, Testing.
  - API RP 53: Blowout Prevention Equipment Systems for Drilling Operations.
  - API RP 59: Recommended Practice for Well Control Operations.
  - API RP 7SC: Recommended Practice for the Safe Use of Gas as a Testing Medium.
  - API Bul E3: Environmental Guidance for P&A.

Many operating companies develop their internal WIMS manuals and standards based on the framework provided by ISO 16,530–1, incorporating the specific technical requirements detailed in NORSOK D-010 and relevant API standards/RPs, tailored to their specific operational context, well types, and local regulatory requirements. The WIMS provides the overarching management framework, while these technical standards define the *how* and *what* of specific integrity practices.

13.4 Integration with Safety Management Systems (SMS)

Well integrity is fundamentally linked to process safety. A major loss of well containment is, by definition, a major accident hazard with potential for catastrophic consequences (Adeyinka et al., 2020). Therefore, the WIMS should not exist in isolation but should be an integral part of, or closely aligned with, the organization's overall Safety Management System (SMS) (also sometimes called Health, Safety, and Environmental Management System—HSEMS or SEMS). A Safety Management System (SMS) is a structured framework designed to systematically manage safety risks in industrial operations. Rooted in principles of accountability, continuous improvement, and compliance, an SMS helps organizations prevent incidents, protect personnel, and ensure regulatory adherence. It integrates policy development, hazard identification, risk management, operational controls, emergency response, and performance monitoring into a cohesive system. The SMS aligns closely with international standards such as ISO 45,001 and API RP 75, and supports a proactive safety culture through clear roles, ongoing evaluation, and documented procedures. Figure 13.3 illustrates the key components of an effective SMS and how they function together to enable safe and reliable operations across the organization (Janicak et al., 2009; Saeed et al., 2018; Sohani & Haugnaess, 2002).

- **Shared Principles**: Many WIMS elements (e.g., policy, risk assessment, MoC, competency, incident investigation, audit) are direct parallels or components of a typical SMS framework.
- **Process Safety Focus**: WIMS provides the specific technical and operational detail required to identify, assess, and manage the major accident hazards associated with wellbore pressure containment, feeding into the broader process safety management framework of the SMS. The SMS provides the overarching structure for managing all major hazards, of which well integrity is a key one.
- **Risk Management Integration**: Risks identified and managed within the WIMS (e.g., risk of casing burst due to SCP, risk of environmental release due to wellhead leak, risk of underground blowout during intervention) should be integrated into the overall corporate risk register managed by the SMS. Risk acceptance criteria should be consistent between the WIMS and SMS.
- **Consistency**: Ensuring consistency in terminology, processes (e.g., MoC workflow, incident reporting hierarchy, audit protocols), and reporting structures between the WIMS and the overarching SMS avoids confusion, duplication of effort, and ensures well integrity receives appropriate focus within the broader safety agenda.
- **Management Oversight**: Integrating well integrity performance and WIMS reviews into broader SMS management review processes ensures it receives appropriate attention at senior leadership levels.

An effective WIMS functions as a specialized component within the company's broader framework for managing operational risks and process safety, ensuring that the specific challenges of maintaining well integrity receive the necessary strategic

Fig. 13.3 Core components of a Safety Management System (SMS), illustrating the integration of leadership commitment, organizational roles, risk management, operational controls, emergency response, monitoring, and continuous improvement. These interconnected elements function as a closed-loop system to ensure safety is planned, implemented, assessed, and enhanced continuously

focus, technical rigor, and resources. As noted by Jaculli et al. (2022), WIMS is a cornerstone of process safety, managing major accident hazards associated with loss of containment, and should be fully integrated with the broader SMS/HSEMS framework (Jaculli et al., 2022).

13.5 Challenges in Implementing and Maturing a WIMS

Implementing and maintaining an effective WIMS is not without challenges, especially for organizations with large, diverse, and aging well portfolios.

- **Data Management**: Achieving universality across assets, full automation of data capture, and effective integration of diverse well data (design, construction, operations, logs, tests) from disparate systems (often historical and siloed) remains a significant challenge (Semwogerere et al., 2025; Yakoot et al., 2021b). Poor data quality and accessibility hinder effective monitoring, analysis, and risk assessment.

- **Technical Complexity**: Applying complex technical standards and understanding intricate degradation mechanisms (corrosion, cement degradation, seal failure) require specialized technical expertise. Bridging knowledge gaps and ensuring consistent application of technical requirements can be difficult.
- **Organizational Integration**: Ensuring seamless integration and clear responsibilities across different functions (drilling, completions, production, reservoir, subsurface, maintenance, HSE, engineering) and potentially multiple service contractors is vital but can be challenging culturally and operationally.
- **Resource Allocation**: Securing adequate resources (personnel, budget for monitoring, inspection, maintenance, and remediation) to support the WIMS objectives, especially in cost-constrained environments.
- **Culture and Competency**: Building and maintaining a strong integrity culture where personnel understand their role, follow procedures, report issues, and feel empowered to stop unsafe work (see Chap. 17). Ensuring a sufficient pool of competent personnel with up-to-date knowledge.
- **Life Extension Decisions**: Applying WIMS principles to justify the life extension of aging wells and structures poses specific challenges in standardizing assessment methodologies and demonstrating acceptable risk over extended periods (Semwogerere et al., 2025).
- **Keeping Pace with Technology**: Integrating new monitoring technologies (fiber optics, AI/ML) and assessment methodologies into established WIMS processes and procedures can lag technological advancements (see Chap. 28).

A mature WIMS moves beyond simply documenting minimum compliance toward leveraging data analytics, predictive capabilities, and continuous learning to optimize integrity management and enable informed decisions, including life extension (Iriogbe et al., 2024; Semwogerere et al., 2025; Yakoot et al., 2021b). Addressing these implementation challenges is key to realizing the full benefits of a systematic approach to well integrity.

References

Adeyinka, A., Tsakporhore, A., Arije, O., Adegoke, O., Aliyev, S., Iwueze, C., & Hulett, R. (2020). Rethinking well integrity for sustainability: Adopting a risk based approach. In *Society of petroleum engineers—SPE Nigeria annual international conference and exhibition 2020, NAIC 2020*. https://doi.org/10.2118/203672-MS

Anunciação Jaculli, M., Choueri, N., da Mata, C. R., Aparecida Silva Leite, A. G., Pelaquim Mendes, J. R., & Colombo, D. (2022). Well safety and integrity evaluation of offshore wells: A review of past, present, and future. *Journal of Petroleum Science and Engineering, 212*. https://doi.org/10.1016/j.petrol.2022.110329

Brekka, L. T., & Vlay, G. J. (1993). Integrated risk management—procedures and tools. *INCOSE International Symposium, 3*(1), 491–498. https://doi.org/10.1002/J.2334-5837.1993.TB01617.X

Daghmouni, H., El Manci, M., Rady, S. A. A., & Shattawa, Y. (2010). Well Integrity Management System WIMS development. In *Society of petroleum engineers—14th Abu Dhabi international petroleum exhibition and conference 2010, ADIPEC 2010* (Vol, 2, pp. 1421–1427). https://doi.org/10.2118/137966-MS

Ghafoor, L. (2023). *Effective risk management using proactive approach.* https://doi.org/10.31219/OSF.IO/5Z7PF

Heydari, M., Lai, K. K., & Zhou, X. (2020). Creating sustainable order fulfillment processes through managing the risk: Evidence from the disposable products industry. *Sustainability, 12*(7), 2871. https://doi.org/10.3390/SU12072871

Janicak, C. A., Professor, A., Program Coordinator, G., Ferguson, L., Professor, C., & Chairperson, D. (2009). *Integrating safety performance measures into the safety management system.* OnePetro. https://dx-doi-org.ezproxy.lib.utexas.edu/

Khan, A., Bareche, M., Al Lawati, A., Abboub, M., & Shekhar, S. (2023). Leveraging the power of data to develop a new well integrity management system. In *Proceedings of the SPE/IADC Middle East drilling technology conference and exhibition.* https://doi.org/10.2118/214592-MS

Maylor, H., Brady, T., Cooke-Davies, T., & Hodgson, D. (2006). From projectification to programmification. *International Journal of Project Management, 24*(8), 663–674. https://doi.org/10.1016/J.IJPROMAN.2006.09.014

Miller, R., & Lessard, D. (2001). Understanding and managing risks in large engineering projects. *International Journal of Project Management, 19*(8), 437–443. https://doi.org/10.1016/S0263-7863(01)00045-X

mypubliclands. (2015). "Oil and Gas (33317868280).jpg." Wikimedia Commons, 16 July 2015. Licensed under CC BY 2.0. https://commons.wikimedia.org/wiki/File:Oil_and_Gas_(33317868280).jpg

Oziegbe Iriogbe, H., Ebeh, C. O., & Onita, F. B. (2024). *Well integrity management and optimization: A review of techniques and tools.* https://doi.org/10.56781/ijsrr.2024.5.1.0041

Ramadhani, F., Nurdin, S., Etuhoko, M. O., Zhi, Y., Mulyono, S., Mintardja, A., & Dawson, M. (2021). An expert software system has accomplished full implementation of well integrity management system WIMS resulting to excellent safety and production sustainability for critical sour HPHT Field in offshore Madura Strait, Indonesia. Society of Petroleum Engineers—SPE/IATMI Asia Pacific Oil and Gas Conference and Exhibition 2021. *APOG, 2021*, 12–14. https://doi.org/10.2118/205697-MS

Saeed, M. R., Chelmis, C., & Prasanna, V. K. (2018). Smart oilfield safety net—an intelligent system for integrated asset integrity management. In *Proceedings—SPE annual technical conference and exhibition* (pp. 24–26). https://doi.org/10.2118/191718-MS

Semwogerere, D., Sangesland, S., Vatn, J., Pavlov, A., & Colombo, D. (2025). Well integrity and late life extension - A current industry state of practice and literature review. In *Geoenergy Science and Engineering* (Vol. 244). Elsevier B.V. https://doi.org/10.1016/j.geoen.2024.213419

Sohani, S., & Haugnaess, T. (2002). *Contractor management by integration into the safety management system.* https://doi.org/10.2118/73836-MS

Wilson, V. (2015). Guidance for the development and implementation of an effective Well Integrity Management System (WIMS). *Proceedings of the Annual Offshore Technology Conference, 2,* 994–1007. https://doi.org/10.4043/25734-MS

Yakoot, M. S., Elgibaly, A. A., Ragab, A. M. S., & Mahmoud, O. (2021b). Well integrity management in mature fields: a state-of-the-art review on the system structure and maturity. In *Journal of petroleum exploration and production* (Vol. 11, Issue 4, pp. 1833–1853). Springer Science and Business Media B.V. https://doi.org/10.1007/s13202-021-01154-w

Yakoot, M. S., Elgibaly, A. A., Ragab, A. M. S., & Mahmoud, O. (2021a). Well integrity management XE "Integrity management" in mature fields: A state-of-the-art review on the system structure and maturity. *Journal of Petroleum Exploration and Production, 11*(4), 1833–1853. https://doi.org/10.1007/S13202-021-01154-W/FIG.S/21

Chapter 14
Risk Assessment and Management in Well Integrity

Abstract This chapter focuses on risk assessment and management as core components of a proactive Well Integrity Management System (WIMS). It highlights how structured risk evaluation enables operators to identify potential hazards, prioritize threats, and implement mitigation strategies across the well lifecycle. The chapter introduces key hazard identification methods such as HAZID and HAZOP, along with structured tools like Failure Modes, Effects, and Criticality Analysis (FMECA) for assessing failure scenarios at a component level. Risk matrices are presented as practical tools for ranking hazards based on likelihood and consequence, while Bow-Tie analysis offers a visual approach for mapping threats, barriers, and outcomes. The chapter also explores the role of Quantitative Risk Assessment (QRA), including Fault Tree and Event Tree Analyses, to provide numerical insights into the likelihood and severity of well integrity failures. It emphasizes the application of risk assessment results for design optimization, inspection planning, remediation prioritization, and demonstrating compliance with ALARP principles. Finally, the chapter introduces emerging data-driven and machine learning approaches that leverage historical well data and real-time monitoring to dynamically predict failure risks and improve decision-making. Through these methodologies, this chapter underscores the importance of integrating both traditional and advanced risk management techniques to enhance well integrity, safety, and sustainability.

© The Author(s), under exclusive license to Springer Nature Switzerland AG 2026
A. Alsubaih and K. Sepehrnoori, *Well Integrity for Oil, Gas, and Subsurface Storage Operations*, https://doi.org/10.1007/978-3-032-15233-6_14

Surface burning of oil at Deepwater Horizon. *Source* David Valentine (n.d.)

Chapter's Objectives

- Explain the fundamental role of risk assessment as a core component of a proactive WIMS.
- Describe common qualitative and semi-quantitative hazard identification techniques (HAZID, HAZOP).
- Detail Failure Modes, Effects, and Criticality Analysis (FMECA) as a structured analysis tool.
- Explain the use of risk matrices for evaluating and prioritizing risk levels.
- Introduce Bow-Tie analysis as a visual tool for understanding barriers and threats.
- Briefly discuss Quantitative Risk Assessment (QRA) applications in well integrity.
- Emphasize how risk assessment results inform integrity management decisions and demonstrate ALARP.
- Introduce the concept of data-driven and machine learning approaches to risk assessment.

14.1 The Role of Risk Assessment in WIMS

Risk assessment and management are fundamental pillars of a proactive Well Integrity Management System (WIMS). Rather than reacting to failures after they occur, risk-based approaches enable operators to systematically identify, evaluate, and mitigate potential threats to well integrity throughout the entire lifecycle—from design and construction to abandonment (Al-Ashhab et al., 2004). These methods incorporate tools such as risk matrices, which classify risks based on likelihood and severity, and probabilistic models, which use operational data, historical incidents, and environmental factors to quantify risk levels.

Modern WIMS frameworks integrate these methods with advanced risk assessment software, allowing operators to model risk scenarios, generate risk profiles for individual wells or entire fields, and visualize vulnerabilities using decision-support dashboards (Samuel et al., 2024). These tools facilitate informed decision-making, resource allocation, and prioritization of mitigation measures.

Comprehensive risk management frameworks also define standardized procedures for hazard identification, risk evaluation, mitigation planning, and ongoing monitoring—ensuring consistency and traceability across operations. By incorporating diagnostic technologies, predictive maintenance, and continuous performance reviews, operators can strengthen well integrity while promoting operational safety, environmental protection, and long-term sustainability. As technologies and methodologies advance, the ability to proactively manage well integrity risks will continue to evolve, enhancing the reliability and resilience of oil and gas assets worldwide (Samuel et al., 2024).

The process of risk assessment typically involves:

1. **Hazard Identification**: Identifying potential undesirable events or situations that could lead to a loss of well integrity (e.g., casing corrosion, cement failure, packer leak, external impact).
2. **Risk Analysis**: Determining the potential causes (threats) of identified hazards and analyzing the potential consequences (impacts on safety, environment, asset, reputation) if the hazardous event occurs. This step also involves analyzing the effectiveness of existing controls or barriers in preventing the event or mitigating its consequences.
3. **Risk Evaluation**: Assessing the level of risk associated with the identified hazard/scenario by combining the likelihood (or frequency) of the event occurring with the severity of its potential consequences (Flynn, 2021). This allows for prioritizing risks.
4. **Risk Treatment**: Determining if the risks are acceptable (meeting predefined acceptance criteria) or require further mitigation measures to reduce them to an acceptable level (e.g., ALARP—As Low As Reasonably Practicable) (Stavert-Dobson, 2016). This involves implementing new or strengthening existing controls.
5. **Risk Monitoring and Review**: Continuously monitoring the effectiveness of controls, collecting new data, and periodically reviewing and updating risk assessments as conditions change, or new information (e.g., incident findings, monitoring data) becomes available.

This systematic process allows organizations to focus resources and attention on the most significant threats to well integrity, prioritize actions (monitoring, inspection, maintenance, remediation), and make informed decisions about design, operation, and expenditure throughout the well lifecycle. As shown in Fig. 14.1, this process begins with the identification of integrity threats, followed by the analysis of their causes and impacts. Risks are then evaluated using matrices or models to determine severity and likelihood. Based on this, operators implement treatment

Fig. 14.1 Visual representation of the five-stage risk management process in well integrity management. Each stage—Identification, Analysis, Evaluation, Treatment, and Monitoring—builds upon the previous to ensure a comprehensive, proactive approach to risk mitigation across the well lifecycle

strategies—ranging from preventive controls to contingency measures—and engage in continuous monitoring to ensure risks remain within acceptable limits over time. This cyclical approach reinforces the core principles of a proactive WIMS, facilitating ongoing improvement and regulatory compliance. Risk assessment techniques, ranging from qualitative HAZIDs to semi-quantitative FMECA and quantitative QRAs, are central to WIMS, enabling prioritization and resource allocation (Yakoot et al., 2021b).

14.2 Hazard Identification Techniques

The first step in risk assessment is to identify what could go wrong–the potential well integrity hazards and failure scenarios. Structured brainstorming techniques involving experienced personnel from relevant disciplines (drilling, completions, production, reservoir, subsurface, process safety, maintenance, inspection, service providers) are commonly used to ensure a comprehensive review.

- **Hazard Identification (HAZID) Study**: A high-level, systematic review, typically conducted during the initial design phase (Concept Select or Front-End Engineering Design—FEED) (CCPS, 2022), to identify potential hazards and major accident events associated with the planned well construction, operation, and abandonment. It uses a structured approach with guidewords (e.g., "Leak," "Release," "Loss of Containment," "Collapse," "Failure," "Corrosion," "Erosion," "Overpressure") applied to different parts of the well system and different lifecycle phases to brainstorm potential hazardous scenarios. The output is typically a list of identified hazards, potential consequences, and initial assessment of their significance, informing subsequent design and risk analysis efforts.
- **Hazard and Operability (HAZOP) Study**: A more detailed, systematic examination, often applied to specific systems (like the wellhead/Xmas tree, or intervention equipment) or detailed operational procedures. It uses specific guidewords (e.g., "No Flow," "More Pressure," "Less Temperature," "Contamination," "Reverse Flow," "Maintenance/Utility Deviation") applied to intended design parameters or procedural steps to identify potential deviations from the intended state, their potential causes, consequences, and existing safeguards (barriers).

While originally developed for process facilities, HAZOP principles can be adapted for analyzing specific well operations, completion designs, or intervention procedures for potential integrity impacts.

These qualitative techniques rely heavily on the collective experience, knowledge, and structured approach of the multidisciplinary team to thoroughly explore potential failure scenarios.

14.3 Failure Modes, Effects, and Criticality Analysis (FMECA)

FMECA is a systematic, bottom-up analysis method used to identify potential failure modes of individual components or systems, analyze their potential effects (consequences), and prioritize them based on their criticality (Signoret & Leroy, 2021). It is often used as a detailed follow-up to HAZID or as a standalone analysis for critical systems or components (e.g., the wellhead, a complex packer system, a specific casing string design).

Steps in FMECA for Well Integrity:

1. **Define the System/Component**: Break down the well or a specific system into key components relevant to integrity (e.g., production tubing string, production packer assembly, A-annulus barrier system, Christmas tree valve assembly, specific casing string and its cement sheath across a zone).
2. **Identify Potential Failure Modes**: For each component or element, list all credible ways it could fail to perform its intended function (e.g., for casing: rupture/burst, collapse, axial separation at connection, leak at connection, corrosion pinhole/leak, wear through; for packer seal: extrusion, chemical degradation, failure to set, leak past slips; for valve: failure to close on demand, leak through bore, stem leak, failure to open).
3. **Identify Potential Causes**: Determine the underlying reasons why each failure mode might occur (e.g., for casing burst: excessive internal pressure exceeding design limits, reduced wall thickness from corrosion/wear; for packer seal extrusion: operating above pressure/temperature rating, improper installation, seal degradation; for valve leak through: erosion damage to seats, obstruction, component wear).
4. **Identify Potential Effects/Consequences**: Describe the local impact on the component/system and the broader impact on the well and objectives if the failure occurs (e.g., loss of primary barrier, SCP in a specific annulus, inter-zonal communication, environmental release, production loss, safety hazard, inability to perform intervention).
5. **Identify Existing Controls/Barriers**: List the measures already in place (design features, operational procedures, monitoring systems, maintenance activities) that are intended to prevent the cause from occurring, prevent the failure mode, or mitigate the consequences if the failure occurs. These are the existing safeguards.

6. **Assess Risk/Criticality**: Evaluate the risk associated with each failure mode considering the likelihood of the failure occurring and the severity of the consequences. This is often done qualitatively using a risk matrix (see Sect. 14.4) or semi-quantitatively by assigning scores for Likelihood (L), Severity of Consequences (S), and potentially Detectability (D) and calculating a Risk Priority Number (RPN = L × S × D) to rank failure modes by criticality.

FMECA is a powerful tool for systematically analyzing potential failures at a component or system level, ensuring that appropriate controls are considered during design and operation, and identifying critical components and failure modes that warrant priority attention in monitoring, inspection, and maintenance plans.

14.4 Risk Matrices

Once potential failure scenarios are identified and analyzed, their associated risk level needs to be evaluated to determine if it is acceptable or requires further action. A **Risk Matrix** is a common tool used for this qualitative or semi-quantitative risk evaluation in well integrity management (Ahmed et al., 2023). To support proactive decision-making, well integrity risk matrices are used to visualize and rank threats across various well components. By evaluating both the frequency and consequence of specific failure scenarios, these matrices guide resource allocation, inspection prioritization, and mitigation planning. Figure 14.2 presents comparative risk matrices for cement, casing, and Xmas tree systems—each a critical element in maintaining zonal isolation and pressure containment throughout the well lifecycle. These structured visual tools are essential in enabling consistency across risk assessments and aligning safety efforts with ALARP principles. By using such matrices, integrity teams can standardize assessments, facilitate ALARP evaluations, and ensure that critical risks are managed systematically and transparently.

- **Structure**: A grid with axes representing **Likelihood** (or Frequency/Probability of the failure occurring) and **Consequence** (or Severity of impact if the failure occurs).
- **Scales**: Each axis has defined categories or levels (e.g., 3 × 3, 4 × 4, or 5 × 5 matrix).
  - **Likelihood Scale**: Defined categories for the probability or frequency of occurrence (e.g., Rare, Unlikely, Possible, Likely, Frequent, or numerical frequencies like events/year). These definitions must be consistent.
  - **Consequence Scale**: Defined categories for the severity of impact across different domains (e.g., Safety/Personnel, Environment, Asset/Cost, Reputation/Political). The highest severity across any domain is typically used to determine the overall consequence category for a scenario. Definitions for each level (e.g., Catastrophic, Major, Moderate, Minor, Negligible) must be clear and consistently applied.

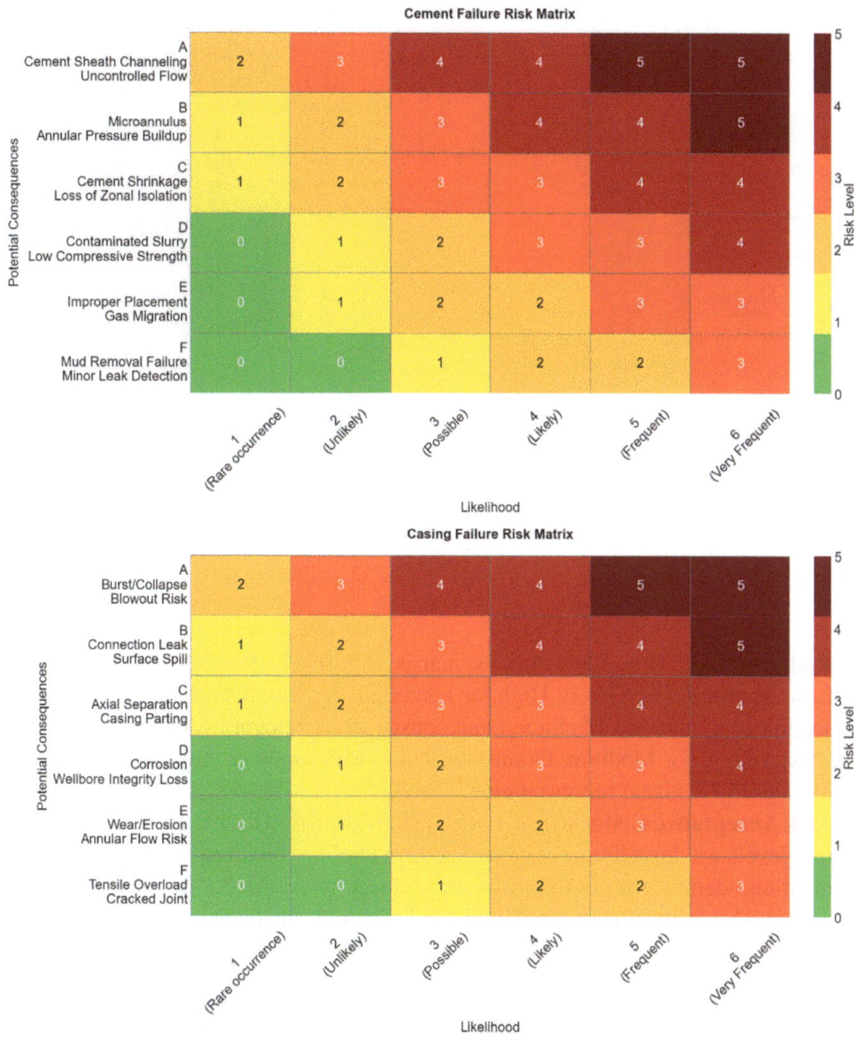

Fig. 14.2 Well Integrity Risk Matrices for Cement, Casing, and Xmas Tree Failures. This set of matrices illustrates the evaluated risk levels associated with failure modes of critical well components. Each matrix ranks failure scenarios by combining the likelihood of occurrence (x-axis) with the severity of potential consequences (y-axis). Cement failure scenarios include channeling, microannuli, and placement issues. Casing failure includes collapse, connection leaks, and corrosion. Xmas tree risks cover valve rupture, seal erosion, and bolt failures. The matrices support risk prioritization and the development of targeted barrier strategies in Well Integrity Management Systems (WIMS)

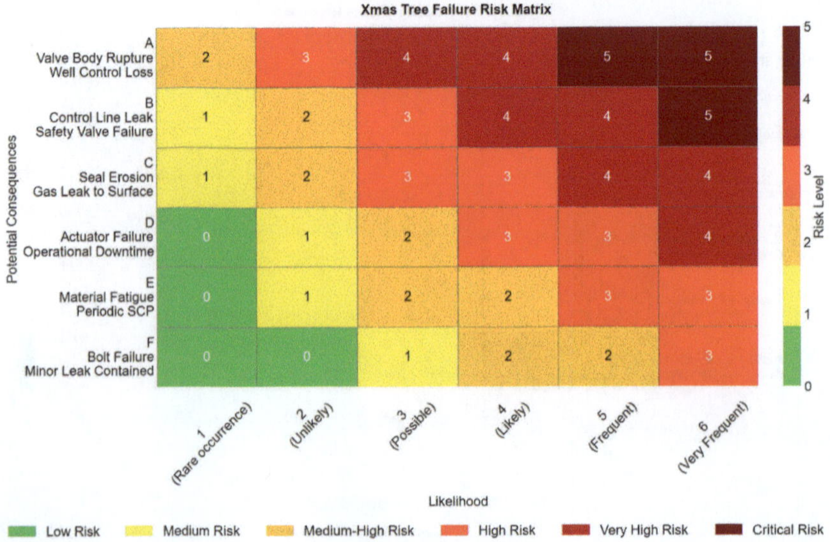

Fig. 14.2 (continued)

- **Risk Ranking**: The cells within the matrix represent different levels of risk (e.g., Low, Medium, High, Very High, or numerical scores), determined by the intersection of the chosen likelihood and consequence categories. Colors (Green for Low, Yellow for Medium, Orange for High, Red for Very High/Intolerable) are often used for visual representation.
- **Risk Acceptance Criteria**: Lines are typically drawn on the matrix defining which risk levels are broadly acceptable ("Acceptable"), which require formal assessment and demonstration that risks are reduced to ALARP ("ALARP Region"), and which are considered intolerable and require immediate action to reduce the risk regardless of cost or feasibility ("Intolerable").

Using the Matrix: For each identified failure scenario (from HAZID, FMECA, incident investigation, or monitoring data), the risk assessment team assigns likelihood and consequence categories based on available data, analysis, models, and collective expert judgment. The corresponding cell on the matrix indicates the risk level and determines the required action (Accept, ALARP assessment/mitigation, or Intolerable/Stop). Risk matrices provide a simple, visual way to prioritize risks but rely heavily on the quality and consistency of subjective judgment in assigning categories. Clear definitions, training, and facilitation are crucial for consistent application.

14.5 Bow-Tie Analysis

The Bow-Tie Diagram is a visual tool that effectively links the potential causes (threats) of an unwanted event (the "Top Event" or "Hazard") to its potential consequences, while explicitly showing the barriers (controls) in place to prevent the event or mitigate its consequences. It provides a holistic view of risk management for a specific hazard (Culwick et al., 2016; Voicu et al., 2018; Zhan et al., 2021). Bow-Tie analysis provides a comprehensive visual framework for understanding how specific failure scenarios in well integrity can be effectively prevented or mitigated. By identifying threats that lead to a central failure event—such as cement debonding, casing collapse, or Xmas tree valve malfunction—and aligning them with control measures, this method clarifies both proactive and reactive risk management strategies. Figure 14.3a–c illustrate Bow-Tie models for three essential well components. These visualizations serve as barrier management tools, highlighting critical safeguards before and after an event, and reinforcing the concept of layered defense within a Well Integrity Management System (WIMS). Their use enables better decision-making, supports ALARP principles, and enhances communication of risk across operational teams.

- **Structure**: Resembles a bow tie shape, read from left to right.
- **Knot (Top Event)**: The central unwanted event where control over the hazard is lost (e.g., "Loss of Containment from "Production Tubing," "Uncontrolled Gas Migration in Annulus," "Cement Sheath Failure").
- **Left Side (Threats/Causes)**: Lines radiating into the knot from the left represent the potential causes or threats that could lead to the central Top Event (e.g., "Tubing Corrosion," "Excessive Internal Pressure," "Poor Primary Cement Job," "Packer Seal Degradation," "External Impact on Wellhead").
- **Right Side (Consequences)**: Lines radiating out from the knot to the right represent the potential undesirable consequences if the central Top Event occurs (e.g., "SCP in "A-Annulus," "Environmental Pollution (Oil Spill/Gas Release)," "Blowout," "Production Loss," "Safety Incident/Fatality").
- **Barriers (Controls)**: Vertical bars (often represented as rectangles) drawn across the lines on both sides of the knot represent the preventative and mitigation barriers or controls in place.
  - **Preventative Barriers (Left Side)**: Located between the Threats and the Top Event. These are controls designed to prevent the threats from causing the central event (e.g., "Corrosion Inhibition Program," "Pressure Relief Valve," "Cement Placement Procedure," "CBL/Ultrasonic Log Verification," "Tubing Material Selection," "Regular Monitoring").
  - **Mitigation/Recovery Barriers (Right Side)**: Located between the Top Event and the Consequences. These are controls designed to reduce the severity or likelihood of the consequences if the central event occurs (e.g., "Annulus Pressure Monitoring and Management," "Emergency Shut Down (ESD)

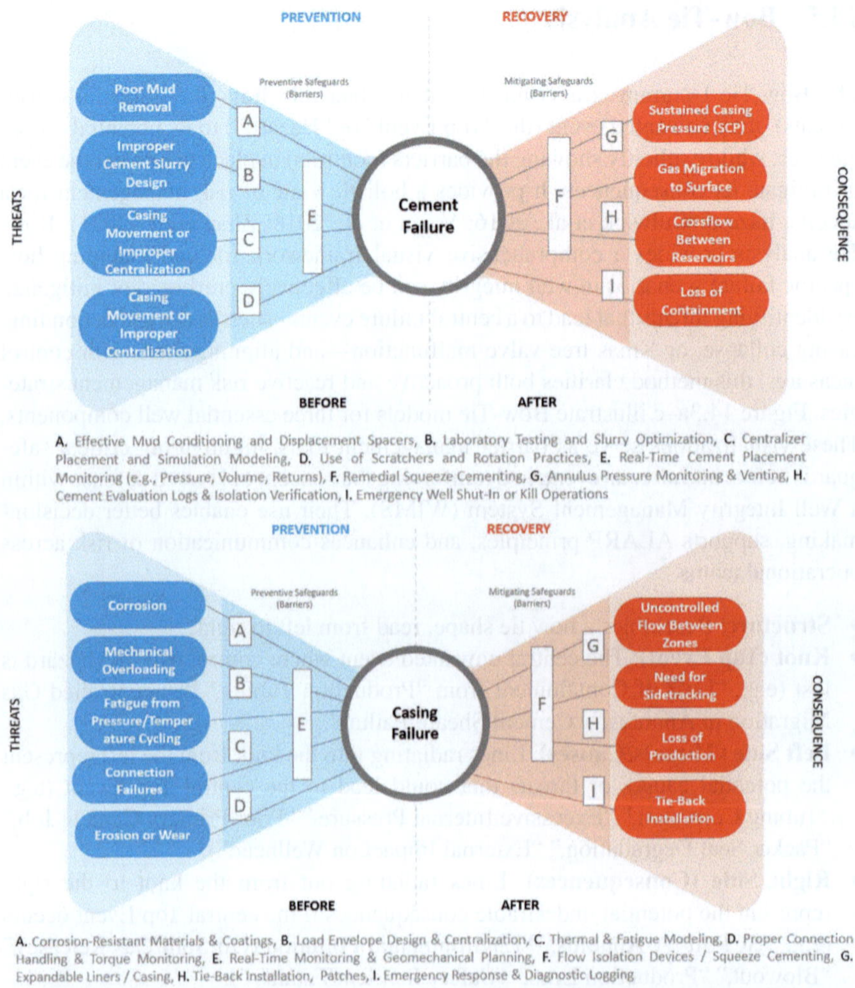

Fig. 14.3 Bow-Tie Diagrams for Cement, Casing, and Xmas Tree Failures in Well Integrity Management. These Bow-Tie diagrams visually depict the relationship between threats, failure events, and consequences for three critical well components: cement sheath, casing, and Xmas tree assemblies. The left side of each diagram identifies primary threats and associated preventive barriers designed to avert failure (e.g., corrosion-resistant coatings, centralization, cement design). The right side illustrates potential consequences if failure occurs, along with mitigation barriers (e.g., diagnostic logging, flow isolation, tie-back installations). These diagrams provide a clear overview of proactive and reactive strategies and serve as vital tools for communicating and managing integrity risks across the well lifecycle

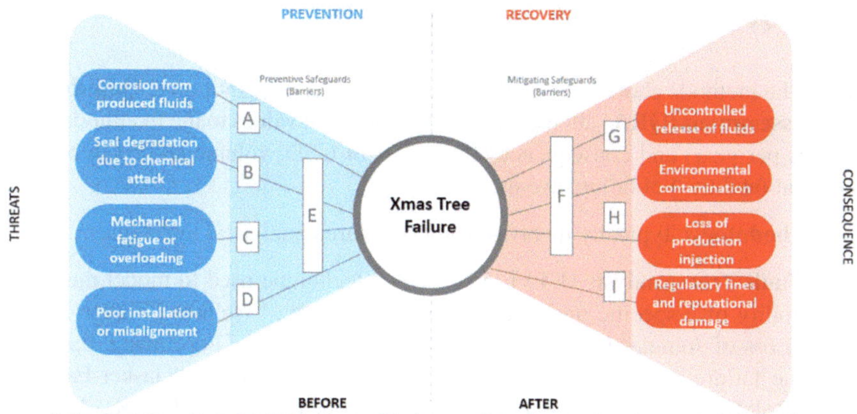

A. Use of corrosion-resistant alloys (CRAs) and compatible elastomers, **B.** Regular preventive maintenance and valve greasing schedule, **C.** Non-destructive testing (NDT) for welds, joints, and structural components, **D.** Proper tree installation procedures with alignment checks and torque specifications, **E.** Functional testing of SCSSVs and actuators prior to commissioning, **F.** Real-time pressure and temperature monitoring at tree outlets **F.** Emergency shutdown system (ESD) activation and remote isolation valves, **G.** Secondary pressure control via BOP (if installed) or kill line activation, **H.** Emergency response plan (ERP) and fire suppression systems, **I.** Mobile well intervention units or rig-based repair

Fig. 14.3 (continued)

System," "Well Control Procedures," "Spill Response Plan," "Remediation Procedures").

- **Degradation Factors and Controls (Optional but Recommended)**: Bow-ties can also incorporate "escalation factors" or "degradation factors"–events or conditions that could impair the effectiveness of a specific barrier (e.g., "Inhibitor Pump Failure," "Pressure Gauge Inaccurate," "Lack of Personnel Competency," "Corrosion Monitoring Data Not Reviewed")—and the controls specifically managing those degradation factors ("Regular Pump Maintenance," "Gauge Calibration Schedule," "Competency Training Program," "Data Review Procedure").

Benefits: Bow-ties provide a clear visual representation of how risks are managed through multiple layers of independent barriers. They are excellent tools for communicating risk, identifying potential weaknesses or single points of failure in barrier systems, assessing the criticality of specific controls, and supporting investigation of incidents (tracing back through the failed barriers). They are widely used in process safety and increasingly applied to well integrity management to visualize and communicate barrier strategies.

14.6 Quantitative Risk Assessment (QRA)

While the methods above are largely qualitative or semi-quantitative, **Quantitative Risk Assessment (QRA)** attempts to assign numerical values to the likelihood (e.g., failures per year) and consequences (e.g., cost in USD, volume spilled in barrels, potential fatalities) of failure events.

- **Methodology**: Typically involves using structured analytical techniques:
  - **Fault Tree Analysis (FTA)**: A top-down technique used to model the probability of a specific undesirable "top event" (e.g., "Wellhead Flange Leak," "Loss of Annular Isolation leading to Surface Release") occurring, based on the failure probabilities of underlying basic events (e.g., "Gasket Failure," "Bolt Corrosion," "Cement Channel") and logical gates (AND, OR). An illustrative example of FTA applied to well integrity is shown in Fig. 14.4, which maps potential leak pathways in a permanently abandoned (PA) well (Lavasani et al., 2015). The left-hand schematic identifies physical leak locations along the wellbore, while the right-hand logic tree connects these locations through AND/OR gates to the central event: "Leak Through PA Well." This dual representation enables risk assessors to estimate the probability of various failure sequences (e.g., plug degradation, casing rupture, annular flow) and calculate the likelihood of the top event using known or assumed failure rates of each contributing component. Such structured analysis supports more accurate risk quantification, facilitates remediation planning, and aligns with regulatory QRA expectations for complex legacy wells.
  - **Event Tree Analysis (ETA)**: A bottom-up technique used to model the probabilities of different outcomes or consequences following an initiating event (e.g., "Loss of Containment from Tubing"). It maps out potential sequences of events, considering whether subsequent barriers or mitigation systems succeed or fail, leading to different endpoints (Shafiee et al., 2019) (e.g., contained SCP, small leak to atmosphere, major blowout).
  - Combining FTA (to calculate the likelihood of the initiating event) and ETA (to calculate the probability of different consequences given the event) allows for calculation of the frequency of different consequence levels (e.g., frequency of a leak > 5 bbl/day, frequency of a fire).
- **Applications in Well Integrity**: QRA is less commonly applied to routine well integrity scenarios for individual wells compared to qualitative/semi-quantitative methods due to data limitations and complexity (Ahmed et al., 2023; Lavasani et al., 2015; Shafiee et al., 2019). However, it may be used for:
  - Comparing the relative risks of different well design concepts or complex completion architectures.
  - Assessing the risk associated with large fields, specific high-consequence scenarios (e.g., subsea blowouts, wells near critical infrastructure), or complex multi-well systems.

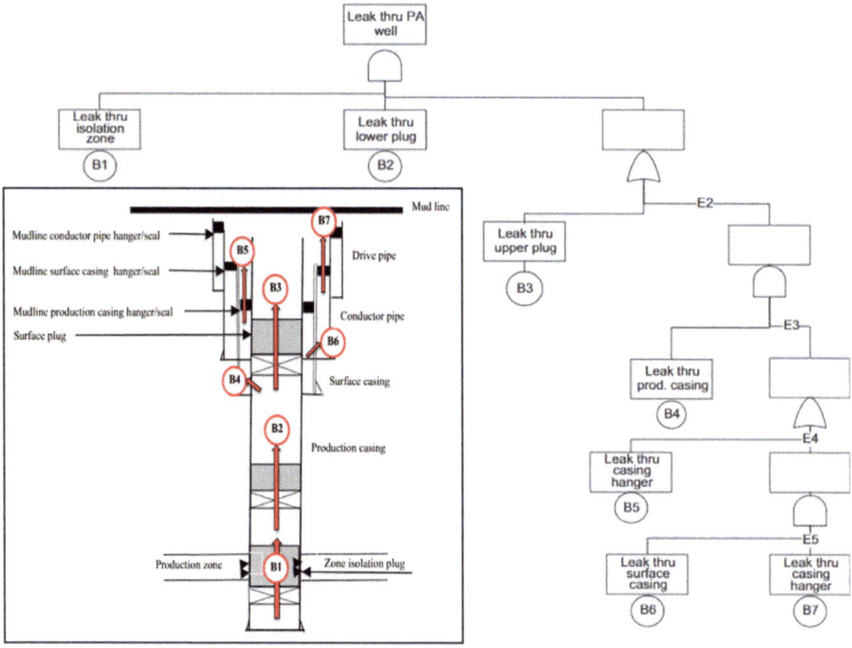

Fig. 14.4 Fault Tree Diagram for Leak Pathways in a Permanently Abandoned (PA) Well. This diagram integrates a physical well schematic (left) with a logical fault tree (right), tracing multiple potential failure paths—from isolation zone breaches to casing hanger leaks—leading to well leakage. It exemplifies the use of FTA in quantitative risk modeling for well integrity (Lavasani et al., 2015)

  - Providing input into overall platform, facility, or field-level QRA studies for regulatory submissions (e.g., Safety Cases).
  - Evaluating the risk reduction benefit of proposed mitigation measures or new technologies quantitatively.

- **Limitations**: Results are highly sensitive to the quality, availability, and applicability of the input failure rate data (which is often generic or requires extensive data collection). Consequence modeling also involves significant assumptions (e.g., weather conditions, ignition probability, dispersion modeling). QRA can be resource-intensive and often yields wide uncertainty ranges.

  For QRA-based Probability of Top Event from Fault Tree Analysis (FTA):
  For **AND gates** (where all basic events must occur simultaneously):

$$P_{\text{Top (AND)}} = \prod_{i=1}^{n} P_i \tag{14.1a}$$

For **OR gates** (where any one of the basic events causes the top event):

$$P_{\text{Top (OR)}} = 1 - \prod_{i=1}^{n}(1 - P_i) \tag{14.1b}$$

where P_{Top} is the probability of the top event (e.g., total system failure), P_i is the probability of individual basic events (e.g., plug failure, casing rupture), and n is the number of independent basic events connected to the same gate

Note: In complex fault trees, these AND and OR gate calculations are used iteratively at different levels to calculate the **overall top event probability**.

For Event Tree Analysis (ETA) Consequences

$$P_{\text{Outcome}} = P_{\text{Initiating}} \times \prod_{i=1}^{m} P_{\text{Barrier}_i} \tag{14.2}$$

where P_{Outcome} is the probability of a specific consequence, $P_{\text{Initiating}}$ is the probability of the initiating event (e.g., tubing leak), P_{Barrier_i} is the probability of barrier i either failing or succeeding depending on the analysis logic, and m is number of barriers considered in the sequence.

14.7 Using Risk Assessment Results

The outputs of risk assessments are not endpoints in themselves but crucial inputs for decision-making and action prioritization within the WIMS:

- **Design Decisions**: Informing critical design choices such as material selection, casing design factors, selection of connection types (API vs. Premium), number and type of barriers required for specific zones, and selection of completion equipment and wellhead components based on assessed risk.
- **Procedural Development**: Identifying steps or activities where risks are highest and ensuring robust operational procedures, checklists, and controls are developed and implemented (e.g., critical steps in cementing, intervention procedures).
- **Prioritizing Actions**: Directing monitoring, inspection, testing, and maintenance efforts towards the components, wells, and failure modes identified as having the highest risk. This forms the basis for Risk-Based Inspection (RBI) strategies (Chap. 12) and differentiated monitoring frequencies (Chap. 8).
- **Remediation Decisions**: Evaluating the risk associated with an observed anomaly (like SCP or a valve leak) to decide on the urgency and type of remediation required (e.g., immediate shut-in vs. continued monitoring with increased frequency, planning a major workover vs. a simple sealant injection).
- **Resource Allocation**: Justifying the allocation of resources (personnel, budget) towards managing the most significant integrity risks across the well portfolio.

- **Demonstrating ALARP**: Providing documented evidence, particularly in goal-setting regulatory regimes (Chap. 16), that hazards have been identified, risks assessed, existing controls evaluated and implemented mitigation measures reduce risks to a level As Low As Reasonably Practicable. Bow-ties are often used to communicate the ALARP argument by showing layers of control.

Regularly reviewing and updating risk assessments based on new operational data, monitoring results, incident findings, changes in technology or operating conditions, or changes in external factors (e.g., regulatory updates, environmental sensitivity) is essential for maintaining their relevance and effectiveness within the WIMS.

14.8 Data-Driven and Machine Learning Approaches to Risk Assessment

Traditional risk assessment relies heavily on historical generic data and expert judgment, which can be subjective and slow to update. The availability of large volumes of well integrity data within WIMS platforms is enabling more data-driven and dynamic approaches.

- **Data Analytics for Likelihood**: Analyzing historical data on component failures, SCP occurrences, test failures, and operational parameters across a well portfolio to derive more specific failure rates or probabilities for use in quantitative or semi-quantitative risk assessments, potentially segmented by well type, age, or operating environment.
- Machine Learning (ML) for Risk Classification and Prediction: Emerging techniques utilize machine learning models trained on historical data (design, construction, operations, monitoring, failures) to provide more dynamic risk assessments and classify well integrity status or predict the likelihood of future failures with high accuracy (Kandel & Baroud, 2024; Murugan, 2023; Salem et al., 2022; Yakoot et al., 2021a).
  - **Predicting Anomaly Development**: Models can be trained to predict the likelihood of SCP developing in a well based on its construction details (e.g., cement bond log quality, casing program), operating history, and initial monitoring data.
  - **Predicting Failure Severity**: Models can potentially predict the likely severity (e.g., high vs. low build-up rate) of an anomaly like SCP once detected.
  - **Dynamic Risk Scoring**: Integrating real-time monitoring data (pressures, temperatures) and operational status into ML models to provide a continuously updated risk score for individual wells or components, reflecting their current condition and operating context.
  - **Root Cause Prediction**: Some models aim to predict the most likely root cause of an integrity anomaly based on the observed symptoms and historical patterns (Salem et al., 2022).

Advantages: Potential for more objective, data-driven assessments; ability to process complex interactions between factors; continuous updates; early warning capabilities.

Challenges: Requires high-quality, well-structured, and extensive historical data; need for domain expertise to select features and validate models; 'explainability' of complex ML models (understanding *why* a risk is predicted) can be a challenge for regulatory acceptance and human trust; managing model drift as conditions change.

These advanced approaches, discussed further in Chap. 31, are enhancing the sophistication of risk assessment within well integrity management, moving towards more predictive and proactive decision-making.

References

Ahmed, A. H., Al-Ali, A., Al Hosani, A., Ali, M. S., & Al Hammadi, H. A. (2023). Digitalization of offshore wells risk management through quantitative, and automated risk assessment techniques. In *Society of Petroleum Engineers—ADIPEC, ADIP, 2023*. https://doi.org/10.2118/216881-MS

Al-Ashhab, J., Afzal, M., & Emenik, C. O. (2004). *Well integrity management system (WIMS)*. https://doi.org/10.2118/88696-MS

Culwick, M. D., Merry, A. F., Clarke, D. M., Taraporewalla, K. J., & Gibbs, N. M. (2016). Bow-tie diagrams for risk management in anaesthesia. *Anaesthesia and Intensive Care, 44*(6), 712–718. https://doi.org/10.1177/0310057X1604400615

Flynn, A. M. (2021). Hazard risk assessment. *Introduction to Environmental Management*, 339–349. https://doi.org/10.1201/9781003171126-44

Hazard Identification. (2022). *Process Safety for Engineers*, 239–262. https://doi.org/10.1002/978 1119831075.CH12

Kandel, R., & Baroud, H. (2024). A data-driven risk assessment of Arctic maritime incidents: Using machine learning to predict incident types and identify risk factors. *Reliability Engineering & System Safety, 243*, Article 109779. https://doi.org/10.1016/J.RESS.2023.109779

Lavasani, S. M., Ramzali, N., Sabzalipour, F., & Akyuz, E. (2015). Utilisation of Fuzzy Fault Tree Analysis (FFTA) for quantified risk analysis of leakage in abandoned oil and natural-gas wells. *Ocean Engineering, 108*, 729–737. https://doi.org/10.1016/J.OCEANENG.2015.09.008

Murugan, M. S. (2023). Large-scale data-driven financial risk management & analysis using machine learning strategies. *Measurement: Sensors, 27*, 100756. https://doi.org/10.1016/J.MEA SEN.2023.100756

Salem, A. M., Yakoot, M. S., & Mahmoud, O. (2022). A novel machine learning model for autonomous analysis and diagnosis of well integrity failures in artificial-lift production systems. *Advances in Geo-Energy Research, 6*(2), 123–142. https://doi.org/10.46690/ager.2022.02.05

Samuel, E., Okem, E. S., Daniel, T., & Akande, D. O. (2024). Revolutionizing U.S. pavement infrastructure: A pathway to sustainability and resilience through nanotechnology and AI Innovations. *World Journal of Advanced Engineering Technology and Sciences, 11*(2), 411–428. https://doi.org/10.30574/WJAETS.2024.11.2.0124

Shafiee, M., Enjema, E., & Kolios, A. (2019). An integrated FTA-FMEA model for risk analysis of engineering systems: A case study of subsea blowout preventers. *Applied Sciences, 9*(6), 1192. https://doi.org/10.3390/APP9061192

Signoret, J. P., & Leroy, A. (2021). Failure mode, effects (and criticality) analysis, FME(C)A. *Springer Series in Reliability Engineering*, 165–172. https://doi.org/10.1007/978-3-030-64708-7_10

Stavert-Dobson, A. (2016). Acceptability and ownership of risk (pp. 39–46). https://doi.org/10.1007/978-3-319-26612-1_3

Valentine, David. (n.d.). "Surface burning of oil at Deepwater Horizon." Wikimedia Commons. Licensed under CC BY-SA 4.0. https://commons.wikimedia.org/wiki/File:Deepwater_Horizon_oil_spill_UC_Santa_Barbara.jpg

Voicu, I., Panaitescu, F. V., Panaitescu, M., Dumitrescu, L. G., & Turof, M. (2018). Risk management with Bowtie diagrams. In *IOP Conference Series: Materials Science and Engineering* (vol. 400, no. 8, p. 082021). https://doi.org/10.1088/1757-899X/400/8/082021

Yakoot, M. S., Elgibaly, A. A., Ragab, A. M. S., & Mahmoud, O. (2021a). Well integrity management in mature fields: a state-of-the-art review on the system structure and maturity. *Journal of Petroleum Exploration and Production, 11*(4), 1833–1853. Springer Science and Business Media B.V. https://doi.org/10.1007/s13202-021-01154-w

Yakoot, M. S., Elgibaly, A. A., Ragab, A. M. S., & Mahmoud, O. (2021b). Well integrity management in mature fields: A state-of-the-art review on the system structure and maturity. *Journal of Petroleum Exploration and Production, 11*(4), 1833–1853. https://doi.org/10.1007/S13202-021-01154-W/FIG.S/21

Zhan, H., Dou, Z., Suo, G. Y., Xuan, S. Q., Tao, J. F., Yang, J. F., Chen, L. C., Li, L. L., Li, R., & Zhou, M. C. (2021). Bow ties of process safety and environmental impact transition from theory to practice. *Bow Ties in Process Safety and Environmental Management: Current Trends and Future Perspectives*, 1–14. https://doi.org/10.1201/9781003140382-1

Chapter 15
Data Management and Key Performance Indicators (KPIs)

Abstract This chapter explores the foundational role of data management and performance measurement in ensuring long-term well integrity across the full life-cycle of oil and gas wells. The chapter emphasizes that effective integrity management is fundamentally dependent on the availability of accurate, complete, and accessible data spanning design, construction, operation, suspension, and abandonment phases. It details the structure and content of Well Integrity Management Systems (WIMS), highlighting how centralized databases capture essential information—such as barrier schematics, cement logs, test results, corrosion monitoring, and intervention records—for diagnostics, risk assessments, and regulatory compliance. Robust data validation and quality control mechanisms are presented as critical safeguards against erroneous entries, ensuring data consistency, reliability, and auditability. The chapter also outlines a systematic framework for defining and tracking Key Performance Indicators (KPIs), categorized into leading and lagging indicators. These KPIs enable proactive monitoring of well health, identification of emerging risks, benchmarking of performance, and continuous improvement in integrity programs. A tiered structure of KPIs is introduced to balance predictive metrics (e.g., test completion rates, maintenance compliance) with outcome-based indicators (e.g., sustained casing pressure, failure events). Furthermore, this chapter highlights the increasing integration of well integrity data with advanced analytics platforms, laying the groundwork for predictive maintenance and AI/ML-based forecasting of integrity failures. It also identifies persistent challenges—such as data silos, legacy systems, inconsistent formats, and lack of standardization—that impede effective data utilization. Addressing these challenges through digital transformation, organizational accountability, and strategic investment in WIMS tools is essential for building a resilient and future-ready integrity management program. Overall, the chapter positions structured data and KPI frameworks as strategic enablers for safer, more sustainable, and cost-effective well operations.

© The Author(s), under exclusive license to Springer Nature Switzerland AG 2026
A. Alsubaih and K. Sepehrnoori, *Well Integrity for Oil, Gas, and Subsurface Storage Operations*, https://doi.org/10.1007/978-3-032-15233-6_15

Oilfield worker at sunset near a drilling rig and pump jack, symbolizing production operations and well lifecycle stages. *Source* The Authors

Chapter's Objectives

- Highlight the critical importance of accurate, complete, and accessible well data for effective integrity management.
- Describe the typical structure and content of a well integrity database or information system.
- Discuss the need for data validation and quality control processes.
- Explain how to define, track, and utilize well integrity Key Performance Indicators (KPIs) for monitoring WIMS performance.
- Outline reporting requirements and performance review processes.
- Connect robust data management to emerging predictive analytics capabilities.
- Discuss common challenges in well integrity data management.

15.1 The Importance of Well Integrity Data

Effective well integrity management is fundamentally data-driven, encompassing the entire lifecycle of a well—from design and construction to operation, suspension, and abandonment. Accurate, complete, and accessible data form the cornerstone of all critical decisions related to well integrity, including material selection, barrier design, risk assessment, anomaly diagnosis, and life extension planning. Without reliable data, integrity management efforts become reactive, disjointed, and less effective, increasing the likelihood of integrity failures and operational disruptions. Robust data management enables the structured analysis of well conditions, supports risk-based prioritization, ensures traceability for compliance, and facilitates continuous improvement through lessons learned. Furthermore, it underpins emerging technologies such as predictive analytics and machine learning, which rely on high-quality historical and real-time data to forecast failures, optimize monitoring, and guide preventative actions. To manage well integrity proactively, data must be systematically captured across all lifecycle phases—design, construction, operations, and diagnostics—and fed into centralized systems for storage, analysis, and decision support. Figure 15.1 illustrates how these data streams are integrated through a structured Well Integrity Management System (WIMS), from initial capture to ultimate use in analytics, performance tracking, and operational decision-making. This systematic data flow ensures timely detection of anomalies, informed risk assessment, optimized interventions, and regulatory compliance. As the oil and gas industry advances toward digital transformation and asset life extension, the value of well integrity data has evolved from a record-keeping necessity to a strategic enabler of safe, cost-effective, and sustainable operations (Semwogerere et al., 2025; Yakoot et al., 2021a). Key reasons why robust data management is critical:

- **Understanding Well Status**: Provides a clear, documented picture of the well's "as-built" construction, components, materials, operational history, and current integrity condition.
- **Risk Assessment**: Accurate data (design details, material certifications, construction reports, operational parameters, monitoring history, failure history) is essential input for meaningful and data-driven risk assessments (Chap. 14).
- **Monitoring and Trend Analysis**: Allows systematic tracking of parameters like annulus pressures, corrosion rates, test results, and operational data over time to identify trends, anomalies, and potential degradation early (Chap. 8).

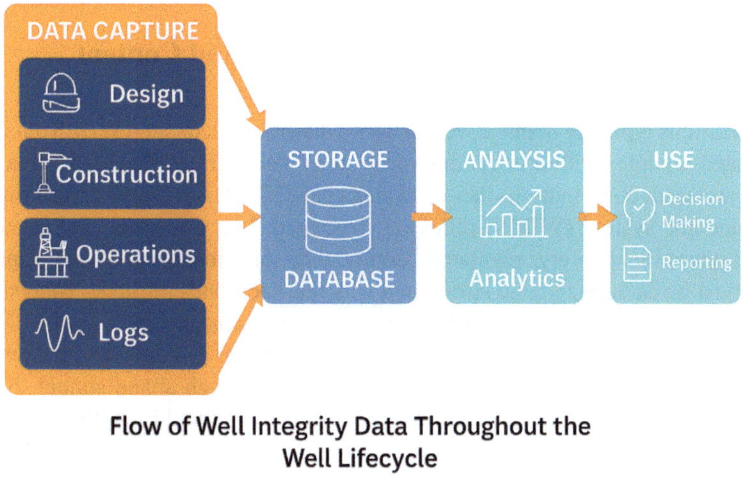

Fig. 15.1 Oilfield worker at sunset near a drilling rig and pump jack, symbolizing production operations and well lifecycle stages. *Source* The Author's

- **Diagnostics**: Comparing current logs or test results against historical and especially baseline data (Chap. 7) is crucial for accurately diagnosing the nature, location, and severity of integrity problems.
- **Compliance and Assurance**: Provides the documented evidence required to demonstrate compliance with regulatory requirements, industry standards, and internal WIMS procedures during audits (Chap. 16).
- **Learning and Improvement**: Analyzing aggregated historical data on failures, near misses, operational anomalies, and interventions helps identify systemic issues, common failure modes, root causes, and inform improvements in design standards, procedures, materials selection, and maintenance strategies.
- **Predictive Analytics**: High-quality, structured, and integrated historical and real-time data is the essential foundation for applying advanced analytics and machine learning techniques to predict potential integrity issues, optimize surveillance, and enable predictive maintenance (Chap. 31).
- **Life Extension Decisions**: Access to comprehensive, reliable data on well history, condition, and degradation rates is paramount for technically justifying the safe life extension of aging wells (Sects. 13.5 and 32.5).

Conversely, poor data management (missing data, inaccurate records, inconsistent formats, information stored in disparate and inaccessible silos) severely hinders effective integrity management, increases operational risks, leads to inefficient decision-making, complicates regulatory compliance, and undermines the ability to learn and improve. Data scarcity, poor quality, and siloed systems remain significant barriers to effective integrity management and optimizing life extension decisions in many organizations.

15.2 Well Integrity Databases: Structure and Content

Organizations typically manage well integrity data using dedicated WIMS databases, integrated modules within larger enterprise asset management (EAM) or data historian systems, or combinations thereof (Khan et al., 2023). An effective Well Integrity Management System (WIMS) relies on structured, lifecycle-wide data capture to ensure operational safety, regulatory compliance, and informed decision-making. Data must be categorized and stored in a way that facilitates retrieval, analysis, and visualization. Table 15.1 outlines key data categories typically managed in a WIMS database, along with examples of the specific data points associated with each category across the well lifecycle—from initial design through abandonment. A well-structured, centralized database or information system should ideally capture key information across the entire well lifecycle (Ramadhani et al., 2021; Wejwittayaklung et al., 2022; Yakoot et al., 2021b):

Table 15.1 Conceptual categorization of WIMS database content. This table summarizes major well integrity data categories with representative examples of the types of data typically captured under each, ensuring comprehensive coverage of integrity-critical information from design to abandonment

WIMS data category	Examples of specific data points
Design	Casing spec, design pressure/temp, cement recipe, barrier diagrams
Construction	Cement placement records, TOC, pressure tests, casing run logs
Operations	Annulus A/B/C pressures, flow rates, corrosion monitoring results
Inspection	Cement evaluation logs (CBL/VDL), leak detection logs, visual checks
Incidents	SCP events, valve leaks, failure investigations, squeeze job results
Maintenance	Wellhead service logs, valve test results, lubrication schedules
Intervention	Temporary barrier used, depth of patch, tool string details
Abandonment	Plug back depth, isolation fluid properties, verification tags

- **Well Identification and Basic Data**
  - Unique Well Identifier (UWI) and other internal identifiers.
  - Location (surface and bottom hole coordinates), field, asset.
  - Well Type (producer, injector, observer, exploration, storage, geothermal, etc.).
  - Current Operational Status (drilling, completing, operating, shut-in, suspended, abandoned).
  - Key Dates (spud date, completion date, first production date, P&A date).
- **Design Data**
  - Basis of Design (BoD) parameters (full P/T/fluid envelopes, design life, etc.).
  - Casing Design (number of strings, sizes, weights, grades, connections, setting depths, design factors, load cases).
  - Tubing Design (size, weight, grade, connections, depth, flow path).
  - Cementing Program Design (slurry types, volumes, additives, design properties).
  - Completion Equipment Design (packer types/settings, DHSV type/depth, wellhead/Xmas Tree specifications including API 6A PSL/PR, material class, pressure/temperature ratings).
  - Initial Well Barrier Schematics (WBS) for planned phases.
- **Construction Data ("As-Built")**
  - Detailed Drilling Reports (mud logs, rig states, events like losses/gains, hole conditions, bit/BHA data).
  - Tubular Running Reports (joint numbers, depths, makeup torque records for connections).

- Cementing Reports (actual slurry properties as mixed, actual pump rates, pressures, volumes pumped, returns, deviations, plug bump pressure, confirmed TOC).
- Cement Evaluation Log Reports (CBL/VDL, ultrasonic)—ideally linked to raw data files.
- Casing and Tubing Material Certificates and Test Reports.
- Completion Installation Reports (equipment depths, setting forces/pressures for packers/hangers, function test results for DHSV/valves).
- Initial Pressure Test Certificates/Records (casing tests, LOT/FIT, tubing test, packer tests, annulus tests, wellhead tests, valve tests)—including test pressure, duration, pressure drop.
- "As-Built" WBS and wellhead/Xmas Tree stack-up drawings.
- Baseline Logs (Caliper, EM Thickness)—ideally linked to raw data files.

• **Operational Data**

- Operating Envelopes (MAASP, MAWOP, MAWIP limits, temperature limits, max flow rates).
- Production/Injection History (rates, pressures, temperatures, fluid composition—water cut, GOR, H_2S, CO_2, salinity, fluid chemistry analysis results).
- Annulus Pressure Monitoring Records (SCP history, bleed-down/build-up test data including volumes/fluid types, monitoring frequency, estimated MAASP/MAWOP calculations).
- Corrosion Monitoring Data (coupon results, probe readings, fluid analysis, inhibitor injection records, scale inhibitor records, biocide treatment records).
- Maintenance Records (planned and unplanned maintenance on wellhead, valves, control systems, etc.).
- Management of Change (MoC) records related to the well.

• **Inspection and Testing Data**

- Records of all periodic integrity tests (valve tests, annulus tests, packer tests) including dates, pressures, durations, results, leak rates, next test due date.
- All Inspection Log Reports (corrosion logs—caliper, EM; Cement evaluation logs run later in life; Leak detection logs—temperature, acoustic; PLT logs)—ideally linked to raw data files.
- Visual Inspection Reports (e.g., wellhead inspection findings).

• **Well Integrity Anomaly/Failure Data**

- Records of all identified SCP occurrences, barrier failures (leaks, non-closures), equipment malfunctions, unexpected pressure changes.
- Detailed anomaly descriptions, date of detection, investigation reports, root cause analyses.
- Remediation actions taken (squeeze jobs, patches, replacements) and records of their execution and effectiveness (post-remediation tests/logs).

- **Intervention Data**
  - Intervention programs and risk assessments.
  - Records of temporary barriers used and verified.
  - Intervention reports detailing activities performed, tools used, any issues encountered, and **post-intervention integrity verification test results**.
- **Suspension Data**
  - Records of steps taken for temporary abandonment (e.g., plugs set, fluids placed).
  - Monitoring performed during suspension.
  - Verification tests performed before returning to service.
- **Abandonment Data**
  - Approved P&A Program/Basis of Design.
  - Records of permanent barrier plugs placed (depth, length, material batch info, placement method).
  - Verification test results for permanent plugs (tagging, pressure test, inflow test, logging).
  - Final P&A schematics and handover documentation.

The database or WIMS software should be designed for easy, structured data entry, automated validation, comprehensive querying, customizable reporting, and ideally, visualization (e.g., linking data to wellbore schematics, displaying trends on graphs, geographical location mapping).

15.3 Data Validation and Quality Control

The principle of "Garbage In, Garbage Out" applies strongly to well integrity data. Inaccurate, incomplete, or inconsistent data leads to flawed analysis, unreliable risk assessments, poor decision-making, and potential compliance issues (Ekseth et al., 2007; Smith & Milanovic, 2008; Wejwittayaklung et al., 2022). Robust data validation and quality control processes are essential throughout the data lifecycle:

- **Standardized Data Entry**: Using consistent templates, forms, and units for data entry reduces errors. Defining mandatory fields.
- **Automated Validation Rules**: Implementing automated checks within the database to flag missing critical data, out-of-range values (e.g., test pressure exceeding component rating), or inconsistencies between related data fields.

- **Verification Processes**: Establishing procedures for verifying the accuracy of critical data entered against source documents (e.g., reviewing original test certificates, comparing data entry to scanned log reports, cross-checking against daily drilling reports).
- **Data Ownership and Responsibility**: Clearly defining who is responsible for collecting, entering, and verifying specific types of data from different sources (e.g., rig personnel for drilling reports, cementing company for placement reports, wireline crew for log reports, production operators for pressure readings).
- **Periodic Data Audits**: Systematically auditing the database content and the data management processes periodically to ensure data accuracy, completeness, and adherence to procedures.
- **Data Cleansing**: Implementing processes to identify and correct erroneous or inconsistent historical data.

Maintaining data quality requires ongoing effort, clear procedures, technological tools, and a culture of diligence throughout the organization.

15.4 Defining, Tracking, and Utilizing Well Integrity KPIs

Key Performance Indicators (KPIs) are measurable metrics used to track the performance and effectiveness of the Well Integrity Management System and the overall health of the well portfolio (Issar & Navon, 2016). They help identify areas of concern, demonstrate progress toward objectives, enable benchmarking, and drive continuous improvement initiatives. Effective KPIs should ideally be SMART: Specific, Measurable, Achievable/Actionable, Relevant, and Time-bound (Mugharbil & Al Khunaizi, 2021; Nicotra et al., 2024a; Sharp et al., 2015).To contextualize performance metrics within a structured integrity framework, industry best practices classify KPIs by severity and temporal relevance. This tiered pyramid (Fig. 15.2) provides a visual model of leading versus lagging indicators. Tier 4 and Tier 3 focus on proactive, preventive measures such as management system performance and early challenges to safety barriers. Tier 2 and Tier 1, conversely, reflect the severity of realized well integrity failures, such as leaks or sustained casing pressure events. Integrating this model with WIMS KPI tracking ensures a balanced approach to both anticipating failures and responding effectively when they occur (Nicotra et al., 2024b).

KPIs can be categorized as Lagging and Leading (Fred, 2009; Stough & De Janeiro, 2010; Vukomanović et al., 2010; Yorio et al., 2020):

- **Lagging KPIs**: Measure outcomes or failures that have already occurred (Nicotra et al., 2024b). They indicate past performance but can highlight areas needing attention.
  - Number/Percentage of wells with active SCP (potentially broken down by annulus, fluid type, build-up rate, or risk level).

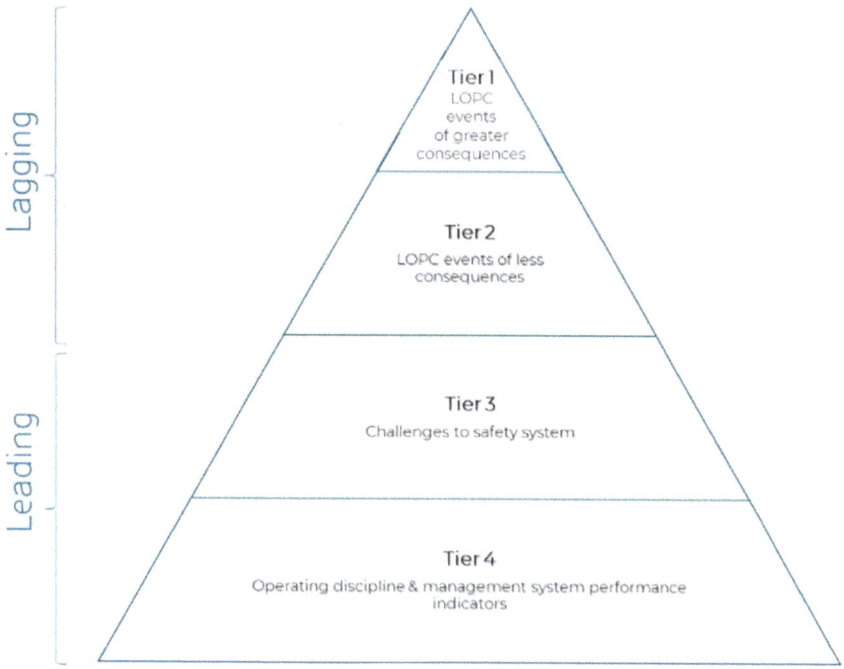

Fig. 15.2 Tiered Framework for Well Integrity Indicators illustrating a hierarchical structure for performance indicators, with Tier 4 and 3 representing leading indicators that measure the robustness of safety systems and management practices, and Tier 2 and 1 representing lagging indicators associated with actual loss of primary containment (LOPC) events (Nicotra et al., 2024a, 2024b)

  - Number of reportable leaks/releases related to well integrity failure (e.g., to atmosphere, sea, ground).
  - Number of integrity-related Non-Productive Time (NPT) events and associated production losses/costs.
  - Number/Percentage of failed integrity tests (e.g., DHSV leak tests, annulus pressure tests, valve integrity tests).
  - Number of wells shut-in or restricted due to integrity issues.
  - Corrosion failure frequency (e.g., number of corrosion-induced leaks per well-year or per specific tubular type).
  - Number of required P&A plugs that failed initial verification tests.

- **Leading KPIs**: Measure proactive activities or prevention efforts designed to manage risk and prevent failures (Nicotra et al., 2024a, 2024b). They indicate the effectiveness of the WIMS processes.

  - Percentage completion of scheduled integrity tests on time (e.g., DHSV tests, annulus tests).
  - Percentage completion of scheduled inspections/logging runs on time (e.g., corrosion logs, cement logs).

- Percentage completion of scheduled preventative maintenance tasks (e.g., wellhead servicing, valve lubrication).
- Number/Percentage of overdue integrity-related actions or recommendations (e.g., from risk assessments, incident investigations, inspections).
- Percentage compliance with defined operating envelopes (e.g., pressure/temperature limits).
- Number of well integrity risk assessments reviewed/updated within the required frequency.
- Percentage of personnel competency assessments completed for integrity-critical roles.
- Number of well integrity-related near misses reported (indicating reporting culture and vigilance).
- Percentage of critical integrity data fields completed in the WIMS database.

Tracking and Utilization

- KPI data should be collected systematically (often automatically from the WIMS database or manually input as required).
- Results should be regularly compiled, analyzed, and reviewed by relevant teams (e.g., asset integrity teams, operations management, senior leadership).
- Trends over time (improving or worsening performance) are often more informative than single data points. Benchmarking against internal targets or industry averages (where comparable data exists) can be valuable.
- Targets should be set for key KPIs, and significant deviations from targets should trigger investigation and action (e.g., root cause analysis for high failure rates, process review for low compliance).
- KPI results provide valuable input for management reviews (Sect. 13.2) and identifying areas needing improvement within the WIMS or operational practices.

Tracking KPIs related to both lagging indicators (failures like SCP, leaks) and leading indicators (test compliance, preventative maintenance completion) allows monitoring of WIMS effectiveness in both preventing failures and executing planned integrity activities (Yakoot et al., 2021b).To effectively monitor the performance and maturity of a Well Integrity Management System (WIMS), organizations rely on a structured set of Key Performance Indicators (KPIs). These indicators provide insights into both proactive risk management activities and reactive event outcomes. Leading indicators measure compliance with planned integrity activities such as testing, inspection, and maintenance, while lagging indicators reflect historical events like leaks, failures, or sustained casing pressure. The combination of both types of KPIs supports a balanced scorecard approach to managing well integrity. Table 15.2 summarizes common KPIs used in the industry, categorized by their leading or lagging nature, and describes the basis of each metric.

Table 15.2 Key Well Integrity Key Performance Indicators (KPIs), categorized as leading and lagging metrics, used to monitor the effectiveness and maturity of a Well Integrity Management System (WIMS)

KPI category	Key Performance Indicator (KPI)	Definition/metric basis
Leading	% of scheduled tests completed on time	Ratio of tests completed vs. scheduled within due period
Leading	% of scheduled inspections completed	Ratio of integrity inspections completed on schedule
Leading	% of preventive maintenance tasks completed	Planned maintenance vs. actual completion rate
Leading	% of integrity risk assessments updated	Number of assessments reviewed vs. required
Leading	% of integrity data fields completed in WIMS	Data completeness score based on mandatory fields
Leading	# of near misses reported (well integrity)	Count of formally logged integrity-related near misses
Lagging	# of wells with active SCP	Count of wells with active sustained casing pressure
Lagging	# of reportable integrity-related leaks	Count of leaks that required regulatory reporting
Lagging	# of failed integrity tests	Total number of failed leak/pressure tests
Lagging	# of wells shut-in due to integrity issues	Wells taken offline due to integrity anomalies
Lagging	Corrosion-induced leak rate (per well-year)	Leaks due to corrosion divided by total active well-years
Lagging	# of failed P&A plug verifications	Number of plugs failing initial post-installation tests

15.5 Reporting and Performance Reviews

Formal reporting and review processes are essential components of the WIMS, ensuring that well integrity status and performance are visible throughout the organization, from the field level to senior management, and that the WIMS remains effective and drives improvement.

- **Routine Reporting**: Regular reports summarizing key well integrity information (e.g., list of wells with active SCP, overdue tests/actions, KPI dashboards, status of critical barriers) for different levels of the organization (e.g., daily/weekly operational handover summaries, monthly asset integrity reports, quarterly business unit reviews) (Smith & Lunt, 2012). Frequency and content depend on the audience and well criticality.

- **Anomaly Reporting**: Clear processes for immediate or rapid reporting of significant integrity anomalies, near misses, and failures (e.g., detection of new SCP, failed valve test, observed leak) to relevant personnel and management, triggering investigation and response workflows.
- **Management Review**: Periodic (e.g., quarterly, semi-annually, or annually) formal reviews of the WIMS by senior management (Islin & Andersen, 2008). These reviews are critical for demonstrating leadership commitment and assessing the overall health and effectiveness of the WIMS. Agenda items typically include:
  - Review of well integrity performance against objectives and KPI targets.
  - Review of results from internal and external audits.
  - Review of findings and lessons learned from incidents and near misses.
  - Review of risk assessment results and status of high-priority risks.
  - Assessment of resource adequacy (personnel, budget, tools) for managing integrity.
  - Review of the status of corrective and preventative actions.
  - Identification of strategic opportunities for improving the WIMS or integrity performance.
  - Minutes and actions from these reviews should be documented and tracked to closure.

Effective reporting ensures transparency and accountability, while formal management reviews ensure that well integrity remains a priority at the highest levels, and that the WIMS is functioning as intended and is continuously improved.

15.6 Connection to Predictive Analytics

The availability of comprehensive, structured well integrity data within a WIMS database provides the essential foundation for applying advanced analytics, particularly Artificial Intelligence (AI) and Machine Learning (ML) techniques (Chap. 31).

- **Data as Fuel**: High-quality, consistent, and integrated historical and real-time data (design parameters, construction specifics, operational profiles, monitoring trends, test results, failure events) is the "fuel" required to train and validate AI/ML models for well integrity applications.
- **Enabling Predictive Capabilities**: By analyzing patterns and correlations within the data that may not be obvious to human operators, AI/ML models can potentially
  - Predict the likelihood and timing of future integrity failures (e.g., SCP development, component wear/failure).
  - Detect subtle anomalies in real-time monitoring data that might indicate an emerging issue (e.g., slight changes in pressure or temperature signatures).
  - Optimize inspection and maintenance schedules based on predicted risk (predictive maintenance, advanced RBI).

– Assist in the interpretation of complex logging data.
– Support more dynamic and accurate risk assessments.

These advanced techniques, discussed further in Chap. 31, hold immense potential to move well integrity management toward a more predictive and proactive state. Investing in robust data management practices, ensuring data quality, and integrating data sources today is key to unlocking the predictive power of AI/ML for well integrity tomorrow.

15.7 Common Challenges in Well Integrity Data Management

To monitor, evaluate, and continuously improve well integrity programs, organizations rely on comprehensive Key Performance Indicators (KPIs) derived from structured datasets. The dashboard illustrated in Fig. 15.3 provides a real-time visualization of key metrics from casing failure events, including service resolution frequency, annual failure counts, time from drilling to failure (from a service perspective), failure status breakdown, and depth distribution by resolution type (Well Casing Failures Report, 2025). These KPIs enable quick assessment of systemic issues, effectiveness of mitigation strategies, and help prioritize high-impact resolutions for operational planning. Integrating such dashboards into WIMS platforms facilitates improved decision-making, cross-asset comparison, and compliance assurance. Despite the clear importance, effective well integrity data management faces common challenges:

- **Data Silos and Integration**: Well data is often generated and stored in disparate systems (drilling databases, production historians, maintenance systems, spreadsheets, paper files) across different departments or contractors. Integrating this data into a single, comprehensive WIMS database is technically and organizationally challenging.
- **Historical/Legacy Data**: Large volumes of historical data exist in old formats (paper charts, scanned images, incompatible databases), making extraction, digitization, and integration difficult and time-consuming.
- **Data Quality and Completeness**: Inconsistent recording practices over time, missing data fields, manual entry errors, and lack of standardized procedures can result in poor data quality, limiting its usefulness.
- **Lack of Standardization**: Variations in terminology, test procedures, and reporting requirements across different assets or even within a single company make data aggregation and comparison difficult.
- **System Usability**: WIMS databases can be complex to use, leading to inconsistent data entry or reluctance by field personnel to utilize the system fully.
- **Ownership and Accountability**: Unclear ownership of data collection and quality can lead to data gaps and inaccuracies.

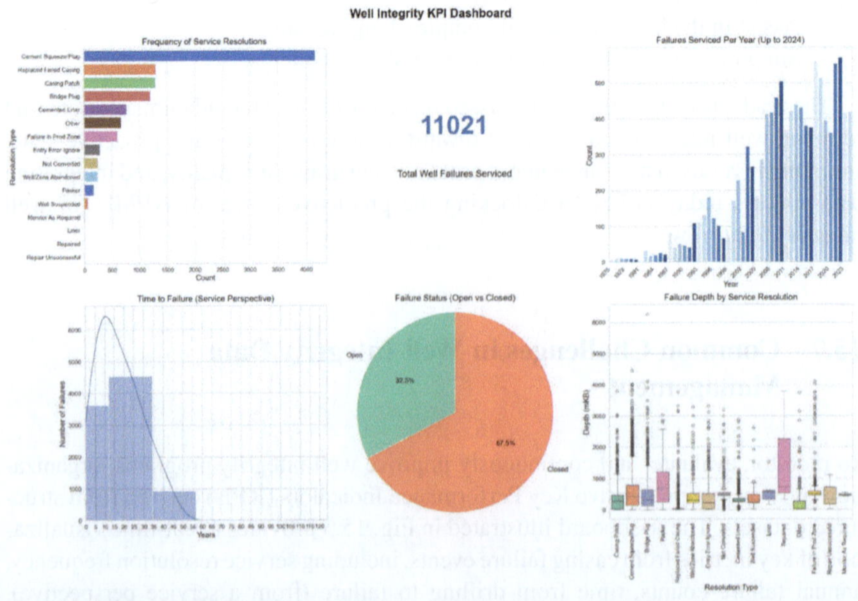

Fig. 15.3 Well integrity KPI dashboard summarizing key performance metrics related to casing failure events. The dashboard includes resolution frequency, failures per year, average time to failure, failure status (open vs. closed), and failure depth distributions by resolution category. These KPIs provide a performance snapshot of the Well Integrity Management System (WIMS) based on historical casing failure data (Well Casing Failures Report, 2025)

- **Cost**: Implementing and maintaining a robust, integrated WIMS data management system requires significant investment in software, hardware, and personnel.

Addressing these challenges through strategic planning, investment in suitable WIMS software, process standardization, training, and clear accountabilities is essential for building the data foundation needed for effective and modern well integrity management. For analyzing well lifecycle performance, the time from startup to first failure is calculated by

$$\textbf{Time to Failure}\ \text{TTF (days)} = \text{Date of Failure} - \text{Date of Well Start of Production} \tag{15.1}$$

References

Ekseth, R., Torkildsen, T., Brooks, A., Weston, J., Nyrnes, E., Wilson, H., & Kovalenko, K. (2007). High-integrity wellbore surveys: Methods for eliminating gross errors. In *SPE/IADC Drilling Conference, Proceedings* (vol. 2, pp. 585–601). https://doi.org/10.2118/105558-MS

Islin, H., & Andersen, T. (2008). The process of management review. *Accreditation and Quality Assurance, 13*(3), 157–160. https://doi.org/10.1007/S00769-008-0380-Z/FIG.S/2

Issar, G., & Navon, L. R. (2016). Key performances indicators (KPI). *Management for Professionals, Part, F327*, 73–76. https://doi.org/10.1007/978-3-319-20699-8_17

Khan, A., Bareche, M., Al Lawati, A., Abboub, M., & Shekhar, S. (2023). Leveraging the power of data to develop a new well integrity management system. In *Proceedings of the SPE/IADC Middle East Drilling Technology Conference and Exhibition.* https://doi.org/10.2118/214592-MS

Manuele, F. A. (2009). Leading & lagging indicators. *American Society of Safety Professionals (ASSP).* https://www.jstor.org/stable/48687557

Mugharbil, M., & Al Khunaizi, M. (2021). Significance of smart and integration system solutions in maintaining well integrity. In *SPE Middle East Oil and Gas Show and Conference, MEOS, Proceedings*. https://doi.org/10.2118/204574-MS

Nicotra, G., Decarli, L., La Rosa, L., Tortora, A., Nocerino, A., Taddei, S., Ponteggia, L., Mangione, A., Bosco, M., & Padovani, R. (2024a). Process safety & asset integrity key performance indicators (PS&AI KPIs): An operative model for strengthening the asset safety barriers. https://doi.org/10.2118/220349-MS

Nicotra, G., Decarli, L., Rosa, L. La, Tortora, A., Nocerino, A., Taddei, S., Ponteggia, L., Mangione, A., Spa, E., Donato Milanese, S., Bosco, M., & Padovani, R. (2024b). Process safety & asset integrity key performance indicators (PS&AI KPIs): An operative model for strengthening the asset safety barriers. *The SPE International Health*, 10–12. https://doi.org/10.2118/220349-MS

Ramadhani, F., Nurdin, S., Etuhoko, M. O., Zhi, Y., Mulyono, S., Mintardja, A., & Dawson, M. (2021). An expert software system has accomplished full implementation of well integrity management system WIMS resulting to excellent safety and production sustainability for critical sour HPHT field in offshore Madura strait, Indonesia. In *Society of Petroleum Engineers—SPE/IATMI Asia Pacific Oil and Gas Conference and Exhibition 2021, APOG 2021* (pp. 12–14). https://doi.org/10.2118/205697-MS

Semwogerere, D., Sangesland, S., Vatn, J., Pavlov, A., & Colombo, D. (2025). Well integrity and late life extension—A current industry state of practice and literature review. In *Geoenergy science and engineering* (vol. 244). Elsevier B.V. https://doi.org/10.1016/j.geoen.2024.213419

Sharp, J., Ersdal, G., & Galbraith, D. (2015). Meaningful and leading structural integrity KPIs. In *Society of Petroleum Engineers—SPE Offshore Europe Conference and Exhibition, OE* (2015). https://doi.org/10.2118/175519-MS

Smith, L. M., & Lunt, G. (2012). Case histories of life cycle well integrity management using iWIT software. In *Society of Petroleum Engineers—SPE Asia Pacific Oil and Gas Conference and Exhibition 2012, APOGCE 2012* (vol. 2, pp. 1287–1295). https://doi.org/10.2118/159441-MS

Smith, L., & Milanovic, D. (2008). The total control of well integrity management. In *Society of Petroleum Engineers—13th Abu Dhabi International Petroleum Exhibition and Conference, ADIPEC 2008* (vol. 1, pp. 145–153). https://doi.org/10.2118/117121-MS

Stough, J., & De Janeiro, R. (2010). *Putting lagging indicators in the rear view mirror with broadly applicable QHSE process-based KPIs.* https://doi.org/10.2118/127188-MS

Vukomanović, M., Radujković, M., & Nahod, M. M. (2010). Leading, lagging and perceptive performance measures in the construction industry. *Organization, Technology & Management in Construction : An International Journal, 2*(1), 103–111. www.constructingexcellence.org.uk

Well Casing Failures Report. (2025). *General well data | Alberta energy regulator*. https://www.aer.ca/data-and-performance-reports/activity-and-data/general-well-data

Wejwittayaklung, K., Nadoon, A., Hnuruang, K., Piron, P., Promwikorn, P., Jirajindakul, P., Chandakaew, R., Singsa-Nga, S., Kunpitaktakun, T., Toempromraj, W., Yokrat, Y., & Konganuntragul, P. (2022). Easy life with automated digitalised analytic tools for well integrity management for unstructured brown field. In *Proceedings of the IADC/SPE Asia Pacific Drilling Technology Conference, APDT*. https://doi.org/10.2118/209836-MS

Yakoot, M. S., Elgibaly, A. A., Ragab, A. M. S., & Mahmoud, O. (2021a). Well integrity management integrity management in mature fields: A state-of-the-art review on the system structure and maturity. *Journal of Petroleum Exploration and Production, 11*(4), 1833–1853. https://doi.org/10.1007/S13202-021-01154-W/FIG.S/21

Yakoot, M. S. E., Salem Ragab, A. M., & Mahmoud, O. (2021b). Multi-class taxonomy of well integrity anomalies applying inductive learning algorithms: Analytical approach for artificial-lift wells. In *Proceedings—SPE Annual Technical Conference and Exhibition* (pp. 21–23). https://doi.org/10.2118/206129-MS

Yorio, P. L., Haas, E. J., Bell, J. L., Moore, S. M., & Greenawald, L. A. (2020). or leading? Exploring the temporal relationship among lagging indicators in mining establishments 2006–2017. *Journal of Safety Research, 74*, 179–185. https://doi.org/10.1016/J.JSR.2020.06.018

Chapter 16
Regulatory Requirements and Industry Standards

Abstract This chapter provides a comprehensive examination of the regulatory frameworks and technical standards that govern well integrity across global oil and gas operations. It begins by differentiating between prescriptive and goal-setting (performance-based) regulatory approaches, highlighting how each influences operator responsibilities and well integrity practices. The chapter underscores that while prescriptive regimes specify detailed technical requirements, goal-setting regimes require operators to demonstrate that risks are reduced to "As Low As Reasonably Practicable" (ALARP), fostering innovation and risk ownership. The evolution of regulatory frameworks is contextualized through lessons learned from major incidents such as Macondo, Montara, and Piper Alpha. Case studies from key jurisdictions—including the U.S. (BSEE), Norway (PSA/Havtil), the UK (HSE/OPRED), Australia (NOPSEMA), and Brazil (ANP)—illustrate diverse regulatory philosophies and their impact on Well Integrity Management Systems (WIMS). These regional models demonstrate how incident-driven reforms have led to increasingly stringent requirements for risk assessment, barrier management, and lifecycle documentation. A detailed review of major industry standards—NORSOK D-010, ISO 16530, and various API specifications—follows, outlining their technical scope, influence, and integration into both regulations and operator WIMS. These standards provide structured guidance on well design, barrier verification, testing, monitoring, and abandonment, serving as critical tools for demonstrating regulatory compliance and engineering excellence. The chapter further explores the Safety Case/Report methodology required in many goal-setting regimes, where operators must explicitly document their identification of major accident hazards, risk assessments, barrier systems, and safety management strategies. This approach shifts the burden of proof from regulators to operators and necessitates robust internal systems for ongoing risk management. Finally, this chapter emphasizes the importance of compliance assurance through internal and external auditing processes. It introduces a systematic audit lifecycle and stresses the value of findings management in driving continuous improvement. The chapter concludes by acknowledging challenges in regulatory alignment with technological innovation, underscoring the need for agile standards and regulatory frameworks that can accommodate emerging risks and advanced tools

© The Author(s), under exclusive license to Springer Nature Switzerland AG 2026

A. Alsubaih and K. Sepehrnoori, *Well Integrity for Oil, Gas, and Subsurface Storage Operations*, https://doi.org/10.1007/978-3-032-15233-6_16

such as AI and digital twins. Overall, this chapter highlights that regulatory compliance, when integrated with industry standards and proactive management systems, forms the cornerstone of sustainable and safe well integrity performance.

Field operators managing surface valves during flaring operations, illustrating gas handling and surface integrity in production fields. *Source* The Authors

Chapter's Objectives

- Provide an overview of the regulatory approach to well integrity in major operating regions, noting varying styles (prescriptive vs. goal-setting).
- Discuss the influence of major incidents on regulatory and standards evolution.
- Conduct a deeper dive into key influential industry standards (NORSOK D-010, ISO 16530, API, NACE), outlining their scope and significance.
- Explain the Safety Case/Report approach used in some regulatory regimes for demonstrating risk management.
- Discuss the importance of compliance assurance, internal auditing, and external verification.
- Highlight areas where standards may lag technological advancements.

16.1 The Regulatory Imperative

The safe construction, operation, and decommissioning of oil and gas wells are underpinned by robust regulatory frameworks and technical standards. These frameworks define not only how integrity is assured throughout the well lifecycle, but also how accountability is maintained between operators, regulators, and the public. In the wake of several high-profile incidents—including Macondo (2010), Montara (2009), and Piper Alpha (1988)—well integrity has moved to the forefront of regulatory agendas across jurisdictions. Today, effective well integrity management is recognized as a non-negotiable expectation, not just a technical discipline.

This chapter explores how regulations and standards collectively serve as the backbone of well integrity governance. Regulations define what must be achieved in terms of safety and environmental protection, while standards provide the detailed "how-to" needed to meet those expectations. Operators are typically required to demonstrate that their Well Integrity Management Systems (WIMS) are aligned with both the letter and the spirit of these instruments.

The regulatory landscape is diverse and continues to evolve. Some jurisdictions, such as the United States and Brazil, tend to rely on prescriptive rules, specifying minimum design requirements, test procedures, and safety thresholds. Others, such as the United Kingdom, Norway, and Australia, adopt a goal-setting or performance-based approach, where operators must demonstrate that their systems and practices reduce risk to "As Low As Reasonably Practicable" (ALARP). Increasingly, many countries employ hybrid models that combine elements of both, balancing enforceability with flexibility to accommodate innovation and site-specific risk profiles.

Complementing the regulatory backbone are industry-developed technical standards. Documents such as NORSOK D-010, ISO 16530, and the wide suite of API specifications and recommended practices form the foundation for global well integrity assurance. These standards detail engineering design criteria, barrier verification protocols, pressure testing methods, cement performance parameters, and abandonment requirements, among others. Their integration into regulatory expectations and internal company systems facilitates a common understanding of best practices across stakeholders.

Importantly, regulatory compliance alone is insufficient in today's risk-informed environment. Regulators increasingly require operators to go beyond minimum standards, mandating a demonstration of effective risk management through tools such as Safety Cases, risk-based verification, and performance monitoring using Key Performance Indicators (KPIs). Audits—both internal and external—are essential for verifying WIMS implementation and identifying systemic weaknesses before they lead to failures.

This chapter begins by contrasting regulatory philosophies (prescriptive vs. goal-setting), then reviews how major incidents have shaped regional regimes. It proceeds to a deep dive into globally influential standards and concludes with an analysis of compliance assurance practices, the role of independent verification, and the implications of emerging technologies and risks that outpace the update cycles of existing standards. Regulatory approaches generally fall into two main categories, or more commonly (Pasman, 2015), a combination thereof:

- **Prescriptive Regulations**: These specify detailed technical requirements, minimum standards, mandatory procedures, testing frequencies, or acceptable practices that operators *must* follow (Pasman, 2015). This approach provides clear, unambiguous requirements and can be easier to enforce. However, it can sometimes be rigid, potentially stifle innovation or adaptation to unique circumstances, and requires regulators to constantly update detailed rules to keep pace with technology and practices. Examples include specific casing design factors, mandatory BOP testing intervals, minimum cement compressive strengths, or specific requirements for P&A plug lengths and materials. API standards are often referenced or incorporated into prescriptive regulations.

- **Goal-Setting (or Performance-Based) Regulations**: These set high-level safety and environmental objectives and outcomes that operators must achieve (Pasman, 2015) (e.g., "maintain two independent barriers," "reduce risks to ALARP," "prevent pollution to the environment," "ensure permanent isolation for abandonment"). The regulations define *what* must be achieved but leave *how* it is achieved to the operator, requiring them to demonstrate (e.g., through Safety Cases or management system documentation) that their chosen management systems, risk assessments, technical solutions, and operational practices are adequate to meet these goals safely and effectively. This approach allows flexibility, encourages operators to use the most appropriate technologies and practices, and fosters a culture of risk ownership. However, it requires robust demonstration and verification processes (like Safety Cases or management system audits) and competent regulators capable of assessing complex technical arguments. Industry standards (like NORSOK D-010 or ISO 16530) often serve as the basis for operators to demonstrate they are meeting goal-setting requirements.

Many modern regulatory regimes, particularly offshore, combine both approaches, setting some mandatory minimum technical requirements (prescriptive elements) while also demanding a comprehensive safety and environmental management system (SMS or SEMS) and a documented demonstration of risk reduction to ALARP (goal-setting elements), of which well integrity is a key component.

16.2 Overview of Major Regulatory Regimes and Incident Influence (Examples)

The evolution of well integrity regulations and standards has been significantly influenced by major incidents and regional operating contexts.

- **United States (Offshore—BSEE)**: The Bureau of Safety and Environmental Enforcement (BSEE), within the Department of the Interior, is the primary regulator for offshore oil and gas operations on the Outer Continental Shelf (OCS). Significant reforms and increased regulatory stringency occurred post-Macondo (2010). Regulations (e.g., 30 CFR Part 250) include a mix of prescriptive requirements (e.g., detailed drilling safety rules, casing design minimums, cementing requirements including cement evaluation, BOP testing protocols, production safety systems including mandated DHSV testing per API RP 14B, lifecycle P&A requirements) and management system elements (e.g., requiring operators to have a Safety and Environmental Management System—SEMS—based on API RP 75) (Maher et al., 2016; Shenold et al., 2018). BSEE focuses heavily on compliance verification through inspections, incident investigations, and enforcement actions.

- **Norway (PSA—Havtil)**: The Petroleum Safety Authority Norway (PSA), now known as Havtil (The Norwegian Offshore Safety and Security Authority), operates a predominantly goal-setting regime based on demonstrating risk reduction to ALARP. Operators must comply with functional requirements laid out in overarching regulations (e.g., Framework Regulations, Management Regulations, Facilities Regulations, Activities Regulations, Technical and Operational Regulations). While specific industry standards like NORSOK D-010 are not legally mandatory in the regulations themselves, compliance with these standards is generally considered the primary and most accepted way for operators to demonstrate they are meeting the functional requirements and achieving ALARP for well integrity and safety. PSA verifies compliance through audits of operator management systems, technical solutions, and offshore facilities. The development of NORSOK D-010 itself was a direct response to incidents and the need for robust, standardized barrier management (Anunciação Jaculli et al., 2022).
- **United Kingdom (Offshore—HSE/OPRED)**: The UK employs a Safety Case regime for offshore operations, overseen primarily by the Health and Safety Executive (HSE) for safety aspects and the Offshore Petroleum Regulator for Environment and Decommissioning (OPRED) for environmental and abandonment aspects. Operators must submit a detailed Safety Case for each installation (demonstrating management of major accident hazards, including well integrity) and obtain acceptance from HSE. Specific well operations require submission of a Well Operations Notification Form (WONF) to HSE and/or an Environment Plan to OPRED. Operators must demonstrate that risks are managed to ALARP and that environmental impacts are acceptable. Compliance with industry guidelines (e.g., from Offshore Energy UK—OEUK, formerly Oil & Gas UK) and standards (like NORSOK D-010, API, ISO) is expected and used to support the ALARP demonstration. The Safety Case approach shifts the onus onto the operator **to** demonstrate safety rather than just comply with a checklist of rules (OEUK Guidelines, 2023).
- **Australia (Offshore—NOPSEMA)**: The National Offshore Petroleum Safety and Environmental Management Authority (NOPSEMA) uses an objective-based regulatory model similar to the UK and Norway. Operators must submit Safety Cases, Well Operations Management Plans (WOMPs), and Environment Plans for assessment and acceptance before conducting activities. These plans must demonstrate that major risks (including well integrity) are managed to ALARP and that environmental impacts are acceptable. NOPSEMA verifies compliance through assessments of these submissions, inspections, and audits.
- **Brazil (ANP)**: The National Agency of Petroleum, Natural Gas and Biofuels (ANP) regulates the industry in Brazil. ANP Resolution No. 46 (2016) specifically addresses operational safety and well integrity management, requiring operators

to implement a Safety Management System for Drilling and Well Intervention and a Well Integrity Management System (WIMS), defining minimum requirements for barrier elements, testing, monitoring, and documentation (*Brazil's ANP Resolution No. 46*, n.d.).

These examples illustrate how different regulators approach well integrity, often incorporating the principles of barrier management, risk assessment, management systems, and verification, with the specific framework and level of prescription varying. The evolution of many of these frameworks has been significantly influenced by major incidents like Piper Alpha and Macondo, highlighting the reactive nature of some regulatory changes but also the drive toward more systematic prevention (Anunciação Jaculli et al., 2022).

16.3 Deep Dive into Key Industry Standards

Industry standards play a pivotal role in establishing a consistent and technically sound foundation for managing well integrity. These standards are either directly incorporated into national regulations or serve as the recognized basis for demonstrating compliance with goal-setting regulatory frameworks. They offer structured guidance across all phases of the well lifecycle—design, construction, operation, monitoring, maintenance, and abandonment—while incorporating decades of industry experience and evolving best practices.

- **NORSOK D-010: Well Integrity in Drilling and Well Operations**
  - Scope: Comprehensive, covering the entire lifecycle from drilling to abandonment.
  - Approach: Relatively prescriptive, providing detailed requirements for well barrier elements, barrier schematics, testing frequencies, monitoring, documentation, and specific operational activities. It defines explicit requirements for barrier acceptance criteria.
  - Key Features: Strong emphasis on the two-barrier principle, clear definitions of barrier elements for different operational phases, specific requirements for verification (testing methods, acceptance criteria), detailed guidance on SCP management, and requirements for P&A barrier placement and verification.
  - Influence: Highly influential globally, often used as a benchmark even where not strictly required by regulation. It is often considered the gold standard for detailed technical integrity requirements and significantly influences regulatory approaches in various jurisdictions (Anunciação Jaculli et al., 2022).
- **ISO16530: Petroleum and Natural Gas Industries—Well Integrity**
  - Scope: Part 1 covers lifecycle governance (the WIMS framework); Part 2 covers the operational phase. (Part 3 on P&A is under development/consideration).

   - Approach: More of a management system framework standard (especially Part 1), outlining the required elements and processes for establishing, implementing, maintaining, and improving a WIMS based on Plan-Do-Check-Act principles. It is less prescriptive on specific technical details compared to NORSOK D-010, allowing more flexibility in implementation, particularly for novel technologies or approaches, provided they can be validated within the WIMS framework.
   - Key Features: Defines well integrity lifecycle phases, emphasizes risk management, outlines requirements for monitoring, testing, maintenance, data management, competency, and continuous improvement during operations (Part 2). Provides a structure for organizing the WIMS, aligning with broader management system concepts (Anunciação Jaculli et al., 2022).

- **API Standards and Recommended Practices (RPs)**: The American Petroleum Institute publishes numerous documents relevant to specific aspects of well integrity. These are widely used globally, particularly where US regulations or practices are influential or where there is no specific international standard. Key examples include

   - API RP 90: Annular Casing Pressure Management (as discussed in Chap. 8), providing detailed guidance on managing SCP.
   - API RP 65-2: Isolating Potential Flow Zones During Well Construction (cementing focus), detailing casing centralization and mud removal practices critical for primary cementing integrity.
   - API Std 65-3: Wellbore Plugging and Abandonment, providing guidelines for P&A procedures and barrier requirements.
   - API Spec 5CT: Casing and Tubing Specification, defining material grades, dimensions, and manufacturing requirements for tubulars (discussed in Chap. 4).
   - API Spec 10A/RP 10B Series: Well Cement Specifications and Testing Procedures, specifying cement classes, additives, and laboratory testing methods (discussed in Chap. 5).
   - API Spec 6A: Wellhead and Christmas Tree Equipment Specification, covering design, materials, pressure ratings, and testing for surface well control equipment (discussed in Chap. 6).
   - API RP 11D1 (ISO 14310): Packers and Bridge Plugs, providing qualification requirements and testing procedures for downhole isolation equipment (discussed in Chap. 6).
   - API RP 14B (ISO 10417): Subsurface Safety Valve Systems—Design, Installation, Operation, outlining requirements and testing procedures for DHSVs (discussed in Chap. 6).
   - API Bul E3: Environmental Guidance for P&A, addressing environmental considerations in abandonment.

Operators typically reference and use a combination of these international and API standards within their own WIMS and internal engineering standards to meet both

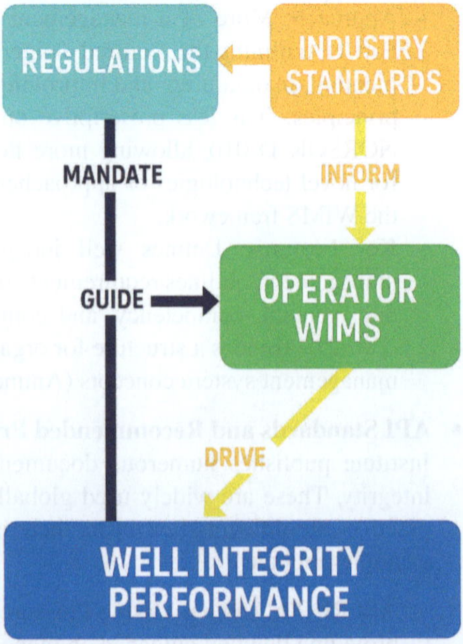

Fig. 16.1 Relationship between regulations, industry standards, operator WIMS, and well integrity performance. Regulations mandate and guide operators; standards inform WIMS design and operation; and WIMS drive real-world integrity outcomes

regulatory obligations and internal governance requirements. The interplay between prescriptive and goal-setting standards, and how they accommodate evolving technology, is a key consideration (discussed further in Chap. 28). Understanding the interplay between regulations, industry standards, and operator-level Well Integrity Management Systems (WIMS) is essential for contextualizing how well integrity performance is governed and achieved across the oil and gas industry. Figure 16.1 illustrates this relationship: regulations provide the mandatory framework and guidance, while industry standards offer detailed technical content and best practices. These standards, in turn, inform the operator's WIMS—an integrated system that drives operational practices and ultimately determines well integrity outcomes. The alignment of these three components is critical for achieving regulatory compliance and sustaining long-term barrier integrity.

16.4 The Safety Case/Report Approach

In goal-setting regulatory regimes like the UK, Norway, and Australia, operators are required to submit a Safety Case (or similar document, e.g., Report on Major Hazards in Norway, WOMP in Australia) to the regulator for acceptance before operations can commence or continue for existing installations (Nelson et al., 1997). This approach shifts the primary responsibility for demonstrating safety from the regulator (via prescriptive rules) to the operator.

Fig. 16.2 Flowchart illustrating the safety case/report process (Hazard Identification → Risk Assessment → ALARP Controls → SMS → Submission → Regulator Acceptance/Audit)

Figure 16.2 illustrates the step-by-step process followed in Safety Case or equivalent regulatory submissions. It begins with hazard identification, followed by structured risk assessment and implementation of ALARP (As Low As Reasonably Practicable) controls. These are then embedded within the operator's Safety Management System (SMS), formally documented, and submitted to the regulator for review, acceptance, and audit.

- Purpose: To demonstrate to the regulator that the operator has a comprehensive understanding of the risks associated with their activities and has implemented adequate measures to control them. Specifically, the Safety Case must demonstrate that the operator has
  - Identified all Major Accident Hazards (MAHs), including those directly related to well integrity failure (e.g., blowout, major leak).
  - Evaluated the associated risks, considering likelihood and potential consequences (safety, environmental, asset, reputation).
  - Implemented adequate control measures (technical barriers, operational procedures, organizational systems) to reduce risks to As Low As Reasonably Practicable (ALARP).
  - Established and implemented a robust Safety Management System (SMS), of which the WIMS is an integral part, to manage these risks effectively throughout the lifecycle.
  - Established adequate emergency response plans for potential MAHs.
- Content Related to Well Integrity: The Safety Case must explicitly and demonstrably address how well integrity MAHs are managed. This typically includes
  - Description of the well-designed basis and the barrier philosophy applied.
  - Well Barrier Schematics (WBS) for critical operational phases, clearly defining primary and secondary barriers.
  - Demonstration of how barriers meet requirements for independence and verifiability, referencing design standards and testing procedures.
  - Summary of risk assessments (HAZID, FMECA, Bow-Tie) related to well integrity failure modes and scenarios (Zhang et al., 2020).
  - Description of the WIMS processes for monitoring, testing, maintenance, anomaly management (e.g., SCP), Management of Change (MoC), and competency assurance.
  - Details of how identified integrity issues (e.g., SCP) are diagnosed, assessed, and managed to maintain risks ALARP.

  - Demonstration that P&A strategies ensure long-term isolation (for existing wells being abandoned or future plans).
- Demonstration of ALARP for well integrity risks through the implemented controls and management system.

The Safety Case is a living document that must be reviewed periodically and updated when significant changes occur (e.g., new wells, major modifications, changes in risk profile, findings from incident investigations or audits). To support regulatory expectations in goal-setting regimes, operators must adopt a systematic, risk-based approach to managing well integrity. The Safety Case or equivalent document provides a structured framework to demonstrate this process. It starts with identifying major accident hazards, assessing associated risks, and implementing controls to reduce those risks to ALARP. These controls are embedded within the operator's Safety Management System (SMS), which governs operational practices. The final step involves submitting the Safety Case to the regulator for review and acceptance, often followed by audits to ensure ongoing compliance (Friend & Friend, 2015).

16.5 Compliance Assurance and Auditing

Regardless of the regulatory style (prescriptive, goal-setting, or mixed), operators must have robust processes in place to ensure ongoing compliance with both external regulations, adopted industry standards, and their own internal WIMS procedures and standards. A structured audit process is critical for verifying that operators are effectively implementing their Well Integrity Management System (WIMS) and maintaining compliance with both regulatory requirements and internal standards. This process not only ensures operational discipline, but also identifies weaknesses, nonconformities, and opportunities for improvement. As illustrated in Fig. 16.3, the audit lifecycle typically includes five key phases: planning, execution, reporting, findings management, and follow-up. Each phase is essential in providing transparency, traceability, and continuous improvement of well integrity practices. This provides crucial assurance to management, regulators, and stakeholders that the WIMS is functioning as intended.

Fig. 16.3 Diagram showing the audit process flow, including Planning → Execution → Reporting → Findings Management → Follow-up

- **Internal Verification**: Routine checks and balances embedded within the operational workflow are the first line of defense. This includes activities like:
  - Requiring sign-offs on critical operational steps (e.g., cement job reports, test certificates, isolation verification).
  - Cross-checking data entry against source documents.
  - Supervisory checks of procedural adherence in the field or control room.
  - Peer reviews of design documentation or operational plans.
- **Auditing**: A formal, systematic process to verify that the WIMS is implemented as planned, is effective in achieving its objectives, and complies with requirements. Audits provide a more structured, independent assessment than routine verification.
  - **Internal Audits**: Conducted by the operator's own personnel (who are independent of the activity being audited) or by third parties on their behalf. They review documentation, interview personnel, and observe activities to check compliance with procedures, standards, and regulatory requirements, verify record keeping, assess competency processes, and identify areas for improvement within the WIMS.
  - **External Audits**: Conducted by regulatory bodies (e.g., BSEE inspections, PSA audits, NOPSEMA audits) or by Independent Verification Bodies (IVBs) often required by regulations (especially in Safety Case regimes). External audits verify compliance with external regulations, adopted standards, and the operator's own commitments (e.g., those documented in the Safety Case or WOMP). They provide an independent assessment of the effectiveness of the WIMS from an external perspective.
- **Findings Management**: A critical system for ensuring that audit findings and other identified non-conformities or areas for improvement are systematically addressed. This involves
  - Documenting the finding clearly.
  - Assigning responsibility for corrective and preventative actions.
  - Setting target completion dates.
  - Tracking progress toward completion.
  - Verifying the effectiveness of implemented actions to ensure the issue is truly resolved and unlikely to recur.

Robust compliance assurance and auditing, supported by an effective findings management system, are essential components of a mature WIMS. They provide confidence to management and regulators that well integrity risks are being adequately controlled and that the organization is committed to continuous improvement.

References

Anunciação Jaculli, M., Choueri, N., da Mata, C. R., Aparecida Silva Leite, A. G., Pelaquim Mendes, J. R., & Colombo, D. (2022). Well safety and integrity evaluation of offshore wells: A review of past, present, and future. *Journal of Petroleum Science and Engineering, 212*. https://doi.org/10.1016/j.petrol.2022.110329

Brazil's ANP Resolution No. 46. (n.d.). Retrieved April 26, 2025, from https://www.legisweb.com.br/legislacao/?id=330646

Friend, M. A., & Friend, K. S. (2015). Aviation Safety Management System (SMS): Applications for general industry. OnePetro. https://dx.doi.org/

Maher, S. T., Long, G. D., Cromartie, R. S., Sutton, I. S., & Steinhilber, M. R. (2016). Paradigm shift in the regulatory application of safety management systems to offshore facilities. *Process Safety Progress, 35*(4), 317–329. https://doi.org/10.1002/PRS.11558

Nelson, S., Tranter, P., Grieve, A., & Krahn, D. (1997). How the world's most comprehensive goal-setting regulatory regime works: A model for the UKCS regulatory system and its unique implications for MODUs. *All Days*. https://doi.org/10.2118/37686-MS

OEUK Guidelines. (2023). Retrieved April 26, 2025, from https://oeuk.org.uk/product/cumulative-risk-guidelines-issue-1/

Pasman, H. (2015). Goal-oriented versus prescriptive regulation. *Risk Analysis and Control for Industrial Processes—Gas, Oil and Chemicals, 407–415*. https://doi.org/10.1016/B978-0-12-800057-1.00010-9

Shenold, C., Teodoriu, C., & Salehi, S. (2018). Consideration of the human element in cementing operations and well integrity. In *Proceedings of the International Conference on Offshore Mechanics and Arctic Engineering—OMAE* (vol. 8). https://doi.org/10.1115/OMAE2018-78275

Zhang, S., Wang, S., Lv, Y., Pan, R., Geng, X., Liu, Y., & Bai, M. (2020). Well integrity analysis and risk assessment for injection wells in CO_2 flooding. In *Offshore Technology Conference Asia 2020, OTCA 2020*. https://doi.org/10.4043/30423-MS

Chapter 17
Human Factors and Safety Culture in Well Integrity

Abstract This chapter explores the critical, yet often underestimated, role of human factors and organizational culture in achieving and sustaining well integrity. While earlier chapters focus on technical systems, materials, and management frameworks, this chapter highlights that well integrity is fundamentally a human-driven endeavor. Drawing from industry incidents such as Macondo and Montara, it demonstrates how even the most robust technical designs can be compromised by procedural deviations, communication failures, poor training, or weak safety culture. The chapter introduces the discipline of Human Factors, emphasizing the dynamic interaction between personnel, tasks, equipment, and environment. It explains how competency development, targeted training, and procedural discipline form the foundation of reliable well integrity performance. It then examines how structured communication—particularly during shift handovers and lifecycle transitions—ensures continuity of safety–critical information and prevents misunderstandings that may jeopardize barrier integrity. A key focus is the Management of Change (MoC) process, which is presented as a vital control mechanism for managing technical, procedural, or organizational changes that could impact integrity. The chapter details how a rigorous MoC system should identify, assess, authorize, and communicate changes before implementation to mitigate new risks. Moreover, the chapter stresses the importance of organizational learning through incident investigation and near-miss reporting. It promotes a "just culture" that encourages transparent reporting without fear of blame, facilitating the root cause analysis of failures and driving continuous improvement. Finally, the chapter defines safety culture as the shared values, behaviors, and attitudes that influence risk management across all organizational levels. It outlines key attributes of a strong safety culture—such as leadership commitment, trust, risk awareness, and learning orientation—and explains how these traits foster proactive behaviors, procedural compliance, and early anomaly detection. This chapter concludes that cultivating human excellence, supported by a mature safety culture and aligned operational systems, is indispensable to well integrity. When personnel are trained, empowered, and supported by clear expectations and strong leadership, they become the most effective defense against integrity failures—transforming human variability from a source of risk into a pillar of operational resilience.

© The Author(s), under exclusive license to Springer Nature Switzerland AG 2026
A. Alsubaih and K. Sepehrnoori, *Well Integrity for Oil, Gas, and Subsurface Storage Operations*, https://doi.org/10.1007/978-3-032-15233-6_17

Oil production engineer illustration. *Source* Mohamed Hassan (2018)

Chapter's Objectives

- Explain the critical influence of human factors on well integrity performance.
- Discuss the importance of competency, training, and procedural discipline.
- Highlight the role of effective communication and handover practices.
- Reinforce the need for robust Management of Change (MoC) processes.
- Emphasize the value of learning from incidents and near misses.
- Define safety culture and its specific relevance to well integrity.

17.1 The Human Element in Well Integrity

While previous chapters focused heavily on technology, materials, physics, and management systems, it is crucial to recognize that people are central to achieving and maintaining well integrity. Human performance is the most variable and, in many cases, the most critical factor in achieving and maintaining well integrity throughout the well lifecycle—from design and construction to production, intervention, and abandonment.

Investigations into major well incidents such as Macondo and Montara have repeatedly shown that technical failures often coincide with, or are precipitated by, human and organizational breakdowns (*Deepwater Horizon—BP Gulf of America Oil Spill|US EPA*, n.d.). These failures include procedural non-compliance, poor communication, misinterpretation of data, inadequate training, or management pressure that leads to compromised safety decisions. The mature integrity systems can be undermined when organizational behaviors or decision-making processes fail to prioritize risk management and safety culture (Jaculli et al., 2022).

This chapter introduces the discipline of Human Factors, which examines how people interact with equipment, systems, tasks, and organizational structures. By understanding these interactions, we can design work environments, tools, and processes that reduce the likelihood of human error while enhancing safety, reliability, and performance. Key areas covered include competency management, procedural discipline, and training systems—all essential to ensuring personnel have the capability and awareness needed for integrity-critical operations. Understanding human factors begins with recognizing the interconnected elements that shape human performance—namely the person, the task, the equipment, and the environment. These components do not function in isolation but interact continuously to influence outcomes. As shown in Fig. 17.1, this relationship is best understood as a dynamic model, where deficiencies in any area can increase the risk of errors, while well-aligned systems can enhance safety and efficiency in well operations.

Communication is another recurring root cause in incident investigations, especially during shift handovers, inter-departmental interfaces, and lifecycle transitions. Failures in communication can lead to missed warning signs or incomplete handovers of critical information. Likewise, an ineffective or absent Management of Change (MoC) process has been a contributing factor in multiple integrity failures, where uncontrolled or undocumented deviations from design, process, or material led to new and unmitigated risks.

A proactive learning culture is vital for improving well integrity outcomes. Organizations must not only investigate incidents and near misses thoroughly, but must also embed the findings into system-wide learning and continuous improvement. However, this is only possible when there is a just culture—one that supports open

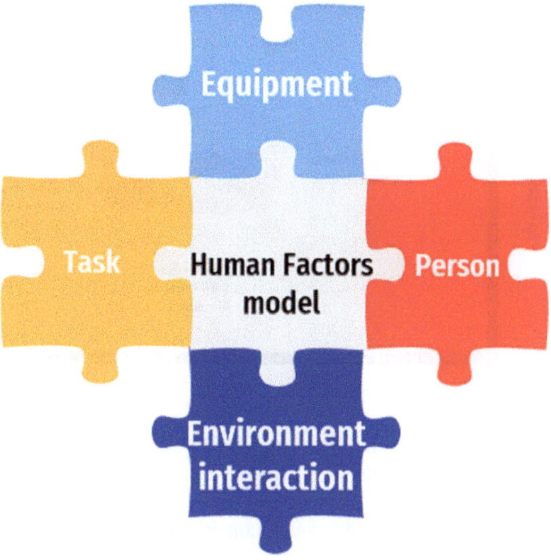

Fig. 17.1 Diagram illustrating the human factors model (Person–Task–Equipment–Environment interaction)

reporting and distinguishes between acceptable human error and negligent or reckless behavior.

Ultimately, safety culture is the foundation on which all other systems rest. A strong culture encourages diligence, accountability, early reporting of anomalies (e.g., SCP), and empowerment to intervene when conditions deviate from safe norms. Conversely, a weak culture—characterized by complacency, blame, or a "production over safety" mindset—can render even the most sophisticated technical safeguards ineffective.

This chapter expands on each of these human and organizational factors, offering both practical strategies and cultural insights to strengthen the integrity of wells not just through design and technology, but through human excellence.

17.2 Competency, Training, and Procedural Discipline

Ensuring that personnel have the capability and discipline to perform their tasks correctly is fundamental (Dieva & Swasti, 2024). Achieving well integrity requires not only technical systems and robust procedures, but also a deliberate progression of human capability and operational integration. Figure 17.2 highlights the structured development path that begins with individual competency, progresses through formal training, and strengthens with procedural discipline, ultimately resulting in consistent and reliable well integrity execution.

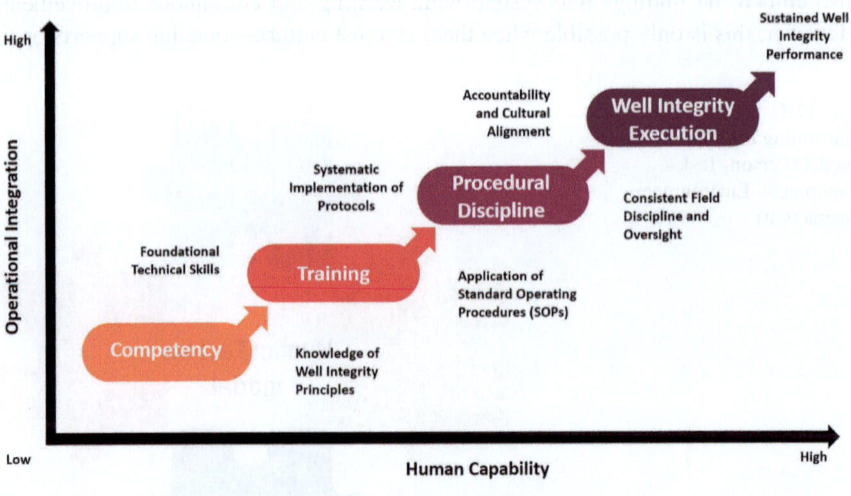

Fig. 17.2 Diagram illustrating the progressive relationship between Competency, Training, Procedural Discipline, and Well Integrity Execution. As human capability increases (x-axis), operational integration (y-axis) strengthens. Each stage builds upon the prior, culminating in sustained well integrity performance through consistent execution and oversight

This layered approach underscores the importance of cultivating foundational technical skills and reinforcing them through structured training programs focused on well integrity principles. As personnel move from training to procedural discipline, emphasis shifts to the systematic implementation of protocols and adherence to Standard Operating Procedures (SOPs). The final stage—well integrity execution—depends on a workforce that is both procedurally disciplined and empowered to maintain field oversight. Sustained performance, therefore, is not a result of isolated skillsets but of integrated human and operational systems aligned toward the same safety and integrity objectives.

- **Competency**: Ensuring that personnel assigned to integrity-critical roles possess the necessary knowledge, skills, experience, and qualifications to perform their tasks safely and effectively (Mahalaxmi et al., 2017). This requires
  - Defining Competency Requirements: Specifying the required competencies for each role (e.g., well integrity engineer, cementing supervisor, valve technician, driller, well operator). This may involve technical knowledge (e.g., understanding cement chemistry, load calculations), practical skills (e.g., operating wellhead valves, performing pressure tests), operational experience (e.g., experience in the specific operating environment or well type), and qualifications (e.g., relevant certifications).
  - Competency Assessment: Processes for verifying that individuals meet these requirements (e.g., through review of qualifications and experience, interviews, written or practical tests, simulator assessments, observation in the field). Assessments should be conducted periodically to confirm ongoing competence.
  - Ongoing Development: Providing targeted training, mentoring, coaching, and development opportunities to maintain and enhance competency, particularly with evolving technology or procedures.
- **Training**: Providing targeted training on well integrity principles, specific procedures, equipment operation, risk awareness, and emergency response relevant to individual roles(Ljubić et al., 2016). Training should cover not only *how* to do a task but also *why* it is important for integrity, emphasizing the consequences of error. Training should include
  - Initial Training: For new personnel or those taking on new roles.
  - Refresher Training: Periodic training to reinforce knowledge and skills.
  - Task-Specific Training: Training on critical procedures (e.g., pressure testing, annulus bleed-down, MoC process).
  - Risk Awareness Training: Ensuring personnel understand the major accident hazards related to well integrity and the critical role of barriers.
- **Procedural Discipline**: Having well-designed, clear, concise, and readily accessible procedures for integrity-critical tasks is essential, but insufficient on its own (Abbas, 2023; Riznic, 2024). Procedural discipline—the consistent adherence to approved procedures by trained and competent personnel—is vital. This requires

- User-Friendly Procedures: Procedures written clearly, logically, and easy to understood and follow in the field environment (e.g., clear language, flowcharts, diagrams). Involvement of end-users in procedure development and review helps ensure practicality.
- Management Expectation: Clear and consistent communication from leadership that adherence to approved procedures for integrity-critical tasks is mandatory and non-negotiable. Leaders should role model adherence.
- Addressing Barriers: Identifying and removing any obstacles in the work environment (e.g., time pressure, resource constraints, conflicting goals like production targets vs. integrity hold points, poorly maintained equipment) that might inadvertently encourage personnel to take shortcuts, skip steps, or deviate from procedures.
- Culture of Intervention: Empowering and encouraging personnel at all levels (e.g., through "Stop Work Authority") to pause or stop a job if they believe a procedure cannot be followed safely or correctly, if conditions change unexpectedly, or if something doesn't seem right ("chronic sense of unease"). This requires a non-retributive environment.

Failures in competency or procedural adherence (e.g., misinterpreting a pressure test, incorrect valve operation, improper cement slurry mixing, failure to follow MoC steps) can directly lead to barrier failures or compromise planned controls.

17.3 Communication and Handover Practices

Effective communication and information transfer between different individuals, teams, shifts, and lifecycle phases are crucial for maintaining a shared, accurate understanding of well status, ongoing operations, potential integrity risks, and required actions (Jones, 2007; Schecter & Mell, 2017). Breakdowns in communication are frequently cited as root causes in incident investigations.

- **Clarity and Accuracy**: Ensuring information (e.g., operational instructions, monitoring data, test results, risk assessments, changes in well conditions, planned activities) is communicated clearly, accurately, and unambiguously, both verbally and in writing. Using standardized terminology, reporting formats, and checklists helps minimize misinterpretation.
- **Shift Handovers**: Structured and disciplined handover processes (verbal briefings supported by documented logs, checklists, and handover sheets) are essential to ensure critical information about well status, ongoing operations, potential integrity concerns, outstanding actions, and known risks is effectively transferred between outgoing and incoming shifts on an asset (e.g., rig, platform).
- **Interface Management**: Establishing clear communication protocols and shared understanding of responsibilities between different departments or companies involved in well operations (e.g., operator well team, drilling contractor, service

companies providing cementing, logging, intervention services). Ensuring alignment on procedures, safety expectations, and risk perception at the interfaces between different contractors or disciplines is vital.

- **Lifecycle Handovers**: Formal and comprehensive processes for transferring well information and integrity status documentation when responsibility shifts between major lifecycle phases (e.g., from the drilling/completions team to the production operations team, or from operations to the abandonment team). This ensures the receiving team fully understands the well's "as-built" condition, operational history, any existing integrity issues (e.g., SCP), past interventions, and known risks, enabling them to manage integrity effectively going forward. Key documentation (design basis, construction records, test history, WBS, risk assessments) must be transferred and accessible.

Breakdowns in communication (e.g., failure to report SCP, misunderstanding handover information, conflicting instructions, lack of access to relevant documentation) can lead to critical information being missed, misinterpreted, or acted upon without full awareness of the integrity implications, resulting in poor decisions and potential integrity failures.

17.4 Management of Change (MoC)

Wells and their operating conditions often change over time, sometimes significantly. Management of Change (MoC) is a formal, systematic system to identify, assess, authorize, and document any temporary or permanent deviations or changes to equipment, procedures, operating conditions, materials, or personnel that could potentially impact safety or well integrity (Chosnek, 2023; "Management of Change," 2022; "Process Safety for Engineers: An Introduction," 2022). Effective Management of Change (MoC) is a cornerstone of well integrity assurance. It ensures that any modifications to equipment, procedures, personnel, or operational parameters are evaluated systematically before implementation. In high-risk environments like oil and gas operations, uncontrolled or undocumented changes have historically contributed to well integrity incidents (Laskar, 2017, 2022). A robust MoC system not only evaluates technical implications but also accounts for hazard analysis, stakeholder coordination, and procedural readiness. The visual representations in Fig. 17.3 illustrate both a linear MoC workflow and a cyclical change management process, emphasizing how structured steps, team roles, and hazard review mechanisms are integral to maintaining safety and regulatory compliance during operational changes. Bypassing MoC is a common factor in incidents.

- **Triggering MoC**: Defining clearly within the WIMS what constitutes a "change" requiring formal MoC review. This includes alterations such as:
  - Changing operating pressure or temperature limits.
  - Modifying fluid injection/production rates significantly.

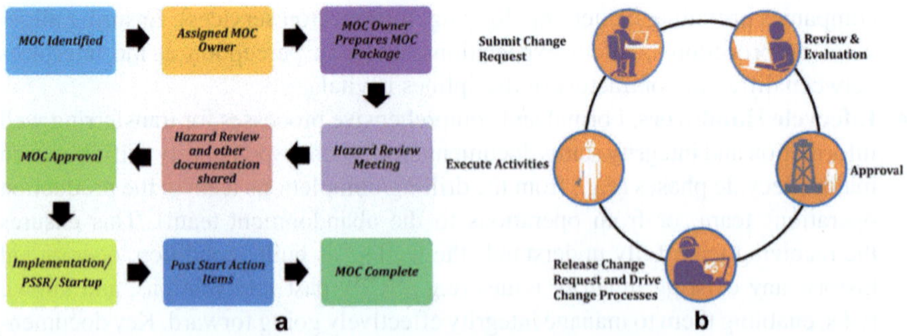

Fig. 17.3 Flowcharts illustrating the Management of Change (MoC) process

- Changing the chemical injection strategy or product.
- Modifying wellhead or downhole equipment.
- Revising an approved critical procedure.
- Introducing new technology or equipment.
- Using different materials than specified.
- Significant organizational or personnel changes impacting critical roles.

- **Risk Assessment**: The core of the MoC process is evaluating the potential impact of the proposed change on well integrity and safety. This involves a multidisciplinary review to determine
  - Does the change introduce new hazards?
  - Does it affect existing barriers (e.g., compromise their effectiveness, introduce new stresses)?
  - Does it create new potential leak paths?
  - Does it invalidate previous design calculations or risk assessments?
  - Are necessary controls in place or need to be implemented?
- **Authorization**: Requiring appropriate levels of technical and management approval before the change is implemented, based on the assessed risk and potential impact on integrity. Approval levels should be defined based on the risk ranking of the change.
- **Documentation**: Documenting the entire MoC process: the change proposal, the risk assessment findings, the technical review, the authorization(s), the implementation plan (including necessary verification steps), and any necessary updates to procedures, drawings, operating envelopes, or the WIMS database.
- **Communication**: Ensuring all relevant personnel (operations, maintenance, integrity team, offshore crew) are made aware of the approved change and any implications for their work *before* the change is implemented.

A robust MoC process is essential for controlling the integrity risks associated with changes and ensuring that potential negative impacts are identified, assessed, and mitigated before they result in failures.

(Left): A linear MoC process flow from identification to completion, incorporating hazard review, documentation, approval, and post-startup verification. (Right): A cyclical model showing continuous steps of submitting change requests, evaluation, approval, process release, and field execution—highlighting the recurring nature of change integration in operational settings (Laskar, 2017, 2022).

17.5 Learning from Incidents and Near Misses

Organizations must actively learn from their own experiences and those of the wider industry to prevent the recurrence of well integrity failures and continuously improve their WIMS. This requires a proactive approach to collecting, investigating, and sharing information about events (Adeyinka et al., 2020; Egbe et al., 2024; Freifeld et al., 2016).

- **Open Reporting Culture**: Fostering an environment where personnel feel safe and are encouraged to report *all* integrity-related incidents, anomalies (like unexpected pressure readings, strange noises), equipment failures, procedural deviations, and near misses without fear of blame or retribution. Near misses, in particular, provide invaluable learning opportunities to address weaknesses *before* significant consequences occur. "Just Culture" principles (as discussed in Sect. 17.6) are essential here.
- **Systematic Investigation**: Implementing structured investigation processes (e.g., Root Cause Analysis—RCA) for all significant integrity incidents and potentially for high-potential near misses. Investigations must aim to thoroughly understand not just the immediate technical cause of an event, but also the underlying human factors, procedural weaknesses, organizational influences, and management system deficiencies that contributed to it. A multidisciplinary team should conduct investigations.
- **Effective Recommendations**: Developing practical, specific, and effective corrective and preventative actions that address the identified root causes, not just the symptoms. Recommendations should focus on fixing systemic issues (e.g., revising a flawed procedure, enhancing training, improving equipment reliability) rather than solely blaming individuals.
- **Sharing Lessons Learned**: Communicating findings and lessons learned broadly across the organization (and often with the wider industry, e.g., via safety alerts, industry forums, JIPs) through safety alerts, revised procedures, training updates, management briefings, and awareness campaigns to ensure the learning translates into tangible improvements in knowledge and practice.
- **Tracking Actions**: Implementing a robust findings management system (as discussed in Sect. 16.5) to monitor the implementation and verify the effectiveness of corrective and preventative actions arising from investigations and reviews.

A strong learning culture transforms incidents from mere failures into opportunities for strengthening the WIMS, improving procedures, enhancing competency, and ultimately preventing similar events from happening again.

17.6 Safety Culture and Well Integrity

Safety Culture refers to the shared values, beliefs, attitudes, and behaviors within an organization regarding safety (Cooper, 2000; Guldenmund, 2000; "Organizational Safety Culture," 2017; Wiegmann et al., 2004). It reflects "the way we do things around here" when it comes to managing risk, particularly when faced with competing pressures like production targets or cost constraints. A positive and mature safety culture is fundamental to achieving sustainable well integrity performance, as it influences individual behaviors and collective decision-making at all levels.

Key attributes of a positive safety culture include.

- **Leadership Commitment**: Visible, consistent, and authentic commitment from leaders at all levels, from senior executives to frontline supervisors. This demonstrates that safety and integrity are core values, not just priorities that can be traded off against production or cost. Leaders actively participate in safety activities, discuss integrity in meetings, allocate necessary resources, and hold themselves and others accountable.
- **Shared Responsibility**: A sense among all personnel, regardless of role or level, that they have a personal responsibility and a collective role to play in ensuring well integrity. This includes looking out for colleagues, raising concerns, and adhering to procedures.
- **Trust and Open Communication**: An environment where people feel psychologically safe to raise concerns, report errors, near misses, or anomalies, and challenge unsafe practices or instructions without fear of reprisal or negative consequences. Information flows freely up, down, and across the organization.
- **Risk Awareness**: A collective understanding among personnel of the major hazards associated with well operations (e.g., pressure, H_2S, dropped objects), the critical function of well barriers, and the potential consequences of integrity failure. This includes developing a "chronic sense of unease"—an alertness to potential deviations or weak signals that might indicate an emerging problem. Resisting complacency is key.
- **Learning Orientation**: An organizational mindset that actively seeks out opportunities to learn from experience (internal incidents, near misses, external industry events), implement lessons learned, and continuously improve procedures, systems, and practices. Failure is viewed as an opportunity to learn and improve the system, not just to blame individuals.
- **Just Culture**: An atmosphere of trust where people are encouraged (and rewarded) for providing safety-related information, but where a clear distinction is made

between acceptable human error (slips, lapses), at-risk behavior (conscious deviation from procedure or accepted practice), and reckless conduct (willful disregard for safety rules), with fair and consistent accountability based on the behavior, not just the outcome.

Relevance to Well Integrity.
A strong safety culture directly supports well integrity by promoting:

- Diligent adherence to procedures for integrity-critical tasks (cementing, testing, monitoring, valve operations).
- Careful execution of work to minimize damage to equipment and barriers.
- Willingness to report anomalies like unexpected SCP, minor leaks, or equipment malfunctions early and accurately.
- Proactive identification and management of risks before they escalate.
- Effective communication and teamwork, especially during complex operations or handovers.
- Commitment to learning from failures and implementing corrective actions.
- Empowerment to intervene and stop unsafe or integrity-compromising work.

Conversely, a weak safety culture (e.g., prioritizing production over procedure, blaming individuals for errors, tolerating shortcuts, ignoring warning signs, poor communication) undermines the effectiveness of even the best technical systems and procedures, significantly increasing the likelihood of well integrity failures and major accidents. Building and sustaining a positive safety culture requires continuous effort, visible leadership engagement, reinforcement through organizational systems and practices, and integration of human factors principles into design and operations. It is the essential human and organizational foundation upon which technical well integrity relies.

References

Abbas, K. (2023). *Quality management for process, polices and procedure.* https://doi.org/10.31219/OSF.IO/KCQ8B

Adeyinka, A., Tsakporhore, A., Arije, O., Adegoke, O., Aliyev, S., Iwueze, C., & Hulett, R. (2020). Rethinking well integrity for sustainability: Adopting a risk based approach. In *Society of Petroleum Engineers—SPE Nigeria Annual International Conference and Exhibition 2020, NAIC 2020.* https://doi.org/10.2118/203672-MS

Anunciação Jaculli, M., Choueri, N., da Mata, C. R., Aparecida Silva Leite, A. G., Pelaquim Mendes, J. R., & Colombo, D. (2022). Well safety and integrity evaluation of offshore wells: A review of past, present, and future. *Journal of Petroleum Science and Engineering, 212.* https://doi.org/10.1016/j.petrol.2022.110329

Chosnek, J. (2023). Managing the risk of MOC. *Process Safety Progress, 42*(S1), S50–S55. https://doi.org/10.1002/PRS.12472;REQUESTEDJOURNAL:JOURNAL:15475913;CSUBTYPE:STRING:SPECIAL;PAGE:STRING:ARTICLE/CHAPTER

Cooper, M. D. (2000). Towards a model of safety culture. *Safety Science, 36*(2), 111–136. https://doi.org/10.1016/S0925-7535(00)00035-7

Deepwater Horizon—BP Gulf of America Oil Spill | US EPA. (n.d.). Retrieved April 19, 2025, from https://www.epa.gov/enforcement/deepwater-horizon-bp-gulf-america-oil-spill

Dieva, D. P., & Swasti, I. K. (2024). Competence, work discipline, and work facilities contribute to improving employee performance. *AJARCDE (Asian Journal of Applied Research for Community Development and Empowerment)*, *8*(3), 93–98. https://doi.org/10.29165/AJARCDE.V8I3.459

Egbe, P. I., Azim, M. K., & Muzairie, T. (2024). Unraveling concentric casing failures: Insights from two decades of investigative endeavors. In *Society of Petroleum Engineers—SPE Symposium—Well Integrity Management 2024*. https://doi.org/10.2118/220588-MS

Freifeld, B. M., Oldenburg, C. M., Jordan, P., Pan, L., Perfect, S., Morris, J., White, J., Bauer, S., Blankenship, D., Roberts, B., Bromhal, G., Glosser, D., Wyatt, D., Rose, K., Freifeld, B. M., Oldenburg, C. M., Jordan, P., Pan, L., Perfect, S., … Rose, K. (2016). *Lawrence Berkeley National Laboratory Recent Work Title Well Integrity for Natural Gas Storage in Depleted Reservoirs and Aquifers* (DOE National Laboratories Well Integrity Work Group Publication Date).

Guldenmund, F. W. (2000). The nature of safety culture: A review of theory and research. *Safety Science, 34*(1–3), 215–257. https://doi.org/10.1016/S0925-7535(00)00014-X

Hassan, M. (2018). Oil rig engineer. Stockvault. Licensed under CC0 (Public Domain). https://www.stockvault.net/photo/255762/oil-rig-engineer

Jones, S. (2007). How to manage knowledge transfer. *Strategic HR Review, 6*(4), 5–5. https://doi.org/10.1108/14754390980000974

Laskar, S. (2017). Implementing an effective management of change MOC system for upstream oil and gas industry. In *Society of Petroleum Engineers—SPE Health, Safety, Security, Environment, and Social Responsibility Conference—North America* (pp. 459–470). https://doi.org/10.2118/184432-MS

Laskar, S. (2022).An integrated management of change Program (MOC) for oil and gas industry through blockchain technology and smart contracts. In *Society of Petroleum Engineers—ADIPEC 2022*. https://doi.org/10.2118/210790-MS

Ljubić, M., Raković, B., Dimitrov, L., & Garvanov, I. (2016). Training of workers as an important safety measure for risk management. In *2016 19th International Symposium on Electrical Apparatus and Technologies, SIELA 2016*. https://doi.org/10.1109/SIELA.2016.7543022

Mahalaxmi, K. R., Priyadharshini, M., & Student, P. G. (n.d.). Implementation of knowledge, skill and ability in an organisation for effective job performance *3*. Retrieved May 24, 2025, from www.ijariie.com

Management of Change. (2022). Process safety for engineers, 383–392. https://doi.org/10.1002/9781119831075.CH18

Organizational Safety Culture. (2017). *Leading reliable healthcare* (pp. 1–24). https://doi.org/10.1201/B21925-1

Process safety for engineers: An introduction. (2022). Process safety for engineers: An introduction, 1–514. https://doi.org/10.1002/9781119831075;JOURNAL:JOURNAL:BOOKS;WGROUP:STRING:PUBLICATION

Riznic, J. (2024). Integrity management of critical systems: Introduction. *Asset Integrity Management of Critical Infrastructure, 77–78*,. https://doi.org/10.1115/1.887738_PT3.I1

Schecter, A., & Mell, J. (2017). The antecedents and implications of information sharing processes in multiteam systems (vol. 2017, no, 1, p. 11860). https://doi.org/10.5465/AMBPP.2017.11860abstract.

Wiegmann, D. A., Zhang, H., Von Thaden, T. L., Sharma, G., & Gibbons, A. M. (2004). Safety culture: An integrative review. *International Journal of Aviation Psychology, 14*(2), 117–134. https://doi.org/10.1207/S15327108IJAP1402_1;REQUESTEDJOURNAL:JOURNAL:HIAP20

Part V
Specialized Topics and Advanced Concepts

The previous parts of this book established the fundamental principles, construction practices, operational monitoring, failure modes, and management systems central to well integrity. Part V delves into more specialized and advanced topics, addressing the unique integrity challenges posed by specific operating environments, well types, and operational phases. This revised Part V includes dedicated chapters on managing integrity in harsh environments like HPHT and deepwater, wells used for energy transition applications (geothermal, EOR, water injection, subsurface storage), unconventional and subsea wells, integrity management during interventions and remediation activities. It also dedicates chapters to the dynamic relationship between established standards and evolving technology, provides statistical context for understanding integrity performance, and concludes with a detailed look at the critical final lifecycle phase: plug and abandonment (P&A). These chapters build upon the foundational knowledge, providing insights into managing integrity under more demanding or complex circumstances and acknowledging the ongoing evolution of the field.

Chapter 18
Integrity Challenges in Harsh Environments

Abstract In this chapter focuses on the structured processes of auditing, performance review, and continuous improvement as essential components of a robust Well Integrity Management System (WIMS). While well integrity relies on sound design, competent execution, and vigilant operations, sustaining performance over time requires periodic evaluation of whether systems, practices, and barriers are functioning as intended. This chapter introduces the concept of integrity assurance as a dynamic, iterative process grounded in evidence-based assessment and organizational learning. The chapter outlines the objectives, methodologies, and types of well integrity audits, distinguishing between internal and external audits, and between system-level audits (e.g., WIMS compliance, barrier verification programs) and operational audits (e.g., well file reviews, surveillance records, field execution). It emphasizes that audits should not merely check for procedural compliance but should actively test the effectiveness of integrity controls in preventing failures. Effective audits identify gaps, inconsistencies, or latent risks that may not be visible during routine operations. Performance reviews are discussed as a complementary activity, integrating audit findings, KPIs, incident data, and operational observations to evaluate the overall health and maturity of the integrity program. The chapter presents structured review cycles—quarterly, annual, and post-event—and outlines tools such as integrity dashboards, bowtie analysis, and maturity models to support high-level evaluations. These reviews are crucial for prioritizing resources, identifying emerging risks, and setting strategic goals for integrity performance. The concept of continuous improvement is introduced not as a separate function, but as the output of a closed-loop integrity management cycle. This includes root cause analysis (RCA) of failures or near misses, implementation of corrective and preventative actions (CAPAs), and verification of effectiveness. Importantly, continuous improvement extends to updating procedures, training content, digital systems, and field practices based on feedback from audits, reviews, and frontline personnel. The chapter concludes by reinforcing that audits and reviews must be perceived not as punitive controls, but as opportunities for learning, improvement, and strengthening organizational resilience. A successful integrity assurance program is the one that evolves with the operating environment, adapts to new challenges, and fosters a culture of accountability, transparency, and proactive risk management.

© The Author(s), under exclusive license to Springer Nature Switzerland AG 2026

A. Alsubaih and K. Sepehrnoori, *Well Integrity for Oil, Gas, and Subsurface Storage Operations*, https://doi.org/10.1007/978-3-032-15233-6_18

Offshore oil production platform during storm conditions illustrating environmental loading and structural integrity challenges. *Source* The Authors

Chapter's Objectives:

- Describe the specific well integrity challenges associated with High-Pressure/High-Temperature (HPHT) wells.
- Discuss integrity considerations unique to deepwater environments.
- Outline the materials and design challenges in highly corrosive environments (H2S, CO_2).
- Explain the impact of geomechanical challenges like salt creep and compaction on well integrity.

18.1 Introduction: Pushing the Boundaries

As the global energy demand continues to rise and conventional hydrocarbon reservoirs mature, the oil and gas industry is increasingly driven to explore and produce resources in technically demanding and environmentally extreme settings—collectively referred to as harsh environments. These include High-Pressure/High-Temperature (HPHT) reservoirs, ultra-deepwater basins, formations containing corrosive gases such as CO_2 and H_2S, and geologically unstable zones like salt formations and compacting reservoirs. Operating in such environments imposes substantial challenges to well integrity due to intensified mechanical, thermal, and chemical stresses, coupled with a significantly narrower margin for operational error.

A failure in design, materials, or execution in these conditions can lead to severe safety incidents, environmental damage, and economic loss. To maintain integrity across the lifecycle of wells in harsh environments, a multidisciplinary approach is required—one that integrates advanced geomechanical modeling, precision material selection, thermally and chemically resistant barrier systems, and operational discipline at every phase. These extreme settings push traditional technologies to their limits: HPHT wells exert triaxial stresses that demand elastomer and seal designs far exceeding standard ratings; deepwater operations require flawless execution due to logistical constraints and the difficulty of intervention; corrosive reservoirs expose wells to hydrogen-induced cracking, CO_2 carbonation, and rapid material degradation; and geomechanical conditions such as salt creep or subsidence introduce long-term structural distortion to casing and cement. This chapter categorizes these integrity threats into five domains: HPHT conditions, deepwater operations, corrosive environments, geomechanical hazards, and the necessary design and operational responses. By understanding the specific risks within each category, operators can deploy targeted mitigation strategies that reinforce barrier performance, regulatory compliance, and long-term well reliability. In such high-risk domains, Well Integrity Management Systems (WIMS) must be especially robust, incorporating predictive analytics, stringent qualification protocols, real-time diagnostics, and a culture rooted in procedural discipline and continuous improvement.

18.2 High-Pressure/High-Temperature (HPHT) Wells

High-pressure and high-temperature (HPHT) wells are defined as those with an expected shut-in surface pressure of at least 690 bar (10,000 psi) and/or temperatures exceeding 150 °C (300 °F) (Paula et al., 2009). As drilling activities expand into deeper and hotter formations, it becomes essential to clearly define the boundary conditions that distinguish standard, HPHT, and ultra-HPHT wells. Fig. 18.1 provides a graphical representation of how static reservoir pressure and temperature define operational regimes. Conventional HPHT wells begin at pressures above 10,000 psi (69 MPa) and temperatures above 150°C, but Ultra-HPHT and HPHT-hc (high complexity) wells push the envelope even further—demanding unconventional materials, specialized tools, and rigorous safety margins. These operational categories are critical for establishing design criteria and equipment ratings, and they guide everything from tubular selection to cementing strategies and pressure control system qualification. These extreme subsurface conditions create favorable environments for fluid compression, making HPHT wells viable candidates for oil and gas storage. However, their development poses considerable technical challenges, particularly during drilling and completion operations (Yuan et al., 2013). HPHT wells are subject to complex coupled interactions between in-situ stress changes, pore pressure variations, and elevated temperatures. These interactions frequently lead to wellbore instability events such as tight holes, stuck pipe incidents, and differential sticking. Thermal expansion under high temperatures reduces the equivalent

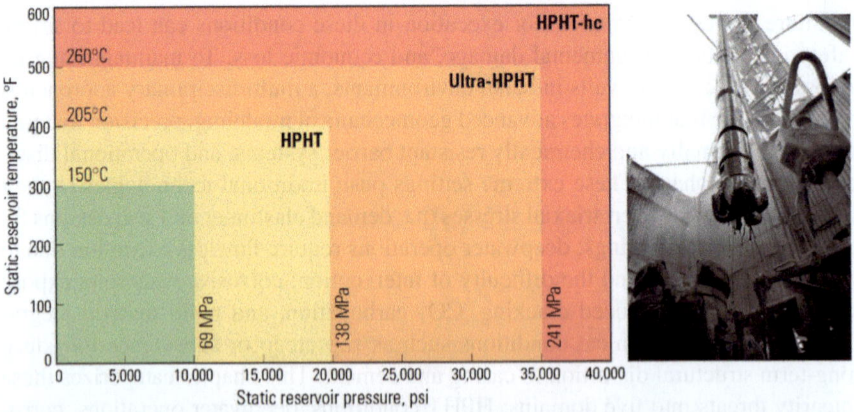

Fig. 18.1 Classification of high-pressure, high-temperature (HPHT) wells based on static reservoir pressure and temperature. The Fig. defines conventional HPHT limits (above 150°C and 10,000 psi) and illustrates how Ultra-HPHT and HPHT-hc conditions exceed these thresholds, with significantly greater engineering and material requirements. Data adapted from Junior et al. (2009) and industry specifications for well design envelopes (SLB, 2025)

circulating density (ECD) of drilling fluids, narrowing the operational window and increasing the risk of formation influx or collapse. Furthermore, the high thermal gradient shortens the thickening time of cement slurries, accelerating early strength development and increasing the potential for post-set cracking (Shadravan & Amani, 2012). Cement rheology, including properties such as plastic viscosity and yield point, also deteriorates under these conditions, affecting pressure management within the wellbore. Without accurate modeling of the pressure profile, both casing and cement sheaths may become structurally compromised, leading to wellbore collapse (Ravi et al., 2003). At temperatures above 450 °F, cement hydration can occur rapidly, often forming porous structures where the tobermorite phase suffers from strength retrogression (Gherardi et al., 2012). Understanding the thermal profile of a well is critical when designing for HPHT conditions. As depth increases, so does the temperature, often exceeding the tolerance limits of standard tools and materials. The illustration in Fig. 18.2 contrasts conventional and HPHT drilling environments, showing the narrowing margin for safe operation as bottomhole temperatures approach or exceed 300°F (150°C). In HPHT settings, equipment such as BHAs becomes limited in function or may fail altogether, while wellbore cement and casing materials face significant risks of degradation or collapse. Such visualizations underscore the importance of temperature control strategies and specialized materials in preserving well integrity under extreme subsurface conditions. Operating under these combined extreme conditions poses severe challenges for well integrity:

- **Material Performance Limitations**:

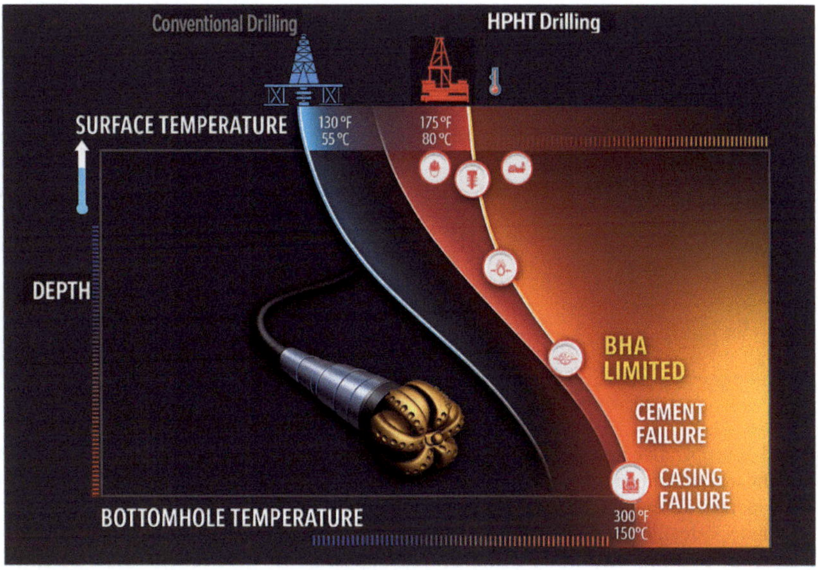

Fig. 18.2 Temperature-depth diagram comparing conventional drilling with HPHT drilling conditions. The illustration highlights surface and bottomhole temperature profiles, as well as potential failure zones for bottomhole assemblies (BHA), cement, and casing. As depth and temperature increase, HPHT wells exhibit limited operational envelopes and elevated risks of structural and thermal failure, necessitating advanced engineering and well temperature control technologies (Drill cool, 2025)

- Steel Strength Derating: The yield strength and tensile strength of conventional carbon and low-alloy steels decrease significantly at HPHT temperatures (as discussed in Chap. 4). Design calculations must use temperature-derated properties, often requiring the use of higher-strength steel grades (e.g., P110, Q125) or thicker walls to meet load requirements. This, in turn, can introduce other issues (e.g., increased susceptibility to SSC if H_2S is present at hardness levels associated with high strength). In HPHT wells, the selection of materials capable of withstanding elevated temperatures and high triaxial stresses is a critical component of well integrity. Most conventional casing steels, elastomers, and cementitious materials suffer from thermal degradation or strength loss beyond 150–200 °C. Fig. 18.3 provide a visual comparison of how different material classes and metallic alloys behave with increasing service temperature. These plots underscore the limitations of standard materials and justify the industry's shift toward high-performance alloys, composites, ceramics, and thermally stable elastomers in HPHT applications.
- Elastomer Failure: Most conventional elastomers degrade rapidly above 300-350°F (150-177°C) due to thermal degradation and chemical attack. HPHT wells demand specialized, expensive, high-grade elastomers (e.g., high-temperature rated FFKMs like Kalrez®, sometimes combined with ceramic

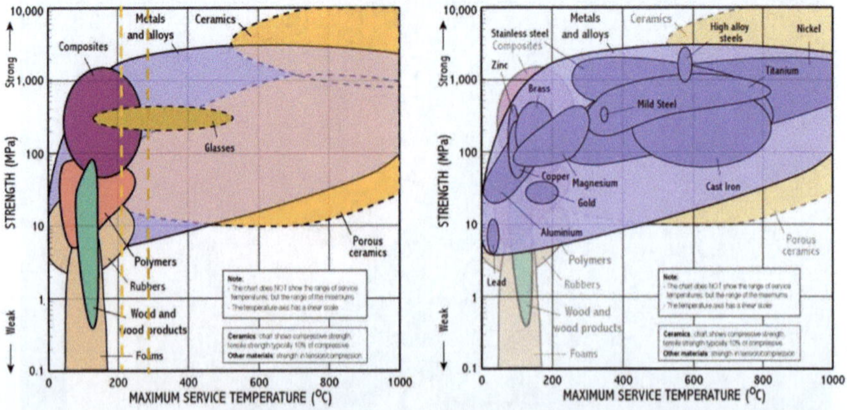

Fig. 18.3 Comparative material performance maps showing the relationship between strength and maximum service temperature. The left plot classifies broad material categories (e.g., polymers, rubbers, metals, ceramics), while the right plot zooms into specific metals and alloys such as mild steel, stainless steel, high-alloy steels, and nickel. These plots highlight the service temperature limitations of conventional oilfield materials and emphasize the need for advanced materials in HPHT environments (materials.eng, 2025)

fillers for added heat resistance) or increased reliance on metal-to-metal (MTM) seals for critical components like packers, hangers, wellheads, and valves. Even advanced elastomers have performance limits, reduced lifespan under extreme conditions, and are highly susceptible to Explosive Decompression (ED) or Rapid Gas Decompression (RGD) in high-pressure gas environments (Patel et al., 2019a, 2019b). Robust qualification testing of elastomers and seals under representative combined HPHT and fluid conditions is essential.

– Electronics Failure: Downhole electronic sensors, control systems, and logging tools have temperature limitations, making real-time monitoring and periodic diagnostics more challenging in very hot wells, although high-temperature electronics are under continuous development.

- **Seal Reliability**: Achieving reliable sealing with elastomers or even metal-to-metal seals under extreme pressure and temperature cycling is extremely difficult. Large differential thermal expansion between metal components (casing, tubing, wellhead) creates significant movements that seals must accommodate without leaking. MTM seals require very high contact stresses and precision manufacturing, and their reliability under extreme cyclic conditions is a focus of ongoing research and qualification efforts.
- **Complex Load Scenarios (Triaxial Stress)**: The combination of high internal pressure (burst loads), high temperature (causing thermal expansion stresses, potentially compression/buckling), and external pressure (collapse loads, geomechanical stress) results in complex triaxial stress states in casing and tubing.

Accurate modeling (e.g., using FEA, considering temperature-derated material properties) and triaxial stress analysis (Chap. 4) are essential for robust tubular design. Annular pressure buildup due to thermal effects in trapped fluids (Annular Pressure Build-up—APB) can be extreme in HPHT wells if annuli are sealed and contain liquid, and must be managed through specific design features (e.g., nitrogen caps, bleed/feed systems) or mitigation techniques to avoid over-pressuring and potential casing collapse or burst.

- **Corrosion**: High temperatures generally accelerate corrosion reactions (CO_2, H_2S) and can alter fluid chemistry. Material selection often necessitates expensive Corrosion Resistant Alloys (CRAs) for production tubing and potentially other components exposed to corrosive fluids. Predicting and monitoring corrosion under combined HPHT conditions and complex fluid chemistries is challenging and may require specialized models and monitoring techniques (Askari et al., 2021).
- **Cementing Challenges**: Designing and placing cement slurries that remain stable and provide long-term zonal isolation at HPHT conditions require specialized additives (e.g., very high amounts of silica flour to prevent strength retrogression at temperatures > 230°F / 110°C, effective high-temperature retarders to control thickening time, gas migration control additives designed for high temperatures) and meticulous execution. Thermal cycling poses a significant threat to long-term cement bond integrity and can induce cracking (Chaps. 10 and 21).
- **Drilling and Completion Complexity**: Managing well control (requiring high-pressure rated BOPs and well control equipment), drilling fluid properties (stability, density control, ECD management), and equipment ratings during drilling and completion operations is significantly more complex, time-consuming, and costly in HPHT environments (Yuan et al., 2013).

HPHT well integrity demands meticulous engineering design based on thorough understanding of material behavior, advanced modeling, rigorous qualification testing of equipment (especially seals and connections), and flawless execution during construction and operation. The reduced operating margin and potentially higher consequences necessitate a stringent WIMS and highly competent personnel.

18.3 Deepwater Well Integrity

Operating in deep water (typically defined as water depths >1000–1500 ft/300–460 m) introduces another set of integrity challenges distinct from onshore or shallow water operations, related to the water depth and the subsea environment (Chap. 20) (korkut, 1998).

- **High External Hydrostatic Pressure:** The deep seawater column exerts significant external pressure on conductor and surface casings, risers, and subsea equipment. This must be considered in the collapse design of these tubulars (Chap. 4), and connections must be rated for this external pressure.
- **Shallow Hazards:** Deepwater environments often feature shallow geohazards not typically encountered onshore or in shallow water, which can impact the structural integrity of the upper well sections(Williams & Pham, 2014). These include:
  - Shallow Water Flow (SWF): Overpressured, unconsolidated sand formations near the seabed that can flow uncontrollably into the wellbore during drilling if not properly managed (e.g., using low-density fluids, specialized drilling techniques, potentially setting a shallow hazard casing). This can lead to loss of structural support for conductor and surface casings and subsea wellheads if surrounding sediments are washed away or destabilized. Requires specialized drilling, casing, and cementing techniques (e.g., blind cementing with quick-setting slurries). Shallow water flows (SWFs) represent a critical hazard in deepwater environments, particularly when unconsolidated, overpressured sands near the mudline are penetrated during drilling. These formations can release large volumes of water and sand into the wellbore, displacing the drilling fluid and threatening well control. If not properly managed, SWFs can erode support around the drive pipe and wellhead system, destabilize surface casing, and pose risks to the riser. The following schematic (Fig. 18.4) visualizes a typical SWF scenario, emphasizing the severity and mechanism of this integrity challenge(Ostermeier et al., 2000; Ren et al., 2018).
  - Gas Hydrates: Ice-like solids formed from water and natural gas (usually methane) under high-pressure and low-temperature conditions near the seabed. Drilling through hydrate zones can cause them to dissociate (melt) due to frictional heating from the drill string or contact with warmer drilling fluids. This releases large volumes of gas, potentially leading to well control issues, destabilizing the formation and wellbore, and creating annular pressure or structural loading on casing. Requires careful temperature management during drilling and cementing across hydrate zones
- **Riser Integrity:** Wells drilled from floating platforms (Semi-submersibles, Drillships, Spars, TLPs) connect the subsea BOP or wellhead on the seabed to the surface facility via a drilling riser (temporary, for drilling) or a production riser (permanent, containing produced fluids). Maintaining the integrity of the riser (against internal pressure, external hydrostatic pressure, tension from surface support, fatigue from vessel motion and wave/current loading) and its connections is critical(Blondeau et al., 2014; Maheshwari et al., 2008; Neidhardt et al., 2017). Production risers contain produced fluids and are part of the primary containment envelope from the seabed to the surface facility (Chap. 20).
- **Subsea Equipment Complexity:** Subsea wellheads, trees, manifolds, and control systems are complex, remotely installed and operated systems located on the seabed (Chap. 20). Access for inspection, maintenance, and intervention is limited, time-consuming, and extremely expensive (requiring specialized vessels, ROVs,

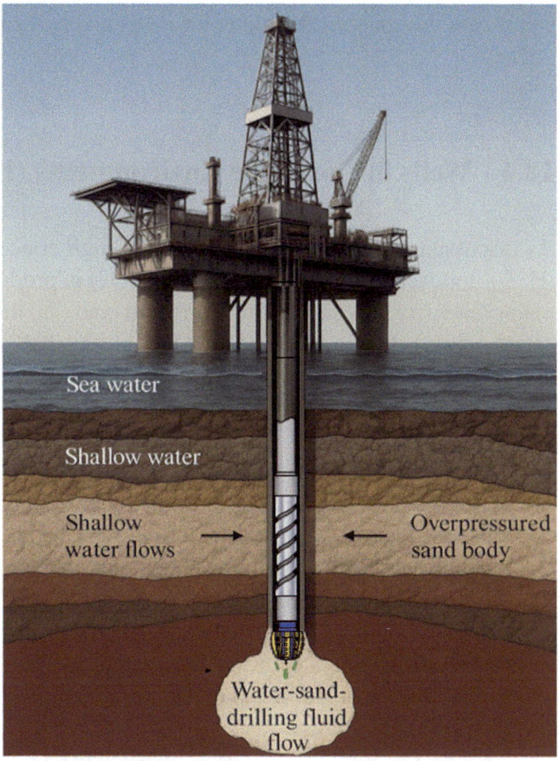

Fig. 18.4 Cross-sectional schematic showing a shallow water flow (SWF) blowout in deepwater drilling. The image illustrates water-sand-drilling fluid escape due to penetration into overpressure unconsolidated sand near the seafloor. These flows, if not controlled, can lead to destabilization around the drive pipe, wellhead, and conductor casing, potentially undermining the structural integrity of the upper well system

or drilling rigs). This places a high premium on getting the initial design and construction right, ensuring long-term reliability through extensive equipment qualification (e.g., API 17D PR2), and relying heavily on remote monitoring capabilities. Seal integrity in subsea connectors (e.g., tree-to-wellhead, flowline connections) is critical and challenging to verify remotely.

- **Logistical Challenges:** Mobilizing specialized rigs and equipment for intervention or repair in deepwater is significantly more complex and costly than onshore or shallow water operations. This reinforces the need for reliable design, robust preventative integrity management, and effective remote diagnostics.
- **Narrow Margin Between Pore Pressure and Fracture Gradient**: Often encountered in deepwater drilling, requiring precise management of drilling fluid density (Equivalent Circulating Density - ECD) to avoid both influx (well control issue) and losses/fracturing (integrity/well control issue at the casing shoe).

Deepwater operations require specialized vessel capabilities, advanced drilling and completion techniques, highly reliable subsea equipment, robust contingency planning, and extensive remote monitoring, all contributing to higher costs and increased complexity for integrity management. Dynamic loading and fatigue are significant design considerations for risers and subsea wellhead connectors (Jaculli, 2020).

18.4 Wells in Corrosive Environments (H_2S, CO_2)

As discussed in Chap. 9, the presence of high concentrations of hydrogen sulfide (H_2S—sour service) or carbon dioxide (CO2) in produced or injected fluids presents major integrity threats due to corrosion and, in the case of H2S, potential for catastrophic brittle fracture (SSC)(Yap et al., 2012). These challenges are often exacerbated by high temperatures or pressures.

- **Material Selection Challenges:**
  - Sour Service (H2S): Strict adherence to NACE MR0175/ISO 15156 is mandatory globally when H2S is present above defined partial pressure thresholds in the presence of water. This often restricts the use of high-strength carbon steels and necessitates specific heat treatments (e.g., for L80, C90, T95 grades) or the use of qualified Corrosion Resistant Alloys (CRAs—e.g., stainless steels, duplex stainless steels, nickel alloys) for production tubing, casing, and wellhead components exposed to sour fluids. Material traceability and stringent quality control during manufacturing are critical to ensure compliance and avoid SSC failures.
  - High CO_2: Requires materials resistant to CO_2 corrosion, potentially involving higher chromium steels (e.g., 13Cr, Super 13Cr) or other CRAs, especially at elevated temperatures and pressures. Accurate corrosion modeling is needed to predict corrosion rates and select the most cost-effective suitable material (Askari et al., 2021). Carbon steel may be usable with robust continuous inhibition programs within defined limits, but CRAs are often required for long-term reliability in severe CO_2 environments.
- **Cost Implications**: CRAs are significantly more expensive than carbon steel, driving up well construction and completion costs substantially in corrosive fields. CRA-lined pipe or clad components offer a compromise but introduce complexities related to weld integrity and liner damage susceptibility during installation or intervention.
- **Seal Compatibility**: Elastomers used in packers, seals, and valves must be carefully selected for resistance to both H2S and CO_2, as well as other fluid components (chlorides, organic acids, inhibitors), at operating temperature and pressure. This often necessitates high-grade FFKM or specialized HNBR/FKM variants (Patel et al., 2019a, 2019b). Compatibility testing under representative conditions is vital.
- **Inhibition Challenges**: While chemical inhibition is used to mitigate internal corrosion, achieving effective inhibition in highly corrosive environments, especially under deposits, in complex flow regimes, or at very high temperatures,

can be difficult and requires continuous monitoring of effectiveness (Chap. 9). Inhibitor selection must consider compatibility with CRAs and elastomers.

- **Cement Degradation:** Both CO_2 (carbonation) and H_2S (acid attack, potentially sulfate formation via oxidation/MIC) can degrade Portland cement over time, necessitating careful slurry design (e.g., use of pozzolans) or consideration of alternative cements for long-term zonal isolation in corrosive environments (Alberdi-Pagola & Fischer, 2023; Kiran et al., 2017b).
- **Safety:** H2S is extremely toxic and flammable. Managing wells with high H_2S concentrations requires extensive safety measures, including specialized detection equipment, breathing apparatus, stringent operating procedures, robust well control protocols, and comprehensive emergency response plans, all intrinsically linked to maintaining primary and secondary containment integrity.

Integrity management in highly corrosive environments relies heavily on upfront material selection based on a thorough understanding of the fluid chemistry and operating conditions throughout the well's life, supplemented by robust corrosion management programs (inhibition, monitoring, inspection).

18.5 Geomechanical Challenges

The interaction between the wellbore tubulars (casing, tubing) and the surrounding rock formations can impose significant and sometimes unpredictable loads on casing and cement, potentially leading to integrity failures, especially in environments with unusual geomechanical behavior (Alsubaih, 2016, 2017, 2018a, 2018b, 2020, 2025).

- **Salt Formations (Creep)**: Thick salt layers are often encountered in sedimentary basins, particularly in regions like the Gulf of Mexico, Brazil, and West Africa. Salt behaves plastically (creeps or flows) under differential stress over time. This creep exerts non-uniform, time-dependent external pressure on the casing string passing through it, which can lead to severe casing deformation (e.g., ovalization, pinching) and potentially collapse if not properly designed for. Cementing effectively within creeping salt is also challenging, as the salt movement can stress or damage the cement sheath. Design requires specialized geomechanical modeling (often FEA) to predict salt behavior and the resulting loads, and often involves using thick-walled casing with high collapse resistance, potentially combined with engineered annulus materials or designs to mitigate stress transfer. Salt formations pose a unique set of geomechanical challenges due to their viscoelastic behavior and solubility. Unlike brittle rocks, salt exhibits time-dependent plastic deformation known as salt creep, which can lead to casing ovalization, compression failure, or collapse if not properly accounted for. Moreover, salt is highly susceptible to chemical leaching by drilling fluids, which can result in borehole enlargement and further mechanical destabilization. Fig. 18.5 presents a visual overview of the integrity risks associated with drilling and casing through salt formations. On the left, it depicts typical problems such as ledges impinging on

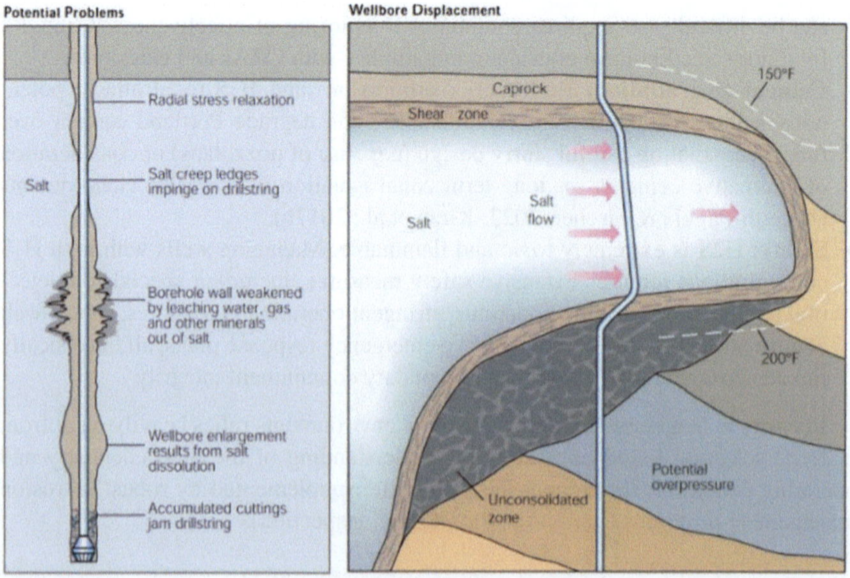

Fig. 18.5 Geomechanical challenges in salt formations. (Left): Common wellbore integrity problems in salt zones, including salt creep, borehole enlargement from dissolution, stress relaxation, and drillstring jamming from accumulated cuttings. (Right): Salt flow-induced wellbore displacement, showing the lateral movement of salt against casing within a surrounding unconsolidated zone and temperature gradients that exacerbate the mechanical stress. These phenomena necessitate specialized casing design and salt behavior modeling for long-term well integrity adapted from (Farmer, 1996)

the drillstring, radial stress relaxation, and dissolution-driven borehole instability. On the right, the diagram illustrates how salt movement can displace the wellbore over time, exerting lateral stress that compromises cement bonds and casing alignment, especially when adjacent to unconsolidated or overpressured formations (Farmer, 1996). Table 18.1 summarizes widely used constitutive relations that mathematically describe the creep strain rate, strain evolution, and damage accumulation in rock salt:

- **Compacting Reservoirs (Subsidence)**: Significant depletion of reservoir pressure during production can cause the reservoir rock matrix to compact (AL-Aamri et al., 2019). This compaction translates to subsidence at the surface and, crucially, imposes high compressive axial loads (due to friction/drag forces as the formation moves relative to the pipe) on the casing strings passing through or near the compacting interval, particularly at interfaces between compacting and non-compacting zones. This can lead to connection failures (compression), buckling, or shear failure of the casing string. Design requires careful assessment of potential compaction/subsidence (e.g., using reservoir simulation integrated with geomechanics) and incorporating features like specialized slip joints in the casing string

Table 18.1 Constitutive relations for creep strain rate, creep strain, and damage evolution in rock salt, relevant to wellbore integrity in salt formations (Ramesh Kumar et al., 2021)

Model	Formulation	Application/ Experiment
Power law	$\dot{\varepsilon}_{cr} = A\exp\left(\frac{-Q}{RT}\right)\sigma^n$	Generic creep behavior under differential stress
Hou/Lux	$\dot{\varepsilon}_{cr} = \frac{3}{2}\left[\frac{1}{\eta_k(\sigma_v)}\left(1 - \frac{\varepsilon_{cr}}{\varepsilon_{max}(\varepsilon_{cr})}\right) + \frac{1}{\eta_m(\sigma_v, T)}\right] \times s_{ij}$	Secondary and tertiary creep in rock salt
MD model	$\dot{\varepsilon}_{cr} = A_1\exp\left(-\frac{Q_1}{RT}\right)\left[\frac{\sigma}{\mu(1-\omega)}\right]^{n_1} + A_2\exp\left(-\frac{Q_2}{RT}\right)\left[\frac{\sigma}{\mu(1-\omega)}\right]^{n_2}$	Steady-state creep incorporating damage evolution
Hoek-brown	$\dot{\varepsilon}_{cr} = \frac{s-D_s\varepsilon_{cr}}{1-D_b}\left(\frac{1}{2G_1} + \frac{1}{2G_2}\left(1 - e^{-G_2\eta}\right)\right) + \frac{1}{2\eta_2}$	Yield criteria and viscoplastic creep in triaxial tests
Kachanov law	$\dot{D} = A\sigma^\nu(1 - D)^\nu$	Uniaxial compression creep damage evolution
Modified Young's modulus	$D = \frac{\varepsilon_f - \varepsilon_0}{\varepsilon_f - \varepsilon_0}$	Stiffness degradation under cyclic loads
Damage and Fatigue variable	$D = k_1 D_c^p + k_2 D_f^p$	Combined creep and fatigue damage

to accommodate movement, or designing casing strings to withstand the anticipated compressive loads (Alsubaih et al., 2020; Kusolsong et al., 2023). Table 18.2 provides a comprehensive summary of the primary approaches and formulas used to evaluate subsidence risks, presenting both empirical and theoretical methods.

- **Unstable Formations (e.g., Reactive Shales, Unconsolidated Sands):** Drilling through formations prone to swelling (shales), sloughing, or collapse requires careful drilling fluid design and casing programs to maintain wellbore stability and ensure successful casing running and cementing, which are foundational for establishing initial integrity (Chap. 5). Failure to manage wellbore stability can lead to stuck pipe, poor cement jobs (channels, voids), or damage to casing during running.
- **Fault Reactivation:** Changes in reservoir pressure or stress field due to production or injection (e.g., waterflooding, CO2 injection, waste injection) can alter the stress state around existing geological faults, potentially causing them to slip (reactivate) (Ben Abdallah et al., 2022). If a wellbore crosses a fault that reactivates and undergoes significant shear displacement, the resulting movement can shear or severely deform the casing string, leading to a complete loss of integrity. Avoiding drilling through potentially active faults or designing casing systems with some degree of compliance or enhanced strength at fault crossings are potential mitigation strategies. Geomechanical modeling is used to assess fault reactivation risk. Horizontal well orientation significantly influences the risk of

Table 18.2 Subsidence prediction approaches and key formulas

Approach	Formula/Equation	Description and key features	Key references/Notes
Empirical methods	– Quadratic: $s = ax^2 + bx + c$ – Linear: $s = ax + b$ – Exponential: $s = ae^{bx}$ – Logarithmic: $\log s = \log a + b\log x$	Extrapolating historical data of subsidence vs. time. Relies on fitting data trends without directly considering causative physical processes. Often used for quick, approximate predictions when historical data is available	(Donald C. Helm, n.d.)
Semi-theoretical approach	– **Wadachi Model**: $\frac{ds}{dt} = k(p_0 - p)$ – **Yamaguchi Model**: $\frac{ds}{dt} = ks_C(p_0 - p)\left(\frac{dp}{dt}\right)\exp[-k(p_0 - p)t]$ – **Subsidence vs Liquid Withdrawal**: $s = aQ + b$ – **Ratio of Subsidence to Liquid Withdrawal**: $s = \frac{C}{m_V}$ – **Subsidence/Head-Decline Ratio**: $\frac{\Delta b}{\Delta p'}$	These models link subsidence to water-level decline or fluid withdrawal. For example, Wadachi's model proposes that the rate of subsidence is proportional to the difference between actual and reference water levels. Others link subsidence directly to cumulative fluid withdrawal or volumetric compressibility, providing semi-empirical insights into subsidence prediction	(Donald C. Helm, n.d.)

(continued)

Table 18.2 (continued)

Approach	Formula/Equation	Description and key features	Key references/Notes
Theoretical approach	– **Compressibility Relationship**: $\Delta H = m_V \Delta p' H_0$ – **Total Subsidence (Multiple Strata)**: $\Delta H = \sum H_{0i} m_{vi} \Delta P'_i$ – **Ground-Water Flow (Aquifer)**: $\frac{\partial^2 h}{\partial x^2} + \frac{\partial^2 h}{\partial y^2} = \frac{S}{T}\frac{\partial h}{\partial t}$ – **Consolidation (Aquitard)**: $\frac{\partial^2 h}{\partial z^2} = \frac{1}{\nu}\frac{\partial h}{\partial t}$ – **Geertsma's Model (Compaction & Overburden)**: $U_z(0,0) =$ $-2(1-\nu)c_m b \Delta p\left(1 - \frac{C}{(1+C)^2}\right)$	These are physically-based models grounded in soil mechanics and groundwater flow. They incorporate compressibility of soils (via consolidation tests), effective stress principles, and the coupling of aquifer/aquitard responses. Geertsma's model further accounts for the effect of overburden on surface subsidence. These models allow for more precise, site-specific predictions but require extensive site data (e.g., soil compressibility, hydraulic parameters)	(Donald C. Helm, n.d.)
Depth-porosity model	– **Athy's Exponential Model**: $n = n_0 e^{-cz}$ – **Dickinson's Logarithmic Model**: $n = n_{ref} - a\ln\frac{z}{z_{ref}}$ – **Specific Storage from Porosity Gradient**: $S'_{skv} = \frac{c}{(1-n)^2(G-1)} n$ (or alternative form using a from Dickinson's model)	Empirical models linking porosity decrease with depth (compaction equilibrium). These relationships provide site-specific estimates of specific storage and compaction potential, especially useful in areas with limited direct subsidence data	(Donald C. Helm, 1983)

(continued)

Table 18.2 (continued)

Approach	Formula/Equation	Description and key features	Key references/Notes
Aquitard-drainage model	– **One-Dimensional Consolidation Equation** (Terzaghi): $\frac{\partial^2 h}{\partial z^2} = \frac{1}{\nu}\frac{\partial h}{\partial t}$ – **Computer-based simulation of residual compaction**	A dynamic simulation approach that predicts time-lagged compaction within aquitards embedded in confined aquifers. Uses consolidation theory to model the delayed response of compressible layers to changing hydraulic head in adjacent aquifers. This method provides time-dependent subsidence predictions, essential for areas with complex geohydraulic systems	(Donald C. Helm, 1983)
Elastic uniaxial compaction	– **Basic compaction formula:** $\Delta h = \frac{(1-2\nu)}{(1-\nu)}\frac{\Delta p}{E}h$ – **Coefficient of uniaxial compaction:** $C_m = \frac{\Delta h}{\Delta p \cdot h}$	Commonly used for **uniaxial compaction** (vertical-only strain) scenarios in reservoirs (e.g., depletion zones). Often used in the petroleum industry for first-order estimates of compaction	("Chapter 12 Reservoir Compaction. Subsidence and Well Problems," 1992)
Nucleus of strain method	– **Geertsma's integral-based model:** Vertical displacement $w(r)$ at the surface is derived by integrating strain "nuclei" throughout the compacting reservoir: $w(r) = \int_V \frac{(1-2\nu)}{8\pi(1-\nu)G}\frac{\Delta p \cdot dV}{r}$ – Analytical disk-shaped solutions use Bessel function integrals A and B	Widely used for predicting **subsidence bowls** above compacting reservoirs; accounts for **three-dimensional** strain distribution. Geertsma's approach has been refined and extended to consider more complex reservoir geometries and overburden heterogeneity	("Chapter 12 Reservoir Compaction. Subsidence and Well Problems," 1992)

(continued)

Table 18.2 (continued)

Approach	Formula/Equation	Description and key features	Key references/Notes
Radial disk-shaped aquifer model	– **Gravanis & Sarris Analytical Model:** For **deep aquifer (D≫R)**: $w(r) = -\frac{(1-2\nu)\alpha Q}{2\pi GD^2}\left(1 + \frac{r^2}{D^2}\right)^{-3/2}$ – **At well location (r = 0):** $w_0 = -\frac{(1-2\nu)\alpha Q}{2\pi GD^2}$ – **Key dimensionless group:** $\frac{r}{D}$ (Viscosity/permeability-independent in deep aquifers)	New **closed-form analytical solutions** for disk-shaped confined aquifers. Provides **time-dependent subsidence profiles** for deep vs. shallow aquifers (radial dependence). Shows that in deep aquifers, subsidence is independent of fluid viscosity and aquifer permeability. Useful for fast, physically based predictions of **radial subsidence** in **homogeneous aquifers**	(Gravanis & Sarris, 2023)
Finite element method (FEM)-based extensions	– **General parametric formulation:** $\Delta h_c = C_1 \Delta p$, Subsidence $U = C_3 \Delta p$, $\Delta\sigma_{\text{vertical}} = C_6 \Delta p$, etc. – C_i factors are tabulated as functions of dimensionless groups: $\frac{h}{r}, \frac{D}{r}, \frac{E_r}{E_c}$	Quick estimation method for **subsidence, compaction, and in-situ stress** changes due to pore pressure changes, accounting for modulus contrast between reservoir and caprock. Faster than full FEM runs while retaining physical insight. Particularly useful in early reservoir evaluation	(Morita et al., 1989)

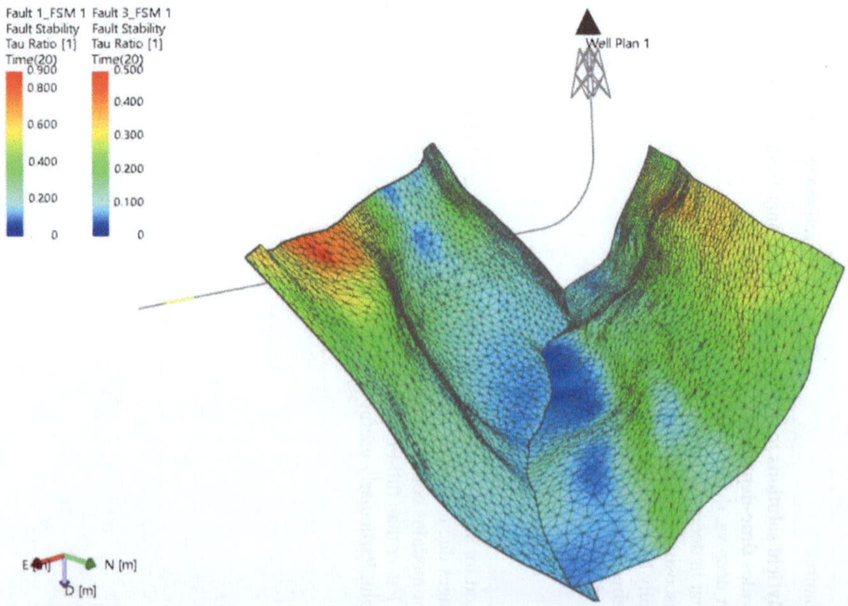

Fig. 18.6 Fault stability map showing Tau ratio distribution on Fault-1 and Fault-3 with a horizontal well azimuth of 90°. At this azimuth, Fault-1 exhibits a maximum τ of 0.9, while Fault-3 shows a maximum τ of 0.5, indicating both faults remain in a stable condition. The analysis demonstrates that extreme azimuths (0° and 90°) reduce the fault slip tendency, enhancing safe injection potential in this geological setting (adapted from (Saber et al., 2023))

fault reactivation during CO_2 injection, especially in faulted reservoirs. The Tau ratio —defined as the ratio of applied shear stress to critical shear strength—serves as a quantitative measure for assessing fault slip potential. Values approaching or exceeding 1 indicate critically stressed or unstable faults. Fig. 18.6 illustrates the Tau ratio distribution on two modeled faults (Fault-1 and Fault-3) for a well azimuth of 90°, as part of a parametric study exploring azimuthal impact. The simulation reveals that at this azimuth, both faults exhibit τ values below 1 (0.9 and 0.5, respectively), signifying stability (Saber et al., 2023). This result supports the strategic placement of injection wells away from fault strike intersections to minimize reactivation risk. Azimuthal sensitivity analyses like this are crucial for optimizing CO_2 sequestration well design and ensuring long-term containment integrity. Table 18.3 compiles the most widely used mathematical criteria and formulas for evaluating the reactivation of geological faults. Each criterion provides insight into the mechanical conditions leading to fault slip, such as shear strength, principal stress interactions, or the influence of pore pressure. The variables involved are fundamental in fault stability analysis, and the references listed provide detailed derivations and applications of these principles.

Table 18.3 Summary of mathematical formulas for fault reactivation

Criterion/Formula name	Mathematical formula	Description and conditions for reactivation	Variable definitions	Key references
1. Mohr-Coulomb Failure Criterion	$\tau \geq C + \mu \cdot \sigma'_n$ where $\sigma'_n = \sigma_n - P_p$	Reactivation occurs when the shear stress (τ) acting on the fault exceeds the shear strength (cohesion C plus frictional resistance $\mu \cdot \sigma'_n$). For pre-existing faults, C is often assumed zero	τ: Shear stress on the fault plane C: Cohesion (or S_0) μ: Coefficient of static friction ($\mu = \tan(\varphi)$) φ: Angle of internal friction σ_n: Normal stress P_p: Pore pressure	(Fossen Frontmatter, 2016; Jaeger et al., 2009)
2. Slip Tendency	$T_s = \frac{\tau}{\sigma'_n} = \frac{\tau}{\sigma_n - P_p}$	A dimensionless ratio comparing the shear stress to the effective normal stress on the fault. Reactivation is likely when $T_s \geq \mu$. Useful for comparing stability of different faults under the same stress field	T_s: Slip tendency (Other variables as above)	(Morris et al., 1996; Zoback, 2007)

(continued)

Table 18.3 (continued)

Criterion/Formula name	Mathematical formula	Description and conditions for reactivation	Variable definitions	Key references
3. Reactivation in Principal Stresses (2D)	$\sigma_n = \frac{\sigma_1+\sigma_3}{2} + \frac{\sigma_1-\sigma_3}{2}\cos(2\theta)$ $\tau = \frac{\sigma_1-\sigma_3}{2}\sin(2\theta)$ Reactivation condition: $\frac{\sigma_1-\sigma_3}{2}\sin(2\theta) \geq C + \mu\left(\frac{\sigma_1+\sigma_3}{2} + \frac{\sigma_1-\sigma_3}{2}\cos(2\theta) - P_p\right)$	Mohr-Coulomb criterion expressed in terms of principal stresses and fault orientation (θ). Helps calculate if a fault with a given orientation will reactivate under a particular stress state and pore pressure	σ_1: Maximum principal stress σ_3: Minimum principal stress θ: Angle between fault normal and σ_1 direction (Other variables as above)	(Jaeger et al., 2009; Zoback, 2007)
4. Critical Pore Pressure for Reactivation	$P_{p,\text{crit}} = \sigma_n - \frac{\tau - C}{\mu}$	Calculates the minimum pore pressure needed to induce slip on a given fault. Key for assessing induced seismicity due to fluid injection (e.g., wastewater disposal, hydraulic fracturing)	$P_{p,\text{crit}}$: Critical pore pressure (Other variables as above)	Zoback (2007)

Addressing geomechanical challenges requires integrating geological and rock mechanics understanding into the well design process, often utilizing advanced modeling techniques (e.g., FEA) to predict formation behavior and its impact on the wellbore, and selecting appropriate casing designs, materials, and cementing strategies to withstand the anticipated loads and movements. Monitoring techniques like strain sensing (DSS) or periodic caliper logging can help detect deformation over time (Chaps. 24 and 30).

Managing well integrity in harsh environments demands a higher level of engineering rigor, specialized technologies, stringent material selection and qualification, robust risk assessment, and meticulous operational discipline due to the increased potential for failure and often higher consequences compared to conventional wells.

References

Ahmed, A., Hayfaa, S., Hussien, A., Maher, H., Madi, S., & Malak, A. (2021). *Extensive hole cleaning optimization to mitigate wellbore instability challenges in shale formations*. OnePetro. https://dx-doi-org.ezproxy.lib.utexas.edu/

AL-Aamri, M., Mahajan, S., & Mukhaini, H. (2019). Geomechanical assessment of compaction related well integrity risks for a large field in sultanate of Oman. In *Society of petroleum engineers—Abu Dhabi international petroleum exhibition and conference 2019, ADIP 2019*. https://doi.org/10.2118/197308-MS

Alberdi-Pagola, P., & Fischer, G. (2023). Review of integrity loss detection and quantification due to cracking in cemented wells. *SPE Journal, 965*. http://onepetro.org/SJ/article-pdf/28/03/965/3120397/spe-214303-pa.pdf

Alsubaih, A. A., & Nygaard, R. (2016). *Shale instability of deviated wellbores in Southern Iraqi fields*. OnePetro. https://dx.doi.org/

Alsubaih, A. A., Dahm, H. H., & Alsubaih, I. (2017). *History matching, sensitivity and uncertainty analysis to improve wellbore stability model predictions*. OnePetro. https://dx-doi-org.ezproxy.lib.utexas.edu/

Alsubaih, A., Albadran, F., Abbood, N., & Alkanaani, N. (2018). *Optimization of the tripping parameters to prevent the surge related wellbore instability*. OnePetro. https://dx-doi-org.ezproxy.lib.utexas.edu/

Alsubaih, A., Albadran, F., & Alkanaani, N. (2018). Mechanical specific energy and statistical techniques to maximizing the drilling rates for production section of mishrif wells in southern Iraq fields. In *Proceedings of the SPE/IADC middle east drilling technology conference and exhibition, 2018-January*. https://doi.org/10.2118/189354-ms

Alsubaih, A. A., Hayfaa, S., Alboshawi, H. S., Alaa, A., Al, lawe, Erfan, M., Sawdi, M., Al, A., & Lubna, H. (2020). *Reservoir geomechanics simulation to optimize reservoir management in Southern Iraq fields*. OnePetro. https://dx.doi.org/

Alsubaih, A., Sepehrnoori, K., & Delshad, M. (2025). Development and optimization of self-healing cement for CO_2 injection and storage wells: Enhancing long-term wellbore integrity in extreme subsurface conditions. *Applied Sciences, 15*(10), 5428. https://doi.org/10.3390/APP15105428

Askari, M., Aliofkhazraei, M., Jafari, R., Hamghalam, P., & Hajizadeh, A. (2021). Downhole corrosion inhibitors for oil and gas production—a review. *Applied Surface Science Advances, 6*. https://doi.org/10.1016/j.apsadv.2021.100128

Ben Abdallah, K., Souissi, S., Hamdi, E., & Pouya, A. (2022). *Geomechanical integrity of fractured reservoirs: hydro-mechanical coupling*. OnePetro. https://dx-doi-org.ezproxy.lib.utexas.edu/

Blondeau, S., Andre, M., Garnier, P., Benmedakhene, S., & Quesnel, J. J. (2014). Riser integrity monitoring system for offshore fields. *All Days*. https://doi.org/10.4043/24791-MS

Chapter 12 Reservoir compaction. Subsidence and well problems. (1992). *Developments in Petroleum Science, 33*(C), 287–317. https://doi.org/10.1016/S0376-7361(09)70198-0

Donald C. Helm. (n.d.). *Guidebook to studies of land subsidence due to ground-water withdrawal*. Retrieved June 6, 2025, from https://wwwrcamnl.wr.usgs.gov/rgws/Unesco/PDF-Chapters/Chapter5.pdf

Drill cool. (2025). *Well temperature control|drill cool systems*. https://drillcool.com/well-temperature-control/

Farmer, P., M. D., P. A., R. J. & W. R. (1996). *Exploring the subsalt. Oilfield Review*. https://www.slb.com/~/media/Files/resources/oilfield_review/ors96/spr96/ors96_salt_p50_64.pdf

Fossen Frontmatter, H. (2016). *Structural geology second edition*. Cambridge University Pres. www.cambridge.org

Gherardi, F., Audigane, P., & Gaucher, E. C. (2012). Predicting long-term geochemical alteration of wellbore cement in a generic geological CO2 confinement site: Tackling a difficult reactive transport modeling challenge. *Journal of Hydrology, 420–421*, 340–359. https://doi.org/10.1016/J.JHYDROL.2011.12.026

Gravanis, E., & Sarris, E. N. (2023). A new analytical method for calculating subsidence resulting by fluid withdrawal from disk-shaped confined aquifers. *Water,* 15, 3175. https://doi.org/10.3390/W15183175

Jaeger, J. C., Cook, N. G. W., & Zimmerman, R. Wayne. (2009). *Fundamentals of Rock mechanics* (1st ed.). Wiley-Blackwell. https://www.dawsonera.com:443/abstract/9781444308914

Kiran, R., Teodoriu, C., Dadmohammadi, Y., Nygaard, R., Wood, D., Mokhtari, M., & Salehi, S. (2017). Identification and evaluation of well integrity and causes of failure of well integrity barriers (A review). *Journal of Natural Gas Science and Engineering, 45*, 511–526. https://doi.org/10.1016/J.JNGSE.2017.05.009

kOrkut, Z. (1998). *Deepwater operations plan guideline*.

Kusolsong, S., Adisornsupawat, K., & Vardcharragosad, P. (2023). Numerical simulation for compaction and subsidence during production period for a large carbonate gas field. *International Petroleum Technology Conference, IPTC 2023*. https://doi.org/10.2523/IPTC-23015-EA

Maheshwari, H., Ruf, W., & Walters, D. (2008). *Riser integrity monitoring techniques and data processing methods*. OnePetro. https://dx.doi.org/

materials.eng. (2025). *Strength—Max. service temperature*. http://www-materials.eng.cam.ac.uk/mpsite/interactive_charts/strength-temp/NS6Chart.html

Morita, N., Whitfill, D. L., Nygaard, O., & Bale, A. (1989). A quick method to determine subsidence XE "Subsidence", reservoir compaction, and in-situ stress induced by reservoir depletion. *Journal of Petroleum Technology, 41*(01), 71–79. https://doi.org/10.2118/17150-PA

Morris, A., Ferrill, D. A., & Henderson, D. B. (1996). Slip-tendency analysis and fault reactivation. *Geology, 24*(3), 275–278. https://doi.org/10.1130/0091-7613(1996)024%3c0275:STAAFR%3e2.3.CO;2

Neidhardt, D., Ziegler, R. F., & Lancaster, J. (2017). Drilling riser integrity assurance for deepwater XE "Deepwater" floating drilling. *Proceedings of the Annual Offshore Technology Conference, 1*, 50–66. https://doi.org/10.4043/27945-MS

Ostermeier, R. M., Pelletier, J. H., Winker, C. D., Nicholson, J. W., Rambow, F. H., & Cowan, K. M. (2000). Dealing with shallow-water flow in the deepwater XE "Deepwater" Gulf of Mexico. *Proceedings of the Annual Offshore Technology Conference, 1*, 75–86. https://doi.org/10.4043/11972-MS

Patel, H., Salehi, S., Ahmed, R., & Teodoriu, C. (2019a). Review of elastomer seal assemblies in oil & gas wells: Performance evaluation, failure mechanisms, and gaps in industry standards. *Journal of Petroleum Science and Engineering, 179*, 1046–1062. https://doi.org/10.1016/j.petrol.2019.05.019

Patel, H., Salehi, S., & Teodoriu, C. (2019). Assessing mechanical integrity of expanding cement. In *Society of petroleum engineers—SPE Oklahoma City Oil and Gas Symposium 2019, OKOG 2019*. https://doi.org/10.2118/195225-MS

Paula, R. R., Ribeiro, P. R., & Santos, O. L. A. (2009). HPHT XE "HPHT" drilling—new frontiers for well safety. *SPE/IADC Drilling Conference, Proceedings, 2*, 1185–1197. https://doi.org/10.2118/119909-MS

Ramesh Kumar, K., Makhmutov, A., Spiers, C. J., & Hajibeygi, H. (2021). Geomechanical simulation of energy storage in salt formations. *Scientific Reports, 11*(1), 19640. https://doi.org/10.1038/S41598-021-99161-8

Ravi, K., Bosma, M., & Hunter, L. (2003). *Optimizing the cement sheath design in HPHT shearwater field*. https://doi.org/10.2118/79905-MS

Ren, S., Liu, Y., Gong, Z., Yuan, Y., Yu, L., Wang, Y., Xu, Y., & Deng, J. (2018). Numerical simulation of water and sand blowouts when penetrating through shallow water flow formations in deep water drilling. *Journal of Ocean University of China, 17*(1), 17–24. https://doi.org/10.1007/S11802-018-3454-5/METRICS

Saber, E., Qu, Q., Aminossadati, S. M., Li, J., & Chen, Z. (2023). A new coupled approach to simulating fault reactivation during CO_2 sequestration. In *Society of petroleum engineers—Asia pacific unconventional resources symposium, APUR 2023*. https://doi.org/10.2118/217287-MS

Shadravan, A., & Amani, M. (2012). HPHT 101—what every engineer or geoscientist should know about high pressure high temperature wells. *Society of Petroleum Engineers—Kuwait International Petroleum Conference and Exhibition 2012, KIPCE 2012: People and Innovative Technologies to Unleash Challenging Hydrocarbon Resources, 2*, 917–943. https://doi.org/10.2118/163376-MS

SLB. (2025). *The defining series: HPHT Wells|SLB*. https://www.slb.com/resource-library/oilfield-review/defining-series/defining-hpht

Williams, M., & Pham, M. (2014). API 17TR12—consideration of external pressure in the design of deepwater XE "Deepwater" subsea equipment. *Proceedings of the Annual Offshore Technology Conference, 5*, 3479–3490. https://doi.org/10.4043/25416-MS

Yap, K. M., Srinivasan, S., & Lagad, V. V. (2012). *Evaluation of a CO2/H2S corrosion prediction model in multiphase oil/gas production systems*. OnePetro. https://dx-doi-org.ezproxy.lib.utexas.edu/

Yuan, Z., Schubert, J., Esteban, U. C., Chantose, P., & Teodoriu, C. (2013). *Casing failure mechanism and characterization under HPHT conditions in South Texas*. https://doi.org/10.2523/IPTC-16704-MS

Zoback, M. D. (2007). Reservoir geomechanics. *Reservoir Geomechanics*, 1–452. https://doi.org/10.1017/CBO9780511586477

Chapter 19
Well Integrity for Unconventional Resources

Abstract This chapter provides a comprehensive evaluation of well integrity challenges specific to unconventional resource development, particularly in shale gas and tight oil plays. The unconventional landscape is marked by long horizontal laterals, multi-stage hydraulic fracturing, and high-density well pads, introducing unique mechanical, thermal, and operational stresses on wellbore systems. The chapter explores the integrity risks posed by repeated high-pressure cyclic loading during fracturing, including casing fatigue, cement debonding, and annular micro-annuli formation, supported by numerical simulations and case studies. It details advanced mechanical models that quantify stress distributions in casing and cement, and presents fracture propagation equations to assess induced pressure effects. Special attention is given to cement sheath performance in horizontal wells, where suboptimal centralization, fluid migration, and thermal cycling pose severe threats to zonal isolation. The design of fracture-resistant and thermally stable cement systems, combined with robust placement techniques and verification methods (e.g., ultrasonic logs, CT imaging), is emphasized as critical for maintaining integrity throughout the well lifecycle. Fracture height containment is addressed through geomechanical modeling, real-time monitoring, and tailored stimulation design to prevent out-of-zone fracture growth, especially toward aquifers. The chapter also discusses long-term integrity threats such as corrosion, casing deformation, and parent–child well interference, all of which can compromise structural barriers if unmitigated. Finally, it outlines abandonment strategies for multi-stage horizontal wells, highlighting the complexity of ensuring permanent isolation in deviated sections. By integrating engineering models, operational practices, and field insights, this chapter provides a framework for adapting conventional well integrity principles to the demanding environment of unconventional resource development.

© The Author(s), under exclusive license to Springer Nature Switzerland AG 2026
A. Alsubaih and K. Sepehrnoori, *Well Integrity for Oil, Gas, and Subsurface Storage Operations*, https://doi.org/10.1007/978-3-032-15233-6_19

Diagram of hydraulic fracturing in the Marcellus Shale. *Source* Al Granberg/ProPublica (2011)

Chapter's Objectives:

- Identify the unique well integrity challenges associated with hydraulically fractured wells in unconventional reservoirs (shale gas/oil, tight formations).
- Discuss the impact of stimulation operations on casing and cement integrity.
- Outline methods for ensuring fracture height containment.
- Address long-term integrity considerations for unconventional wells.

19.1 Introduction: The Unconventional Landscape

The development of unconventional resources, primarily shale gas and tight oil, has revolutionized energy production in recent decades. These resources are typically extracted from low-permeability rock formations using a combination of horizontal drilling (long lateral sections) and high-volume, multi-stage hydraulic fracturing (Boak & Kleinberg, 2020). While sharing fundamental integrity principles with conventional wells (barrier philosophy, WIMS, basic equipment), the specific nature of unconventional well construction (long horizontal sections, high-rate drilling) and operation (high-pressure cyclic fracturing of multiple stages) introduces unique well integrity challenges that must be carefully managed to ensure safe and environmentally sound operations. The scale and density of development in some unconventional plays also amplify potential cumulative impacts and require careful management of interactions between wells.

Unconventional wells differ significantly in geometry, loading cycles, and operational intensity compared to conventional vertical wells. Long lateral wellbores intersect multiple zones, are subject to hundreds of high-pressure fracture stages, and often operate in tightly spaced multi-well pads. These factors impose complex mechanical and thermal stresses on the well casing, cement sheath, and annular barriers—frequently over prolonged periods (Patel & Salehi, 2019). Repeated high-pressure injection and rapid pressure bleed-off cycles during stimulation can degrade cement bonding and initiate casing fatigue, especially when coupled with long-term production-induced cycling, erosion, or corrosive reservoir fluids.

From cementing in horizontal sections with suboptimal centralization, to frac hits between parent and child wells, to abandonment complexity at the end of life, unconventional development introduces risks that demand proactive design and monitoring strategies. Moreover, the cumulative environmental exposure and regulatory scrutiny of unconventional operations further emphasize the need for robust zonal isolation, fracture containment, and lifecycle barrier assurance.

This chapter explores the integrity challenges unique to unconventional wells across five key dimensions: (1) stimulation-related mechanical stress, (2) cement design and placement issues, (3) fracture height containment strategies, (4) long-term fatigue, corrosion, and well interactions, and (5) specialized plug and abandonment considerations. Through technical examples and best practices, we aim to outline how industry can adapt conventional well integrity principles to this high-demand, high-risk operational domain.

19.2 Hydraulic Fracturing and Integrity Impacts

Hydraulic fracturing involves pumping large volumes of fluid (primarily water) mixed with proppant (sand, ceramic particles, or bauxite) into the wellbore at high pressure to create and propagate fractures in the target formation (e.g., shale, tight sand), enhancing permeability and allowing hydrocarbons to flow at economic rates. This process, particularly the high-pressure cycling, imposes significant stresses on the wellbore components below the isolation packer or plug, notably the production casing string and the surrounding cement sheath. Modern hydraulic fracturing operations in unconventional reservoirs rely on highly efficient, large-scale well pad designs that enable multiple wells to be drilled and stimulated from a single surface location. This not only reduces surface disruption but also maximizes lateral coverage within the shale formation. As shown in Fig. 19.1, the combination of dense horizontal well architecture and repeated high-pressure stimulation stages creates a complex network of induced fractures. Each well can consist of dozens of individually stimulated stages, increasing the mechanical and thermal loads on the casing, cement sheath, and surrounding formation. The visual integration of surface infrastructure and subsurface fracture propagation emphasizes the multidisciplinary engineering required to manage operational integrity and prevent zonal isolation failures or unintended fracture growth.

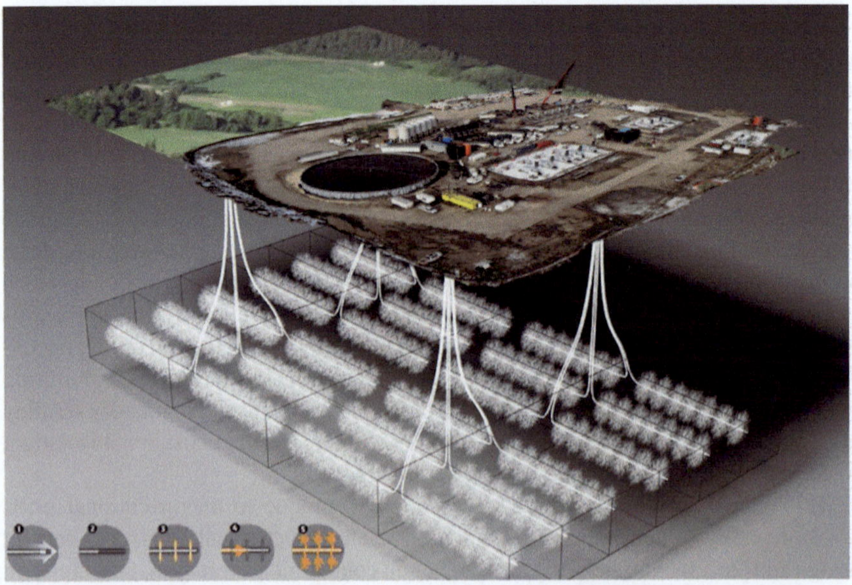

Fig. 19.1 Integrated 3D conceptual illustration of a multi-well unconventional hydraulic fracturing pad. The surface view shows an actual industrial fracking site with equipment for fluid storage, pumping, monitoring, and chemical injection. The subsurface section displays multiple horizontal wellbores and associated induced fracture networks extending laterally from each well. At the base, step-by-step icons summarize the key stages of the fracking process—from horizontal drilling to fracture propagation—highlighting the mechanical complexity and environmental footprint of high-density unconventional development

- **High Cyclic Pressures**: Fracturing operations typically involve repeated cycles of very high injection pressure (fracturing pressure, potentially exceeding 10,000 psi/69 MPa) followed by rapid pressure bleed-off. This cyclic pressure loading inside the production casing (or the temporary frac string used for stimulation) imposes significant fatigue stress on the casing body, connections, and crucially, the surrounding cement sheath (Guo et al., 2018; Liu et al., 2018; Zheng et al., 2024b).
  - Casing Fatigue: Requires careful selection of casing connections with proven fatigue resistance designed for cyclic loading applications (Hamilton & David Pattillo, 2019c; Liu et al., 2018; Zara et al., 2021). Fatigue life analysis should be part of the casing design phase, especially for wells anticipating many frac stages (some wells have over 50–100 stages) or potential future re-fracturing operations. Connection makeup quality is critical to achieve design fatigue life and sealability under cyclic stress. Hydraulic fracturing imposes severe internal pressure loads on the production casing, especially when combined with repeated cycling over multiple stimulation stages. The abrupt pressure fluctuations generate dynamic mechanical stresses—including radial expansion and axial elongation—that interact with pre-existing manufacturing or

operational imperfections. Figure 19.2 presents a numerical analysis of Von Mises stress distribution along a casing segment under internal pressures of 0, 60, and 120 MPa. As seen in the progression, stress concentrations shift and intensify with increasing pressure, particularly around geometric discontinuities such as bends or ovality zones. The maximum stress in the 120 MPa case surpasses 700 MPa, indicating the potential for localized yield or fatigue damage (Yang et al., 2023). This simulation underscores the critical need for fatigue-resistant casing materials and the importance of accounting for fracture-induced pressure loads in unconventional well design.

  - Cement Fatigue/Debonding: The pressure cycling causes the casing to repeatedly expand radially outwards under high internal pressure and contract when pressure is bled off. This movement is highly detrimental to the cement bond at the casing-cement interface, potentially initiating or propagating micro-annuli, and can induce fatigue cracking within the cement sheath itself, especially in conventional brittle cement formulations (Kiran, 2917, Alberdi-Pagola, 2023). This mechanical degradation over many cycles can compromise zonal isolation between frac stages along the lateral or from the annulus to other formations (Chap. 10).

- **Erosion**: Flowback of fracturing fluid containing proppant (sand or ceramic particles) immediately after stimulation or during subsequent production can cause severe erosion damage to the inside of the production casing, connections, perforations, or completion equipment near the perforations if not managed carefully (e.g., by controlling flowback rates, using erosion-resistant materials or flow control devices, implementing sand control measures where possible) (ElSebaee et al., 2020; Vincent et al., 2005). Erosion rates are highly dependent on flow velocity and proppant concentration.
- **Wellhead Pressure Limitations**: Surface wellhead equipment and the permanent Christmas tree must be rated to withstand the maximum anticipated treating pressures during fracturing, which can often exceed the well's normal operational pressure rating (Renpu, 2011b). Temporary, high-pressure frac heads or frac trees with higher pressure ratings (e.g., 15,000 psi or higher) are often installed above the permanent wellhead assembly during stimulation to manage these extreme pressures safely. The transition piece and seals between the frac head/tree and the permanent wellhead are critical temporary barriers.

Understanding the interplay between hydraulic fracturing pressures and well integrity is critical for predicting and preventing casing and cement failures, as well as for ensuring long-term well performance in unconventional wells. High-pressure, cyclic hydraulic fracturing induces significant mechanical stresses—including hoop, axial, and shear stresses—on the casing and cement sheath. Repeated pressure fluctuations, temperature changes, and mechanical loads can degrade these components over time. In addition to standard casing stress and fatigue formulas, advanced formulations such as the net pressure scaling, fracture width scaling, and tip pressure drop (Eqs. 19.1–19.3) are used to capture the complex mechanical interactions during

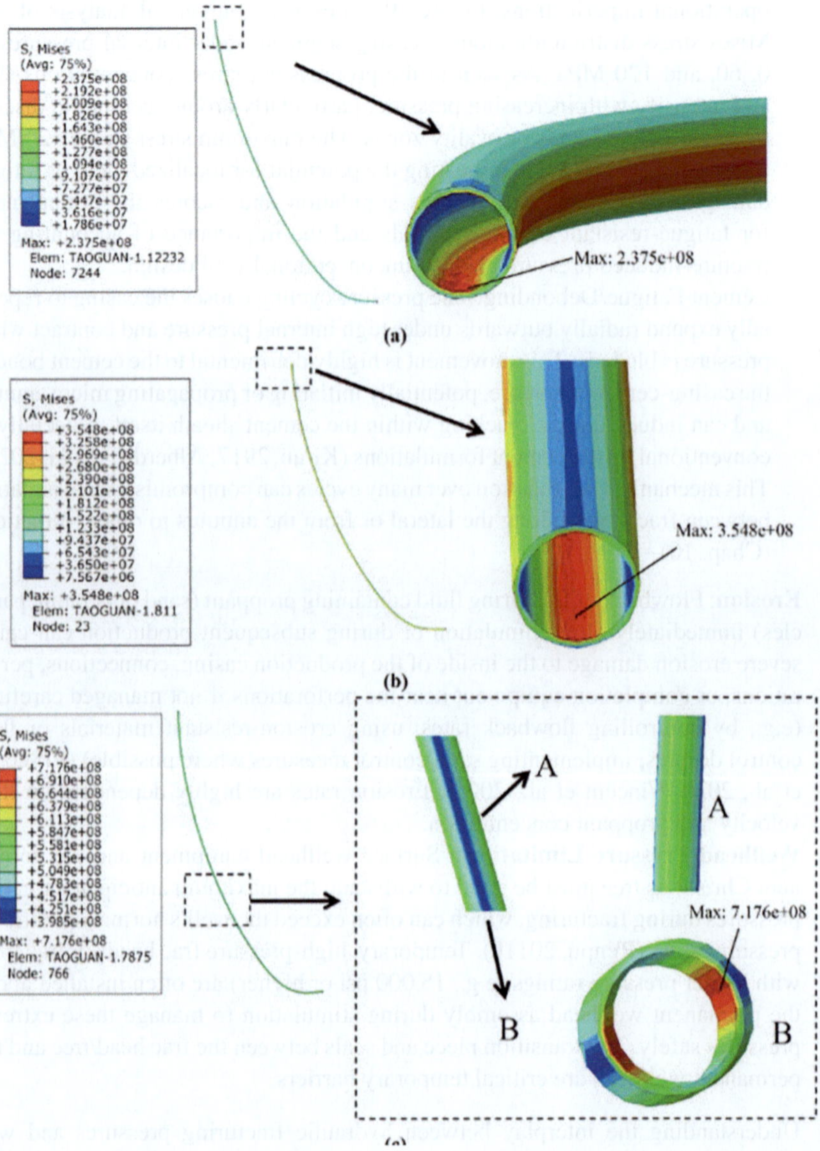

Fig. 19.2 Finite element simulation of casing stress distribution under varying internal pressures during hydraulic fracturing. **a** Stress profile under zero internal pressure, showing residual geometry-induced stress. **b** Stress concentration at 60 MPa internal pressure indicates the initial onset of hoop stress buildup. **c** Stress amplification at 120 MPa, demonstrating significant triaxial stress development and plastic deformation near casing ovalization zones. The maximum Von Mises stress observed exceeds 700 MPa, indicating potential integrity risk if not properly mitigated by design (Yang et al., 2023)

hydraulic fracturing. These equations consider the role of fracture toughness, fluid viscosity, injection rates, and mechanical properties, providing a more comprehensive framework for evaluating fracture-driven integrity risks. This section presents the key mathematical formulations used to evaluate these effects.

The net fracturing pressure is expressed as:

$$p_{\text{net}} = (p - \sigma_c) \propto \left[\frac{E'^4}{h^4} \left(\frac{\mu q x_f}{E'} \right) + \frac{K_{\text{lc-app}}^4}{h^4} \right]^{\frac{1}{4}} \tag{19.1}$$

where p_{net} is net pressure (Pa), p is fluid pressure (Pa), σ_c is confining stress (Pa), E' is plane strain Young's modulus (Pa), h = fracture height (m), μ is fluid viscosity (Pa·s), q is injection rate (m^3/s), x_f is fracture half-length (m), $K_{\text{lc-app}}$ is apparent fracture toughness (Pa·m^0.^{5}),

The corresponding fracture width scaling during hydraulic fracturing is:

$$w \propto \left[\left(\frac{\mu q x_f}{E'} \right) + \frac{K_{\text{lc-app}}^4}{E'^4} \right]^{\frac{1}{4}} \tag{19.2}$$

where:

- w = fracture width (m)
- (Other variables as above)

The pressure drop at the fracture tip region is estimated by:

$$\Delta p^* \propto \frac{E' C h_p}{h^2} \tag{19.3}$$

where:

- Δp^* = pressure drop (Pa)
- C = fluid leak-off coefficient (m/s^0.^{5})
- h_p = fracture tip height (m)
- (Other variables as above)

The casing string's mechanical response to internal and external pressure gradients during hydraulic fracturing can be described using classical thick-walled cylinder elasticity solutions (Lamé solutions). These formulations consider the casing as a thick-walled tube subjected to hydrostatic pressures from the wellbore fluids and the surrounding cement sheath, providing essential insights for assessing casing integrity under high-pressure fracturing and long-term cyclic loading.

Using the **Lamé solutions** for thick-walled cylinders, the radial, tangential, and axial stresses in the casing walls under pressure gradients are given by:

$$\sigma_{rr,\text{csg}} = \frac{p_w r_w^2 - p_{\text{cem}} r_{\text{csg}}^2}{r_{\text{csg}}^2 - r_w^2} - \frac{r_w^2 r_{\text{csg}}^2 (p_w - p_{\text{cem}})}{\left(r_{\text{csg}}^2 - r_w^2\right) r^2} \tag{19.4}$$

$$\sigma_{\theta\theta,\text{csg}} = \frac{p_w r_w^2 - p_{\text{cem}} r_{\text{csg}}^2}{r_{\text{csg}}^2 - r_w^2} + \frac{r_w^2 r_{\text{csg}}^2 (p_w - p_{\text{cem}})}{\left(r_{\text{csg}}^2 - r_w^2\right) r^2} \tag{19.5}$$

$$\sigma_{zz,\text{csg}} = \nu\left(\sigma_{rr,\text{csg}} + \sigma_{\theta\theta,\text{csg}}\right) \tag{19.6}$$

where:

- p_w = pressure inside the wellbore (Pa)
- p_{cem} = pressure outside casing (Pa)
- r_w = wellbore inner radius (m)
- r_{csg} = outer radius of casing (m)
- ν = Poisson's ratio of steel
- r = radial position (m)

19.3 Cement Integrity and Zonal Isolation in Multi-stage Fracturing

Achieving effective zonal isolation with the primary cement job behind the production casing or liner is particularly critical in long horizontal wells designed for multi-stage hydraulic fracturing. This cement sheath is a fundamental barrier for preventing communication between zones and ensuring fracture containment.

- **Isolating Frac Stages**: Mechanical isolation devices like bridge plugs (often composite, dissolvable, or drillable types) or swellable/mechanical packers are used sequentially to isolate individual stages in the lateral during fracturing. After each stage is fractured and isolated, the next stage is treated (Almulhim et al., 2023; Davis & McCrady, 2008). However, the integrity of the cement sheath in the annulus behind the casing provides the ultimate, long-term barrier preventing fracture fluid and pressure from communicating *between stages* along the lateral *outside the casing* or migrating vertically outside the target zone via the annulus. Poor cement bonding or channeling (unfilled sections in the annulus) can lead to inefficient stimulation (e.g., fracturing into a previously fractured stage via a channel, or fluid loss to unwanted zones behind the pipe) or, more critically, out-of-zone fracture growth and potential communication with overlying aquifers or underlying formations.
- **Challenges in Horizontal Cementing**: Achieving good mud removal, ensuring adequate casing centralization, and obtaining uniform cement placement in long, deviated, and horizontal sections (often thousands of feet in length) is inherently more difficult and complex than in vertical wells (Chap. 5). Gravity effects can cause cement particles to settle out in slurries, leading to density variations, and can

cause the casing to lie on the low side of the hole if not adequately centralized. This results in a non-uniform annular gap, making effective mud displacement difficult and potentially leaving mud channels on the low side. Techniques like extensive pipe rotation and/or reciprocation (moving the pipe up and down) if feasible and mechanically sound, specialized turbulent-flow or effective laminar flow spacer systems designed for horizontal geometries, optimized cement slurry rheology (e.e., low viscosity for turbulent flow potential, stable suspension properties), and stringent quality control during mixing and pumping are crucial for improving mud displacement and achieving a continuous, competent cement sheath.

- **Cement System Design**: Cement slurries for unconventional wells must be designed not only for successful placement in challenging horizontal geometries but also to withstand the severe mechanical stresses and cyclic loading imposed by fracturing. Thermal cycling induced by fracturing operations and subsequent shut-ins can significantly degrade the mechanical integrity of the cement sheath, particularly in horizontal wells subjected to repeated stimulation. The alternating radial expansion and contraction of the casing exerts fatigue stress on the cement, promoting the formation of microcracks and interfacial debonding (Vralstad et al., 2015). Figure 19.3 illustrates the evolution of such damage through high-resolution X-ray computed tomography. Initially isolated micro-defects (left) grow and connect under thermal cycling (right), forming continuous pathways that can allow fluid migration along the annulus. This damage mechanism highlights the importance of thermally stable cement systems and the need for advanced monitoring and modeling to ensure long-term zonal isolation in unconventional wells. This often involves designing for:

  - Improved Mechanical Properties: Enhanced flexibility (lower Young's modulus, making the cement less brittle) and toughness/ductility (using additives like polymers or fibers) to better resist cracking and debonding under cyclic loading from fracturing and subsequent production (Al-Yami et al., 2018b; Segre et al., 2006; Wray et al., 2009).
  - Good Bonding: Emphasis on achieving a strong, durable bond to both casing and formation is paramount to resist debonding during pressure cycling, crucial for maintaining zonal isolation along the lateral and vertical sections. Surface preparation (e.g., casing cleaning), adequate centralization, and effective displacement are key operational requirements for achieving this bond. During hydraulic fracturing operations in unconventional wells, the cement sheath integrity is highly sensitive to pressure cycling and the resulting mechanical stresses. Micro-annuli and interfacial debonding at the casing-cement interface can significantly compromise zonal isolation. To better understand these integrity challenges under different fracturing pressures, the morphology of the cement sheath was examined across varying pressure conditions. Figure 19.3 illustrates the microstructural evolution of the cement sheath at incremental internal pressures (70, 80, and 90 MPa), highlighting the progressive development of micro-annuli and debonding zones that can lead to pathways for fluid migration and integrity failure (Wang et al., 2024).

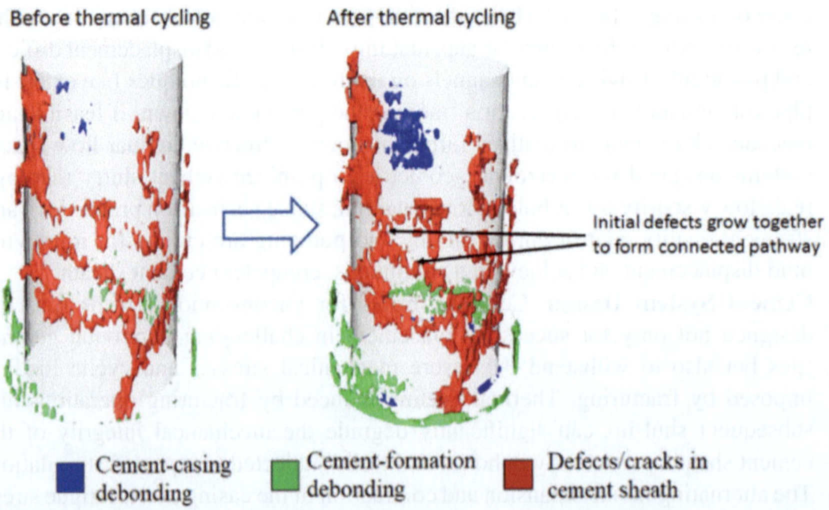

Fig. 19.3 3D X-ray CT visualization of internal defect evolution in the cement sheath before and after thermal cycling. Red regions represent internal cracks, green indicates cement-formation debonding, and blue shows cement-casing debonding. Post-cycling image (right) reveals the growth and coalescence of isolated micro-defects into connected fracture pathways, compromising zonal isolation. Image resolution: ~200 μm (adapted from (Vralstad et al., 2015))

– Zero or Minimal Free Fluid: Preventing free water separation in the slurry is essential to avoid creating weak points or channels, especially on the high side of highly deviated or horizontal sections where water can separate from the solid cement.

Cement sheath mechanical behavior can be quantitatively analyzed using plane axisymmetric, linear-elastic constitutive relations that describe the radial, tangential, and axial stress and strain responses of the cement sheath after setting. These equations are crucial for predicting potential micro-annulus formation, casing de-bonding, and cement fatigue cracking due to cyclic fracturing pressures. The following formulations describe these fundamental mechanical responses.

After cement setting, radial and axial stresses within the cement sheath under plane axisymmetric, linear-elastic behavior can be expressed as:

$$\begin{aligned}
\varepsilon_{rr,\text{cem}} &= \frac{\sigma_{rr,\text{cem}}}{E_\text{cem}} - \frac{\nu_\text{cem}}{E_\text{cem}}\left(\sigma_{\theta\theta,\text{cem}} + \sigma_{zz,\text{cem}}\right) + \varepsilon_\text{free} \\
\varepsilon_{\theta\theta,\text{cem}} &= \frac{\sigma_{\theta\theta,\text{cem}}}{E_\text{cem}} - \frac{\nu_\text{cem}}{E_\text{cem}}\left(\sigma_{rr,\text{cem}} + \sigma_{zz,\text{cem}}\right) + \varepsilon_\text{free} \\
\varepsilon_{zz,\text{cem}} &= \frac{\sigma_{zz,\text{cem}}}{E_\text{cem}} - \frac{\nu_\text{cem}}{E_\text{cem}}\left(\sigma_{\theta\theta,\text{cem}} + \sigma_{rr,\text{cem}}\right) + \varepsilon_\text{free}
\end{aligned} \tag{19.7}$$

where:

- $\varepsilon_{rr,\text{cem}}, \varepsilon_{\theta\theta,\text{cem}}, \varepsilon_{zz,\text{cem}}$ = radial, tangential, and axial strains
- $\sigma_{rr,\text{cem}}, \sigma_{\theta\theta,\text{cem}}, \sigma_{zz,\text{cem}}$ = radial, tangential, and axial stresses
- $E_\text{cem}, \nu_\text{cem}$ = Young's modulus and Poisson's ratio of cement

- $\varepsilon_{\text{free}}$ = free strain due to thermal or chemical effects

  The radial displacement of the cement sheath, u_r, is governed by:

$$\frac{\partial^2 u}{\partial r^2} + \frac{\partial u}{r \partial r} - \frac{u}{r^2} = 0 \tag{19.8}$$

with general solution:

$$u_r = Xr + \frac{Y}{r} \tag{19.9}$$

where X and Y are integration constants determined by boundary conditions.

The resulting stress components in the cement sheath are:

$$\sigma_{rr,\text{cem}} = \frac{E_{\text{cem}}}{1+\nu_{\text{cem}}}\left(\frac{X - \frac{Y}{r^2}}{1-2\nu_{\text{cem}}}\right) + \frac{2X\,\nu_{\text{cem}} E_{\text{cem}}}{(1+\nu_{\text{cem}})(1-2\nu_{\text{cem}})} + \frac{E_{\text{cem}}\varepsilon_{\text{free}}}{1-2\nu_{\text{cem}}} \tag{19.10}$$

$$\sigma_{\theta\theta,\text{cem}} = \frac{E_{\text{cem}}}{1+\nu_{\text{cem}}}\left(\frac{X + \frac{Y}{r^2}}{1-2\nu_{\text{cem}}}\right) + \frac{2X\,\nu_{\text{cem}} E_{\text{cem}}}{(1+\nu_{\text{cem}})(1-2\nu_{\text{cem}})} + \frac{E_{\text{cem}}\varepsilon_{\text{free}}}{1-2\nu_{\text{cem}}} \tag{19.11}$$

$$\sigma_{zz,\text{cem}} = \frac{2\nu_{\text{cem}} E_{\text{cem}} X}{(1+\nu_{\text{cem}})(1-2\nu_{\text{cem}})} + \frac{E_{\text{cem}}\varepsilon_{\text{free}}}{1-2\nu_{\text{cem}}} \tag{19.12}$$

- **Verification**: Running cement evaluation logs (especially advanced ultrasonic tools providing azimuthal coverage—Chap. 12) along the entire cemented interval, particularly the horizontal section, is critical to verify cement placement, bonding quality, and zonal isolation *before* initiating fracturing operations. Interpreting logs in horizontal wells can be challenging due to tool centralization issues and wellbore rugosity. Remedial cementing (squeeze jobs—Chap. 26) may be required if significant defects (channels, voids, poor bond) are found in critical zones. Distributed Temperature Sensing (DTS) run during or after cementing can also help confirm top of cement and potentially identify areas of poor hydration or flow (Fig. 19.4).

19.4 Fracture Height Containment

A key objective during hydraulic fracturing is to confine the created fractures within the target reservoir zone and prevent them from growing vertically upwards into overlying formations, particularly shallow aquifers containing potable water, or downwards into water-bearing zones or other unintended formations. A major challenge in hydraulic fracturing operations is ensuring that fracture growth remains confined within the intended reservoir interval. Inadequate containment can lead to upward

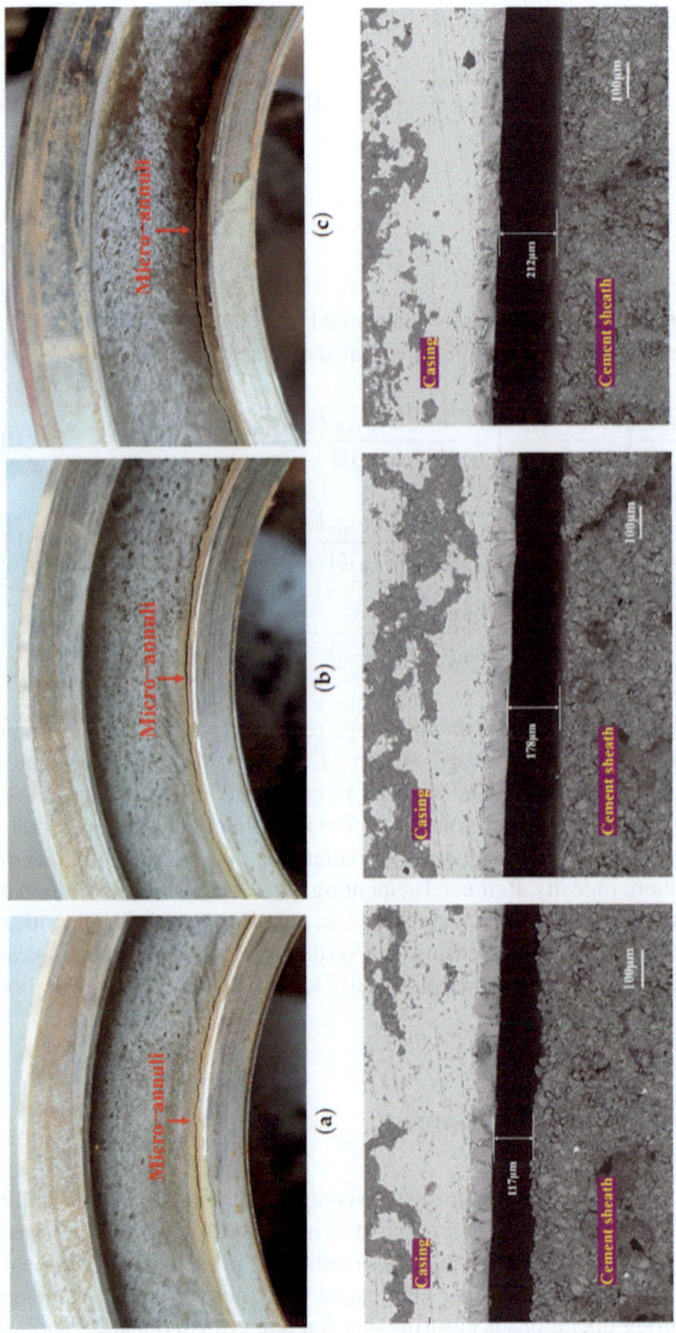

Fig. 19.4 Morphology of cement sheaths under different internal pressures during hydraulic fracturing: **a** 70 MPa, **b** 80 MPa, and **c** 90 MPa. The images show the casing-cement interface, where increasing internal pressure promotes the formation and widening of micro-annuli, compromising the zonal isolation crucial for well integrity in unconventional wells (Wang et al., 2024)

or downward fracture propagation into non-target formations, including freshwater aquifers or water-bearing zones, thereby compromising well integrity and environmental safety. Figure 19.5 illustrates a simulated propped fracture profile, where the fracture has initiated at ~8200 ft and exhibits growth into adjacent layers above and below. The proppant concentration map shows the highest density localized within the pay zone, but low-concentration tails extend into surrounding strata. Accurate characterization of in-situ stress contrasts, elastic properties, and natural barriers (e.g., shale breaks or tight layers) is essential to limit this vertical migration (Kolawole et al., 2020). This visualization underscores the importance of tailored treatment design and real-time fracture monitoring to ensure effective zonal isolation and stimulation containment. This environmental containment objective is directly linked to well integrity and geological understanding. This relies on several factors:

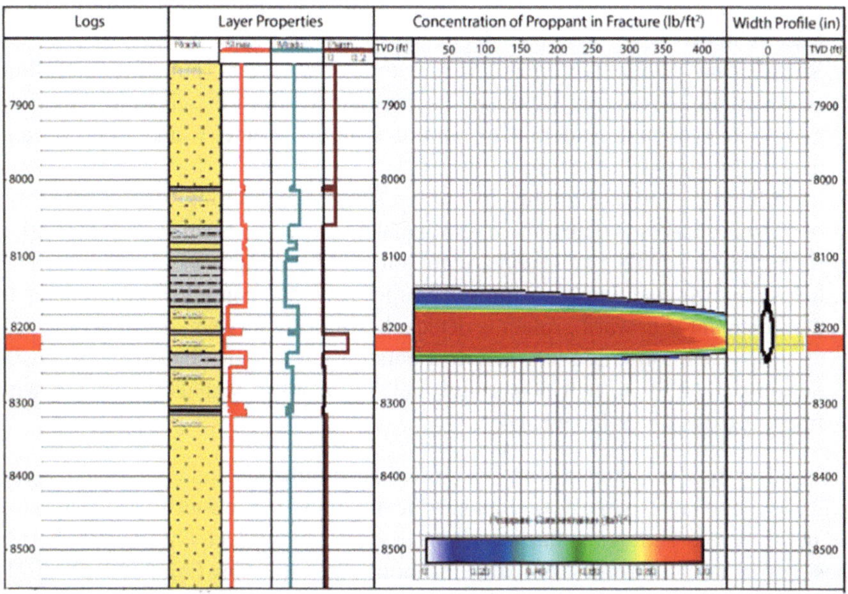

Fig. 19.5 Simulated profile of fracture height growth and proppant distribution in a hydraulically fractured well. Left: Log and formation layer properties, including gamma ray, Poisson's ratio, Young's modulus, and minimum horizontal stress (σ). Center: Color map representing proppant concentration (lb/ft^2) in the fracture plane, with peak distribution near the perforation cluster. Right: Fracture width profile (inches) along the vertical axis. The red band at 8200 ft indicates the primary fracture zone, with vertical growth extending into surrounding layers. Containment risks arise when proppant or fracture width extends significantly beyond the target formation (Kolawole et al., 2020)

- **Geological Barriers**: Utilizing natural stress contrasts or impermeable rock layers (e.g., thick shales, evaporites, unfractured dense rock) above and below the target zone that act as natural barriers to vertical fracture growth. Careful geological characterization, geomechanical modeling (based on logging and core data), and in-situ stress measurements (e.g., LOTs/FITs—Chap. 7, diagnostic fracture injection tests—DFITs) are needed to identify these barriers and predict fracture behavior.
- **Cement Integrity**: The integrity of the cement sheath in the annulus behind the casing provides the primary wellbore barrier preventing fractures from directly migrating upwards or downwards *outside the casing* along the wellbore path from the target zone into other formations. A well-placed, well-bonded, and competent cement sheath that extends continuously across the geological barrier formations above and below the target zone is essential. Defects (channels, voids, debonding) in the cement sheath can provide low-stress pathways for undesired fracture propagation ("frac hits" behind pipe) or subsequent fluid migration outside the intended fracture network.
- **Stimulation Design**: Optimizing the treatment parameters (pumping rates, fluid viscosity, fluid volumes, proppant concentration, injection schedule) based on geomechanical understanding to control fracture geometry and limit vertical height growth within the target zone.
- **Monitoring**: Using techniques like micro-seismic monitoring (deploying arrays of seismic sensors to detect micro-earthquakes caused by fracture growth) during fracturing to map the extent and orientation of created fractures in real-time. This helps confirm containment within the target zone and can allow adjustments to the treatment if upward growth is detected. Tiltmeter surveys on the surface can also provide information on fracture orientation and height. Distributed Acoustic Sensing (DAS) can also potentially pick up acoustic signals associated with fracture growth near the wellbore.

Ensuring fracture containment is a critical environmental and operational aspect of unconventional resource development, directly linked to robust well design (casing set points across geological barriers), construction quality (cement integrity across these barriers), and careful stimulation planning and monitoring (Ahmed et al., 2021; Alsubaih et al., 2020).

Understanding the temporal development of fractures is critical for assessing both the efficiency of stimulation and its implications for long-term well integrity. The proposed fracture propagation model offers a mathematical framework to reconstruct fracture geometry based on operational data recorded during hydraulic fracturing operations.

At each stage of fracturing, slurry volumes V_{si} are injected into the formation. Accounting for the fluid loss volume V_{li}, the effective fracture volume V_{Fi} for the ith interval can be obtained using a material balance:

$$V_{Fi} = V_{si} - V_{li} \tag{19.13}$$

Using the classic leakoff equation, the lost volume V_{li} can be modeled as:

$$V_{li} = C_L A t^{\frac{1}{2}} \tag{19.14}$$

where C_L is the leakoff coefficient, A is the fracture area, and t is the duration of exposure.

Assuming that the fracture grows proportionally in all dimensions, we define a fracture expansion coefficient α_i such that:

$$\frac{L_i}{L} = \left(\frac{A_i}{A}\right)^{\frac{1}{2}} = \left(\frac{V_{Fi}}{V_F}\right)^{\frac{1}{3}} = \alpha_i \tag{19.15}$$

Here, L_i, A_i, and V_{Fi} represent the fracture's linear dimension, area, and volume at the end of the ith period, respectively.

Substituting into the material balance and applying recurrence for multiple stages, we obtain the cubic recurrence relationship:

$$V_F \alpha_i^3 + C_L A \sqrt{\frac{\Delta t_i}{2}} \alpha_i^2 = r_i, i = 1, ..., N \tag{19.16}$$

The term r_i on the right-hand side is defined by the operational injection schedule and previous increments:

$$r_i = V_{si} - C_L \left[\sum_{j=1}^{i-1} \Delta A_j \sqrt{\sum_{k=j+1}^{i} \Delta t_k + \frac{1}{2} \Delta t_j} + \sum_{j=1}^{i-1} \Delta A_j \sqrt{\frac{\Delta t_j}{2}} \right] \tag{19.17}$$

The incremental area increase per stage is given by:

$$\Delta A_j = A_j - A_{j-1} = A\left(\alpha_j^2 - \alpha_{j-1}^2\right) \tag{19.18}$$

Equations (19.15), (19.16), and (19.17) collectively form a nonlinear system used to iteratively solve for fracture geometry development over time. These cubic recurrence equations allow reconstruction of fracture growth using only surface injection data and fluid properties, providing a computationally efficient and physics-consistent method to assess induced fracture evolution.

19.5 Long-Term Integrity Considerations

While initial stimulation operations pose immediate, severe cyclic loading challenges, the long-term integrity of unconventional wells during their production life also requires attention.

- **Cumulative Fatigue**: The initial high-pressure cycles during fracturing represent a significant portion of the well's fatigue life loading (potentially hundreds or thousands of cycles), but subsequent production cycles (e.g., pressure changes from shut-in to flowing, flow rate fluctuations, artificial lift cycling) also contribute to cumulative fatigue damage on the casing, connections, and cement sheath over many years. Long-term fatigue performance needs continued consideration in design and operational management.
- **Corrosion**: Produced fluids from unconventional reservoirs can still be corrosive (containing CO2, H2S, saline water, and organic acids). While initial production may be dominated by frac water flowback, native reservoir fluids will eventually dominate. Corrosion management strategies (appropriate material selection for production tubing, chemical inhibition, monitoring) remain important, although fluid compositions and corrosion rates can vary significantly depending on the specific formation and location.
- **Wellbore Deformation**: Potential for formation movements over time (e.g., shale creep, minor reservoir compaction) could impose loads on the casing, potentially leading to deformation or damage, especially in long horizontal sections or areas with complex geomechanics.
- **Parent–Child Well Interactions**: Infill drilling (child wells) close to existing "parent" wells in multi-well pads is common in unconventional development. Stimulating the new child well can lead to unintended pressure communication ("frac hits") with nearby parent wells. This can occur via the reservoir or, potentially, via unsealed pathways (natural fractures, compromised cement, mechanical damage during drilling/completion) behind pipe in the parent well. Frac hits can potentially damage the integrity of the parent well (e.g., overpressure casing, cause deformation, displace equipment) if not managed properly (e.g., by protecting the parent wellbore with pressure, temporary plugs, or strategic shut-in during child well fracturing). Managing these interactions is a critical aspect of multi-well pad integrity management. In tightly spaced multi-well pad developments, stimulation of a new (child) well can result in unintended hydraulic communication with an existing (parent) well—an event commonly referred to as a frac hit. These interactions can compromise the structural integrity of the parent well by inducing pressure spikes, reactivating fractures, or shearing annular seals and cement. Figure 19.6 illustrates a numerical simulation of a frac hit scenario where closely spaced child well stages generate a fracture network that intercepts the parent well (Joslin et al., 2020). The visualization highlights how insufficient spacing and high-pressure stimulation can create shear pathways, elevate inter-well pressures, and increase the likelihood of integrity failure in the offset well. This underscores the need for careful geomechanical modeling, parent well monitoring, and pre-stimulation mitigation strategies such as shut-in pressure control or mechanical isolation.
- **Abandonment**: Permanently plugging and abandoning long horizontal wells with multiple fractured stages presents unique challenges in ensuring effective isolation of all potential leak paths along the lateral section and vertically through the cemented annulus. Ensuring plug placement and verification in horizontal

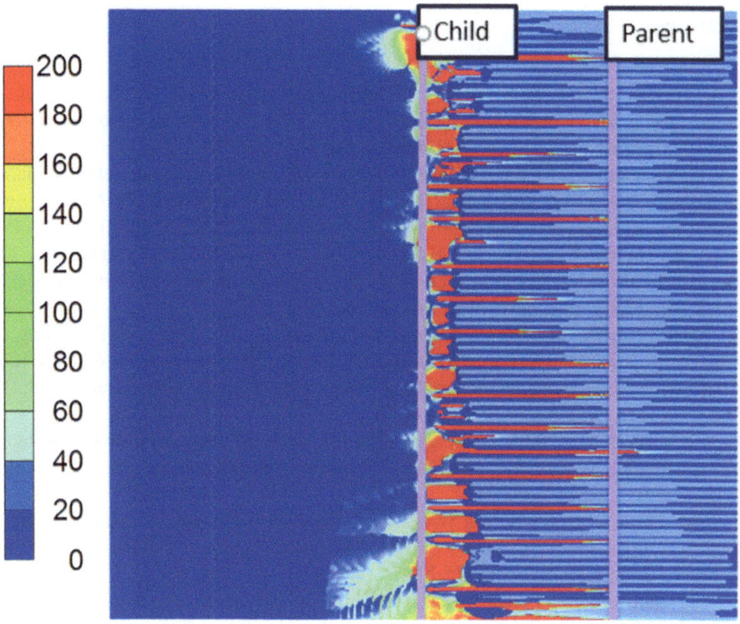

Fig. 19.6 Simulated hydraulic communication during a "frac hit" from a child well to an adjacent parent well. Fracture propagation from the child well (left) with 50-ft stage spacing results in significant stress perturbation and fluid migration toward the previously completed parent well (right). The color scale represents stress or pressure intensity, with high values (red/yellow) indicating active fracture zones. The interaction may lead to shear failure or mechanical disruption of cement and casing in the parent well, especially in the absence of zonal isolation or pressure management protocols (Joslin et al., 2020)

sections, dealing with potential debris or restrictions, and confirming long-term sealing across multiple casing strings and annuli can be more difficult than in vertical wells (Chap. 27). Tailored P&A strategies and verification methods are needed.

Managing well integrity in unconventional plays requires adapting conventional practices to address the specific challenges of horizontal drilling, multi-stage high-pressure hydraulic fracturing, cyclic loading effects, managing multi-well pad interactions, and the unique geology of these reservoirs, with a particular focus on robust cementing practices (especially in horizontal sections), managing fatigue, and ensuring effective long-term zonal isolation.

References

Ahmed, A., Hayfaa, S., Ali, J., Sarah, A. A., Abdulrahman, A., Madi, S., & Malak, A. (2021). *Hydraulic fracture design to improve well injectivity in Southern Iraqi limestone reservoir.* OnePetro. https://dx-doi-org.ezproxy.lib.utexas.edu/

Almulhim, B. K., Ashby, S., Alobaid, H., & Alyousef, A. (2023). Evaluation and study of dissolvable Frac plugs on horizontal well fracturing, plug and Perf completions. In *Proceedings—SPE annual technical conference and exhibition, 2023-October*. https://doi.org/10.2118/214841-MS

Alsubaih, A. A., Hayfaa, S., Alboshawi, H. S., Alaa, A., Al, lawe, Erfan, M., Sawdi, M., Al, A., & Lubna, H. (2020). *Reservoir Geomechanics simulation to optimize reservoir management in Southern Iraq Fields*. OnePetro. https://dx.doi.org/

Al-Yami, A. S., Alqam, M. H., Riefky, A., & Shafqat, A. U. (2018). Self healing durable cement; development, lab testing, and field execution. In *Proceedings of the SPE/IADC Middle East drilling technology conference and exhibition, 2018-January*. https://doi.org/10.2118/189397-MS

Boak, J., & Kleinberg, R. (2020). Shale gas, tight oil, shale oil and hydraulic fracturing. *Future Energy: Improved, Sustainable and Clean Options for Our Planet*, 67–95. https://doi.org/10.1016/B978-0-08-102886-5.00004-9

Davis, T., & McCrady, D. (2008). Using Swellable packers to provide annular isolation for multistage fracture treatments. *Proceedings—SPE Annual Technical Conference and Exhibition, 4*, 2240–2250. https://doi.org/10.2118/115775-MS

ElSebaee, M., Alekseev, A., Plyashkevich, V., Yudin, A., AlSomali, A., & Chertov, M. (2020). Novel trends in fracturing Proppant Flowback control. In *Proceedings—SPE annual technical conference and exhibition, 2020-October*. https://doi.org/10.2118/201308-MS

Granberg, A. (2011). Hydraulic Fracturing Marcellus Shale. Wikimedia Commons. Licensed under CC BY-ND 3.0. https://commons.wikimedia.org/wiki/File:Hydraulic_Fracturing_Marcellus_Shale.jpg

Guo, X., Li, J., Yu, Y., & Liu, G. (2018). Influence of cement defection on casing stress under cyclic loading during multi-stage hydro-fracturing. *Lecture Notes in Mechanical Engineering*, 1039–1054. https://doi.org/10.1007/978-981-13-0107-0_99/FIG.S/14

Hamilton, K., & David Pattillo, P. (2019). Developing an evaluation method for casing connections used in hydraulically fractured wells. In *Society of petroleum engineers—SPE hydraulic fracturing technology conference and exhibition 2019, HFTC 2019*. https://doi.org/10.2118/194369-MS

Joslin, K., Thomas, A., Tran, D., Shrivastava, V., & Erdle, J. (2020). Assessment of Frac-hit criteria in parent child well configurations. *Springer Series in Geomechanics and Geoengineering*, 3703–3718. https://doi.org/10.1007/978-981-15-2485-1_337/FIG.S/12

Kolawole, O., Wigwe, M., Ispas, I., & Watson, M. (2020). How will treatment parameters impact the optimization of hydraulic fracturing XE "hydraulic fracturing" process in unconventional reservoirs? *SN Applied Sciences, 2*(11), 1–21. https://doi.org/10.1007/S42452-020-03707-W/FIG.S/21

Liu, Z., Samuel, R., Gonzales, A., & Kang, Y. (2018). Analysis of casing fatigue failure during multistage fracturing operations. In *Society of petroleum engineers—Abu Dhabi international petroleum exhibition and conference 2018, ADIPEC 2018*. https://doi.org/10.2118/193189-MS

Patel, H., & Salehi, S. (2019). Development of an advanced finite element model and parametric study to evaluate cement sheath barrier. *Journal of Energy Resources Technology, Transactions of the ASME, 141*(9). https://doi.org/10.1115/1.4043137/725906

Renpu, W. (2011). Wellhead assembly. *Advanced Well Completion Engineering*, 572–616. https://doi.org/10.1016/B978-0-12-385868-9.00007-5

Segre, N., Galves, A. D., Rodrigues, J. A., & Joekes, I. (2006). Tire rubber-cement composites: Effect of slag on properties. *Cerâmica, 52*(324), 283–292. https://doi.org/10.1590/S0366-69132006000400012

Vincent, M. C., Miller, H. B., Milton-Taylor, D., & Kaufman, P. B. (2005). Erosion XE "Erosion" by sand and ceramic Proppants during slurry injection and Proppant Flowback. *Journal of Petroleum Technology, 57*(03), 35–37. https://doi.org/10.2118/0305-0035-JPT

Vralstad, T., Skorpa, R., Opedal, N., & De Andrade, J. (2015). Effect of thermal cycling on cement sheath integrity: Realistic experimental tests and simulation of resulting leakages. *Society of Petroleum Engineers—SPE Thermal Well Integrity and Design Symposium.* https://doi.org/10.2118/178467-MS

Wang, Y., Feng, Y., Wang, Y., & Feng, Y. (2024). Experimental and numerical simulation investigation of cement sheath integrity during multi-stage fracturing in offshore tight oil reservoir. *Journal of Marine Science and Engineering, 12*(5), 814. https://doi.org/10.3390/JMSE12050814

Wray, B., Bedford, D., Leotaud, L., & Hunter, B. (2009). The application of high-density elastic cements to solve HPHT XE "HPHT" challenges in South Texas: The success story. *Proceedings—SPE Annual Technical Conference and Exhibition, 1*, 135–151. https://doi.org/10.2118/122762-MS

Yang, Z., Sun, R., & Liu, Z. (2023). Casing tensile stress distribution in the bending section of a well during hydraulic fracturing XE "hydraulic fracturing." *Journal of Petroleum Exploration and Production Technology, 13*(5), 1361–1378. https://doi.org/10.1007/S13202-023-01609-2/FIG.S/19

Zara, E. M., Grittini, S. J., Jordan, G. R., Perello, L., & Sanders, M. (2021). Premium connections fatigue assessment methodology fit for multi-stage hydraulic fracturing operations. In *Proceedings—SPE annual technical conference and exhibition, 2021-September*. https://doi.org/10.2118/206286-MS

Zheng, D., Ozbayoglu, E., Miska, S., & Baldino, S. (2024). *An experimental study on the influence of cyclic loading and eccentric casing geometry on cement sheath integrity*. https://doi.org/10.56952/ARMA-2024-0305

Chapter 20
Subsea Well Integrity

Abstract This chapter presents an in-depth examination of the unique integrity challenges associated with subsea well systems, which are critical to deepwater oil and gas operations. Unlike surface wells, subsea wells place key control and structural components—such as the wellhead, Christmas tree, connectors, and flowlines—on the seabed, often at extreme depths exceeding 3000 m. This remote configuration introduces significant engineering, operational, and monitoring complexities. Subsea environments impose high hydrostatic pressures, aggressive corrosion conditions, dynamic loading, and limited accessibility, necessitating a robust design philosophy focused on reliability, remote surveillance, and advanced materials. The chapter begins by detailing the primary components of subsea production systems, including wellheads, tubing hangers, horizontal and vertical trees, connectors, manifolds, risers, and umbilicals. The interdependencies between these systems demand precise engineering and adherence to stringent industry standards (e.g., API 17D, DNVGL-ST-F101). Emphasis is placed on the integrity of metal-to-metal seals, load-bearing structures, and the hydraulic/electrical subsystems that control flow and ensure safety barriers. Monitoring and intervention challenges are a central theme. Unlike topside wells, subsea assets rely heavily on sensor networks and Subsea Control Modules (SCMs) for pressure, temperature, and valve diagnostics. Intervention requires costly, weather-dependent deployment of Remotely Operated Vehicles (ROVs) or Light Well Intervention Vessels (LWIVs), making proactive integrity assurance vital. The chapter outlines failure modes in subsea connectors and valves, including fatigue, erosion, seal degradation, and control system failure, and describes best practices for inspection, maintenance, and design for redundancy. Special sections explore the integrity of flowlines and risers, which serve as vital production conduits. Topics include internal corrosion mitigation, external cathodic protection, hydrate and wax management, and structural assessments under thermal, axial, and environmental loads. The criticality of riser integrity for preventing hydrocarbon releases is emphasized. Finally, the chapter introduces an integrated Subsea Integrity Management (SIM) framework aligned with global regulatory guidelines (e.g., PSA Norway, HSE UK). It emphasizes lifecycle management, digital twin implementation, risk-based inspection (RBI), and cross-functional coordination. Through historical incident analysis and evolving safety standards, the chapter underscores how subsea well

© The Author(s), under exclusive license to Springer Nature Switzerland AG 2026
A. Alsubaih and K. Sepehrnoori, *Well Integrity for Oil, Gas, and Subsurface Storage Operations*, https://doi.org/10.1007/978-3-032-15233-6_20

integrity has become a multidisciplinary domain demanding advanced engineering, remote operability, and real-time integrity governance.

Gas kick migration behavior in deepwater drilling modeled using CFD-DEM. *Source* American Society of Mechanical Engineers (ASME, Xu et al., 2021)

Chapter's Objectives

- Describe the main components of subsea production systems relevant to integrity (wellheads, trees, manifolds, flowlines).
- Highlight the unique challenges of monitoring and accessing subsea wells.
- Discuss the integrity management aspects of subsea wellheads and connectors.
- Outline integrity considerations for subsea Christmas trees and control systems.
- Briefly touch upon riser and flowline integrity management

20.1 Introduction: The Subsea Environment

As offshore oil and gas operations venture into increasingly deeper waters and more complex reservoirs, subsea production systems have become a cornerstone of modern field development. In contrast to conventional offshore wells where the wellhead and Christmas tree are located on fixed platforms above the sea surface, subsea wells place these critical control elements directly on the seabed—often at depths exceeding several thousand meters. This configuration eliminates the need for surface structures at each well location and allows multiple wells to tie back to a central floating facility or platform, thereby improving field economics and enabling development of otherwise inaccessible reservoirs.

However, this subsea configuration introduces unique integrity challenges that distinguish it significantly from surface well architecture. Once installed, subsea well components are largely inaccessible without the use of Remotely Operated Vehicles (ROVs), Light Well Intervention Vessels (LWIVs), or full drilling rigs—resources that are costly to mobilize and subject to weather constraints and logistical delays (Karlsen et al., 2013; Kumar Singhal et al., 2008; McLean et al., 2020). Therefore, the emphasis in subsea integrity shifts heavily toward designing for reliability, pre-emptive risk mitigation, and robust remote monitoring and control. The ability to predict, detect, and respond to anomalies becomes a critical determinant of well performance and safety.

The subsea environment itself poses extreme external loads and aggressive exposure conditions. Hydrostatic pressures can exceed 20 MPa (3000 psi) at deepwater depths, necessitating the use of high-strength materials and precision-engineered seals to ensure structural and containment integrity. Seawater, with its high salinity and oxygen content, accelerates corrosion and material degradation, further complicating long-term integrity assurance. In addition, dynamic environmental forces such as seabed subsidence, current-induced vibration, and thermal expansion of flowlines must be accounted for in structural design and fatigue assessments. Offshore and subsea well integrity has evolved in direct response to a series of significant accidents and operational failures over the past four decades. These events have exposed vulnerabilities in platform structures, well control systems, subsea components, and regulatory oversight. In turn, they have prompted governments, industry bodies, and operators to implement increasingly stringent safety legislation, integrity guidelines, and monitoring protocols. Table 20.1 outlines a chronological sequence of key incidents and regulatory actions that have influenced offshore oil and gas integrity practices, particularly in the North Sea and global offshore basins. Understanding this timeline provides critical context for current offshore well design, risk mitigation strategies, and lifecycle integrity programs.

This chapter provides a comprehensive overview of the key elements contributing to subsea well integrity. It begins by examining the architecture and critical components of a typical subsea production system—including wellheads, trees, tubing hangers, connectors, control modules, and flowlines—each of which plays a vital role in maintaining barrier integrity and operability over the field life. We then address the monitoring and access limitations inherent in subsea systems and how advanced sensor networks, communication systems, and ROV technology are deployed to overcome them.

Further sections focus on the structural, sealing, and corrosion challenges associated with subsea wellhead assemblies and connectors, followed by the complexities of managing subsea tree valves, control actuators, and umbilicals. The chapter concludes with an overview of flowline and riser integrity, highlighting how failures in these segments can compromise overall system integrity and environmental safety.

With subsea field developments trending toward deeper, higher-pressure, and higher-temperature environments—often exceeding traditional equipment qualification limits—the demand for enhanced engineering rigor, materials innovation, and digital integrity surveillance is higher than ever. The subsea domain is thus a testbed for some of the most advanced integrity strategies in the oil and gas industry, and effective well integrity management here is a direct function of robust system integration, proactive design, and real-time operational vigilance.

20.2 Subsea Production System Components

The complexity of subsea production systems lies not only in their components but in how they are interconnected and managed as an integrated unit on the seafloor (Li et al., 2022). Fig. 20.1 provides a schematic representation of a complete subsea containment and production infrastructure, illustrating how key elements such as the subsea containment assembly, manifold, risers, umbilical, and flying leads connect to floating support vessels and offloading systems at the surface. It demonstrates the spatial and functional relationships between critical integrity elements and emphasizes the need for coordinated design and monitoring across all subsea assets to ensure long-term containment and performance. Understanding the key components of a subsea production system is essential for discussing integrity:

- **Subsea Wellhead**: Similar in principle to a surface wellhead, it is installed on the seabed and supports the weight of the casing strings via casing hangers. It provides housing for casing hangers and their seals and incorporates features for

Table 20.1 Offshore oil and gas incident timeline and associated regulatory and integrity milestones. This timeline captures key industry accidents (e.g., Alexander L. Kielland, Piper Alpha, Deepwater Horizon) and their role in driving regulatory reforms, well integrity inspection programs (KP1–KP4), and safety guidance publications issued by entities such as HSE, the Petroleum Safety Authority Norway, and Oil & Gas UK. The events underscore how failures in structural, mechanical, and operational systems have catalyzed continuous improvement in offshore integrity management (Wood, 2020)

Year	Incident/integrity development
1980	Alexander L. Kielland disaster—Catastrophic structural failure of a semi-submersible accommodation rig in the North Sea
1981	Official investigation report on the Alexander L. Kielland accident published
1985	Norwegian petroleum act enacted, reinforcing regulatory oversight
1988	Piper alpha disaster—A major explosion and fire on a UK North sea platform
1990	Cullen inquiry report issued, recommending safety reforms for the UK offshore sector
1992	UK offshore safety act introduced to strengthen offshore safety regulations
1993	Wellhead oil leakage in the UK sector (~13.5 m^3) due to mechanical failure during shutdown
1996	Gas leak from subsea XMT (~41.6 tonnes) attributed to a dropped object incident in the UK sector
1996	Hydrogen-induced stress cracking (HISC) identified as the root cause of a subsea manifold leak in the UK sector during final commissioning
1996	HSE pipeline safety regulations implemented in the UK
2000	HSE launches key programme 1 (KP1) targeting hydrocarbon release (HCR) reductions
2003	Operational error causes oil release (~30 m^3) from a subsea manifold on the Norwegian continental shelf (NCS)
2003	Major oil spill (500–800 m^3) on the NCS due to rupture between a manifold and production line
2004	Well control failure on the NCS led to barrier breach and gas breakthrough at the seabed
2004	Petroleum safety authority Norway (PSA) established, taking over safety responsibilities from the Norwegian petroleum directorate
2004	HSE key programme 3 (KP3) launched to address asset integrity
2006	HSE releases asset integrity toolkit to guide integrity management
2007	KP3 asset integrity report published by HSE
2009	HSE introduces asset integrity KPIs to measure and improve performance
2010	Deepwater Horizon blowout in the Gulf of Mexico—One of the worst offshore oil disasters
2010	Well control incident on the NCS involving gas release on the platform deck
2010	HSE key programme 4 (KP4) initiated for ageing infrastructure and life extension assessment
2012	Oil & gas UK guidance published for managing ageing and life extension of UKCS installations

(continued)

Table 20.1 (continued)

Year	Incident/integrity development
2012	UK sector well control incident leads to gas leak lasting 51 days
2013	Human error leads to valve misconfiguration on the NCS, causing ~2.5 tonnes of oil to spill
2014	Oil & gas UK issues guidelines on ageing and life extension
2014	HSE'S KP4 programme continues, focusing on ageing offshore assets
2016	Well workover incident on the NCS results in gas release through the drill rig riser
2016	Oil & gas UK releases guidelines for managing life extension of subsea pipelines and risers
2016	Oil & gas UK issues well Lifecycle integrity guidelines
2017	Flexible pipe integrity management guidance and best practices published by oil & gas UK

interfacing with the subsea Blowout Preventer (BOP) stack during drilling and the subsea Christmas Tree (XT) during production. Designed for installation using specialized running tools from a drilling rig. Governed by standards like API Spec 17D and API Spec 17TR8 (HPHT).

- **Tubing Hanger**: Lands inside the subsea wellhead (for vertical trees) or directly into the tree body (for horizontal trees). It supports the weight of the production tubing string, seals the A-annulus (tubing-by-production casing) at the top, and provides conduits for essential downhole functions (e.g., SCSSV hydraulic control line, chemical injection lines, electrical cables for ESPs, fiber optic lines). Critical seals around these penetrations are required.
- **Subsea Christmas Tree (XT)**: The assembly of valves, spools, chokes, and fittings installed on top of the subsea wellhead, controlling flow from the well and providing access for interventions. Significantly more complex, automated, and robustly designed than surface trees due to remote operation, the harsh subsea environment, and high-pressure/high-temperature requirements. Contains critical barrier valves (master valves, wing valves, annulus valves), chokes, and connection hubs for flowlines and control umbilical's.
  - Vertical Tree: Valves are stacked vertically above the tubing hanger. Access for intervention typically requires removing the tree.
  - Horizontal Tree: The tubing hanger lands in the tree body beside the main bore valves, allowing larger bore access into the tubing below the tree valves for through-tree interventions (wireline, coiled tubing) without removing the main tree valves. Often preferred for wells requiring frequent intervention. The subsea Christmas tree (XT) is a critical component of the subsea production system, comprising a complex assembly of valves and fittings that control flow from the well and provide intervention access. There are two primary designs: vertical (conventional) and horizontal. Vertical trees stack valves directly above the tubing hanger, requiring the removal of the valve block for through-tubing

intervention. In contrast, horizontal trees position the tubing hanger beside the main valves, allowing intervention without disturbing the primary flow path (Bai & Bai, 2010). Figure 20.2 illustrates the key mechanical differences between these two configurations.

- **Connectors**: Specialized hydraulic connectors are used to join the subsea tree to the wellhead (tree connector), flowlines to the tree or manifold (flowline connector), and umbilical (umbilical connector). These connectors must provide reliable pressure containment and structural integrity under high pressure, temperature, and external loads (e.g., flowline thermal expansion forces), often using metal-to-metal seals remotely actuated by ROVs or dedicated hydraulic systems. Qualification standards (e.g., API 17D) and testing are crucial.
- **Subsea Manifolds**: Structures located on the seabed that commingle flow from multiple production wells or distribute injection fluids to multiple injection wells before routing them through larger export flowlines towards the processing facility (Bai & Bai, 2010). They contain valves, chokes, and connection points for wells, flowlines, and umbilical.
- **Flowlines and Risers**: Pipelines transporting produced fluids from the subsea tree/manifold along the seabed (flowlines) and vertically through the water column to the surface facility (risers) (Duan et al., 2022). Flowlines and risers are part of the overall production containment system. Risers can be rigid steel pipes, flexible composite pipes, or hybrid systems (e.g., SCR–Steel Catenary Risers). Their integrity is subject to internal pressure, external pressure, axial loads, bending, fatigue, and environmental forces (waves, currents).
- **Control System (Umbilicals and SCMs)**: Umbilicals run from the surface facility (platform, FPSO) to the subsea wellheads and manifolds, providing hydraulic power for valve actuation, electrical power for sensors and controls, chemical injection pathways (e.g., hydrate inhibitors, corrosion inhibitors), and communication lines (electrical or fiber optic) (Mudrak, 2016). A Subsea Control Module (SCM) is mounted on each tree (or manifold), receiving commands and power from the umbilical and translating them into actions (e.g., operating valves) and transmitting sensor data back to the surface. The integrity of the umbilical and SCM is vital for operating, monitoring, and controlling the subsea well remotely.

20.3 Challenges of Subsea Monitoring and Access

The remote, underwater location and significant water depth (ranging from hundreds to thousands of meters) impose unique and significant challenges for monitoring and accessing subsea wells for integrity management purposes compared to surface wells.

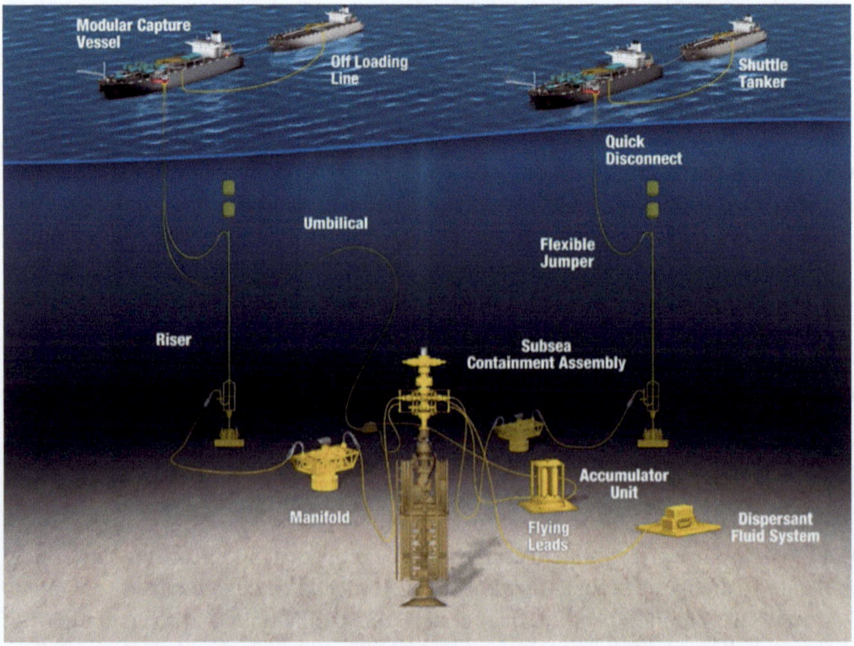

Fig. 20.1 Subsea production and containment system layout, showing the modular capture vessel, shuttle tanker, risers, umbilical, flexible jumpers, subsea containment aassemblies manifold, flying leads, and supporting equipment such as the dispersant fluid system and accumulator unit (Koto & Junaidi, 2014)

- **Limited Direct Access**: Unlike surface wells, routine visual inspection by personnel, manual intervention, or easy access for conventional logging is impossible. All monitoring and intervention must be done remotely (via sensors and the control system) or using specialized vessels, equipment, and robots (ROVs, AUVs–Autonomous Underwater Vehicles). This makes problem diagnosis and simple repairs significantly more difficult, time-consuming, and expensive.
- **Reliance on Remote Monitoring**: Integrity surveillance depends heavily on permanently installed sensors on the subsea equipment (e.g., pressure and temperature sensors in annuli, leak detection sensors around connectors, valve position sensors, flow meters) transmitting data to the surface via the control system/umbilical (Chap. 30). Reliability and accuracy of these sensors, the control system, and the umbilical are paramount for detecting integrity anomalies. Annulus pressure monitoring capabilities, while critical (Chap. 8), may be limited depending on how annuli are configured and whether sensors are installed below the mudline.
- **Intervention Complexity and Cost**: Intervening in a subsea well (e.g., for logging, complex diagnostics, changing out equipment, plug setting for P&A) requires mobilization of expensive specialized assets like drilling rigs, light well intervention vessels (LWIVs–for simpler tasks like wireline), or specialized installation vessels equipped with heavy lift capability and ROVs (Farquhar et al.,

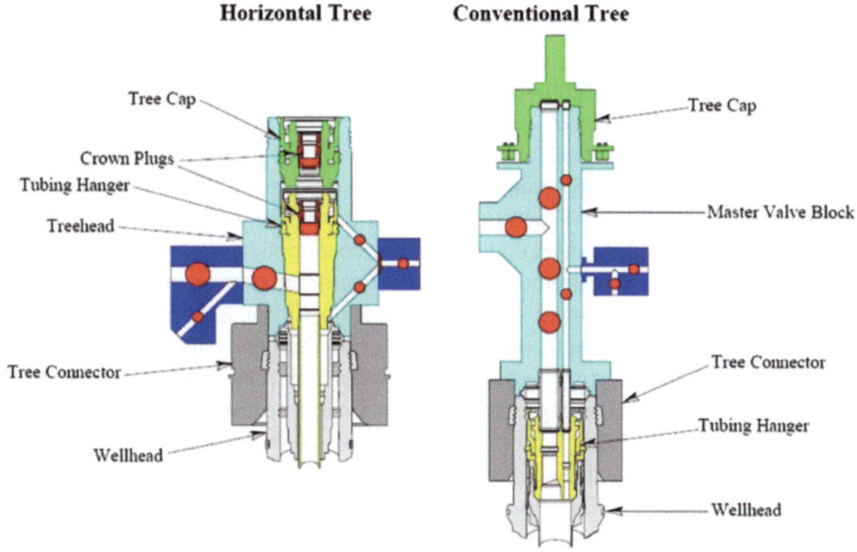

Fig. 20.2 Comparison of a horizontal subsea tree (left) and a conventional vertical tree (right), showing key structural differences in valve arrangement and tubing hanger placement. Horizontal trees allow easier intervention access without removing tree valves, whereas conventional trees require removal of the master valve block for tubing access (Bai & Bai, 2010)

2019; Loov et al., 2015). Weather conditions can significantly impact intervention windows and operational time. This high cost places a high premium on ensuring long-term reliability through robust design and extensive equipment qualification (e.g., API 17 series PR2 testing), and on designing for ease of intervention where possible (e.g., using horizontal trees).

- **ROV Dependence**: Remotely Operated Vehicles (ROVs), operated from surface vessels, are essential for numerous tasks: visual inspection (often primary method), operating valves (via torque tools), manipulating components (e.g., operating connectors, clearing debris), assisting with installation or intervention tasks (e.g., connecting hydraulic lines, guiding tools), and deploying small sensors or monitoring equipment. Reliability and capabilities of ROVs and their tooling are critical for subsea operations and integrity management.

20.4 Subsea Wellhead and Connector Integrity

Ensuring the integrity of the subsea wellhead assembly on the seabed and the connectors joining it to the tree and subsequent riser systems is critical for overall containment.

- **External Pressure**: Subsea wellhead housings and connectors must be designed to withstand the high external hydrostatic pressure exerted by the deep seawater column (Chap. 18). Collapse resistance of components and tubulars is a key design parameter.
- **Seal Integrity**: Hanger seals (casing and tubing) supporting the tubular strings inside the wellhead, and connector seals (e.g., tree-to-wellhead connector seal, flowline connector seals) are critical pressure barriers. Metal-to-metal (MTM) seals are standard for primary containment in these applications due to the need for high reliability under extreme internal and external pressures, temperatures, and the difficulty of replacement subsea (Singeetham, 1990). Ensuring clean sealing surfaces during installation (which is performed remotely using running tools from the surface) and proper actuation/preload of connectors to energize the MTM seals is crucial for achieving initial integrity. Leak detection systems (e.g., monitoring pressure in a test port or drain line) may be incorporated into connector designs. Elastomer seals may be used as secondary seals or in less critical applications, requiring careful selection for subsea conditions (temperature, pressure, fluid compatibility, long-term aging in seawater).
- **Structural Loads**: The subsea wellhead must support the weight of the entire casing string, the tubing string, and the subsea tree assembly. It must also withstand significant bending moments induced by drilling risers, production risers (especially SCRs), or thermal expansion/movement of flowlines connected to the tree (Stewart & Andueza, 2020). Fatigue analysis is important, especially for connections experiencing cyclic loading (e.g., from floating vessel motion and wave/current loading on risers, or thermal cycling of flowlines). The interface with the seabed foundation is also crucial for structural integrity.
- **Corrosion Protection**: Subsea equipment relies on multiple layers of protection against external corrosion from seawater: careful material selection (often extensive use of CRAs or cladding in wetted areas), protective coatings (e.g., fusion-bonded epoxy, polyurethane), and active cathodic protection systems (typically sacrificial anodes attached to the wellhead structure or frame, or potentially impressed current systems) (Britton & Taylor, 2017; Okyere, 2019). Regular ROV inspection is used to monitor the condition of coatings and anodes, ensuring CP systems remain effective. Internal corrosion is managed by the integrity of the inner tubulars and cement.

20.5 Subsea Christmas Tree and Control System Integrity

The subsea Christmas tree is a highly complex and critical assembly responsible for controlling flow at the seabed, providing isolation barriers, and enabling remote well operation. It includes multiple valves (e.g., production master valves, annulus valves, swab valves), flowlines, sensors, and connectors designed for long-term deployment in harsh subsea environments. As shown in Fig. 20.3, these trees integrate both hydraulic and manual actuation systems, temperature/pressure sensors, and mechanical lockdown systems to ensure well control and structural integrity under extreme pressures and temperatures.

- **Valve Integrity**: Subsea tree valves (e.g., master valves, wing valves, annulus valves, swab valve) must provide extremely high reliability for both sealing (when closed) and operation (opening/closing) over decades in a harsh, inaccessible environment. Extensive qualification testing according to standards like API 17D PR2 (involving rigorous pressure, temperature, and endurance cycling) is mandatory. The gate and seat sealing mechanisms (or ball/seat) are critical barrier elements and must resist erosion (from high velocity flow or solids) and corrosion (Chap. 11). Regular testing (e.g., monitoring leak rates or pressure build-up across closed valves) is crucial, often performed remotely via the control system.
- **Actuator Reliability**: Hydraulic or electric actuators for subsea tree valves must function reliably when commanded by the subsea control system. Failure to open or close on demand can compromise well control or containment. Redundancy (e.g., dual master valves, fail-safe design—closing on loss of power/signal) is built into critical functions.
- **Control System Integrity**: The Subsea Control Module (SCM) and the umbilical connecting it to the surface are vital for controlling the well and receiving sensor data. The SCM contains complex electronics and hydraulics operating in a subsea housing. Its reliability is critical; failure often requires retrieval and replacement using a vessel (typically an LWIV or specialized vessel). The hydraulic integrity of control fluid lines and the electrical/fiber optic integrity of communication lines within the umbilical are equally important. Leaks in hydraulic lines or failures in electrical/fiber optic conductors can disable well control, safety functions (e.g., SCSSV control), or monitoring capabilities. Protecting the umbilical from damage (e.g., trawl boards, dropped objects) is essential. Fiber optic lines enable high-bandwidth data transmission for advanced sensors (Chaps. 24, 30).
- **Choke Integrity**: Subsea chokes controlling flow rates are subjected to severe erosion, especially in wells with high flow rates or sand production. They must be designed with hardened materials (e.g., tungsten carbide trims) or allow for remote change-out of internal components (trims) using ROVs. Choke body integrity under pressure is also critical.
- **Annulus Access**: Subsea trees provide access ports to the A-annulus (and potentially outer annuli via specialized subsea annulus vent valves–SSAVVs) for monitoring pressure (via permanently installed sensors), fluid injection (e.g., hydrate inhibition chemicals), or intervention. The integrity of these access paths and their associated valves is crucial. Monitoring annular pressures remotely is a key aspect of subsea integrity surveillance (similar to surface wells, Chap. 8), although full SCP diagnosis (bleed-down/build-up) is more complex subsea and requires specific procedures.

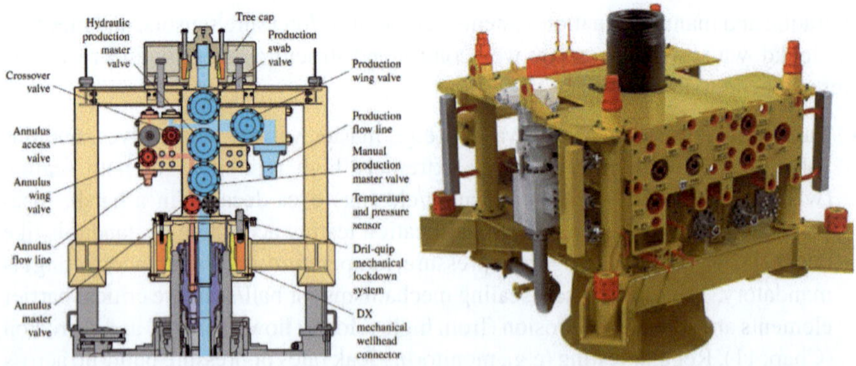

Fig. 20.3 Detailed schematic (left) and physical layout (right) of a subsea horizontal Christmas tree, showing key barrier valves, flow paths, control interfaces, and the mechanical lockdown connector. These systems are critical for flow control, safety, and remote operability in deepwater environments (Bai & Bai, 2010)

20.6 Riser and Flowline Integrity Management

While technically distinct from the wellbore itself (which typically ends at the subsea tree isolation valve), the flowlines along the seabed and risers connecting the subsea tree/manifold to the surface facility are part of the overall production containment system. The riser system acts as a critical conduit between the subsea blowout preventer (BOP) and the surface drilling or production facility. It must accommodate not only the mechanical transfer of fluids but also house essential control and communication lines. These include choke and kill lines for pressure control, hydraulic booster lines for valve actuation, and multiplex cables (Mux) for signal transmission to and from subsea control pods. Figure 20.4 highlights the configuration of these integrated systems at a riser connection, showing how operational and safety-critical functions are maintained across the water column. Their integrity management is closely linked to well integrity, as a failure can lead to uncontrolled releases of hydrocarbons or injection fluids to the environment. Key aspects include:

- **Internal Corrosion/Erosion**: Managing metal loss from produced fluids, which may be corrosive (CO2, H2S, water, organic acids) or contain solids (sand). Often requires continuous corrosion inhibition injection at the subsea tree or manifold via the umbilical, and potentially intelligent pigging for inspection (if the line is designed for pigging, involving launching and receiving specialized internal inspection tools).
- **External Corrosion**: Protecting against seawater corrosion using external coatings (e.g., FBE) and cathodic protection systems (sacrificial anodes or ICCP). One of the most critical elements in ensuring the long-term integrity of subsea risers and flowlines is protecting against external corrosion caused by continuous exposure to seawater. Impressed current cathodic protection (ICCP) systems are widely used in offshore environments to achieve this. These systems operate by

applying a direct current from an external power source (typically located topside on the platform), which flows through a submerged anode and forces the steel structure to become a cathode. This process inhibits the electrochemical reactions responsible for corrosion (Britton & Taylor, 2017). Unlike sacrificial anode systems, ICCP provides adjustable protection levels and is suitable for larger or higher-value installations with long design lives. Figure 20.5 demonstrates how ICCP is applied to an offshore platform structure, highlighting the current path, protective circuit, and electrochemical reactions involved in maintaining structural integrity in aggressive subsea environments.

- **Structural Integrity**: Ensuring resistance to internal pressure, external hydrostatic pressure (for risers), axial tension (risers), bending, and fatigue (especially for dynamic risers subjected to vessel motion, waves, and currents). Protection against dropped objects (e.g., from vessels), anchor drag, or seabed hazards (e.g., mudslides, seismic activity) is also necessary.
- **Flow Assurance (Hydrates/Wax)**: Preventing blockages from hydrate formation (ice-like solids from water and gas under high pressure/low temperature) or wax deposition (from crude oil) during flow. Often managed through insulation, heating (e.g., pipe-in-pipe designs, electric heating), or chemical injections (e.g., methanol or MEG for hydrates, wax inhibitors via the umbilical). Blockages can lead to overpressure situations or require complex, risky remediation (e.g., chemical dissolution, pigging).

Subsea well integrity management demands a comprehensive, system-level approach, integrating the wellbore, tree, controls, flowlines, and risers. It focuses

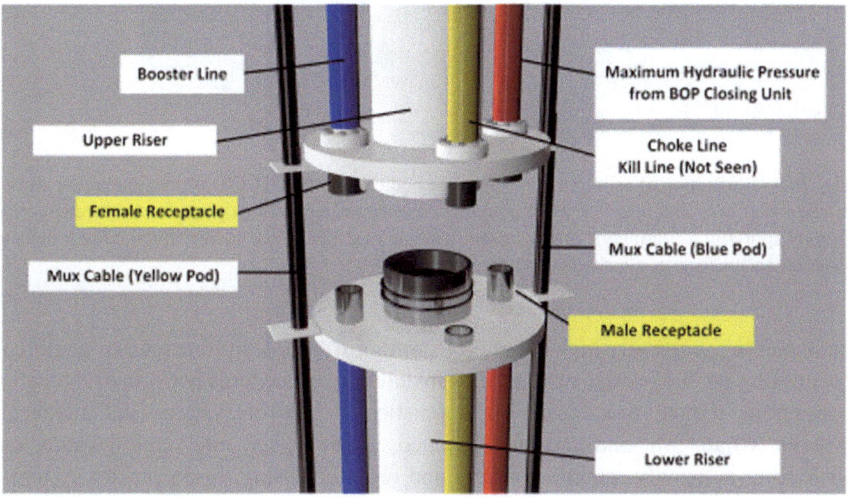

Fig. 20.4 Detailed illustration of a subsea riser connection showing critical control and communication lines. The system includes choke and kill lines, booster lines, hydraulic control pressure from the BOP closing unit, and multiplex cables (Mux) used to transmit control signals between surface and subsea equipment. The male and female receptacles enable modular, secure connection between riser sections (Raabe & Jortner, 2022)

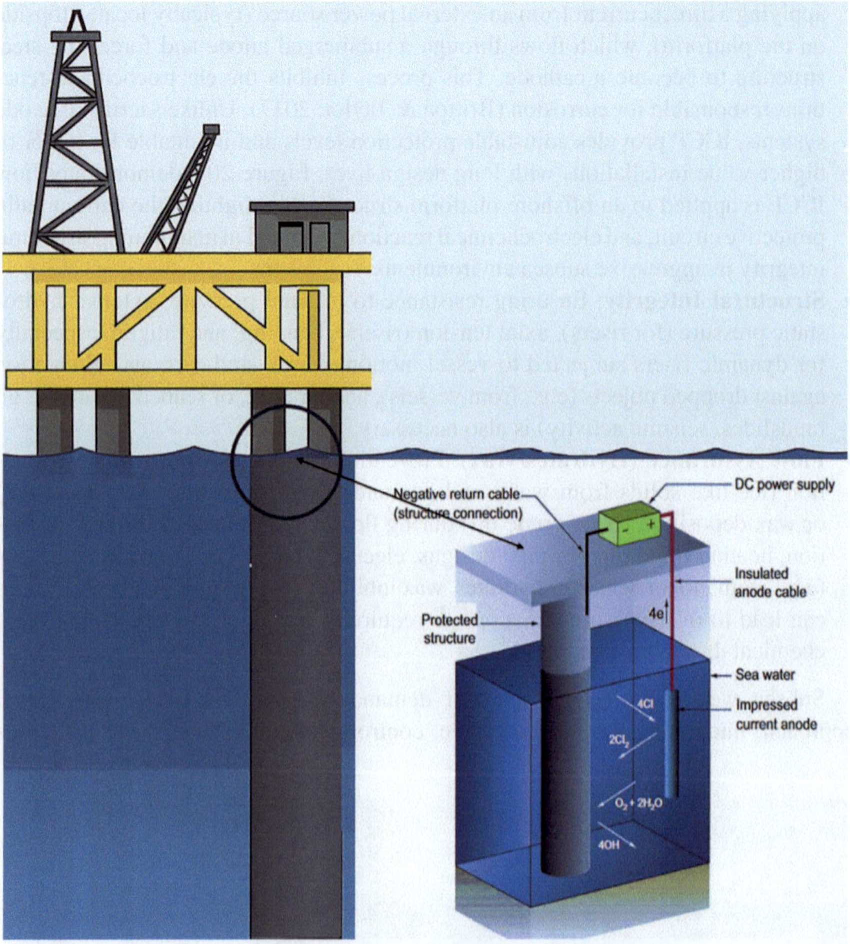

Fig. 20.5 Illustration of an impressed current cathodic protection (ICCP) system integrated into an offshore platform structure. The system protects submerged steel components from seawater corrosion by applying a continuous direct current from an external power source, forcing the protected structure to function as a cathode (Britton & Taylor, 2017)

on achieving extremely high reliability through robust design, extensive qualification testing (API 17 series), meticulous installation procedures performed remotely, continuous remote monitoring via sophisticated control systems and advanced sensors (Chap. 30), and carefully planned, albeit costly, intervention strategies. Inspection is heavily reliant on ROVs and remote sensing. In deepwater systems, physical inspection and maintenance of risers and flowlines are performed using ROVs, which can manipulate connectors, perform valve operations, and carry sensors for corrosion and pressure monitoring. Figure 20.6 illustrates a subsea production

Fig. 20.6 Remotely operated vehicles (ROVs) deployed around a subsea production system, inspecting and maintaining flowlines, riser connections, and control units. ROV-based intervention is critical for subsea integrity management, especially in the absence of direct human access (Park & Kim, 2022)

system where ROVs provide vital support for real-time integrity assessments and operational adjustments.

20.7 Integrated Subsea Integrity Management (SIM) Framework

Effective subsea well integrity management extends beyond individual components and encompasses a **system-wide integrity strategy** aligned with recognized regulatory and operational frameworks. According to PSA (2020), a **Subsea Integrity Management (SIM)** approach must ensure hydrocarbon containment between the wellhead and the topside Emergency Shutdown Valve (ESDV), addressing the entire lifecycle—from design to decommissioning (Wood, 2020).

Key aspects to incorporate in the SIM framework include:

- **Auditable Elements**: Operators must maintain documentation and systems for threat assessment, operational data trends, lifetime extension planning, and change management. Table 20.1 from the PSA guideline outlines 19 integrity assurance requirements that serve as auditable elements.
- **System Subdivision and Interfaces**: Equipment must be categorized by pressure-retaining vs. non-retaining, and their interface standards must be clearly defined (e.g., API 17D, DNVGL-ST-F101).

- **Lifecycle Management**: SIM should be embedded in all phases including feasibility, FEED, manufacturing, installation, operation, and eventual decommissioning.
- **Digital Tools and Data Governance**: A centralized data management system (CMS/CMMS) must store LCI (Life Cycle Information), operational records, inspection data, and configuration changes. GIS technology and digital twins are increasingly used to manage, visualize, and analyze this data for anomaly detection and proactive risk mitigation.
- **Risk-Based Inspection (RBI)**: RBI methodologies should identify equipment-specific degradation mechanisms (e.g., corrosion, fatigue, HISC) and align inspection frequency and scope accordingly.
- **Organizational and Technical Interfaces**: Roles, responsibilities, and interdisciplinary communication channels must be established through RACI matrices. These interfaces ensure coordinated integrity actions across asset teams.

Monitoring, Emergency Preparedness, and Performance Evaluation: KPI-based integrity tracking, emergency response planning (ALARP), and internal/external audits are mandatory for continuous improvement. PSA's framework regulations (e.g., §§17, 19–23) emphasize this loop.

References

Bai, Y., & Bai, Q. (2010). Subsea wellheads and trees. In *Subsea Engineering Handbook* (pp. 703–761). https://doi.org/10.1016/B978-1-85617-689-7.10022-6

Britton, J., & Taylor, M. L. (2017). Advancements in cathodic protection of offshore structures. In *Trends in Oil and Gas Corrosion Research and Technologies: Production and Transmission* (pp. 593–612). https://doi.org/10.1016/B978-0-08-101105-8.00025-5

Duan, M., Zhang, Y., & Jia, Z. (2022). Design of pipelines and risers. In *Encyclopedia of Ocean Engineering* (pp. 319–327). https://doi.org/10.1007/978-981-10-6946-8_272

Farquhar, R., Rodriguez, J., & MacPherson, R. (2019). Evolution of Bacchus subsea well interventions using rig and light well intervention vessels. In *Society of Petroleum Engineers-Spe/Icota Well Intervention Conference and Exhibition 2019*. https://doi.org/10.2118/194270-MS

Karlsen, O. E., Morrison, B., & Maciel, P. (2013). Improved recovery rate in brownfield subsea wells using riserless light well intervention. In *Proceedings of the Annual Offshore Technology Conference* (Vol. 2, pp. 813–827). https://doi.org/10.4043/24367-MS

Koto, J., & Junaidi, A. K. (2014). Analysis of vortex-induced vibration of riser using Spalart-Almaras model. *Jurnal Teknologi (Sciences and Engineering), 69*(7), 9–15. https://doi.org/10.11113/JT.V69.3260

Kumar Singhal, A., Gerito, T., Bedada, D., & Alemu, O. (2008). An analysis of micro finance institutions (with respect to loan repayment). International Journal of Advances in Engineering and Management (IJAEM), 2(1), 1192. 10.35629/5252-45122323.

Li, J., Chang, Y., Shi, J., Liu, X., Chen, G., Zhang, N., Guan, Q., & Dai, Y. (2022). Probability prediction approach of fatigue failure for the subsea wellhead using Bayesian regularization artificial neural network. *Journal of Marine Science and Engineering, 10*(11), 1627. https://doi.org/10.3390/JMSE10111627

Loov, R., Locken, D., McLnnes, N., & Chippett, E. (2015). Light well intervention using conventional Slickline and electric line off the East Coast of Canada. In *Society of Petroleum Engineers-Coiled Tubing and Well Intervention Conference and Exhibition 2015* (pp. 694–704). https://doi.org/10.2118/173687-MS

McLean, D. L., Parsons, M. J. G., Gates, A. R., Benfield, M. C., Bond, T., Booth, D. J., Bunce, M., Fowler, A. M., Harvey, E. S., Macreadie, P. I., Pattiaratchi, C. B., Rouse, S., Partridge, J. C., Thomson, P. G., Todd, V. L. G., & Jones, D. O. B. (2020). Enhancing the scientific value of industry remotely operated vehicles (ROVs) in our oceans. In *Frontiers in Marine Science* (Vol. 7). https://doi.org/10.3389/FMARS.2020.00220

Metal-to-Metal Annulus Packoff for a Subsea Wellhead System. (1990).

Mudrak, C. (2016). Subsea Production Systems—A Review of Components, Maintenance and Reliability. https://pure.unileoben.ac.at/en/publications/subsea-production-systems-a-review-of-components-maintenance-and-.

Okyere, M. S. (2019). Corrosion protection for the oil and gas industry: Pipelines, subsea equipment, and structures. In *Corrosion Protection for the Oil and Gas Industry*. https://doi.org/10.1201/9780429056451

Park, J., & Kim, J. (2022). Robust underwater localization using acoustic image alignment for autonomous intervention systems. *IEEE Access, 10*, 58447–58457. https://doi.org/10.1109/ACCESS.2022.3179430

Raabe, G., & Jortner, S. (2022). Reference: Subsea equipment. In *Universal Well Control* (pp. 547–574). https://doi.org/10.1016/B978-0-323-90584-8.00003-4

Stewart, B., & Andueza, A. (2020). ASME sec VIII div. 3 and API TR 17TR8 verification and validation methodology for subsea connectors. In *American Society of Mechanical Engineers, Pressure Vessels and Piping Division (Publication) Pvp* (Vol. 5). https://doi.org/10.1115/PVP2020-21784

Xu, H., Zhang, H., & Wang, J. (2021). Numerical study of gas kick migration behavior in deepwater drilling using a coupled CFD-DEM method. In *Proceedings of the ASME 2021 40th International Conference on Ocean, Offshore and Arctic Engineering (OMAE2021)* (Vol. 2: Offshore Technology, V002T02A029). American Society of Mechanical Engineers. https://doi.org/10.1115/OMAE2021-62013

Wood. (2020). Guideline to Subsea Integrity Management-Wellhead to Topside Esdv. www.woodplc.com

Chapter 21
Geothermal Well Integrity

Abstract This chapter presents a comprehensive exploration of well integrity challenges specific to geothermal energy systems, emphasizing the extreme thermal and chemical environments that differentiate these wells from conventional oil and gas operations. Geothermal wells are routinely exposed to sustained high temperatures (>200 °C) and aggressive fluid chemistries rich in CO_2, H_2S, and dissolved minerals. These conditions impose severe stress on casings, cement sheaths, seals, and surface equipment, demanding specialized material selection, robust design, and tailored operational practices. The chapter begins by characterizing geothermal resources and their system configurations—including hydrothermal, Enhanced Geothermal Systems (EGS), and closed-loop designs—highlighting the diverse and often hostile downhole conditions. Design challenges include thermal stress cycling, strength degradation of materials, and cement stability. Thermal-mechanical modeling is introduced to predict casing expansion, annular pressure buildup, and cement cracking. Innovative cement formulations incorporating silica, elastomers, or self-healing properties are reviewed to improve resilience. The chapter also details the use of premium casing connections, thermal-resistant seals, and large-diameter designs essential for high-flow geothermal applications. Drilling and cementing geothermal wells introduce additional difficulties, such as poor mud displacement in high-angle holes, logging tool temperature limits, and slurry instability during placement. Operationally, corrosion, mineral scaling, and thermal fatigue represent persistent integrity threats. The chapter outlines monitoring strategies using Distributed Temperature and Strain Sensing (DTS/DSS), annular pressure surveillance, and temperature-compensated wellhead design to mitigate risks. Field case studies and figures demonstrate how thermal stress reversals, material degradation, and cyclic fatigue compromise casing and cement over time. Abandonment of geothermal wells introduces further complexity due to residual heat, chemically degraded primary barriers, and the difficulty of verifying annular sealing in high-temperature zones. The chapter concludes by framing geothermal well integrity as a high-stakes, long-term commitment requiring an integrated approach—from design to plug and abandonment—based on rigorous thermal analysis, tailored materials, and advanced monitoring technologies. It emphasizes the need for continued innovation to support geothermal's critical role in a decarbonized energy future.

© The Author(s), under exclusive license to Springer Nature Switzerland AG 2026
A. Alsubaih and K. Sepehrnoori, *Well Integrity for Oil, Gas, and Subsurface Storage Operations*, https://doi.org/10.1007/978-3-032-15233-6_21

Newberry Geothermal Lease Project (geothermal well site). *Source* Bureau of Land Management/U.S. federal government (n.d.)

Chapter's Objectives

- Characterize geothermal systems and their relevance to well integrity.
- Detail the design challenges posed by high temperatures and unique fluid chemistries.
- Discuss specific construction and cementing issues in geothermal wells.
- Outline operational integrity management, focusing on corrosion, scaling, and thermal cycling.
- Address P&A considerations for long-term integrity in high-temperature environments.

21.1 Introduction: Harnessing Earth's Heat

Geothermal energy is a renewable energy source derived from the natural heat of the Earth's interior. It represents one of the most promising pathways to decarbonize global energy systems due to its reliability, minimal surface footprint, and capacity to provide base-load power with low greenhouse gas emissions (Abesser et al., 2023; Min, 2018; Stober & Bucher, 2021). Geothermal systems harness subsurface heat through injection and production wells that circulate fluid to and from a hot geological formation—commonly hot sedimentary aquifers, volcanic zones, or engineered reservoirs in enhanced geothermal systems (EGS). The subsurface environment accessed by geothermal wells is vastly different from conventional oil and gas targets. Temperatures often exceed 200 °C (392 °F) and can reach or surpass 400 °C (752 °F) in volcanic regions or deep EGS projects (Eckstein, 2025; Khodayar & Björnsson, 2024)s. In addition to heat, geothermal fluids typically exhibit aggressive chemistries—including high salinity, elevated concentrations of dissolved gases (e.g., CO_2, H_2S), and significant mineral content—which pose continuous threats to well integrity. These conditions accelerate corrosion, degrade elastomers, promote mineral scaling, and challenge traditional cementing materials and casing designs.

Unlike oil and gas production, geothermal operations often involve sustained injection or cyclic injection/production processes, creating substantial thermal gradients within the wellbore. The repeated expansion and contraction of casing and cement during start-up, shut-down, and load-following operations generate complex thermal-mechanical stresses (Zhou et al., 2018). These stresses can lead to fatigue in metallic components and cracking or debonding of the cement sheath, compromising zonal isolation and raising the risk of subsurface leakage or surface safety hazards (Roselli et al., 2024). The mechanical integrity of geothermal wells must also account for large-diameter casing strings needed to deliver the high flow rates required for economic energy conversion. This increases the complexity of achieving complete and durable cement coverage in the annulus—especially when combined with difficult borehole geometries, fractured zones, or lost circulation intervals. Verification of cement placement and bond quality is further hindered by the temperature limitations of most logging tools. The thermal and chemical environment of geothermal wells presents one of the harshest settings for well integrity worldwide. Unlike conventional hydrocarbon wells, geothermal systems experience extreme thermal gradients between injected and formation fluids, and repeated heating and cooling cycles that stress both the casing and the cement sheath. The result is a spectrum of failure mechanisms—ranging from casing buckling and fatigue cracking to cement debonding, radial fractures, and chemical degradation. As illustrated in Fig. 21.1, understanding and mitigating these failure modes is central to achieving reliable, long-term geothermal well performance. Without targeted design and proactive integrity management, the sustainability and safety of geothermal operations can be severely compromised.

Moreover, many geothermal projects are in tectonically active regions, adding another dimension of geomechanical uncertainty. Fault reactivation, induced seismicity, and natural fractures introduce additional risks for well deformation or breach, particularly in EGS systems where hydraulic stimulation is used to create permeability in low-porosity formations. As geothermal deployment expands globally and technologies evolve—from flash steam and binary systems to closed-loop and hybrid configurations—there is a growing emphasis on understanding and improving the long-term integrity of geothermal wells. Ensuring the structural and hydraulic isolation of subsurface formations is essential for safe operations, maintaining system performance, minimizing environmental risks (e.g., brine or gas leaks), and achieving regulatory compliance.

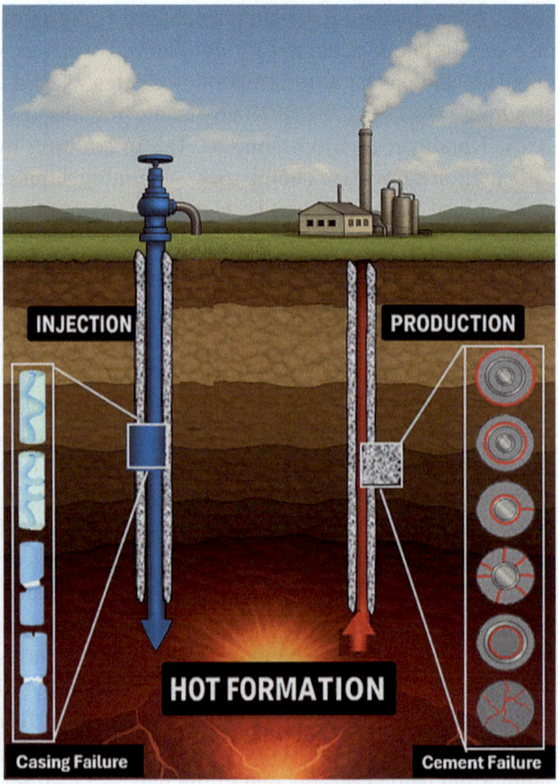

Fig. 21.1 Common well integrity failure modes in geothermal wells. Casing failures (left) include thermal fatigue cracking, collapse, and deformation, while cement failures (right) include debonding, micro-annuli formation, radial cracking, and loss of zonal isolation. These issues are driven primarily by extreme temperatures, thermal cycling, and corrosive fluid chemistries

This chapter provides a comprehensive overview of geothermal well integrity challenges, spanning from initial design and construction through long-term operation and eventual abandonment. It emphasizes the critical role of material selection, thermal modeling, cement performance, and robust monitoring in addressing the unique stressors encountered in geothermal environments. Special attention is also given to Enhanced Geothermal Systems (EGS), which present the most extreme conditions in terms of temperature, pressure, and rock mechanics—offering both tremendous potential and formidable technical challenges. While sharing basic construction principles with oil and gas wells, geothermal wells present a unique and often severe set of well integrity challenges primarily driven by:

- Extreme High Temperatures: Reservoir temperatures can range from moderate (for direct heating) to very high (>300 °C/570 °F or even >400 °C/750 °F in some volcanic or enhanced systems), significantly exceeding typical oil and gas HPHT conditions (Feder, 2021). This places extreme demands on materials and cement.
- Severe Thermal Cycling: Significant temperature differences between injected fluids (often cooler surface water) and hot reservoir rock, combined with production/injection cycles (e.g., intermittent injection/production, startup/shutdown),

induce extreme thermal stresses and fatigue loading on well components and cement. This is a primary integrity challenge (Feng et al., 2020).
- Unique Fluid Chemistries: Geothermal fluids (brines) are often highly saline and can contain dissolved gases (CO_2, H_2S, NH_3), silica, heavy metals, and other species leading to aggressive corrosion and scaling issues, requiring specialized materials (Khasani et al., 2021).
- Variable Formation Types: Ranging from porous sedimentary rocks to fractured crystalline basement rock, impacting drilling and completion strategies.
- System Types: Integrity considerations vary between vapor-dominated (steam production), liquid-dominated (hot water/brine production), and Enhanced Geothermal Systems (EGS–requiring hydraulic stimulation of hot, dry rock, similar to unconventional fracturing but at higher temperatures).

Maintaining well integrity under these conditions is critical for safe, efficient, and long-term operation of geothermal power plants or direct-use facilities, minimizing risks of environmental contamination (e.g., brine leaks), loss of well capacity, or safety hazards (e.g., steam leaks). The diversity of geothermal energy systems—ranging from shallow sedimentary hydrothermal to deep superhot dry rock and engineered closed-loop designs—requires tailored integrity strategies. Figure 21.2 illustrates how geothermal wells differ significantly from water and oil & gas wells in terms of depth, target formations, and system configuration, highlighting the complexity of geothermal well design and monitoring.

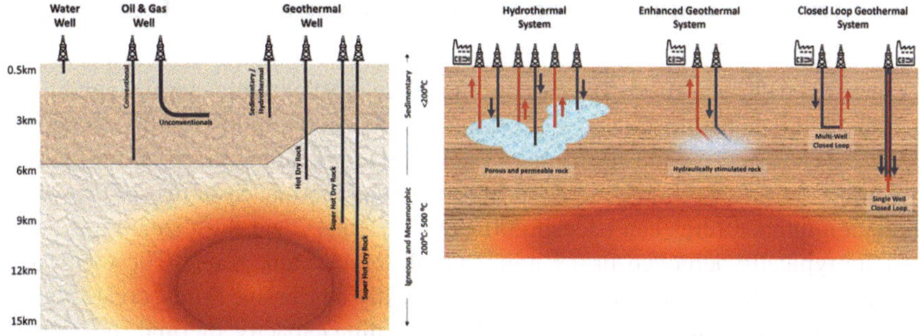

Fig. 21.2 Comparison of well depths and subsurface targets for water, oil & gas, and different types of geothermal systems. The right panel shows key geothermal system configurations: hydrothermal, enhanced geothermal systems (EGS), and closed-loop systems. Each presents distinct challenges for maintaining well integrity, including depth, temperature, rock type, and fluid flow configuration (Izadi & Freitag, 2025)

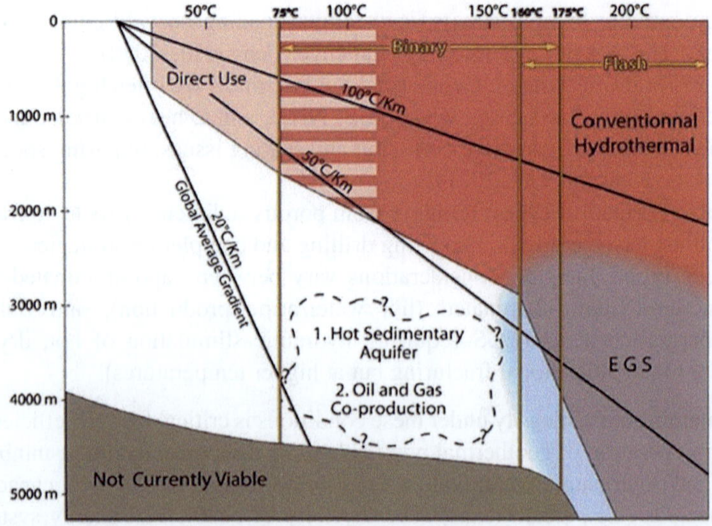

Fig. 21.3 Schematic depth-temperature classification of geothermal resources. The diagram outlines thermal gradients, resource viability zones, and system applications including direct use, binary cycles, flash steam systems, and Enhanced Geothermal Systems (EGS). Also shown are geothermal co-production opportunities from oil and gas fields and hot sedimentary aquifers. (Adapted from EGS)

21.2 Design Challenges for High Temperatures and Fluids

Geothermal wells are often exposed to some of the most extreme thermal conditions encountered in subsurface energy development. As shown in Fig. 21.3, geothermal resource types span a wide spectrum of depth and temperature ranges, from shallow direct-use systems (<100 °C) to deep Enhanced Geothermal Systems (EGS) where formation temperatures can exceed 200–300 °C. Binary and flash steam technologies are typically deployed in the mid- to high-temperature zones. These elevated temperatures, often in combination with chemically aggressive fluids, create significant design constraints on material durability, sealing capacity, thermal stress resistance, and long-term integrity of casings, cements, and elastomeric components. Understanding the thermal regime and resource classification at the outset is essential for appropriate well design, particularly for managing casing stress, cement formulation, and seal system selection in both injection and production wells. The extreme temperature and corrosive fluid environment dictates many specialized design choices:

- Material Selection:
  - Casing/Tubing: Requires careful selection considering significant strength derating at high temperatures (as discussed in Chap. 4). While carbon steels are used in moderate temperature sections, thermal fatigue resistance becomes

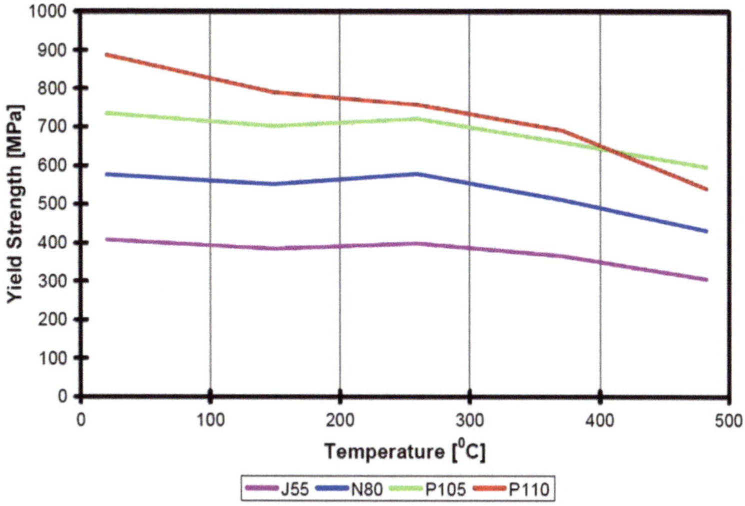

Fig. 21.4 Yield strength degradation with increasing temperature for different casing materials commonly used in geothermal and oil and gas wells (J55, N80, P105, and P110). Thermal exposure above 300 °C significantly reduces strength, particularly for high-strength grades like P110, necessitating careful material selection for geothermal applications (Allahvirdizadeh, 2020)

critical in the hot zones. Corrosion resistance to specific geothermal fluid chemistry (salinity, low pH, H2S, CO2) may necessitate expensive stainless steels or other CRAs (e.g., duplex stainless steels, nickel alloys), significantly impacting cost (Maury Fernandez et al., 2024). Thermal expansion coefficients must be carefully managed in string design. Connection performance under severe thermal cycling is a major concern, often requiring robust premium connections with proven high-temperature, fatigue-resistant, and sealing capability (often MTM seals designed for thermal applications). Geothermal well designs must account for thermal derating of casing materials, as strength loss becomes pronounced at elevated temperatures. As shown in Fig. 21.4, commonly used grades such as J55 and P110 experience measurable declines in yield strength above 200–300 °C, which can compromise structural capacity if not properly accounted for in design (Allahvirdizadeh, 2020).

– Cement Formulations: Portland cement requires very high percentages of silica flour (typically 35–40% or more) to prevent strength retrogression (conversion of stable C-S-H to weaker phases) at temperatures above ~110 °C (230 °F). To further evaluate the performance of cement systems at elevated temperatures, (Fig. 21.5) the impact of average silica particle size on compressive strength and permeability must be considered. While the silica content stabilizes the cement matrix at high temperatures, its particle size distribution significantly affects both strength development and sealing quality. Finer silica particles enhance pozzolanic reactivity and densify the matrix but may lead to higher permeability if poorly graded. Conversely, coarser particles

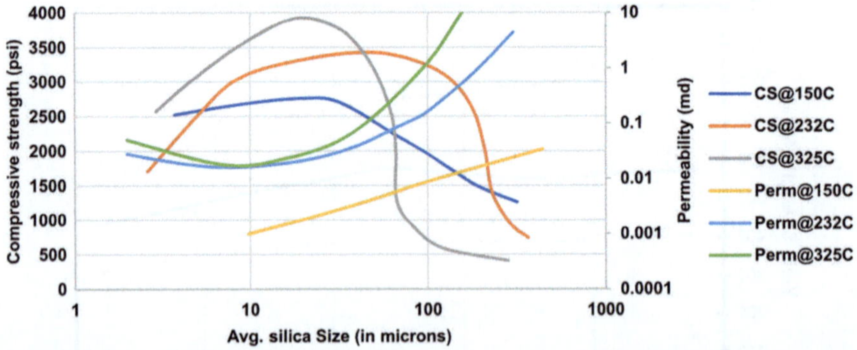

Fig. 21.5 Effect of average silica particle size on compressive strength (psi) and permeability (md) of geothermal cement systems at 150 °C, 232 °C, and 325 °C. Results show the temperature-dependent trade-off between strength and sealing performance, highlighting the need for particle size optimization to balance durability and integrity under geothermal well conditions (Arbad et al., 2022)

reduce early strength gain but may improve long-term performance under certain thermal regimes. The relationship is highly temperature-dependent and optimized through targeted formulation design. Specialized high-temperature retarders are needed to manage thickening time. Achieving good bonding, low permeability, and long-term stability under extreme thermal cycling and potential chemical attack (e.g., CO2, low pH brines) requires careful slurry design and potentially non-Portland-based cements (e.g., calcium aluminates, geopolymers) in very aggressive, high-temperature environments (Kiran et al., 2017). Flexibility (low Young's modulus) and resistance to thermal cracking are highly desirable properties to withstand thermal stresses. One effective way to modify the mechanical properties of cement is to incorporate elastomer particles. When added to cement slurries, elastomers lower the Young's modulus, making the cement more elastic (Al-Yami et al., 2018). To address these concerns, Pyatina & Sugama (2020) developed Thermal Shock Resistant Cement (TSRC) composites with self-healing properties, which demonstrated an 86% recovery in compressive strength (CS) after a five-day healing period, compared to only 36% recovery in Portland cement under similar conditions.

Recent advancements in self-repairing polymer-cement formulations have been explored, including Calcium Aluminate Phosphate (CaP) cement and Sodium Silicate-Activated Slag (SSAS) cement (Sugama, 2007). These materials offer cost-effective solutions, as they incorporate industrial by-products from coal combustion and steel manufacturing. To enhance cement resistance in acidic environments, Achang & Radonjic (2020) recommended incorporating 5% BWOC olivine microparticles, which act as sacrificial materials available for carbonation.

  - Elastomers and Seals: Standard elastomers fail rapidly at typical geothermal temperatures. Packers, expansion joints, and wellhead seals in hot zones rely heavily on high-performance fluoroelastomers (FFKM, sometimes combined with ceramic fillers for added heat resistance) or, more commonly and reliably, metal-to-metal (MTM) sealing technology designed and qualified for extreme temperatures and severe thermal cycling (Patel & Salehi, 2019). Qualification standards need to rigorously test seals under representative combined loads and temperatures.

- **Thermal Stress Analysis**: This is perhaps the most critical aspect of geothermal well mechanical design. Rigorous modeling (often Finite Element Analysis–FEA) is required to predict and manage stresses induced by temperature changes:

  - Casing/Tubing Movement: Significant axial expansion and contraction during heat-up and cool-down cycles (potentially hundreds of feet of movement in deep, hot wells) (Huber & Wijewickreme, 2014; Yohana, 2022; Yohana & Rahyang, 2022). Designs must accommodate this movement (e.g., using specialized expansion joints in tubing strings, or designing floating casing hangers) or the tubulars must be designed to withstand the induced high compressive stresses (buckling risk is significant) and tensile stresses (connection failure risk).
  - Annular Pressure Buildup (APB): Trapped fluids in sealed annuli heating up can generate extreme pressures due to thermal expansion if not properly managed (as discussed in Chap. 18). Engineered annular fluids (e.g., with controlled compressibility, phase change properties), bleed systems, or rupture disks designed for high temperature and cyclic service may be used.
  - Stresses on Connections: Thermal cycling places severe fatigue loads on casing and tubing connections, requiring fatigue-rated premium connections (Maruyama et al., 1990). Thermal cycling in geothermal wells induces significant mechanical stress fluctuations on casing and couplings. As the well heats up during production, the steel expands, building axial compressive stresses. When cooling occurs—such as during injection or shutdown—contraction induces axial tensile stresses. Stresses on Cement: Differential expansion/contraction between casing, cement, and formation induces high shear and tensile stresses at the interfaces (causing debonding) and within the cement sheath (causing cracking), particularly during rapid cool-down events (injection).

- **Large Diameter Designs**: Geothermal wells often require large diameters (e.g., 9–5/8" or larger production casing/liners, potentially up to 13-3/8") to achieve high flow rates needed for economic power generation (King et al., 1995). This increases material costs, complicates casing running and cementing (larger annular volumes, higher risk of poor displacement, heavier pipe), and magnifies the impact of thermal expansion forces.

- **Wellhead Design**: Must be capable of handling high surface pressures and temperatures, accommodate significant thermal expansion and contraction of casing strings (thermal wellheads often allow controlled upward movement of hangers via screw mechanisms), and provide reliable sealing (often MTM seals) under cyclic conditions.

In geothermal systems, wells are subjected to repeated heating and cooling cycles due to fluid injection, production, and operational shut-ins. These **thermal and pressure cycles** impose mechanical stresses on the casing, cement, and surrounding formations, potentially compromising long-term well integrity. This chapter provides a detailed mathematical derivation of the thermal conduction and mass balance equations that govern these effects, followed by a discussion of their implications for casing collapse, cement failure, and zonal isolation.

The one-dimensional transient heat conduction equation in porous rock is:

$$\frac{\partial T}{\partial t} = \kappa \frac{\partial^2 T}{\partial x^2} \tag{21.1}$$

where T is Temperature (K or °C), t is Time (s), κ is Thermal diffusivity (m^2/s), and x is Spatial coordinate (m).

Discretizing Eq. (21.1) using forward difference in time and central difference in space gives:

$$\frac{T_i^{n+1} - T_i^n}{\Delta t} = \kappa \frac{T_{i+1}^n + T_{i-1}^n - 2T_i^n}{\Delta x^2} \tag{21.2}$$

Define the **Fourier number**:

$$F = \frac{\kappa \Delta t}{\Delta x^2} \tag{21.3}$$

Then:

$$T_i^{n+1} = F\left(T_{i+1}^n + T_{i-1}^n\right) + (1 - 2F)T_i^n \tag{21.4}$$

Stability condition: $F < 0.5$

To improve stability, the implicit form is:

$$\frac{T_i^{n+1} - T_i^n}{\Delta t} = \kappa \frac{T_{i+1}^{n+1} + T_{i-1}^{n+1} - 2T_i^{n+1}}{\Delta x^2} \tag{21.5}$$

Which can be expressed as:

$$a_i T_{i-1}^{n+1} + b_i T_i^{n+1} + c_i T_{i+1}^{n+1} = d_i \tag{21.6}$$

The general mass balance in 3D for compressible fluids is:

$$\phi\frac{\partial\rho}{\partial t} + \left(\frac{\partial(\rho u)}{\partial x} + \frac{\partial(\rho v)}{\partial y} + \frac{\partial(\rho w)}{\partial z}\right) = 0 \tag{21.7}$$

In integral form:

$$\frac{\partial A_m}{\partial t} + \nabla \cdot F_m + \dot{q}_m = 0 \tag{21.8}$$

where:

- $A_m = \phi(S_v\rho_v + S_l\rho_l)$: Fluid mass in control volume
- F_m: Mass flux
- $\dot{q}_m$: Source term (injection/production).

Thermal stresses induced in the casing due to this temperature change are given by:

$$\sigma_{th} = E \cdot \alpha_t \cdot \Delta T \tag{21.9}$$

where:

- E: Young's modulus of casing material
- α_t: Linear thermal expansion coefficient
- ΔT: Local change in temperature over time.

In geothermal wells, prolonged exposure to internally sourced heat can increase ΔT, causing significant **hoop, axial, and radial stress changes**, eventually leading to **failure of wellbore integrity barriers**.

21.2.1 Geothermal Gradient and Transient Wellbore Temperature Model

To refine the estimation of geothermal gradients and transient wellbore temperature behavior in geothermal wells, (Al Saedi et al., 2018, 2019a, 2019b; Al Saedi & Kabir, 2020) proposed a comprehensive transient heat conduction model. This approach is especially valuable for assessing temperature profiles during cyclic injection and production phases, which are common in geothermal operations. The model captures the dynamic thermal interactions between injected fluids and surrounding formations by accounting for the exponential integral function.

The **geothermal gradient** can be expressed as:

$$g_T = \frac{T_{ei} + \left(\frac{Q}{4\pi k} \cdot \mathrm{Ei}\left(\frac{r^2 C_e \rho_e}{4k_e t}\right) - T_s\right)}{D_{TV}} \tag{21.10}$$

where:

- g_T = geothermal gradient (°F/ft)
- T_{ei} = initial formation temperature (°F)
- Q = heat flow rate (Btu/hr)
- k = thermal conductivity of the formation (Btu/hr · ft·°F)
- Ei(·) = exponential integral function
- r = radial distance from the wellbore (ft)
- c_e = specific heat of formation (Btu/lbm·°F)
- ρ_e = density of the formation (lbm/ft^3)
- t = elapsed time (hr)
- T_s = surface temperature (°F)
- D_{TV} = true vertical depth (ft)

The corresponding **transient wellbore temperature profile** is given by:

$$T(r,t) = T_{ei} + \frac{Q}{4\pi k} \cdot \mathrm{Ei}\left(\frac{r^2 c_e \rho_e}{4k_e t}\right) \tag{21.11}$$

where $T(r, t)$ is the transient temperature is at a given radial distance r and time t.

These formulations provide a more precise understanding of the temperature variations encountered in geothermal wells, essential for the accurate prediction of thermal stresses, material selection, and long-term integrity management.

In geothermal well design and integrity analysis, the accurate calculation of hydrostatic pressure as a function of depth is crucial due to the strong dependency of fluid density on both pressure and temperature. Unlike conventional wells, geothermal systems experience significant thermal gradients, requiring integration techniques that account for variable fluid properties.

$$dz = f(P,T)\,dP = \frac{1}{g\rho(P,T)}\,dP \tag{21.12}$$

where:

- dz: Incremental depth (m)
- P: Pressure (Pa)
- T: Temperature (K or °C)
- $\rho(P, T)$: Density of the fluid, a function of pressure and temperature (kg/m^3)
- g: Gravitational acceleration (≈ 9.81 m/s^2).

21.3 Construction and Cementing Issues

Drilling and cementing geothermal wells present specific operational challenges intensified by the environment:

- Drilling Fluids: Must remain stable (rheology, fluid loss control, suspension of weighting materials) at very high bottom-hole temperatures (Mohamed et al., 2019). Lost circulation in fractured formations is a common problem. Air drilling or aerated fluids (potentially exacerbating oxygen corrosion risk) are sometimes used. Managing mud properties is key for subsequent cementing success.
- Cement Placement: Achieving complete mud removal and uniform cement placement in large-diameter, potentially deviated holes at high temperatures is difficult. Proper centralization (using an adequate number and spacing of centralizers), pipe movement (rotation/reciprocation if feasible), effective spacer fluids, and optimized slurry rheology (dispersed flow) are crucial. Managing slurry temperature during mixing and pumping (e.g., using chiller units) is important to control thickening time at depth. Managing potential lost circulation zones during cementing may require specialized slurries or stage cementing tools.
- Cement Verification: Running cement evaluation logs (CBL/VDL, ultrasonic) can be challenging due to severe temperature limitations of standard logging tools. Specialized high-temperature tools (often more expensive and with performance limitations) are required. Interpreting logs can be complicated by fractured formations, high-density cements, or variability in set cement properties at high temperatures. Obtaining accurate baseline logs after cementing is critical for future comparison and integrity assessment.

21.4 Operational Integrity Management

Maintaining integrity during the long operational life of a geothermal well requires diligent management of several key threats and ongoing monitoring.

- Corrosion: Geothermal brines can be highly corrosive due to high salinity, low pH (from dissolved CO2, H2S, and potentially organic acids), presence of chlorides, and high temperatures, which accelerate reaction rates. This can affect carbon steel casing, tubing, wellheads, and surface equipment. Material selection (CRAs) is the primary defense in high-risk sections, but corrosion monitoring (e.g., downhole coupons, specialized probes if temperature allows, fluid analysis, periodic caliper/EM logging) is still necessary, especially in sections where carbon steel is used. Chloride stress corrosion cracking (CSCC) can be a significant risk for certain stainless steels (e.g., 13Cr) at elevated temperatures and chloride levels. Geothermal brines can be highly corrosive due to high salinity, low pH (from dissolved CO_2, H_2S, and potentially organic acids), presence of chlorides, and elevated temperatures—all of which accelerate reaction rates and increase the risk of casing, tubing, and wellhead degradation. In particular, wells operating in volcanic regions or with high CO_2 and H_2S concentrations are prone to severe localized corrosion. Figure 21.6 shows an actual liner pipe retrieved from the KJ-39 geothermal well, exhibiting extreme corrosion damage with multiple penetration points and near-total wall loss (Klapper et al., 2019). This highlights the urgent need for proper material selection (e.g., use of corrosion-resistant alloys),

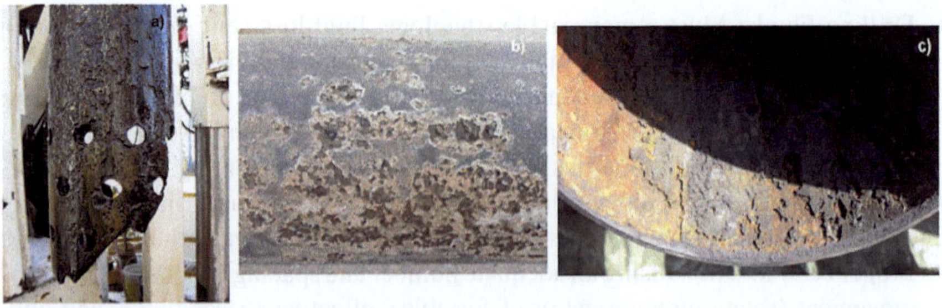

Fig. 21.6 **a** Heavy corrosion of the liner pipe in the KJ-39 geothermal well. The liner exhibits severe wall loss and perforations due to exposure to aggressive geothermal brine conditions (Kruszewski & Wittig, 2018), Outside (**b**) and Inside (**c**) of production tubing after 6-month exposure to geofluids (Klapper et al., 2019)

continuous fluid monitoring, and protective measures like corrosion inhibitors and barrier coatings in high-risk geothermal wells.

- Scaling: Changes in temperature and pressure as geothermal fluids flow up or down the wellbore, or mixing of incompatible fluids (e.g., injected water with formation water), can cause dissolved minerals (primarily silica, calcite, and sometimes sulfides) to precipitate and form hard scale deposits inside tubing, casing, valves, and surface equipment (Matoorian & Malaieri, 2022). Scaling reduces flow rates (injectivity/producibility), can seize or block valves, hinders logging/intervention access, and can potentially hide underlying corrosion or exacerbate MIC if deposits harbor bacteria. Scale management involves chemical scale inhibitor injection (where effective and economical at temperature), operational controls (managing pressure/temperature drop, blending incompatible fluids before they enter the well), and periodic mechanical (milling, scraping) or chemical (acidizing) cleanouts (which carry intervention risks and could potentially damage well integrity).
- Managing Thermal Cycling Effects: This remains a primary concern during operation.
  - Fatigue Assessment: Monitoring operational cycles (number of startups, shutdowns, injection/production switches) and potentially using permanently installed strain sensors (DSS) to assess cumulative fatigue damage on casing, connections, and wellhead components.
  - Connection Integrity: Connections are particularly vulnerable to leaks developing under repeated thermal stress. Monitoring annulus pressures for SCP is essential, although interpretation is complicated by thermal effects on annular fluid pressure. Figure 21.7 illustrates this thermal stress behavior, where the system transitions from elastic compression to potential plastic deformation and eventually to axial tension under cooling. These thermally induced stress reversals can lead to fatigue, casing distortion, or failure at the coupling if not properly accounted for in material selection and connection design.

Yield point exceedance due to repeated cycling poses a major integrity risk in high-temperature wells.

– Cement Bond Degradation: Thermal cycling continues to stress cement bonds and induce cracking in the cement sheath, potentially opening leak paths over time. Many of these issues originate from cement degradation due to mechanical, thermal, or chemical stresses (Lambrescu et al., 2024). At high temperatures, the cement experiences thermal fluctuation in geothermal wells, leading to the development of micro-annuli and micro-cracks due to thermal stresses. Also, cement strength deteriorates either during placement or post-hardening, with the situation worsening in the presence of CO_2 or H_2S, which chemically interact with cement and accelerate its degradation. Previous studies have shown that above the critical temperature range of 104–160 °C, cement strength declines as temperature and aging increase (Knowles et al., 2024).

Research by Buntoro & Rubiandini (2001) examined the effect of neat magnesium oxide (MgO) on the compressive and shear strength of API Class G HSR-type cement. Southon (2005) analyzed casing failure modes, focusing on trapped fluid expansion in the casing-to-casing annulus, suggesting that while tie-back liners could mitigate this issue, they introduce other operational challenges. Proper construction and completion techniques were identified as crucial in minimizing such risks. Regular annulus pressure monitoring is critical, although interpretation is complicated by thermal effects on annular fluids. Permanently installed Distributed Temperature Sensing (DTS) is highly

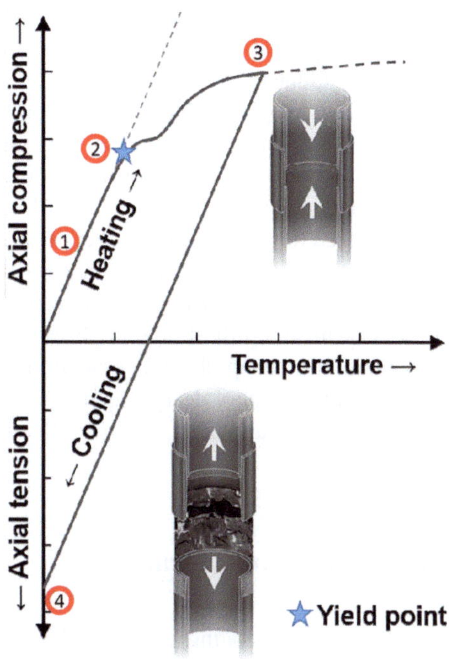

Fig. 21.7 Thermal stress evolution in casing connections during heating and cooling. Heating induces axial compression up to the yield point, beyond which plastic deformation or buckling may occur. Cooling after thermal cycling introduces tensile stresses, risking failure at the casing-coupling interface

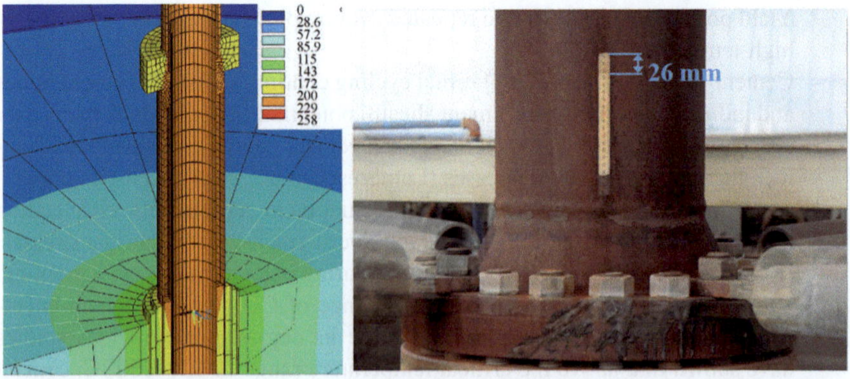

Fig. 21.8 Simulated and observed thermal expansion of a geothermal wellhead. (Left) The modeled temperature distribution nine days after well discharge from HE-46 shows axial thermal gradients. (Right) Field measurement confirms a 26 mm expansion in the master valve spool due to thermal loading. (Jónsson et al., 2017)

valuable for tracking thermal profiles and identifying anomalies potentially related to leaks or flow behind pipe (Chap. 24).

- Monitoring and Logging: Requires logging tools and permanently installed sensors rated for the specific well temperature range. Access for logging may be limited by scale, wellbore geometry, or temperature. Continuous monitoring using permanently installed sensors (DTS, DSS, downhole pressure/temperature gauges) is highly valuable but expensive and technologically challenging at the highest temperatures.
- Wellhead Integrity: Regular inspection and maintenance are needed, focusing on seal integrity, lubrication of thermal movement mechanisms, and managing potential corrosion or scaling at the surface. Geothermal wells experience intense thermal gradients not only downhole but also at the surface wellhead due to prolonged flow testing or production. As heat transfers from the produced fluid to the steel wellhead, thermal expansion can cause measurable axial displacement, especially when thermal growth is unmitigated. Figure 21.8 demonstrates this effect with an uplift of 26 mm observed on the RN-32 well during extended flow testing (Jónsson et al., 2017). Such thermal movement must be accounted for in wellhead and spool design, as it can impact sealing integrity, stress distribution across flanges and bolts, and overall structural stability.

21.5 P&A Considerations

Permanently abandoning geothermal wells presents unique and demanding challenges for achieving long-term isolation.

- High Temperatures: Placing stable, long-term barrier materials (cement or alternatives) that can withstand residual high temperatures (potentially still elevated even after the well is shut-in) and potential chemical attack from geothermal fluids is critical. Standard Portland cement must be formulated with high silica content for temperature stability, but its long-term durability in highly aggressive geothermal brines or under potential future changes in fluid chemistry or temperature is uncertain. Material qualification for specific geothermal P&A conditions is needed. Temperature limitations of placement tools and verification logs must be considered.
- Annular Sealing: Ensuring effective sealing across large diameter annuli, potentially containing thermally degraded and cracked primary cement, can be difficult. Techniques like section milling to remove existing casing and cement and place a qualified permanent barrier directly against the formation (rock-to-rock plug) may be necessary but are challenging at depth and temperature.
- Long-Term Stability: Demonstrating that P&A barriers will remain effective over the very long term in a potentially corrosive, high-temperature, and geomechanically active environment is a significant challenge, similar to but distinct from CCS P&A (discussed in Chap. 24). Research into alternative barrier materials (e.g., specific geopolymers, ceramics) with demonstrated long-term stability in geothermal conditions is ongoing.
- Verification: Developing reliable methods to verify the hydraulic sealing integrity of permanent barriers at high temperatures, especially across the full annulus, remains a challenge.

Geothermal well integrity requires a specialized approach heavily focused on managing extreme temperatures and their consequences (thermal stresses, cycling fatigue, material limitations), alongside aggressive fluid chemistry (corrosion, scaling). Robust design, meticulous construction, and vigilant operational management considering these specific factors are essential for harnessing geothermal energy safely and sustainably over the long term.

References

Achang, M., & Radonjić, M. (2020). Olivine–Portland cement as a solution to wellbore integrity issues in CO_2-rich geothermal reservoirs. *Geothermal Resources Council*. Retrieved January 30, 2025, from https://grc2020.mygeoenergynow.org/olivine-portland-cement-solution-wellbore-integrity-issues-co2-rich-geothermalreservoirs

Abesser, C., Gonzalez Quiros, A., & Boddy, J. (2023). *Evidence Report Supporting the Deep Geothermal Energy White Paper: The Case for Deep Geothermal Energy–Unlocking Investment at Scale in the UK*. https://evidencehub.northeast-ca.gov.uk/downloads/668/nel1435a-geothermal-white-paper-report-v12.pdf

Al Saedi, A. Q., Flori, R. E., & Kabir, C. S. (2018). New analytical solutions of wellbore fluid temperature profiles during drilling, circulating, and cementing operations. *Journal of Petroleum Science and Engineering, 170*, 206–217. https://doi.org/10.1016/J.PETROL.2018.06.027

Al Saedi, A. Q., Flori, R. E., & Kabir, C. S. (2019a). Estimating the initial-formation temperature and flowing-temperature gradient with transient-temperature analysis: Applications in gas reservoirs. *Journal of Natural Gas Science and Engineering, 66*, 126–137. https://doi.org/10.1016/J.JNGSE.2019.03.008

Al Saedi, A. Q., Flori, R. E., & Kabir, C. S. (2019b). Exploring alteration of near-wellbore geothermal gradient during fluid circulation and production. *Journal of Petroleum Science and Engineering, 177*, 909–920. https://doi.org/10.1016/J.PETROL.2019.02.013

Al Saedi, A. Q., & Kabir, C. S. (2020). Estimation of heat-flow rate from fluid-flow rate. *Journal of Petroleum Science and Engineering, 191*, 107203. https://doi.org/10.1016/J.PETROL.2020.107203

Allahvirdizadeh, P. (2020). A review on geothermal wells: Well integrity issues. *Journal of Cleaner Production, 275*, 124009. https://doi.org/10.1016/J.JCLEPRO.2020.124009

Al-Yami, A. S., Alqam, M. H., Riefky, A., & Shafqat, A. U. (2018). Self healing durable cement; development, lab testing, and field execution. In *Proceedings of the SPE/IADC Middle East Drilling Technology Conference and Exhibition, 2018-January*. https://doi.org/10.2118/189397-MS

Arbad, N., Emadi, H., & Watson, M. (2022). A comprehensive review of geothermal cementing from well integrity perspective. *Journal of Petroleum Science and Engineering, 217*, 110869. https://doi.org/10.1016/J.PETROL.2022.110869

Buntoro, A., & Rubiandini, R. R. (2001). *The Effect of Neat Magnesium Oxide (Mgo) as Expanding Additive with Burning Temperature 1200 O C and 1300 O C on Cement Shear Bond Strength at High Pressure and Temperature*.

Eckstein, G. (2025). Who owns the heat? Property rights in geothermal energy. In *University of Illinois Law Review, 2025*. https://scholarship.law.tamu.edu/facscholar/2200

Feder, J. (2021). Geothermal well construction: A step change in oil and gas technologies. *Journal of Petroleum Technology, 73*(01), 32–35. https://doi.org/10.2118/0121-0032-JPT

Feng, G., Wang, X., Kang, Y., & Zhang, Z. (2020). Effect of thermal cycling-dependent cracks on physical and mechanical properties of granite for enhanced geothermal system. *International Journal of Rock Mechanics and Mining Sciences, 134*, 104476. https://doi.org/10.1016/J.IJRMMS.2020.104476

Huber, M., & Wijewickreme, D. (2014). *Response of Buried District Heating Pipelines under Relative Axial Movements*. https://doi.org/10.1115/IPC2014-33530

Izadi, G., & Freitag, H. C. (2025). Resource assessment and management for different geothermal systems (hydrothermal, enhanced geothermal, and advanced geothermal systems). *Geothermal Energy Engineering: Technology Transfer with the Oil and Gas Industry 2025*, pp. 23–72. https://doi.org/10.1016/B978-0-443-21662-6.00003-7

Jónsson, M. Þ., Pálsson, H., & Karlsdóttir, S. N. (2017). *Structural Analysis of the Casings in High-Temperature Geothermal Wells the Deep Roots of Geothermal Systems (Drg) Final Meeting 2017*. Reykjavik.

Khasani, K., Utami, P., & Budiarto, R. (2021). Corrosion in geothermal facilities: Their causes, effects, mitigation, and worldwide cases. In *AIP Conference Proceedings* (Vol. 2338). https://doi.org/10.1063/5.0066755

Khodayar, M., & Björnsson, S. (2024). Conventional geothermal systems and unconventional geothermal developments: An overview. *Open Journal of Geology, 14*(02), 196–246. https://doi.org/10.4236/OJG.2024.142012

King, T. R., Freeston, D. H., & Winmill, R. L. (1995). *A Case Study of Wide Diameter Casing for Geothermal Systems*. https://doi.org/10.2172/895960

Kiran, R., Teodoriu, C., Dadmohammadi, Y., Nygaard, R., Wood, D., Mokhtari, M., & Salehi, S. (2017). Identification and evaluation of well integrity and causes of failure of well integrity barriers (a review). *Journal of Natural Gas Science and Engineering, 45*, 511–526. https://doi.org/10.1016/J.JNGSE.2017.05.009

Klapper, H. S., Schorling, P., Lause, B., Mok, W., van Zonneveld, W., Castricum, A., & Moeller, M. (2019). *Addressing Integrity Issues in Downhole Equipment by Corrosion Inhibition in Sweet Geothermal Brines*. OnePetro. https://dx.doi.org/

Knowles, J. S., Johnson, C. R., Liapunova, N., & Syslak, H. (2024). Specialized carbon dioxide self-healing cement system engineered for carbon capture and storage well in Norwegian continental shelf. In *Offshore Technology Conference Asia, OTCA 2024*. https://doi.org/10.4043/34674-MS

Kruszewski, M., & Wittig, V. (2018). Review of failure modes in supercritical geothermal drilling projects. *Geothermal Energy, 6*(1), 1–29. https://doi.org/10.1186/S40517-018-0113-4/TABLES/2

Lambrescu, I., Abid, K., & Teodoriu, C. (2024). Cement-formation Debonding due to temperature variation in geothermal Wells: An intensive numerical simulation assessment. *Energies, 17*(19), 4953. https://doi.org/10.3390/EN17194953

Maruyama, K., Tsuru, E., Ogasawara, M., Inoue, Y., & Peters, E. J. (1990). An experimental study of casing performance under thermal cycling conditions. *SPE Drilling Engineering, 5*(02), 156–164. https://doi.org/10.2118/18776-PA

Matoorian, R., & Malaieri, M. (2022). Flow Assurance Management in Geothermal Production Wells. In *Society of Petroleum Engineers-SPE Thermal Well Integrity and Production Symposium 2022, TWID 2022*. https://doi.org/10.2118/212144-MS

Maury Fernandez, D., Emadi, H., Hussain, A., & Thiyagarajan, S. R. (2024). A holistic review on wellbore integrity challenges associated with underground hydrogen storage. In *International Journal of Hydrogen Energy* (Vol. 57, pp. 240–262). Elsevier Ltd. https://doi.org/10.1016/j.ijhydene.2023.12.230

Min, H. S. (2018). Carbon Nanotube Wires and Cables: Review.

Mohamed, A., Salehi, S., & Ahmed, R. (2019). Rheological properties of drilling fluids containing special additives for geothermal drilling applications. In *Proceedings* (p. 46).

Newberry Geothermal Lease Project (14513387233) (n.d.). Wikimedia Commons. Public domain (U.S. Government work). https://commons.wikimedia.org/wiki/File:Newberry_Geothermal_Lease_Project_(14513387233).jpg

Patel, H., & Salehi, S. (2019). Development of an advanced finite element model and parametric study to evaluate cement sheath barrier. *Journal of Energy Resources Technology, Transactions of the ASME*, 141(9). https://doi.org/10.1115/1.4043137/725906.

Pyatina, T., & Sugama, T. (2020). Cements with supplementary cementitious materials for high-temperature geothermal wells. *Geothermics, 86*, 101840. https://doi.org/10.1016/J.GEOTHERMICS.2020.101840

Roselli, C., Deng, Y., Chen, L., Wang, K., Wu, J., Corina, A. N., & Moghadam, A. (2024). The sealing performance of cement sheaths under thermal cycles for low-enthalpy geothermal Wells. *Energies, 17*(1), 239. https://doi.org/10.3390/EN17010239

Southon, J. A. (2005). Geothermal Well Design, Construction and Failures.

Stober, I., & Bucher, K. (2021). Geothermal energy: From theoretical models to exploration and development. In *Geothermal Energy: From Theoretical Models to Exploration and Development* (pp. 1–390). https://doi.org/10.1007/978-3-030-71685-1/COVER

Sugama, T. (2007). *Advanced Cements for Geothermal Wells*. https://doi.org/10.2172/909955

Yohana, E. (2022). Tubing movement analysis in completion process at well X. *International Journal of Multidisciplinary Research and Analysis, 05*(10) https://doi.org/10.47191/IJMRA/V5-I10-20

Yohana, E., & Rahyang, I. S. (2022). Tubing movement analysis in completion process at well X. *International Journal of Multidisciplinary Research and Analysis, 05*(10), 2706–2716. https://doi.org/10.47191/IJMRA/V5-I10-21

Zhou, H., Amodio, A., Rodriguez, A. B., Boylan, N., & Deeks, A. (2018). Recent advances in well modeling: Axial interaction between structural casings and soil under cyclic thermal loads. In *Proceedings of the Annual Offshore Technology Conference* (Vol. 3, pp. 1545–1564). https://doi.org/10.4043/28932-MS

Chapter 22
Well Integrity in Enhanced Oil Recovery (EOR)

Abstract This chapter provides a comprehensive analysis of well integrity in the context of Carbon Capture and Storage (CCS), where long-term containment of CO_2 in deep geological formations is critical for climate mitigation. CCS wells, often repurposed from oil and gas operations or newly drilled in saline aquifers and depleted reservoirs, face unique integrity risks stemming from the chemical reactivity of supercritical CO_2, elevated pressures, thermal gradients, and potential reactivation of pre-existing faults or fractures. Ensuring robust zonal isolation and preventing CO_2 migration are paramount for environmental and regulatory compliance. The chapter begins by examining the geochemical environment created by CO_2 injection, highlighting degradation mechanisms affecting Portland cement (e.g., carbonation, leaching, and strength reduction) and steel casings (e.g., corrosion under acidic conditions). It introduces experimental and field-based evidence on CO_2-induced damage to wellbore materials and evaluates long-term stability using reactive transport models. Self-healing cements, pozzolanic additives, and CO_2-resistant coatings are discussed as emerging technologies for improving barrier performance. Mechanically, CO_2 injection alters the stress regime in the wellbore and surrounding formation, potentially leading to thermal fracturing, cement debonding, and micro-annulus formation. Coupled thermo-hydro-mechanical (THM) simulations are presented to assess risk scenarios for leakage pathways, particularly in legacy wells with incomplete or deteriorated primary barriers. Field experiences from projects such as Sleipner, Quest, and Illinois Basin–Decatur are used to illustrate lessons in well design, monitoring, and remediation. The chapter emphasizes the importance of risk-based integrity assessments, including pre-injection wellbore diagnostics, cement bond evaluation, annular pressure testing, and long-term pressure/temperature monitoring. It introduces the concept of the "smart injection field," incorporating fiber-optic sensing, distributed acoustic measurements, and digital twin technologies to enable real-time integrity management. Plugging and abandonment of CCS wells pose additional integrity challenges, requiring the use of permanent barriers that resist CO_2 attack for hundreds to thousands of years. Regulatory frameworks (e.g., EPA Class VI, ISO 27914) are reviewed, along with evolving best practices for long-term liability and post-injection site care. By integrating geomechanical, geochemical, and operational insights, this chapter offers a holistic framework

© The Author(s), under exclusive license to Springer Nature Switzerland AG 2026

A. Alsubaih and K. Sepehrnoori, *Well Integrity for Oil, Gas, and Subsurface Storage Operations*, https://doi.org/10.1007/978-3-032-15233-6_22

for designing, operating, and decommissioning CCS wells with verifiable integrity, positioning CCS as a viable pillar of global decarbonization strategies.

EOR production wellhead system with surface flow lines and control valves, illustrating well integrity and flow management. *Source* The Authors

Chapter's Objectives

- Provide an overview of EOR methods and their specific integrity challenges for both injectors and producers.
- Detail the integrity management requirements for Thermal EOR (steam injection).
- Discuss integrity considerations for Chemical EOR (polymer, surfactant, ASP).
- Outline integrity aspects specific to Gas EOR (CO_2, N_2, hydrocarbon gas).

22.1 Introduction: Pushing Reservoir Limits

Enhanced Oil Recovery (EOR) represents the third stage in the lifecycle of hydrocarbon production, following primary recovery—driven by natural reservoir pressure—and secondary recovery, typically implemented through waterflooding. Despite the success of primary and secondary methods, a substantial portion of the original oil in place (OOIP)—often more than 50%—remains trapped due to capillary forces, poor mobility, or unfavorable sweep efficiency. EOR techniques aim to mobilize this residual oil by altering the reservoir's physical, chemical, or thermodynamic conditions.

EOR encompasses a wide array of processes including thermal (e.g., steam injection), chemical (e.g., polymer, surfactant, or alkaline-surfactant-polymer flooding), and gas-based methods (e.g., CO_2, N_2, or hydrocarbon gas injection). Each method modifies key parameters such as interfacial tension, fluid viscosity, phase behavior, or pressure distribution to enhance oil flow and displacement efficiency. The selection of an EOR technique is contingent on reservoir characteristics, oil viscosity, depth, temperature, and economic feasibility.

While EOR provides a means to maximize recovery and extend the productive life of mature assets, it introduces complex and often severe challenges to well integrity. These challenges are not only operational but also deeply material and chemistry dependent. EOR fluids—whether hot steam, reactive chemicals, or high-pressure gases—interact differently with the wellbore environment compared to traditional injection or production operations. For instance, steam injection leads to rapid thermal cycling, inducing mechanical fatigue, while CO_2 injection may cause aggressive corrosion and cement degradation due to acid formation. Similarly, chemical floods can lead to elastomer degradation, cement destabilization, and formation plugging if not properly managed.

Integrity risks extend to both injector and producer wells. Injectors are subjected to the direct impact of high-pressure, high-temperature, or chemically active fluids, while producers encounter altered reservoir fluids that may include emulsions, gas breakthroughs, or high-viscosity slugs. Moreover, EOR processes can exacerbate pressure and temperature gradients, accelerate wear and corrosion, and induce formation deformation or compaction, all of which pose risks to zonal isolation, tubular stability, and barrier functionality.

Therefore, the success of EOR projects hinges not only on the optimization of reservoir recovery mechanisms but also on the deployment of a robust and tailored well integrity management strategy. This requires the integration of high-performance materials, rigorous quality assurance during construction, real-time monitoring technologies, and adaptive operational practices. Well designs must anticipate long-term exposure to aggressive environments, with special attention to elastomer seal performance, cement resilience, tubular metallurgy, and wellhead configuration.

As EOR becomes increasingly vital in global energy strategies—especially in heavy oil, deepwater, and mature field redevelopment—understanding and mitigating

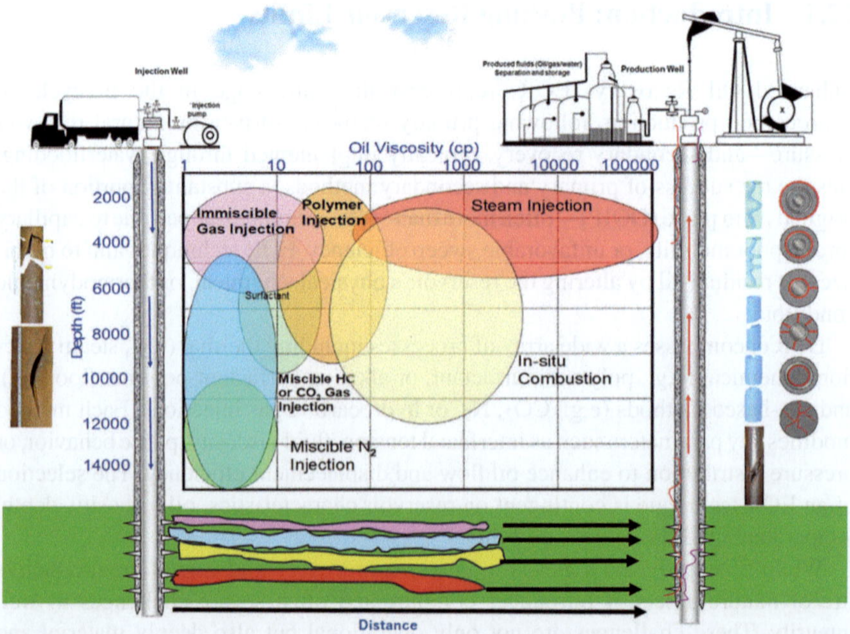

Fig. 22.1 Showing the operating depth and oil viscosity range for various Enhanced Oil Recovery (EOR) methods (steam, gas, polymer, surfactant), overlaid with wellbore configurations for injectors and producers. The injector (left) delivers EOR fluids (e.g., steam, polymer, CO_2) into the reservoir, while the producer (right) recovers the mobilized oil. The Figure also shows possible well integrity failure mechanisms including casing deformation, corrosion, cement debonding, and packer seal failure, highlighting the need for robust integrity management tailored to each EOR type

the integrity risks associated with each EOR process is critical. This chapter addresses these issues in detail, structured by the EOR method, to provide engineers and asset managers with the insights necessary to maintain safe and sustainable operations throughout the extended life of EOR wells. Figure 22.1 provides a comprehensive visualization of how different EOR technologies operate across various reservoir conditions—defined by oil viscosity and depth—while also illustrating typical injection and production well designs. It highlights the fluid flow directions, major failure risks along the wellbore (e.g., casing deformation, chemical attack, thermal cycling effects), and key components impacted by EOR operations. This visual serves as a reference point for the detailed integrity discussions that follow for thermal, chemical, and gas-based EOR methods. Table 22.1 provides a comparative summary of the major EOR technologies and the specific well integrity challenges associated with each. These methods—thermal, gas, chemical, and waterflooding—employ different fluids to enhance oil recovery by mobilizing residual hydrocarbons or maintaining reservoir pressure. However, these fluids also introduce aggressive thermal, chemical, or mechanical environments that can compromise wellbore integrity if not properly

Table 22.1 Summary of EOR Methods and Well Integrity Challenge Overview of common Enhanced Oil Recovery (EOR) techniques, detailing the types of injected fluids and the primary integrity risks they pose to both injection and production wells. Each method introduces unique thermal, chemical, or mechanical conditions that can compromise tubulars, seals, cements, and downhole components if not properly managed through material selection, operational control, and surveillance

EOR method	Injected fluid	Primary integrity challenges (Injectors)	Primary integrity challenges (producers)
Thermal (Steam)	Steam, hot water	– Severe thermal stress and casing deformation (up to 300°C)- Corrosion from high-temperature fluids- Cement degradation due to thermal cycling	– Thermal expansion affecting casing and wellhead components- Sand production from reservoir heating- Corrosion from hot produced fluids
Gas Injection	CO_2, natural gas, nitrogen	– CO_2 corrosion (carbonic acid in wet environments)- High-pressure-induced casing or tubing leaks- Gas migration or breakthrough affecting isolation	– Gas coning or fingering- Corrosion from produced CO_2 or sour gases- Scale or asphaltene deposition
Chemical	Polymers, surfactants, alkalis (ASP)	– Chemical-induced corrosion or elastomer degradation- Injectivity loss from polymer plugging or biofouling- Cement degradation from high-pH fluids	– Emulsion formation reducing production efficiency- Scale or polymer precipitation- Handling and separation of produced chemicals
Waterflooding	Brine, treated seawater, recycled water	– Corrosion from oxygen-rich or saline water- Injectivity loss due to scaling or microbial fouling- Long-term cement degradation from persistent water exposure	– Water breakthrough reducing oil cut- Corrosion from produced brine- Scale deposition in tubing and surface equipment

mitigated. Injectors are directly exposed to the injected fluids under elevated pressure and temperature conditions, while producers may encounter altered formation fluids, breakthrough of injection media, or unintended chemical interactions. Understanding these risks is essential for designing resilient wells and implementing an effective integrity management program tailored to the selected EOR approach.

22.2 Thermal EOR Integrity (Steam Injection)

Thermal EOR, primarily using steam injection (Cyclic Steam Stimulation—CSS, Steam Assisted Gravity Drainage—SAGD, Steam Flooding), is used for heavy oil and bitumen recovery. It involves injecting high-temperature steam (often > 300°C/570°F, potentially up to 350°C/660°F in some designs) into the reservoir to reduce oil viscosity and improve mobility. This process creates significant integrity challenges for injection wells and for nearby production wells operating in the heated zone.

- **Extreme Thermal Cycling and High Temperatures:** This is the defining integrity challenge, similar to geothermal wells but often involving more rapid and extreme temperature fluctuations. Figure 22.2 demonstrates the cyclic mechanical response of L80 casing material under thermal loading, highlighting how rapid temperature swings during steam injection can drive plastic strain accumulation and stress reversal (Jueren Xie, 2008). This behavior underpins the critical need for thermally robust casing designs and material selection in CSS and SAGD operations.
  - Injectors (CSS): Subjected to severe and rapid temperature swings between high-temperature steam injection periods, potentially followed by cooler soak periods, and then hot production phases. This induces massive, rapid thermal stresses and fatigue loading on tubulars, connections, and cement.
  - Producers (SAGD/Steam Flood): Operate continuously at high temperatures (potentially up to steam temperature in SAGD producers) due to contact with the heated reservoir fluids. Producers in steam floods may also experience significant temperature changes.
- **Impacts on Well Components:**
  - Casing/Tubing: Severe thermal expansion and contraction require specialized designs to accommodate significant axial movement (e.g., using expansion joints in production tubing, designing floating casing hangers in the wellhead) (Wu et al., 2008). Tubulars must be designed to withstand high induced compressive stresses (buckling risk is significant) and tensile stresses (connection failure risk) when movement is constrained (Chang et al., 2009). Connection integrity under severe cyclic thermal fatigue is critical; robust premium connections with proven high-temperature, fatigue-resistant, and gas-tight sealing capability (often MTM seals designed for thermal applications) are essential. Strength derating at high temperatures must be applied in design calculations (as discussed in Chap. 4). External casing corrosion can be accelerated by high temperatures if corrosive species are present in the annulus or formation.
  - Cement Sheath: Extreme thermal cycling is highly damaging to cement bonds (debonding at interfaces, especially casing-cement) and can cause cracking (radial and longitudinal) within the sheath, leading to progressive loss of zonal isolation (critical for containing steam within the target zone, preventing

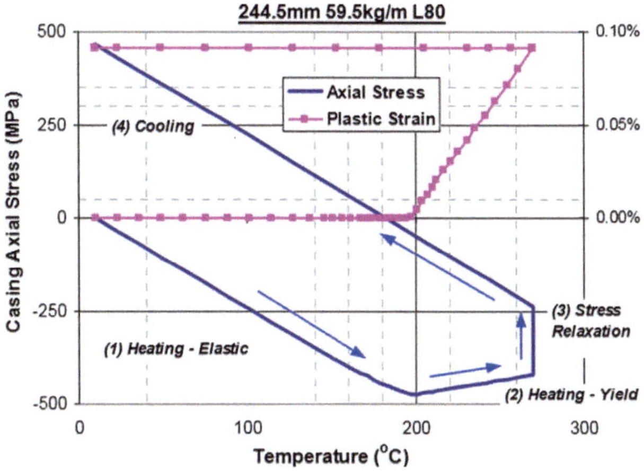

Fig. 22.2 Thermal Stress and Plastic Strain Behavior of L80 Casing During Steam Injection Cycles Axial stress and plastic strain response of L80 casing (244.5 mm, 59.5 kg/m) subjected to cyclic thermal loading from 0°C to 300°C. The graph shows four distinct stages: (1) Elastic heating, (2) Thermal yielding, (3) Stress relaxation, and (4) Cooling-induced stress reversal. This illustrates how repeated thermal cycling in steam injection operations can push casing materials into plastic deformation, contributing to integrity loss, buckling, or failure (Jueren Xie, 2008)

surface heave/steam breakthrough) and potential SCP (Medeiros de Souza et al., 2018). High percentages of silica flour (often 35–40% or more) are mandatory to prevent strength retrogression at high temperatures. As shown in Fig. 22.3, poor cement integrity—represented by high cement void—leads to significantly higher casing strain and curvature during thermal cycles. These mechanical deformations increase the risk of casing ovalization, buckling, and failure, emphasizing the importance of optimized cement slurry designs and placement techniques in thermal EOR wells. Flexible or thermally resilient cement systems designed to accommodate strain and resist thermal cracking are highly (Eric Broni-Bediako et al., 2016).

- Packers/Seals: Require specialized high-temperature elastomers (FFKM with appropriate fillers) or, more commonly and reliably, Metal-to-Metal (MTM) seals designed explicitly for thermal cycling and expansion. Specialized thermal packers designed to accommodate significant tubing movement while maintaining a seal are often used (Surve et al., 2024).
- Wellheads: Specialized thermal wellheads designed to handle high temperatures and high thermal expansion forces are necessary. They often allow controlled upward movement of casing strings via screw mechanisms while maintaining seal integrity. Seals are typically high-temperature elastomers or MTM.

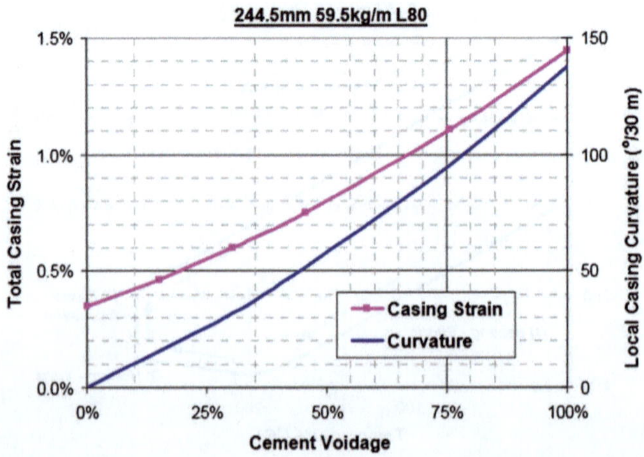

Fig. 22.3 Impact of Cement Voidage on Casing Strain and Curvature during Thermal Loading Graph showing the increase in total casing strain and local curvature with increasing cement voidage for L80 casing (244.5 mm, 59.5 kg/m). As cement voidage increases—indicating poor cement placement or debonding—the casing is subjected to higher mechanical strain and curvature, especially during thermal cycling. This highlights the critical need for high-quality cement placement and void minimization to preserve wellbore mechanical integrity during steam injection operations (Jueren Xie, 2008)

- **Operational Integrity Management:**
  - Monitoring: Tracking temperature profiles (Distributed Temperature Sensing—DTS is widely used and critical for steam chest management), casing strain (Distributed Strain Sensing—DSS), and annulus pressures is crucial for understanding thermal loading, monitoring steam conformance, and detecting leaks or casing deformation (Chap. 24). Annulus pressure management must account for significant thermal effects (APB) and may involve specialized bleed/feed or nitrogen cap systems.
  - Casing Inspection: Periodic inspection logging (caliper, EM thickness) is important to assess casing condition (deformation, particularly ovalization or buckling, internal/external corrosion, wear) after thermal cycles.
  - Sand Control: Heavy oil production often involves significant sand production as viscosity is reduced, requiring effective sand control completions (e.g., screens, gravel packs, slotted liners) to prevent erosion damage to tubulars and equipment and maintain wellbore stability.
  - Artificial Lift: High temperatures and viscous fluids pose challenges for artificial-lift systems (e.g., progressing cavity pumps—PCPs, electrical submersible pumps—ESPs) used in producers, requiring specialized high-temperature rated equipment (Al Jabri et al., 2024).

Thermal EOR well integrity centers on managing the extreme mechanical stresses induced by high temperatures and severe thermal cycling, requiring specialized materials, designs, construction practices, and rigorous monitoring. As shown in Fig. 22.4, casing integrity failures in thermal EOR operations peaked during periods of intense CSS and SAGD development (late 1990s to early 2010s). The data also reveal that most failures were addressed using interventions such as casing patching, liner installation, or cement squeezing—each indicating mechanical damage to well components, likely caused by thermal expansion, poor cement bonding, or pressure cycling. Understanding these patterns is essential to inform proactive well designs and failure prevention strategies.

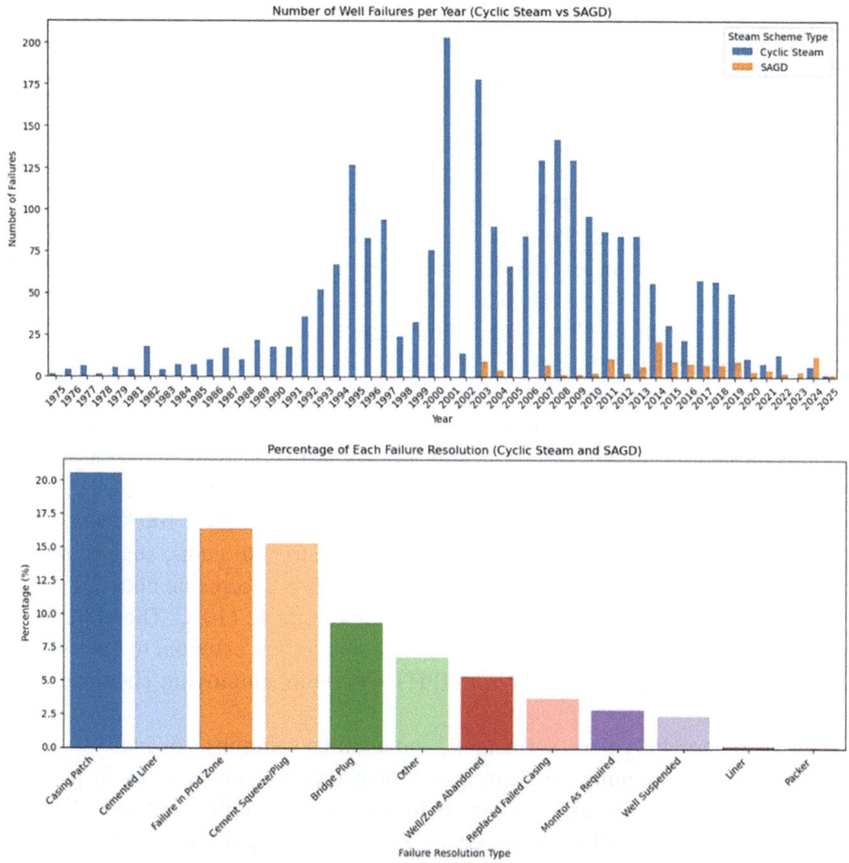

Fig. 22.4 Failure Trends and Remediation Practices in Cyclic Steam and SAGD Wells. (Top) Historical timeline of well integrity failures in thermal EOR wells (CSS and SAGD) from 1975 to 2025, showing the frequency of incidents by year. (Bottom) Distribution of failure resolution strategies, where casing patching, cemented liners, and cement squeezing dominate as the most common remediation approaches. These trends underscore the persistent challenges associated with thermal cycling and the importance of robust well design and monitoring in steam injection environments

22.3 Chemical EOR Integrity (Polymer, Surfactant, Alkaline-Surfactant-Polymer—ASP)

Chemical EOR involves injecting chemicals into the reservoir to alter fluid properties or rock-fluid interactions. This includes injecting polymers (to increase water viscosity and improve sweep efficiency), surfactants (to reduce interfacial tension between oil and water/rock), or combinations (Alkaline-Surfactant-Polymer—ASP).

- **Chemical Compatibility:** This is the primary concern for injected fluids interacting with wellbore materials.
  - Tubulars/Equipment: Assessing potential corrosion caused by injected chemicals or their byproducts, especially if oxygen ingress occurs (e.g., with injection water sources) or if bacterial activity is involved (MIC with biopolymers). Compatibility with existing coatings or inhibitors is important. Dissolution of certain chemicals can also impact pH, affecting corrosion.
  - Cement: While it is widely recognized that high-pH alkaline agents used in ASP flooding can degrade Portland cement over time, recent experimental studies have shown that the severity of degradation is strongly temperature-dependent. (Alias et al., 2016) investigated the impact of polymer and surfactant chemical EOR agents on Class G cement under elevated temperatures (100–130°C) and high-pressure conditions (2000 psi). Their findings indicated that while minimal mass loss (<1%) occurred even after 21 days of exposure, significant reductions in compressive strength were observed at higher temperatures. For example, exposure to surfactant and polymer at 130°C resulted in compressive strength losses of up to 53%, attributed to thermal-induced breakdown of the C-S–H gel and strength retrogression. In contrast, exposures at 100°C and 110°C showed negligible impact or even temporary strength gains, suggesting that cement compatibility with chemical EOR fluids is acceptable under moderate conditions but becomes increasingly vulnerable above 110°C. Therefore, laboratory testing under field-representative conditions is essential to assess long-term durability, particularly for HPHT reservoirs employing chemical EOR strategies.
  - Elastomers: Polymers, surfactants, and alkaline solutions can cause swelling, softening, hardening, or degradation of certain elastomer seals in packers, valves, and wellheads, leading to loss of sealing ability (Chandrasekaran, 2010). Material selection based on extensive compatibility testing for the specific chemical mixture, concentration, temperature, and pressure is vital.
  - Formation Rock: Injected chemicals can react with reservoir minerals or native clays, potentially causing dissolution (leading to subsidence or integrity issues near the wellbore) or precipitation (causing plugging of perforations or the near-wellbore formation) (Dai, 2018).
- **Injectivity Management**

  - Plugging Risk: High viscosity polymer solutions can be difficult to inject and prone to plugging perforations or the near-wellbore formation if not properly designed (considering polymer shear degradation) or if injection water quality (solids, bacterial biomass) is poor (requiring stringent filtration requirements) (Garrepally et al., 2020). Surfactants can create stable emulsions that hinder injection or production. ASP flooding can cause significant scale precipitation or fines migration. Plugging increases injection pressures, potentially exceeding equipment ratings or formation fracture pressure limits, which can compromise integrity.
  - Biological Fouling: Biopolymers can be susceptible to bacterial degradation in injection systems, requiring biocide treatment of injection water to prevent reservoir souring (H2S production via SRB) and wellbore or formation plugging (Spark & Mutch, 2002).

- **Flow Assurance (Producers):** Handling produced fluids containing residual chemicals, potentially viscous emulsions, or altered scales from the chemical flood requires careful management in producer wells and surface facilities to prevent plugging or equipment damage.
- **Zonal Isolation:** Ensuring injected chemical slugs effectively sweep the target zone requires good initial zonal isolation (cement integrity) to prevent channeling behind pipe or via fractures into unwanted zones (thief zones) or directly to nearby producers through unintended pathways (Ibukun et al., 2024). Poor conformance wastes chemicals, reduces recovery efficiency, and can exacerbate integrity risks (e.g., if chemicals channel to shallower, unprotected intervals). Monitoring injection profiles (PLT) and produced fluid composition helps assess conformance and identify potential out-of-zone communication due to integrity failures.

Chemical EOR integrity focuses heavily on ensuring chemical compatibility between injected fluids and all wellbore/reservoir components, managing injectivity (preventing plugging that increases pressure), and maintaining zonal isolation to control fluid movement.

22.4 Gas EOR Integrity (CO_2, N_2, Hydrocarbon Gas)

Injecting gases (e.g., CO_2, nitrogen (N_2), or hydrocarbon gas like methane or associated gas) aims to miscibly or immiscibly displace oil, reduce viscosity, or maintain reservoir pressure.

- **CO_2 Injection:** (Significant overlap with CCS discussed in Chap. 24, applied here to EOR context)

  - Corrosion: As discussed previously (Chaps. 9, 18), wet CO_2 is highly corrosive to carbon steel due to carbonic acid formation. This is a major concern for injectors and for producers that experience CO_2 breakthrough, especially with increasing water cut (Bakar et al., 2021). Injectors and the lower section

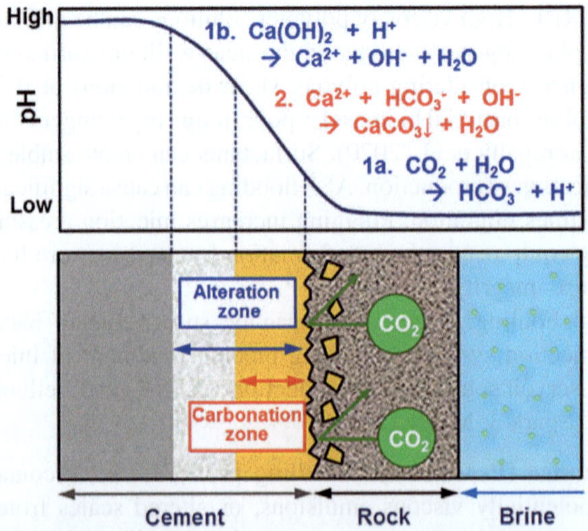

Fig. 22.5 Schematic of Cement–CO_2 Interaction Showing Carbonation and Alteration Zones. Illustration of the chemical interaction between injected CO_2 and well cement, highlighting the formation of two distinct zones: the outer carbonation zone (where CO_2 reacts with cement to form $CaCO_3$) and the inner alteration zone (where pore structure and chemical composition are modified). The upper plot shows the associated drop in pH due to CO_2 dissolution and acid–base reactions, which progressively degrade the cement's microstructure over time (Mito et al., 2015)

of producers require careful material selection (CRAs are often needed, particularly with high water cut), robust continuous corrosion inhibition programs, or specialized internal coatings/liners.

– Cement Degradation: CO_2 attack (carbonation) degrades Portland cement over long exposure times, increasing porosity/permeability and reducing strength, potentially compromising zonal isolation (Abid et al., 2015; Bakar et al., 2021). Requires use of CO_2-resistant blended cements (high pozzolan content, e.g., fly ash, slag, silica fume) or qualified alternative materials for zones exposed to CO_2. When CO_2 is injected during EOR or CCS operations, it dissolves in formation water and forms carbonic acid, leading to a significant drop in pH. This acidic environment initiates a series of chemical reactions with the hydrated cement phases. As shown in Fig. 22.5, these interactions generate two chemically distinct regions in the cement matrix: a carbonation zone, where calcium from cement reacts with bicarbonate to form calcium carbonate ($CaCO_3$), and an alteration zone, where further decalcification and structural transformation occur. Over time, these zones compromise the physical and mechanical integrity of the cement sheath, increasing the risk of zonal isolation failure(Mito et al., 2015).

– Elastomer Performance: Elastomer seals (in packers, valves, wellheads) exposed to high-pressure or dense-phase CO_2 can experience significant swelling, degradation, and Explosive Decompression (ED/RGD) risk (Daou

et al., 2014; O. M. Davies et al., 1999). Requires use of specialized, CO_2-resistant, high-grade elastomers (e.g., specific FFKM, HNBR grades) and potentially MTM seals in critical applications.
- Hydrates: CO_2 can form hydrates under certain P/T conditions if free water is present, potentially causing blockages in tubulars or surface lines.
- Reservoir Sour Gas: If native reservoir fluids contain H_2S, CO_2 injection can alter geochemistry and fluid properties, potentially exacerbating existing sour service challenges.

- **Nitrogen (N_2) or Hydrocarbon Gas Injection**
  - High Pressure Management: Often involves injecting at high pressures (sometimes exceeding formation fracture pressure to create controlled fractures in tight reservoirs), requiring appropriately rated tubulars, packers, wellheads, and connections (Mungan, 2000). Pressure control and monitoring systems must be robust.
  - Material Compatibility: Generally, less corrosive than CO_2 unless H_2S is present in the injected or native gas/fluid, which requires sour service materials according to NACE MR0175/ISO 15,156. Compatibility of elastomers and non-metallic seals with the injected gas type and pressure (ED risk) is important.
  - Thermal Effects: Injection of high-pressure gas can cause significant cooling due to Joule-Thomson expansion near the injection point or downhole restrictions, potentially affecting material toughness or creating hydrate risks if water is present.

- **General Gas EOR Considerations**
  - Zonal Isolation: Critical to prevent injected gas from bypassing oil and channeling directly to producers via high-permeability streaks or behind-pipe leaks (poor cement). Requires good primary cementing and potentially conformance control techniques (e.g., using gels or foams). Monitoring injection profiles and gas breakthrough at producers helps assess conformance.
  - Gas Breakthrough Monitoring: Tracking the arrival of injected gas at producer wells is important for reservoir management and potentially adjusting operations, but also for identifying potential out-of-zone communication indicative of integrity failure.
  - Well Control: Standard well control precautions apply, especially when handling high-pressure gas injection or production.

Each EOR method imposes a unique set of integrity risks that must be understood and proactively managed through appropriate design, material selection, construction quality, operational surveillance, and maintenance practices tailored to the specific process. The complexity and potential severity of EOR environments often necessitate higher levels of technical rigor and investment in integrity management compared to conventional production.

References

Abid, K., Gholami, R., Choate, P., & Nagaratnam, B. H. (2015). A review on cement degradation under CO2-rich environment of sequestration projects. *Journal of Natural Gas Science and Engineering, 27*, 1149–1157. https://doi.org/10.1016/J.JNGSE.2015.09.061

Al Jabri, J., Rubiano, E., & Skoczylas, P. (2024). Extending the limits of high-efficiency artificial lift method in high downhole temperatures: a successful field trial of the mechanically bonded progressing cavity pump system. *Society of Petroleum Engineers—SPE Middle East Artificial Lift Conference and Exhibition 2024, MEAL 2024*. https://doi.org/10.2118/221540-MS

Alias, N. S., Ridha, S., & Yerikania, U. (2016). Effect of chemical EOR XE "EOR" on class G oilwell cement integrity under elevated conditions. *Procedia Engineering, 148*, 1066–1073. https://doi.org/10.1016/J.PROENG.2016.06.593

Bakar, M. A. A., Mohamad, W. A. W., Wahi, M. W. M., & Amir, M. S. E. (2021). Material selection and corrosion rate analysis for CO2 injection well: a case study of K1 field CO2 sequestration project. *International Petroleum Technology Conference, IPTC, 2021*,. https://doi.org/10.2523/IPTC-21818-MS

Chandrasekaran, V. C. (2010). Swelling aspects of rubber related to seal performance. *Rubber Seals for Fluid and Hydraulic Systems, 79–90*,. https://doi.org/10.1016/B978-0-8155-2075-7.10008-5

Chang, J., Ivory, J., & Rajan, R. S. V. (2009). Cyclic steam-solvent stimulation using horizontal wells. *Canadian International Petroleum Conference 2009, CIPC 2009*. https://doi.org/10.2118/2009-175

Dai, C. (2018). Formation damage during chemical flooding. *Formation Damage during Improved Oil Recovery: Fundamentals and Applications, 275–304*,. https://doi.org/10.1016/B978-0-12-813782-6.00007-5

Daou, F., De Miranda, C. R., De Oliveira, J. L., Engelke, B., Borman, C., Le Roy-Delage, S., & Lungwitz, B. (2014). Swelling of elastomers in CO2 environment: Testing methodology and experimental data. *SPE Latin American and Caribbean Petroleum Engineering Conference Proceedings, 1*, 409–424. https://doi.org/10.2118/169277-MS

Davies, O. M., Arnold, J. C., & Sulley, S. (1999). The mechanical properties of elastomers in high-pressure CO2. *Journal of Materials Science, 34*(2), 417–422. https://doi.org/10.1023/A:1004442614090/METRICS

Eric Broni-Bediako, E., friday Joel, O., Ofori-Sarpong Akuffo, G., Broni-Bediako, E., Joel, O. F., & Ofori-Sarpong, G. (2016). Oil well cement additives: a review of the common types. *Oil Gas Res, 2*, 1. https://doi.org/10.4172/ogr.1000112

Garrepally, S., Jouenne, S., Leuqeux, F., & Olmsted, P. D. (2020). Polymer Flooding—Towards a better control of polymer mechanical degradation at the near wellbore. *Proceedings—SPE Symposium on Improved Oil Recovery, 2020-August*. https://doi.org/10.2118/200373-MS

Ibukun, M., Elyan, E., Amish, M., Njuguna, J., & Oluyemi, G. F. (2024). A review of well life cycle integrity challenges in the oil and gas industry and its implications for sustained casing pressure (SCP). In *Energies* (Vol. 17, Issue 22). Multidisciplinary Digital Publishing Institute (MDPI). https://doi.org/10.3390/en17225562

Jueren Xie, Y. L. (2008). *Analysis of Casing Deformations in Thermal Wells*. https://www.researchgate.net/publication/308709003_Analysis_of_Casing_Deformations_in_Thermal_Wells

Medeiros de Souza, W. R., Bouaanani, N., Martinelli, A. E., & Bezerra, U. T. (2018). Numerical simulation of the thermomechanical behavior of cement sheath XE "cement sheath" in wells subjected to steam injection. *Journal of Petroleum Science and Engineering, 167*, 664–673. https://doi.org/10.1016/J.PETROL.2018.04.023

Mito, S., Xue, Z., & Satoh, H. (2015). Experimental assessment of well integrity for CO2 geological storage: Batch experimental results on geochemical interactions between a CO2–brine mixture and a sandstone–cement–steel sample. *International Journal of Greenhouse Gas Control, 39*, 420–431. https://doi.org/10.1016/J.IJGGC.2015.06.007

Mungan, N. (2000). Enhanced oil recovery XE "enhanced oil recovery" with high pressure nitrogen injection. *SPE/AAPG Western Regional Meetings, 253–256*,. https://doi.org/10.2118/62547-MS

Spark, I., & Mutch, K. (2002). *Microbial Monitoring to Minimise Corrosion and Formation Damage in the Oil Industry.* https://doi.org/10.2118/78572-MS

Surve, K., Cai, M., Yun, J., Zolfaghari, A., Krishnamoorti, R., & Bhowmick, A. K. (2024). Effect of extreme environments on aging of fluoroelastomers. *Polymer Degradation and Stability, 227*, Article 110875. https://doi.org/10.1016/J.POLYMDEGRADSTAB.2024.110875

Wu, J., Knauss, M. E., & Kritzler, T. (2008). Casing Failures in Cyclic Steam Injection Wells. *Society of Petroleum Engineers - IADC/SPE Asia Pacific Drilling Technology Conference 2008 - "Drilling Technology to Access Future Energy Demand and Environmental Challenges,"* 26–32. https://doi.org/10.2118/114231-MS

Chapter 23
Integrity Management for Water Injection Wells

Abstract This chapter explores the multifaceted integrity challenges associated with water injection wells—critical components of secondary recovery operations aimed at maintaining reservoir pressure and enhancing hydrocarbon production. These wells must reliably inject large volumes of water over extended timeframes, often under high pressure and in chemically aggressive environments. The chapter outlines the key well designs and operational principles governing water injection, emphasizing the central role of injection water quality in sustaining well performance and structural integrity. The chapter first addresses the detrimental effects of poor water quality, highlighting mechanisms such as internal corrosion from oxygen, CO_2, and H_2S; microbiologically influenced corrosion (MIC) by sulfate-reducing bacteria; and inorganic scaling caused by mixing incompatible water chemistries or temperature/pressure shifts. These phenomena can significantly reduce injectivity, damage tubulars, and compromise zonal isolation. Effective mitigation strategies—such as water filtration, deaeration, biocide, and scale inhibitor treatments—are examined in detail. Injectivity and conformance are then analyzed from an integrity perspective. Plugging, casing corrosion, and poor zonal isolation can impair water sweep efficiency and result in out-of-zone injection, undermining reservoir management objectives. Monitoring techniques such as Production Logging Tools (PLT), Distributed Temperature Sensing (DTS), and acoustic diagnostics are presented as essential tools for detecting water flow paths and identifying integrity failures. A detailed geomechanical discussion follows, focused on pressure containment and casing stress evolution during high-pressure injection. The chapter presents evidence from numerical simulations and field data that excessive injection pressures, particularly near faults or in weak formations, can cause casing deformation, tensile failure, or loss of cement bond. Historical trends from major fields (e.g., Daqing and Pubei) are used to illustrate the correlation between injection pressure and casing failure incidence. Finally, the chapter examines long-term integrity management strategies for water injectors. These include cumulative corrosion assessment using periodic logging, fatigue evaluation under cyclic pressure conditions, and cement degradation from persistent chemical exposure. Integrity considerations during well abandonment (P&A) are discussed, highlighting the challenges of sealing altered zones and dealing with residual pressure build-up. By integrating chemical, mechanical,

© The Author(s), under exclusive license to Springer Nature Switzerland AG 2026
A. Alsubaih and K. Sepehrnoori, *Well Integrity for Oil, Gas, and Subsurface Storage Operations*, https://doi.org/10.1007/978-3-032-15233-6_23

and operational factors, this chapter provides a comprehensive framework for safeguarding the integrity of water injection wells throughout their lifecycle. The insights presented are critical for optimizing injection performance, protecting subsurface environments, and ensuring regulatory compliance in mature field development.

Wellhead and production manifold in a desert field, illustrating surface equipment integrity and flow control systems in water injection well. *Source* The Authors

Chapter's Objectives

- Describe the purpose and common configurations of water injection wells.
- Detail the impact of injection water quality on corrosion, scaling, and plugging.
- Discuss the importance of injectivity and conformance for reservoir management and integrity.
- Outline pressure containment requirements and monitoring strategies.
- Address long-term integrity considerations for water injectors.

23.1 Introduction: Supporting Production

Water injection wells play a vital role in maintaining reservoir pressure and enhancing oil recovery in mature fields. Their primary function is to inject water into targeted reservoir formations, displacing hydrocarbons toward production wells and sustaining pressure above the bubble point to maximize hydrocarbon recovery (Abdeli et al., 2024). This process, known as secondary oil recovery or waterflooding, is a cornerstone of field development strategies in many oil fields worldwide.

Water injection has been implemented extensively because it is a cost-effective and proven method to enhance recovery efficiency and extend the economic life of a reservoir. By replacing the volume of produced fluids with injected water, these wells ensure that reservoir pressure remains sufficiently high to drive hydrocarbons toward the producing wells. In addition, water injection can help improve sweep efficiency—ensuring that more of the reservoir's oil is mobilized and produced, rather than bypassed. While the fundamental concept of water injection is relatively simple, the practical implementation of water injection wells introduces a range of unique challenges and integrity considerations. These wells must handle large volumes of injection water—often sourced from a variety of origins, including surface water (such as seawater or river water), aquifers, or treated produced water. As a result, maintaining the quality of this water is essential to prevent issues like corrosion, scaling, and plugging, which can compromise well integrity and injectivity. Furthermore, the integrity of injection wells is crucial for achieving the intended reservoir management goals. Poor injectivity due to plugging, or poor conformance due to water channeling into unintended zones, can reduce the effectiveness of waterflooding operations and ultimately impact hydrocarbon recovery. Even worse, integrity failures in these wells—such as casing corrosion, cement sheath degradation, or loss of zonal isolation—can lead to environmental incidents, including unintended water migration to sensitive formations or even to the surface. Water injection wells must also be designed and operated to safely contain the high injection pressures required for effective displacement of fluids in the reservoir. Failure to control injection pressures can lead to fracturing of the reservoir or overlying formations, with serious implications for wellbore integrity, environmental protection, and regulatory compliance (Bringedal et al., 2006). In addition to these immediate operational concerns, long-term integrity management of water injection wells is essential. Over the course of decades-long field lifespans, injection wells are subject to cumulative damage mechanisms—including corrosion, scale build-up, and fatigue. These mechanisms can progressively degrade well components and lead to complex repair or abandonment challenges if not managed proactively. As a result, the integrity of water injection wells is a multifaceted challenge that requires a holistic approach—integrating robust water treatment practices, precise pressure control, monitoring of injectivity and conformance, and diligent inspection and maintenance practices. By addressing these factors, operators can ensure that water injection wells contribute effectively to reservoir management goals while protecting the environment and upholding the highest standards of well integrity and safety. Figure 23.1 demonstrates the fundamental process of water injection and production in a typical waterflood operation. Injection wells deliver treated water into the reservoir, displacing oil and maintaining reservoir pressure to enhance hydrocarbon recovery. The produced hydrocarbons are lifted to the surface using a beam pumping unit in the production well. This integrated approach ensures a continuous energy supply within the reservoir and promotes efficient oil displacement and extraction.

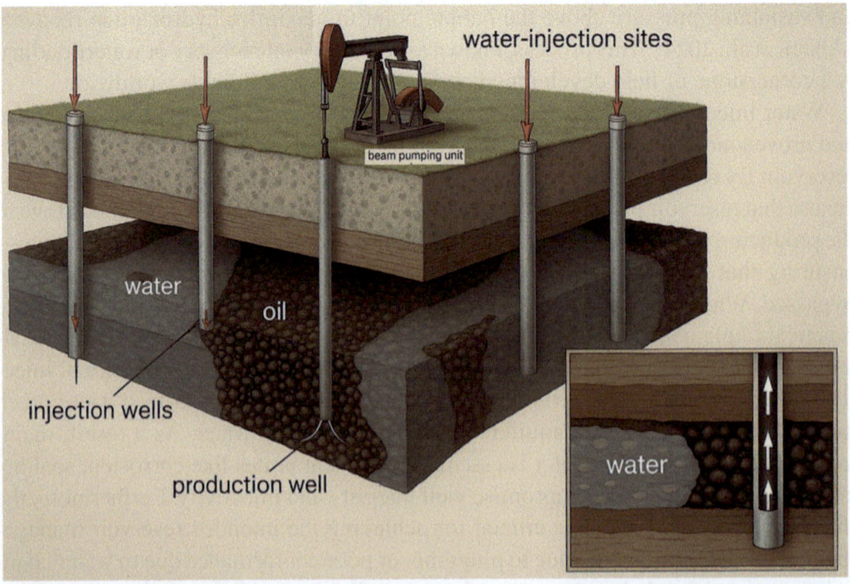

Fig. 23.1 Illustration of water injection wells and production wells in an enhanced oil recovery system. The Figure highlights the flow of injected water into the reservoir formation to maintain pressure and mobilize hydrocarbons toward the production well, which is completed with a beam pumping unit for hydrocarbon lifting

23.2 Impact of Injection Water Quality

The physical and chemical properties of the injected water are arguably the most significant factors influencing the integrity and performance of water injection wells. Source water can be surface water (seawater, river water, lake water), aquifer water (groundwater), or produced water (process water separated from produced hydrocarbons, often treated and re-injected—PWRI). Water quality requires careful management through treatment processes.

- **Corrosion:** Injection water can be a major source of internal corrosion in carbon steel tubulars.
  - Dissolved Oxygen: This is often the primary culprit for severe internal pitting corrosion in water injectors using surface water sources (sea, river, aquifer) if not properly treated. Even trace amounts of oxygen (<10–20 ppb) can cause aggressive pitting corrosion in carbon steel tubing and casing. Effective deaeration (e.g., vacuum towers, gas stripping using nitrogen or fuel gas) and chemical oxygen scavenger treatment (e.g., sodium bisulfite, ammonium bisulfite) are essential but require constant monitoring and control (Al Khamis et al., 2014;

Bhatkar et al., 2023) High water velocities increase the rate of oxygen transport to the metal surfaces, leading to faster corrosion. Furthermore, velocity-induced removal of protective films exacerbates the issue, leaving the material exposed to corrosive substances like dissolved oxygen, acidity (low pH), and high temperatures. These factors significantly contribute to the corrosive environment in water injection systems (Gordon & Survey, 2016). Elevated water velocities can accelerate the transport of oxygen to the metal surface, thus enhancing the corrosion rate. Additionally, high velocities can mechanically remove protective scales or films that inhibit corrosion, further increasing the degradation of the well components.

- CO_2/H_2S: If present in the source water, dissolved from the reservoir, or generated by bacterial activity, these components cause acid corrosion of carbon steel similar to producer wells (Vignes et al., 2008).
- Salinity (Chlorides): High chloride levels, common in seawater or produced water, exacerbate corrosion, particularly pitting, and potentially contribute to stress corrosion cracking in susceptible materials.
- Microbiologically Influenced Corrosion (MIC): Injection water systems provide ideal environments for bacterial growth (e.g., Sulfate-Reducing Bacteria—SRB, Acid-Producing Bacteria—APB, slime formers) if not properly treated with biocides. SRB can metabolize sulfates to produce H2S, leading to sour corrosion and pitting, often occurring under biofilms or deposits where biocides cannot easily penetrate (Askari et al., 2021). Regular biocide treatment (continuous or batch) and maintaining cleanliness are usually required. Figure 23.2 highlights the most common corrosion and cracking processes that compromise the structural integrity of water injection well tubulars. The illustration shows how internal and external corrosion, hydrogen-induced cracking, flow-induced erosion, and stress corrosion cracking manifest in the wellbore environment. These damage mechanisms can arise due to poor water quality, high injection pressures, microbiologically influenced corrosion (MIC), and mechanical fatigue—each posing significant risks to the casing and tubing's long-term performance. Proper understanding of these mechanisms is crucial for designing effective water treatment, selecting corrosion-resistant materials, and planning ongoing integrity monitoring and maintenance strategies.

- **Scaling**
  - Compatibility Issues: Mixing injected water (with one ionic composition) with native formation water (with a different ionic composition) in the reservoir or wellbore can lead to precipitation of mineral scales when solubility limits are exceeded. Common scales include calcium carbonate ($CaCO_3$), calcium sulfate ($CaSO_4$—anhydrite, gypsum), and barium sulfate ($BaSO_4$—barite) (Askari et al., 2021). This reduces injectivity by plugging perforations and the near-wellbore formation and can hinder logging or intervention access. Scale prediction modeling based on water chemistries and compatibility testing is crucial before injection begins.

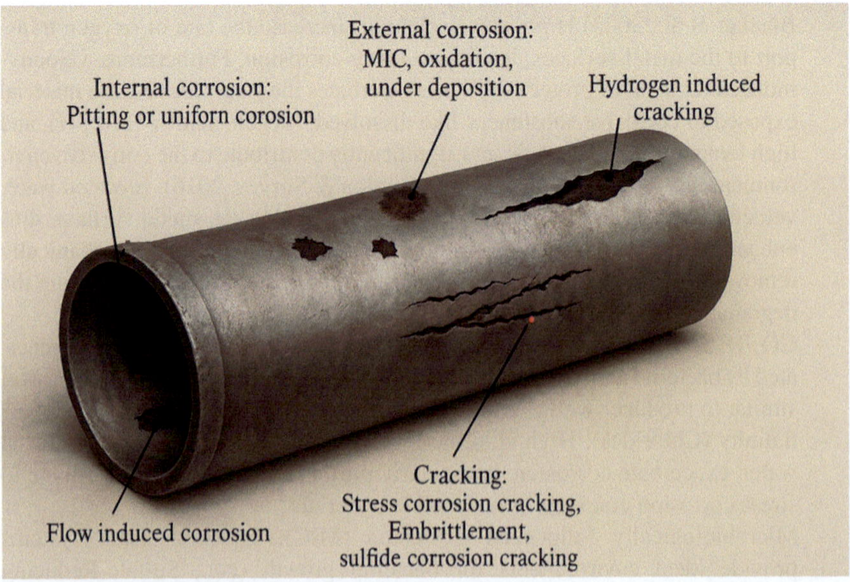

Fig. 23.2 Illustration of typical corrosion and cracking mechanisms in water injection well tubulars, showing internal corrosion, external corrosion, hydrogen-induced cracking, flow-induced corrosion, and stress corrosion cracking. Understanding these damage mechanisms is essential for managing the long-term integrity of water injection wells

- Temperature/Pressure Changes: Changes in temperature and pressure as water flows down the wellbore (especially if cooling occurs) can also shift solubility limits and cause scaling.
- Corrosion Products: Iron oxides/sulfides generated by corrosion can themselves form scales or deposits, further exacerbating corrosion (under-deposit corrosion) and plugging.
- Mitigation: Scale inhibitor chemicals are often added to injection water (continuous injection) to prevent crystal growth. Periodic scale removal treatments (chemical using acids or chelants, or mechanical using coiled tubing/wireline tools) may be needed, which carry intervention risks and could potentially damage well integrity.

- **Solids and Particulates**
  - Source: Suspended solids from the source water (e.g., sand, silt), corrosion products (e.g., iron oxides/sulfides), bacterial biomass, precipitated scales, or fines migrated from the formation.
  - Impact: Solids can cause erosion damage at high flow velocities (though less common than in producers unless injection rates are very high)(Vignes et al., 2008). High velocity can cause mechanical scouring of the well's internal surfaces, particularly if solid particles are present in the water. Cavitation, which occurs when pressure drops lead to bubble formation and collapse, can

also cause material damage (Speight, 2015). To mitigate erosion, engineers may design wells with additional erosion allowances or use materials that offer higher resistance to wear and tear. Primarily, however, they lead to plugging of perforations and the formation face, significantly reducing injectivity and increasing pressure requirements. This can force the operator to inject at pressures exceeding the formation fracture pressure to maintain rates, leading to unintended out-of-zone injection (fracturing) or potential damage to the caprock.

- Mitigation: Effective filtration of injection water to stringent quality specifications (e.g., particle size limits, turbidity) is crucial before injection. Produced water re-injection (PWRI) systems require particularly robust treatment to remove oil, solids, and bacteria. To manage corrosion and erosion effectively, several mitigation strategies can be implemented. Using corrosion-resistant alloys (CRAs) is one option to reduce corrosion damage (Metikoš-Huković et al., 2011). Reducing water velocity can also significantly decrease both the rate of corrosion and erosion (Li et al., 2022). Furthermore, filtering the water to remove solid particles can prevent mechanical scouring, while corrosion inhibitors can help prevent the formation of corrosion products, thus extending the lifespan of the wellbore components.

Managing injection water quality through appropriate treatment (deaeration, filtration, biocide treatment, scale/corrosion inhibitors) is fundamental to maintaining the integrity, injectivity, and performance of water injection wells.

23.3 Injectivity and Conformance

Maintaining the well's ability to accept the desired water volume at acceptable pressures (injectivity) and ensuring the injected water enters the intended reservoir zone(s) and sweeps oil effectively toward producers (conformance) are critical for reservoir management and production optimization. Well integrity problems directly impair these objectives:

- **Maintaining Injectivity:** The ability to inject the desired water volume at acceptable pressures is key to sweep efficiency. Well integrity problems directly impair injectivity:
  - Plugging: Perforation or near-wellbore plugging by solids, scale, or biomass increases injection pressure, potentially exceeding equipment ratings or requiring injection at pressures that fracture the formation.
  - Corrosion Damage: Severe internal corrosion can lead to restrictions or casing collapse, hindering injection flow paths (Saintier et al., 2019).
  - Wellbore Fill: Accumulation of debris or settled solids can block flow paths.

- Remediation: Interventions to restore injectivity (e.g., acid stimulation, reperforating, mechanical cleanouts, scale removal treatments) must be carefully planned and executed to avoid damaging well integrity (e.g., creating communication behind pipe via micro-annuli, casing corrosion from incompatible acid, mechanical damage during cleanout).

- **Conformance Control:** Ensuring the injected water enters the intended reservoir zone(s) and sweeps oil effectively toward producers is critical for reservoir management. This relies heavily on well integrity and monitoring:
  - Zonal Isolation: Effective cement bonding and packer seals are required to prevent injected water from channeling behind casing into unwanted, non-productive zones (thief zones), or overlying/underlying formations (e.g., gas caps, water aquifers), or directly to nearby producers through unintended pathways. Poor conformance wastes injection water, bypasses oil (reducing recovery efficiency), and can exacerbate integrity risks (e.g., injecting into a permeable shallow zone lacking robust outer casing and cement barriers).
  - Monitoring Injection Profiles: Periodic logging (using Production Logging Tools—PLT, Distributed Temperature Sensing—DTS, or Distributed Acoustic Sensing—DAS) is used to determine where the injected water is actually going downhole along the perforated interval and potentially in the annulus, verifying conformance and identifying potential out-of-zone injection or channeling due to integrity failures (e.g., cement channels, casing leaks below the packer) (Chap. 24).

23.4 Pressure Containment

Injection pressures are a critical integrity parameter that must be carefully managed.

- **Managing Injection Pressure:** Water injection in oilfields, while beneficial for maintaining reservoir pressure, can lead to significant well integrity challenges, particularly casing damage. The process of water injection can cause imbalance in the injection-production pressure difference, resulting in formation uplift and tensile deformation of the casing. This deformation occurs due to the redistribution of stresses in the surrounding rock and wellbore, especially under high-pressure conditions. Numerical simulations have shown that high water injection pressures can result in casing tensile failure as a result of axial stress build-up, further exacerbated by fluid–solid interactions between the casing, cement, and reservoir (Liu & Yuan, 2022). In regions like Daqing, a significant number of casing failures are associated with high-pressure water injection, which induces stress concentrations that lead to tensile damage, particularly near fault zones or areas with formation dislocations (Zhang et al., 2023).Water is typically injected at pressures high enough to displace fluids and maintain reservoir pressure but below the pressure that would fracture the reservoir cap rock or overlying formations (the fracture gradient). Exceeding the fracture pressure can lead to:

  - Loss of injected water out of zone into unintended formations.
  - Potential communication with other wells or formations through induced or reactivated fractures.
  - In severe cases, the broaching of injected water to the surface (especially for shallow injectors).
  - This requires accurate knowledge of formation fracture gradients (often determined via step-rate tests, DFITs, or minifracs) and reliable surface pressure control and monitoring, including establishing Maximum Allowable Wellhead Injection Pressure (MAWIP) limits. Exceeding MAWIP should trigger automated shutdowns.

- **Tubing and PackerIntegrity:** The tubing string and packer typically form the primary barrier containing injection pressure and protecting the production casing from direct exposure to high-pressure injection water. Leaks in the tubing or packer allow high-pressure injection water to enter the A-annulus, potentially over-pressuring the production casing beyond its design limits or causing external corrosion. Regular monitoring of A-annulus pressure is essential (as discussed in Chap. 8), as significant pressure build–up can indicate primary barrier failure.
- **Casing and Cement Integrity:** The production casing string and its cement sheath form the secondary barrier. Failures (corrosion holes, connection leaks, cement channels) can allow injection water to leak into outer annuli or surrounding formations, bypassing the reservoir. Outer annulus pressure monitoring is required to detect such leaks. Long-term exposure of cement to potentially aggressive injection water chemistry (e.g., with high CO_2/H_2S or low pH) needs consideration regarding potential long-term degradation.Water injection can induce casing deformation due to differential pressures between strata, especially when faults are present. The presence of faults can significantly impact casing integrity as the injection fluid seeps into fault zones, triggering shear failure along the fault plane, which in turn affects the well casing. Studies in fields such as the Sabei Development Zone have highlighted that casing damage is most concentrated in areas where the formation pressure differs drastically across blocks, particularly when high-pressure water injection leads to excessive pressure differences (Ding et al., 2023). These conditions can cause not only mechanical failure of the casing but also compromises in the overall production efficiency, making it essential for operators to carefully monitor injection pressures and adjust them to avoid triggering casing damage in vulnerable areas (Guo et al., 2015). The casing damage in oilfields like Pubei has been observed to increase with water injection pressure, where the most common forms are deformation and dislocation. This increase in damage is closely tied to the pressure and the geological conditions such as faults and dip angles, which exacerbate the deformation of the casing.Furthermore, casing damage from high-pressure water injection can be caused by factors like channeling, where water surpasses fracture pressure in poorly cemented areas, leading to large immersion zones and strata stress imbalances (Han et al., 2006). This issue is most pronounced when high-pressure water injection results in localized slipping of formation layers, causing shear damage

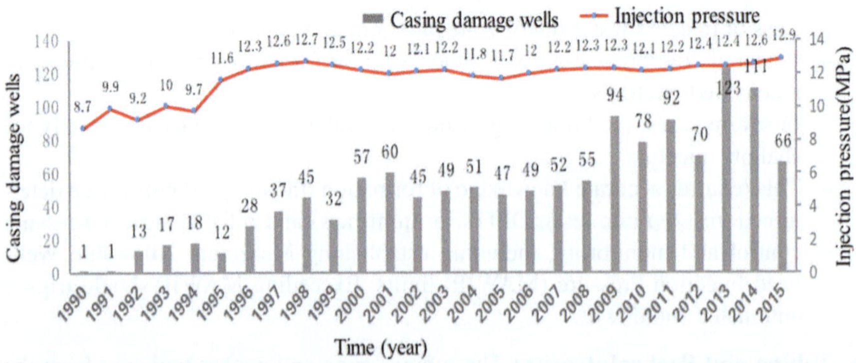

Fig. 23.3 The relationship between injection pressure (red line, right axis) and the number of casing damage wells (gray bars, left axis) over time from 1990 to 2015. The data highlights how sustained high injection pressures correlate with a steady increase in the number of casing damage incidents, reflecting the importance of pressure control in injection well integrity management (Yao, 2021)

in the casing due to relative movement between formation interfaces. In Daqing, the effect of increasing water injection pressure on casing damage was evident, with a direct correlation between higher injection pressures and increased casing failure (Han et al., 2006). Figure 23.3 presents a clear visual representation of the long-term relationship between injection pressure and casing damage incidents in water injection wells. The increasing number of casing failures over the years closely tracks with elevated injection pressures, underscoring the critical role of precise pressure management in preserving well integrity. This trend also highlights the cumulative impact of sustained high-pressure injection on casing integrity, emphasizing the need for proactive monitoring and mitigation measures.

23.5 Long-Term Integrity Management

Water injection wells often operate for decades as a field matures, requiring a long-term perspective on integrity that considers cumulative effects. Over the long operational life of water injection wells, mechanical stresses—including those from fault reactivation—can compromise casing and cement integrity. Figure 23.4 illustrates the stress distribution in the casing and cement sheath during water injection scenarios, showing how fault plane reactivation can trigger shear failure and structural damage (Tuo et al., 2024).

- **Cumulative Corrosion:** Assessing the total internal metal loss over the well's life through periodic logging (caliper, EM thickness) or modeling is essential, especially if water treatment has been inconsistent or challenges arise (e.g., SRB excursions). Determining remaining structural capacity is critical for continued safe operation or future interventions/abandonment.

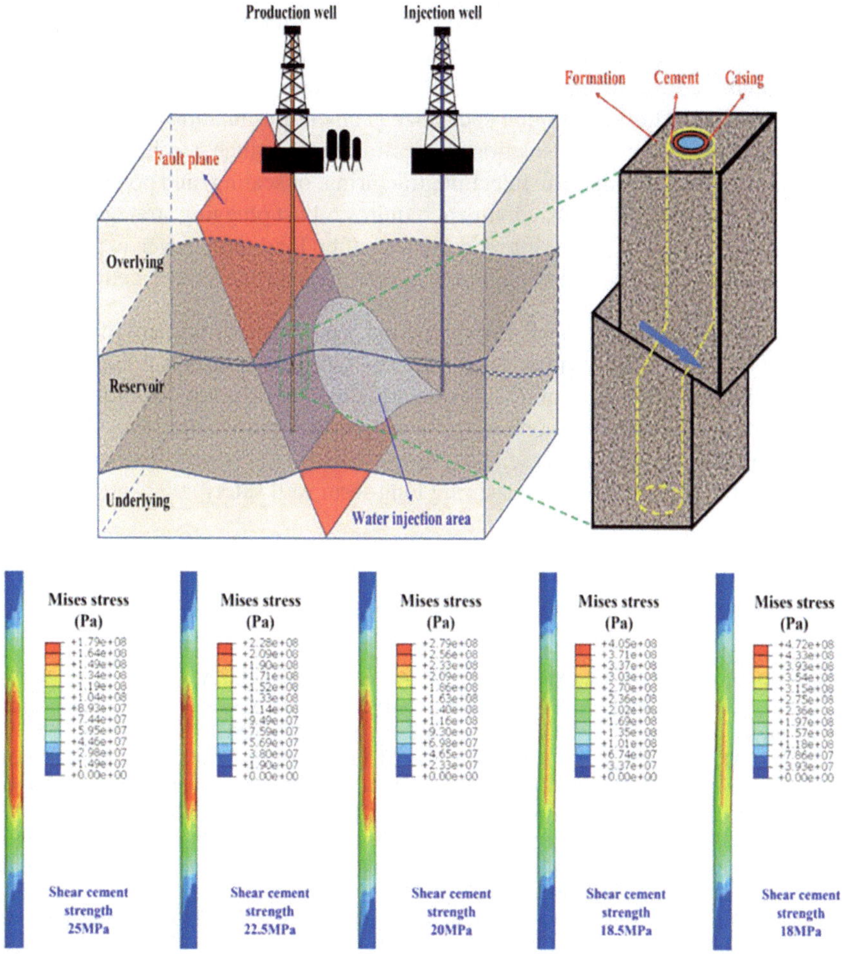

Fig. 23.4 Numerical simulation results showing Mises stress distributions (Pa) around the casing and cement sheath in an injection well affected by fault reactivation. The shear cement strength thresholds indicate the vulnerability of the casing and cement to mechanical shear failure as the fault plane is reactivated during water injection (Tuo et al., 2024)

- **Scaling:** Long-term scale build-up needs continuous monitoring and management to prevent blockages, maintain injectivity, and avoid forcing injection pressures too high.
- **Fatigue:** While less severe than thermal EOR or gas storage, pressure cycling associated with intermittent injection or shutdowns can contribute to fatigue damage, particularly in connections or wellhead components over decades of service.

- **Cement Durability:** Evaluating the potential long-term impact of large volumes of injection water (potentially undersaturated with respect to cement minerals, or containing corrosive species like CO_2/H_2S/bacteria) on cement sheath integrity and bonding. This may require specialized cement formulations and monitoring.
- **P&A:** Ensuring effective isolation of injection zones (which may have been significantly altered by long-term injection, fracturing, or scaling) and potentially pressurized aquifers during abandonment requires robust plugging strategies, considering the condition of casing and cement after long-term water exposure and potential scaling issues that might hinder plug placement or verification.

Maintaining the integrity of water injection wells requires diligent management of injection water quality (treatment), careful control of injection pressures below fracture limits, verification of zonal isolation and conformance (monitoring injection profiles), and ongoing monitoring for corrosion, scaling, and leaks throughout their operational life. It is a crucial aspect of maximizing reservoir recovery while protecting the environment and ensuring long-term well safety.

References

Abdeli, D. Z., Yskak, A. S., Shilanbayev, B. A., & Baluanov, B. A. (2024). Enhanced oil recovery of deposits by maintaining a rational reservoir pressure. *Naukovyi Visnyk Natsionalnoho Hirnychoho Universytetu, 4*, 48–54. https://doi.org/10.33271/NVNGU/2024-4/048

Al Khamis, M. N., Al Khalewi, F., Al Hanabi, M., Al Yateem, K., Al Qatari, A., Al Muailu, H., & Aramco, S. (2014). A comprehensive approach of well integrity surveillance. *International Petroleum Technology Conference.* https://doi.org/10.2523/IPTC-17465-MS

Askari, M., Aliofkhazraei, M., Jafari, R., Hamghalam, P., & Hajizadeh, A. (2021). Downhole corrosion inhibitors for oil and gas production – a review. *Applied Surface Science Advances, 6*,. https://doi.org/10.1016/j.apsadv.2021.100128

Bhatkar, S. A., Badhe, N. G., Adhatrao, A. P., & Wadgaonkar, V. S. (2023). Optimizing produced water quality for enhancing oil recovery. *Materials Today: Proceedings, 77*, 81–85. https://doi.org/10.1016/J.MATPR.2022.10.101

Bringedal, B., Morud, S. A., Hall, N. A., & Huseman, G. (2006). Online water-injection optimization and prevention of reservoir damage. *Proceedings - SPE Annual Technical Conference and Exhibition, 5*, 3089–3094. https://doi.org/10.2118/102831-MS

Ding, L., Chen, W., Han, C., Geng, H., & zhang, q. (2023). research on a typical casing failure during drilling of cement plugs in ultradeep wells. In *SPE Journal.* http://onepetro.org/SJ/article-pdf/28/06/2753/3332024/spe-215837-pa.pdf

Gordon, R. P., & Survey, M. G. (2016). *Corrosive Water-Facts, Common Questions, and Resources.* http://water.usgs.gov/nawqa/pubs/gw_corrosivity/

Guo, Y., Blanford, M. L., & Candella, J. D. (2015). Evaluating the risk of casing failure caused by high-density perforation: A 3D finite-element-method study of compaction-induced casing deformation in a deepwater reservoir, Gulf of Mexico. *SPE Drilling and Completion, 30*(2), 141–151. https://doi.org/10.2118/170618-PA

Han, H., Dusseault, M., Xu, B., & Peng, B. (2006). *Simulation of Tectonic Deformation and Large Area Casing Shear Mechanisms—Part B: Geomechanics.* OnePetro. https://dx.doi.org/

Li, Z., Li, G., Li, H., He, L., & Wang, X. (2022). Mechanisms of casing deformation during hydraulic fracturing XE "Hydraulic Fracturing" in the horizontal wells of the weirong shale field. *ACS Omega, 7*(11), 9796–9807. https://doi.org/10.1021/acsomega.2c00044

Liu, S. shan, & Yuan, W. wei. (2022). Numerical simulation of casing tensile failure in water injection wells. *Springer Series in Geomechanics and Geoengineering*, 119–132. https://doi.org/10.1007/978-981-19-2149-0_11/FIG.S/14

Metikoš-Hukovič, M., Babić, R., Škugor Rončević, I., & Grubač, Z. (2011). Corrosion resistance XE "Corrosion resistance" of copper–nickel alloy under fluid jet impingement. *Desalination, 276*(1–3), 228–232. https://doi.org/10.1016/J.DESAL.2011.03.056

Saintier, N., El May, M., Odemer, G., Hénaff, G., Bosch, C., Feaugas, X., & Couvant, T. (2019). Corrosion and hydrogen fatigue at different scales. *Mechanics—Microstructure—Corrosion Coupling: Concepts, Experiments, Modeling and Cases, 385–411*,. https://doi.org/10.1016/B978-1-78548-309-7.50017-X

Speight, J. G. (2015). Fouling as a result of corrosion. *Fouling in Refineries, 413–452*,. https://doi.org/10.1016/B978-0-12-800777-8.00017-6

Tuo, Y., Yu, H., Lian, Z., Lin, T., & Wang, H. (2024). Impact of fault slip induced by water injection XE "water injection" on casing deformation: A numerical study. *Petroleum Science and Technology*. https://doi.org/10.1080/10916466.2024.2356734

Vignes, B., Tonning, S. A., & Aadnoy, B. S. (2008). Injection wells with integrity challenges on the NCS. *Society of Petroleum Engineers—13th Abu Dhabi International Petroleum Exhibition and Conference, ADIPEC 2008, 3*, 1328–1340. https://doi.org/10.2118/118101-MS

Yao, H. tian. (2021). Cause analysis and prevention of casing damage in X Oilfield. *Springer Series in Geomechanics and Geoengineering*, 1125–1131. https://doi.org/10.1007/978-981-16-0761-5_105/TABLES/1

Zhang, N., Wang, P., Li, J., Ma, W., Zhang, X., Zhang, H., Jiang, C., Huang, W., Feng, X., & Liu, S. (2023). A study on the mechanism of casing deformation and its control strategies in shale oil hydraulic fracturing. *Processes, 11*(8). https://doi.org/10.3390/pr11082437

Chapter 24
Integrity for Subsurface Storage Wells (CCS, Gas Storage, H_2 Storage)

Abstract As global energy systems shift toward sustainability and decarbonization, subsurface storage of fluids—including carbon dioxide (CO_2), natural gas (NG), and hydrogen (H_2)—has emerged as a critical component of climate mitigation and energy reliability strategies. This chapter provides a comprehensive analysis of the well integrity challenges associated with Carbon Capture and Storage (CCS), Underground Natural Gas Storage (NGS), and Underground Hydrogen Storage (UHS), highlighting the stringent requirements for long-term containment, cyclic loading resilience, and material compatibility unique to each application. For CCS, well integrity is paramount to ensuring millennia-scale containment of CO_2, necessitating specialized focus on corrosion resistance in steel components, degradation of cement under carbonic acid attack, elastomer compatibility with supercritical CO_2, and rigorous abandonment and monitoring protocols. Advanced materials and sealing systems must be validated under reservoir-specific geochemical and geomechanical conditions. Cement micro-annuli, legacy well reactivation, and induced seismicity are explored as critical risks requiring predictive modeling and long-term MMV (Monitoring, Measurement, and Verification). In NGS applications, wells are subject to frequent and rapid pressure and temperature cycling, which can induce mechanical fatigue in casing, cement, and wellhead equipment. Deliverability assurance and leak prevention are central to operational and economic success. The chapter discusses cyclic fatigue management, pressure testing, and corrosion control strategies, alongside field data highlighting the primary causes of well failures in storage environments. The chapter also explores the emerging frontier of UHS, where the unique physical and chemical properties of hydrogen—including its low molecular weight, high diffusivity, and tendency to induce embrittlement—pose unprecedented integrity threats to steel casings and cement sheaths. Mechanisms such as hydrogen embrittlement (HE), hydrogen-induced cracking (HIC), and hydrogen blistering (HB) are examined in depth, along with mitigation strategies through alloy selection, coatings, and seal optimization. Salt cavern-specific geomechanical challenges and the impact of cyclic operations on casing design and cement integrity are also detailed. Through detailed case studies, laboratory data, corrosion rate modeling, and analytical leakage equations, this chapter equips engineers, researchers, and policymakers with a deep understanding of the technical complexities governing well

© The Author(s), under exclusive license to Springer Nature Switzerland AG 2026

A. Alsubaih and K. Sepehrnoori, *Well Integrity for Oil, Gas, and Subsurface Storage Operations*, https://doi.org/10.1007/978-3-032-15233-6_24

integrity in subsurface storage applications. The insights provided form the basis for designing robust wells, managing long-term risks, and advancing safe and efficient storage of critical energy and climate-related fluids.

Interior view of Çankırı Salt Cave. *Source* Murat Özsoy (n.d.)

Chapter's Objectives

- Emphasize the critical role and heightened requirements for well integrity in subsurface storage applications.
- Detail the specific integrity challenges for Carbon Capture and Storage (CCS) wells, focusing on long-term containment and material compatibility.
- Discuss integrity management for Natural Gas Storage wells, focusing on cyclic loading and deliverability.
- Outline the emerging integrity considerations for Underground Hydrogen Storage (UHS) wells, including material embrittlement and sealing small molecules.

24.1 Introduction: Security and Reliability in Storage

Subsurface geological formations are increasingly used not just for producing resources but also for storing substances—either permanently (like CO_2 in Carbon Capture and Storage—CCS) or temporarily as part of energy supply management

(Natural Gas Storage—NGS, and potentially Underground Hydrogen Storage—UHS). Wells provide the essential access for injecting and potentially withdrawing these stored substances. However, the integrity requirements for storage wells are often even more stringent than for production wells due to:

- **Containment Assurance**: The primary objective is secure storage, making any leakage potentially a failure of the system's core purpose, with significant environmental (CCS, UHS, potentially NGS) or economic (NGS) consequences (Alsulaiman, 2024; Schultz et al., 2014). This demands extremely low leak rates, often aiming for near-zero over very long periods.
- **Long Timescales (CCS, potentially UHS)**: CCS requires containment assurance for centuries or millennia, demanding exceptional long-term durability of well barriers under evolving subsurface conditions (Krevor et al., 2023). UHS may also involve multi-year storage durations.
- **Cyclic Operations (NGS, UHS)**: Frequent and sometimes rapid injection and withdrawal cycles, driven by energy demand fluctuations, impose severe fatigue loading on well components and cement (Wood, 2024)
- **Unique Fluid Properties**: Stored substances (supercritical CO_2, pure hydrogen) have unique interactions with wellbore materials (corrosion, degradation, embrittlement, permeation) that differ from traditional hydrocarbons or water (Abraham et al., 2025).
- **Public and Regulatory Scrutiny**: Storage projects, particularly CCS and UHS linked to climate mitigation and future energy systems, often face heightened public awareness and demand extremely high levels of safety and environmental assurance, requiring robust demonstration of containment integrity.

Well integrity is therefore the absolute cornerstone of safe, reliable, and publicly acceptable subsurface storage. The technical integrity challenges require adapting and extending existing oil and gas well integrity knowledge and practices.

24.2 Carbon Capture and Storage (CCS) Well Integrity

CCS involves capturing CO_2 emissions from industrial sources and injecting it deep underground into suitable geological formations (e.g., saline aquifers, depleted oil/gas reservoirs) for permanent storage. Wells (injectors, monitoring wells, and potentially legacy wells penetrating the storage complex) are critical potential leakage pathways that must be secured.

- **Long-Term Containment (Primary Goal)**: The paramount objective is to ensure containment of the injected CO_2 for hundreds to thousands of years. The WIMS must be designed with this perpetual perspective in mind, going beyond typical operational lifespans.
- **Material Durability versus Wet CO_2**: This is the dominant technical challenge for long-term integrity.

- Corrosion: Steel and steel alloys commonly used in wellbore components—such as casing, tubing, wellheads, blowout preventers, and subsurface safety valves—are inherently vulnerable to corrosion when exposed to CO_2-rich environments during enhanced oil recovery (EOR) or storage operations. Under high CO_2 pressure conditions, the corrosion rate of carbon steel can be significant; however, over extended periods, it can decrease as protective $FeCO_3$ layers develop (Choi et al., 2013). Cui et al., (2006) observed that corrosion rates decline with increasing temperature due to the passivation effect of corrosion products like α-FeOOH and $FeCO_3$ on the steel surface. At elevated temperatures, these protective films become denser and more uniform, enhancing corrosion resistance. The decrease in CO_2 solubility in water at higher temperatures also contributes to a reduction in carbonic acid concentration, further mitigating corrosion. Choi et al., (2010) studied the effects of oxygen in CO_2-saturated water and water-saturated CO_2 environments and found that oxygen presence accelerates corrosion by inhibiting $FeCO_3$ protective film formation. Similarly, William Carey et al., (2010) highlighted that in scenarios where micro-annuli exist between cement and carbon steel, steel tends to be more reactive than cement when CO_2 and brine flow along this interface. Beltrán-Jiménez et al., (2021) further emphasized that cement contamination can exacerbate casing corrosion. Shaikhah et al., (2022) explored how engineered composite layers can enhance corrosion and erosion resistance in casing steel. By incorporating carboxyl-functionalized polystyrene (PSCOOH) microspheres, they successfully manipulated the nucleation and growth of $FeCO_3$ corrosion products, forming robust composite layers that offered improved mechanical stability and corrosion protection. Additionally, recent work by Shaikhah et al., (2024) demonstrated that the inclusion of 100 ppm of poly (allylamine hydrochloride) (PAH) significantly reduces corrosion rates, indicating the promise of tailored chemical interventions in mitigating corrosion challenges in CO_2 injection wells. Legacy wells within the storage complex made of carbon steel are a major concern if they are exposed to wet CO_2, as they represent potential failure points that need to be identified, evaluated, and potentially remediated or re-abandoned to CCS standards (Alsubaih et al., 2024, 2025). Figure 24.1 provides a conceptual illustration of a subsurface storage complex, highlighting the interactions between active CO_2 injection wells, abandoned wells, and orphan wells that may compromise long-term containment integrity. Figure 24.1 also emphasizes the potential role of satellite monitoring and wellhead instrumentation in ensuring storage security. Understanding these subsurface and surface features is critical to addressing the unique well integrity challenges posed by carbon capture and storage (CCS) operations. Figure 24.2 presents corrosion rate data and reaction mechanisms for steel casing materials under supercritical CO_2 conditions. It highlights how temperature, pressure, and casing steel grade influence corrosion behavior in CO_2-rich environments, and how electrochemical reactions—such as anodic dissolution, cathodic reduction, and iron carbonate scale formation—affect long-term well integrity. This information is crucial for selecting appropriate

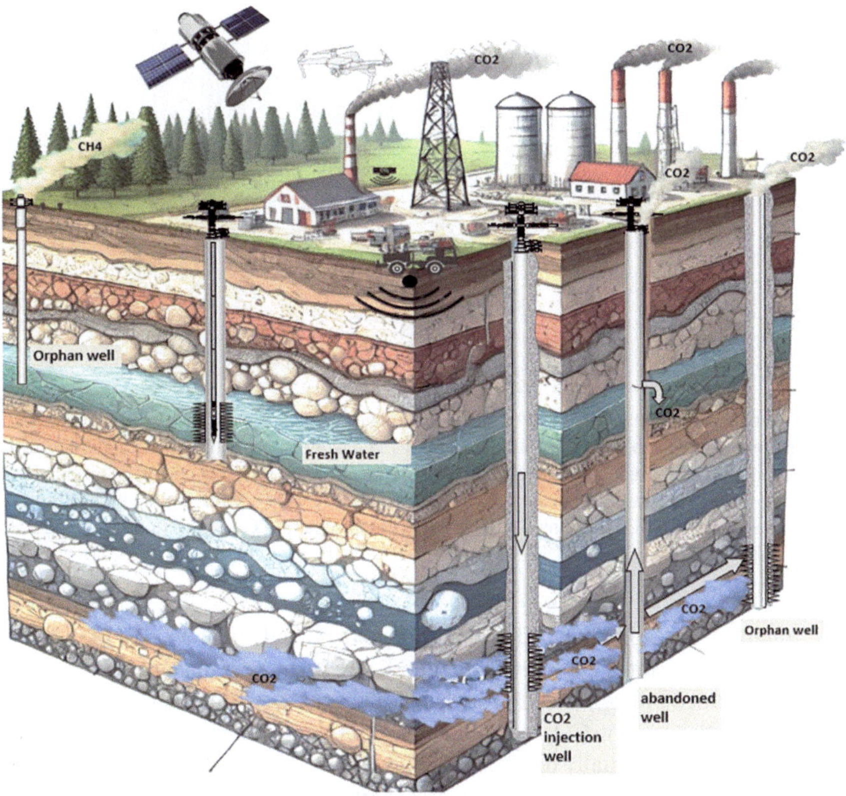

Fig. 24.1 Schematic illustration of a subsurface storage complex, showing: • Active CO_2 injection wells, • Abandoned wells within the storage complex, • Orphan wells intersecting storage formations, and • Surface monitoring infrastructure (e.g., satellite and drone-based observations)

casing materials and implementing corrosion management strategies in these wellbore environments.

– Cement Degradation: Cement degradation in CCS wells occurs primarily due to reactions between carbonic acid—formed when CO_2 dissolves in water—and the cement matrix, leading to increased porosity and reduced strength. Two key conditions for calcium carbonate dissolution are undersaturated formation water in calcium and carbonate ions, and the presence of flow to transport dissolved species and bring fresh water to the cement. The extent of degradation depends on the formation mineralogy and the water's ability to remain undersaturated as CO_2 solubility increases. Experimental studies have illuminated these processes in detail. Connell et al., (2015) examined cement degradation at the sandstone–cement interface, showing that pressure-driven flow within cement channels can accelerate degradation significantly. Their model indicated that even narrow channels can propagate degradation over years, with

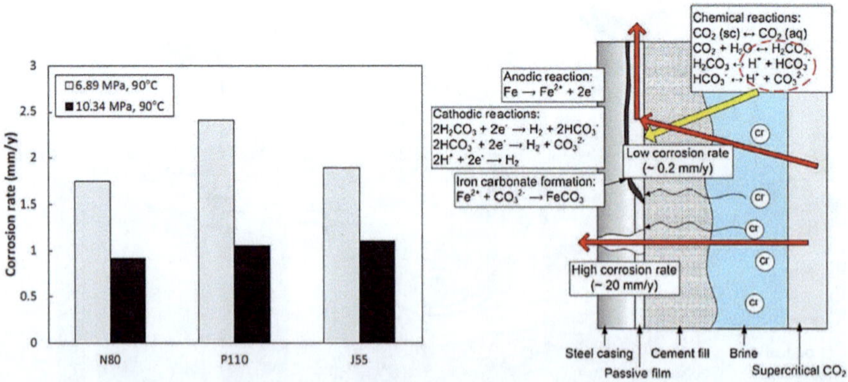

Fig. 24.2 Corrosion behavior of steel casing materials in supercritical CO_2 environments. (Left) Comparison of corrosion rates for N80, P110, and J55 casing steels at 6.89 MPa and 10.34 MPa (90 °C), showing increased rates with higher pressure. (Right) Schematic of corrosion mechanisms, illustrating anodic and cathodic reactions, iron carbonate ($FeCO_3$) formation, and differences in corrosion rates across well sections—emphasizing the impact of CO_2 interactions with steel and cement in the wellbore (Choi et al., 2013)

migration speed influenced by pressure gradient and initial width. For instance, a 0.01 cm channel could erode up to 50 m in eight years under specific conditions. Once the erosion front breached the seal, flow rates surged, indicating a potential loss of CO_2 containment. Complementing this, Rimmelé et al., (2008) experimentally demonstrated how porosity changes in cement samples exposed to CO_2-rich water evolve over time. They found that carbonic acid, formed by dissolved CO_2, reacts with calcium hydroxide in Portland cement to produce calcium carbonate and water. Additionally, CO_2 interacts with calcium silicate hydrates (C-S–H), forming calcium carbonate and a silica gel. Excess CO_2 can dissolve the newly formed calcium carbonate, generating calcium bicarbonate, which may react further to form new calcium carbonate within the cement matrix. Their observations showed that millimeter-sized calcium carbonate grains precipitate onto the cement surface shortly after exposure. The carbonated zone that forms includes calcium carbonate, silica gel, and calcium-depleted C-S–H phases, bounded by a thin, low-porosity carbonation front (about 50–200 μm wide) rich in calcium carbonates. Ahead of this front lies a dissolution front—a high-porosity region where portlandite and C-S–H dissolve, facilitating further degradation. Thomas et al., (2005) also highlighted that when carbonic acid infiltrates the annulus, it lowers the pH, rapidly compromising cement integrity and raising the risk of leakage. While dry supercritical CO_2 can slightly enhance cement strength and reduce permeability, the presence of water—enabling carbonic acid formation—makes this interaction the dominant and most aggressive degradation pathway. Figure 24.3 provides advanced log-based insights into the initiation of micro-annulus (debonding) along the casing–cement interface in CCS wells. These micro-annuli represent potential pathways for CO_2 leakage and are often linked to changes in stress, temperature, or chemical environments. By analyzing acoustic impedance, flexural attenuation, and solid–liquid–gas mapping logs, alongside gamma ray and bond index data, engineers can pinpoint the precise depths and extents of debonding, enabling more targeted remediation and long-term integrity assurance.

- Elastomer Performance: Elastomer seals (in packers, valves, wellheads) exposed to high-pressure or dense-phase CO_2 can experience significant swelling, degradation, and the risk of Explosive Decompression (ED/RGD). This swelling is driven by CO_2 sorption, as documented by (Dubois et al., 2018), who measured significant CO_2 uptake in various elastomers—up to 30% by weight in some cases—using in-situ FTIR microscopy. The study highlighted that swelling is directly proportional to CO_2 uptake, with typical swelling levels ranging from 15 to 30% at 15 MPa and 50 °C for common elastomers. This swelling can alter elastomer mechanical properties, leading to compromised sealing performance and increased risk of ED upon pressure drops. Additionally, elastomer chemical compatibility varies: materials like EPDM, SBR, and EVM exhibit substantial swelling due to their high chain mobility and affinity for CO_2. To ensure reliable performance in CCS applications, it is necessary to use specialized, CO_2-resistant, high-grade elastomers (such as specific FFKM and HNBR grades) and potentially Metal-to-Metal (MTM) seals in critical applications. Proper material selection and validation

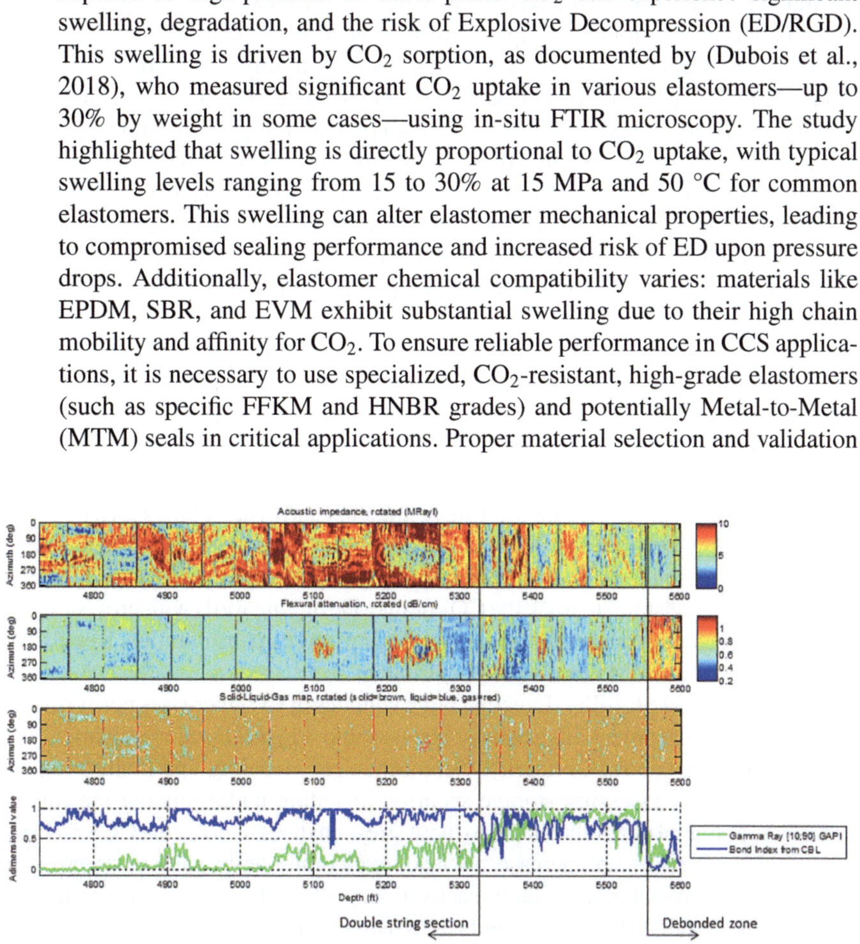

Fig. 24.3 Advanced logging data revealing micro-annulus initiation in CCS wells. The logs show variations in acoustic impedance, flexural attenuation, and solid–liquid–gas mapping that correlate with the formation of micro-annuli (debonded zones). The bond index from CBL and gamma ray data confirm these compromised zones, typically manifesting at interfaces of significant stress or chemical reactions. Identifying these early is crucial for mitigating CO_2 leakage risks and preserving long-term storage security

under representative CO_2-rich conditions are crucial to mitigating these risks and maintaining long-term well integrity.

- **Well Design for Long-Term Performance**: Designs must minimize potential leak paths (e.g., simplify completions), utilize materials expected to remain stable for centuries under storage conditions (pressure, temperature, geochemistry), and account for potential long-term stresses (e.g., due to pressure changes in the reservoir, geomechanical effects).
- **Rigorous P&A for Permanence**: P&A strategies for CCS wells (and any legacy wells within the storage complex) must provide extremely high confidence in long-term, perpetual sealing. This often involves multiple, qualified permanent barrier plugs (as discussed in Chap. 27) made from qualified CO_2-resistant materials, placed at strategic locations to seal all potential migration pathways (inside casing and all annuli). Verification testing must be stringent, and techniques like section milling to ensure effective annular sealing (rock-to-rock barriers) are often considered necessary to bypass potentially degraded original cement and provide a seal directly against the formation. Abandoning legacy wells within or near the storage complex to modern CCS P&A standards is often a crucial prerequisite for storage project approval.
- **Comprehensive Monitoring, Measurement, and Verification (MMV)**: Extensive MMV programs are required *during* CO_2 injection and, critically, potentially for decades or centuries *post-injection* to provide assurance of containment and detect any leakage early (Lee et al., 2016; Newmark et al., 2010; Plampin et al., 2014; Siirila et al., 2012; Terzi et al., 2014). This includes downhole pressure/temperature monitoring in injection and observation wells, potential fluid sampling, wellhead integrity checks, near-surface/soil gas monitoring, atmospheric monitoring, groundwater monitoring, and potentially periodic geophysical surveys (e.g., seismic) to track the CO_2 plume. Designing wells with provisions for long-life sensors or access for future monitoring is a key consideration. Regulatory requirements for MMV are typically stringent.
- **Geomechanical Stability**: Injecting large volumes of CO_2 increases pore pressure in the reservoir, potentially altering stresses and triggering micro-seismicity or compromising cap rock integrity. Geomechanical modeling and monitoring are crucial parts of assessing storage security, including understanding the role of wells as potential conduits related to these stress changes. During the CCS process, excessive pore pressure increase can lead to alteration of the state of stresses and consequently rock failure or fault reactivation, potentially causing seismic activity (White & Foxall, 2016). If this induced seismicity is severe, it can be significant enough to be felt by humans. Even though most seismic events associated with CO_2 injection are micro-seismic, there are instances of larger magnitude events (Yang et al., 2023).

CCS well integrity demands a long-term perspective, focusing on material science, robust P&A with qualified materials, comprehensive MMV, and careful management of geomechanical effects to provide the necessary assurance of permanent CO_2 containment. Geologic carbon storage (GCS) represents a cornerstone of global climate mitigation strategies by permanently sequestering carbon dioxide (CO_2) in deep geological formations. However, achieving secure and long-lasting CO_2 containment is challenged by potential leakage pathways, including natural fractures, insufficient caprock integrity, and, most critically, legacy wells that were drilled and abandoned before modern sealing practices. These wells, often featuring compromised casings and cement sheaths, can provide high-permeability pathways for brine and CO_2 migration into overlying freshwater aquifers, driven by injection-induced pressure gradients. As summarized in Table 24.1, a variety of analytical and empirical equations have been developed to quantify leakage behaviors through these complex systems, including models of gas velocity, liquid velocity, momentum balance, and drift-flux dynamics, as well as Darcy-based flow through micro-annuli, fractures, and cement. These formulations, drawn from foundational studies and modern risk assessments, underpin site-specific evaluations and highlight key factors—such as micro-annulus formation, brine salinity, reservoir permeability, and mechanical degradation of well barriers—that critically influence leakage onset and severity. By integrating these equations with advanced numerical simulations, reduced-order modeling, and data-driven predictive frameworks, stakeholders can better quantify brine migration risks, design robust well integrity programs, and optimize monitoring and remediation strategies to ensure the long-term safety and efficacy of geologic CO_2 storage operations.

In wells used for Carbon Capture and Storage (CCS), steel casing and tubing are often exposed to CO_2-rich environments. This leads to the formation of carbonic

Table 24.1 Analytical and empirical equations for well integrity assessment in subsurface storage wells

Mathematical expression	Reference
$\overline{Q}_i^\alpha = \sum_{j=1}^{N} \left\{ \frac{\left(f_i^\alpha \xi_{i,j}^\alpha - g_i^\alpha \xi_{i+\alpha,j}^\alpha\right)}{\lambda_j} \left[\left(\sum_{k=1}^{N} \xi_{k,j} \underset{k}{\bar{Q}} \right) + c_j^L \right] \right\}$	(Javandel et al., 1988)
$\underset{d}{\bar{Q}} = \frac{\frac{1}{v} K_0 (R_d \sqrt{v})}{K_0 (r_{ld} \sqrt{v}) + \frac{1}{T_d} K_0 \left(r_{ld} \sqrt{\frac{v}{\eta_{ad}}} \right) + \frac{1}{a_d}}$	Das and Hassanzadeh (2023)
$q' = \frac{1}{l} - \frac{4}{\pi} \sum_{n=1}^{\infty} \left\{ \begin{array}{c} \left[\frac{J_1(\alpha_n a_1')}{J_1(\alpha_n)} \right]^2 \left[\frac{1+\exp(2\alpha_n \gamma_1)}{1-\exp(2\alpha_n \gamma_1)} \right] \beta_{1n}' + \\ \left[\frac{J_1(\alpha_n a_2')}{J_1(\alpha_n)} \right]^2 \left[\frac{1+\exp(2\alpha_n \gamma_2)}{1-\exp(2\alpha_n \gamma_2)} \right] \beta_{2n}' \end{array} \right\}$	(Avc, 1992)
$Q_w(t) = K_{\text{well}} \pi r_w^2 \frac{h_{\text{top}}(x_w) - h(x_w,t)}{B(x_w)}$	(Nordbotten et al., 2004)
$q_1 = \frac{q a_t \text{Ei}\left(b_t \frac{R^2}{t_1}\right)}{\Omega\left(1 + a_t \text{Ei}\left(b_t \frac{r_w^2}{t_1 - t_b}\right)\right) + a_u \text{Ei}\left(b_u \frac{r_w^2}{t_1 - t_b}\right)}$	(Qiao et al., 2021)
$Q = \left(\frac{h^3 \cos\alpha}{12\mu}\right)\left(\frac{\Delta P}{L}\right) W$	Trudel and Frigaard (2023)
$Q = \left(\frac{\pi R_c \Delta P}{6\mu L}\right) \delta R^3$	Ford et al., (2017)
$Q = k_{\text{eff}} A \frac{\psi_L - \psi_T}{L}$	WLAT-MSWM Method (2021)

acid, accelerating corrosion rates before protective films can form. One widely used empirical correlation for estimating initial corrosion rates in such conditions is based on the partial pressure of CO_2 and temperature.

$$logCR\ (\text{mpy}) = 8.78 - \frac{2320}{T\ (^{\circ}K)} - 0.00555 \cdot T\ (^{\circ}C) + 0.67log\left(\text{PP}_{CO_2}\right) \qquad (24.1)$$

where CR (mpy) is Corrosion rate in mils per year (1 mpy = 0.001 in/year), T (K) is Temperature in Kelvin, T (°C) is Temperature in Celsius, and $PP_{(CO_2)}$ (psi): Partial pressure of CO_2 in psi.

24.3 Natural Gas Storage Well Integrity

Underground storage of natural gas (primarily methane) in depleted reservoirs, aquifers, or salt caverns is essential for balancing seasonal energy supply and demand by injecting gas during periods of low demand and withdrawing it during high demand. Storage wells experience frequent and often rapid cycles of injection and withdrawal, imposing significant stress.

- **Cyclic Loading Fatigue**: This is a primary integrity concern, distinguishing NGS wells from conventional production wells (Su et al., 2022). Repeated pressure cycles (from low withdrawal pressures to high injection pressures, potentially daily or weekly) and associated temperature fluctuations impose significant fatigue stress on:
  - Casing and Tubing Connections: Premium connections with proven fatigue resistance are often required and need to be designed and qualified for the anticipated number and severity of cycles.
  - Cement Sheath and Bonds: Cyclic loading from pressure and temperature variations can degrade cement bonds (leading to micro-annuli) and potentially cause fatigue cracking in the cement sheath itself (Zheng et al., 2024). This can compromise zonal isolation, allowing gas migration.
  - Wellhead Equipment: Valves and seals experience frequent pressure cycles, requiring robust designs and potentially more frequent maintenance/ replacement cycles.
  - Fatigue life assessment and management should be an integral part of the design and ongoing integrity management program for NGS wells.
- **Deliverability Assurance**: High injection and withdrawal rates ("deliverability") are crucial for meeting peak energy demand and operational flexibility. Well integrity issues like corrosion damage, scale, restrictions, or perforation plugging can impair deliverability. Maintaining wellbore condition is important for economic performance, but interventions to restore deliverability must be performed without compromising integrity.

- **Leak Prevention**: Preventing leakage of valuable working gas or cushion gas (the base gas volume needed to maintain sufficient pressure for withdrawal) is economically critical and environmentally necessary (methane is a potent greenhouse gas). Requires robust, verifiable barriers (tubing, packer, casing, cement, wellhead) and diligent monitoring. Small leaks, even if not posing an immediate safety hazard, result in significant economic loss over time.
- **Frequent Testing**: Regular functional and leak testing of safety–critical valves (e.g., SSVs, wellhead valves) and pressure testing of annuli is standard practice due to the frequent operational changes and criticality of preventing leakage.
- **Monitoring**: Continuous annulus pressure monitoring is common and often required. Gas detection systems at the surface are typically employed as a primary method for identifying leaks. Downhole monitoring (e.g., pressure/temperature) may also be used.
- **Wellhead Growth/Movement**: Significant temperature changes during high-rate injection (potentially cooling due to Joule–Thomson expansion) and withdrawal (potentially heating) can cause axial movement of casing/tubing strings. Requires thermal wellhead designs that can accommodate this movement while maintaining seal integrity.
- **Corrosion**: While typically injecting dry gas into a depleted (and hence potentially dry) reservoir, cushion gas may interact with native formation water. Water vapor can also condense during pressure/temperature changes, creating a potential for CO_2/H_2S corrosion if these components are present in the stored gas or native formation fluids. MIC can also occur in some storage environments. Corrosion management strategies may be required.

Gas storage well integrity focuses on managing fatigue from severe cyclic operations, ensuring high deliverability, and preventing leaks through robust barrier design, material selection (especially connections), frequent testing, and diligent monitoring. Figure 24.4 provides an overview of well leakage events in underground natural gas storage (UNGS) facilities, highlighting trends in reported events over time, as well as the primary components and causes of well failures. This dataset includes events reported to PHMSA (Pipeline and Hazardous Materials Safety Administration) and historic data from the literature. Notably, a significant increase in reported events after 2017 aligns with updated PHMSA reporting requirements, reflecting improved data capture rather than necessarily increased leakage rates. The data reveal that casing and wellhead failures are among the most frequently reported failure modes, and highlight the particular challenges associated with maintaining well integrity in cyclically operated storage wells (Lackey et al., 2024). Erosion and flow behavior in natural gas storage wells under different gas production volumes and particle velocities (Fig. 24.5). The top panel shows how the maximum erosion rate increases with higher gas production and particle injection rates. The middle panel visualizes the erosion distribution across a multi-perforated pipe system, highlighting areas of concern at different mass flow rates. The bottom panel demonstrates how varying particle velocities (from 8 to 16 m/s) influence particle impact zones and erosion

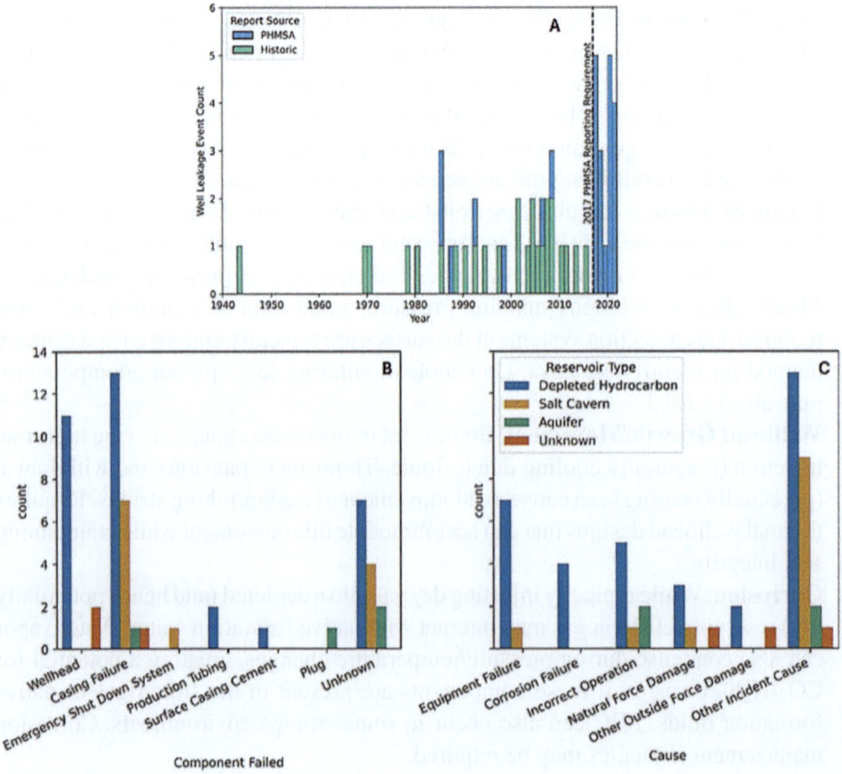

Fig. 24.4 Historical and component-specific distribution of well leakage events in underground natural gas storage (UNGS) wells **a** Temporal distribution of leakage events from historic data and PHMSA reports, noting the 2017 PHMSA reporting requirement as a key milestone **b** Component-specific failure counts, showing that casing and wellhead failures dominate **c** Leakage caused by reservoir type, highlighting the range of failure mechanisms including corrosion, incorrect operation, and natural forces (Lackey et al., 2024)

concentration, emphasizing risks to wellbore components under high-cycle loading (Zhu et al., 2022).

24.4 Underground Hydrogen Storage (UHS) Well Integrity (Emerging)

Storing pure hydrogen underground (e.g., in salt caverns, depleted reservoirs, aquifers) is being explored as a method for large-scale, long-duration energy storage to buffer intermittent renewable energy supplies. As an emerging application, the specific integrity challenges are being actively researched and understood. Hydrogen

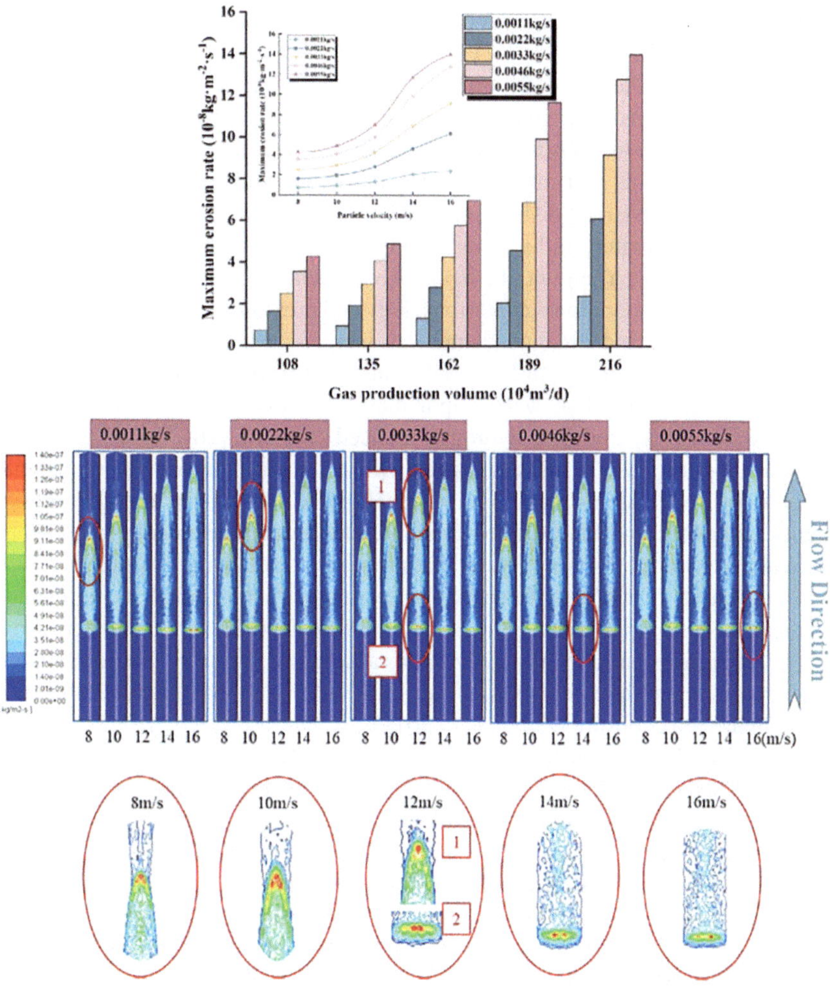

Fig. 24.5 Illustrates how particle-laden gas flow can induce localized erosion in natural gas storage wells, especially under high production volumes and particle velocities, underscoring the need for erosion-resistant materials and smart flow path design in cyclic gas injection environments (Zhu et al., 2022)

injection into geological formations for underground hydrogen storage (UHS) introduces unique and complex integrity challenges for well components, particularly steel casing. Due to hydrogen's small atomic radius, low molecular weight, and high diffusivity, it can permeate metallic materials more rapidly than other gases such as methane (Maury Fernandez et al., 2023). This characteristic makes conventional casing materials susceptible to several hydrogen-related degradation mechanisms including hydrogen embrittlement (HE), hydrogen-induced cracking (HIC), and hydrogen blistering (HB), each of which can critically undermine the structural integrity of the well.

Hydrogen embrittlement (HE) refers to the degradation of mechanical properties such as ductility, tensile strength, and fracture toughness resulting from hydrogen atom diffusion into steel. Once adsorbed on the metal surface, molecular hydrogen dissociates into atoms that diffuse through the lattice and accumulate at trapping sites like grain boundaries, dislocations, or non-metallic inclusions, initiating crack nucleation and premature failure—especially in high-strength steels ($\geq$1000 MPa) (Dubois et al., 2018; Young et al., 2011). This mechanism is highly dependent on environmental and operational conditions. Elevated pressures and temperatures, long exposure durations, and stress concentrations increase susceptibility (C. Zhou et al., 2019). Moreover, hydrogen sulfide (H_2S) and other impurities exacerbate embrittlement, whereas some like water vapor may have less effect (Dwivedi & Vishwakarma, 2018). By contrast, hydrogen-induced cracking (HIC) does not require externally applied stress to initiate. Instead, HIC results from internal pressure build-up due to hydrogen atoms recombining into H_2 molecules within internal trap sites such as manganese sulfide (MnS) inclusions. The pressure eventually ruptures the surrounding material, forming cracks that propagate through the casing. The likelihood and severity of HIC are influenced by steel microstructure, impurity segregation, and the presence of high-angle grain boundaries (Laadel et al., 2022; Song & Curtin, 2011). Crack propagation is typically intergranular and driven by the formation of micro-cracks at multiple neighboring trap sites, which may coalesce under stress or further hydrogen exposure (Zhou et al., 2019).

Hydrogen blistering (HB) is a surface-level manifestation of internal hydrogen pressure. When recombination occurs near the steel surface, the resulting H_2 build-up can lift surface layers, creating dome-like blisters. Though often mistaken for HIC, blistering specifically occurs in low-strength steels or regions with soft zones and often coexists with HIC (Laadel et al., 2022). Due to the similarity in nucleation and propagation mechanisms, HB is considered a near-surface variant of HIC.

Understanding the distinct mechanisms of HE, HIC, and HB is critical, as they are often confused. HE typically occurs in high-strength steels under tensile stress and can sometimes be reversed through hydrogen removal and heat treatment. In contrast, HIC and HB result from molecular hydrogen build-up and cause permanent damage. HIC occurs in the bulk material, while HB affects surface layers. The interaction of hydrogen with steel microstructure remains complex and highly dependent on the specific environment, material grade, and mechanical loading regime (Song & Curtin, 2011; Van Der Schijff et al., 2021).

Preventing casing failure due to hydrogen exposure involves material selection, barrier technologies, and operational control. Alloys with low hydrogen diffusivity, refined grain structure, and reduced inclusion content offer improved resistance. Thermal treatments can stabilize austenitic microstructures, and metallic coatings may act as diffusion barriers if free from defects. Advanced lab techniques such as thermal desorption spectrometry (TDS), hydrogen microprint techniques (HMT), and electrochemical noise analysis can identify and quantify hydrogen damage, though in-field detection remains limited (Dwivedi & Vishwakarma, 2018). More research is needed to develop field-deployable monitoring tools and to understand hydrogen's interaction with well cements, which remains an underexplored topic despite its significance to zonal isolation. Hydrogen poses severe and multifaceted risks to casing integrity in UHS applications. Its ability to penetrate steel, alter mechanical properties, and induce failure through both stress-driven and spontaneous mechanisms necessitates a rigorous approach to material qualification, monitoring, and operational management. As hydrogen storage scales up globally, understanding and mitigating these failure mechanisms will be essential to ensure long-term well integrity and safety. In salt cavern storage, salt dissolution leads to the creation of cavities and non-uniform well diameters. Additionally, salt creep results in reduced well diameter and challenges in running casing pipes. Water and brine facilitate corrosion and damage to drilling pipes. The significant depth of caverns subjects casing pipes to high lateral pressures. Furthermore, the lack of wall straightness and non-uniform well diameter induces bending stresses. The low density and creep behavior of salt cause displacement at the salt-formation interface and exert shear forces on casing pipes (Nagumo & Takai, 2019). High pressure within the caverns creates conditions for gas leakage through the cemented space around the casing. The combination of these conditions necessitates high knowledge and experience for proper drilling operations. Arguably, designing casing pipes is the most critical part of drilling operations. This involves appropriate geometric design considering specific conditions at each depth and the use of materials with high resistance and consistent quality, especially for the inner casing wall. In addition, creep creates a plastic transition layer in the formation. The pressure from this plastic transition layer increases the force exerted on the interface between the cement and the formation, which can threaten the well's integrity throughout its lifespan. The geomechanical effects of plastic creep formations, although not detectable from surface observations, can have detrimental consequences for cement. An important parameter in well designation is the well resistance against the stresses from external or internal forces. This ensures the safety and preservation of the well during the project lifespan. Compared to methane-storage salt caverns, the small size of hydrogen molecules results in a higher rate of dispersion and diffusion. This presents challenges for the stability and preservation of the well. The high affinity of hydrogen to combine with mineral elements in rocks and fluids in the well leads to erosion and corrosion of the well walls and infrastructure due to the pressure and temperature resulting from these reactions. Considering the small size of hydrogen, it tends to escape from the cavities around the well. Furthermore, the reaction between hydrogen inside the well and sulfur in the earth creates water and corrosive fluids, which degrade the casing used in the well. Sand and fine rock particles resulting from pressure changes, pore openings, and stresses applied to the well walls lead to erosion and penetration into the well (Atrens et al., 2018). Figure 24.6 illustrates the multifaceted security risk mechanisms associated with underground hydrogen storage (UHS) and natural hydrogen extraction wells. The Figure highlights various well integrity challenges—ranging from hydrogen embrittlement, metal corrosion, and hydrogen-induced fractures to cement degradation and elastic failures. It also shows how hydrogen's unique properties, such as its high permeability and reactivity with metal and cement materials,

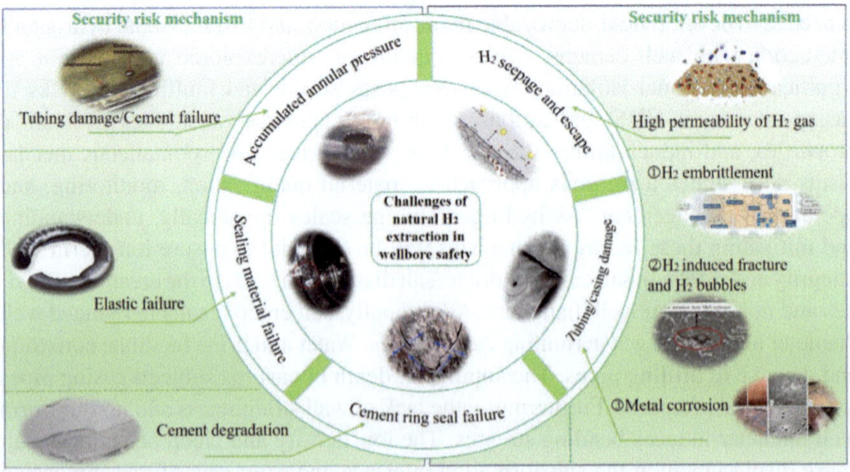

Fig. 24.6 Security risk mechanisms for hydrogen storage and extraction wells. The schematic highlights key integrity risks associated with hydrogen in underground storage and extraction scenarios, including: • Accumulated annular pressure, • H_2 seepage and escape through well materials, • Hydrogen embrittlement and hydrogen-induced fractures/bubbles in steel casings, • Metal corrosion exacerbated by hydrogen reactions, • Elastic and sealing material failures in packers and wellhead seals, and • Cement degradation and ring seal failures in cement sheaths (Sun et al., 2025)

create complex interactions that can compromise well safety and storage efficiency. Understanding these risk pathways is essential for developing robust well designs, selecting appropriate materials, and implementing effective integrity management practices in UHS projects. Injecting hydrogen into empty salt dome cavities is ideal for energy storage and carbon mitigation. However, ensuring the safety and integrity of the well is essential to prevent any environmental challenges and is crucial for environmental health. The mechanism of underground hydrogen storage (UHS) as well as the cyclic injection-production through the well creates pressure cycles that can lead to erosion and corrosion of the well.

- **Material Compatibility (Hydrogen Embrittlement—HE)**: This is a major potential integrity threat unique to hydrogen storage. Hydrogen atoms can readily diffuse into the crystal lattice of steel and cause embrittlement, significantly reducing fracture toughness and potentially leading to sudden, brittle failure at stresses below the yield strength, especially under tensile stress (applied or residual) and cyclic loading.
  – Susceptibility: Higher-strength steels, welds, and hard zones are generally more susceptible to HE. Temperature, pressure (hydrogen partial pressure), hydrogen purity, and the presence of impurities (e.g., H_2O, H_2S) influence HE risk.

  - Mitigation: Requires careful screening and selection of steels (potentially lower-strength grades or specialized alloys resistant to HE), rigorous material qualification testing under representative UHS conditions, and potentially protective internal coatings or liners (e.g., CRA liners, non-metallic liners) to prevent hydrogen contact with the load-bearing steel casing/tubing. NACE/ISO standards (e.g., ISO 19880–3 for hydrogen fueling stations, standards being developed for pipelines and storage) are being developed specifically for hydrogen service and need to be adapted and applied to well materials.

- **Sealing Hydrogen (Small Molecule Permeation)**: Hydrogen is the smallest molecule, allowing it to permeate more readily through materials compared to methane, CO_2, or water. This poses a challenge for maintaining tight seals over time.
  - Elastomers: Standard elastomers may exhibit high permeation rates and potential degradation/swelling in hydrogen. Specialized elastomers (potentially based on specific HNBR, FKM, FFKM variants) or non-elastomeric seals (e.g., MTM seals, polymers with very low permeability like PEEK or specialized composites) are likely required for reliable long-term sealing in packers, valves, and wellheads. Qualification and testing under high-pressure hydrogen are critical.
  - Cement: Concerns exist about hydrogen diffusion through the pore structure of conventional Portland cement over long timescales, potentially leading to inventory loss or contamination of overlying formations. Research into low-permeability cement formulations or alternative barrier materials is ongoing.

- **Cyclic Loading**: Similar to NGS, UHS for energy balancing involves frequent cycling of pressure and temperature, posing significant fatigue risks that are potentially exacerbated by hydrogen embrittlement effects on steel. Fatigue design and assessment need to consider HE impacts.
- **Corrosion and Biogeochemical Reactions**: While pure hydrogen is generally not corrosive to steel, contaminants (e.g., H_2O, H_2S, CO_2) in the stored hydrogen or native formation fluids can cause corrosion. Certain subsurface microorganisms can potentially consume stored hydrogen (acting as a food source), leading to loss of inventory and potentially generating corrosive byproducts (e.g., H_2S via SRB activity). Understanding and mitigating these biogeochemical reactions is necessary for both integrity and storage efficiency. This MIC risk can affect both steel and cement integrity.
- **Purity Maintenance**: Preventing contamination of the high-purity stored hydrogen by leakage of other formation fluids (e.g., formation water, native gas) into the wellbore or storage formation is critical for many end uses (e.g., fuel cells require very high hydrogen purity). Requires extremely high integrity and verifiable sealing of all well barriers.

UHS well integrity requires addressing fundamental material science challenges related to hydrogen embrittlement and permeation, alongside managing cyclic loading, specific corrosion/MIC risks, and ensuring exceptionally high sealing performance for a small molecule. Significant research and qualification work are needed to establish reliable designs, materials, and operational practices, leveraging but adapting the knowledge base from oil and gas and NGS storage. This also applies to potential wells for natural hydrogen extraction, which may encounter similar challenges regarding hydrogen-material interactions (Sun et al., 2025).

Managing well integrity for subsurface storage demands a tailored, risk-based approach focusing on the specific risks posed by the stored fluid, the operational mode (cycling versus permanent storage), and the required duration of containment, often necessitating advanced materials, rigorous verification, and comprehensive monitoring programs beyond standard oil and gas practice.

References

Abraham, J. J., Carvero, A., Teodoriu, C., AlMujalhem, M., & Amani, M. (2025). Evaluating potential near wellbore integrity issues in cement and casing layers during subsurface hydrogen storage and production. *GOTECH*. https://doi.org/10.2118/224759-MS

Alsubaih, A., Sepehrnoori, K., & Delshad, M. (2024). Environmental impacts of orphaned and abandoned wells: Methane emissions, and implications for carbon storage. *Applied Sciences, 14*(24), 11518. https://doi.org/10.3390/APP142411518

Alsubaih, A., Sepehrnoori, K., & Delshad, M. (2025). Development and optimization of self-healing cement for CO_2 injection and storage wells: Enhancing long-term wellbore integrity in extreme subsurface conditions. *Applied Sciences, 15*(10), 5428. https://doi.org/10.3390/APP15105428

Alsulaiman, A. (2024). Review of hydrogen leakage along the supply chain: Environmental impact, mitigation, and recommendations for sustainable deployment. https://www.econstor.eu/handle/10419/306643

Atrens, A., Venezuela, J., Liu, Q., Zhou, Q., Verbeken, K., Tapia-Bastidas, C., Gray, E., Christien, F., & Wolski, K. (2018). Electrochemical and mechanical aspects of hydrogen EmbrittlementEvaluation of martensitic steels. *Encyclopedia of Interfacial Chemistry: Surface Science and Electrochemistry*, 201–225. https://doi.org/10.1016/B978-0-12-409547-2.13770-9

Avc, C. B. (1992). Flow occurrence between confined aquifers through improperly plugged boreholes. *Journal of Hydrology, 139*(1–4), 97–114. https://doi.org/10.1016/0022-1694(92)90197-4

Beltrán-Jiménez, K., Anwar, I., Gebremariam, K. F., Kragset, S., Gardner, D., Skadsem, H. J., & Stormont, J. C. (2021). Assessment of corrosion in the interface casing–cement and its effect on the leakage potential. *SPE/IADC Drilling Conference, Proceedings*. https://doi.org/10.2118/204080-MS

Choi, Y. S., Young, D., Nešić, S., & Gray, L. G. S. (2013). Wellbore integrity and corrosion of carbon steel in CO_2 geologic storage environments: A literature review. *International Journal of Greenhouse Gas Control, 16*. https://doi.org/10.1016/j.ijggc.2012.12.028

Choi, Y. S., Nesic, S., & Young, D. (2010). Effect of impurities on the corrosion behavior of CO_2 transmission pipeline steel in supercritical CO_2-water environments. *Environmental Science and Technology, 44*(23), 9233–9238. https://doi.org/10.1021/ES102578C/SUPPL_FILE/ES102578C_SI_001.PDF

Connell, L., Downa, D., Lua, M., Hay, D., Heryantoa, D. (2015). An investigation into the integrity of wellbore cement in CO_2 storage wells: Core flooding experiments and simulations. *International Journal of Greenhouse Gas Control.*

Cui, Z. D., Wu, S. L., Zhu, S. L., & Yang, X. J. (2006). Study on corrosion properties of pipelines in simulated produced water saturated with supercritical CO_2. *Applied Surface Science, 252*(6), 2368–2374. https://doi.org/10.1016/J.APSUSC.2005.04.008

Das, A. K., & Hassanzadeh, H. (2023). Analytical solutions for brine invasion across caprock and leakage through an abandoned well in CO_2 storage sites. *Journal of Hydrology, 625*, Article 130046. https://doi.org/10.1016/J.JHYDROL.2023.130046

Dubois, J., Grau, E., Tassaing, T., & Dumon, M. (2018). On the CO2 sorption and swelling of elastomers by supercritical CO_2 as studied by in situ high pressure FTIR microscopy. *The Journal of Supercritical Fluids, 131*, 150–156. https://doi.org/10.1016/J.SUPFLU.2017.09.003

Dwivedi, S. K., & Vishwakarma, M. (2018). Hydrogen embrittlement in different materials: A review. *International Journal of Hydrogen Energy, 43*(46), 21603–21616. https://doi.org/10.1016/J.IJHYDENE.2018.09.201

Ford, E. P., Moeinikia, F., Lohne, H. P., Arild, O., Majoumerd, M. M., & Fjelde, K. K. (2017). Leakage calculator for plugged and abandoned wells. *Society of Petroleum Engineers–SPE Bergen One Day Seminar 2017*. https://doi.org/10.2118/185890-MS

Javandel, I., Tsang, C. F., Witherspoon, P. A., & Morganwalp, D. (1988). Hydrologic detection of abandoned wells near proposed injection wells for hazardous waste disposal. *Water Resources Research, 24*(2), 261–270. https://doi.org/10.1029/WR024I002P00261

Krevor, S., de Coninck, H., Gasda, S. E., Ghaleigh, N. S., de Gooyert, V., Hajibeygi, H., Juanes, R., Neufeld, J., Roberts, J. J., & Swennenhuis, F. (2023). Subsurface carbon dioxide and hydrogen storage for a sustainable energy future. *Nature Reviews Earth & Environment*, *4*(2), 102–118. https://doi.org/10.1038/s43017-022-00376-8

Laadel, N. E., El Mansori, M., Kang, N., Marlin, S., & Boussant-Roux, Y. (2022). Permeation barriers for hydrogen embrittlement prevention in metals–A review on mechanisms, materials suitability and efficiency. *International Journal of Hydrogen Energy, 47*(76), 32707–32731. https://doi.org/10.1016/J.IJHYDENE.2022.07.164

Lackey, G., Mundia-Howe, M., & Pekney, N. J. (2024). Underground natural gas storage facility operations and well leakage events in the United States. *Geoenergy Science and Engineering, 234*, Article 212630. https://doi.org/10.1016/J.GEOEN.2023.212630

Lee, K. K., Lee, S. H., Yun, S. T., & Jeen, S. W. (2016). Shallow groundwater system monitoring on controlled CO_2 release sites: A review on field experimental methods and efforts for CO_2 leakage detection. *Geosciences Journal, 20*(4), 569–583. https://doi.org/10.1007/S12303-015-0060-Z/METRICS

Maury Fernandez, D., Emadi, H., Thiyagarajan, S., Hussain, A., & Ispas, I. (2023). Effects of hydrogen on class H well cement's properties under geological storage conditions. In *57th US Rock Mechanics/Geomechanics Symposium*. https://doi.org/10.56952/ARMA-2023-0735

Nagumo, M., & Takai, K. (2019). The predominant role of strain-induced vacancies in hydrogen embrittlement of steels: Overview. *Acta Materialia, 165*, 722–733. https://doi.org/10.1016/J.ACTAMAT.2018.12.013

Newmark, R. L., Friedmann, S. J., & Carroll, S. A. (2010). Water challenges for geologic carbon capture and sequestration. *Environmental Management, 45*(4), 651. https://doi.org/10.1007/S00267-010-9434-1

Nordbotten, J. M., Celia, M. A., & Bachu, S. (2004). Analytical solutions for leakage rates through abandoned wells. *Water Resources Research, 40*(4), 4204. https://doi.org/10.1029/2003WR002997

Özsoy, Murat (n.d.). "Çankırı Salt Cave, YERALTI TUZ ŞEHRİ – Murat Özsoy 01." Wikimedia Commons. Licensed under Public Domain/CC0. https://commons.wikimedia.org/wiki/File:%C3%87ANKIRI_Salt_Cave,_YERALTI_TUZ_%C5%9EEHR%C4%B0_-_Murat_%C3%96zsoy_01.jpg

Plampin, M., Illangasekare, T., Sakaki, T., & Pawar, R. (2014). Experimental study of gas evolution in heterogeneous shallow subsurface formations during leakage of stored CO_2. *International Journal of Greenhouse Gas Control, 22*, 47–62. https://doi.org/10.1016/J.IJGGC.2013.12.020

Qiao, T., Hoteit, H., & Fahs, M. (2021). Semi-analytical solution to assess CO_2 leakage in the subsurface through abandoned wells. *Energies, 14*(9), 2452. https://doi.org/10.3390/EN1409 2452

Rimmelé, G., Barlet-Gouédard, V., Porcherie, O., Goffé, B., & Brunet, F. (2008). Heterogeneous porosity distribution in Portland cement exposed to CO2-rich fluids. *Cement and Concrete Research, 38*(8–9), 1038–1048. https://doi.org/10.1016/j.cemconres.2008.03.022

Schultz, R. A., Summers, L. E., Lynch, K. W., & Bouchard, A. J. (2014). Subsurface containment assurance program–key element overview and best practice examples. *Proceedings of the Annual Offshore Technology Conference, 2*, 1575–1586. https://doi.org/10.4043/24851-MS

Shaikhah, D., Cowe, B., Taleb, W., Neville, A., Mohamed-Said, M., & Barker, R. (2022). In-situ Synthesis of Corrosion Product-polymer composites on carbon steel for enhanced erosion-corrosion protection in CO_2 corrosion environments. *CONFERENCE, 2022*, 1–17. https://doi.org/10.5006/C2022-17717

Shaikhah, D., Taleb, W., Mohamed-Said, M., Cowe, B., & Barker, R. (2024). Augmentation of polymer-FeCO3 Microlayers on carbon steel for enhanced corrosion protection in hydrodynamic CO_2 corrosion environments. *ACS Omega, 9*(29), 31745–31753. https://doi.org/10.1021/ACSOMEGA.4C02616/ASSET/IMAGES/LARGE/AO4C02616_0006.JPEG

Siirila, E. R., Navarre-Sitchler, A. K., Maxwell, R. M., & McCray, J. E. (2012). A quantitative methodology to assess the risks to human health from CO_2 leakage into groundwater. *Advances in Water Resources, 36*, 146–164. https://doi.org/10.1016/J.ADVWATRES.2010.11.005

Song, J., & Curtin, W. A. (2011). A nanoscale mechanism of hydrogen embrittlement in metals. *Acta Materialia, 59*(4), 1557–1569. https://doi.org/10.1016/J.ACTAMAT.2010.11.019

Su, D., Li, Z., Wu, X., Li, J., Sun, J., & Zheng, G. (2022). Cement sheath integrity evaluation under multiple cyclic loading using mechanical equivalent experiment for gas storage wells in Eastern China. In *Proceedings of the International Conference on Offshore Mechanics and Arctic Engineering*–OMAE, Vol. 10. https://doi.org/10.1115/OMAE2022-80440

Sun, B., Zhang, M., Sun, Q., Zhong, J., & Shao, G. (2025). Review on natural hydrogen wells safety. *Nature Communications 16*(1), 1–14. https://doi.org/10.1038/s41467-024-55773-y

Terzi, K., Aggelopoulos, C. A., Bountas, I., & Tsakiroglou, C. D. (2014). Effects of carbon dioxide on the mobilization of metals from aquifers. *Environmental Science and Technology, 48*(8), 4386–4394. https://doi.org/10.1021/ES405032D/SUPPL_FILE/ES405032D_SI_001.PDF

Thomas, D. C., Benson, S. M., Scherer, G. W., Celia, M. A., Prrvost, J.-H., Bachu, S., Bruant, R., Duguid, A., Fuller, R., Gasda, S. E., Radonjic, M., & Vichit-Vadakan, W. (2005). Leakage of CO_2 through abandoned wells: Role of corrosion of cement. www.eub.gov.ab.ca

Trudel, E., & Frigaard, I. A. (2023). Stochastic modelling of wellbore leakage in British Columbia. *Journal of Petroleum Science and Engineering, 220*, Article 111199. https://doi.org/10.1016/J.PETROL.2022.111199

Van Der Schijff, O. J., Budiansky, N., & Latanision, R. M. (2021). Hydrogen damage and embrittlement. *Failure Analysis and Prevention*, 518–532. https://doi.org/10.31399/ASM.HB.V11.A0006784

White, J. A., & Foxall, W. (2016). Assessing induced seismicity risk at CO_2 storage projects: Recent progress and remaining challenges. *International Journal of Greenhouse Gas Control, 49*, 413–424. https://doi.org/10.1016/j.ijggc.2016.03.021

William Carey, J., Svec, R., Grigg, R., Zhang, J., & Crow, W. (2010). Experimental investigation of wellbore integrity and CO2–brine flow along the casing–cement microannulus. *International Journal of Greenhouse Gas Control, 4*(2), 272–282. https://doi.org/10.1016/J.IJGGC.2009.09.018

Wood, D. A. (2024). Well integrity for underground gas storage relating to natural gas, carbon dioxide, and hydrogen. *Sustainable Natural Gas Drilling: Technologies and Case Studies for the Energy Transition*, 551–576. https://doi.org/10.1016/B978-0-443-13422-7.00019-2

Yang, B., Wang, H., Wang, B., Yi, Y., Zhao, C., & Tian, G. (2023). Effect of supercritical CO_2-Water/Brine-rock interaction on microstructures and mechanical properties of tight sandstone. *Transport in Porous Media, 149*(1), 87–115. https://doi.org/10.1007/s11242-022-01834-z

Young, D., Sevre, A. K., Smith, P. E., Al-Madlouh, M., & Gjelstad, G. (2011). Effects of rapid gas decompression on Swellable rubber and common oilfield rubber compounds. In *Proceedings of the SPE/IADC Middle East Drilling Technology Conference and Exhibition*, pp. 738–744. https://doi.org/10.2118/148584-MS

Zheng, D., Ozbayoglu, E., Miska, S., & Baldino, S. (2024). An experimental study on the influence of cyclic loading and eccentric casing geometry on cement sheath integrity. https://doi.org/10.56952/ARMA-2024-0305

Zhou, C., Ye, B., Song, Y., Cui, T., Xu, P., & Zhang, L. (2019). Effects of internal hydrogen and surface-absorbed hydrogen on the hydrogen embrittlement of X80 pipeline steel. *International Journal of Hydrogen Energy, 44*(40), 22547–22558. https://doi.org/10.1016/J.IJHYDENE.2019.04.239

Zhu, X., Peng, P., Jing, J., & Song, W. (2022). Research on casing erosion in reservoir section of gas storage wells based on gas-solid two-phase flow. *Engineering Failure Analysis, 142*, Article 106688. https://doi.org/10.1016/J.ENGFAILANAL.2022.106688

Chapter 25
Well Intervention and Integrity

Abstract This chapter explores the critical intersection between well intervention operations and well integrity management across the well lifecycle. Well interventions—including wireline, coiled tubing, and workover operations—are essential for restoring productivity, acquiring downhole data, maintaining wellbore components, and preparing for abandonment. However, these operations inherently compromise the established barrier envelope of the well, introducing elevated risks to its mechanical and hydraulic integrity. The chapter begins by identifying the drivers behind well interventions, from production enhancement to integrity remediation and regulatory compliance. It then introduces the **temporary barrier philosophy**, emphasizing the principle of maintaining at least two independent, tested barriers between the reservoir and the environment during any intervention. The various forms of temporary barriers—including mechanical plugs, fluid columns, pressure control equipment (PCE), and cement plugs—are discussed alongside their verification requirements and integration into barrier schematics. A detailed review of **well control systems** used during interventions is presented, highlighting the configurations and functions of wireline and slickline PCEs, coiled tubing BOPs, and workover/snubbing stacks. The importance of surface BOP systems, shear rams, lubricators, and injectors is emphasized, with attention to proper functionality, regular testing, and emergency response procedures. The chapter also addresses the **potential for mechanical and chemical damage** to well components during intervention activities. These include casing and tubing wear, elastomer degradation, scale deposition, and thermal or pressure-induced failures. Case studies and data illustrate how intervention tools and fluids—if poorly selected or executed—can compromise the well's barrier system. The Swiss cheese model is applied to visualize how multiple, seemingly minor weaknesses can align to cause integrity loss. Finally, the chapter outlines the **critical need for post-intervention integrity verification** before returning a well to service. This includes reinstating permanent barriers, pressure testing critical components (e.g., packers, downhole safety valves), and documenting the verified integrity status. Trends in equipment failures, particularly for blowout preventers and safety valves, are examined to stress the importance of post-job testing and record keeping. Overall, this chapter emphasizes that successful well interventions require meticulous planning, the use of reliable and verified temporary barriers, strict adherence to well

© The Author(s), under exclusive license to Springer Nature Switzerland AG 2026

A. Alsubaih and K. Sepehrnoori, *Well Integrity for Oil, Gas, and Subsurface Storage Operations*, https://doi.org/10.1007/978-3-032-15233-6_25

control procedures, damage prevention strategies, and rigorous post-intervention verification to preserve the well's long-term integrity.

BO shifting tool (wireline tool used in well operations). *Source* J. E. Simpson (n.d.)

Chapter's Objectives

- Explain why well interventions are performed and the associated integrity risks.
- Detail the concept of temporary barriers during interventions.
- Discuss maintaining well control during different intervention types (wireline, coiled tubing, workover).
- Emphasize the importance of verifying integrity after intervention operations.

25.1 The Need for Well Intervention

Well intervention refers to any operation performed on or inside a well after its initial completion to diagnose problems, restore or enhance production/injection, perform maintenance, or gather data (Aihevba, 2004). These operations are necessary to optimize reservoir performance and extend the well's productive life, but they inherently involve manipulating or bypassing the well's established barrier system, introducing significant risks to well integrity. Figure 25.1 illustrates the lifecycle of a typical well intervention, highlighting the process from problem identification through execution to verification and return to service. Common reasons for intervention include (Aihevba, 2004; Carragher & Talapatra, 2025; Mohamed Hanafi et al., 2024; Oatey et al., 2021; Solli, 2011):

- Data Acquisition: Running logging tools (production logging, integrity logs like caliper or EM thickness), taking fluid samples, performing pressure surveys.
- Well Surveillance/Testing: Performing integrity tests (e.g., leak tests of valves, annulus tests), checking valve operation.
- Production Enhancement: Performing stimulations (acidizing, hydraulic fracturing—re-fracs), scale removal, paraffin removal, perforating new zones, changing out artificial-lift equipment (e.g., gas lift valves, ESPs), installing flow control devices.
- Integrity Remediation: Setting plugs/patches to isolate leaks (e.g., in tubing or casing), performing cement squeezes to repair annular isolation, replacing failed downhole components like packers or DHSVs, and fishing lost equipment that might obstruct barriers.
- Routine Maintenance: Changing out inserts in wellhead/Xmas tree valves, performing scale squeezes or inhibitor treatments.

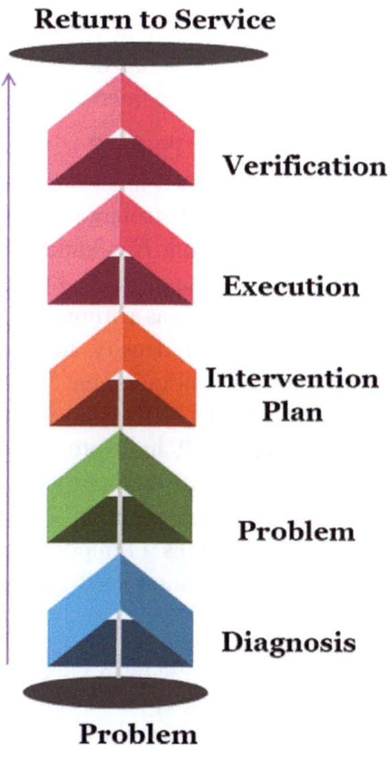

Fig. 25.1 Lifecycle of a well intervention operation, from problem identification and diagnosis, through planning and execution of the intervention, to verification of integrity and return to service. The process emphasizes the critical need to maintain and test well barriers at each stage to ensure long-term integrity

- Preparing for Abandonment: Setting initial plugs, cutting/pulling tubing or casing sections, and preparing for segment milling.

While necessary for asset management, interventions inherently involve risks to well integrity because they often require temporarily defeating or manipulating the well's permanent barrier system under pressure and introduce potential for damaging existing components.

25.2 Maintaining Barriers During Intervention: The Temporary Barrier Philosophy

The fundamental principle of well integrity—maintaining at least two independent and tested barriers between any potential pressure source and the environment—remains paramount during intervention operations (*NORSOK D-010*, 2021). However, since the intervention itself often requires accessing the primary production conduit (tubing) or annulus, the permanent barriers (like the Xmas tree valves or DHSV) may need to be opened or removed. To manage this risk, the concept of temporary barriers is employed.

- **Temporary Barrier Element**: A component or fluid column deployed specifically for the intervention operation to provide a verifiable pressure seal, which is intended to be removed or cease functioning as a barrier after the operation. Unlike permanent barriers designed for long-term life, temporary barriers are for the duration of the intervention. Examples include:
  - Mechanical Plugs: Wireline-set or coiled-tubing-set bridge plugs, straddle packers, or test tools set inside the tubing or casing below the working area. These provide isolation from lower zones.
  - Fluid Column: A column of fluid (e.g., kill-weight mud, heavy brine) in the wellbore with sufficient hydrostatic pressure to overbalance the formation pressure. This acts as a primary or secondary barrier, but requires careful monitoring of fluid level, density, and potential losses or gains (Costa et al., 2024).
  - Intervention Equipment Seals: The seals and valves within the pressure control equipment (PCE) used at surface (e.g., wireline BOP rams, coiled tubing stripper/BOP, lubricator valves, shear/seal rams in workover BOPs). These act as critical temporary barriers against pressure from within the wellbore during the intervention activity.
  - Cement Plugs (Temporary): Sometimes used for longer interventions or specific isolation needs, designed to be easily drilled out later (Achang et al., 2020).

- **Verification:** Crucially, temporary barriers must also be tested and verified after installation (e.g., pressure testing a mechanical plug from above, inflow testing from below if possible, pressure testing the surface PCE stack) to confirm their

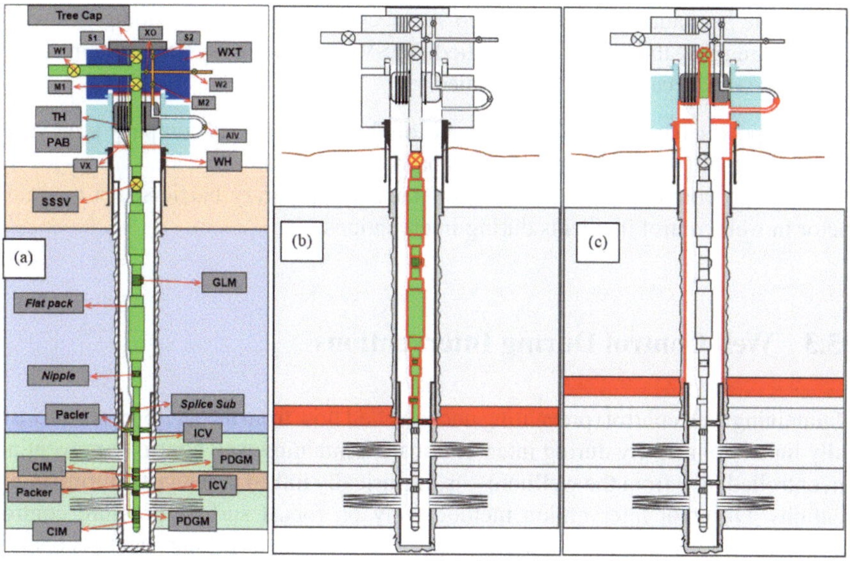

Fig. 25.2 Completion schematic and barrier configurations **a** Completion schematic showing key components such as the wellhead (WH), packer, tubing hanger (TH), master valves (M1, M2), wing valves (W1, W2), swab valves (S1, S2), annulus valves (AIV), inflow control valves (ICV), chemical injection mandrel (CIM), permanent downhole gauge mandrel (PDGM), gas lift mandrel (GLM), and the surface-controlled subsurface safety valve (SSSV) **b** Primary Barrier Integrated Set (BIS) providing initial well control during interventions; and **c** Secondary BIS as an additional protective layer, maintaining overall well integrity throughout intervention activities (Da Fonseca et al., 2013)

sealing integrity *before* relying on them and proceeding with activities that might compromise other barriers (Cully et al., 2022; Daohmareeyor et al., 2022; Raaen & Fjær, 2020).

- **Two Barrier Requirement:** The intervention plan must explicitly demonstrate using Well Barrier Schematics (WBS) how two independent, tested barriers (which may be a combination of permanent and temporary barriers, or two temporary barriers) will be always maintained between the pressure source (e.g., reservoir pressure, annulus pressure) and the external environment throughout each phase of the intervention operation. Figure 25.2 illustrates the completion schematic of a typical well system, along with the primary and secondary Barrier Integrated Sets (BIS). These BIS examples highlight how well integrity is maintained through mechanical and fluid-based barriers in different well intervention phases (Da Fonseca et al., 2013).

For example, during a wireline operation inside the tubing below an open tree:

- Possible Primary Barrier: The Wireline Pressure Control Equipment (PCE) stack (stuffing box and wireline BOP) closed around the wireline cable.

- Possible Secondary Barrier: A deep-set mechanical plug in the tubing below the work area, OR the closed and verified DHSV if its control system remains intact and its sealing capability is confirmed.

(The specific barrier definition depends on the operation and company policy, but the principle of two independent, tested barriers remains).

Failure to correctly install, verify, or manage temporary barriers is a common factor in well control incidents during interventions.

25.3 Well Control During Interventions

Maintaining well control (preventing uncontrolled flow from the wellbore) is intrinsically linked to integrity during interventions. Maintaining well control—preventing uncontrolled flow from the wellbore—is intrinsically linked to integrity during interventions. Different intervention methods rely on robust surface pressure control equipment (PCE) and well barriers to provide the crucial primary and secondary barriers. Figure 25.3 illustrates the arrangement of primary and secondary well barriers during two live well intervention scenarios: wireline interventions and hydraulic workover operations (Crumpton, 2018). These schematics emphasize the multiple layers of barriers, including shear seal BOPs, coiled tubing injectors, snubbing safety heads, and annular preventers, that work together to ensure well control and maintain overall well integrity throughout the intervention. Different intervention methods utilize different surface pressure control equipment (PCE) to provide the crucial upper temporary barrier.

- **Wireline/Slickline:** Uses a PCE stack assembled above the wellhead/Xmas tree (Ashena, 2025), typically consisting of:
  - Stuffing Box/Grease Injection Head: The uppermost sealing element, designed to seal around the moving wireline or slickline cable using packing elements or grease injection pressure. Requires careful monitoring and management of grease pressure or packing compression.
  - Wireline Blowout Preventer (BOP): A critical safety device below the stuffing box. Contains rams designed to seal around the wireline (pipe rams), cut the wireline in an emergency (shear rams), and potentially seal the open hole if the wireline is removed (blind rams). Acts as a crucial temporary barrier if the stuffing box fails. Must be regularly function and pressure tested according to standards (e.g., API RP 53, API Spec 16ST).
  - Lubricator Sections: Pressurized riser sections placed between the BOP and the Xmas tree, used to house the toolstring so it can be introduced into or retrieved from the pressurized wellbore. Valves between sections and at the Christmas tree connection provide further barriers.

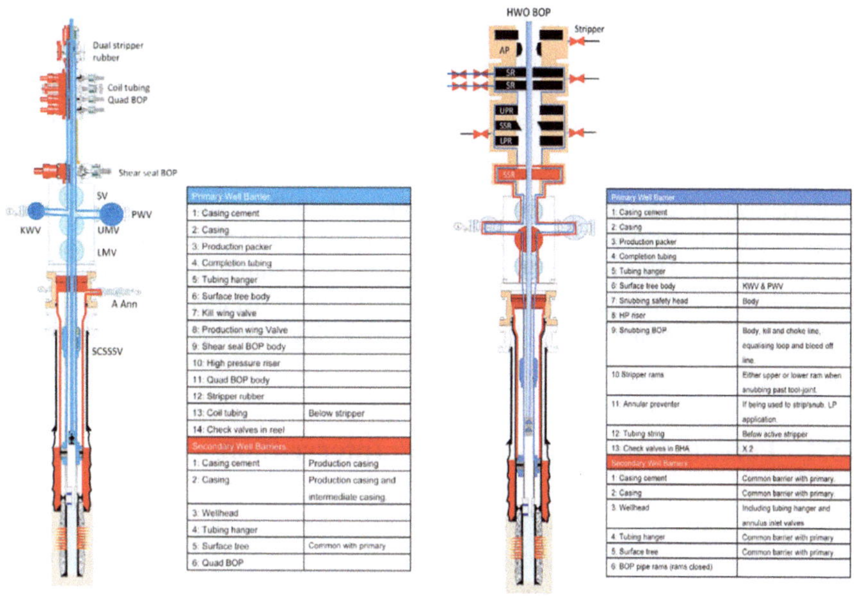

Primary Well Barrier	
1: Casing cement	
2: Casing	
3: Production packer	
4: Completion tubing	
5: Tubing hanger	
6: Surface tree body	
7: Kill wing valve	
8: Production wing Valve	
9: Shear seal BOP body	
10: High pressure riser	
11: Quad BOP body	
12: Stripper rubber	
13: Coil tubing	Below stripper
14: Check valves in reel	
Secondary Well Barriers	
1: Casing cement	Production casing
2: Casing	Production casing and intermediate casing
3: Wellhead	
4: Tubing hanger	
5: Surface tree	Common with primary
6: Quad BOP	

Primary Well Barrier	
1: Casing cement	
2: Casing	
3: Production packer	
4: Completion tubing	
5: Tubing hanger	
6: Surface tree body	KWV & PWV
7: Snubbing safety head	Body
8: HP riser	
9: Snubbing BOP	Body, kill and choke line, equalising loop and bleed off line.
10 Stripper rams	Either upper or lower ram when snubbing past tool-joint.
11: Annular preventer	If being used to strip/snub. LP application.
12: Tubing string	Below active stripper
13: Check valves in BHA	X 2
Secondary Well Barriers	
1: Casing cement	Common barrier with primary.
2: Casing	Common barrier with primary.
3: Wellhead	Including tubing hanger and annulus inlet valves
4: Tubing hanger	Common barrier with primary
5: Surface tree	Common barrier with primary
6: BOP pipe rams (rams closed)	

Fig. 25.3 Schematic of primary and secondary well barrier configurations during live well interventions. (Left) Wireline intervention barriers, showing shear seal BOPs, coiled tubing, and tree valves. (Right) Hydraulic workover unit (HWO BOP) arrangement, depicting multiple rams, annular preventers, and tubing interfaces. These barriers are essential for well control during interventions (Crumpton, 2018)

  - Shear Seal Valve (SSV) or Master Valve: Often the uppermost valve on the Xmas tree is part of the PCE stack, providing the interface and a potential lower temporary barrier.

- **Coiled Tubing (CT):** Uses a similar PCE stack, but adapted for the continuous coiled tubing string:

  - Stripper/Packer Assembly: The uppermost sealing element seals around the moving coiled tubing pipe, often using hydraulic pressure to energize packing elements. Subject to significant wear.
  - Coiled Tubing BOP: Below the stripper. Contains specialized rams (pipe rams to seal around the CT pipe, slip rams to grip and hold the CT pipe if tensile load is lost, shear rams to cut the CT pipe in an emergency, and blind rams). Provides well control if the stripper fails (Hadzihafizovic, 2024). Must be regularly function and pressure tested.
  - Injector Head: The unit that grips and pushes/pulls the coiled tubing string into/out of the well, mounted above the PCE stack.

- **Workovers/Snubbing**: Involve running jointed pipe (drill pipe, tubing) into a live well, often after removing the Xmas tree. Requires a larger Workover BOP stack similar to a drilling BOP, but often configured differently and potentially designed for higher pressures. May include annular preventers (to seal around varying pipe sizes or the open hole), multiple sets of pipe rams, and shear/blind rams. Snubbing operations involve using a hydraulic jack (snubbing unit) to push pipe into the well against pressure. Requires highly skilled crews, robust equipment, and rigorous procedures.

Regardless of the intervention method, the surface PCE stack acts as a critical temporary barrier system. Its reliable function, verified through regular testing (e.g., per API RP 53, API Spec 16ST), is paramount. Procedures must be in place for responding to leaks or failures in the PCE or downhole temporary barriers, often involving well shut-in, pumping kill fluid, or activating emergency shear rams.

25.4 Potential for Damage During Intervention

Intervention operations themselves, while necessary, can inadvertently damage permanent well components if not planned and performed carefully. Such damage can create new integrity issues or compromise existing barriers.

- **Mechanical Damage**: Tools run into the well (logging tools, perforating guns, milling assemblies, wireline toolstrings, coiled tubing ends) can cause wear, scraping, or impact damage to the inside of tubing, casing, completion components (e.g., DHSV flappers, packer bores), or wellhead sealing surfaces, particularly in deviated or narrow sections, or when running heavy or rigid tools (Campos et al., 2024). Setting mechanical tools forcefully or jarring operations can impose significant stress on casing or cement bonds. Intervention operations, while essential for well performance, can inadvertently introduce mechanical damage to well components. Figure 25.4 illustrates how the use of different coiled tubing (CT) grades influences the creation of groove damage in the wellbore. Higher-grade CT was associated with more wells experiencing groove damage, highlighting the importance of selecting appropriate materials and managing the number of CT runs to minimize wear.
- **Corrosion/Scale Introduction:** Intervention fluids (e.g., workover fluids, completion fluids, stimulation chemicals) if not properly selected, filtered, inhibited, or treated (e.g., for oxygen or bacteria content) can introduce corrosive agents or incompatible fluids leading to scale deposition, formation damage, or accelerated corrosion of tubulars and equipment. Compatibility testing of fluids with well materials is important.
- **Pressure Effects**: Applying excessive pressure during intervention activities (e.g., pressure testing, pumping stimulation fluids, setting hydraulic tools) can potentially overstress and damage downhole components (tubulars, packers) or formations (causing fractures). Ensuring injected volumes and pressures are within design limits is crucial.

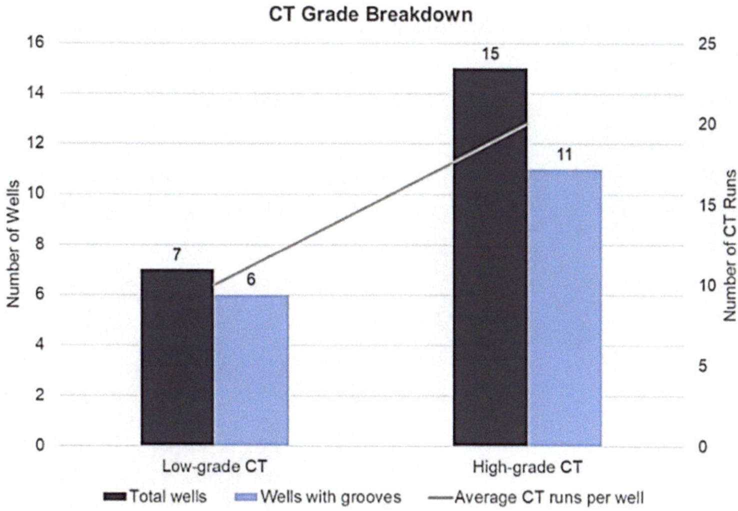

Fig. 25.4 Number of wells and occurrence of groove damage associated with different coiled tubing (CT) grades. The left bars represent total wells and wells with grooves for low-grade CT, while the right bars show the same data for high-grade CT. The line indicates the average number of CT runs per well, showing a clear correlation between CT grade, number of runs, and wellbore groove damage (Campos et al., 2024)

- **Temperature Effects**: Pumping large volumes of cool fluid into a hot well can induce thermal shock and stresses, potentially damaging cement or casing (similar to thermal cycling discussed in Chaps. 10and 21).
- **Perforating Operations**: While designed to create connections to the formation, perforating can impact casing and cement integrity. High-perforating shock can sometimes damage nearby cement bonds or even cause micro-cracks. Back-phasing perforating guns help reduce casing deformation.

Careful planning, selection of appropriate tools for the wellbore geometry and condition, adherence to operating limits (e.g., jarring limits, maximum tool OD vs. drift diameter, pressure limits), managing intervention fluid quality, and use of experienced, competent personnel minimize the risk of causing damage during intervention. Intervention operations themselves, while necessary, can inadvertently damage permanent well components if not planned and performed carefully. Such damage can create new integrity issues or compromise existing barriers. The Swiss cheese model of barrier defense (Fig. 25.5) illustrates how the alignment of weaknesses in multiple defense layers—ranging from mechanical and procedural to human decision-making—can lead to a loss of well integrity.

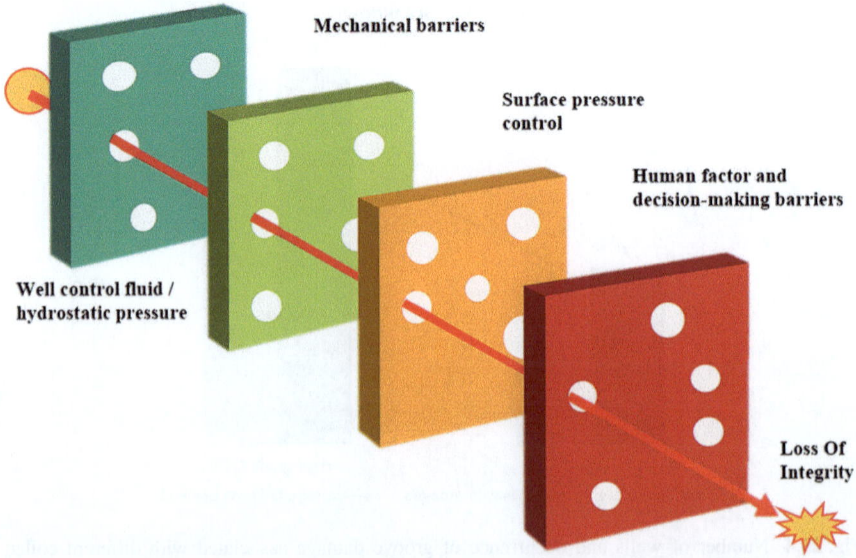

Fig. 25.5 The Swiss cheese model of barrier defense in well interventions. Each "slice" represents a layer of defense: well control, fluid/hydrostatic pressure, mechanical barriers, surface pressure control, and human factors. Holes in the slices represent weaknesses or failures in these barriers. When these weaknesses align, a clear path forms, leading to a well integrity failure (loss of integrity)

25.5 Integrity Verification Post-intervention

After the intervention is complete, temporary barriers are removed (if applicable), and permanent components are reinstated, it is essential to re-establish and verify the integrity of the well's permanent barrier system *before* returning the well to normal operation or shut-in status. Failing to perform adequate post-intervention verification is a significant risk that can leave undetected integrity issues, potentially leading to failures later. Historical failure data can illustrate how critical this process is. Figure 25.6 shows the number of failures per year for Downhole Safety Valves (DHSV) and Blowout Preventers (BOP), highlighting how equipment reliability fluctuates over time (Vignes, 2011). This typically involves:

- **Reinstating Permanent Barriers**: Ensuring all permanent components that were manipulated or bypassed (e.g., tubing, packer, DHSV, Xmas tree valves) are correctly re-installed or back in their normal operational configuration.
- **Pressure Testing**: Performing a series of relevant integrity tests to confirm that the permanent primary and secondary barriers are intact and sealing effectively *after* the disturbance of the intervention operation. These tests should repeat key initial verification tests (discussed in Chap. 7), such as:

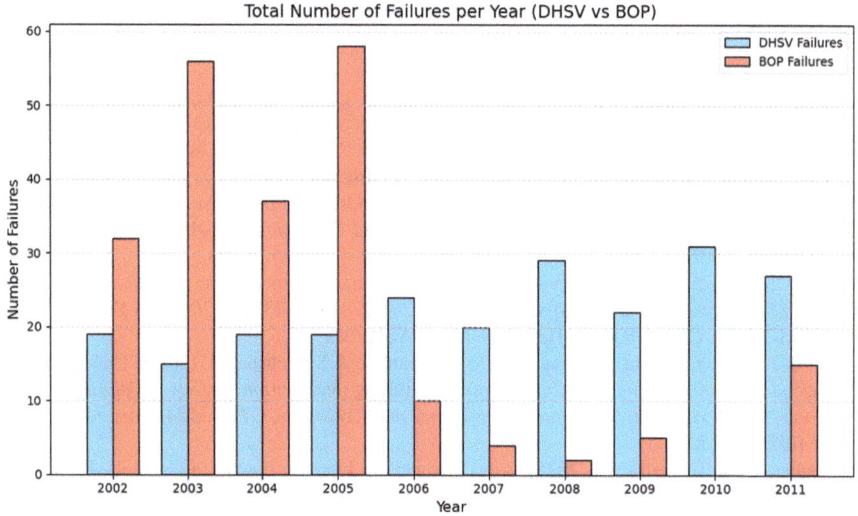

Fig. 25.6 Total number of failures per year for Downhole Safety Valves (DHSV) and Blowout Preventers (BOP) from 2002 to 2011, highlighting the importance of rigorous integrity verification and continuous monitoring to ensure safe and reliable operations (Vignes, 2011)

- Tubing string test (confirming tubing/connection integrity).
- Packer seal test (confirming isolation of annulus from below).
- DHSV functional and leak test (confirming operability and sealing of the critical downhole safety valve).
- Wellhead and annulus seal tests (confirming sealing of surface barriers).

  The specific tests depend on the nature of the intervention and the barriers affected.

- **Documentation**: Recording all intervention activities, all barrier tests performed (both temporary barriers during the job and permanent barriers upon completion), any anomalies observed, and the final verified integrity status of the well. This becomes part of the well's permanent history data (Chap. 15).
- **Handover**: The formal handover process back to the production/injection operations team must explicitly confirm that the well's integrity status has been verified to meet the required standards for the subsequent operational phase.

Skipping or inadequately performing post-intervention verification tests can leave undetected integrity issues, potentially leading to SCP, leaks, or well control incidents later. Meticulous planning, robust well control practices, careful execution to avoid damage, and thorough verification testing upon completion are fundamental for managing well integrity throughout intervention activities.

References

Achang, M., Yanyao, L., & Radonjic, M. (2020). A review of past, present, and future technologies for permanent plugging and abandonment of wellbores and restoration of subsurface geologic barriers. *Environmental Engineering Science, 37*(6), 395–408. https://doi.org/10.1089/EES.2019.0333/ASSET/IMAGES/EES.2019.0333_FIGURE5.JPG

Aihevba, L. O. (2004). Well intervention management in Oman fields. In *Proceedings of the IADC/SPE Asia Pacific Drilling Technology Conference and Exhibition*, 451–459. https://doi.org/10.2118/88025-MS

Ashena, R. (2025). Wireline operations and well control setup. *The Practical Handbook of Well Control*, 289–295. https://doi.org/10.1201/9781003473770-12/WIRELINE-OPERATIONS-WELL-CONTROL-SETUP-RAHMAN-ASHENA

Campos, J., Keong, A., Liu, Z., Tomczak, D., Kleppan, T., & Soerensen, M. (2024). Understanding mechanisms of wellbore abrasion during coiled tubing intervention. *Society of Petroleum Engineers–SPE/ICoTA Well Intervention Conference and Exhibition, CTWI 2024*. https://doi.org/10.2118/218335-MS

Carragher, P., & Talapatra, D. (2025). Advancements in bismuth-based sealing tools: Enhancing efficiency and integrity in well interventions. *Society of Petroleum Engineers–SPE/ICoTA Well Intervention Conference and Exhibition, CTWI, 2025*, 25–26. https://doi.org/10.2118/224020-MS

Costa, W. R. P. da, Nóbrega, K. C., Costa, A. C. A., Nascimento, R. C. A. de M., Souza, E. A. de, Oliveira, T. A. de, Barros, M., & Amorim, L. V. (2024). *A Liquid Well Barrier Element for Temporary Plug & Abandonment Operations: A Breakthrough Approach*. https://doi.org/10.20944/PREPRINTS202408.0572.V1

Crumpton, H. (2018). Coiled tubing well control. *Well Control for Completions and Interventions*, 483–566. https://doi.org/10.1016/B978-0-08-100196-7.00010-5

Cully, R., Franklin, C. M., Lewis, K., & Mesa, C. (2022). A new approach to assuring well integrity and performance. *Proceedings of the Annual Offshore Technology Conference*. https://doi.org/10.4043/32004-MS

Daohmareeyor, T., Lock, D. L. W., & Wattanasuwan, R. (2022). The use of an organic Crosslinked polymer sealant as a barrier to retrieve stuck coiled tubing from a live high pressure well after over a year: Case study from offshore Vietnam. *Proceedings of the IADC/SPE Asia Pacific Drilling Technology Conference, APDT, 2022-August*. https://doi.org/10.2118/209856-MS

Da Fonseca, T. C., Miura, K., & Mendes, J. R. P. (2013). Well integrity analysis applied to workover prediction. *Proceedings of the Annual Offshore Technology Conference, 2*, 832–839. https://doi.org/10.4043/24369-MS

Hadzihafizovic, D. (2024). *Well Intervention Pressure Control*. https://papers.ssrn.com/abstract=5031370

Mohamed Hanafi, M. M., Ismail, N., Anoar, I., Ong, V., Hamzah, S., Khairul Anuar, K., Singh, K., Cavalleri, C., Yong, N., Zailani, M. F. I., Ramondenc, P., & Hofacker, M. (2024). Intervention synergy improving production enhancement efficiency for remote offshore well intervention operation. *Society of Petroleum Engineers–SPE/ICoTA Well Intervention Conference and Exhibition, CTWI 2024*. https://doi.org/10.2118/218364-MS

NORSOK D-010. (2021). https://online.standard.no/en/norsok-d-010-2021

Oatey, M., Duff, F., Emslie, N., Christie, S., Rikabi, R., Henderson, P., Singh, K., Kumar, A., Agrawal, G., Dutta, S., & Bajwa, H. (2021). Enhancing production through well interventions using end-to-end evaluation methodology. *SPE Middle East Oil and Gas Show and Conference, MEOS, Proceedings*. https://doi.org/10.2118/204540-MS

Raaen, A. M., & Fjær, E. (2020). Pressure testing of barrier integrity. In *Proceedings of the International Conference on Offshore Mechanics and Arctic Engineering–OMAE*, Vol. 11. https://doi.org/10.1115/OMAE2020-18713

Simpson, J. E. (n.d.). "BO Shifting Tool." Wikimedia Commons. Licensed under CC BY 2.0. https://commons.wikimedia.org/wiki/File:BO_Shifting_Tool.jpg

Solli, T. E. (2011). Workover/Well intervention and regulatory challenges. *Proceedings of the Annual Offshore Technology Conference, 2*, 942–948. https://doi.org/10.4043/22586-MS

Vignes, B. (2011). *Contribution to well integrity and increased focus on well barriers from a life cycle aspect (2010). Birgit Vignes. 26 Citations* (Faculty of Science and Technology University of Stavanger). https://scispace.com/papers/contribution-to-well-integrity-and-increased-focus-on-well-n5cyhzz40j

Chapter 26
Well Integrity Remediation Techniques

Abstract This chapter presents a comprehensive overview of well integrity remediation techniques aimed at restoring barrier functionality when failures occur during a well's operational life. While preventative design and maintenance are foundational, issues such as sustained casing pressure (SCP), casing corrosion, cement sheath failure, packer leaks, and surface valve damage can still arise and compromise zonal isolation or environmental safety. Remediation becomes essential to extend the well's productive life and mitigate operational and regulatory risks. The chapter begins with the remediation of **cement sheath failures**, emphasizing squeeze cementing as the most widely used method. Techniques such as continuous and hesitation squeeze methods are detailed, including slurry design considerations for penetration into micro-annuli and cracks. Case studies and laboratory data are used to highlight the effectiveness and limitations of various cement formulations, including microfine and chemical sealants. For **casing leak repair**, the chapter outlines the application of expandable casing patches and liners to seal localized damage or reinforce weakened sections. Straddle packer assemblies are described as temporary diagnostic tools, while full-length liners are recommended for extensive damage. Cement squeezing and epoxy-augmented expandable systems are also discussed. **Leaks from downhole packers and tubing/casing connections** are addressed through workovers involving component retrieval and replacement. Elastomeric expansion sleeves and internal sleeves may serve as patching alternatives where geometry permits. However, in many cases, major intervention is necessary due to limited access or component failure. **Wellhead and valve leaks**, though more accessible, still require precise isolation, pressure management, and seal replacement strategies. Temporary fixes such as sealant injection are discussed, alongside definitive solutions like gasket replacement, valve packing changeouts, and full valve replacements. Emphasis is placed on operational safety and verification during surface interventions. The final section outlines **critical planning considerations** for any remediation effort. These include accurate diagnostics, risk assessment, well kill procedures, equipment selection, fluid compatibility, verification testing, and economic analysis. Diagrams and field case data throughout the chapter illustrate the complexities of repair operations

© The Author(s), under exclusive license to Springer Nature Switzerland AG 2026
A. Alsubaih and K. Sepehrnoori, *Well Integrity for Oil, Gas, and Subsurface Storage Operations*, https://doi.org/10.1007/978-3-032-15233-6_26

and reinforce the necessity for experienced personnel, proper tooling, and a systematic approach. Collectively, this chapter provides a field-ready guide to selecting and implementing effective remediation strategies tailored to the specific integrity challenge, supporting safe and sustainable well operation across its lifecycle.

Onshore drilling rig and crane operation during rig-up phase, illustrating surface well construction and infrastructure mobilization. *Source* The Authors

Chapter's Objectives

1. Describe common reasons for well integrity remediation.
2. Outline techniques for repairing cement sheath failures (e.g., squeeze cementing).
3. Discuss methods for repairing casing leaks (patches, liners).
4. Explain approaches for addressing leaking packers or connections.
5. Cover remediation options for wellhead and valve leaks.
6. Discuss general considerations for planning and executing remediation.

26.1 Introduction: Addressing Integrity Failures

Despite best efforts in design, construction, and preventative maintenance programs, well integrity issues can still arise during the operational life of a well. Monitoring and diagnostic techniques (Chaps. 8 and 12) may reveal problems like Sustained Casing Pressure (SCP), leaking valves, packer failures, or casing corrosion that require intervention to restore the integrity of the well barriers. This chapter outlines common remediation techniques used to repair failed components and re-establish containment, thereby extending the well's life and mitigating risks. Remediation is often complex, expensive, and carries its own risks, highlighting the value of preventing failures in the first place through robust initial design, construction quality, and proactive surveillance.

26.2 Remediation of Cement Sheath Failures

Failures in the primary cement sheath (e.g., channels, debonding, cracks) leading to SCP, inter-zonal flow, or surface leaks are common integrity issues requiring remediation. The most common technique is squeeze cementing, but others exist.

- **Squeeze Cementing**: Squeeze cementing involves injecting a cement slurry at elevated pressures into the annulus or any fractures and micro-annuli within the existing cement sheath or surrounding formations. It is the most used technique for repairing cement sheath failures due to its similarities with primary cementing operations and the availability of materials in the field (Izadi et al., 2024; Yousuf et al., 2021). However, some key differences in slurry properties exist compared to conventional primary cement jobs. Specifically, the viscosity of the squeeze cement slurry is typically lower, with densities ranging between 1380 and 1500 kg/m^3 (11.5–12.5 lbm/gal), to facilitate penetration through narrow fractures (Meek & Harris, 1993). When densities are too high, slurry penetration can be hindered. In the continuous method, slurry is continuously injected until the final squeeze pressure is reached, with periodic re-pressurization if pressure falls. In the hesitation squeeze, slurry is injected in stages and paused to enhance penetration through differential pressure. The injectivity factor, calculated as the injection pressure divided by injection rate, guides the cement type selection. For injectivity factors below 2000, conventional Class G cement may be used, while factors above 2000 require microfine cement for effective zone penetration (Asyri & Mulia, 2018).
  - Objective: To plug the leak path and re-establish a hydraulic seal in the annulus.
  - Procedure (Simplified):
    1. Diagnose the leak path and location (using logs, pressure analysis).

2. Isolate the zone needing squeeze, typically by setting mechanical plugs (e.g., wireline-set bridge plug) inside the casing below the suspected leak interval.
3. Perforate the casing at or near the suspected leak depth/interval to gain direct access from inside the casing to the annulus.
4. Establish Injectivity: Pump fluid (water, brine, or acid) down tubing or drill pipe and out through the perforations to confirm communication with the leak path in the annulus and determine how easily fluid can be injected. Acid washes may be used to clean the leak path or break down residual mud/filter cake.
5. Pump Cement Slurry: Mix and pump a small volume of specialized "squeeze slurry" (typically low viscosity to penetrate small channels, low fluid loss to prevent premature dehydration against permeable formations, potentially with rapid gel strength development or slight expansion) down tubing or drill pipe and out through the perforations into the leak path in the annulus.
6. Apply Squeeze Pressure: After pumping the planned slurry volume, apply surface pressure to force the cement into the leak channels until a predetermined final squeeze pressure is reached, indicating the channels are plugged (or until a maximum allowable surface pressure is hit to prevent casing burst or formation fracture). Pressure is held for a specified "squeeze" time.
7. Reverse Out/Wait on Cement: Any excess slurry remaining inside the casing/tubing above the perforations is circulated out to the surface. Wait for the squeezed cement in the leak path to set (WOC—Waiting On Cement).
8. Drill Out and Test: Drill out the hardened cement inside the casing (at the perforations and above the plug) and pressure test the squeezed interval from inside the casing to verify that the leak path has been sealed.

- Challenges: Locating the exact leak path can be difficult, impacting perforation placement. Achieving effective penetration and sealing of very fine channels or micro-annuli requires careful slurry design and execution. There is a risk of plugging the perforations needed for production or fracturing the formation if squeeze pressures are too high. Success rates can be variable, and multiple squeeze attempts may be necessary to fully seal a leak path (Ma et al., 2018). Field data show that conventional cement slurries often struggle with microfractures (<150 μm aperture) (Wu B et al., 2020). For example, in the New Hope Field (USA), an average of 5.33 squeeze jobs were needed to seal an unproductive perforation, yet some zones remained unsealed. Laboratory tests (Ewert et al., 1990) demonstrated that microfine cement slurries could seal a 40/60 gravel pack, whereas conventional slurries failed. Additional experiments (Meek & Harris, 1993) confirmed that micro-cement slurries could penetrate slots as narrow as 0.01" (250 μm) without bridging or forming filter cakes.
- **Other Cement Remediation**:

- Perforate and Circulate (Top Fill): In some cases where a channel is suspected higher up the annulus (e.g., above the cement top), perforating the casing below the channel and circulating large volumes of cement up the annulus and out through the perforations above might be attempted. However, achieving a reliable, continuous seal this way is often difficult and depends on the ability to establish circulation.
- Resins or Chemical Sealants: Various resin systems or other chemical sealants are sometimes used as alternatives or supplements to cement for sealing small leaks or cracks in the cement sheath (Wu B et al., 2020). They may have better penetration properties than cement slurries in very fine pathways, but can have temperature limitations and higher cost. Figure 26.1 illustrates the comparison of cement bond log responses before and after a remedial cement squeeze job. Such cement bond logs (CBL) are crucial for diagnosing annular cement sheath integrity, locating potential leak paths, and verifying the effectiveness of cement remediation. In this example, the log tracks (amplitude, travel time, and attenuation) show significant improvements post-remediation, indicating successful sealing of the previously leaking intervals.

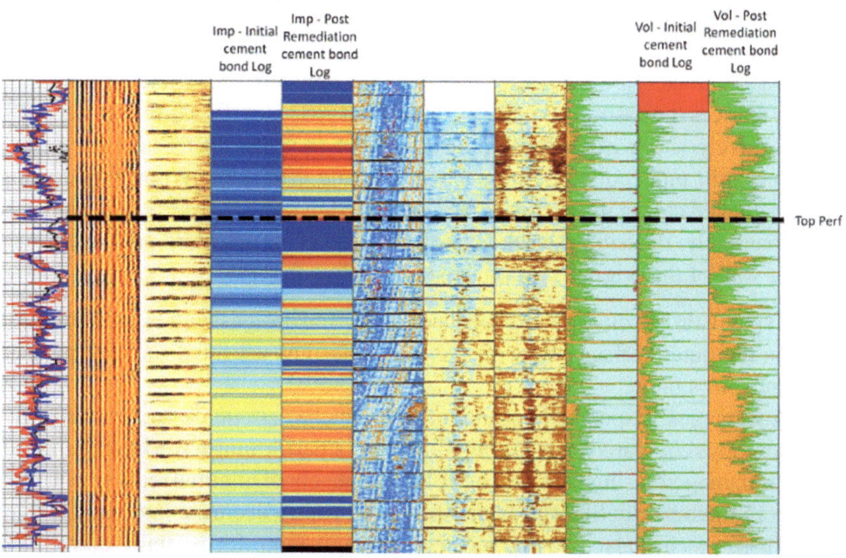

Fig. 26.1 Cement bond logs before and after remedial cement job. The left-side logs represent the initial cement bond quality, while the right-side logs show the improved cement bonding after the remedial squeeze operation. The enhanced amplitude and attenuation responses confirm the establishment of a hydraulic seal across the treated interval (Alvarado et al., 2024)

26.3 Remediation of Casing Leaks

Leaks through the casing wall itself (e.g., corrosion pinholes, cracks, connection leaks, wear damage, splits) require different repair methods, primarily involving placing a patch or liner inside the damaged section.

- **Casing Patch**: Another method used to repair or reinforce damaged casing or tubing is the installation of an expandable casing or liner (patching) (Chustz et al., 2005; Daigle et al., 2000; Durst & Ruzic, 2009; Shahreyar et al., 2023; Storaune & Winters, 2005). In this method, a corrugated liner or patch is precisely positioned across the damaged interval within the casing or tubing. Once in place, an expander assembly tool is run upward to apply mechanical forces that radially expand the liner, pressing it tightly against the existing casing wall. The compressive hoop stress developed during expansion creates sufficient friction to anchor the patch securely, forming hydraulic and gas-tight seals between the patch and the original casing or liner. To enhance bonding, the patch's external surface can be coated with epoxy resin, which extrudes into holes and cavities in the casing wall during expansion, acting as an additional sealing agent. After expansion and anchoring, the expander tool is removed, and a pressure test is performed to confirm the patch's integrity and effectiveness. This process does not significantly affect the internal diameter of the casing; for instance, a standard 1/8-inch (3.18 mm) thick liner reduces the casing diameter by only about 7.62 mm (Edvardsen et al., 2018). A relatively short, expandable tubular section (a sleeve) runs into the well on tubing, drill pipe, coiled tubing, or wireline and is accurately positioned across the damaged casing interval. It is then expanded radially outwards (mechanically, hydraulically, or using thermal/explosive methods) to create a seal against the inner wall of the original casing above and below the leak.
  - Types: Can be solid expandable tubes or layered designs with elastomeric seals on the outer surface of the sleeve. Some designs are rated for high pressure and temperature.
  - Advantages: Expandable casing or liner patches are quick and cost-effective solutions compared to installing a full liner or performing a workover to replace the damaged casing section. They offer the benefit of minimal reduction in wellbore diameter, unlike full liners that significantly decrease internal diameter. These patches also provide localized strengthening for partially corroded or weakened casing and completion equipment. Additionally, they can be used to fully replace a damaged casing interval by integrating a tie-back system, which creates a hydraulic and gas-tight connection between the new and existing casing while preserving the well's internal diameter.
  - Limitations: Despite their advantages, expandable casing or liner patches have several limitations. Their effective length is typically limited to a few hundred feet, which may restrict their use for longer sections of damage. Their success largely depends on the condition of the casing's inner wall, which must be relatively smooth above and below the damaged area to ensure a proper seal.

They are not suitable for severely damaged, collapsed, or heavily pitted casing sections where a good seal cannot be achieved. Accurate depth correlation and precise positioning of the patch are also essential for ensuring the success of the repair.

- **Cement Squeeze (for Casing Leaks)**: If the leak is through a hole or crack in the casing wall that communicates with the annulus, squeeze cementing (as described in 26.2) can sometimes be used to plug the hole from the outside. This often requires placing a temporary plug inside the casing across the leak to contain the squeeze pressure and prevent cement from entering the production bore.
- **Straddle Packer Assembly**: Setting inflatable or mechanical packers on a tubing or coiled tubing string, positioned above and below the leak point inside the casing, can sometimes isolate a leak temporarily for diagnostic purposes (e.g., confirming the depth and leak rate of a casing hole). However, this is rarely a permanent solution and restricts access below the assembly. Figure 26.3 displays the key components of a straddle packer assembly, a temporary isolation system used during well interventions to seal off specific intervals of the wellbore. Straddle packer assemblies typically consist of a bottom packer, anchoring slips, and a top packer. These components work together to provide a reliable seal and support during remediation or testing operations. The images below show a typical configuration of the bottom packer, anchoring slips, and top packer used in these downhole isolation systems.
- **Casing Liner/Cladding**: For extensive damage or long sections of corrosion or predicted failure (e.g., long section of severe corrosion or wear), installing a complete inner liner string (cemented or sealed with packers top and bottom) inside the existing casing may be necessary (Jabs et al., 2004). This is a major workover operation, requires running and potentially cementing a new casing string inside the old one, significantly reduces the wellbore diameter, and is very expensive. CRA cladding applied during manufacturing (preventative) is not a remediation technique itself.

The choice of casing repair method depends on the nature and extent of the damage, its location, the wellbore geometry, the required pressure rating of the repair, and economic considerations. Figure 26.2 presents borehole casing deformation and thickness logs, captured before and after the patch installation at the most-damaged interval. These logs offer valuable insights into the effectiveness of the patch and the restoration of casing integrity. The left panels show the initial casing condition with significant deformation and wall thinning, while the right panels depict the improved casing geometry and thickness post-remediation. These logs are critical for verifying successful patch placement and evaluating the effectiveness of well remediation efforts.

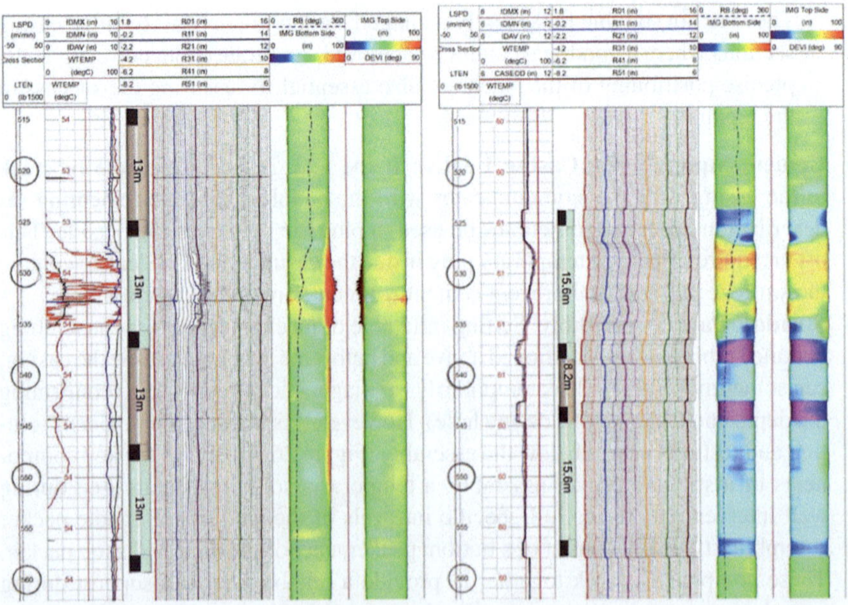

Fig. 26.2 Casing deformation and thickness logs first patch (left) and after (right) the third patch installation in the most-damaged interval. The images highlight the deformation zones, areas of wall thinning, and the post-remediation improvements achieved by the patch system (Shahreyar et al., 2023)

Fig. 26.3 Components of a straddle packer assembly: bottom packer (left), anchoring slips (middle), and top packer (right). These elements are integral to isolating specific wellbore zones for remediation or diagnostic purposes (Gadelhak et al., 2021)

26.4 Remediation of Leaking Packers or Connections

Leaks through downhole components like production packers or tubing/casing connections often require significant interventions.

- **Leaking Packer:**
  - Sealant Injection (Limited Use): Injecting viscous sealant materials down control lines or adjacent conduits designed for chemical injection has occasionally been attempted to stop packer leaks but is generally considered unreliable for achieving a long-term, high-integrity seal for typical downhole packer leak paths (e.g., past seal elements) (Burgess et al., 2022).
  - Replacement: The most common and reliable solution for a leaking production packer is to perform a workover operation to retrieve the production tubing string and the leaking packer (if it is a retrievable type) and replace it with a new, functional packer. If the packer is a permanent type, it must be milled out before running a new completion. This is a significant and costly operation, requiring a rig or heavy workover unit.
- **Leaking Tubing/Casing Connections:**
  - Internal Patch/Sleeve: Expandable sleeves or patches (like casing patches) can sometimes be set across leaking tubing or casing connections if the geometry allows and if the leak is localized to a single or few connections.
  - Tubing Replacement: If multiple tubing connections are leaking significantly, or if the leaks are spread out, replacing the entire production tubing string (or the section containing the leaks) during a workover is often the only reliable long-term solution.
  - Casing Connection Leaks: Extremely difficult to repair reliably from inside. May require internal casing patches, full inner liners, or potentially specialized expandable casing systems if the leak is severe and accessible. Squeeze cementing might be attempted from inside if communication to the annulus through the connection leak is confirmed.

Remediation of downhole packer and connection leaks often involves major well interventions (workovers), highlighting the importance of selecting high-integrity components and ensuring quality installation during initial completion. Figure 26.4 presents a labeled test chart from the successful second tubular expansion test of an elastomeric expansion system. This graph illustrates the relationship between pressure (red line) and flow (blue line) throughout the expansion process (Chustz et al., 2005).

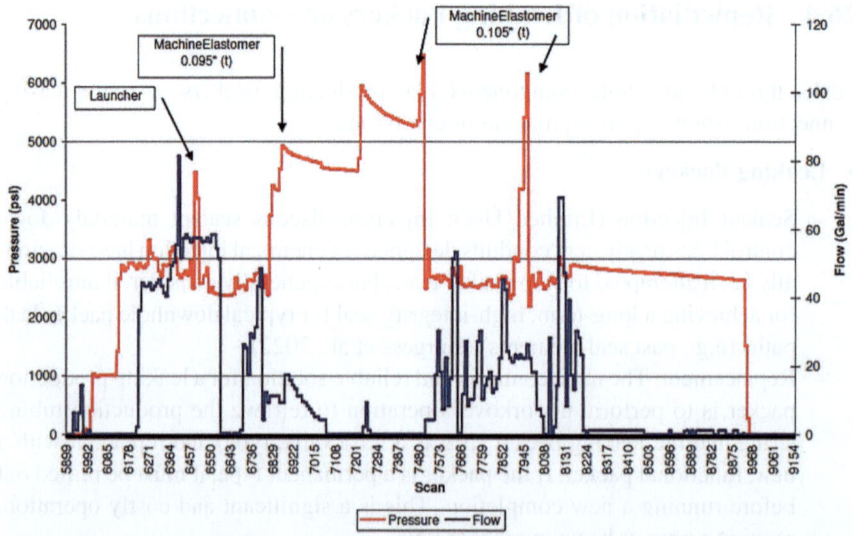

Fig. 26.4 Labeled test chart of the successful second surface expansion test, highlighting the launcher, elastomer expansion points (0.095" and 0.105" thickness), and system performance during the expansion cycle (Chustz et al., 2005)

26.5 Remediation of Wellhead and Valve Leaks

Surface leaks from wellhead or Christmas tree components are generally more accessible than downhole leaks but still require careful planning and execution due to the proximity to pressurized hydrocarbons. Figure 26.5 presents photographs of a well B tubing head flange connection showing signs of leakage (Yuan et al., 2022a, 2022b). These images document the visible external corrosion, fluid leakage, and associated damage on the tubing head. Such photographic documentation is essential for verifying wellhead integrity and identifying leak-prone areas, especially in wellhead and valve leak remediation planning.

- **Leaking Wellhead/Flange Seals:**
  - Retightening Bolts: May temporarily stop minor leaks from flanged connections, but often indicates a deeper issue with the gasket or flange surfaces. Requires following proper torque procedures and patterns.
  - Sealant Injection: Some wellhead flanges and valve bodies have injection ports allowing specialized sealant/packing material (e.g., injectable packing, grease) to be pumped in to temporarily seal minor leaks around gaskets, valve seats, or stem packings (Yuan et al., 2022b). This is often a short-term or temporary fix to manage a leak until a more permanent repair can be planned.
  - Gasket/Seal Replacement: The definitive, permanent solution involves safely isolating and depressurizing the affected wellhead/tree section (often requiring

Fig. 26.5 Tubing head flange connection showing leakage, and fluid accumulation. These images highlight the visual cues of compromised wellhead integrity and emphasize the need for targeted repair strategies (Yuan et al., 2022a, 2022b)

setting downhole plugs or killing the well to remove pressure from above).(Yuan et al., 2022a, 2022b) This allows the faulty equipment (e.g., Xmas tree, tubing head spool) to be safely removed, the damaged ring gasket or seal to be replaced, the seal surfaces repaired if necessary, and the equipment re-installed and pressure tested. This is a significant operation requiring lifting equipment and meticulous adherence to procedures.

- **Leaking Valve Stem Packing:**
  - Tightening the Packing Gland: May compress the packing further and stop minor leaks temporarily, but over-tightening can damage the stem.
  - Injecting Packing/Sealant: Pumping additional packing material through injection ports (if available) into the stem packing area can provide a temporary seal.
  - Replacing Packing: Requires isolating the valve (or well section), safely bleeding pressure from the packing area, removing the actuator/handwheel, and replacing the stem packing set. It can sometimes be done online with specialized techniques and equipment if downstream isolation exists, but often requires well shut-in and depressurization.
- **Leaking Valve (Through-Bore)**: When a valve fails to provide a seal in the closed position (leakage through the bore).

- Greasing/Cycling: Injecting heavy valve grease through lubrication fittings and cycling the valve (opening/closing) may dislodge debris or improve sealing temporarily for minor seat leaks.
- Downstream Isolation: Relying on a downstream valve that is sealing correctly (if available and verified, e.g., the second master valve if one is leaking, or an ESD valve). This is not a repair but a management strategy relying on redundancy.
- Insert Valve (for some gate valves): Installing a specialized insert valve assembly inside the body of the leaking valve (e.g., some gate valves) during a workover/intervention can restore sealing capability without removing the faulty valve body.
- Valve Replacement: The ultimate solution involves safely isolating the valve (requiring upstream isolation, often via downhole plugs or killing the well, and potentially downstream isolation), removing the faulty valve, and installing a new one. This is a major intervention, especially for critical master valves on the Xmas tree.

Remediation of surface equipment requires specialized technicians and adherence to strict safety procedures due to the proximity to pressurized hydrocarbons.

26.6 General Considerations for Remediation

Several factors are crucial when planning and executing any well integrity remediation operation:

- **Accurate Diagnosis**: Thoroughly diagnosing the exact leak path, failure mechanism, and location is crucial before attempting remediation. Using the wrong technique based on an incorrect diagnosis can be ineffective, waste time and money, or even worsen the problem or introduce new issues. This often involves reviewing historical data (logs, tests, operations), monitoring trends, and running diagnostic logs (Chap. 12).
- **Risk Assessment**: Remediation operations themselves carry significant risks (well control during intervention, personnel safety, potential for further damage to the well). A thorough risk assessment and detailed, risk-mitigated procedures must be developed and followed for the specific remediation activity.
- **Well Kill versus Live Intervention**: Many remediation techniques require "killing" the well first (filling with kill-weight fluid with sufficient hydrostatic pressure to overbalance formation pressure) to safely remove wellbore pressure from above the issue location. Some techniques (like certain patches or sealant injections) might be attempted under live well conditions using appropriate pressure control equipment (Chap. 25), but this increases complexity and risk and requires highly specialized procedures and personnel.

- **Tooling and Equipment**: Selecting the appropriate tools (e.g., perforating guns, milling assemblies, setting tools, patches, squeeze tools) and equipment (e.g., workover rig, coiled tubing unit, wireline unit, high-pressure pumping equipment) suitable for the downhole conditions and the specific remediation task.
- **Fluid Design**: Designing specialized fluids for the remediation (e.g., squeeze cement slurry properties, acid types and concentrations for stimulation/cleanup, killing fluids, displacement fluids) is critical for success and avoiding damage.
- **Verification**: After any remediation operation, thorough testing (pressure testing of the repaired section, inflow testing, running logs to assess bond or leak paths) is essential to verify that the repair has been successful and that integrity has been restored before returning the well to service.
- **Economics**: Remediation costs can be substantial (potentially millions of dollars for complex workovers). The decision to repair versus shut-in and abandon a well often involves a careful economic assessment comparing the estimated cost and likelihood of success of the remediation against the remaining value of the well's production or injection capability.

Well integrity remediation requires specialized expertise, tools, and meticulous planning and execution to safely and effectively restore barrier integrity when failures occur.

References

Achang, M., Yanyao, L., & Radonjic, M. (2020). A review of past, present, and future technologies for permanent plugging and abandonment of wellbores and restoration of subsurface geologic barriers. *Environmental Engineering Science, 37*(6), 395–408. https://doi.org/10.1089/EES.2019.0333/ASSET/IMAGES/EES.2019.0333_FIG.5.JPG

Aihevba, L. O. (2004). Well intervention management in Oman fields. In *Proceedings of the IADC/SPE Asia Pacific Drilling Technology Conference and Exhibition*, pp. 451–459. https://doi.org/10.2118/88025-MS

Alvarado, P., Ramasamy, S., Valecillos, J. C., Arnone, M., Kotara, B., Spreafico, D., Booker, D., Christensen, C., Thain, J., & Brinkley, R. (2024). Managed pressure cementing in deepwater environment, evolving from saving the well with perfectly executed challenging primary cementing operations to bring back wells to productive life with state-of-the-art remedial cementing jobs on controlled pressure manner. In: *Society of Petroleum Engineers–SPE/IADC Managed Pressure Drilling and Underbalanced Operations Conference and Exhibition, MPDU*. https://doi.org/10.2118/221446-MS

Ashena, R. (2025). Wireline operations and well control setup. *The Practical Handbook of Well Control*, pp. 289–295. https://doi.org/10.1201/9781003473770-12/WIRELINE-OPERATIONS-WELL-CONTROL-SETUP-RAHMAN-ASHENA

Asyri, M., & Mulia, K. (2018). *Study of microfine cement use on squeeze cementing operations in plug and abandonment work*. https://doi.org/10.1051/e3sconf/20186703001

Burgess, A. T., Hosli, C., & Whittaker, K. (2022). A return to production: Subsea packer leak remediation with pressure-activated sealant. In *Society of Petroleum Engineers–SPE 2022 Symposium Compilation, SYM 2022*. https://doi.org/10.2118/210607-MS

Campos, J., Keong, A., Liu, Z., Tomczak, D., Kleppan, T., & Soerensen, M. (2024). Understanding mechanisms of wellbore abrasion during coiled tubing intervention. In *Society of Petroleum*

Engineers–SPE/ICoTA Well Intervention Conference and Exhibition, CTWI. https://doi.org/10.2118/218335-MS

Carragher, P., & Talapatra, D. (2025). Advancements in bismuth-based sealing tools: enhancing efficiency and integrity in well interventions. In *Society of Petroleum Engineers–SPE/ICoTA Well Intervention Conference and Exhibition, CTWI 2025*, pp. 25–26. https://doi.org/10.2118/224020-MS

Chustz, M., Mason, D., Schober, J., De Lucia, F., Butterfield, C., & Didykenventure, V. (2005a). Expandable liner installation avoids sidetracking following production casing failure in the Gulf of Mexico. In *SPE/IADC Drilling Conference, Proceedings*, pp. 107–113. https://doi.org/10.2118/91923-MS

Costa, W. R. P. da, Nóbrega, K. C., Costa, A. C. A., Nascimento, R. C. A. de M., Souza, E. A. de, Oliveira, T. A. de, Barros, M., & Amorim, L. V. (2024). *A Liquid Well Barrier Element for Temporary Plug & Abandonment Operations: A Breakthrough Approach*. https://doi.org/10.20944/PREPRINTS202408.0572.V1

Crumpton, H. (2018). Coiled tubing well control. *Well Control for Completions and Interventions*, 483–566. https://doi.org/10.1016/B978-0-08-100196-7.00010-5

Cully, R., Franklin, C. M., Lewis, K., & Mesa, C. (2022). A new approach to assuring well integrity and performance. *Proceedings of the Annual Offshore Technology Conference*. https://doi.org/10.4043/32004-MS

Da Fonseca, T. C., Miura, K., & Mendes, J. R. P. (2013). Well integrity analysis applied to workover prediction. *Proceedings of the Annual Offshore Technology Conference, 2*, 832–839. https://doi.org/10.4043/24369-MS

Daigle, C. L., Campo, D. B., Naquin, C. J., Cardenas, R., Ring, L. M., & York, P. L. (2000). Expandable Tubulars: Field examples of application in well construction and remediation. In *SPE Reservoir Engineering (Society of Petroleum Engineers), A*, 139–152. https://doi.org/10.2118/62958-MS

Daohmareeyor, T., Lock, D. L. W., & Wattanasuwan, R. (2022). The use of an organic crosslinked polymer sealant as a barrier to retrieve stuck coiled tubing from a live high pressure well after over a year: Case study from offshore Vietnam. In *Proceedings of the IADC/SPE Asia Pacific Drilling Technology Conference, APDT*. https://doi.org/10.2118/209856-MS

Durst, D. G., & Ruzic, N. (2009). Expandable tubulars facilitate improved well stimulation and well production. In *SPE Western Regional Meeting 2009–Proceedings*, 255–269. https://doi.org/10.2118/121147-MS

Edvardsen, L., Gawel, K., & Cerasi, P. (2018). *Project acronym and title: SECURe-Subsurface Evaluation of Carbon capture and storage and Unconventional risks Report on Remediation Strategies for Tubings and Casings*.

Ewert, D. P., Almond, S. W., & Bierhaus, W. M. (1990). *Small Particle Size Cement*. https://doi.org/10.2118/90-117

Gadelhak, A. M., Al-Badi, M., Al-Bairaq, A., Al Mheiri, E., Al-Hosani, A. H., Ahmed, Z., Ahmed, S. U. B., Ahmed, M., Abedelkhalik, W. O., Naser, S. H., El-Shater, H., Al Assar, W., Ross, S., & Duncan, B. (2021). Lead application to cure sap wells by deploying straddle packer, success story. In: *Society of Petroleum Engineers–Abu Dhabi International Petroleum Exhibition and Conference, ADIP*. https://doi.org/10.2118/207840-MS

Hadzihafizovic, D. (2024). *Well Intervention Pressure Control*. https://papers.ssrn.com/abstract=5031370

Izadi, M., Trudel, E., & Frigaard, I. A. (2024). Risk-based analysis of squeeze cementing operations. *Geoenergy Science and Engineering, 234*, Article 212687. https://doi.org/10.1016/J.GEOEN.2024.212687

Jabs, M., Harmon, C. E., Harmon, S. C., & Harmon, R. E. (2004). New expandable cladding technique enables extended length casing repair. In *Proceedings of the Drilling Conference*, 985–988. https://doi.org/10.2118/87212-MS

Ma, C., Deng, J., Wu, R., Yu, B., Zhou, Y., & Lin, S. (2018). A novel self-healing spacer fluid for sustained casing pressure mitigation. *Journal of Petroleum Science and Engineering, 163*, 722–730. https://doi.org/10.1016/J.PETROL.2017.11.027

Meek, J. W., & Harris, K. (1993). Repairing casing leaks using small-particle-size cement. *SPE Production & Facilities, 8*(01), 45–50. https://doi.org/10.2118/21972-PA

Mohamed Hanafi, M. M., Ismail, N., Anoar, I., Ong, V., Hamzah, S., Khairul Anuar, K., Singh, K., Cavalleri, C., Yong, N., Zailani, M. F. I., Ramondenc, P., & Hofacker, M. (2024). Intervention synergy improving production enhancement efficiency for remote offshore well intervention operation. In *Society of Petroleum Engineers - SPE/ICoTA Well Intervention Conference and Exhibition, CTWI*. https://doi.org/10.2118/218364-MS

NORSOK D-010. (2021). https://online.standard.no/en/norsok-d-010-2021

Oatey, M., Duff, F., Emslie, N., Christie, S., Rikabi, R., Henderson, P., Singh, K., Kumar, A., Agrawal, G., Dutta, S., & Bajwa, H. (2021). Enhancing production through well interventions using end-to-end evaluation methodology. In *SPE Middle East Oil and Gas Show and Conference, MEOS, Proceedings*. https://doi.org/10.2118/204540-MS

Raaen, A. M., & Fjær, E. (2020). Pressure testing of barrier integrity. In *Proceedings of the International Conference on Offshore Mechanics and Arctic Engineering–OMAE*, pp. 11. https://doi.org/10.1115/OMAE2020-18713

Shahreyar, R., Kelly, M., Albouy, C., Saltel, B., & Le Pouezard, B. (2023). How expandable casing patches can restore casing integrity and enable the sidetracking of an old well. *International Petroleum Technology Conference, IPTC 2023*. https://doi.org/10.2523/IPTC-22810-MS

Solli, T. E. (2011). Workover/Well intervention and regulatory challenges. *In Proceedings of the Annual Offshore Technology Conference, 2*, 942–948. https://doi.org/10.4043/22586-MS

Storaune, A., & Winters, W. J. (2005). *Versatile Expandables Technology for Casing Repair.* https://doi.org/10.2118/92330-MS

Vignes, B. (2011). *Contribution to well integrity and increased focus on well barriers from a life cycle aspect (2010), Birgit Vignes, 26 Citations* (Faculty of Science and Technology University of Stavanger). https://scispace.com/papers/contribution-to-well-integrity-and-increased-focus-on-well-n5cyhzz40j

Wu, B., Arjomand, E., Tian, W., Dao, B., & Yan, S. (2020). Sealant technologies for remediating cement-related oil and gas well leakage a state-of-the-art literature review. *Commonwealth Scientific and Industrial Research Organisation.*

Yousuf, N., Olayiwola, O., Guo, B., & Liu, N. (2021). A comprehensive review on the loss of wellbore integrity due to cement failure and available remedial methods. *Journal of Petroleum Science and Engineering, 207*, Article 109123. https://doi.org/10.1016/J.PETROL.2021.109123

Yuan, J., bo, Zheng, X., Vo, M., Wang, Y., Zhang, J., & Chen, S. (2022b). Effective application of wellhead integrity reparation by pressure activated sealant technology. *Springer Series in Geomechanics and Geoengineering*, pp. 5713–5720. https://doi.org/10.1007/978-981-19-2149-0_521/FIG.S/6

Yuan, J., bo, Zheng, X., Vo, M., Wang, Y., Zhang, J., & Chen, S. (2022a). Effective application of wellhead integrity reparation by pressure activated sealant technology. *Springer Series in Geomechanics and Geoengineering*, pp. 5713–5720. https://doi.org/10.1007/978-981-19-2149-0_521/FIG.S/6

Chapter 27
Plug and Abandonment (P&A) for Permanent Integrity

Abstract This chapter addresses the final stage of a well's life—**plug and abandonment (P&A)**—focusing on the permanent restoration of subsurface and surface barriers to prevent fluid migration, environmental contamination, and future liability. Proper P&A operations are critical for long-term integrity, regulatory compliance, and public trust, especially in the context of increasing attention to orphaned and legacy wells. The chapter begins by introducing the **regulatory and environmental drivers** for well abandonment, referencing international standards (e.g., NORSOK D-010, API RP 17W, ISO 16530-1) and country-specific regulations. The importance of permanent barriers to isolate permeable formations, prevent crossflow, and ensure caprock sealing is emphasized. Next, the chapter outlines the **design philosophy of permanent barriers**, including the placement of mechanical plugs (bridge plugs, cement retainers) and the formulation of abandonment-grade cement slurries with long-term durability. Topics covered include:

- Minimum barrier length and placement relative to formation tops and casing shoes.
- Use of dual barriers across critical zones (e.g., open-hole intervals, perforated sections, surface casing shoe).
- Cement evaluation tools (e.g., ultrasonic logs, radial bond logs) to verify placement and zonal isolation.

Operational methods for **well section milling**, **cutting and pulling casing strings**, and **setting cement plugs** are explained, along with the conditions requiring section milling versus casing perforation and squeeze. The use of **resin-based sealants**, **bismuth alloys**, and **geopolymers** as novel P&A materials is introduced to address long-term chemical degradation concerns. The chapter also explores **challenges in abandoning complex wells**, such as those with sustained casing pressure (SCP), inaccessible annuli, or old completion hardware. Risk-based approaches and adaptive designs are discussed to address uncertain well conditions. The chapter further reviews **offshore P&A practices**, which require severing wells below the mudline and restoring seabed integrity, often with specialized subsea tooling. Finally, the chapter stresses the importance of **P&A documentation, regulatory reporting**, and **post-abandonment monitoring**, including the use of surface markers, cement

© The Author(s), under exclusive license to Springer Nature Switzerland AG 2026

A. Alsubaih and K. Sepehrnoori, *Well Integrity for Oil, Gas, and Subsurface Storage Operations*, https://doi.org/10.1007/978-3-032-15233-6_27

tagging, and environmental surveillance to ensure that abandonment is permanent and verifiable. Through diagrams, case studies, and global best practices, this chapter equips engineers with the knowledge and tools needed to execute effective and compliant P&A operations that protect both the environment and stakeholder interests in perpetuity.

Abandoned gas well on refuge lands in the Lower Rio Grande Valley National Wildlife Refuge *Source* Steve Hillebrand/U.S. Fish and Wildlife Service (n.d.)

Chapter's Objectives

- Define the objectives and principles of permanent well abandonment.
- Discuss regulatory requirements and industry guidelines for P&A.
- Describe common permanent barrier materials (cement and alternatives).
- Outline techniques for placing and verifying permanent plugs.
- Address the significant challenge of ensuring long-term P&A integrity.
- Introduce emerging rigless P&A technologies.

27.1 Introduction: The Final Lifecycle Stage

Plug and Abandonment (P&A) represents the final, crucial phase of a well's lifecycle, where the responsibility of the operator transitions from active well management to ensuring permanent environmental and public safety. Unlike routine operational or temporary well suspensions, which anticipate future re-entry or ongoing monitoring,

P&A operations are designed to leave the well in a permanently secure state that requires no further human intervention. The goal is to ensure that no fluids—hydrocarbons, formation water, injected fluids like CO_2—can migrate to the surface, into groundwater aquifers, or across geological formations. The two major differences between the production period and abandonment period are addressed as: mechanical loading scenario, and downhole environment (Lecolier et al., 2006).

The scale and importance of P&A are significant and growing. As global hydrocarbon production matures, the number of wells reaching the end of their productive lives continues to accelerate, with tens of thousands of wells already requiring permanent abandonment worldwide. This includes not only conventional oil and gas wells but also wells drilled for carbon storage, geothermal projects, and other purposes. Managing this growing inventory of inactive or low-productivity wells is a critical aspect of responsible resource stewardship and environmental protection.

The primary challenge in P&A lies in replicating the natural geological barriers that the wellbore originally penetrated—barriers that took millions of years to form and stabilize. P&A involves constructing multiple, carefully designed impermeable barriers—often referred to as "plugs"—at strategic locations within the wellbore to prevent fluid migration. These barriers must be built using materials that can endure the extreme and evolving downhole conditions for centuries or even millennia, facing chemical, thermal, and geomechanical challenges far beyond the expected life of a producing well.

Modern P&A practices are heavily influenced by evolving regulatory requirements, environmental stewardship, and a rapidly advancing technical toolkit. Regulations worldwide increasingly demand thorough risk assessments, robust permanent barrier designs, comprehensive documentation, and verification of barrier integrity. The move toward using advanced materials beyond traditional Portland cement—such as geopolymers, resins, bismuth alloys, and even reactive thermite-based barriers—reflects a growing awareness of the limitations of older approaches in guaranteeing permanent well integrity. Moreover, emerging rigless P&A technologies are reshaping how operators approach well abandonment, offering safer and more cost-effective solutions, particularly for simpler well architectures.

Well integrity failures are a key driver for permanent well abandonment and highlight the significant environmental and safety risks associated with aging well infrastructure worldwide. Documenting and analyzing the rates and distribution of these failures helps identify regions and well types most susceptible to long-term integrity issues and informs targeted P&A strategies. Figure 27.1 presents a global overview of well integrity failures, emphasizing the magnitude of the challenge and the variability of failure rates across different jurisdictions and geological contexts.

Ultimately, effective P&A is not just about sealing a well; it is about upholding a fundamental environmental and social responsibility. Failure of abandoned wells can lead to groundwater contamination, greenhouse gas emissions, and long-term liabilities that extend far beyond the operator's financial and regulatory obligations. In this chapter, we will explore the objectives and principles of P&A, review key regulatory frameworks, discuss the selection and placement of permanent barrier materials, examine verification challenges, and introduce emerging rigless P&A

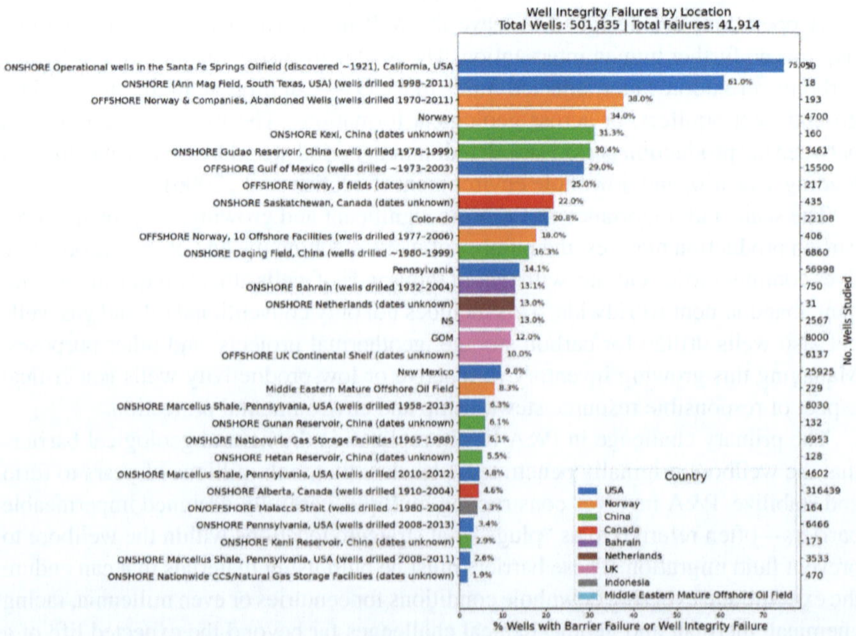

Fig. 27.1 The percentage of wells with documented barrier or integrity failures across various global regions and well types. The total number of wells studied (501,835) and total failures (41,914) are indicated, providing a clear overview of the global well integrity challenge. The Figure also highlights differences in failure rates between countries and regions, emphasizing the need for tailored P&A strategies based on specific well construction practices and geological settings

technologies. Together, these topics form the foundation for permanently securing wells and ensuring that past energy production activities do not pose a risk to future generations.

27.2 Objectives and Principles of P&A

The primary objective of P&A is permanent zonal isolation—re-establishing the natural geological containment that was breached by drilling the well (Hovda et al., 2020). This involves placing multiple impermeable barriers (plugs) at strategic locations within the wellbore to seal all potential pathways for fluid flow, both inside casing and in the annuli. Key principles guiding P&A design and execution include:

- **Isolate Porous/Permeable Zones**: Prevent flow from or into any formation containing hydrocarbons, high pressure, or water (especially potable water aquifers). This requires placing plugs across the full width of these zones (Abiabhar et al., 2022).

- **Prevent Vertical Migration**: Seal the full wellbore cross section (inside the innermost casing and *all* annular spaces) to prevent fluids migrating vertically toward the surface or shallower zones. This includes sealing across casing shoes, casing stubs, and any sections of potentially poor primary cement.
- **Use Permanent Materials**: Employ barrier materials demonstrated to have long-term physical and chemical stability and sealing properties under the specific downhole conditions for potentially geological timescales (hundreds to thousands of years) (Carpenter, 2023; Vissa & Radonjic, 2020).
- **Verify Barrier Placement and Integrity**: Confirm that barriers are placed at the correct depth, have the required dimensions (length, diameter), and most critically, achieve a reliable, long-term hydraulic seal across the full wellbore cross section.
- **Remove Surface Equipment (Optional but common)**: Typically involves cutting casing strings below ground level (onshore) or seabed (offshore) and removing the wellhead assembly. This eliminates surface obstruction and potential future liability points (e.g., surface wellhead corrosion or damage). The cutting processes can be performed using explosives, chemicals, mechanical cutters, or abrasive cutters (Khalifeh & Saasen, 2020).
- **Restore Geological Integrity**: The ultimate goal is to restore the sealing function of the overburden and caprock formations that confined reservoir fluids before drilling. P&A plugs act as engineered seals within the wellbore to replace this function.

The two-barrier principle (Chap. 2), while adapted, remains relevant. P&A designs often aim for multiple, independent barriers at key isolation points, providing redundancy for long-term assurance.

27.3 Regulatory Requirements and Guidelines

P&A operations are strictly regulated worldwide due to the potentially severe and long-lasting environmental consequences of failure (e.g., groundwater contamination, surface hydrocarbon/GHG emissions) (Khalifeh & Saasen, 2020; Razak et al., 2024). Regulations typically specify minimum requirements for:

- **Number and Location of Plugs**: Mandating specific locations for permanent barriers (e.g., across reservoir zones, above high-pressure zones, across defined aquifer boundaries, at surface casing shoe, across casing stubs, shallow plug near surface/seabed) (Ahmed et al., 2024). The number and placement are determined by well type, geology, and regulatory requirements.
- **Plug Length**: The minimum required length for each permanent barrier (e.g., commonly ranging from 100 to 500 ft/30–150 m or more, depending on location, regulation, and plug purpose). Longer plugs provide greater assurance but add cost and complexity.

- **Barrier Material**: Often specifying approved materials (typically Portland cement with specific properties, e.g., density, strength, permeability, resistance to downhole fluids like CO_2) or requiring rigorous justification and qualification for alternative materials to demonstrate long-term performance equivalence or superiority to cement (Sect. 27.4).
- **Annular Sealing**: Increasingly, regulations emphasize the need to ensure barriers seal not just inside the innermost casing string but also across the entire wellbore cross section, including potentially poorly cemented annuli (Morales-Monsalve et al., 2022). This may require specific techniques like perforating casing to squeeze cement into annuli, or section milling to remove sections of existing casing and cement and place permanent barriers directly against the formation ("rock-to-rock" or "environmental" plugs) to ensure a seal across all potential annular pathways.
- **Verification Requirements**: Specifying mandatory methods for verifying plug placement (e.g., tagging the top of the set plug with tubing or wireline to confirm depth and sufficient length above a base) and sealing integrity (e.g., mandatory pressure testing of the plug from above, and potentially inflow testing from below if feasible). Acceptance criteria for pressure tests (e.g., maximum allowable pressure drop over a defined time) are specified.
- **Documentation**: Requiring detailed P&A programs, step-by-step procedures, records of all activities performed (including dates, times, materials used, volumes pumped, pressures, test results), and final P&A reports and handover documentation confirming compliance with the approved program.

Key guidance documents influencing or adopted by national regulations include API Std 65-3: Wellbore Plugging and Abandonment, and guidelines from organizations like Oil & Gas UK (OEUK) in the UK and NORSK Olje & Gass 081 Guideline for Plugging and Abandonment in Norway. The developing ISO 16530-3 standard is also expected to provide a global framework for P&A lifecycle management. Regulatory requirements are continuously evolving, driven by lessons learned and the need for greater long-term environmental protection. Regulatory frameworks for well abandonment vary widely across jurisdictions, reflecting differences in geological conditions, historical practices, and environmental priorities. Understanding these variations is crucial for designing P&A programs that meet both regulatory requirements and long-term integrity goals. Figure 27.2 summarizes the number of U.S. states that specify different plug placement requirements for surface casing, oil and gas-bearing strata, and water bearing strata, as well as the requirement for a surface plug. This regulatory diversity underscores the need for adaptable and robust P&A strategies tailored to regional contexts. Barrier placement requirements for plug and abandonment (P&A) vary widely across countries and regulatory frameworks, reflecting differences in environmental protection goals, well construction practices, and geological risks. This diversity underscores the importance of understanding local regulatory requirements when planning and executing permanent well abandonment. Understanding the global regulatory landscape for petroleum activities is essential for ensuring compliance and alignment with national and regional

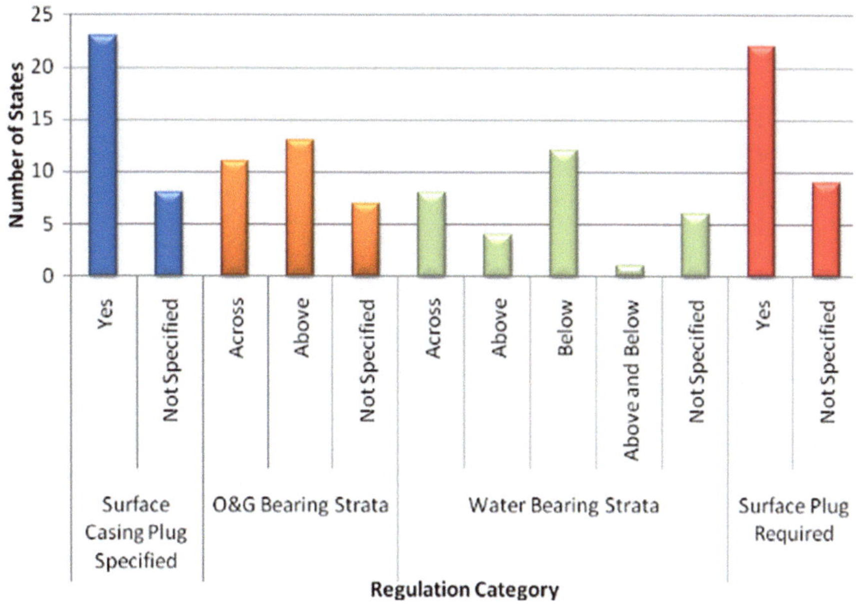

Fig. 27.2 Number of U.S. states specifying different plug placement regulations for surface casing, oil and gas-bearing strata, water bearing strata, and surface plug requirements. The figure highlights the variability of state-level regulations and the importance of regulatory compliance for effective long-term well integrity (National Petroleum Council [NPC], 2011)

standards during plug and abandonment (P&A) operations. Different countries have their regulatory agencies responsible for overseeing well integrity, abandonment practices, and environmental stewardship. Figure 27.3 provides a global overview of these key regulatory authorities, highlighting the diversity of governance frameworks that shape P&A activities and influence best practices. Table 27.1 summarizes these requirements for key jurisdictions, highlighting the specified barrier lengths in both open-hole and cased-hole scenarios, as well as the emphasis on multiple, independent barriers for high-risk zones.

27.4 Permanent Barrier Materials

The selection of appropriate permanent barrier materials is critical for ensuring long-term P&A integrity. The material must be impermeable, physically and chemically stable under downhole conditions for potentially geological timescales, and capable of forming a reliable seal across the wellbore cross section (Vissa & Radonjic, 2020). Permanent well abandonment requires cementing materials that can perform under diverse and challenging subsurface conditions. Over the years, a wide range of cement types has been developed to address specific needs, including expansion to counteract

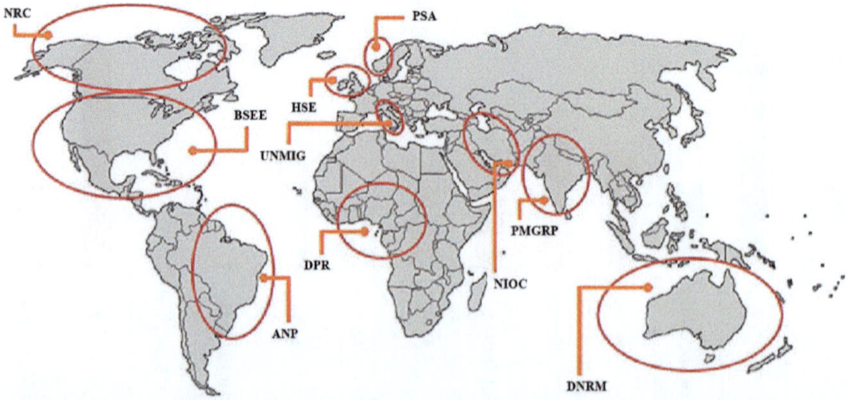

Fig. 27.3 Global Petroleum regulatory authorities world map highlighting the main regulatory authorities that oversee petroleum activities and well integrity in their respective countries or regions. Examples include BSEE (USA), HSE (UK), DPR (Nigeria), ANP (Brazil), PNG (India), and DNRM (Australia), among others (Khalifeh & Saasen, 2020)

shrinkage, resistance to high temperatures, or stability in permafrost environments. Table 27.2 provides an overview of these cement types, highlighting their composition, typical field applications, and key performance considerations essential for ensuring long-term well integrity in P&A operations.

- **PortlandCement**: The most widely used material for P&A plugs globally due to its availability, relatively low cost, established placement techniques, and known short-to-medium term sealing properties.
  - Requirements: P&A cement slurries must be designed for the specific downhole environment (temperature, pressure, fluid chemistry). Key properties for P&A cement include long-term stability (e.g., using silica flour at high temperatures), low permeability, good bonding to casing and formation, resistance to downhole fluids (CO_2, H_2S, brines), minimal shrinkage upon setting, and potentially some ductility or slight expansion to ensure a lasting seal and mitigate micro-annuli formation. Additives are used to achieve these properties (as discussed in Chap. 5).
  - Limitations: Concerns exist about the very long-term (100s–1000s of years) durability and sealing performance of Portland cement under evolving downhole conditions. Potential failure mechanisms include chemical degradation (carbonation by CO_2, sulfate attack, acid attack, leaching), shrinkage leading to micro-annuli, and potential brittleness leading to cracking under geomechanical stress or residual thermal/pressure cycling. Achieving a reliable hydraulic seal across annular spaces with potentially poor original cement or complex geometries remains challenging with standard cementing techniques.

Table 27.1 International barrier requirements for permanent well abandonment

Country/ regulation	Barrier requirements in open hole	Barrier requirements in cased hole	References
United Kingdom/ OGUK Guidelines (2015)	Two independent plugs, each at least 100 ft (30.4 m) MD of good-quality cement (commonly 500 ft). For a single continuous barrier, 200 ft of good cement (typically 800 ft)	Two independent plugs of at least 100 ft of good cement each (commonly 500 ft). Annulus must contain at least 100 ft of quality cement for each plug	Liversidge et al. (2006), Øksnes (2017)
Norway/ NORSOK D-010 (2013)	100 m MD of good-quality cement; at least 50 m above any inflow or leak point. Plugs transitioning from open hole to casing extend 50 m above and below casing shoe	50 m MD if set on a mechanical/cement plug foundation; otherwise 100 m MD. Two independent barriers are required for hydrocarbon-bearing or overpressured zones	Liversidge et al. (2006), Øksnes (2017)
Denmark/DEA guidelines	Minimum 50 m of good cement above and below each zone. In open holes below the deepest casing, plugs extend 50 m above and below casing shoe	Perforated zones isolated with cement plugs extending 50 m below and above topmost and bottommost perforations. For liners, 50 m above and below liner top	–
Alberta & British Columbia/Energy Regulator's Guidelines	Level 1 wells (acid gases, high-pressure gradients, or BOP > 10,000 psi): permanent bridge plug within 15 m above interval, 30 m cement above/below. Level 2 wells: permanent bridge plug within 15 m above, 15 m cement above/below	For perforated cased holes, 60 m cement plug for level 1 wells; 30 m for level 2 wells. Minimum 15 m cement above reservoirs from TVD; additional plugs to 15 m above deepest set casing shoe. Permanent bridge plug capped with 8 m cement acceptable	Trudel et al. (2019)
Northwest Territories/ OROGO (2016)	Not addressed in OROGO 2016 guidelines into two levels. Level 1 wells contain acid gases, have reservoirs of at least 0.8psi/foot pressure gradient, or require Blow out preventer equipment rated >10,000 pPsi. All other wells are classified as level 2	Cased wells with cemented liners require a permanent bridge plug within 15 m above the interval	

(continued)

Table 27.1 (continued)

Country/ regulation	Barrier requirements in open hole	Barrier requirements in cased hole	References
United States/ Pennsylvania DEP (2018)	50 ft below and 50 ft above each water, oil, or gas-bearing zone	50 ft below and 100 ft above each water, oil, or gas-bearing zone, or mechanical plug set 20 ft above the zone	Pennsylvania (2018)
United States/ Texas Railroad Commission (2021)	Cement plug across productive zone extending 50 ft below and above the zone, plus an extra 10% slurry volume per 1000 ft MD	At least 100 ft of plug material, with an additional 10% slurry volume per 1000 ft MD	RRC Texas (2025)

- **Alternative Barrier Materials**: Research and field application of non-Portland cement materials are growing rapidly, aiming to overcome some of cement's limitations and provide greater confidence in long-term isolation, particularly in challenging environments or for CCS.
  - Geopolymers: Inorganic binders formed by reacting aluminosilicate precursors (like fly ash or slag) with alkaline activators. It can offer high strength, very low permeability, good chemical resistance (especially to acids/CO_2), good bonding, and high-temperature stability. Offer potential for better long-term durability than Portland cement in some environments, but long-term downhole performance validation is still being evaluated (Chap. 23).
  - Epoxy/Resins: Polymer-based materials that can offer good bonding, chemical resistance, and low permeability. Can sometimes penetrate smaller leak paths better than cement but may have temperature limitations, are typically more rigid (susceptible to cracking), and are more expensive.
  - Bismuth Alloys: Metal alloys that melt at relatively low temperatures (~100–300 °C) can be placed as a liquid and solidify upon cooling or contact with cooler formations, providing a high-density, impermeable, non-shrinking, and corrosion-resistant seal. Offer potential for excellent long-term durability as they are metallic barriers but are expensive and require specialized placement techniques (heating elements, controlled cooling). Still under development and field trial for broader P&A use.
  - Thermite: Exothermic reactions generating very high temperatures can potentially melt surrounding materials (casing, formation rock) locally to form a fused, impermeable, vitrified barrier. Still largely experimental and requires significant technical development and validation for reliable field application as a permanent barrier.
  - Compacted Clay (Bentonite): Used in some shallow onshore P&A applications, relying on swelling properties in the presence of water to create a low-permeability seal. Suitability for high-pressure or complex downhole

Table 27.2 Summary of cement types for permanent plug and abandonment applications

Cement type	Main composition/ feature	Typical applications	Key considerations
Pozzolanic-Portland cement	Blended cement with pozzolanic materials (e.g., fly ash)	Strength and water-tightness improvement in P&A plugs	Reactive siliceous materials hydrate with lime and water
Pozzolan-lime cement	Siliceous materials (fly ash), hydrated lime, and chemical activator	High-temperature wells (>140 °F)	Slow reaction at low T; additives can accelerate/ retard
Resin/plastic cement	API class A/B/G/ H cement mixed with liquid resins and catalyst	Small-volume plugging in open holes, squeezing perforations	Used in the temperature range 60–200 °F
Gypsum cement	API Class A/G/H cement with 8–10% gypsum (hemihydrate or dihydrate)	Lost circulation zones; rapid setting and early strength	Water solubility of hardened cement limits its use in wet environments
Expanding cement	Portland cement with expansive additives (e.g., Na_2SO_4, lime)	Mitigating shrinkage and micro-annuli in the wellbore	Expansion compensates for wellbore stress changes; improves bond strengths
Calcium aluminate cement	High-alumina cements from bauxite and limestone	High-temperature, corrosive environments; rapid hardening	Low silica content; flash setting if blended with Portland cement
Latex cement	API class A/G/H cement with latex (e.g., polyvinyl acetate, styrene)	Gas migration control, fluid-loss prevention, and improved bonding	Sensitive to temperature, mechanical energy, and salts; use anionic surfactants
Cement for permafrost	Blended cements (e.g., gypsum-cement blends, refractory cements)	Wells in sub-freezing/ permafrost zones	Mitigations include calcium chloride, alcohols; challenges with rapid setting
Diesel oil cement	API class A/B/G/ H cement with diesel oil and surface-active agent	Water control in producing wells or during drilling	Unlimited pumping time until water contact; low fluid-loss protects zones

environments is limited, and long-term performance depends on continuous hydration.

The selection of permanent barrier material depends on the specific well conditions (temperature, pressure, fluid chemistry), regulatory requirements, the required longevity (especially for CCS), and cost considerations. Rigorous qualification and verification of the long-term performance and sealing capability of any barrier material are critical, especially for alternatives where field history is limited. Joint Industry Projects (JIPs) are actively working on standardizing qualification protocols for alternative P&A materials. Selecting materials for permanent well barriers requires careful consideration of their physical, chemical, and mechanical performance under wellbore conditions. Industry guidelines classify barrier materials into specific groups based on composition and functionality, each with mandatory qualification tests to ensure long-term integrity. Table 27.3 categorizes these materials, provides representative examples, and outlines the critical tests used to verify their suitability for P&A operations.

27.5 Plug Placement and Verification Techniques

Placing permanent barriers effectively and reliably and verifying their integrity are crucial steps in the P&A process.

- **Placement Techniques**: Methods used to get the barrier material into the correct location and geometry.
  - Balanced Plug Cementing: The most common method for setting cement plugs inside casing. A carefully calculated volume of cement slurry is pumped down the drill pipe or tubing to the desired depth, balanced by displacing fluid (e.g., mud, brine) pumped down the annulus (if accessible) or up inside the pipe itself, leaving a planned length of cement across the isolation zone. Requires accurate volume calculations, precise depth control, and effective displacement.
  - Dump Bailer: A wireline tool used to transport and dump small volumes of cement slurry, sand, or other materials (like bismuth pellets) at a specific depth, typically on top of a mechanical plug or fill. Often used for setting smaller plugs or placing materials where circulation is not possible.
  - Perforate and Squeeze/Circulate: To seal annuli, the casing may be perforated at specific depths, and cement squeezed into the annulus or circulated up behind the pipe if possible. This can help establish a seal in potentially poorly cemented annuli.
  - Section Milling: Removing a section of casing entirely (using specialized milling tools run on drill pipe or coiled tubing) allows a cement plug (or other material) to be placed directly against the exposed formation. This provides a more robust annular seal ("rock-to-rock" plug), bypassing the original cement sheath and sealing across the full wellbore cross section, which is increasingly

Table 27.3 Material grouping and mandatory qualification tests for permanent well barriers (Chukwuemeka et al., 2023)

Type	Material	Examples	Mandatory qualification tests
A	Cement and ceramic (setting materials)	Portland cement, pozzolanic cement, slag cement, phosphate cement, hardening ceramic geopolymers	Nitrogen permeability, dry mass, expansion/ shrinkage during and after hardening (API RP 10B-5 ring test), differential thermal expansion (ASTM E228), tensile strength (ASTM C496), UCS (API RP 10B-2)
B	Grouts	Sand or clay mixtures, barite plug, bentonite, calcium carbonate, and other inert materials	Nitrogen permeability, dry mass, shrinkage during and after hardening, differential thermal expansion (ASTM E228), UCS (API RP 10B-2)
C	Thermosetting polymers and composites	Resins (epoxy, polyesters, vinyl esters), including fiber reinforcements	Diffusion coefficient, dry mass, expansion/ shrinkage during and after hardening (API RP 10B-5 ring test), differential thermal expansion (ASTM E228), tensile strength (ISO 527-1), UCS (API RP 10B-2), shear bond strength, thermal decomposition (ASTM D1712)
D	Thermoplastic polymers and composites	Polyethylene, polypropylene, polyamide, PTFE, PEEK, PVS, PVD, and fiber-reinforced variants	Diffusion coefficient, dry mass, expansion/ shrinkage during and after hardening, differential thermal expansion (ASTM E228), tensile strength (ISO 527-1), UCS (API RP 10B-2), shear bond strength, thermal decomposition (ASTM D1712)
E	Elastomeric polymers and composites	Natural rubber, neoprene, nitrile, EPDM, FKM, silicone rubber, polyurethane, PUE, and fiber-reinforced swellable rubbers	Dry mass, expansion/shrinkage during and after hardening (API RP 10B-5 ring test), differential thermal expansion (ASTM E228), tensile strength (ISO 527-1), shear bond strength, thermal decomposition (ASTM D1712)
F	Formation	Claystone, shale, salt	Nitrogen permeability, dry mass, and creep (ASTM C512-10)
G	Gels	Polymer gels, polyacrylamides, starch- and silicate-based gels, clay/silt mixtures	Testing for major transport mechanism (permeability or diffusion coefficient), dry mass, expansion/shrinkage during and after hardening (API RP 10B-5), differential thermal expansion (ASTM E228), tensile strength (ISO 527-1), shear bond strength, thermal decomposition (ASTM D1712)
H	Glass		Dry mass, expansion/shrinkage during and after hardening (API RP 10B-5), differential thermal expansion (ASTM E228), thermal shock, UCS (API RP 10B-2)

(continued)

Table 27.3 (continued)

Type	Material	Examples	Mandatory qualification tests
I	Metals	Steel and other alloys, including bismuth-based materials	Dry mass, expansion/shrinkage during and after hardening (API RP 10B-5 ring test), differential thermal expansion (ASTM E228), creep (ISO 204), tensile strength (ISO 527-1), UCS (API RP 10B-2), shear bond strength, thermal decomposition (ASTM D1712)
J	Modified in-situ materials	Casing or formation materials chemically or thermally modified in-situ	Nitrogen permeability, dry mass, and creep (ASTM C512-10)

mandated for critical isolation zones or annuli with suspected poor primary cement. Requires specialized tools and significant rig time.
  - Other Techniques: Placement methods for alternative materials (e.g., heated lines for bismuth alloys, specialized pumping systems for resins/geopolymers).

- **Verification Techniques**: Confirming the plug is in the right place and, most importantly, provides a hydraulic seal. Regulatory requirements often mandate specific verification methods and acceptance criteria (e.g., maximum allowable pressure drop during a test).
  - Tagging Plug Top: Running tubing, drill pipe, wireline, or coiled tubing tools to physically tag the top of the set plug confirms its depth and verifies that the minimum required plug length is present above a base. However, tagging provides no information about the quality of the plug or its sealing capability.
  - Weight Testing: Applying weight onto the plug with tubing or drill pipe provides some indication of its load-bearing capacity and confirms it has set to a certain strength, but does not confirm a hydraulic seal.
  - Pressure Testing (Above Plug): Applying pressure from the surface onto the top of the plug is the most common method used to verify sealing integrity. The pressure is held for a specified duration (e.g., 30 min to several hours), and any pressure decline ("leak off") is monitored and compared against acceptance criteria (e.g., less than 5–10% drop). Requires careful interpretation as pressure can potentially leak off elsewhere in the wellbore above the plug if there are secondary issues.
  - Inflow Testing (Below Plug): Reducing hydrostatic pressure above the plug (e.g., by displacing to a lighter fluid) and monitoring for pressure build-up from below provides a more direct test of sealing against formation pressure, but is often difficult to achieve safely or practically, especially if there is no significant pressure below the plug or if the wellbore cannot be safely lightened.

- Logging: Cement evaluation logs (CBL/Ultrasonic) can assess the bonding quality within the plug interval, but don't directly measure hydraulic sealing capability under pressure. Specialized logging techniques (e.g., acoustic monitoring during testing, potential for dedicated P&A verification logs) are being developed for better plug verification.

Achieving reliable, long-term verification, especially for annular seals achieved via section milling or squeezing, remains a significant technological challenge in P&A. The type of plug used during well abandonment has evolved over time, reflecting changes in regulatory standards, industry practices, and technical understanding of long-term well integrity. Analyzing the historical use of different plug types—particularly cement plugs, capped mechanical plugs (BP), and combinations—provides insights into the legacy risks posed by older abandonment practices and informs strategies for re-abandonment or integrity assessments. Figure 27.4 illustrates these trends in plug placement across abandonment years, revealing the dominance of cement plugs throughout the decades and the limited adoption of alternative or combined plug approaches in many older wells. The length of the cement plug is a critical design parameter influencing the long-term hydraulic sealing performance of abandoned wells. Increasing plug length improves the contact area and bonding strength, thereby reducing potential leakage rates. Figure 27.5 illustrates the relationship between plug length (inside casing, for combined barriers 1 and 2) and the calculated leakage rate. The clear inverse relationship shown emphasizes the importance of ensuring sufficient plug length for robust and permanent zonal isolation.

27.6 The Challenge of Long-Term P&A Integrity

Potential leakage pathways in permanently abandoned wells are critical considerations for evaluating long-term well integrity. Despite the placement of cement or other barrier materials, cracks within the cement, micro-annuli along the casing-cement or formation-cement interfaces, and poorly bonded interfaces can all serve as potential flow paths for fluids. Figure 27.6 provides a schematic cross-sectional view of a plugged and abandoned well, highlighting these primary leakage pathways, including migration through cement cracks, up the cement sheath, and along casing or formation interfaces. Recognizing these pathways is fundamental to improving plug design and verifying permanent abandonment integrity. Ensuring abandoned wells remain safely sealed for centuries or millennia presents significant technical, societal, and financial challenges. The primary uncertainty lies in the very long-term performance of barrier materials and their interaction with the evolving subsurface environment.

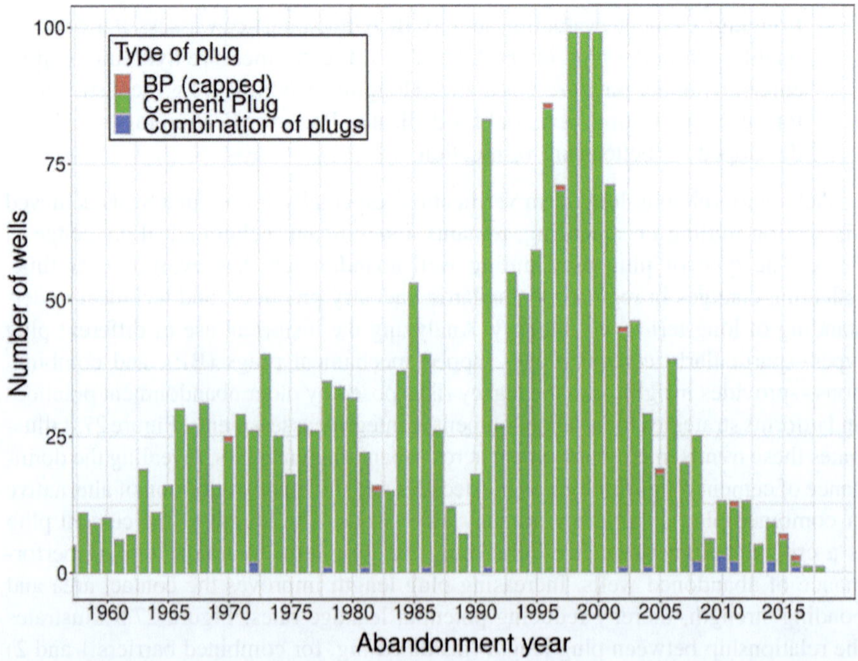

Fig. 27.4 Number of wells abandoned each year (from 1960 to 2015) and the types of plugs used: BP (capped mechanical bridge plugs), cement plugs, and combinations of plug types. This figure highlights the dominance of cement plugs historically, the evolution of abandonment practices over the decades, and the relatively limited use of combination plugs in older wells (Trudel et al., 2019)

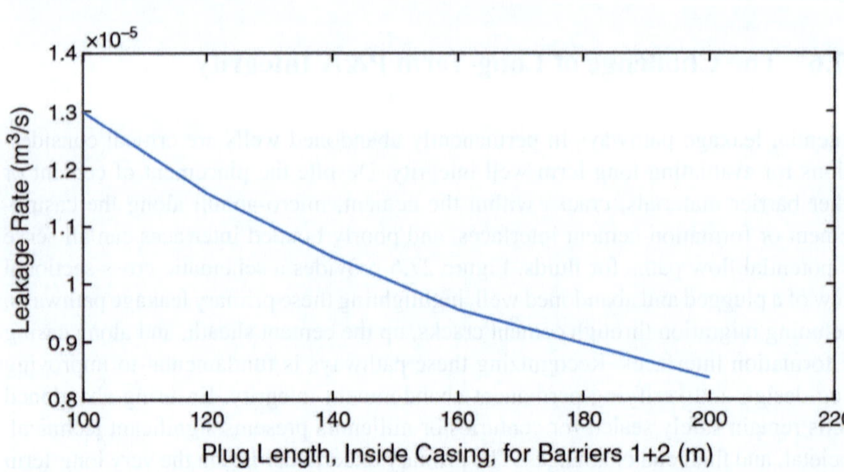

Fig. 27.5 Effect of plug length on leakage rate in abandoned wells: the decrease in leakage rate (m^3/s) with increasing plug length (m) inside casing, for combined barriers 1 and 2. This trend underscores the importance of appropriate plug design and length for long-term well integrity (Moeinikia et al., 2018)

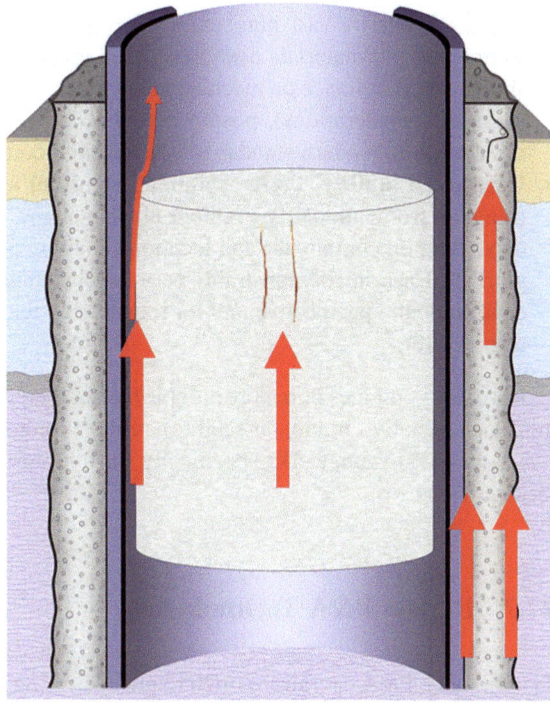

Fig. 27.6 Schematic illustration of leakage pathways in an abandoned well cross-sectional diagram showing potential fluid migration routes in a plugged and abandoned well: through cracks in the cement plug, micro-annuli at the casing-cement interface, and along the cement-formation interface. The figure highlights the need for robust plug placement, thorough verification, and the use of durable barrier materials to mitigate these leakage risks (Vrålstad et al., 2019)

- **Material Degradation**: Uncertainty about the long-term (100s–1000s of years) physical and chemical stability of cement and alternative materials under evolving downhole conditions (e.g., changing temperatures, pressures, pore water chemistry as surrounding formations re-equilibrate, potential for slow diffusion of corrosive species like CO_2). Degradation mechanisms (carbonation, leaching, sulfate attack) can increase permeability over time.
- **Geomechanical Changes**: Long-term geological processes like subsidence (due to regional aquifer depletion), compaction of reservoir rock (even after abandonment), tectonic activity, or salt creep could potentially impose significant stresses on abandoned wells, leading to casing deformation or damage to plugs/cement sheaths.
- **Leakage Pathways**: Micro-annuli or defects not detected or considered insignificant during initial verification could become significant leak paths over long timescales as degradation occurs. The complex interaction between cement, multiple casing strings, and the formation creates numerous potential leak paths that must be sealed.
- **Verification Limitations**: Current verification methods primarily confirm short-term sealing at the time of abandonment. Extrapolating this to long-term performance involves significant uncertainty. Non-invasive, long-term monitoring methods for abandoned wells are generally lacking, and perpetual monitoring is typically not feasible or intended.

- **Legacy Wells**: A vast number of older wells globally were abandoned using techniques and materials now considered inadequate by modern standards (e.g., shorter plugs, reliance on mechanical plugs, limited annular sealing, no long-term material considerations), posing potential future leakage risks. Re-abandoning these wells to modern standards is technically complex and extremely costly.
- **Cost and Liability**: P&A operations represent a significant cost for operators (ranging from hundreds of thousands to many millions of dollars per well, depending on complexity and location), totaling potentially hundreds of billions globally. Ensuring demonstrably permanent abandonment is critical from both an environmental perspective and for transferring the long-term liability away from the operator.

Research continues into more durable barrier materials, improved placement techniques (especially for annular sealing), reliable long-term verification methods, and risk assessment frameworks specifically for long-term P&A integrity and potential leakage scenarios.

27.7 Rigless P&A Technologies

Traditionally, P&A operations often require a drilling rig or workover unit to provide the necessary heavy lifting capability, well control, and pipe handling for plug placement, section milling, and equipment removal. However, mobilizing and operating these assets, particularly offshore, is expensive. Rigless P&A techniques utilize intervention vessels (e.g., Light Well Intervention Vessels—LWIVs) or specialized modular equipment deployed from smaller vessels, platforms, or land, aiming to reduce costs and logistical complexity for suitable wells.

- **Methods**: Rigless P&A employs advanced wireline, coiled tubing, or specialized downhole tools to perform tasks traditionally requiring a rig:
  - Setting plugs (mechanical, cement via dump bailer or specialized pumps).
  - Perforating casing.
  - Cutting and pulling tubing and potentially shallow casing sections (using specialized cutting tools and tension jacks).
  - Section milling (using specialized downhole motors and mills run on CT or drill pipe).
  - Cementing operations (using specialized downhole mixing units, high-pressure pumping spreads on vessels, or coiled tubing for placement).
  - Wellhead severing and recovery (using specialized cutting tools and lifting equipment).

- **Benefits**: Potential for significant cost savings (rig rates vs. LWIV rates, reduced logistics), reduced personnel on board (especially offshore), faster operations for certain plug types, and the ability to access platforms with limited deck space or load capacity. This is particularly attractive for abandoning large numbers of simpler wells.
- **Limitations**: May have limitations on the depth, complexity, wellbore angle (horizontal sections can be challenging), required pressure ratings, or type of P&A operations that can be performed reliably compared to a full rig. Section milling of deeper or larger diameter casing strings may still require a rig due to power and debris handling requirements. Requires specialized equipment and highly competent personnel.

Rigless P&A is a growing area of development, driven by the need for more cost-effective solutions to address the increasing number of wells requiring abandonment globally. Its applicability depends on the specific well characteristics, regulatory requirements, and the maturity of the available technology.

Effective and permanent P&A is the final, critical act of responsible well stewardship, ensuring that wells do not become future sources of environmental contamination or safety hazards. It requires rigorous planning, execution, verification (where possible), and increasingly, detailed consideration of very long-term material performance and degradation. Methane emissions from abandoned wells represent a significant environmental concern, with emission rates varying by location, well type, and abandonment practices. Understanding these emission profiles is essential for prioritizing mitigation efforts and assessing the long-term environmental liabilities associated with legacy wells. Figure 27.7 presents a comparative analysis of methane emission rates and the number of abandoned wells—both unplugged and plugged—across various states and regions in the United States, with a specific focus on Pennsylvania's historical trends. The data highlight the disproportionate contributions of unplugged wells to methane emissions and underscore the need for robust P&A practices that permanently seal wells and eliminate pathways for methane leakage.

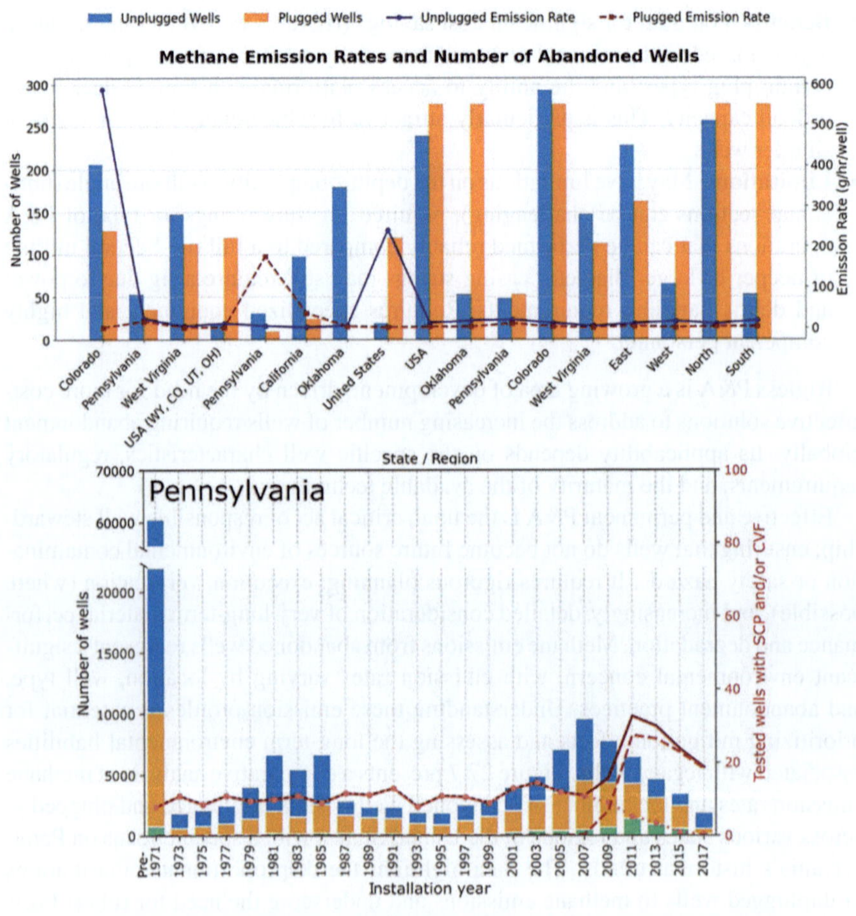

Fig. 27.7 Methane emission rates and abandoned well statistics. Top: Comparative methane emission rates and the number of unplugged versus plugged abandoned wells across selected states and regions. Bottom: Historical abandonment data for Pennsylvania, illustrating the number of wells abandoned by installation year and the percentage of tested wells showing sustained casing pressure (SCP) or casing vent flow (CVF), indicators of ongoing methane emissions. Together, these plots underscore the persistent integrity challenges of legacy well infrastructure and the environmental consequences of incomplete well sealing

References

Abiabhar M. A., Abu Bakar, N. N. E., Aizat Haidi Hod, M., Kumar, A. K., Lau, C. H., & Thuzar, M. (2022). Abandonment of wells under the new norm—Sacrificial tubing approach. In *Offshore Technology Conference Asia, OTCA 2022*. https://doi.org/10.4043/31369-MS

Ahmed, A. H., Srivastava, M., Hosani, A. Al, & Ali, M. S. (2024). Protecting shallow aquifer during entire well lifecycle through integral formations isolation. In *Society of Petroleum Engineers—SPE Water Lifecycle Management Conference and Exhibition, WLMC 2024*. https://doi.org/10.2118/218954-MS

Carpenter, C. (2023). Studies quantify cement sealing performance under downhole conditions. *Journal of Petroleum Technology, 75*(05), 101–103. https://doi.org/10.2118/0523-0101-JPT

Chukwuemeka, A. O., Oluyemi, G., Mohammed, A. I., & Njuguna, J. (2023). Plug and abandonment of oil and gas wells—A comprehensive review of regulations, practices, and related impact of materials selection. *Geoenergy Science and Engineering, 226*, Article 211718. https://doi.org/10.1016/J.GEOEN.2023.211718

Hillebrand, Steve (U.S. Fish and Wildlife Service). (n.d.). "Abandoned gas well 2". Wikimedia Commons. Public domain (U.S. federal government work). https://commons.wikimedia.org/wiki/File:Abandoned_gas_well_2.png

Hovda, L., Phadke, A., Mueller, D., Stevens, J., Ryan, A., & Nevøy, H. (2020). Best practice for cementing and zonal isolation using the jet-type perforate, wash and cement technique. In *Society of Petroleum Engineers—SPE Asia Pacific Oil and Gas Conference and Exhibition 2020, APOG 2020.* https://doi.org/10.2118/202397-MS

Khalifeh, M., & Saasen, A. (2020). Tools and techniques for plug and abandonment (pp. 213–247). https://doi.org/10.1007/978-3-030-39970-2_8

Lecolier, E., Rivereau, A., Ferrer, N., Audibert, A., & Longaygue, X. (2006). Durability of oilwell cement formulations aged in H_2S-containing fluids. In *SPE/IADC Drilling Conference, Proceedings* (pp. 861–869). https://doi.org/10.2118/99105-MS

Liversidge, D., Taoutaou, S., & Agarwal, S. (2006). Permanent plug and abandonment solution for the North Sea. In *Proceedings—SPE Asia Pacific Oil and Gas Conference and Exhibition 2006: Thriving on Volatility* (Vol. 1, pp. 91–95). https://doi.org/10.2118/100771-MS

Moeinikia, F., Ford, E. P., Lohne, H. P., Arild, Ø., Majoumerd, M. M., & Fjelde, K. K. (2018). Leakage calculator for plugged-and-abandoned wells. *SPE Production & Operations, 33*(04), 790–801. https://doi.org/10.2118/185890-PA

Morales-Monsalve, C. B., Fontoura, S. A. B., Rosero-Argote, S., Lomba, R. F., Machado, L., Carrera, S., & Ezechiello, I. (2022). Numerical simulation of wellbore closure due to shale creep: Potential for annular sealing barrier formation. In *56th U.S. Rock Mechanics/Geomechanics Symposium.* https://doi.org/10.56952/ARMA-2022-0575

National Petroleum Council. (2011). Plugging and abandonment of oil and gas wells (Paper #2-25; Working document of the NPC North American Resource Development Study). https://www.npc.org/Prudent_Development-Topic_Papers/2-25_Well_Plugging_and_Abandonment_Paper.pdf

Øksnes, A. L. B. (2017). Permanent plugging and abandonment—An identification and discussion of technologies and the differences in UKCS and NCS regulations.

Pennsylvania, D. (2018). History of well plugging and risks associated with orphan & abandoned wells and improperly plugged wells 2018 information for potential CFA OAWP grant applicants.

Razak, M. N. A., Hamid, A. A., Zawawi, A., Khalid, M., Shah, S. M., Ezani, F. S., & Sellapan, S. (2024). Holistic approach in dealing with accelerated plug and abandonment (P&A) project to mitigate gas leak. In *Proceedings of the IADC/SPE Asia Pacific Drilling Technology Conference, APDT*. https://doi.org/10.2118/219612-MS

RRC Texas. (2025). *RRC current rules.* https://www.rrc.texas.gov/general-counsel/rules/current-rules/

Trudel, E., Bizhani, M., Zare, M., & Frigaard, I. A. (2019). Plug and abandonment practices and trends: A British Columbia perspective. *Journal of Petroleum Science and Engineering, 183*, Article 106417. https://doi.org/10.1016/J.PETROL.2019.106417

Vissa, S. V. K., & Radonjic, M. (2020). Designing wellbore plugging and abandonment materials based on nature's hydraulic barrier materials: A solution to prevent hydrocarbon leakage over time. In *Society of Petroleum Engineers—SPE International Conference and Exhibition on Health, Safety, Environment, and Sustainability 2020, HSE and Sustainability 2020.* https://doi.org/10.2118/199528-MS

Vrålstad, T., Saasen, A., Fjær, E., Øia, T., Ytrehus, J. D., & Khalifeh, M. (2019). Plug & abandonment of offshore wells: Ensuring long-term well integrity and cost-efficiency. *Journal of Petroleum Science and Engineering, 173*, 478–491. https://doi.org/10.1016/J.PETROL.2018.10.049

Chapter 28
Bridging the Gap: Standards Versus Technology and Emerging Challenges

Abstract This chapter explores the evolving tension between well integrity standards and the rapid pace of technological innovation in the oil and gas industry. While standards—such as those from ISO, API, and NORSOK—form the bedrock of regulatory compliance and safe practices, their development cycles are often slow and consensus driven. In contrast, emerging technologies in monitoring, artificial intelligence, alternative barrier materials, and advanced modeling are progressing rapidly, frequently outpacing the guidance provided by existing standards. This disparity gives rise to two critical gaps: one where cutting-edge technologies lack formal recognition within regulatory frameworks, and another where well-established standards pose challenges that current technologies have yet to fully address. The chapter identifies key areas where technology leads, including distributed fiber-optic sensing, AI-driven predictive maintenance, non-Portland P&A materials, and high-fidelity simulations. Simultaneously, it highlights pressing needs such as reliable long-term abandonment verification, improved cement evaluation, MIC detection, leak sensitivity, and cost-effective remediation—all areas where technology still lags behind regulatory expectations. To bridge these gaps, the chapter proposes adaptive mechanisms such as pilot projects, Joint Industry Projects (JIPs), performance-based standards, regulatory flexibility, and industry-wide knowledge sharing. These tools enable safe deployment of novel technologies while preserving the reliability and robustness embedded in existing standards. A lifecycle integrity model is presented to demonstrate how technology and regulation must co-evolve to ensure sustained integrity across all well phases. Ultimately, the chapter emphasizes the importance of enabling responsible innovation through collaboration, transparent validation, and adaptive standardization. By proactively managing the interface between technology and standards, the industry can achieve enhanced safety, environmental stewardship, and operational excellence in the face of evolving energy challenges and stakeholder expectations.

© The Author(s), under exclusive license to Springer Nature Switzerland AG 2026

A. Alsubaih and K. Sepehrnoori, *Well Integrity for Oil, Gas, and Subsurface Storage Operations*, https://doi.org/10.1007/978-3-032-15233-6_28

Engineer check the well status *Source* The Authors

Chapter's Objectives

- Discuss the inherent tension between established standards and rapidly evolving technology in well integrity.
- Identify specific areas where technology offers capabilities potentially ahead of standard practices or regulatory frameworks.
- Highlight known integrity challenges where technology is still striving to meet the needs defined by standards.
- Consider mechanisms for bridging the gap, such as pilot projects, Joint Industry Projects (JIPs), and adaptable regulatory approaches.

28.1 Introduction: The Pace of Change

Industry standards (like ISO, API, NORSOK) and regulations provide an essential foundation for well integrity, codifying best practices, minimum requirements, and lessons learned from decades of experience. They promote consistency, provide benchmarks for design and operation, and form the basis for regulatory compliance. However, the process of developing and updating these standards is inherently collaborative and deliberate, often involving extensive discussion, testing, and consensus-building among stakeholders (operators, service companies, regulators, researchers).

This process is necessarily cautious and can take several years. In contrast, technology development, driven by innovation, computational power, market demands, and research breakthroughs, can advance much more rapidly.

This disparity in pace creates an inevitable gap: situations where emerging technologies offer new capabilities for monitoring, analysis, intervention, or barrier materials that are not yet fully reflected, accommodated, or even explicitly permitted in established standards or regulations. Conversely, there are also situations where standards define clear integrity objectives (often driven by the need to prevent known failure modes or achieve long-term goals like permanent abandonment) that current established technology struggles to meet reliably, cost-effectively, or with sufficient long-term confidence. Figure 28.1 a conceptual illustration of how a gap can emerge between technology development and the evolution of standards and regulations. The red and blue cars represent the relative progress of technology and standards, respectively, moving along diverging paths over time. This gap illustrates the need for mechanisms to align innovation with formalized practices and regulatory expectations. Recognizing and proactively addressing this gap is essential for sustaining robust well integrity management. In the next sections, we explore examples of where this gap is most evident and how the industry and regulators navigate this evolving interface. This chapter explores some key areas where this gap exists and discusses how the industry and regulators navigate this dynamic interface to enable responsible innovation and continuous improvement.

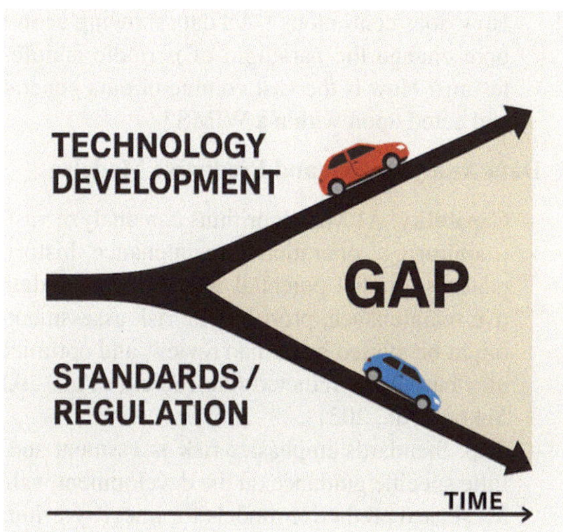

Fig. 28.1 Illustration of the growing gap between the pace of technology development and the evolution of standards and regulatory frameworks. The red and blue cars metaphorically represent technology and standards, respectively, highlighting how they diverge over time without intentional efforts to bridge the gap

28.2 Areas Where Technology Leads Standards

In several areas, technological capabilities have emerged or are maturing that offer potentially significant improvements in well integrity management, but their integration into standard practices or regulatory frameworks may lag behind their demonstrated potential.

- **Advanced Monitoring (Fiber Optics, Wireless)**:
  - Capability: Distributed Fiber Optic Sensing (DTS/DAS/DSS—Chap. 30) provides continuous, real-time data along the entire wellbore, offering unprecedented spatial and temporal coverage for detecting thermal anomalies, acoustic signatures (potentially indicating leaks), or strain changes. Wireless sensors offer the potential for monitoring pressure/temperature in previously inaccessible sealed annuli or deep downhole locations without penetrations.
  - Gap: While standards mandate monitoring (e.g., annulus pressure monitoring, periodic logging), they generally don't prescribe *how* beyond conventional gauges/transmitters at the surface or traditional logging tools. Specific guidance on qualifying, installing, interpreting the complex, high-volume data from, and integrating DTS/DAS/wireless systems into routine integrity management frameworks (e.g., defining specific alarm thresholds for leak detection, required actions based on sensor readings, acceptable levels of confidence for anomaly detection) is still evolving within standards committees and operator practices. How does continuous DAS data showing acoustic signatures around the wellbore change the paradigm of periodic annulus pressure monitoring or leak testing? How is the vast volume of data generated by these systems managed and acted upon within a WIMS?

- **Data Analytics, AI, and Predictive Models**:
  - Capability: AI/ML algorithms can analyze vast datasets from various sources (monitoring, operations, maintenance history, logs) to identify complex patterns, predict potential failures or degradation before they occur (predictive maintenance, probabilistic risk assessment), detect subtle anomalies that might be missed by human review, and optimize inspection and testing schedules based on predicted risk (Salem et al., 2022; Semwogerere et al., 2025; Yakootet al., 2021).
  - Gap: Standards emphasize risk assessment and data management but provide little specific guidance on the development, validation, and deployment of AI/ML-based predictive models for integrity-critical decisions. Questions remain regarding the level of rigor required for model validation, managing model drift over time as conditions change, ensuring transparency ("explainable AI"—understanding *why* a prediction is made), and defining acceptable reliance on predictive alerts or AI-driven recommendations versus traditional surveillance methods and engineering judgment. Regulatory acceptance of decisions based

primarily on AI predictions (e.g., extending inspection intervals based on a low predicted failure probability) is still an open area.

- **Alternative P&A Materials**:
  - Capability: Materials like bismuth alloys, geopolymers, and advanced resins offer potential advantages over Portland cement for long-term P&A barriers in specific conditions (e.g., higher durability in CO_2, better resistance to thermal cycling, non-shrinking properties, potentially better bonding to varied surfaces) (Todorovic et al., 2023).
  - Gap: Most standards and regulations are heavily based on the historical use and verification procedures for Portland cement plugs. While some performance-based frameworks allow for alternatives, the qualification process (demonstrating long-term equivalence or superiority to cement for centuries) can be lengthy, expensive, and lacks standardized, widely accepted protocols across all jurisdictions. This can hinder the adoption of potentially more durable or effective materials despite promising laboratory and field trial results. JIPs are actively working on developing qualification guidelines.
- **Advanced Modeling and Simulation**:
  - Capability: Sophisticated Finite Element Analysis (FEA) and Computational Fluid Dynamics (CFD) tools allow for detailed modeling of complex stress states (e.g., triaxial stresses in HPHT wells, geomechanical loading from salt creep or compaction, thermal stresses in geothermal wells) and fluid dynamics (e.g., erosion prediction, transient multiphase flow effects on corrosion, detailed cement placement simulation) beyond the scope of simpler analytical calculations often referenced in older prescriptive standards.
  - Gap: Standards often specify minimum design factors based on simpler analytical load cases and standardized API connection performance tables. While advanced modeling is used by operators for complex or critical designs (e.g., HPHT casing design, thermal well design), incorporating its outputs formally into compliance demonstrations or standardized design workflows across the industry can be inconsistent. Verifying the accuracy and assumptions of complex models remains crucial.
- **Complex Well Designs and Completions**:
  - Capability: Technologies enable highly complex wells (e.g., long horizontal laterals, multilateral wells with multiple branches, intelligent completions with numerous downhole control valves, sensors, and flow lines).
  - Gap: Applying traditional barrier definitions and verification methods (e.g., pressure testing specific annuli, defining two independent barriers) can be challenging, ambiguous, or physically impossible in these complex geometries. Standards and WIMS procedures need further evolution to provide clear guidance on defining and verifying barrier envelopes in multilateral junctions, across numerous downhole components, or within highly instrumented completions.

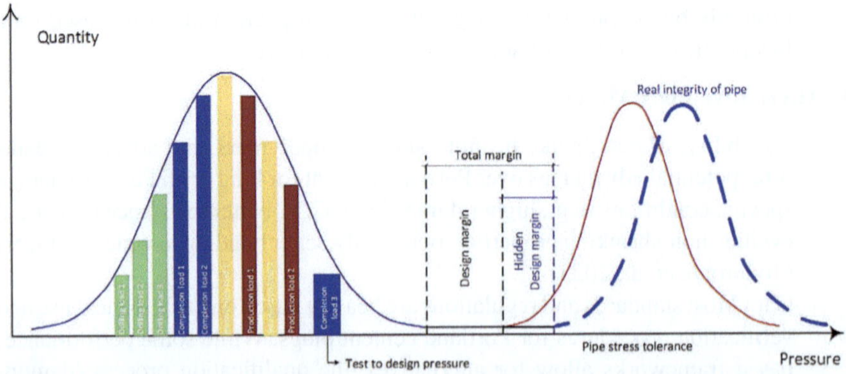

Fig. 28.2 Real integrity of pipe and the associated total margin to loads, highlighting the difference between design margin (from codes and standards) and hidden margin (not fully captured by standard tests). This underscores the need for advanced models, testing, and continuous performance monitoring to ensure that actual performance aligns with—or exceeds—design expectations (Brechan et al., 2019)

Another dimension of the gap between standards and technology emerges in understanding the real integrity of materials and the hidden margins that may not be fully accounted for in standard design practices. Figure 28.2 Illustration of the real integrity of a pipe compared to the standardized design margins. The figure highlights how real-world capacity (blue dashed line) can differ significantly from the assumed design capacity (red line), and how hidden margins—undocumented in standards—can exist but are not always captured or verified. This gap underscores the value of advanced testing, real-time monitoring, and continuous performance evaluation in well integrity management. Such insights reveal the importance of incorporating field-derived data, advanced probabilistic models, and continuous integrity assessment tools to bridge the gap between the standardized design assumptions and the actual performance envelope. As operators push the boundaries of well complexity and performance, these enhanced approaches ensure a more robust and realistic understanding of true well integrity.

28.3 Areas Where Technology Lags Standards/Needs

Conversely, there are well-understood integrity requirements defined by standards (often driven by lessons learned from past failures or long-term environmental concerns) (Brechan et al., 2019) where current technology still faces significant limitations in reliably or cost-effectively meeting the desired objective, or where the 'gold standard' remains challenging to achieve consistently in practice.

- **Long-Term P&A Verification**:
  - Need: Standards require permanent isolation and verification of P&A barriers to prevent long-term fluid migration.
  - Gap: Current verification methods (tagging plug top, pressure testing from above, occasional inflow testing, cement evaluation logs) primarily assess short-term sealing integrity and structural properties at the time of abandonment. Technology for reliably and non-destructively verifying the *hydraulic seal quality* across the entire wellbore cross section (including annuli and interfaces) and predicting its performance over geological timescales (centuries) remains a major technological gap. Developing non-invasive or long-lived monitoring methods for abandoned wells is also challenging. This is a critical gap given the increased focus on permanent abandonment assurance, especially for CCS.
- **Cement Evaluation Limitations**:
  - Need: Standards (e.g., API RP 65-2, NORSOK D-010) require evaluation of primary cement jobs to confirm zonal isolation before critical steps like pressure testing casing shoes or perforating.
  - Gap: Even advanced ultrasonic logging tools have limitations in definitively detecting very small micro-annuli (especially if liquid-filled), reliably evaluating bond in gas-filled or highly attenuated slurries (lightweight, foam, high-additive cements), or accurately assessing cement properties far behind thick or multiple casing strings. Distinguishing different materials behind casing can be challenging. Achieving consistently accurate, unambiguous cement evaluation across all well types and conditions remains a technological aspiration.
- **MIC Prediction and Real-Time Detection**:
  - Need: Standards and risk assessments recognize Microbiologically Influenced Corrosion (MIC) as a significant integrity threat in certain environments (e.g., water injection, produced water handling, potentially some storage or abandonment scenarios).
  - Gap: Reliably predicting where and when MIC will occur based on fluid chemistry, flow conditions, and biofilm formation remains difficult. Furthermore, detecting and quantifying active MIC in-situ and in real time using sensors is technologically challenging. Current monitoring methods (e.g., coupons, probes) provide averaged or localized data, making proactive, targeted mitigation difficult beyond broad-spectrum biocide treatments.
- **Leak Detection Sensitivity and Reliability**:
  - Need: Early detection of small leaks is critical for preventing escalation to major failures (driven by lessons from incidents).
  - Gap: While advanced technologies like DAS/DTS offer high sensitivity, reliably distinguishing very small leak signals from background noise, normal operational fluctuations, or other benign phenomena under all conditions, and

avoiding false alarms, remains an ongoing challenge requiring sophisticated signal processing, interpretation, and validation. False positives can lead to unnecessary intervention costs; false negatives mean delayed detection.

- **Cost-Effective Downhole Remediation:**
  - Need: Standards require remediation of integrity failures posing unacceptable risks.
  - Gap: Many reliable downhole remediation techniques (especially those requiring major workovers like packer/tubing replacement or section milling for P&A) remain very expensive and time consuming. There is a continuous need for more cost-effective, reliable, and less intrusive technologies for diagnosing and repairing downhole integrity issues (e.g., long-term sealing of micro-annuli, repairing casing leaks in challenging environments). While rigless intervention advances cost-effectiveness for some tasks, limitations remain.

28.4 Bridging the Gap: Mechanisms for Progress

Effectively managing the dynamic interface between standards and technology is crucial for continuous improvement in well integrity. This holistic view emphasizes that managing the gap between technology and standards isn't a one-time task but a continuous process throughout the well's life cycle. Figure 28.3 illustrates how well integrity must be actively maintained and verified at every stage, with adaptive risk assessment and data-driven decision-making at the core of this integrated approach (Brechan et al., 2019). This model underscores the need to bridge the gap between technology, standards, and regulatory frameworks by adopting a lifecycle approach that leverages innovative technologies, performance-based standards, and continuous learning from field data and lessons learned. Several mechanisms help bridge the gap:

- **Pilot Projects and Field Trials**: Operators and service companies test and validate new technologies (e.g., alternative P&A materials, advanced monitoring systems, novel remediation tools) in pilot wells or controlled field trials before wider deployment. This generates performance data, identifies limitations, and builds confidence for potential standardization and regulatory acceptance.
- **Joint Industry Projects (JIPs)**: Collaborative efforts between multiple operators, service companies, research institutions, and sometimes regulators to fund research, develop new technologies, collect and share data (often anonymized), and develop best practices or qualification guidelines for emerging areas (e.g., long-term P&A material qualification, validation of AI models for integrity, defining requirements for fiber-optic systems). JIPs pool resources and knowledge to address shared challenges.

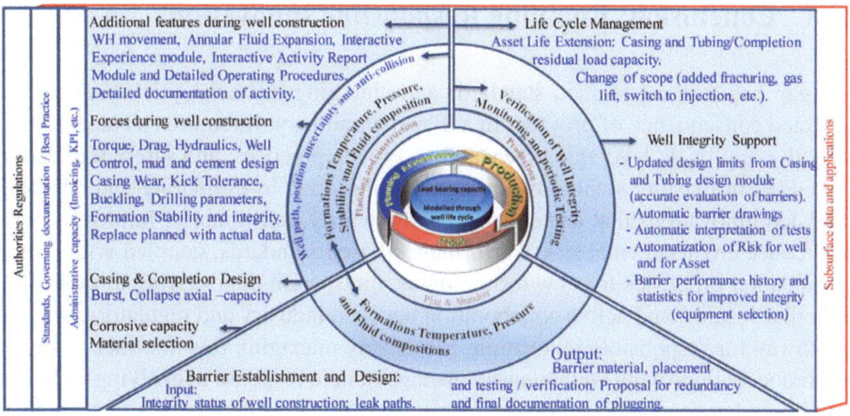

Fig. 28.3 A comprehensive life cycle well integrity model that integrates technical, regulatory, and operational considerations across all phases of a well's life—from design and construction to operation, plug and abandonment, and post-abandonment monitoring. The model highlights the importance of holistic planning, robust barrier design and verification, advanced modeling and testing, and continuous risk management to ensure long-term integrity and compliance with evolving standards and technological capabilities (Brechan et al., 2019)

- **Performance-Based Standards and Regulations**: Goal-setting regulations and standards that define *what* must be achieved (e.g., "demonstrate two verifiable barriers," "ensure permanent isolation over geological timescales," "reduce risks to ALARP") rather than prescribing *how* it must be achieved, allow operators the flexibility to propose and utilize innovative technological solutions, provided they can rigorously demonstrate effectiveness and safety (often via Safety Cases or similar submissions). This encourages innovation while maintaining a safety envelope.
- **Updating Standards**: Standards committees actively work to incorporate new knowledge, lessons learned from incidents, and technological advancements into revised versions of standards, although this consensus-driven process necessarily takes time. Industry feedback and active participation in these committees (e.g., API, ISO, NACE, NORSOK) are crucial for ensuring standards remain relevant and incorporate validated technologies.
- **Industry Forums and Knowledge Sharing**: Conferences, workshops, technical papers, and online forums provide platforms for sharing experiences, successes, failures, and lessons learned with new technologies and approaches, accelerating learning and adoption across the industry.
- **Regulatory Adaptation**: Regulators may engage with industry through various mechanisms (e.g., workshops, JIP participation, technical committees). They may issue interim guidance, approve specific field trials, or adapt requirements based on demonstrated performance and rigorous risk assessment of new technologies before formal standards are updated.

28.5 Conclusion: Enabling Responsible Innovation

The gap between established standards and emerging technology is a natural and expected consequence of progress in a dynamic industry. It requires a balanced and proactive approach: leveraging the stability, proven practices, and safety foundation embedded in existing standards while remaining open to evaluating and adopting validated innovations that can enhance well integrity performance, improve safety, and reduce environmental risk. Performance-based standards, coupled with robust qualification processes for new technologies, transparent data sharing through JIPs and other forums, and active collaboration between industry and regulators, provide a pathway for responsible innovation. Effectively managing this interface is key to continuously improving well integrity management in response to evolving technical challenges, environmental requirements (especially for energy transition applications), and societal expectations. The goal is not simply to adopt new technology for its own sake, but to strategically deploy innovations that address critical integrity challenges and deliver demonstrable improvements in safety, environmental protection, and operational efficiency.

References

Brechan, B., Sangesland, S., & Dale, S. (2019). Well integrity—Managing the risk using accurate design factors. In *Society of Petroleum Engineers—SPE Norway One Day Seminar 2019*. https://doi.org/10.2118/195604-MS

Salem, A. M., Yakoot, M. S., & Mahmoud, O. (2022). A novel machine learning model for autonomous analysis and diagnosis of well integrity failures in artificial-lift production systems. *Advances in Geo-Energy Research, 6*(2), 123–142. https://doi.org/10.46690/ager.2022.02.05

Semwogerere, D., Sangesland, S., Vatn, J., Pavlov, A., & Colombo, D. (2025). Well integrity and late life extension—A current industry state of practice and literature review. In *Geoenergy science and engineering* (Vol. 244). Elsevier B.V. https://doi.org/10.1016/j.geoen.2024.213419

Todorovic, J., Stroisz, A. M., Duda, M. I., Agofack, N., Lange, T., Nilsen, N. I., Aas, P. G., Sefidroodi, H., & Ringe, T. (2023). Durability of polymer-based wellbore sealants in a corrosive environment. In *Society of Petroleum Engineers—SPE EuropEC—Europe Energy Conference Featured at the 84th EAGE Annual Conference and Exhibition, EURO 2023*. https://doi.org/10.2118/214340-MS

Yakoot, M. S., Ragab, A. M. S., & Mahmoud, O. (2021). Machine learning application for gas lift performance and well integrity. In *Society of Petroleum Engineers—SPE Europe Featured at 82nd EAGE Conference and Exhibition, EURO 2021*. https://doi.org/10.2118/205134-MS

Chapter 29
Statistical Insights Into Well Integrity Performance

Abstract This chapter investigates the critical role of statistical analysis in understanding and managing well integrity performance across the oil and gas industry. It emphasizes how aggregated data on integrity failures—such as sustained casing pressure (SCP), barrier degradation, leaks, and plug and abandonment (P&A) issues—can uncover trends, inform risk assessments, benchmark performance, and guide proactive integrity strategies. However, it also acknowledges the significant challenges that hinder consistent, global data collection and interpretation. These include varying definitions of failure events, inconsistent reporting requirements, data confidentiality, under-reporting of minor issues, and a lack of normalization against well count, age, or complexity. The chapter presents illustrative statistics, offering conceptual estimates on SCP prevalence (10–40% depending on region and well age), component-level failure patterns (e.g., casing, cement, elastomer seals), and the scale of the global P&A backlog. It discusses how failures vary by depth and time since drilling, supported by figures showing the frequency of intervention types, depth distributions, and failure onset timelines. Regional disparities are also explored, linking integrity trends to basin maturity, environmental severity, well types, and regulatory practices. Through visualizations such as histograms, heatmaps, and boxplots, the chapter illustrates how age-dependent failure patterns and intervention needs can shape long-term monitoring and maintenance strategies. It further underscores the need for transparency, standardized reporting formats, and collaborative data sharing initiatives—such as Joint Industry Projects (JIPs)—to improve the quality and comparability of well integrity datasets. These improvements would enable more effective application of advanced analytics, machine learning models, and probabilistic risk assessments. In conclusion, the chapter advocates for a data-driven approach to well integrity management, grounded in carefully interpreted statistics. While recognizing the current limitations, it emphasizes that even imperfect data, when normalized and contextualized, provides valuable insights. Greater industry-wide cooperation in standardizing definitions, enhancing reporting practices, and sharing anonymized integrity data is essential for improving reliability, reducing risks, and sustaining well integrity throughout the well lifecycle.

© The Author(s), under exclusive license to Springer Nature Switzerland AG 2026

A. Alsubaih and K. Sepehrnoori, *Well Integrity for Oil, Gas, and Subsurface Storage Operations*, https://doi.org/10.1007/978-3-032-15233-6_29

Top view of a drill pipe connection showing internal threads before make-up, illustrating well construction and integrity inspection steps. *Source* The Authors

Chapter's Objectives

- Discuss the importance of statistics for understanding well integrity performance and risks.
- Highlight the challenges in collecting consistent and comprehensive global well integrity data.
- Present illustrative statistics (where available or conceptually typical) on SCP, barrier failures, leaks, and P&A.
- Discuss potential regional variations and the limitations of available data.
- Emphasize the value of data sharing and standardized reporting.

29.1 Introduction: The Power and Pitfalls of Statistics

Statistics provide a powerful lens through which to view the overall performance of well integrity management across the industry or within specific regions, companies, or asset types. Analyzing aggregated data on the prevalence of specific issues (like sustained casing pressure), failure rates for components or barriers, common causes of failure, and trends over time can provide valuable insights. These insights can help to:

- Identify systemic weaknesses or common failure modes requiring focused attention and improvement efforts (e.g., consistent issues with cementing in a specific formation, higher failure rates for a certain equipment type, or increasing SCP prevalence with well age).
- Benchmark performance against industry peers or historical trends within an organization, facilitating continuous improvement.
- Provide quantitative input for risk assessments, such as generic failure frequencies for specific components or barrier types, while noting these must be used cautiously and tailored to specific wells and environments.
- Track the effectiveness of improvement initiatives, technology adoption, or revised operational practices over time by observing whether key failure indicators decrease.
- Inform regulatory focus, policy development, and resource allocation for interventions, inspections, or technology qualification processes.
- Support transparent communication and accountability within an organization and with stakeholders, such as regulators or joint industry project (JIP) collaborators.

However, obtaining reliable, consistent, and comprehensive well integrity statistics on a global or even regional scale is fraught with significant challenges. Data is often fragmented across different sources and databases, and definitions of integrity events can vary widely between operators, standards, and regulatory bodies. Reporting requirements also differ significantly by jurisdiction—some require detailed reporting of all SCP occurrences above a minimal threshold, while others only require reports if a leak to the environment occurs.

Adding to the challenge, data on well integrity issues is frequently considered commercially sensitive or proprietary by operators, limiting the extent to which it is shared outside of anonymized, aggregated studies or collaborative industry initiatives. Even within companies, data can be stored in disparate systems, ranging from modern digital well integrity management platforms to older paper records or spreadsheets, making consistent aggregation and analysis difficult.

Compounding these challenges is the reality of under-reporting—particularly for minor incidents or near-miss events that might be managed internally without entering formal reporting systems. These events, though often low in immediate consequence, can provide crucial early signals of potential broader issues if captured systematically. A culture of transparent reporting and learning (as discussed in Chap. 17) is essential for ensuring a full picture of integrity performance.

Normalization of statistics is another critical consideration. Direct comparisons of raw numbers of integrity failures across regions, operators, or periods are rarely meaningful without accounting for context: the number of active wells, the age and condition of those wells, the complexity of the well types and completions, the severity of the operating environment, and the production volumes involved.

Normalized metrics—such as failures per well-year, per completion, or per meter drilled—are far more informative for meaningful performance benchmarking.

Despite these pitfalls, well integrity statistics—when interpreted carefully—remain invaluable. They can reveal trends over time within an organization or region, help identify leading indicators of emerging risks, and support proactive management of well integrity. They also provide a crucial foundation for data-driven risk assessments, continuous improvement initiatives, and regulatory oversight. In the following sections, we will explore the challenges of data collection and comparability, provide illustrative statistics (where available or conceptually typical), and highlight the regional and operational factors that shape integrity performance. Finally, we will discuss the importance of data transparency and standardization as foundational pillars for more robust, globally comparable integrity performance insights.

29.2 Challenges in Data Collection and Comparability

Several factors make it difficult to collect and compare well integrity data consistently across the industry:

- **Varying Definitions**: What constitutes an "integrity failure," an "SCP incident," or a "reportable leak" can differ significantly between operators, regulatory bodies, and industry standards. For example, the threshold pressure for reporting SCP may vary.
- **Reporting Requirements**: Regulatory requirements for reporting well integrity issues vary significantly by jurisdiction. Some may require detailed reporting of all SCP occurrences above a minimal threshold, while others may only require reporting above certain pressure thresholds or only if a leak to the environment occurs. Some events might be managed internally and not enter any external reporting system.
- **Data Confidentiality**: Operators often consider specific failure rate data for their wells, components, or service providers to be commercially sensitive or proprietary, limiting public disclosure or sharing outside of anonymized, aggregated studies (e.g., JIPs).
- **Data Quality and Consistency**: Ensuring the accuracy, completeness, and consistency of data collected across different assets, regions, operational teams, and periods within a single company, let alone across the industry, is challenging. Data may be stored in disparate databases, spreadsheets, or paper records, hindering aggregation and analysis (as discussed in Chap. 15). Legacy data may use outdated formats or definitions.
- **Under-Reporting**: Minor issues, anomalies, or near misses that are managed internally without significant consequence may not always be captured in formal reporting systems, potentially skewing statistics toward more severe events. A culture of open reporting (Chap. 17) is needed to capture a fuller picture.

- **Normalization**: Comparing raw numbers of failures between different regions, companies, or time periods is often meaningless without normalization. Factors like the number of active wells in the inventory, the age distribution of the well stock, the complexity of the wells (e.g., HPHT, deepwater, unconventional), the operating environment severity (e.g., presence of H2S, high temperature), or production volumes should be considered when comparing failure rates (e.g., failures per well-year, failures per completion, failures per meter drilled).

These challenges mean that publicly available, globally comparable well integrity statistics are scarce, fragmented, and often require careful scrutiny of methodologies and definitions.

29.3 Illustrative Statistics (Conceptual and Order-of-Magnitude)

Given the limitations in comprehensive, comparable global data, the following provides conceptual examples and potential orders of magnitude based on information sometimes shared in industry forums, regional regulatory reports, or specialized studies. These figures are illustrative and require verification against specific, credible data sources for any formal use.

- **Sustained Casing Pressure (SCP)**:
  - Prevalence: Studies in mature basins (e.g., Gulf of Mexico, North Sea) have indicated that a significant percentage of wells experience measurable SCP in at least one annulus at some point in their life. Figures ranging from 10% to over 40% have been cited in different studies/regions, depending on the definition used, the age of the well stock, and the geological/operational context. This highlights SCP as a widespread issue.
  - Sources: Analysis of investigated SCP cases often points to cement quality issues (poor zonal isolation, debonding, channeling) as a primary cause, particularly for outer annuli (B, C) (Ibukun et al., 2024). Tubing/packer leaks are common sources for A-annulus SCP (tubing-casing annulus).
  - Severity: Most reported SCP incidents involve relatively low pressures that are managed through monitoring or bleeding. However, a small percentage may involve higher pressures or increasing trends requiring more significant intervention or well shut-in due to risk concerns (e.g., risk exceeding casing ratings—MAASP/MAWOP).
- **Barrier Failure Statistics (Component Level)**:
  - Data Scarcity: Reliable, publicly available statistical failure rate data for specific downhole components (e.g., production packers, DHSVs, specific casing connection types, wellhead seals) is very limited due to commercial

sensitivity and data collection challenges. JIPs sometimes collect anonymized, aggregated data for specific equipment types to analyze reliability.
  - Qualitative Trends: Experience suggests that elastomeric seals in packers and DHSVs are common failure points, particularly in harsher environments (HPHT, corrosive) or after long service lives, due to degradation mechanisms (Chap. 11). Connection leaks (casing/tubing) are also recognized failure modes, especially for standard API connections or if installation quality is poor (Goodman et al., 2021). Catastrophic casing body failure (burst/collapse) is relatively rare if designed and operated correctly within limits, but corrosion-induced leaks (pinholes, splits) are more common in aging wells or corrosive service (Chap. 9). Cement sheath failure (leading to SCP or crossflow) is a significant and often recurring problem.

- **Leak/Release Statistics**:
  - Regulatory Reports: National regulators (e.g., BSEE in the US, PSA/Havtil in Norway, NOPSEMA in Australia) publish annual statistics on reported hydrocarbon spills or releases. While not all releases are due to wellbore integrity failure (many relate to pipelines, process equipment, surface facilities), well-related leaks (e.g., from wellheads, subsea trees, or surface casing/annuli) are a component of these statistics.
  - Severity: Most reported well integrity-related leaks are small and contained, but rare, large-scale uncontrolled releases (like blowouts or major spills) dominate the cumulative volume released and public attention, highlighting the high-consequence nature of integrity failure. Preventing these high-consequence events is a primary focus of integrity management.
  - Trends: Some regulatory reports may suggest trends in reported incidents over time, although these trends can be influenced by changes in reporting requirements, definitions, or inspection/monitoring intensity as much as actual underlying performance changes.

- **Plug and Abandonment (P&A) Statistics**:
  - Well Inventory: Estimates suggest hundreds of thousands, potentially millions, of offshore and onshore wells globally will eventually require P&A. The number of wells reaching the end of their design life is increasing rapidly, creating a significant backlog of inactive wells.
  - Activity Levels: The number of wells being abandoned annually is significant but often lags behind the rate at which wells become inactive, leading to a growing inventory of idle wells awaiting P&A.
  - Costs: P&A costs are substantial, ranging from hundreds of thousands (simple onshore) to many millions of dollars (complex deepwater, HPHT, highly deviated) per well(Saxena et al., 2019). The total global P&A liability is estimated in the hundreds of billions or even trillions of dollars.
  - Failure/Re-Abandonment: Data on the failure rate of previously abandoned wells or the number requiring re-abandonment due to leakage is extremely scarce and challenging to collect but anecdotal evidence and studies on

observed methane emissions from some abandoned wells suggest it is a non-negligible issue, particularly for older wells abandoned to lower standards (Jackson, 2014). This underscores the challenge of ensuring permanent isolation.

In addition to numerical prevalence and failure rates, another valuable perspective is the distribution of intervention or repair types undertaken to address well integrity issues. This provides insights into which resolutions are most implemented, shedding light on the practical challenges operators face in managing well integrity over time. Figure 29.1 shows the relative frequency of different reported resolution types based on a dataset of interventions logged in well integrity management systems. It demonstrates the range of common interventions, from cement squeezes and casing repairs to more specialized interventions like packer installation or well suspension. Beyond understanding the frequency of various well integrity interventions, analyzing the depth distribution of these failures can reveal important insights into the typical zones or depth ranges where specific integrity challenges tend to occur. Figure 29.2 presents boxplots illustrating the top depth of failures associated with different resolution types. Such visualizations can help highlight patterns—such as whether certain interventions tend to be required more frequently at specific depths—providing valuable context for planning surveillance programs, optimizing equipment selection, or tailoring barrier designs to likely failure zones. Understanding not only how often different integrity failures occur, but when they tend to occur relative to well drilling is critical for integrity planning and surveillance. Figure 29.3 presents a series of histograms showing the distribution of years between initial drilling and the first reported failure for various intervention types. This temporal perspective provides insights into well age effects, typical degradation timelines, and when specific failure modes become more prevalent—informing proactive integrity management strategies and resource allocation.

29.4 Regional Variations

Reported statistics on well integrity issues, their prevalence, and dominant failure modes can vary significantly by region due to several factors:

- **Basin Maturity/Age**: Older, more mature basins (e.g., North Sea, parts of the Gulf of Mexico) have a higher proportion of aging wells (30+ years old), increasing the prevalence of age-related issues like corrosion, cement degradation, and SCP compared to newer exploration areas.
- **Operating Environment**: Harsh environments (HPHT, deepwater, arctic, highly corrosive sour gas fields) inherently pose greater technical challenges and may exhibit different failure modes or potentially higher failure rates if not managed with specialized materials and techniques (discussed in Chaps. 18, 21, 22, 23 and 24).

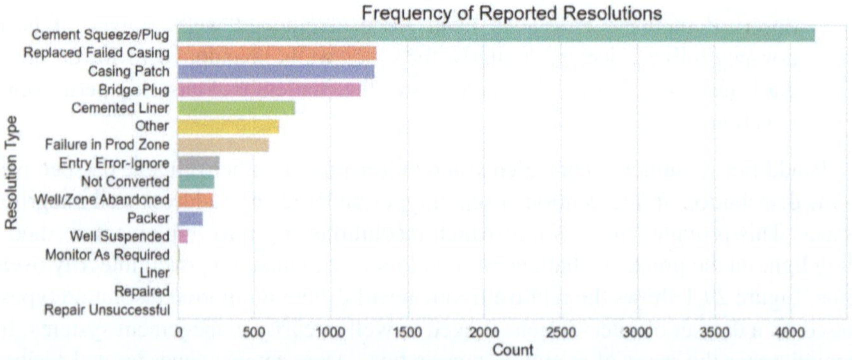

Fig. 29.1 Frequency of reported resolution types for well integrity interventions. The dataset highlights the prevalence of cement squeeze/plug jobs as the most common integrity repair method, followed by casing repairs (e.g., casing patch, replaced failed casing) and bridge plug installations. Less frequent interventions include packer placement, well suspension, and rare instances of repair failures. Such data helps prioritize improvement efforts and technology development in well integrity management

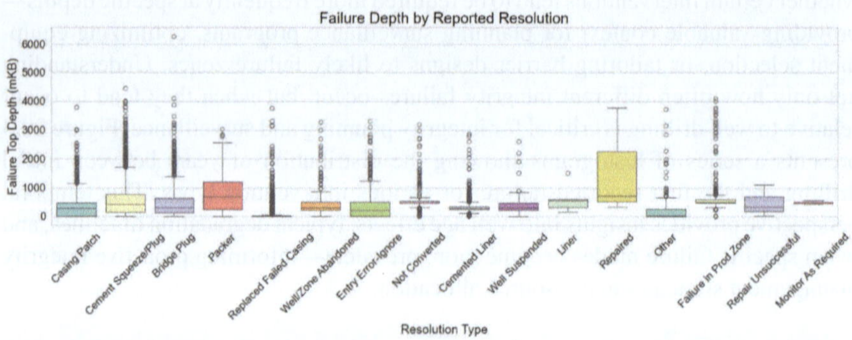

Fig. 29.2 Boxplot distribution of failure top depths (measured in meters below Kelly bushing) by reported resolution type. Each box shows the median depth, interquartile range, and outliers for failures addressed by different interventions (e.g., cement squeeze, casing patch, bridge plug). This data helps illustrate how well integrity issues span a range of depths and may vary systematically between resolution types—information that can inform more effective well integrity surveillance and remediation strategies

- **Regulatory Regime and Reporting Culture**: Differences in regulatory requirements for monitoring, reporting, and intervention thresholds, coupled with variations in company safety and reporting culture, can significantly influence the number and type of integrity issues that are formally recorded and reported.
- **Dominant Well Types**: Regions dominated by unconventional wells (Chap. 19) face different integrity challenges (e.g., cyclic stress from fracturing, horizontal cementing) compared to regions with conventional vertical wells. Similarly, areas

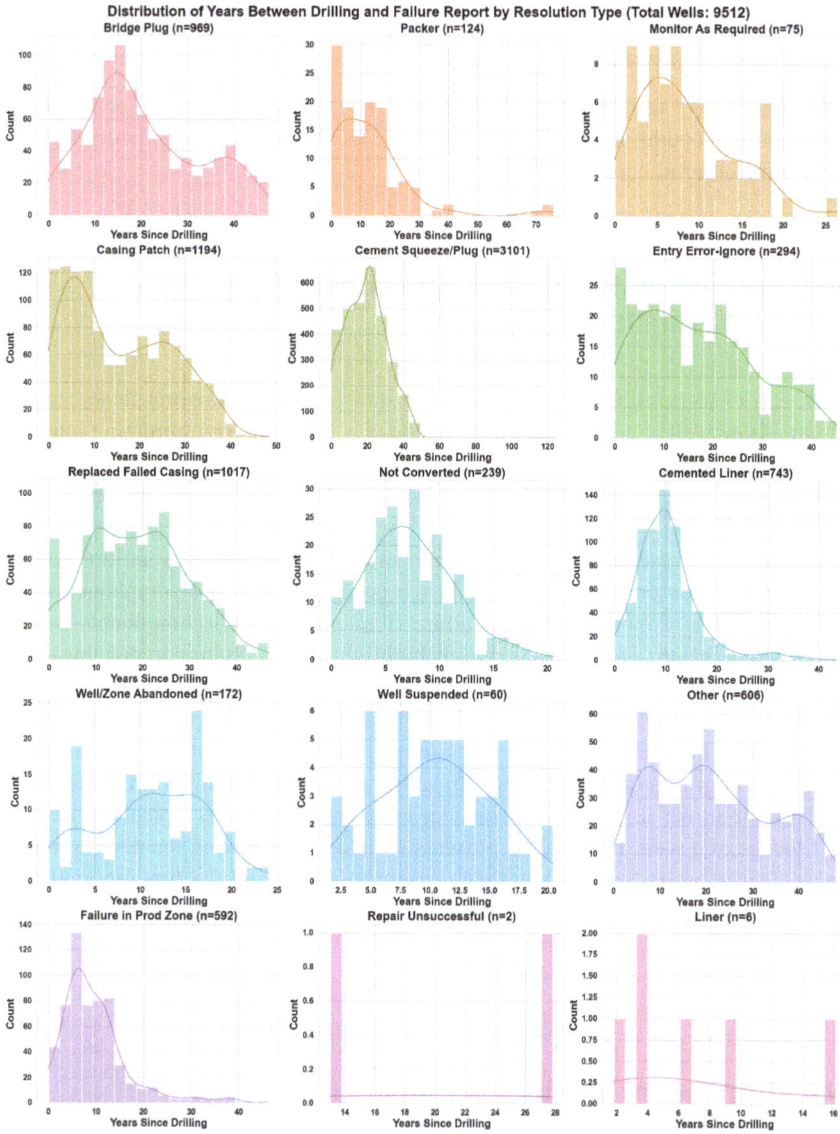

Fig. 29.3 Distribution of the number of years between well drilling and first reported failure for different resolution types (total wells: 9694). The plots highlight typical timeframes when certain failures tend to emerge—such as bridge plug issues clustering early, while cement squeeze/plugs are often required throughout the well's life. This analysis provides insights into the temporal patterns of well integrity challenges and can guide inspection scheduling, lifecycle integrity plans, and technology targeting for critical well age windows

with significant EOR (Chap. 22), water injection (Chap. 23), or storage (Chap. 24) activities will see statistics influenced by the specific risks of those operations.

Benchmarking well integrity performance should ideally compare performance within similar operating contexts (age, environment, well type, regulatory regime) to draw meaningful conclusions.

29.5 Age-Dependent Failure Patterns and Component Vulnerabilities

While earlier sections have focused on broad statistics and failure prevalence, understanding how well age influences failure modes is essential for optimizing integrity management. Certain well components are more prone to failure as wells mature due to cumulative mechanical stress, environmental exposure, and evolving operational conditions.

Figure 29.4 presents a heatmap visualization of well integrity failure counts, cross-tabulated by component and well age group. Figure 29.4 complements this with a bar chart showing the total failure frequency of well components broken down by age group. These visualizations highlight those certain components—like gas lift valves (GLVs), casing, and cement—that tend to show peak failure counts in specific age ranges, such as the 10–19-year mark. These patterns are crucial for informing lifecycle-based risk assessments, surveillance planning, and targeted equipment upgrades or interventions. In addition to understanding failure types and depth distributions, it is crucial to consider when integrity issues typically arise over the lifetime of a well. Figure 29.5 presents a histogram showing the distribution of the number of years between the final drilling of a well and the first reported well integrity failure. This data provides important insight into the age-dependent reliability of well systems, helping to inform proactive integrity management and the development of long-term monitoring strategies.

29.6 Limitations and Interpretation

When encountering well integrity statistics, it is crucial to approach them with a critical eye and understand their limitations:

- **Understand the Source and Definitions**: Always question where the data came from, who collected it, the methodology used, and precisely how key terms like "failure," "SCP," or "leak" were defined.
- **Consider the Population**: What specific group of wells does the data represent (e.g., all wells in a region, only offshore wells, wells operated by a specific company, wells of a certain age or type)? Is the sample representative?

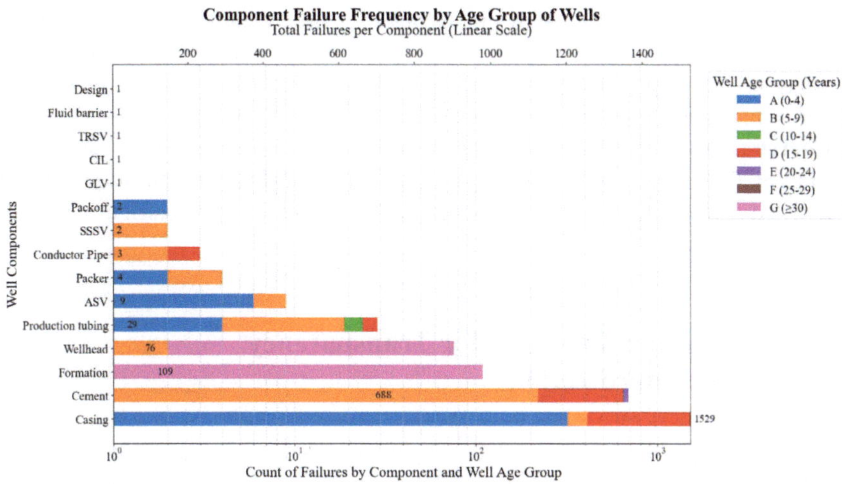

Fig. 29.4 Log-scale bar chart of well component failure frequency by well age group. Colors represent different well age groups, showing how failures cluster for certain components at particular life stages

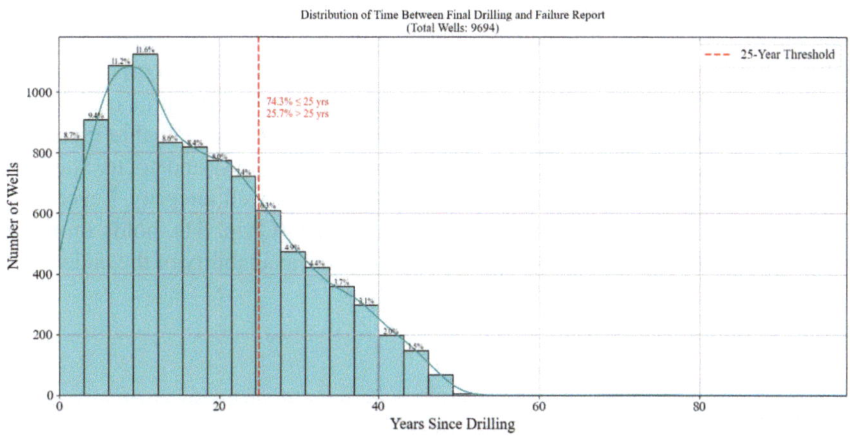

Fig. 29.5 Distribution of the number of years between final drilling and first reported failure for 9,694 wells. The histogram shows a clear peak in failures within the first 10–20 years of a well's life, after which the failure frequency gradually declines. This highlights the critical early lifecycle period for focused integrity monitoring and emphasizes the importance of long-term surveillance to capture later-onset issues

- **Look for Normalization**: Are failure rates presented per well-year, per completion, per meter drilled, or just as raw counts? Normalized data is generally more useful for comparison.
- **Be Aware of Bias**: Recognize potential biases in data collection or reporting (e.g., under-reporting of minor issues, focus on specific failure types).

- **Focus on Trends**: Often, analyzing trends over time within a consistent dataset (even if the absolute numbers have limitations) is more informative than interpreting single-year data points in isolation. Trends can indicate whether performance is improving or worsening.
- **Correlation versus Causation**: Statistics can show correlations between factors (e.g., SCP increasing with well age) but do not inherently prove causation without detailed investigation (e.g., root cause analysis of the failures contributing to the trend).

29.7 Conclusion: The Need for Transparency and Standardization

While challenging to obtain consistently, well integrity statistics are vital for understanding industry performance, identifying systemic risks, supporting risk assessments, and driving targeted improvements in design, procedures, and management systems. Greater transparency and standardized data sharing across the industry (e.g., through anonymized industry-wide databases, standardized reporting formats, or JIPs for data collection and analysis) would be highly beneficial for accelerating learning and improving overall performance. Furthermore, standardization of definitions and reporting criteria related to well integrity failures (e.g., SCP classifications, leak severity levels) would significantly improve the quality and comparability of data globally. Until then, available statistics must be interpreted with a critical eye, understanding their context and limitations, but still utilized as valuable indicators to guide risk management efforts, inform strategic decisions, and focus attention on critical areas for continuous improvement in ensuring well integrity. Leveraging data effectively, including through advanced analytics (Chap. 31), holds immense potential, but it relies fundamentally on the quality and availability of the underlying statistics.

References

Goodman, M. A., Mitchell, R. F., & Kalil, I. A. (2021). API connection leak equation extended with dependence on axial force and backup pressure. *SPE Drilling & Completion, 36*(02), 338–351. https://doi.org/10.2118/194059-PA

Ibukun, M., Elyan, E., Amish, M., Njuguna, J., & Oluyemi, G. F. (2024). A review of well life cycle integrity challenges in the oil and gas industry and its implications for sustained casing pressure (SCP). In *Energies* (Vol. 17, Issue 22). Multidisciplinary Digital Publishing Institute (MDPI). https://doi.org/10.3390/en17225562

Jackson, R. B. (2014). The integrity of oil and gas wells. In *Proceedings of the National Academy of Sciences of the United States of America, 111*(30), 10902. https://doi.org/10.1073/PNAS.1410786111

Saxena, A., Chaudhary, S., Adarsh, D., Anshul, K., Nihalani, D., & Tewari, P. (2019). Challenges in HPHT well: A case study. In *Society of Petroleum Engineers—SPE Oil and Gas India Conference and Exhibition 2019, OGIC 2019*. https://doi.org/10.2118/194633-MS

Chapter 30
Advanced Monitoring and Sensing Technologies

Abstract This chapter presents a comprehensive overview of the advanced monitoring and sensing technologies transforming well integrity surveillance and management. As the limitations of conventional methods—such as periodic pressure readings and discrete logging—become increasingly evident in the face of complex well designs and dynamic operational conditions, the industry is transitioning toward continuous, real-time monitoring systems that deliver greater sensitivity, spatial coverage, and diagnostic capability. At the forefront of this transformation is **fiber-optic sensing**, with Distributed Temperature Sensing (DTS), Distributed Acoustic Sensing (DAS), and Distributed Strain Sensing (DSS) technologies turning the entire wellbore into a continuous sensor array. These systems provide high-resolution data streams for early leak detection; thermal and mechanical stress monitoring; and real-time evaluation of cement curing, flow dynamics, and structural deformation. Complementary technologies like **wireless downhole sensors** are also emerging, offering untethered data acquisition in annuli and deep subsurface zones previously unreachable without compromising barrier integrity. The chapter also explores **advancements in conventional logging tools**, including high-resolution ultrasonics, multi-frequency electromagnetic logs, and through-tubing diagnostics. These innovations improve the accuracy of cement evaluation, corrosion mapping, and leak detection, ensuring the continued relevance of traditional logging in modern well integrity programs. Finally, the chapter highlights the growing role of **drones and robotics** for surface and structural inspections. These platforms enhance safety, reduce inspection costs, and expand coverage for wellheads, risers, and platform structures—critical components of overall well integrity. Through illustrative figures and real-world case studies, the chapter emphasizes the importance of integrating multiple data sources—fiber optics, wireless telemetry, periodic logs, and robotic inspection—into a unified well integrity monitoring strategy. This integrated approach enables earlier anomaly detection, supports predictive maintenance, reduces costly interventions, and enhances decision-making throughout the well's lifecycle. As the industry navigates more challenging environments and energy transition applications, advanced sensing technologies offer a robust foundation for ensuring safe, reliable, and sustainable well operations.

© The Author(s), under exclusive license to Springer Nature Switzerland AG 2026
A. Alsubaih and K. Sepehrnoori, *Well Integrity for Oil, Gas, and Subsurface Storage Operations*, https://doi.org/10.1007/978-3-032-15233-6_30

Wellhead of Obenor 2 oil well, Arnegard, North Dakota, July 2013. *Source* Bureau of Land Management (BLM)/U.S. federal government (2013)

Chapter's Objectives

- Provide a detailed overview of fiber-optic sensing technologies (DTS, DAS, DSS) and their application to well integrity.
- Discuss the potential of wireless downhole sensors.
- Highlight advancements in conventional logging tool technology.
- Mention the role of drones and robotics in surface and structural inspection.

30.1 Introduction: Beyond Conventional Monitoring

Traditional well integrity monitoring has primarily relied on surface pressure measurements from gauges or transmitters and periodic downhole logging runs, as covered in earlier (Chaps. 8 and 12). While these methods are valuable, they only provide discrete, infrequent snapshots of well conditions, often missing transient events or localized anomalies that can jeopardize well integrity over time.

The past decade has witnessed a dramatic evolution in monitoring and sensing technologies. Driven by the need for more proactive, predictive, and continuous integrity surveillance, operators and service companies have increasingly embraced advanced sensing systems that provide real-time or near-real-time data across entire well sections. These cutting-edge technologies, including fiber-optic sensing, wireless downhole sensors, enhanced conventional logging tools, and the integration of drones and robotics, represent a significant paradigm shift in well integrity management.

By capturing a wealth of high-resolution data streams from multiple sources, these technologies enable earlier detection of well integrity anomalies, more precise localization of leaks or structural weaknesses, and a deeper understanding of downhole conditions under dynamic operational stresses. They also support more informed, data-driven decision-making, reducing reliance on costly and intrusive interventions and ultimately improving well performance and safety. This chapter delves into the core principles and applications of these advanced monitoring and sensing technologies. It begins with an in-depth discussion of fiber-optic sensing technologies—highlighting how they transform an entire wellbore into a continuous sensor network. It then explores the promise and challenges of wireless downhole sensors for barrier integrity and annulus pressure monitoring. The chapter also examines how conventional logging tools are advancing to offer improved measurements and diagnostics. Finally, it considers the emerging role of drones and robotic inspection systems in surface integrity assurance, extending the well integrity paradigm beyond the subsurface. Together, these technologies form the foundation of a comprehensive, integrated well integrity monitoring strategy, poised to meet the challenges of modern energy production and storage operations.

30.2 Fiber-Optic Sensing

Permanently installed fiber-optic cables, typically strapped to the outside of casing or tubing strings during completion or deployed inside capillary tubes, enable distributed sensing. This revolutionary approach treats the entire length of the fiber-optic cable as a continuous linear sensor, providing data at thousands of points along its path. The technology relies on analyzing laser light backscattered along the fiber, which is affected by external conditions. A laser pulse is launched into the sensing fiber, where it interacts with the silica-based core's crystalline structure, generating spontaneous scattering. The glass structure of the fiber responds to external influences like thermal and pressure changes (including noise and sound), which induce oscillations in the core. Some of the scattered photons are captured by the fiber's guided modes and propagate back toward the interrogator, where they are detected by a fast photodetector. This scattering results from the interaction between the laser light and density fluctuations within the fiber, producing a backscattered spectrum composed of Rayleigh, Brillouin, and Raman bands. The three primary fiber-optic measurement techniques used today are Distributed Temperature Sensing (DTS), Distributed Acoustic Sensing (DAS), and Distributed Strain Sensing (DSS). DAS relies on Rayleigh scattering, DTS uses Raman scattering, and DSS utilizes Brillouin scattering. Variations in the backscattered light can thus be correlated to acoustic and thermal changes along the fiber. Distributed sensing systems operate through optical time domain reflectometry (OTDR), which locates local losses in the fiber. By knowing the propagation time of a pulse at a particular wavelength along a fiber with a given refractive index, the interaction's position can be pinpointed. When applied to sensing, this technique allows the corresponding measurement perturbations along the fiber to be identified (Feo et al., 2020a, 2020b). Figure 30.1 illustrates the fundamental interactions between the launched laser light and the fiber core's silica structure, highlighting the mechanisms of backscattering that enable distributed sensing. It shows how Rayleigh, Brillouin, and Raman backscattered signals are generated and captured, with distinct dependencies on temperature, strain, or both. These interactions form the scientific basis for the three main distributed sensing techniques—DTS, DAS, and DSS—that deliver real-time insights into downhole wellbore conditions. Key technologies include the following:

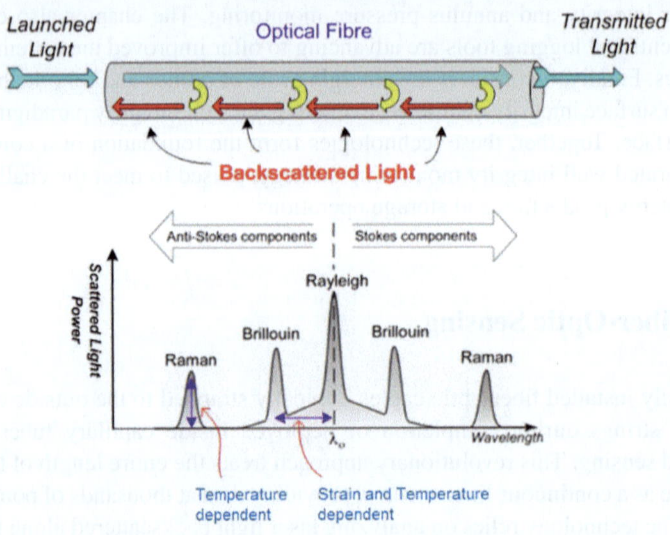

Fig. 30.1 Schematic of backscattering phenomena in optical fiber sensing. Launched light propagates through the fiber, with some photons backscattered by interactions with density and structural variations in the silica core. The scattered light spectrum includes Rayleigh, Brillouin, and Raman peaks, each sensitive to different physical parameters. Rayleigh scattering is used in Distributed Acoustic Sensing (DAS), Brillouin scattering for Distributed Strain Sensing (DSS), and Raman scattering for Distributed Temperature Sensing (DTS). The figure also highlights how these scattering components relate to temperature and strain dependencies (Soga et al., 2008)

- **Distributed Temperature Sensing (DTS)**
  - Principle: Measures temperature at thousands of points along the fiber-optic cable. Based on analyzing backscattered laser light (Raman scattering), whose intensity ratio at different wavelengths varies predictably with temperature (Ekechukwu & Sharma, 2021).
  - Applications in Well Integrity: Leak detection, cement cure monitoring, flow profiling, and thermal stress monitoring.
  - Advantages: Continuous temperature profile along the entire instrumented section, real-time or frequent data updates, relatively mature technology with field-proven reliability in many environments.
- **Distributed Acoustic Sensing (DAS)**
  - Principle: Measures dynamic strain variations (vibrations or acoustic energy) along the fiber caused by acoustic waves (sound propagating through the wellbore or formation). Based on analyzing coherent backscattered laser light (Rayleigh scattering). Effectively turns the fiber into a continuous array of microphones or hydrophones (Rashid et al., 2025).
  - Applications in Well Integrity: Leak detection, flow monitoring, well operations monitoring, and seismic monitoring.
  - Advantages: High sensitivity to dynamic events, continuous spatial coverage, broad frequency response, and potential for multiple simultaneous applications (leak detection, flow, seismic).
- **Distributed Strain Sensing (DSS)**
  - Principle: Measures changes in axial strain (elongation or compression) along the fiber. Based on analyzing Brillouin or Rayleigh backscattering, whose frequency/phase shifts are sensitive to both strain and temperature. Requires careful coupling of the fiber to the structure being monitored (casing/tubing) to transfer strain effectively (Liu et al., 2024).
  - Applications in Well Integrity: Structural integrity monitoring (deformation, buckling), pressure change detection, and fracture monitoring.
  - Advantages: Provides direct measurement of mechanical strain, crucial for assessing structural integrity.
- **Installation and Challenges**: Fiber-optic cables can be strapped to the outside of casing/tubing during installation using specialized clamps, pumped inside small-diameter capillary tubes, or integrated into specialized completion components or coiled tubing. Challenges include protecting the fiber and couplings during installation in the harsh downhole environment, ensuring long-term survivability and integrity of the fiber and any feed-throughs at packers or wellheads, managing the large volumes of data generated, and developing robust interpretation algorithms and workflows to translate raw data into actionable integrity insights. Defining standard interpretation guidelines and alarm thresholds for these complex data streams is an ongoing effort.

Fiber-optic sensing offers a paradigm shift from periodic logging to continuous, real-time or near-real-time surveillance of the entire instrumented wellbore section, enabling earlier detection of integrity anomalies, providing richer diagnostic information about the nature and location of issues, and supporting proactive decision-making within the WIMS. Figure 30.2 highlights the versatility and depth of information provided by permanently installed fiber-optic sensing systems in modern well integrity and production monitoring. From seismic profiling and hydraulic fracture analysis to flow monitoring, casing leak detection, and gas lift optimization, these distributed sensors enable continuous, high-resolution measurements along the entire length of the wellbore. The figure demonstrates how the optical fiber network captures dynamic and static data, transforming the wellbore into a rich data source for integrity management and operational optimization. Figure 30.3 illustrates the application of Raman-based fiber-optic Distributed Temperature Sensing (DTS) in identifying inflow points within a wellbore. The temperature profile, measured along the fiber-optic cable, enables detection of fluid entry and production zones, as indicated by distinct deviations in the wellbore temperature gradient. This approach provides a continuous, high-resolution temperature log that enhances understanding

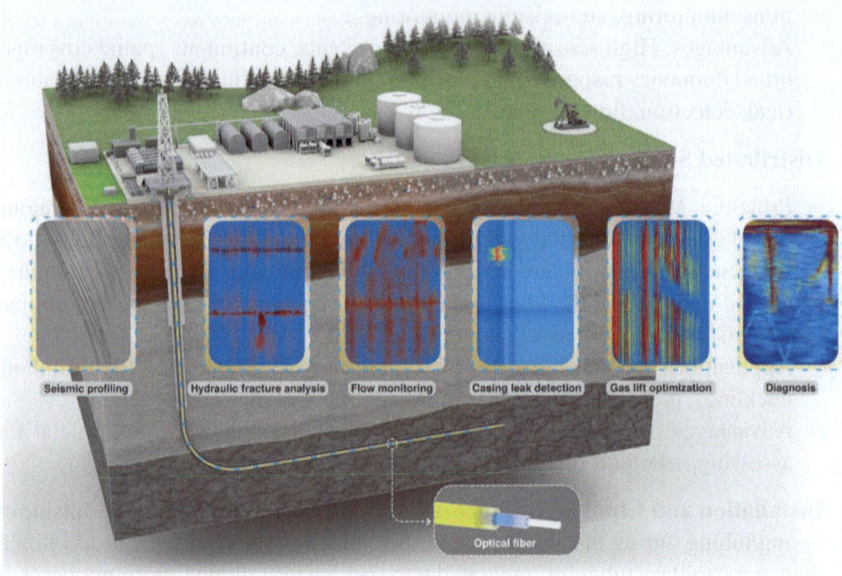

Fig. 30.2 Schematic illustrating the multi-functional applications of fiber-optic sensing in wellbore integrity and performance monitoring. Key applications include seismic profiling, hydraulic fracture analysis, flow monitoring, casing leak detection, gas lift optimization, and diagnostic analyses. The optical fiber (shown in yellow and blue) is permanently deployed along the wellbore, enabling real-time distributed sensing and providing a paradigm shift from conventional discrete measurements to continuous, integrated well integrity monitoring (Ashry et al., 2022)

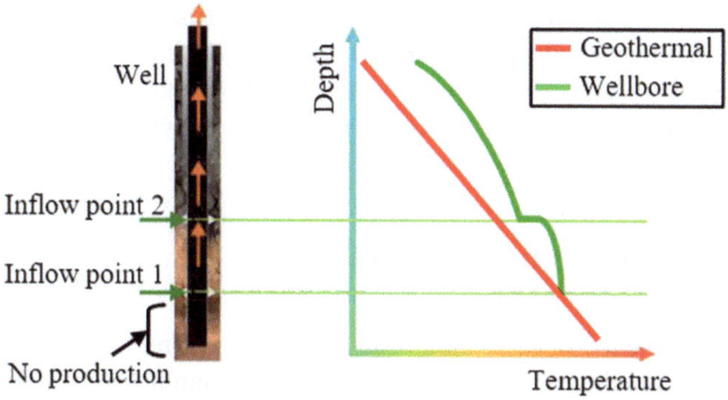

Fig. 30.3 Example of using Raman-based fiber-optic DTS to locate inflow points in a wellbore. The temperature profile reveals two inflow points (green lines), while the red and green curves represent geothermal and wellbore temperature gradients, respectively. The optical fiber enables continuous and accurate detection of dynamic temperature changes associated with fluid entry (Ashry et al., 2022)

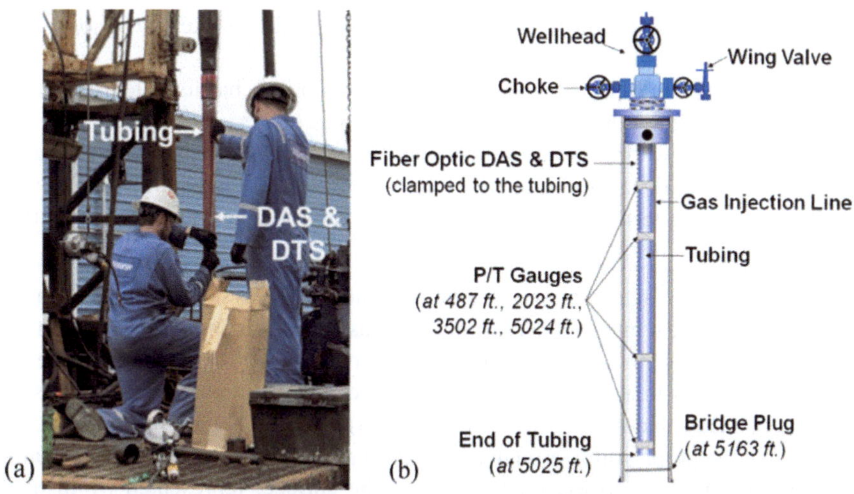

Fig. 30.4 Real-world application of fiber-optic DAS and DTS systems in well integrity monitoring. **a** Field technicians are preparing the tubing with fiber-optic cables clamped securely in place. **b** Schematic of a typical well completion showing the fiber-optic cables along the tubing string, surface wellhead equipment, P/T gauges at various depths, gas injection lines, and bridge plugs. The clamped fiber-optic cables provide continuous, real-time data streams for temperature, strain, and acoustic variations along the entire wellbore (Ekechukwu & Sharma, 2021)

of well performance and integrity (Ashry et al., 2022). Figure 30.4 offers a practical illustration of fiber-optic Distributed Acoustic Sensing (DAS) and Distributed Temperature Sensing (DTS) installation in a real-world well completion. Panel (a) shows field technicians attaching the fiber-optic cables to the tubing before deployment. Panel (b) provides a schematic view of the well completion with fiber-optic cables clamped to the tubing and connected to surface equipment, alongside critical components like pressure/temperature (P/T) gauges, gas injection lines, and bridge plugs. This setup underscores the engineering complexity and integration challenges of deploying fiber-optic sensing systems for continuous well integrity monitoring.

30.3 Wireless Downhole Sensors

Traditional permanent downhole monitoring systems often rely on wired connections (electric lines, hydraulic control lines, capillary tubes) running from the surface to the sensors downhole. These lines involve penetrations through packers, hangers, and wellheads—which themselves represent potential leak paths. Wireless technologies aim to overcome this limitation by allowing downhole sensors to communicate data to the surface without a physical wire or tube connection.

- **Principle**: Sensors (measuring pressure, temperature, flow, vibration, etc.) installed downhole collect data and communicate it to the surface using various non-wired methods: acoustic telemetry, electromagnetic (EM) telemetry, or within intelligent completions.
- **Applications in Well Integrity**: Annulus monitoring, barrier integrity monitoring, and potential long-term P&A monitoring.
- **Challenges**: Powering the downhole sensors, achieving reliable data transmission over long distances through complex and variable media, sensor survivability and long-term drift in harsh downhole environments, and cost compared to conventional methods.

While still evolving and facing technical hurdles, wireless sensing holds promise for providing critical integrity data from locations previously impossible to monitor directly, improving the understanding of barrier status, and reducing reliance on wired penetrations that can introduce potential leak paths. Figure 30.5 presents an example of a Distributed Acoustic Sensing (DAS) waterfall plot, capturing dynamic acoustic energy propagating through the wellbore. The plot depicts how gas migrates through the annulus and eventually reaches the surface, represented by distinct acoustic signals that change over time and depth. DAS offers continuous, real-time monitoring of dynamic events such as gas migration, flow anomalies, and potential integrity threats, making it a powerful tool for proactive well integrity surveillance (Feo et al., 2020a, 2020b). Figure 30.6 illustrates a multi-dimensional analysis of pressure, strain, and fracture width evolution during wellbore stimulation, highlighting the capabilities of Distributed Strain Sensing (DSS). The center panel shows the spatial–temporal distribution of strain along the sensing fiber, revealing dynamic changes in response to stimulation events. The integration of strain data with pressure logs and fracture width measurements provides a holistic view of wellbore deformation processes, essential for assessing well integrity during fracturing operations.

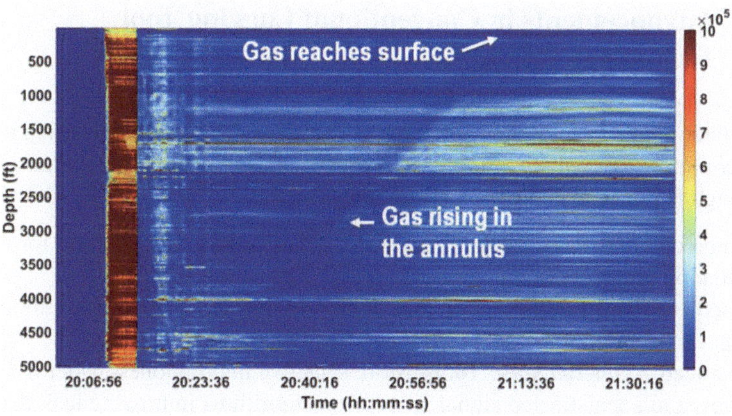

Fig. 30.5 DAS waterfall plot (frequency range 10–50 Hz) for Trial-2, illustrating gas migration in the well annulus and eventual surface emergence. Color intensities represent the magnitude of acoustic energy, showing the powerful diagnostic capabilities of real-time distributed sensing (Feo et al., 2020a, 2020b)

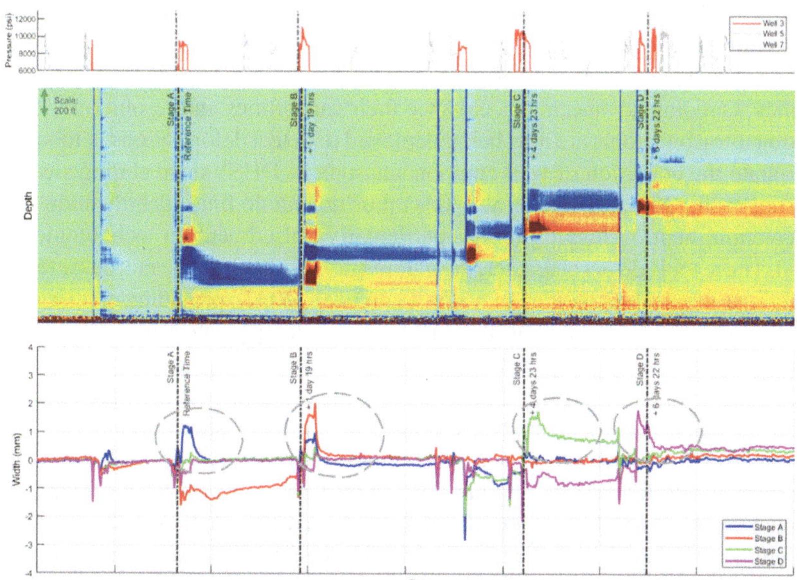

Fig. 30.6 Multi-panel view of well stimulation monitoring using DSS. (Top) Pressure distribution along the wellbore during stimulation. (Center) Spatio-temporal distribution of temperature-compensated strain (blue: compressive, red: tensile, green: small magnitude). The y-axis represents distance along the fiber, while the x-axis is time. (Bottom) Time series of fracture width changes, highlighting the dynamic strain responses of the wellbore to stimulation events. Dotted vertical lines mark the initiation of each stimulation stage, providing crucial insights into well integrity and deformation during hydraulic fracturing (Saw et al., 2023)

30.4 Advancements in Conventional Logging Tool Technology

While fiber optics and wireless are newer paradigms, conventional wireline logging tools continue to evolve, offering improved measurements and diagnostics for assessing the physical condition of well components.

- **Enhanced Ultrasonic Tools**: Improved resolution and interpretation for cement evaluation and corrosion mapping.
- **Pulsed Neutron Logging**: Provides behind-casing fluid information to identify channels or migration.
- **Advanced Acoustic/Noise Tools**: More sensitive microphones, wider frequency ranges, and sophisticated signal processing algorithms improving leak detection and characterization.
- **High-Resolution Calipers**: More detailed mapping of internal pipe diameter and defects.
- **Combined Toolstrings**: Integrating multiple measurements onto a single toolstring for improved efficiency and correlation.

Conventional logging remains a vital part of periodic integrity assessment, providing direct measurements of pipe dimensions, cement presence, and flow anomalies. Ongoing advancements enhance their capabilities and resolution. High-resolution ultrasonic tools are now being deployed to detect deformations in tubulars and evaluate the condition of well components such as DHSVs and sliding sleeves (Littleford et al., 2022; Pehlke et al., 2024). For downhole fluid identification and leak detection, neutron logging tools—particularly pulsed-neutron technologies—are used. These tools detect characteristic gamma rays emitted when neutrons interact with target atoms such as oxygen in water (Galli et al., 2022), providing precise data on fluid type and location behind the casing. For detecting internal casing corrosion, caliper logs remain a key method. However, new electromagnetic flux logging tools can now measure corrosion in the outer casing, even across multiple concentric casing strings—advancing behind-casing integrity diagnostics (Zeghlache et al., 2021). Imrie et al. (2021) also showed that combinations of through-tubing logging tools can precisely locate leaks. Figure 30.7 showcases a comparative example of traditional cement bond logging (CBL) and distributed fiber-optic sensing (DFOS) temperature profiling. It demonstrates how fiber-optic DTS can reveal dynamic temperature variations along the wellbore, providing continuous, real-time insights that complement static CBL data. The integrated approach helps identify cement integrity issues, zonal isolation effectiveness, and potential fluid migration pathways more comprehensively.

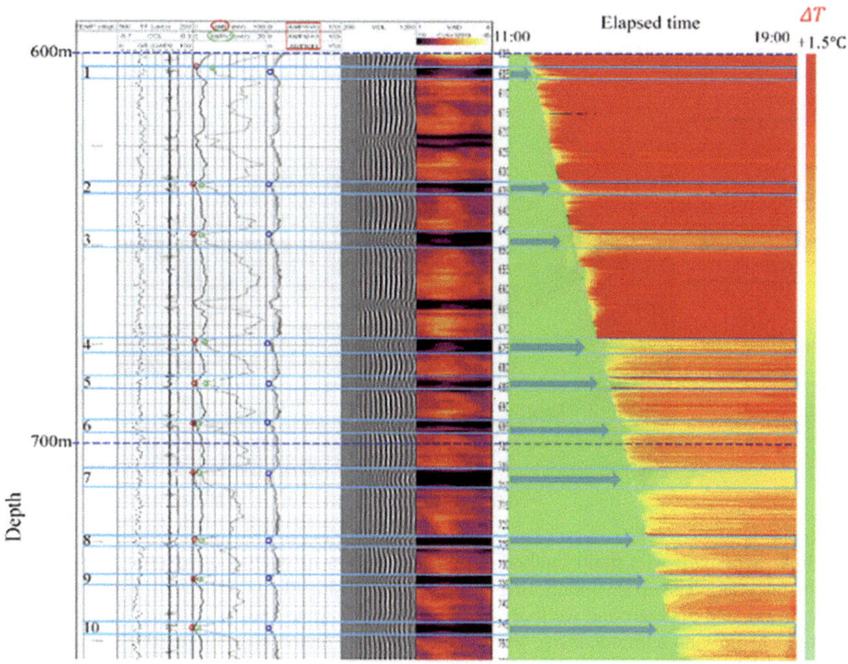

Fig. 30.7 Comparison of radial cement bond log (CBL) measurements and distributed fiber-optic sensing (DFOS) temperature profiles (adapted from Sun et al., 2018). The figure highlights how DFOS profiles (right panel) provide continuous temperature data along the wellbore, complementing discrete CBL measurements (left panel). This combined approach enhances the understanding of cement sheath integrity, fluid movement, and thermal behavior in the well (Sun et al., 2018)

30.5 Drones and Robotics in Inspection

While the focus is often on downhole integrity, managing overall well integrity also involves surface facilities and structures (wellheads, Christmas trees, production manifolds, platforms, risers). Drones (Unmanned Aerial Vehicles—UAVs) and robotic crawlers/ROVs are increasingly used for safer, more efficient, and often more comprehensive inspection of these surface and structural assets, particularly in hard-to-reach or hazardous areas. These technologies enhance the efficiency, safety, and coverage of surface integrity monitoring and inspection programs, complementing downhole integrity management by ensuring the integrity of the systems connected to the wellbore. Figure 30.8 illustrates a comprehensive, multi-method monitoring system that integrates data from various technologies and subsystems. It highlights the complementary roles of aerospace monitoring (e.g., InSAR satellites, UAVs), surface monitoring (e.g., visual inspections, drones), and underground (wellbore) monitoring (e.g., micro-seismic, VSP, and fiber-optic sensors). This multi-tiered approach provides a holistic view of well and reservoir integrity, extending beyond

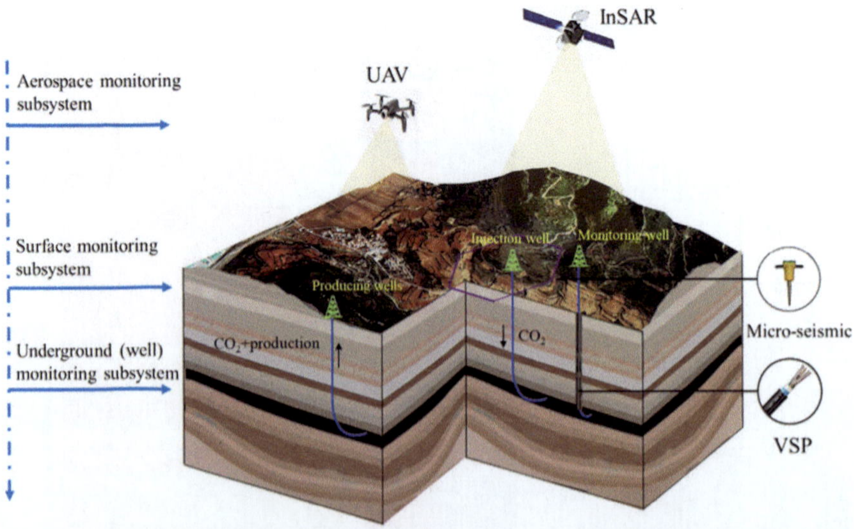

Fig. 30.8 Schematic of a multi-method monitoring system for well and reservoir integrity. The system integrates aerospace monitoring (UAVs, InSAR satellites), surface monitoring (drones and sensors), and underground wellbore monitoring (fiber optics, micro-seismic, vertical seismic profiling) to provide comprehensive surveillance of CO_2 injection wells, producing wells, and surrounding formations

individual measurements to encompass the entire system—surface to subsurface—enabling proactive and data-driven management of complex wells and reservoirs. The future of well integrity monitoring lies in integrating data from multiple advanced sources—continuous fiber optics, targeted wireless sensors, high-resolution periodic logs, and robotic inspections—providing a more complete and timelier picture of the well's health.

References

Ashry, I., Mao, Y., Wang, B., Hveding, F., Bukhamsin, A., Ng, T. K., & Ooi, B. S. (2022). A review of distributed fiber-optic sensing in the oil and gas industry. *Journal of Lightwave Technology, 40*(5), 1407–1431. https://doi.org/10.1109/JLT.2021.3135653

Ekechukwu, G. K., & Sharma, J. (2021). Well-scale demonstration of distributed pressure sensing using fiber-optic DAS and DTS. *Scientific Reports, 11*(1), 1–18. https://doi.org/10.1038/S41598-021-91916-7 (SUBJMETA = 1075, 1107, 166, 4077, 4082, 624, 639; KWRD = APPLIED+OPTICS,ENGINEERING, FOSSIL+FUELS, OPTICAL+TECHNIQUES)

Feo, G., Sharma, J., Kortukov, D., Williams, W., & Ogunsanwo, T. (2020). Distributed fiber optic sensing for real-time monitoring of gas in riser during offshore drilling. *Sensors, 20*(1), 267. https://doi.org/10.3390/S20010267

Feo, G., Sharma, J., Santos, O., Williams, W., & Ogunsanwo, T. (2020b). Multiphase flow characterization and modeling using distributed fiber optic sensors to prevent well blowout. *Optics InfoBase Conference Papers*. https://doi.org/10.1364/ES.2020.EM3C.5

Galli, G., Pirrone, M., Pirola, N., Parodi, L., Eni, Guiducci, M., & Kozlowski, M. (2022). Exploring the downhole waterways: Identification of the sneaky path of water through the well completion. https://doi.org/10.30632/SPWLA-2022-0060

Imrie, A., Kozlowski, M., Torky, O., & Wijaya, A. A. (2021). Full well corrosion insight—Case studies in the added value of electromagnetic thickness measurements during well interventions. In *SPE/ICoTA Well Intervention Conference and Exhibition 2021, CTWI 2021*. https://doi.org/10.2118/204431-MS

Littleford, T., Simpson, G., & Halpenny, M. (2022). High-resolution acoustic imaging and multi-axis robotics: Three-dimensional downhole obstruction evaluation in real time. In *Society of Petroleum Engineers—SPE Hydraulic Fracturing Technology Conference and Exhibition, HFTC 2022* (pp. 1–3). https://doi.org/10.2118/209142-MS

Liu, L., Guo, X., & Wang, X. (2024). Integrated wellbore-reservoir-geomechanics modeling for enhanced interpretation of distributed fiber-optic strain sensing data in hydraulic-fracture analysis. *Journal of Rock Mechanics and Geotechnical Engineering, 16*(8), 3136–3148. https://doi.org/10.1016/J.JRMGE.2023.09.027

Obenor 2 oil well–wellhead–Arnegard North Dakota–2013-07-04. (2013). Wikimedia Commons. Public domain (U.S. government work). https://commons.wikimedia.org/wiki/File:Obenor_2_oil_well_-_wellhead_-_Arnegard_North_Dakota_-_2013-07-04.jpg

Pehlke, T., Simpson, G., Roberts, A., Maxwell, T., & Chu, J. (2024). High-resolution acoustic imaging deployed for mapping permanent fiber optic lines. In *Society of Petroleum Engineers—SPE Hydraulic Fracturing Technology Conference and Exhibition, HFTC 2024*. https://doi.org/10.2118/217819-MS

Rashid, A., Tackie-Otoo, B. N., Latiff, A. H. A., Otchere, D. A., Jamaludin, S. N. F., & Asfha, D. T. (2025). Research advances on distributed acoustic sensing technology for seismology. *Photonics, 12*(3), 196. https://doi.org/10.3390/PHOTONICS12030196

Saw, J., Luo, L., Correa, J., Soga, K., Zhu, X., Ajo-Franklin, J., Kerr, E., & Bohn, R. (2023). Hydraulic fracture stage identification and size estimation using distributed strain and temperature sensing. In *SEG Technical Program Expanded Abstracts, 2023-August* (pp. 469–473). https://doi.org/10.1190/IMAGE2023-3915606.1

Soga, K., Mohamad, H., Bennett, P. J., & Soga Hisham Mohamad Peter Bennett, K. J. (2008). Distributed fiber optics strain measurements for monitoring geotechnical structures. In *International Conference on Case Histories in Geotechnical Engineering* (Vol. 11, pp. 55–67). https://scholarsmine.mst.edu/icchge/6icchge/session14/4

Sun, Y., Xue, Z., & Hashimoto, T. (2018). Fiber optic distributed sensing technology for real-time monitoring water jet tests: Implications for wellbore integrity diagnostics. *Journal of Natural Gas Science and Engineering, 58*, 241–250. https://doi.org/10.1016/J.JNGSE.2018.08.005

Zeghlache, M. L., Bazaid, A., Hill, F., & Guergueb, N. (2021). Comprehensive casing corrosion inspection using multi-frequency array EM technology. In *Society of Petroleum Engineers—SPE International Oilfield Corrosion Conference and Exhibition 2021, OFCS 2021* (pp. 16–17). https://doi.org/10.2118/205050-MS

Chapter 31
Digitalization, AI, and Predictive Analytics in Well Integrity

Abstract This chapter explores how digital transformation, artificial intelligence (AI), and predictive analytics are reshaping the landscape of well integrity management. With the proliferation of real-time sensors, fiber-optic monitoring systems, and advanced logging tools, the oil and gas industry is generating vast volumes of data across all phases of the well lifecycle. Harnessing this "Big Data" presents both immense opportunities and complex challenges. When managed effectively, it enables a shift from reactive and scheduled integrity practices toward proactive, predictive strategies that can significantly improve safety, reduce costs, and extend well life. The chapter highlights the core applications of **AI and machine learning** in well integrity, including predictive maintenance, corrosion forecasting, early anomaly detection, dynamic risk assessment, and automated interpretation of complex logs. These techniques empower operators to anticipate failures and intervene before they escalate into costly or high-risk events. Central to this transformation is the deployment of **Digital Twins**—dynamic, virtual replicas of physical wells that integrate static and real-time data, physics-based models, and AI outputs into interactive dashboards. Digital twins facilitate enhanced diagnostics, scenario simulations, predictive modeling, and optimized operations, offering a holistic and adaptive framework for integrity management. To support these technologies, robust **integrated data platforms** are essential. The chapter underscores the importance of breaking down data silos, ensuring data quality, and using intuitive visualization tools to transform complex outputs into actionable insights for engineers and decision-makers. Moreover, the chapter introduces the concept of **Smart Fields**, where real-time control, automation (e.g., SCADA and PLCs), and AI-driven optimization converge to enable adaptive production and well integrity assurance. These smart systems can dynamically respond to integrity threats, such as SCP, annular pressure build-up, or thermal cycling, and even integrate geomechanical models to manage risks in high-pressure injection scenarios like CO_2 or H_2 storage. Through case studies, system architectures, and maturity models, this chapter presents a forward-looking vision where digitalization is not simply a technological trend but a strategic imperative for well integrity. By integrating AI, digital twins, and smart field technologies, operators can unlock new levels of operational intelligence, resilience, and sustainability in an increasingly complex energy environment.

© The Author(s), under exclusive license to Springer Nature Switzerland AG 2026

A. Alsubaih and K. Sepehrnoori, *Well Integrity for Oil, Gas, and Subsurface Storage Operations*, https://doi.org/10.1007/978-3-032-15233-6_31

Drilling bit at the rig floor during drilling operations, representing the integration of real-time data acquisition for predictive well integrity monitoring. *Source* The Authors

Chapter's Objectives

- Discuss the opportunities and challenges presented by Big Data in well integrity.
- Explain the potential applications of Artificial Intelligence (AI) and Machine Learning (ML) for predictive integrity management.
- Describe the concept and application of Digital Twins for wells.
- Highlight the role of integrated data platforms and visualization tools.

31.1 The Digital Transformation and Big Data

The oil and gas industry is undergoing a significant digital transformation, driven by increased sensor deployment, connectivity, and computing power. This transformation is generating vast amounts of data—often referred to as "Big Data"—that are highly relevant to well integrity management. Data sources include the following:

- Real-time sensor data (surface and downhole pressure, temperature, flow rates, vibration, data from fiber optics like DTS, DAS, DSS).
- Periodic inspection and logging data (ultrasonic scans, EM thickness logs, caliper logs, cement evaluation logs).
- Operational history (production/injection rates and volumes, pressure/temperature trends, shutdowns, startups, operational events, intervention history).

- Testing records (valve functional tests, leak tests, annulus pressure tests, pressure test charts).
- Design and construction data (basis of design, casing/tubing specifications, cement slurry design, placement records, initial test results, "as-built" documentation).
- Maintenance records (preventative and corrective maintenance performed).
- Failure and anomaly reports (SCP occurrences, leaks, equipment malfunctions, investigation reports).
- External data (geological models, seismic data, environmental data).

Opportunities: This wealth of data, if properly managed and analyzed, offers unprecedented opportunities to move from reactive (addressing issues after they occur) or purely preventative (scheduled maintenance/inspection) integrity management toward more predictive and prescriptive approaches. Analyzing data can allow operators to anticipate failures before they occur, optimize operations to minimize integrity risks, and recommend the most effective actions. Leveraging this data is key to improving efficiency, reducing costs, and enhancing safety and environmental performance.

Challenges: Despite the potential, managing well integrity in Big Data presents significant challenges:

- Data Volume and Velocity: Handling the sheer volume and speed of data generated, especially from high-frequency, continuous sensors like fiber optics. Requires robust data infrastructure and processing capabilities.
- Data Variety: Integrating and making sense of diverse data types (structured numerical data from databases, unstructured text reports, log files, images, videos). Requires data cleansing, standardization, and contextualization.
- Data Quality and Silos: Ensuring data accuracy, completeness, and consistency. Data is often stored in disparate systems and legacy databases ("data silos"), making it difficult to access, integrate, and analyze holistically (as discussed in Chap. 15).
- Analytical Expertise: Requiring personnel skilled in data science, AI/ML techniques, and deep domain knowledge (well integrity engineering, operations) to develop, validate, interpret, and apply analytical models effectively.
- Cybersecurity: Protecting sensitive operational and integrity data from cyberthreats.

Figure 31.1 presents a conceptual overview of how digital twin technology, real-time monitoring, and AI-based adaptability are integrated into a single operational ecosystem. Although developed for the context of space exploration, the core principles—real-time status detection, simulation of environmental adaptability, remote task planning, and energy management—parallel the ambitions of digital transformation in the energy industry. In well integrity management, similar frameworks of integrated data platforms, digital twins, AI/ML algorithms, and advanced visualization tools enable a holistic view of well performance, early anomaly detection, and proactive decision-making for safer and more efficient operations.

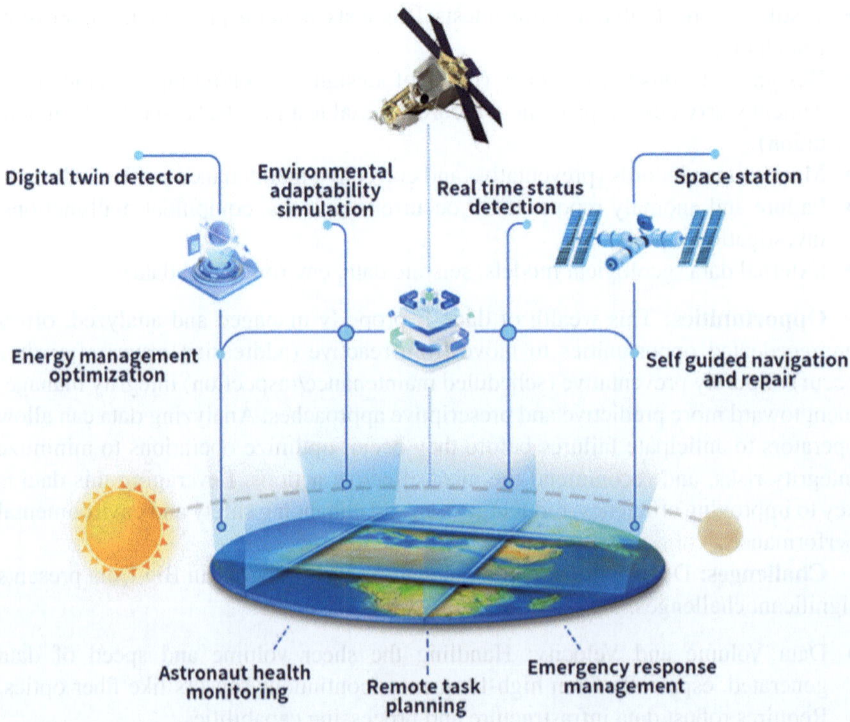

Fig. 31.1 Conceptual illustration of a fully integrated digital system combining digital twins, real-time monitoring, simulation capabilities, and adaptive control—demonstrating a model that parallels digital transformation efforts in well integrity management. The figure underscores the importance of combining data streams, predictive modeling, and human decision-making support to manage complex operational environments effectively

31.2 Artificial Intelligence (AI) and Machine Learning (ML) Applications

AI/ML algorithms, particularly machine learning, are well suited to identifying complex patterns, correlations, and anomalies in large, multivariate datasets that may not be apparent through traditional analysis methods or simple thresholds. This makes them powerful tools for well integrity management. Key applications include the following:

- **Predictive Maintenance**
  - Component Failure Prediction: Training ML models on historical component failure data, real-time sensor readings (e.g., pressure, temperature, vibration, cycling counts), and operational parameters to predict the Remaining Useful Life (RUL) of critical components like valves, packers, or ESPs. This allows

maintenance or component replacement to be scheduled proactively before failure occurs, reducing unplanned downtime and intervention costs.
  - Corrosion Rate Prediction: Developing data-driven models that incorporate real-time operational data (pressure, temperature, flow rate), fluid properties (composition, water cut, pH), and potentially sensor data (e.g., from corrosion probes) to provide more accurate, dynamic predictions of corrosion rates compared to static physics-based models.

- **Anomaly Detection**
  - Early Leak Detection: Using ML algorithms to analyze subtle changes or deviations in pressure trends (e.g., SCP build-up signatures that don't fit expected thermal patterns), temperature profiles (DTS), acoustic signals (DAS), or combinations of sensor data that deviate from established normal operating behavior patterns. This can potentially indicate an incipient leak or barrier degradation earlier than conventional threshold-based alarms, allowing for timely intervention.
  - Equipment Malfunction Detection: Identifying abnormal operating signatures of equipment (e.g., pumps, valves, actuators) that may precede outright failure.
- **Risk Assessment Improvement:** Incorporating historical failure data, real-time monitoring inputs, and predictive model outputs into ML models to provide more dynamic and potentially more accurate assessments of well integrity risk levels for individual wells or assets.
- **Optimizing Inspectionand Testing:** Using predictive models to forecast the likelihood of degradation or failure in specific wells or components based on their predicted risk. This allows for optimization of inspection logging and integrity testing frequencies based on predicted risk rather than fixed time intervals (moving toward a truly risk-based inspection—RBI—approach on steroids, as discussed in Chap. 12).
- **Automated Log Interpretation:** Training AI models (particularly using image recognition and pattern recognition algorithms) to assist in the interpretation of complex logs like ultrasonic cement evaluation images (identifying channels, bond quality variations) or multi-finger caliper data (quantifying pitting, wear). This can potentially improve consistency and speed of interpretation.

Implementation Considerations: Requires high-quality, appropriately labeled training data, careful feature engineering (selecting the right input variables and representations), appropriate algorithm selection, rigorous model validation against real-world outcomes, continuous model retraining and monitoring (managing model drift), ensuring transparency ("explainable AI"—understanding *why* a prediction is made) for building trust among engineers and operators, and seamless integration with existing operational workflows and WIMS processes.

31.3 Digital Twins for Wells

A digital twin is a dynamic virtual representation of a physical asset or system, in this case, an individual well. It integrates multiple sources of data and models to mirror the well's current structure, historical behavior, and dynamic condition in near-real-time. Digital twin technology has emerged as a transformative tool in well integrity management, integrating data streams, real-time monitoring, and physics-based simulations to provide a comprehensive view of well health. For example, the use of digital twins in underground gas storage (UGS) wells allows operators to simulate the evolution of leaks, evaluate intervention strategies without physically entering the field, and optimize maintenance plans to reduce environmental risks and operational costs (Abdo et al., 2023). This approach leverages real-time data from sensors and monitoring tools, feeding it into dynamic models that can replicate the behavior of the well system. By incorporating predictive analytics, digital twins can estimate future operating conditions and failure probabilities, enabling proactive maintenance and intervention planning (Straw et al., 2022).

Furthermore, digital twins not only enhance integrity diagnostics but also enable scenario-based planning and risk assessment. Recent studies highlight how digital twins can be integrated with real-time monitoring of subsea assets, identifying degradation mechanisms and assessing structural integrity using advanced simulations. By combining high-fidelity physics models, data-driven algorithms, and artificial intelligence, these digital twin frameworks offer predictive insights into well integrity status, guide inspection scheduling, and support adaptive decision-making in well integrity management (Anjos et al., 2020; Busollo et al., 2021). This data-driven, physics-informed digital twin approach underscores the future of well integrity management, where digital systems will increasingly drive operational excellence and environmental stewardship.

- **Components:** A well-designed digital twin typically combines
  - Static Data: Well-designed specifications, construction details ("as-built"), geological models, material properties.
  - Dynamic Data: Real-time or near-real-time sensor readings (pressure, temperature, flow, DTS, DAS, DSS), operational data, and periodic inspection/test results.
  - Physics-Based Models: Engineering models simulating wellbore flow, heat transfer, stress distribution in tubulars and cement, corrosion processes, and geomechanical interactions.
  - Data Analytics/AI Models: Predictive models for failure, anomaly detection algorithms, and root-cause analysis helpers (as described above).
  - Visualization Interface: User-friendly dashboard displaying key parameters, model outputs, alerts, KPI status, and often a 3D visualization of the wellbore schematic overlaid with relevant data (e.g., pressure profiles, temperature anomalies, corrosion maps).

- **Applications in Well Integrity**

- Enhanced Visualization and Situation Awareness: Provides an integrated, intuitive view of the well's status and integrity barriers in real time, helping engineers and operators quickly grasp complex information.
- Improved Diagnostics: Combining real-time data with physics models and AI allows for more sophisticated analysis to better understand the root cause of observed anomalies (e.g., simulating different leak scenarios to see which best matches observed DTS/DAS signatures or pressure build-up patterns).
- Scenario Planning ("What-if" Analysis): Allowing engineers to simulate the potential impact of planned operational changes (e.g., increasing flow rate, changing injection parameters, different shutdown/startup sequences) on integrity parameters (e.g., thermal stresses, corrosion rates, potential for scale formation) *before* implementing them in the real well.
- Predictive Integrity Management: Integrating AI/ML predictions directly into the twin to forecast the well's future condition (e.g., predicted time to failure for a component, likelihood of SCP escalation) and automatically trigger alerts or recommend maintenance/intervention actions.
- Optimized Operations: Using the twin to identify optimal operating parameters (e.g., flow rate, injection pressure) that maximize performance while respecting integrity constraints and minimizing long-term degradation.
- Training and Simulation: Providing a realistic simulation environment for training personnel on well integrity diagnosis and response.

Digital twins offer a powerful platform for integrating diverse data streams and models; transforming raw data into actionable insights; providing a holistic view; and enabling more informed, proactive, and optimized integrity management decisions. Developing and maintaining accurate digital twins requires significant investment in data infrastructure, modeling capabilities, integration efforts, and ongoing validation. Figure 31.2 illustrates the progressive capacity levels of digital twins, ranging from simple standalone models to fully autonomous systems. These levels highlight how the integration of connectivity, analytics, and decision support evolves to enhance well integrity management.

Digital twins in well integrity management integrate a range of data sources, engineering models, and advanced visualization tools to provide a holistic, real-time understanding of the well's condition and potential integrity risks. Figure 31.3 illustrates a comprehensive digital twin architecture tailored for well integrity management. It highlights the workflow from data acquisition (downhole sensors, operational data, and fluid models) through to advanced modeling (transient flow models, structural analyses), real-time dashboards, and actionable alerts. This integrated approach enables proactive decision-making and enhanced integrity assurance across the well lifecycle.

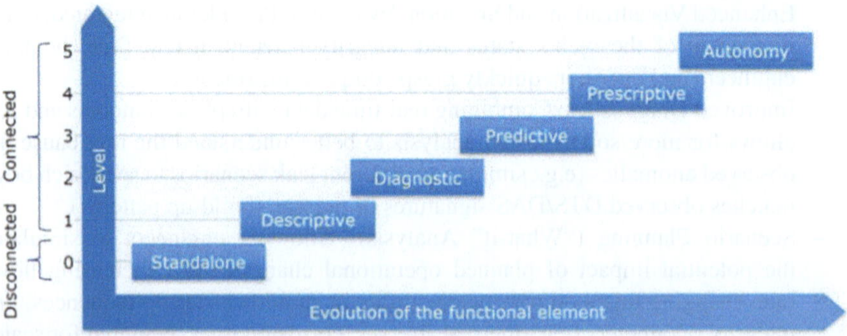

Fig. 31.2 Digital twin capacity levels. It shows the evolution of digital twin capabilities across five progressive levels: standalone, descriptive, diagnostic, predictive, prescriptive, and autonomous. These levels highlight how digital twins integrate increasing levels of connectivity, data analytics, and decision support to enhance real-time well integrity monitoring and proactive management (DNV, 2021)

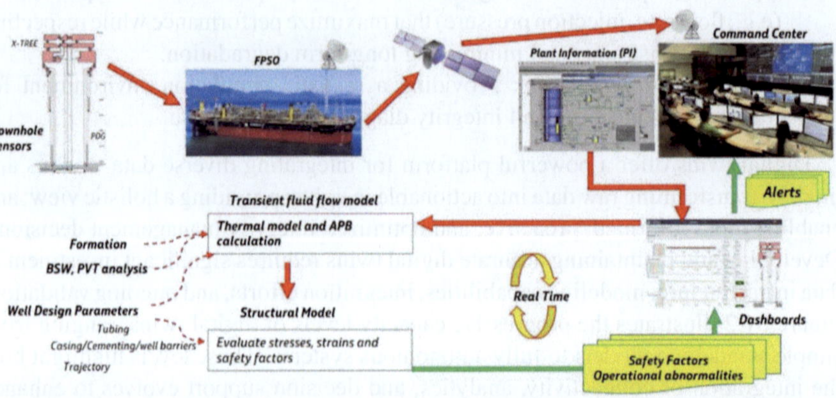

Fig. 31.3 Digital twin architecture for well integrity management. The diagram illustrates the integration of real-time data from downhole sensors and plant information systems into a digital twin environment. It includes transient fluid flow models, thermal and APB (annular pressure build-up) calculations, structural modeling for stress and strain evaluation, and safety factors analysis. The results feed into dashboards and alert systems within the command center to ensure proactive well integrity management (Aranha et al., 2025)

31.4 Integrated Data Platforms and Visualization

Underpinning AI/ML applications and digital twins is the fundamental need for robust, integrated data platforms. As noted, data silos are a major challenge (Chap. 15). Integrated platforms, often cloud-based, aim to solve this by.

- **Breaking Down Silos:** Aggregating data from various sources (real-time historians, operational databases, maintenance systems, regulatory reports, spreadsheets, log files, service company reports) into a single, accessible, unified location.
- **Ensuring Data Quality:** Incorporating tools for data cleansing, validation, standardization, and contextualization to improve accuracy and consistency.
- **Facilitating Analysis:** Providing tools for querying, analyzing, and performing complex computations on the integrated data.
- **Supporting Applications:** Serving as the scalable and reliable foundation upon which AI/ML models, digital twins, advanced visualization tools, and other analytical applications can be built, deployed, and run.

Effective visualization tools are also critical, transforming complex data and model outputs into intuitive dashboards, trend charts, interactive wellbore schematics overlaid with data (e.g., temperature curves, pressure gauges in annuli, corrosion maps), and potentially 3D models. Good visualization helps engineers and managers quickly grasp the well's integrity status, identify anomalies, understand the outputs of complex analytical models, and communicate insights effectively.

The successful adoption of digitalization, AI, and predictive analytics holds immense potential to significantly enhance well integrity performance, improve safety, reduce environmental impact, optimize operational efficiency, and extend asset life. However, realizing this potential requires a strong foundation of high-quality, well-managed data; appropriate technological infrastructure; and a workforce equipped with the necessary digital and domain skills to leverage these new tools effectively. Bridging the gap between technology potential and standardized practice (Chap. 28) is also key.

31.5 Smart Field

A smart field, also known as an intelligent field, is an advanced oil and gas field that integrates digital technologies to enable real-time monitoring, automation, and optimization of production. Smart fields utilize data-driven decision-making, leveraging artificial intelligence, machine learning, and predictive analytics to enhance operational efficiency and minimize risks (Arif & Al Senani, 2020a). One of the core technologies enabling smart fields is the Supervisory Control and Data Acquisition (SCADA) system, which provides remote monitoring and control of field operations. SCADA systems collect real-time data from sensors and field devices, allowing for continuous performance optimization and seamless integration with reservoir models and analytics platforms (Prati et al., 2015). Another crucial component is the Programmable Logic Controller (PLC), which automates control processes and enhances operational reliability by processing real-time data from field sensors to execute optimized production strategies (Sehr et al., 2021). Additionally, Inflow Control Valves (ICVs) and downhole chokes play a pivotal role in regulating fluid flow within the wellbore, facilitating dynamic adjustments in response to reservoir conditions. These components enable enhanced production efficiency and improved reservoir management by optimizing flow rates and preventing water or gas breakthrough (Al-Subaiei et al., 2019). The integration of these technologies establishes a comprehensive digital oil field framework, streamlining operations and ensuring greater efficiency in reservoir monitoring and production optimization. A key example of digitalization in the industry is the first fully remote-operated oil field in the United Arab Emirates, which integrates sensors, engineering models, and advanced algorithms to facilitate unmanned operations and enhance efficiency (Arif & Al Senani, 2020b). The transition toward Energy 5.0 also emphasizes human-centric approaches, incorporating collaborative artificial intelligence and co-bots to ensure that technological advancements align with broader societal objectives.

The benefits of smart fields arise from a comprehensive understanding and precise control of operational parameters. This is facilitated through high-resolution measurements of pressure, temperature, flow rate, and fluid composition at multiple points along the wellbore. Additionally, advanced geophysical imaging techniques, such as four-dimensional (4D) seismic monitoring, enable real-time observation of fluid dynamics within the petroleum-bearing formation, as well as associated changes in reservoir properties. The transition from conventional binary (on/off) control systems to continuously adjustable inflow control valves (ICVs) and variable frequency pumps further enhances operational efficiency, enabling adaptive reservoir management and optimized production strategies (Redutskiy, 2017).

31.6 Smart Field Applications in Well Integrity

Beyond enhancing production efficiency, smart field technologies are proving indispensable for managing well integrity across all phases of the well lifecycle—from drilling and completion to production, injection, and abandonment. Real-time monitoring of key parameters such as pressure, temperature, flow rates, and fluid chemistry—enabled by high-resolution sensors and SCADA systems—serves as an early warning system for well integrity issues. Continuous data streams allow for detecting deviations from baseline operating conditions that may signal integrity threats like sustained casing pressure (SCP), annular pressure build-up, or abnormal thermal cycling, which can lead to cement sheath degradation and casing damage.

Dynamic control capabilities of smart fields, including programmable logic controllers (PLCs) and downhole inflow control valves (ICVs), provide a means to respond proactively to these signals. For instance, automated control loops can modulate injection or production rates to mitigate abrupt pressure or temperature fluctuations—fluctuations known to accelerate mechanical fatigue, casing collapse, or thermal stresses at casing–cement interfaces (Arif & Al Senani, 2020b). This real-time adaptation not only preserves the structural integrity of downhole equipment but also extends the lifespan of key wellbore barriers.

Moreover, smart fields integrate advanced data analytics and artificial intelligence (AI) to identify complex, multi-parameter signatures that indicate emerging well integrity concerns. Machine learning algorithms, trained on historical data from similar wells, can predict the risk of issues like cement debonding, corrosion under thermal and pressure cycling, or chemical attack by formation fluids (Sehr et al., 2021; Al-Subaiei et al., 2019). Such predictive insights enable targeted interventions before failures escalate, reducing costly workovers and environmental risks.

Importantly, the continuous, high-resolution monitoring characteristic of smart fields also supports regulatory compliance and environmental stewardship. For example, it enables better documentation of annular pressure trends and ensures that any unexpected behavior (e.g., SCP exceeding thresholds) is captured and addressed promptly. In injection wells used for carbon capture and storage (CCS), hydrogen storage, or geothermal energy, enhanced oil recovery smart field data can also be integrated with geomechanical models to track pressure and stress changes, ensuring the well and surrounding formation remain within safe operating limits. For example, ensuring the long-term integrity of storage sites remains a significant challenge due to risks such as cap rock leakage, orphan well pathways, and reservoir pressure-induced geomechanical instabilities (Alsubaih et al., 2024; Guglielmi et al., 2021; Khakim et al., 2012). Excessive pore pressure increases can alter the in-situ stress state, potentially triggering rock failure, fault reactivation, and seismic activity. While most induced seismicity associated with CO_2 injection is micro-seismic, instances of larger magnitude events have been reported. Seismic events related to CO_2 injection have been observed at various injection sites, with earthquake rates and magnitudes increasing as reservoir pressure rises (Huang et al., 2018; Simpson, 1976; White & Foxall, 2016; Zhao et al., 2022). Studies indicate that CO_2 plume migration can weaken faults, leading to reactivation due to localized pressure build-up (Vilarrasa et al., 2017). Additionally, micro-seismic events are more frequent during operational flow changes, such as shutdowns, rather than during steady-state injection. The fault reactivation or changes in reservoir pressure can cause significant damage to the casing, a phenomenon also highlighted in Chap. 23. Figure 23.3 presents a clear visual representation of the long-term relationship between injection pressure and casing damage incidents in water injection wells. The figure shows that as injection pressures increase, the number of casing failures rises correspondingly, underscoring the critical role of precise pressure management in preserving well integrity. This trend illustrates how sustained high-pressure injection can cumulatively degrade casing performance and compromise the well's mechanical barriers. The data emphasize the importance of continuous monitoring and proactive intervention strategies to mitigate the risks associated with injection-induced stresses and to maintain safe and reliable operations throughout the well's lifecycle.

The holistic, integrated view offered by smart field data—combined with robust visualization and digital twin technologies—enables operators to identify potential integrity threats early, quantify risks, and deploy remedial measures with confidence. This proactive approach not only enhances operational safety and performance but also underpins the long-term sustainability and resilience of well assets in increasingly challenging environments. As smart field technologies evolve, their role in well integrity assurance is set to expand further, supporting data-driven integrity management frameworks that can adapt to the unique challenges of each well and reservoir system.

References

Abdo, E., Baronio, E., Mauro, S., Troise, M., & Salamina, L. (2023). Case study of the use of a digital XE "Digital" twin for leak detection and quantification in underground gas storage wells. *SPE Journal, 28*(05), 2415–2424. https://doi.org/10.2118/214357-PA

Al-Subaiei, D., Al-Hamer, M., Al-Zaidan, A., Chetri, H., & Nawaz, M. S. (2019). Intelligent digital oilfield implementation: production optimization using north kuwait integrated digital oil field NK KwIDF. *Society of Petroleum Engineers—Abu Dhabi International Petroleum Exhibition and Conference 2019, ADIP 2019.* https://doi.org/10.2118/197811-MS

Alsubaih, A., Sepehrnoori, K., & Delshad, M. (2024). Environmental impacts of orphaned and abandoned wells: methane emissions, and implications for carbon storage. *Applied Sciences, 14*(24), 11518. https://doi.org/10.3390/APP142411518

Anjos, J. L., Aranha, P. E., Martins, A. L., Oliveira, F. L., Joségoncalves, C., Silva, D. R., Dudek, C. L., & Lima, C. B. (2020). Digital twin for well integrity with real time surveillance. In: *Proceedings of the Annual Offshore Technology Conference, 2020-May.* https://doi.org/10.4043/30574-MS

Aranha, P. E., De Souza, C. O., Ferrarezi, L. F., Valore, F., De Oliveira, F. L., Dos Santos, M. V., Donatti, C. N., Mendes, H. U. S., Dearaújo, J. P. N., & Souza Neto, S. L. (2025). Look ahead: optimization of well operations using real time digital XE "Digital" twin for well integrity. *Proceedings of the Annual Offshore Technology Conference.* https://doi.org/10.4043/35617-MS

Arif, M., & Al Senani, A. M. (2020b). Digitalization in oil and gas industry—a case study of a fully smart field in United Arab Emirates. In: *Society of Petroleum Engineers—Abu Dhabi International Petroleum Exhibition and Conference 2020, ADIP 2020.* https://doi.org/10.2118/203461-MS

Arif, M., & Al Senani, A. M. (2020a). Digitalization in oil and gas industry—a case study of a fully smart field in united arab emirates. In: *Society of Petroleum Engineers—Abu Dhabi International Petroleum Exhibition and Conference 2020, ADIP 2020.* https://doi.org/10.2118/203461-MS

Busollo, C., Mauro, S., Nesci, A., Scimmi, L. S., & Baronio, E. (2021). Development of a digital twin for well integrity management in underground gas storage fields. In: *Proceedings - SPE Annual Technical Conference and Exhibition, 2021-September*, 21–23. https://doi.org/10.2118/206252-MS

DNV. (2021). *DNV-RP-F116 Integrity management of submarine pipeline systems.* https://www.dnv.com/energy/standards-guidelines/dnv-rp-f116-integrity-management-of-submarine-pipeline-systems/

Guglielmi, Y., Nussbaum, C., Cappa, F., De Barros, L., Rutqvist, J., & Birkholzer, J. (2021). Field-scale fault reactivation experiments by fluid injection highlight aseismic leakage in caprock analogs: Implications for CO2 sequestration. *International Journal of Greenhouse Gas Control, 111*, Article 103471. https://doi.org/10.1016/J.IJGGC.2021.103471

Huang, R., Zhu, L., Encarnacion, J., Xu, Y., Tang, C. C., Luo, S., & Jiang, X. (2018). Seismic and geologic evidence of water-induced earthquakes in the three gorges reservoir region of China. *Geophysical Research Letters, 45*(12), 5929–5936. https://doi.org/10.1029/2018GL077639

Khakim, M. Y. N., Tsuji, T., & Matsuoka, T. (2012). Geomechanical modeling for InSAR XE "InSAR" -derived surface deformation at steam-injection oil sand fields. *Journal of Petroleum Science and Engineering, 96–97*, 152–161. https://doi.org/10.1016/J.PETROL.2012.08.003

Prati, T. J., Farines, J. M., & De Queiroz, M. H. (2015). Automatic test of safety specifications for PLC XE "PLC" programs in the Oil and Gas Industry. *IFAC-PapersOnLine, 48*(6), 27–32. https://doi.org/10.1016/J.IFACOL.2015.08.005

Redutskiy, Y. (2017). Conceptualization of smart solutions in oil and gas industry. *Procedia Computer Science, 109*, 745–753. https://doi.org/10.1016/J.PROCS.2017.05.435

Sehr, M. A., Lohstroh, M., Weber, M., Ugalde, I., Witte, M., Neidig, J., Hoeme, S., Niknami, M., & Lee, E. A. (2021). Programmable logic controllers in the context of industry 4.0. *IEEE Transactions on Industrial Informatics, 17*(5), 3523–3533. https://doi.org/10.1109/TII.2020.3007764

Simpson, D. W. (1976). Seismicity changes associated with reservoir loading. *Engineering Geology, 10*(2–4), 123–150. https://doi.org/10.1016/0013-7952(76)90016-8

Straw, M., Mistry, R., Lewis, S., Straw, N., Aglave, R., & Kolak, D. (2022). Predictive digital XE "Digital" twin for performance and integrity. *Proceedings of the Annual Offshore Technology Conference, 11*,. https://doi.org/10.4043/31863-MS

Vilarrasa, V., Makhnenko, R. Y., & Laloui, L. (2017). Potential for fault reactivation XE "Fault Reactivation" due to CO2 injection in a semi-closed Saline Aquifer. *Energy Procedia, 114*, 3282–3290. https://doi.org/10.1016/J.EGYPRO.2017.03.1460

White, J. A., & Foxall, W. (2016). Assessing induced seismicity risk at CO2 storage projects: Recent progress and remaining challenges. *International Journal of Greenhouse Gas Control, 49*, 413–424. https://doi.org/10.1016/J.IJGGC.2016.03.021

Zhao, R., Xue, J., & Deng, K. (2022). Modelling seismicity pattern of reservoir-induced earthquakes including poroelastic stressing and nucleation effects. *Geophysical Journal International, 232*(2), 739–749. https://doi.org/10.1093/GJI/GGAC361

Chapter 32
Future Directions and Concluding Remarks

Abstract This final chapter provides a comprehensive synthesis of the book's insights and casts a forward-looking perspective on the evolving landscape of well integrity. As the energy industry faces mounting challenges—from aging well infrastructure and increasingly complex operating conditions to the global imperative for low-carbon technologies—well integrity management must continuously adapt and improve. The chapter highlights key areas shaping the future of integrity assurance, including ongoing research in advanced barrier materials, evolving plug and abandonment (P&A) practices, and the critical role of well integrity in enabling energy transition applications such as carbon capture and storage (CCS), underground hydrogen storage (UHS), and geothermal energy. Emerging material innovations—such as self-healing cements, flexible and CO_2-resistant systems, corrosion-resistant alloys, smart materials, and advanced elastomers—promise to enhance the durability and functionality of well barriers under increasingly harsh downhole conditions. Simultaneously, P&A strategies are evolving to ensure permanent, verifiable isolation with a growing focus on alternative materials, rigless technologies, improved annular sealing methods, and risk-based abandonment design. The chapter also reaffirms that technological innovation must be matched by unwavering organizational discipline, risk-based thinking, and a strong culture of safety and vigilance. It emphasizes that well integrity is not a static achievement but a dynamic, long-term commitment requiring continuous monitoring, learning, and improvement. The principles of multi-barrier design, lifecycle integrity management, and proactive risk mitigation remain foundational as the industry expands its scope to support emerging energy systems. In conclusion, this chapter summarizes the essential takeaways from across the book—ranging from technical principles and operational practices to human and organizational factors—while reinforcing that the pursuit of flawless well integrity is an ongoing journey. In a rapidly evolving energy context, ensuring robust, safe, and sustainable well performance demands innovation, integration, and a steadfast commitment to excellence from all stakeholders involved.

© The Author(s), under exclusive license to Springer Nature Switzerland AG 2026

A. Alsubaih and K. Sepehrnoori, *Well Integrity for Oil, Gas, and Subsurface Storage Operations*, https://doi.org/10.1007/978-3-032-15233-6_32

Chapter's Objectives

- Highlight research and development in novel materials for enhanced integrity.
- Discuss evolving strategies and materials for plug and abandonment.
- Consider the application of well integrity principles to energy transition technologies.
- Reiterate the enduring importance of vigilance, safety culture, and continuous improvement.
- Summarize the key takeaways of the book.

32.1 The Continuous Pursuit of Enhanced Integrity

The discipline of well integrity stands at the intersection of engineering innovation, regulatory compliance, environmental stewardship, and operational excellence. Far from being a static or narrowly defined field, well integrity management has evolved into a dynamic, multifaceted discipline that adapts to the shifting challenges of modern energy operations. Today, it is driven by the convergence of several critical forces: the need to safely extend the productive life of aging assets, the push to explore and operate in more complex and extreme environments, the imperative to meet ever more stringent environmental and regulatory standards, and the growing awareness of the long-term liabilities and risks associated with subsurface activities—especially in the era of energy transition.

This chapter serves as both a synthesis and a forward-looking reflection. It builds on the specialized topics and advanced practices discussed in the preceding chapters, uniting them into a cohesive narrative of continuous improvement and adaptation. It underscores how the principles of well integrity must be reimagined and reinforced to keep pace with emerging challenges: from permanent storage of CO_2 in carbon capture and storage (CCS) projects, to the unique stresses of geothermal wells, to the novel requirements of underground hydrogen storage (UHS). These evolving frontiers highlight those traditional approaches—while foundational—must be complemented by robust materials research, refined operational strategies, and a steadfast culture of safety and accountability.

In this chapter, we highlight cutting-edge research in advanced materials that promise to extend the lifespan and reliability of well barriers—such as self-healing cements, corrosion-resistant alloys, and smart materials that respond to stress or damage. We examine how evolving plug and abandonment (P&A) practices are leveraging innovative materials and risk-based approaches to achieve permanent isolation while managing cost and complexity. We also explore the essential role of well integrity in enabling the secure deployment of low-carbon energy technologies critical to the energy transition.

Beyond technical advancements, this chapter reaffirms that the human element—embodied in a culture of vigilance, organizational learning, and unwavering commitment to safety—remains paramount. As new challenges arise, complacency can be the most insidious threat to well integrity. A proactive, systematic, and continuous approach to improvement is the cornerstone of maintaining the integrity of wells, ensuring not only operational reliability but also societal trust and environmental protection.

This concluding chapter synthesizes key insights from the entire book, articulating how the core principles of well integrity must evolve and be reinforced in the face of future demands. It affirms that well integrity is not a static endpoint, but an ongoing journey—one requiring continuous innovation, collaboration across disciplines, and an unyielding dedication to excellence. As the energy landscape transforms, the principles and practices of well integrity will remain vital to safeguarding both people and the planet. This chapter looks at some key areas shaping the future of well integrity management, building on the specialized topics and advanced concepts discussed in Part 5.

32.2 Novel Materials Research

Developing materials with improved performance, durability, and multi-functional capabilities under downhole conditions remains a key focus area for enhancing long-term well integrity.

- **Advanced Cement Systems:** Research continues to focus on developing cement formulations with enhanced properties for long-term isolation. This includes:
  - Self-Healing Cements: Incorporating materials (e.g., expansive minerals, microencapsulated healing agents like polymers or minerals, bacterial processes) that can automatically react to seal small cracks or micro-annuli that form over time due to stress or degradation, thereby restoring low permeability and enhancing long-term zonal isolation reliability.
  - Flexible/Ductile Cements: Formulations with lower stiffness (Young's modulus) and greater toughness/ductility (often achieved using polymers like latex or fibers) to better withstand cyclic stresses (pressure, temperature, geomechanical loads) without cracking or debonding, particularly relevant for HPHT, thermal, or geomechanically active wells.
  - CO_2-Resistant Cements: Blended cements (using high volumes of fly ash, slag, or silica fume) or alternative binders (geopolymers, calcium aluminate cements, advanced calcium silicate-based cements) designed for improved long-term stability and lower permeability when exposed to CO_2, crucial for CCS and CO_2 EOR applications.
  - Low Permeability & High Bonding Cements: Developments aimed at inherently lower permeability and stronger, more durable bonds to casing and formation to minimize leak path potential.

- **High-Performance Elastomers and Seals:** Continued development of elastomers with higher temperature ratings, improved chemical resistance (especially to aggressive fluids like H_2S, CO_2, specific EOR chemicals, aggressive inhibitors), better explosive decompression (ED/RGD) resistance, and lower compression set for more reliable and longer-lasting sealing in increasingly harsh and cyclic environments. Advancements in metal-to-metal (MTM) seal design and materials for extreme pressure, temperature, and thermal cycling conditions are also ongoing, including exploring novel geometries and surface treatments. Qualification standards need to evolve to match the capabilities and risks of these materials under realistic combined conditions.
- **Advanced Corrosion Resistant Alloys (CRAs) and Coatings:** Research into more cost-effective CRAs, improved manufacturing processes (e.g., for clad pipes or corrosion-resistant liners), and better understanding of their long-term performance limits (e.g., susceptibility to environmentally assisted cracking, localized corrosion) in specific downhole environments. Development of advanced internal and external coatings, including nanomaterial-enhanced composites and functionalized polymers, offering improved barrier properties, increased resistance to damage/disbonding, or potential self-healing functionalities. Qualification and reliability verification of these advanced coatings are key.
- **Smart Materials:** A longer-term research area involves materials incorporating embedded sensors or functionalities, such as materials that change properties (electrical resistance, color) in response to damage, stress, or specific chemical environments, potentially enabling self-monitoring components or providing earlier indication of degradation.

32.3 Evolving Plug and Abandonment (p&A) Strategies and Materials

Ensuring demonstrably permanent and cost-effective abandonment is a major focus area due to the vast number of wells requiring P&A globally and the concerns over long-term liability and environmental impact.

- **Alternative P&A Barrier Materials:** Increased research, testing, and field trials of non-cementitious materials (geopolymers, bismuth alloys, advanced resins, potentially other engineered barriers) aiming for enhanced long-term durability, better bonding (especially in potentially challenging annular spaces), lower shrinkage, greater resistance to specific downhole fluids (like CO_2), and potentially easier placement or verification compared to traditional Portland cement. Regulatory acceptance and robust long-term performance validation remain key hurdles for widespread adoption.

- **Improved Annular Sealing Techniques:** Significant focus remains on reliably sealing the entire wellbore cross section, not just inside the innermost casing. This includes advancements in section milling technology (Chap. 27), perforate-and-circulate methods, and placement techniques for both cement and alternative materials that can effectively fill and seal annular spaces, especially in challenging geometries or when original cement is degraded. Techniques for cleaning annuli prior to plug placement are also evolving.
- **Enhanced Verification Methods:** Development of logging tools or other non-intrusive techniques capable of reliably verifying the hydraulic sealing integrity of permanent barriers (both cement and alternatives) across the full wellbore cross section, moving beyond simple tagging and pressure testing. This includes research into advanced acoustic, ultrasonic, or other geophysical methods that can "see" behind pipe and assess seal quality, and potentially methods for monitoring long-term stability (though perpetual monitoring is typically not the goal). This is a major technology gap.
- **Risk-Based P&A Design:** Increasingly, efforts are being made to tailor P&A requirements (number, type, location, length of plugs, verification methods) based on well-specific risks (e.g., proximity to valuable aquifers, presence of high-pressure/corrosive zones, condition of existing barriers) rather than relying solely on a one-size-fits-all prescriptive approach. This aims to optimize cost while ensuring safety commensurate with risk (requires robust risk assessment methodologies and regulatory acceptance).
- **Rigless P&A Advancement:** Continued development of specialized tools and intervention vessel capabilities to perform a broader range of P&A operations (including milling, complex plug setting, equipment removal) without requiring expensive drilling rigs, contributing to cost-effectiveness for abandoning large numbers of wells.

32.4 Well Integrity for Energy Transition Applications

As the global energy landscape evolves, applying well integrity principles and knowledge to emerging energy technologies is crucial for their safe and successful deployment. This requires leveraging the decades of experience gained in oil and gas while addressing the unique challenges of these new applications (as discussed in Chaps. 21, 22, 23, 24).

- **CCS Well Integrity:** Developing industry standards and best practices specifically for CCS well design (CO_2 compatibility, long-term stress management), materials selection (CO_2-resistant steels, cements, seals), monitoring (long-term surveillance needs), and P&A (demonstrating permanence) to ensure secure geological storage of CO_2 is a critical ongoing effort. Learning from oil and gas experience with CO_2 EOR is valuable but needs adaptation for permanent storage timescales.

- **Hydrogen Storage Well Integrity:** Significant research and development are required to establish integrity practices for wells used in Underground Hydrogen Storage (UHS). This includes addressing fundamental challenges like hydrogen embrittlement of steels, reliable sealing of small hydrogen molecules (permeation), cement compatibility, management of cyclic loading fatigue potentially exacerbated by hydrogen, and understanding potential biogeochemical reactions. Tailored material qualification standards and operational guidelines are needed.
- **Geothermal Well Integrity:** Advancements in materials (cements, elastomers, casing) capable of withstanding sustained extreme temperatures and severe thermal cycling common in geothermal applications are essential. Improved methods for designing for and managing thermal stresses are key.
- **Natural Hydrogen Extraction:** While nascent, potential future extraction of natural hydrogen will require applying well integrity principles, addressing similar challenges to UHS regarding hydrogen-material interaction, potentially combined with high pressures and temperatures.

Leveraging the established knowledge base from oil and gas well integrity while actively researching and adapting practices to address the unique technical, material, and long-term challenges of these new energy applications is essential for the success of the energy transition and ensuring the responsible use of the subsurface.

32.5 The Enduring Importance of Vigilance, Culture, and Continuous Improvement

While technological advancements offer exciting possibilities for the future of well integrity, it is crucial to reiterate that technology alone is not a panacea. The fundamental pillars of effective well integrity management, rooted in robust systems and human performance, remain timeless and indispensable.

- **Technical Rigor:** Adherence to sound engineering principles in design, construction, operation, intervention, and abandonment remains non-negotiable. This includes accurate load calculations, proper material selection based on anticipated conditions throughout the lifecycle, robust barrier design, and meticulous planning.
- **Barrier Philosophy:** Consistent application of the multi-barrier principle (independent, verifiable barriers) is the cornerstone of well design and operation and must be rigorously applied in all phases, including interventions (temporary barriers) and P&A (permanent barriers).
- **Lifecycle Perspective:** Managing integrity proactively from initial geological concept and design through construction, operation, intervention, and into permanent abandonment is essential. Decisions made early have lasting impacts, and integrity is a dynamic, time-dependent condition that evolves over decades.

- **Risk Management:** Systematically identifying, assessing, and mitigating risks associated with potential integrity failures is fundamental to prioritizing effort and ensuring resources are focused on the most significant threats (Chap. 14). This should be a continuous process, informed by data and learning.
- **Vigilance and Monitoring:** Maintaining alertness to potential deviations, unexpected conditions, and weak signals that might indicate an emerging integrity problem ("chronic sense of unease") is crucial. Diligent monitoring of well conditions (pressures, temperatures, flow, logs) and recognizing when something is abnormal allows for early detection and intervention (Chapters 8, 12, 30).
- **Safety Culture:** Fostering an organizational environment where integrity is prioritized as a core value, communication is open and trusted, procedures are followed, individuals feel empowered to stop unsafe work, and learning from experience (incidents and near misses) is ingrained in the DNA of the organization (Chap. 17). Human factors remain critical; technology cannot compensate for a weak culture.
- **Continuous Improvement:** The WIMS (Part 4) provides the framework for this. Driven by performance monitoring (KPIs), audits, management reviews, and lessons learned (internal and external), continuous improvement in procedures, standards, competencies, and technology adoption must be an ongoing process to adapt to evolving challenges and maintain high performance. Addressing organizational challenges, improving data availability and integration, standardizing life extension approaches, and fostering knowledge transfer are key aspects of this.

Complacency is the enemy of well integrity. A long period without incidents can breed a false sense of security. Maintaining the required level of vigilance, discipline, and commitment is a perpetual challenge requiring active reinforcement.

32.6 Concluding Remarks: Summary of Key Principles

This book has journeyed through the multifaceted world of well integrity in the oil and gas industry and related subsurface applications. It has covered the foundational principles, the engineering and operational practices across the lifecycle, the technical threats and failure modes, the management systems required for effective oversight, and the advanced technologies shaping the future.

Key takeaways include:

- Well integrity is paramount for safety, environmental protection, economic performance, and societal acceptance. It is fundamentally about maintaining containment and control of subsurface fluids and pressures throughout the well lifecycle, from design to permanent abandonment.
- The two-barrier philosophy (independent, verifiable barriers) is the cornerstone of well design and operation, providing crucial redundancy against failure.

- Integrity must be managed proactively across all lifecycle phases. Decisions in early phases (design, construction quality) have lasting impacts, and integrity is a dynamic, time-dependent condition that evolves over decades.
- Understanding potential failure modes (corrosion, cement degradation, seal failure, mechanical overload, fatigue) is crucial for designing preventative measures, implementing effective monitoring, and planning remediation.
- Robust monitoring (annulus pressure, logging, advanced sensing) and diagnostics are essential for detecting degradation and emerging issues early.
- A systematic Well Integrity Management System (WIMS), incorporating clear policy, defined organization, risk assessment, standardized procedures, competent personnel, robust data management, assurance processes, and learning mechanisms, is vital for effective, consistent, and auditable integrity management across an organization.
- Human factors and a positive safety culture are indispensable elements, underpinning the effectiveness of technical systems, procedures, and management frameworks.
- Specialized challenges exist in harsh operating environments (HPHT, deepwater, corrosive), unconventional plays, subsea operations, and particularly for energy transition applications like Geothermal, EOR, Water Injection, and Subsurface Storage (CCS, NGS, UHS), requiring tailored technical solutions and management practices.
- Permanent plug and abandonment (P&A) present significant technical challenges for ensuring long-term isolation over geological timescales, requiring durable materials, reliable placement techniques, and verifiable seals.
- Technological advancements (fiber optics, wireless sensors, AI/ML, digital twins, novel materials, rigless P&A methods) offer exciting opportunities to enhance predictive and proactive integrity management, improve efficiency, and address current limitations, provided they are rigorously validated and integrated effectively.

Ensuring the integrity of the millions of wells penetrating the Earth's crust is a complex but non-negotiable responsibility. It requires a blend of sound engineering, diligent execution, vigilant monitoring, adaptive management, continuous learning, and an unwavering organizational commitment to safety and environmental stewardship. As the energy landscape evolves, the principles and practices of well integrity will remain fundamental to the safe and sustainable utilization of subsurface resources, whether for traditional hydrocarbon production or new energy applications. The journey toward flawless well integrity is ongoing, demanding continuous effort, collaboration, and innovation from all involved.

Index

© The Editor(s) (if applicable) and The Author(s), under exclusive license to Springer Nature Switzerland AG 2026

A. Alsubaih and K. Sepehrnoori, *Well Integrity for Oil, Gas, and Subsurface Storage Operations*, https://doi.org/10.1007/978-3-032-15233-6

MIX
Papier aus verantwortungsvollen Quellen
Paper from responsible sources
FSC® C105338

If you have any concerns about our products,
you can contact us on
ProductSafety@springernature.com

In case Publisher is established outside the EU,
the EU authorized representative is:
Springer Nature Customer Service Center GmbH
Europaplatz 3, 69115 Heidelberg, Germany

Printed by Libri Plureos GmbH
in Hamburg, Germany